Tumor Suppressor Genes

METHODS IN MOLECULAR BIOLOGY™

John M. Walker, Series Editor

234. **p53 Protocols**, edited by *Sumitra Deb and Swati Palit Deb, 2003*
233. **Protein Kinase C Protocols**, edited by *Alexandra C. Newton, 2003*
232. **Protein Misfolding and Disease:** *Principles and Protocols*, edited by *Peter Bross and Niels Gregersen, 2003*
231. **Directed Evolution Library Creation**: *Methods and Protocols*, edited by *Frances H. Arnold and George Georgiou, 2003*
230. **Directed Enzyme Evolution**: *Screening and Selection Methods*, edited by *Frances H. Arnold and George Georgiou, 2003*
229. **Lentivirus Gene Engineering Protocols**, edited by *Maurizio Federico, 2003*
228. **Membrane Protein Protocols**: *Expression, Purification, and Characterization*, edited by *Barry S. Selinsky, 2003*
227. **Membrane Transporters:** *Methods and Protocols,* edited by *Qing Yan, 2003*
226. **PCR Protocols, Second Edition,** edited by *John M. S. Bartlett and David Stirling, 2003*
225. **Inflammation Protocols,** edited by *Paul G. Winyard and Derek A. Willoughby, 2003*
224. **Functional Genomics:** *Methods and Protocols,* edited by *Michael J. Brownstein and Arkady Khodursky, 2003*
223. **Tumor Suppressor Genes:** *Volume 2: Regulation, Function, and Medicinal Applications,* edited by *Wafik S. El-Deiry, 2003*
223. **Tumor Suppressor Genes:** *Volume 2: Regulation, Function, and Medicinal Applications,* edited by *Wafik S. El-Deiry, 2003*
222. **Tumor Suppressor Genes:** *Volume 1: Pathways and Isolation Strategies,* edited by *Wafik S. El-Deiry, 2003*
221. **Generation of cDNA Libraries:** *Methods and Protocols*, edited by *Shao-Yao Ying, 2003*
220. **Cancer Cytogenetics:** *Methods and Protocols,* edited by *John Swansbury, 2003*
219. **Cardiac Cell and Gene Transfer:** *Principles, Protocols, and Applications,* edited by *Joseph M. Metzger, 2003*
218. **Cancer Cell Signaling:** *Methods and Protocols,* edited by *David M. Terrian, 2003*
217. **Neurogenetics:** *Methods and Protocols,* edited by *Nicholas T. Potter, 2003*
216. **PCR Detection of Microbial Pathogens:** *Methods and Protocols,* edited by *Konrad Sachse and Joachim Frey, 2003*
215. **Cytokines and Colony Stimulating Factors:** *Methods and Protocols,* edited by *Dieter Körholz and Wieland Kiess, 2003*
214. **Superantigen Protocols,** edited by *Teresa Krakauer, 2003*
213. **Capillary Electrophoresis of Carbohydrates,** edited by *Pierre Thibault and Susumu Honda, 2003*
212. **Single Nucleotide Polymorphisms:** *Methods and Protocols,* edited by *Piu-Yan Kwok, 2003*
211. **Protein Sequencing Protocols, Second Edition,** edited by *Bryan John Smith, 2003*
210. **MHC Protocols,** edited by *Stephen H. Powis and Robert W. Vaughan, 2003*
209. **Transgenic Mouse Methods and Protocols,** edited by *Marten Hofker and Jan van Deursen, 2002*
208. **Peptide Nucleic Acids:** *Methods and Protocols,* edited by *Peter E. Nielsen, 2002*
207. **Recombinant Antibodies for Cancer Therapy:** *Methods and Protocols,* edited by *Martin Welschof and Jürgen Krauss, 2002*
206. **Endothelin Protocols,** edited by *Janet J. Maguire and Anthony P. Davenport, 2002*
205. ***E. coli* Gene Expression Protocols,** edited by *Peter E. Vaillancourt, 2002*
204. **Molecular Cytogenetics:** *Protocols and Applications,* edited by *Yao-Shan Fan, 2002*
203. ***In Situ* Detection of DNA Damage:** *Methods and Protocols,* edited by *Vladimir V. Didenko, 2002*
202. **Thyroid Hormone Receptors:** *Methods and Protocols,* edited by *Aria Baniahmad, 2002*
201. **Combinatorial Library Methods and Protocols,** edited by *Lisa B. English, 2002*
200. **DNA Methylation Protocols,** edited by *Ken I. Mills and Bernie H. Ramsahoye, 2002*
199. **Liposome Methods and Protocols,** edited by *Subhash C. Basu and Manju Basu, 2002*
198. **Neural Stem Cells:** *Methods and Protocols,* edited by *Tanja Zigova, Juan R. Sanchez-Ramos, and Paul R. Sanberg, 2002*
197. **Mitochondrial DNA:** *Methods and Protocols,* edited by *William C. Copeland, 2002*
196. **Oxidants and Antioxidants:** *Ultrastructure and Molecular Biology Protocols,* edited by *Donald Armstrong, 2002*
195. **Quantitative Trait Loci:** *Methods and Protocols,* edited by *Nicola J. Camp and Angela Cox, 2002*
194. **Posttranslational Modifications of Proteins:** *Tools for Functional Proteomics,* edited by *Christoph Kannicht, 2002*
193. **RT-PCR Protocols,** edited by *Joe O'Connell, 2002*
192. **PCR Cloning Protocols, Second Edition,** edited by *Bing-Yuan Chen and Harry W. Janes, 2002*
191. **Telomeres and Telomerase:** *Methods and Protocols,* edited by *John A. Double and Michael J. Thompson, 2002*
190. **High Throughput Screening:** *Methods and Protocols,* edited by *William P. Janzen, 2002*
189. **GTPase Protocols:** *The RAS Superfamily,* edited by *Edward J. Manser and Thomas Leung, 2002*
188. **Epithelial Cell Culture Protocols,** edited by *Clare Wise, 2002*
187. **PCR Mutation Detection Protocols,** edited by *Bimal D. M. Theophilus and Ralph Rapley, 2002*
186. **Oxidative Stress Biomarkers and Antioxidant Protocols,** edited by *Donald Armstrong, 2002*
185. **Embryonic Stem Cells:** *Methods and Protocols,* edited by *Kursad Turksen, 2002*
184. **Biostatistical Methods,** edited by *Stephen W. Looney, 2002*
183. **Green Fluorescent Protein:** *Applications and Protocols,* edited by *Barry W. Hicks, 2002*
182. **In Vitro Mutagenesis Protocols, Second Edition**, edited by *Jeff Braman, 2002*
181. **Genomic Imprinting:** *Methods and Protocols,* edited by *Andrew Ward, 2002*
180. **Transgenesis Techniques, Second Edition:** *Principles and Protocols,* edited by *Alan R. Clarke, 2002*
179. **Gene Probes:** *Principles and Protocols,* edited by *Marilena Aquino de Muro and Ralph Rapley, 2002*
178. **Antibody Phage Display:** *Methods and Protocols,* edited by *Philippa M. O'Brien and Robert Aitken, 2001*
177. **Two-Hybrid Systems:** *Methods and Protocols,* edited by *Paul N. MacDonald, 2001*
176. **Steroid Receptor Methods:** *Protocols and Assays,* edited by *Benjamin A. Lieberman, 2001*
175. **Genomics Protocols**, edited by *Michael P. Starkey and Ramnath Elaswarapu, 2001*
174. **Epstein-Barr Virus Protocols,** edited by *Joanna B. Wilson and Gerhard H. W. May, 2001*
173. **Calcium-Binding Protein Protocols, Volume 2:** *Methods and Techniques,* edited by *Hans J. Vogel, 2001*
172. **Calcium-Binding Protein Protocols, Volume 1:** *Reviews and Case Histories,* edited by *Hans J. Vogel, 2001*
171. **Proteoglycan Protocols,** edited by *Renato V. Iozzo, 2001*
170. **DNA Arrays:** *Methods and Protocols,* edited by *Jang B. Rampal, 2001*
169. **Neurotrophin Protocols,** edited by *Robert A. Rush, 2001*
168. **Protein Structure, Stability, and Folding,** edited by *Kenneth P. Murphy, 2001*
167. **DNA Sequencing Protocols,** *Second Edition,* edited by *Colin A. Graham and Alison J. M. Hill, 2001*

METHODS IN MOLECULAR BIOLOGY™

Tumor Suppressor Genes

Volume 1
Pathways and Isolation Strategies

Edited by

Wafik S. El-Deiry, MD, PhD

Departments of Medicine, Genetics, and Pharmacology,
University of Pennsylvania School of Medicine, Philadelphia, PA

Humana Press Totowa, New Jersey

999 Riverview Drive, Suite 208
Totowa, New Jersey 07512

www.humanapress.com

This publication is printed on acid-free paper. ∞
ANSI Z39.48-1984 (American Standards Institute)
Permanence of Paper for Printed Library Materials.

Production Editor: Tracy Catanese

Cover illustration: Fig. 3B from Volume 2, Chapter 5, "*In Situ* Hybridization in Cancer and Normal Tissue," by Shrihari Kadkol, Jeremy Juang, and Tzyy-choou Wu.

Cover design by Patricia F. Cleary

For additional copies, pricing for bulk purchases, and/or information about other Humana titles, contact Humana at the above address or at any of the following numbers: Tel.: 973-256-1699; Fax: 973-256-8341; E-mail: humana@humanapr.com; or visit our Website: humanapress.com

Printed in the United States of America. 10 9 8 7 6 5 4 3 2 1

Library of Congress Cataloging-in Publication Data

Tumor suppressor genes / edited by Wafik S. El-Deiry.
p. ; cm. -- (Methods in molecular biology ; v. 222)
Includes bibliographical references and index.
Contents: v. 1. Pathways and isolate strategies -- v. 2. Regulation, function, and medicinal application.
ISBN 0-89603-986-2 (v. 1 : alk. paper); 1-59259-328-3 (v. 1 : e-book) -- ISBN 0-89603-987-0 (v. 2 : alk. paper);1-59259-329-1 (v. 2 : e-book)
1. Antioncogenes--Laboratory manuals. I. El-Deiry, Wafik S. II. Methods in molecular biology (Totowa, N.J.) ; v. 222-223.
[DNLM: 1. Genes, Tumor Suppressor--physiology--Laboratory Manuals. 2. Neoplasms--genetics--Laboratory Manuals. 3. Neoplasms--therapy--Laboratory Manuals. 4. Signal Transduction--physiology--Laboratory Manuals. 5. Tumor Suppressor Proteins--physiology--Laboratory Manuals. QZ 25 T925 2003]
RC268.43 .T82 2003
616.99'4042--dc21

2002027501

Preface

It has become clear that tumors arise from excessive cell proliferation and a corresponding reduction in cell death. Tumors result from the successive accumulation of mutations in key regulatory target genes over time. During the 1980s, a number of oncogenes were characterized, whereas from the 1990s to the present, the emphasis shifted to tumor suppressor genes (TSGs). It has become clear that oncogenes and tumor suppressor genes function in the same pathways, providing positive and negative growth regulatory activities. The signaling pathways controlled by these genes involve virtually every process in cell biology, including nuclear events, cell cycle, cell death, cytoskeletal, cell membrane, angiogenesis, and cell adhesion effects. Tumor suppressor genes are mutated in hereditary cancer syndromes, as well as somatically in nonhereditary cancers. In their normal state, TSGs control cancer development and progression, as well as contribute to the sensitivity of cancers to a variety of therapeutics. Understanding the classes of TSGs, the biochemical pathways they function in, and how they are regulated provides an essential lesson in cancer biology. We cannot hope to advance our current knowledge and to develop new and more effective therapies without understanding the relevant pathways and how they influence the present approaches to therapy. Moreover, it is important to be able to access the powerful tools now available to discover these genes, as well as their links to cell biology and growth control.

The scope of this two volume work, *Tumor Suppressor Genes*, *Volume 1: Pathways and Isolation Strategies* and *Volume 2: Regulation, Function, and Medicinal Applications*, is broad in the sense that it covers all the known tumor suppressor pathways and provides key information on the road to their discovery, analysis, and uses in cancer therapeutics. The aim of the first volume, *Pathways and Isolation Strategies*, is to educate the reader about known TSGs and the relevance of the biochemical pathways they regulate to human cancer. The reader has an opportunity in Volume 1 to access state-of-the-art protocols that have been successful in the identification of TSGs in the past, and that can be applied to isolate novel TSGs. With a novel TSG in hand, the reader has an opportunity in Volume 2, *Regulation, Function, and Medicinal Applications*, to explore the cell biology and biochemical function of the encoded protein, as well as its physiological role in vivo. Finally, in Volume 2, the reader is exposed to strategies for the use of information on TSGs to develop diagnostic and therapeutic strategies for cancer.

The two volumes of *Tumor Suppressor Genes* bring together many of the world's experts in the identification and characterization of TSGs. The work is intended to become the core reference and compilation of the emerging pathways and the growing number of molecules that suppress cancer. Importantly, it should also serve as a wide-ranging source of protocols useful in understanding and characterizing the function of

TSGs. One of the challenges facing cancer researchers and clinicians is to bring forward and develop active therapeutics. This book, by example, puts forward highly useful paradigms for rational drug design, based on our dramatic new understanding of molecular pathogenesis.

Tumor Suppressor Genes thus lays down a firm and timely foundation for understanding cancer. In this age of expression profiling and proteomics, there has already been revealed a remarkable complexity and interrelatedness of seemingly diverse processes and signal transduction pathways. For the student, this book provides a reference to the basics concerning the identity of the major TSGs and the signaling pathways they use to inhibit tumors. For the investigator, it provides not only a critical update, but also an extremely useful compendium of newly assembled research protocols, including both classical methods and state-of-the-art techniques. For the translational scientist, the book provides fertile ground for the development of therapeutic strategies based on understanding the mechanisms of action and appreciating the existing preclinical data. One of the criticisms of an effort leading to such a book is that the field is moving very quickly and material is likely to be outdated. However, with many of the world's leading experts providing a comprehensive overview of all tumor suppressing pathways, along with their detailed protocols, we believe we have provided an invaluable resource for continued learning and discovery. Finally, *Tumor Suppressor Genes* provides a bridge to those interested in translational research by giving examples of the rationale for many of the most promising manipulations that may lead to novel therapeutic agents.

Tumor Suppressor Genes is targeted at a broad audience including medical and graduate students, postdoctoral fellows, physician scientists, academics, and principal investigators. The text provides critical information to rapidly gain appreciation of important TSGs in human cancer, as well as modern methods for their discovery, analysis, and clinical application. The reader is enabled to learn the background and then access the literature in which studies designed to define their biology and biochemistry have been performed on known TSGs. Details of protocols with examples of their previous uses allow the researcher to apply current technologies to novel or known genes whose role has not yet been defined.

In summary, *Tumor Suppressor Genes* is a comprehensive compilation of the known tumor suppressing pathways, and the key molecular approaches to their discovery, analysis, and clinical applications. It should be of enduring value to students at all levels of experimental biology and medicine, as well as those clinicians who want to better understand the molecular biology of cancer.

Wafik S. El-Deiry, MD, PhD

Contents

Contents of the Companion Volume

Volume 2: Regulation, Function, and Medicinal Applications

Contributors

MAMOUN AHRAM • *Molecular Biosciences, Pacific Northwest National Laboratory, Battelle, Richland, WA*

TERESA ACOSTA ALMEIDA • *Department of Pathology, Institute of Cancer Genetics, Columbia University, New York, NY*

DARIO C. ALTIERI • *Department of Cancer Biology and the Cancer Center, University of Massachusetts Medical School, Worcester, MA*

SALLY A. AMUNDSON • *National Cancer Institute, National Institutes of Health, Bethesda, MD*

ANDREW M. ARSHAM • *Abramson Family Cancer Research Institute and Howard Hughes Medical Institute, University of Pennsylvania School of Medicine, Philadelphia, PA*

JENNIFER J. ASCAÑO • *Division of Gastroenterology, University of Virginia Health System, Charlottesville,* VA

EVANS C. BAILEY • *Department of Biochemistry and Molecular Genetics, The University of Alabama at Birmingham, Birmingham, AL*

ALBERT S. BALDWIN, JR. • *Lineberger Comprehensive Cancer Center, University of North Carolina, Chapel Hill, NC*

VIMLA BAND • *Department of Radiation Oncology, New England Medical Center, Boston, MA*

BRYAN L. BETZ • *Department of Pathology and Laboratory Medicine, University of North Carolina, Chapel Hill, NC*

KAPIL BHALLA • *Department of Interdisciplinary Oncology, Moffitt Cancer Center and Research Institute, Tampa, FL*

SIHAM BIADE • *Department of Pharmacology, Fox Chase Cancer Center, Philadelphia, PA*

MICHAEL BIRRER • *Biomarkers Branch, Division of Clinical Sciences, National Cancer Institute, Rockville, MD*

MIKHAIL V. BLAGOSKLONNY • *Medicine Branch, National Institutes of Health, Bethesda, MD*

KATHY BOON • *Duke University Medical Center, Durham, NC*

RAINER K. BRACHMANN • *Departments of Medicine and Biological Chemistry, University of California at Irvine, Irvine, CA*

JAMES M. BUGNI • *McArdle Laboratory for Cancer Research, University of Wisconsin Medical School, Madison, WI*

JUDITH CAMPISI • *Life Sciences Division, Lawrence Berkeley National Laboratory, Berkeley, CA, and Buck Institute for Age Research, Novato, CA*

XINBIN CHEN • *Department of Cell Biology, University of Alabama at Birmingham, Birmingham, AL*

SANJEEV DAS • *Department of Microbiology and Cell Biology, Indian Institute of Science, Bangalore, India*

PAUL DENT • *Department of Radiation Oncology, Virginia Commonwealth University, Medical College of Virginia, Richmond, VA*

THEODORE L. DEWEESE • *Johns Hopkins Oncology Center, Baltimore, MD*

DAVID T. DICKER • *Departments of Medicine, Genetics, and Pharmacology, Howard Hughes Medical Institute, University of Pennsylvania School of Medicine, Philadelphia, PA*

MICHAEL DOHN • *Department of Cell Biology, University of Alabama at Birmingham, Birmingham, AL*

LAWRENCE A. DONEHOWER • *Division of Molecular Virology, Baylor College of Medicine, Houston, TX*

NORMAN R. DRINKWATER • *McArdle Laboratory for Cancer Research, University of Wisconsin Medical School, Madison, WI*

ANNIE DUTRIAUX • *Department of Molecular Biology, Cell Biology, and Biochemistry, Brown University, Providence, RI*

WAFIK S. EL-DEIRY • *Departments of Medicine, Genetics, and Pharmacology, University of Pennsylvania School of Medicine, Philadelphia, PA*

MICHAEL R. EMMERT-BUCK • *Pathogenetics Unit, Laboratory of Pathology, National Cancer Institute, National Institutes of Health, Bethesda, MD*

GREG H. ENDERS • *Departments of Medicine and Genetics, University of Pennsylvania School of Medicine, Philadelphia, PA*

ERVIN EPSTEIN, JR. • *Department of Dermatology, UCSF Medical Center, San Francisco, CA*

KERI FAIR • *Section of Hematology/Oncology, University of Chicago, Chicago, IL*

ANDREW P. FEINBERG • *Johns Hopkins University School of Medicine, Baltimore, MD*

ARMANDO FELSANI • *CNR, Istituto di Neurobiologia e Medicina Molecolare, Rome, Italy*

PAUL B. FISHER • *Department of Urology and Pathology, College of Physicians and Surgeons, Columbia University, New York, NY*

ALBERT J. FORNACE, JR. • *National Cancer Institute, National Institutes of Health, Bethesda, MD*

ADI F. GAZDAR • *Hamon Cancer Center, UT Southwestern Medical Center, Dallas, TX*

AMATO J. GIACCIA • *Department of Radiation Oncology, Stanford University School of Medicine, Stanford, CA*

DONNA GILFOR • *Department of Radiation Oncology, Medical College of Virginia, Virginia Commonwealth University, Richmond, VA*

ANTONIO GIORDANO • *Sbarro Institute for Cancer Research and Molecular Medicine, Temple University, Philadelphia, PA*

Eyal Gottlieb • *Department of Cancer Biology, Abramson Family Cancer Research Institute, University of Pennsylvania, Philadelphia, PA*

William M. Grady • *Division of Gastroenterology, Vanderbilt University Medical Center, Nashville, TN*

Susan F. Grammer • *Biotechwrite: Biomedical and Science Communications, Kalamazoo, MI*

Stephen Grant • *Department of Medicine/Hematology, Medical College of Virginia, Virginia Commonwealth University, Richmond, VA*

Andrei V. Gudkov • *Department of Molecular Biology, Lerner Research Institute, Cleveland Clinic Foundation, Cleveland, OH*

Ruth Hemmer • *Department of Molecular Biology, Cell Biology, and Biochemistry, Brown University, Providence, RI*

Meenhard Herlyn • *The Wistar Institute, Philadelphia, PA*

Tyler Jacks • *Center for Cancer Research and Howard Hughes Medical Institute, Massachusetts Institute of Technology, Cambridge, MA*

Lesley J. Jardine • *Department of Cancer Cell Biology, Harvard School of Public Health, Boston, MA*

Stephen N. Jones • *Department of Cell Biology, University of Massachusetts Medical Center, Worcester, MA*

Jeremy Juang • *Department of Pathology, The Johns Hopkins Medical Institutions, Baltimore, MD*

Shrihari Kadkol • *Department of Pathology, The Johns Hopkins Medical Institutions, Baltimore, MD*

William G. Kaelin • *Dana-Farber Cancer Institute, Boston, MA*

Kum Kum Khanna • *Queensland Institute of Medical Research, Brisbane, Queensland, Australia*

Adi Kimchi • *Department of Molecular Genetics and Virology, the Weizmann Institute of Science, Rehovot, Israel*

Takahiko Kobayashi • *Hokkaido University Medical Hospital, Sapporo, Japan*

Elena Komarova • *Department of Molecular Biology, Lerner Research Institute, Cleveland Clinic Foundation, Cleveland, OH*

Smita Lakhotia • *Department of Microbiology and Cell Biology, Indian Institute of Science, Bangalore, India*

Fredrick S. Leach • *Departments of Urology and Molecular and Human Genetics, Baylor College of Medicine, Houston, TX*

Michelle M. Le Beau • *Section of Hematology/Oncology, University of Chicago, Chicago, IL*

Cristina T. Lewis • *Pfizer Global Research and Development, San Diego, CA*

Iuri D. Louro • *Division of Hematology/Oncology, Department of Medicine, The University of Alabama at Birmingham, Birmingham, AL*

Timothy K. MacLachlan • *CuraGen Corporation, Branford, CT*

Lee V. Madrid • *Lineberger Comprehensive Cancer Center, University of North Carolina, Chapel Hill, NC*

CARL G. MAKI • *Department of Cancer Cell Biology, Harvard School of Public Health, Boston, MA*

JAMES J. MANFREDI • *Derald H. Ruttenberg Cancer Center, Mount Sinai School of Medicine, New York, NY*

FRANCESCA M. MARASSI • *The Burnham Institute, La Jolla, CA*

SANFORD D. MARKOWITZ • *Howard Hughes Medical Institute, and Department of Medicine and Ireland Cancer Center, Case Western Reserve University and University Hospitals of Cleveland, Cleveland, OH*

MICHAEL R. MATTERN • *Department of Oncology Research, GlaxoSmithKline, King of Prussia, PA*

MATTHEW MAURER • *Derald H. Ruttenberg Cancer Center, Mount Sinai School of Medicine, New York, NY*

E. ROBERT MCDONALD III • *Departments of Medicine, Genetics, and Pharmacology, Howard Hughes Medical Institute, University of Pennsylvania School of Medicine, Philadelphia, PA*

MARGARET E. MCLAUGHLIN • *Center for Cancer Research and Howard Hughes Medical Institute, Massachusetts Institute of Technology, Cambridge, MA;*
Department of Pathology, Brigham and Women's Hospital, Boston, MA

PATRICE J. MORIN • *Laboratory of Biological Chemistry, National Institute on Aging, National Institutes of Health, Baltimore, MD*

MAUREEN E. MURPHY • *Department of Pharmacology, The Fox Chase Cancer Center, Philadelphia, PA*

KIICHIRO NAKAJIMA • *Peptide Institute Co., Ltd., Osaka, Japan*

MARK A. NELSON • *Department of Pathology, College of Medicine, University of Arizona, Tucson, AZ*

WILLIAM G. NELSON • *Johns Hopkins Oncology Center, Baltimore, MD*

RAMADEVI NIMMANAPALLI • *Department of Interdisciplinary Oncology, Moffitt Cancer Center and Research Institute, Tampa, FL*

SUSAN NOZELL • *Department of Cell Biology, University of Alabama at Birmingham, Birmingham, AL*

PATRICK M. O'CONNOR • *Pfizer Global Research and Development, San Diego, CA*

MICHAEL OHH • *Dana-Farber Cancer Institute, Boston, MA*

JUN-ICHI OKANO • *Division of Gastroenterology, Department of Medicine, University of Pennsylvania School of Medicine, Philadelphia, PA*

MICHAEL S. O'REILLY • *Department of Radiation Oncology and Cancer Biology, M. D. Anderson Cancer Center, University of Texas, Houston, TX*

MARCO G. PAGGI • *Center for Experimental Research, Regina Elena Cancer Institute, Rome, Italy*

RAYMOND A. PAGLIARINI • *Boyer Center for Molecular Medicine, Yale University School of Medicine, New Haven, CT*

NICKOLAS PAPADOPOULOS • *Department of Pathology, Institute of Cancer Genetics, Columbia University, New York, NY; Present address: GMP Genetics, Waltham, MA*

RAMON PARSONS • *Department of Pathology, Columbia University, New York, NY*
STEVEN M. POWELL • *Division of Gastroenterology and Hepatology, University of Virginia Health System, Charlottesville, VA*
NADINE PUGET • *Beth Israel Deaconess Medical Center and Harvard Medical School, Boston, MA*
HUA QIAN • *Department of Surgery, Shanghai Second Medical University, Shangai, P. R. China*
LIANG QIAO • *Department of Radiation Oncology, Medical College of Virginia, Virginia Commonwealth University, Richmond, VA*
ANTHONY G. QUINN • *Experimental Medicine, AstraZeneca R&D Charnwood, Leicestershire, UK*
ANA T. QUIÑONES • *Yale University School of Medicine, New Haven, CT*
TIMOTHY R. REBBECK • *Department of Biostatistics and Epidemiology, University of Pennsylvania School of Medicine, Philadelphia, PA*
MARK REDSTON • *Department of Pathology, Brigham and Women's Hospital, Boston, MA*
M. STACEY RICCI • *Departments of Medicine, Genetics, and Pharmacology, Howard Hughes Medical Institute, University of Pennsylvania School of Medicine, Philadelphia, PA*
GREGORY J. RIGGINS • *Duke University Medical Center, Durham, NC*
GAVIN P. ROBERTSON • *Department of Pharmacology, Penn State College of Medicine, Hershey, PA*
IGOR B. RONINSON • *Department of Molecular Genetics, University of Illinois, Chicago, IL*
JACK A. ROTH • *Department of Thoracic and Cardiovascular Surgery, University of Texas M. D. Anderson Cancer Center, Houston, TX*
J. MICHAEL RUPPERT • *Department of Medicine, University of Alabama at Birmingham, Birmingham, AL*
ANIL K. RUSTGI • *Division of Gastroenterology, Department of Medicine, University of Pennsylvania School of Medicine, Philadelphia, PA*
LINDA SARGENT • *Health Effects Laboratory Division, Toxicology and Molecular Biology Branch, Genetic Susceptibility Team, NIOSH, Morgantown, WV*
KAPAETTU SATYAMOORTHY • *The Wistar Institute, Philadelphia, PA*
RALPH SCULLY • *Beth Israel Deaconess Medical Center, and Harvard Medical School, Boston, MA*
JOHN M. SEDIVY • *Department of Molecular Biology, Cell Biology, and Biochemistry, Brown University, Providence, RI*
JULIAN A. SIMON • *Program in Molecular Pharmacology, Fred Hutchinson Cancer Research Center, Seattle, WA*
M. CELESTE SIMON • *Abramson Family Cancer Research Institute and Howard Hughes Medical Institute, University of Pennsylvania School of Medicine, Philadelphia, PA*
LAURA SIMPSON • *Institute of Cancer Genetics, College of Physicians and Surgeons, Columbia University, New York, NY*

LORRAINE SNYDER • *Division of Gastroenterology, Department of Medicine, University of Pennsylvania School of Medicine, Philadelphia, PA*
KUMARAVEL SOMASUNDARAM • *Department of Microbiology and Cell Biology, Indian Institute of Science, Bangalore, India*
LI-KUO SU • *Department of Molecular and Cellular Oncology, M.D. Anderson Cancer Center, The University of Texas, Houston, TX*
KATSUYUKI TAMAI • *Cyclex Co., Ltd., Nagano, Japan*
YOICHI TAYA • *National Cancer Center Research Institute, Tokyo, Japan*
CRAIG B. THOMPSON • *Department of Cancer Biology, Abramson Family Cancer Research Institute, University of Pennsylvania, Philadelphia, PA*
EDWARD C. THORNBORROW • *Derald H. Ruttenberg Cancer Center, Mount Sinai School of Medicine, New York, NY*
MEREDITH UNGER • *Abramson Family Cancer Research Institute, University of Pennsylvania Cancer Center, Philadelphia PA*
JENNY VARLEY • *CR-UK Department of Cancer Genetics, Paterson Institute of Cancer Research, Manchester, UK*
ARVIND K. VIRMANI • *Hamon Cancer Center, UT Southwestern Medical Center, Dallas, TX*
NARENDRA WAJAPEYEE • *Department of Microbiology and Cell Biology, Indian Institute of Science, Bangalore, India*
TING WANG • *Department of Genetics, Washington University School of Medicine, St. Louis, MO*
VALERIE M. WEAVER • *Department of Pathology, The Institute for Medicine and Engineering, University of Pennsylvania, Philadelphia, PA*
ASHANI T. WEERARATNA • *Cancer Genetics Branch, National Human Genome Research Institute, National Institutes of Health, Bethesda, MD*
WENYI WEI • *Department of Molecular Biology, Cell Biology, and Biochemistry, Brown University, Providence, RI*
BERNARD E. WEISSMAN • *Department of Pathology and Laboratory Medicine, University of North Carolina, Chapel Hill, NC*
BRYAN R. G. WILLIAMS • *Department of Cancer Biology, Lerner Research Institute, Cleveland, OH*
AMY WILLIS • *Department of Cell Biology, University of Alabama at Birmingham, Birmingham, AL*
GEN SHENG WU • *Department of Pathology, Karmanos Cancer Institute, Wayne State University School of Medicine, Detroit, MI*
TZYY-CHOOU WU • *Department of Pathology, The Johns Hopkins Medical Institutions, Baltimore, MD*
TIAN XU • *Boyer Center for Molecular Medicine, Yale University School of Medicine, New Haven, CT*
TIMOTHY J. YEN • *Fox Chase Cancer Center, Philadelphia, PA*
KUMIKO YOSHIZAWA-KUMAGAYE • *Peptide Institute, Co., Ltd., Osaka, Japan*

Zhong Yun • *Division of Radiation and Cancer Biology, Department of Radiation Oncology, Stanford University School of Medicine, Stanford, CA*
Bin-Bing S. Zhou • *Incyte Genomics, Newark, DE*

I

Known Tumor Suppressor Genes and Pathways

1

Growth Control by the Retinoblastoma Gene Family

Marco G. Paggi, Armando Felsani, and Antonio Giordano

1. Introduction

1.1. The Retinoblastoma Gene Family

The Retinoblastoma family consists of three genes, *RB*, *p107,* and *Rb2/p130*, all fundamental in the control of important cellular phenomena, such as cell cycle, differentiation, and apoptosis. The "founder" and the most investigated gene of the family is *RB*, which is considered the prototype for the tumor suppressor genes ***(1,2)***. The other two genes, *p107* and *Rb2/p130*, and the proteins they code for, p107 and pRb2/p130, respectively, clearly reflect a high degree of structural and functional similarity to the *RB* gene product, pRb ***(3,4)***. The *RB* family proteins were disclosed initially by investigators working on viral oncoproteins. In particular, a set of proteins associated with the Adenovirus 5 E1A oncoprotein was identified, and the bands representing the most abundant ones were named p60, p105, p107, p130, and p300, in keeping with their apparent molecular mass, as determined by SDS-PAGE ***(5)***. The subsequent characterization of these proteins identified p105 as the product of the *RB* gene ***(6)***. Later, genes encoding p107 ***(7,8)*** and pRb2/p130 ***(9–11)*** were cloned using different strategies.

In essence, the structure of the three RB family gene products, pRb, p107, and pRb2/p130, consist of (a) an N-terminal portion, (b) a "pocket structure" subdivided into domain A, spacer, and domain B, and (c) a C-terminal portion, a.k.a. domain C ***(3)***. The pocket functional domains A and B are the most conserved among the three RB family members and are responsible for most of the interactions involving either some endogenous proteins, such as those of the E2F family ***(3,4,12–14)***, or viral oncoproteins, i.e., Adenovirus 5 E1A, SV40 large T, and Human Papillomavirus E7 ***(6,7,15–20)***.

A common relevant biological feature shared by the three members of this family is the ability to control the cell cycle ***(8,21–23)***. In fact, they negatively modulate the transition between the G1 and S phases, utilizing mechanisms mostly related to inactivation of transcription factors, such as those of the E2F family, that promote the cell's entrance into the S phase ***(3,4,12–14)***. The role of the RB family proteins as key negative cell cycle regulators is mainly modulated by posttranslational modifications, the most important one being phosphorylation. In fact, the *RB* gene product is a well-known substrate for either kinase or phosphatase activity, thus undergoing extensive and regular changes in its

From: *Methods in Molecular Biology, Vol. 222: Tumor Suppressor Genes: Pathways and Isolation Strategies*
Edited by: Wafik S. El-Deiry © Humana Press Inc., Totowa, NJ

phosphorylation status throughout the cell cycle. In asynchronous cells, pRb is present at various degrees of phosphorylation, which is well depicted as a microheterogeneous pattern typically evident in SDS-PAGE analysis: the more the molecule is phosphorylated, the more slowly it migrates. For this reason the apparent molecular mass of pRb ranges between 105 and 115 kDa, when estimated by SDS-PAGE ***(6,24–25)***. Canonically, pRb is hyperphosphorylated (inactive) in proliferating cells, while it is underphosphorylated (active) in quiescent or differentiating cells. In this second form, however, it shows enhanced affinity for the nuclear compartment ***(26)***. At a single cell level, pRb is underphosphorylated in G0 and early G1. In late G1, the protein becomes phosphorylated at the restriction point and phosphorylation increases in S phase and at the G2–M transition. The protein is found again to be underphosphorylated when the cell has completed the mitotic process ***(25,27–29)***. With a closer look at the phases of pRb phosphorylation during the cell cycle, we can argue that D-type cyclin-dependent kinases might be responsible for early G1, cyclin E/cdk2 for mid/late G1, and cyclin A and cyclin B/Cdc2 for G2/M phosphorylation of pRb. After the cycle has been completed, the cells that have been generated by the mitotic process again display underphosphorylated pRb, because of the specific phosphatase activities. The stringent timing of the activity of the cyclin/kinase complexes on pRb is guaranteed further by the cdk inhibitors (*see* **ref. *30*** for a review). Also, p107 undergoes similar modifications during the cell cycle, but its pattern in SDS-PAGE analysis usually appears less heterogeneous. The major complex responsible for its phosphorylation is the cyclin D1/cdk4 complex ***(31–33)***. As far as pRb2/p130 is concerned, it displays evident cell cycle-related changes in phosphorylation, coupled with an extensive microheterogeneity in SDS-PAGE migration pattern ***(34,35)***. It has been found associated with cyclin A and cyclin E ***(10)*** and with cdk2 ***(11)***. From a functional point of view, cyclins A, D-type and E overexpression rescue pRb2/p130-mediated growth arrest in SAOS-2 human osteosarcoma cells ***(36)***.

During the last decade, evidence has been gathered, indicating that, in addition to the cell cycle, the *RB* family regulates a wide spectrum of complex biological phenomena, such as differentiation, embryonic development, and apoptosis (*see* **refs. *37–40*** and refs. therein).

1.2. Retinoblastoma Proteins and Cancer

The crucial role of the "RB pathway" and of all three RB proteins in cell cycle regulation ***(41)*** is profoundly linked to cancer transformation and/or progression. For several decades, cancer development has been ideally related to loss of control in the cellular processes that regulate the cell cycle.

RB complies with all the prerequisites to be considered a bona-fide tumor suppressor gene. In fact, its mutations or deletions are shared by several malignancies and, in addition, exogenous expression of wild-type *RB* in *RB*-defective cancer cells promptly reverts main characteristics of the neoplastic phenotype (*see* **refs. *3,13,14*** and refs. therein).

In light of the significant structural and functional similarities among the three members of the *RB* family, one can argue that also *p107* or *Rb2/p130*, known to be potent cell cycle inhibitors, could also act as tumor suppressors.

Cytogenetically, *Rb2/p130* maps to the region 16q12.2-13, an area repeatedly altered in human cancers ***(10,11,42)***, whereas *p107* maps to the human chromosome region 20q11.2, a locus not frequently found to be involved in human neoplasms ***(7)***. To date,

Table 1
Cancer Types in Which Different Impairments of *Rb2/p130* or *p107* Genes Have Been Found

Type of cancer	Kind of alteration
Rb2/p130 LOH, downregulation, mutation, or functional inactivation	
Breast cancer	LOH of region 16q12.2 ***(42)***
Ovarian carcinoma	LOH of region 16q12.2 ***(42)***
Prostatic carcinoma	LOH of region 16q12.2 ***(42)***
Small-cell lung cancer	Low to undetectable expression of gene product ***(59)***; point mutation ***(60)***
Endometrial cancer	Low expression of gene product ***(61)***
Choroidal melanoma	Low expression of gene product ***(62)***
Non-Hodgkin lymphoma	Low expression of gene product ***(63)***
Vulvar cancer	Low expression of gene product ***(64)***
Mesothelioma	SV40-mediated functional inactivation ***(65)***
Burkitt lymphomas EBV+	Point mutations at NLS in 56% of cases ***(63)***
Non-small-cell lung cancer	Point mutation ***(66)***
Nasopharyngeal carcinoma	Point mutation ***(67)***
p107 mutation	
B-cell lymphoma	Intragenic deletion ***(68)***

there is only one recent report of deletion or functional inactivation for *p107* in human tumors, while mutations of the *Rb2/p130* gene have been often detected in human cancers (*see* **Table 1**).

The three RB proteins are localized mainly in the nuclear compartment of the cell ***(7,24,34)***. For pRb and pRb2/p130, extranuclear localization has been associated with genetic mutations involving the nuclear localization signal (NLS) of the protein ***(43–45)***. Such mutations do not make it possible to reach an acceptable intranuclear localization, strongly impairing the ability of these proteins to interfere with nuclear transcription factors.

This chapter focuses on the current methods employed to assess the phosphorylation status of the RB family proteins. This approach is currently used to estimate the tendency of a cellular population to proceed in the cell cycle. In addition, a coprecipitation assay, able to validate RB protein functional status, is also described.

2. Materials

2.1. Instrumentation

1. A large gel electrophoresis apparatus, such as the Bio-Rad Protean II xi (cat. no. 165-1811), complete with glass plates, spacers and combs, able to accommodate gels with a size of 16 × 16 cm (W × L).
2. A large cell for protein electrophoretic transfer, such as the Bio-Rad Trans-Blot cell (cat. no. 170-3939).
3. A small gel electrophoresis apparatus, such as the Bio-Rad Mini-Protean 3 (cat. no. 165-3301), complete with glass plates, spacers, and combs, for minigels with a size of 8 × 7 cm (W × L).

4. A small cell for protein electrophoretic transfer, such as the Bio-Rad Mini Trans-Blot transfer cell (cat. no. 170-3930).
5. A power supply for gel electrophoresis (500-V/0.5-A output) and one for electrophoretic transfer (200-V/2-A output).

2.2. Reagents

2.2.1. Western Blot Sample Preparation

1. *Lysis buffer* for detection of retinoblastoma proteins by Western blot: 50 m*M* Tris-HCl, pH 7.40, 5 m*M* EDTA, 250 m*M* NaCl, 50 m*M* NaF, 0.1% Triton X-100, 0.1 m*M* Na_3VO_4. Store at 4°C for up to 6 mo. Immediately prior to use, add: 1 m*M* phenylmethylsulfonyl fluoride (PMSF) (*see* **Note 1**) and 10 μg/mL leupeptin.
2. *SDS lysis buffer:* 125 m*M* Tris-HCl (pH 6.8), 4 % SDS, 20% glycerol.
3. β-Mercaptoethanol (*see* **Note 1**), Sigma cat. no. M-3148.
4. *Laemmli sample buffer* 1× : 62.5 m*M* Tris-HCl (pH 6.8), 2% SDS, 10% glycerol, 5% β-mercaptoethanol.
5. Bromophenol blue, Bio-Rad cat. no. 161-0404 saturated and filtered solution. Add 10 μL for each milliliter of sample buffer.
6. BCA-200 protein assay kit, Pierce cat. no. 23226.

2.2.2. Immunoprecipitations/Co-immunoprecipitations

1. *IPPT lysis buffer*: 50 m*M* Tris-Cl, pH 7.40, 5 m*M* EDTA, 150 m*M* NaCl, 50 m*M* NaF, 1% Nonidet P-40, 0.1 m*M* Na_3VO_4. Store at 4°C for up to 6 mo. Immediately prior to use, add: 1 m*M* PMSF (*see* **Note 1**) and 10 μg/mL leupeptin.
2. *Immobilized protein G*: when dealing with mouse antibodies (from Pierce, Sigma, or Pharmacia).
3. *Immobilized protein A:* when dealing with rabbit antibodies (from Pierce, Sigma, or Pharmacia).

2.2.3. Phosphatase Treatment

1. *MES* buffer: 100 m*M* MES (2-[N-morpholino] ethanesulfonic acid), pH 6.0, containing 1 m*M* of PMSF (*see* **Note 1**).
2. *Potato acid phosphatase preparation:* An aliquot of potato acid phosphatase in ammonium sulfate suspension (Roche, cat. no. 108 197) is pelletted by centrifugation and resuspended in an equal volume of the MES buffer.

2.2.4. SDS-PAGE

1. Acrylamide 30% stock solution (*see* **Note 1**) (29:1 acrylamide–bisacrylamide, Bio-Rad cat. no. 161-0124)
2. Resolving gel buffer: Tris-HCl 1.5 *M*, pH 8.8.
3. Stacking gel buffer: Tris-HCl 0.5 *M*, pH 6.8.
4. TEMED (*see* **Note 1**) (Bio-Rad cat. no. 161-0801).
5. Ammonium persulfate (Bio-Rad cat. no. 161-0700), 10% solution.
6. SDS-PAGE running buffer 1×: 25 m*M* Tris, glycine 192 m*M*, SDS 0.1%, pH 8.3.

2.2.5. Immunoblotting and Detection

1. 3MM paper, Whatman cat. no. 3030917.
2. *Transfer membrane*: polyvinyldifluorene (PVDF) membrane roll, Immobilon, Millipore cat. no. IPVH00010.

3. *PVDF transfer buffer:* 10 m*M* CAPS (3-[cyclohexilamino] 1-propanesulfonic acid), Sigma cat. no. C-2632, pH 11.0, 10 *M* NaOH, 20% methanol. This buffer cannot be reused.
4. Nonfat dry milk (NFDM): Carnation, or Bio-Rad cat. no. 170-6404.
5. *TBS* 10×: 0.2 *M* Tris-HCl, pH 7.60, 1.37 *M* NaCl.
6. *TBS-T*: TBS 1 × plus 0.5% Tween-20.
7. Mouse monoclonal anti-pRB antibody, clone G3-245, PharMingen cat. no. 14001A.
8. Rabbit polyclonal anti-p107 antibody (SC-318, Santa Cruz Biotechnology, Santa Cruz, CA).
9. Rabbit polyclonal anti-pRb2/p130 antibody (SC-317, Santa Cruz Biotechnology, Santa Cruz, CA).
10. Purified rabbit IgG (Pierce, Sigma, Bio-Rad).
11. Purified mouse IgG (Pierce, Sigma, Bio-Rad).
12. Anti-mouse peroxidase-conjugated antibody, Bio-Rad cat. no. 170-6516.
13. Anti-rabbit peroxidase-conjugated antibody, Bio-Rad cat. no. 170-6515.
14. ECL chemiluminescence kit, Amersham cat. no. RPN-2106.
15. Biomax MR film, 18 × 24 cm, Kodak.

2.2.6. In Vitro Transcription and Translation of ^{35}S-Labeled E1A Protein

1. E1A 12S plasmid containing the HindIII–BamHI fragment of pMTEB12S ***(46)*** cloned in the pSP64 vector cut with the same enzymes (courtesy of Dr. E. Moran, Cold Spring Harbor Laboratory, Cold Spring Harbor, NY).
2. EcoRI restriction enzyme, with appropriate buffer.
3. Diethylpyrocarbonate (DEPC) (*see* **Note 1**), Sigma cat. no. D 5758.
4. *DEPC treatment:* In order to destroy any nuclease activity, water and solutions are treated *in glass bottles* with 0.1% DEPC, shaken vigorously, and incubated with the cap not completely sealed overnight under a fume hood (*see* **Note 1**). Then DEPC is decomposed into ethanol and CO_2 by autoclaving the solutions for 15 min (liquid cycle).
5. Bidistilled H_2O, DEPC-treated.
6. 3 *M* NaAcetate, DEPC-treated.
7. *Transcription buffer* 5×*:* 200 m*M* Tris-HCl pH 7.5, 30 m*M* MgCl2, 10 m*M* spermidine, 50 m*M* NaCl.
8. *rNTP mix*: A solution of 2.5 mmol each of ATP, GTP, CTP, and UTP, pH 7.0, in nuclease-free DEPC-treated water.
9. SP6 RNA polymerase, Promega cat. no. P1081.
10. RNase-free DNase RQ1, Promega cat. no. M6101.
11. Dithiothreitol (DTT) 100 m*M*, Sigma cat. no. D 9779.
12. TE solution, Tris-HCl 10 m*M* pH 8.0, EDTA 1 m*M*.
13. Chloroform:isoamyl alcohol mixture (20:1 v/v) (*see* **Note 1**).
14. TE saturated bidistilled phenol, pH 8 (*see* **Note 1**), for DNA extraction.
15. Water-saturated bidistilled phenol (*see* **Note 1**), for RNA extraction.
16. Agarose, gel electrophoresis grade.
17. TBE (10×) electrophoresis buffer: 90 m*M* Tris-borate, 2 m*M* EDTA.
18. Amino acid mixture (1 m*M* each) (minus methionine) (Amersham or Promega).
19. L-[^{35}S]-Methionine (*see* **Note 1)**, in-vitro translation grade, Amersham cat. no. AG1594 or NEN cat. no. NEG009T, with a specific activity of about 1000 Ci/mmol (40 TBq/mmol).
20. Rabbit reticulocyte lysate system, nuclease treated, Promega cat. no. L4960.
21. Ribonuclease inhibitor, recombinant RNAsin, Promega cat. no. N2511.
22. 96% ethanol.
23. 70% ethanol.
24. E1A-specific M73 mouse monoclonal antibody (Oncogene Research Products, cat. no. DP11L-100UG).

Table 2
Antibodies Recognizing the Different Products of the *Rb* Gene Family

	Western blot	Immunoprecipitation
pRb	G3-245, mouse monoclonal 0.5–1 μg/mL	C36, mouse monoclonal 1 μg/sample
pRb2/p130	SC-317, rabbit polyclonal 0.5 μg/mL	SC-317, rabbit polyclonal 1 μg/sample
p107	SC-318 rabbit polyclonal 0.25–0.5 μg/mL	SC-318 rabbit polyclonal 1 μg/sample

3. Methods

Due to their crucial role in modulating key cellular processes, the RB family proteins have been extensively investigated in several cellular models and tissues. As affirmed before, their degree of phosphorylation is a valid hallmark for their functional status, given that their phosphorylated forms are the inactive ones in inhibiting the cell cycle; consequently their functional status is at least as important as their degree of expression (quantification). As specified under **Subheading 1.**, the easiest method for assessing RB family protein phosphorylation is to check their mobility pattern in SDS-PAGE. The subsequent Western blot analysis can thus provide information about two fundamental issues: amount and phosphorylation of the RB family proteins. This technique, however, is not expected to give a reliable answer to another key question: "Are these proteins functional?" It is possible, in fact, that minor structural mutations can totally impair protein function. In the case of the RB family proteins, major functional impairments are due to mutations, often point mutations, in the A/B pocket region, or in the C-terminal moiety of these proteins. A straightforward correlation has been often demonstrated between RB protein function and their ability to interact physically with key cellular regulators ***(3,47)***. For this reason, we developed a method in which the ability of pRb to bind to the Adenovirus E1A oncoprotein was representative of the functional integrity of the protein ***(48)***. This method has been also extended to p107 and pRb2/p130, and has been shown to be as sensitive to discriminate the pRb impaired function also in the H209 human lung carcinoma cell line ***(49)***, where the point mutation (Cys 706 to Phe) makes pRb unable to bind to E1A ***(48)***. Obviously, a negative coprecipitation test can be considered necessary and sufficient to state that a determined cell type has a nonfunctional pRb molecule, or does not have one at all. On the other hand, a positive test is only necessary, but not sufficient, to identify a wild-type pRb.

Immunoprecipitation of the RB family proteins is also discussed. This technique is able to provide a high purification yield for any protein for which a good antibody is obtainable. It is essentially used for biochemical studies, such as the analysis of posttranslational modifications. Suitable commercial antibodies for each of the three RB family members are indicated (**Table 2**). They have been successfully employed, for example, to immunoprecipitate pRb or pRb2/p130 and to demonstrate that a specific enzymatic activity (phosphatase) was able to modify the electrophoretic mobility of the purified

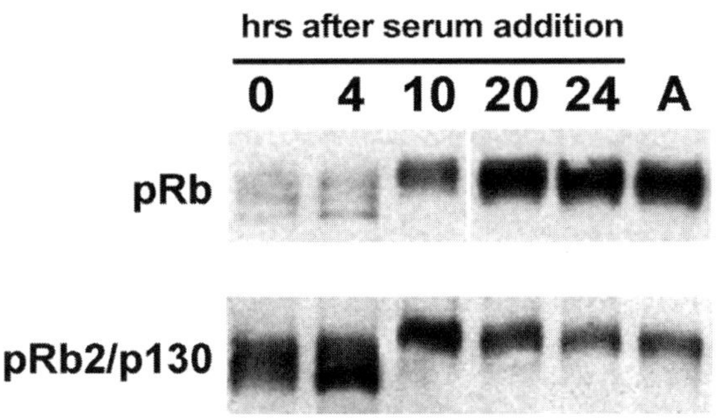

Fig. 1. Western blot showing phosphorylation kinetics of pRb and pRb2/p130 in T98G human malignant glioma cell line. Cells were arrested in G0 by serum deprivation for 48 h, then serum was added and the synchronized cells were allowed to proceed through the cell cycle and collected after 4, 10, 20, and 24 h. A, asynchronous cell population.

protein, thus attributing the microheterogeneous electrophoretic pattern to phosphorylation ***(25,34)***.

3.1. Quantitative and Phosphorylation Analysis

3.1.1. Cultured Cells and Tumor Samples

Mammalian cells are cultured according to standard procedures. In order to display a complete phosphorylation pattern, cells should not be more than 60–70% confluent. Alternatively, phosphorylation kinetics can be easily shown in selected cell types by arresting cells in G0, for example by serum deprivation for 48 h, and then allowing synchronized cells to proceed through the cell cycle by serum addition, and collecting them at specific time points (**Fig. 1**).

While cultured cells are a homogeneous population and represent the optimal condition for the study of RB family protein patterns, these biochemical approaches can also be extended to human tumor samples selected for a very low contamination of nontumor cells. For example, acute myeloid leukemia cells are easily prepared from peripheral blood by Lymphoprep™ (Nycomed Pharma AS, Norway), a one-step technique, to >80% purity ***(50)***. In addition, these samples can also be stored in liquid nitrogen using standard cell cryopreservation procedures.

3.1.2. Cell Lysis

1. Collect cells ($1–10 \times 10^6$), wash them in PBS, pellet in microfuge, and completely discard the supernatant. At this step, you can store dry cell pellets at –80°C.
2. Add 25–100 µL of lysis buffer to the cell pellet. Resuspend cells by gentle pipetting and incubate on ice for 30 min.
3. Centrifuge in a microfuge at 4°C for 10 min at 10,000 *g*. Recover the supernatant (cell lysate).
4. Using 5 µL of cell lysate, measure the total protein content of the samples, by means of the BCA-200 protein assay kit, according to the manufacturer's instructions. Adjust samples to equal amounts of total protein.
4. Add to the lysate an equal volume of 2 × Laemmli sample buffer. Heat at 95°C for 5 min to completely denature the proteins.
5. Load the sample on the gel made as indicated under **Subheading 3.1.4.**

Alternatively, to exhaustively extract the RB family proteins from terminally differentiated cells, where they show enhanced affinity for the nuclear compartment ***(26,51)***, a stronger extraction procedure is recommended, following **step 1** described above.

2. Add 25–100 μL of 2 × SDS sample buffer, resuspend cells by pipetting, then incubate at 95°C for 5 min to completely lyse the cell pellet.
3. Centrifuge in a microfuge at room temperature for 10 min at 10,000 *g* and discard the pellet.
4. Measure the total protein content of the samples, using 5 μL of cell lysate and the BCA-200 protein assay kit, according the manufacturer's instructions. Adjust samples to equal amounts of total protein.
5. Add to the samples 10% β-mercaptoethanol (final concentration). Heat at 95°C for 5 min to completely denature the proteins.
6. Load the sample on the gel made as indicated under **Subheading 3.1.5.**

3.1.3. Immunoprecipitation

1. Collect cells (1–5 × 10^6), wash them in PBS, pellet in microfuge, and completely discard the supernatant.
2. Add 0.5 ml of IPPT lysis buffer to the cell pellet, resuspend cells, and incubate on ice for 30 min to achieve a complete lysis.
3. Spin down in microfuge at the maximum speed for 10 min.
4. Transfer the supernatant to a fresh tube, add 5 μg/pellet of normal IgG (mouse or rabbit IgG, when mouse or rabbit antibodies, respectively, are involved in the immunoprecipitation as primary antibodies) and incubate for 30 min in ice.
5. Using a cut micropipet tip, add to the IgG-treated cell lysate 20 μl of immobilized protein G, or immobilized protein A for mouse or rabbit primary antibodies, respectively, and incubate for 30 min at 4°C.
6. Spin down for 10 s. Carefully transfer the supernatant (precleared lysate) without taking any immobilized protein G or A particle (spin down again and repeat, if necessary).
7. Add to the precleared lysate 1 μg of the relevant antibody (see Table 2), and incubate with rocking for 1 h at 4°C.
8. Using a cut micropipet tip, add 20 μL of immobilized protein G or immobilized protein A for mouse or rabbit antibodies, respectively, and rock for 1 h at 4°C.
9. Spin down 5–10 s in microfuge and discard supernatant.
10. Resuspend the pellet in 1 mL of IPPT lysis buffer, and wash, vortexing briefly. Spin down 5–10 s in microfuge and discard the supernatant.
11. Repeat step 10 twice.
12. Add to each dry pellet 30–40 μL of Laemmli's sample buffer. Heat samples at 95°C (*see* **Note 2**) for 5 min, add 0.5 μL of bromophenol blue saturated solution and spin them down. The samples are now ready for SDS/PAGE.

3.1.4. Phosphatase Treatment

In order to clearly identify the alterations induced by phosphorylation in the pattern of migration of the RB family protein members, the immunoprecipitated samples can be subjected to a dephosphorylation step prior to the electrophoretic analysis.

1. The immunocomplexes containing the RB family protein members, bound to immobilized protein G/A and washed, as indicated in the previous paragraph (end of step 11), are equilibrated in MES buffer.

2. Each sample is resuspended in 60 µL of MES buffer and split in two aliquots, to give a control and a phosphatase treated experimental sample.
3. Add 0.5 units of potato acid phosphatase prepared as indicated under **Subheading 2.**, to each experimental sample.
4. Incubate the control and the experimental sample for 15 min at 37°C (*see* **Note 3**).
5. Stop the reaction by adding 1 volume of sample buffer 2 × to each sample, heat samples at 95°C for 5 min, add 0.5 mL of bromophenol blue saturated solution and spin them down. The samples are now ready for SDS/PAGE.

3.1.5. Electrophoresis Analysis on SDS-PAGE

The gel system used has been described by Laemmli ***(52)***. To resolve the RB family proteins a gel with an acrylamide concentration of 6.5% is adequate.

3.1.5.1. Preparation of the Resolving Gel

In a 50-mL disposable tube, add 4.3 mL acrylamide 30% stock solution, 10.3 mL H_2O, 5 mL 1.5 *M* Tris-HCl pH 8.8, 0.2 mL 10% SDS. Mix gently, avoiding the generation of air bubbles. This amount of resolving gel mixture is sufficient to prepare one 16 × 16-cm gel, 0.75 cm thick. Add 0.2 mL of 10% ammonium persulfate and 8 µL of TEMED just prior to pouring the mixture, since these reagents promote and catalyze the formation of the gel. Mix again and pour the resolving gel mixture into the assembled gel plates, leaving a 4.0-cm space at the top for the stacking gel. Gently overlay the gel surface with a sufficient amount of water-saturated 2-butanol, and allow the gel to polymerize in the appropriate stand for at least 30 min, in order to obtain a perfectly horizontal gel surface. Usually, an evident change in the diffraction at the interface acrylamide/2-butanol indicates the completion of the polymerization. Then remove the 2-butanol overlay and rinse the gel surface with distilled water. Drain the residual water well, and then fill the upper part of the plates with the freshly prepared staking gel mixture.

3.1.5.2. Preparation of the Stacking Gel

In a 10-mL tube, add 1.00 mL of acrylamide 30% stock solution, 4.10 mL of H_2O, 0.75 mL of 0.5 *M* Tris-HCl pH 6.8, 0.06 mL of 10% SDS, 0.06 ml of ammonium persulfate, and 6 µL of TEMED. Mix gently, pour the stacking gel solution to the top of the plates, and insert the comb immediately, avoiding the formation of air bubbles.

After the stacking gel has polymerized, remove the comb and rinse the wells with water gently to remove the unpolymerized acrylamide. The distance from the top of the resolving gel and the bottom of the wells should be about 1.5 cm.

Mount the gels in the gel running apparatus, fill it with the running buffer, and load the heat-denatured samples. As migration standards, prestained gel markers can be used, such as the Life-Technology Benchmark Prestained Protein Ladder (cat. no. 10748-010).

Run at 18 mA (constant current) for each 0.75-mm-thick gel for about 3.5 h, or until the bromophenol blue stain reaches the bottom of the gel.

3.1.5. Immunoblotting and Detection

1. At the end of the electrophoresis run, dismount the gel from the running apparatus, open the plates carefully, and soak the gel in PVDF transfer buffer for 10 min.
2. Cut from a PVDF-membrane roll a rectangle with the same dimension of the gel.

3. Soak the cut PVDF hydrophobic membrane in 100% methanol for 5 min, then equilibrate the membrane in PVDF transfer buffer for at least 10 min. Do not allow the membrane to dry out before the electrophoretic transfer.
4. Mount the gel and the membrane in the Bio-Rad transfer apparatus, according to the instructions of the manufacturer, fill the apparatus with the PVDF transfer buffer and start the electrophoretic transfer at 70 V at 4°C for 2 h for regular gels or for 1 h for minigels.
5. After the transfer has been completed, in order to saturate nonspecific protein binding, incubate the membrane in a tray containing TBS-T plus 5% NFDM, for 60 min at room temperature.
6. Replace the blocking solution with TBS-T plus 3% NFDM containing the appropriate dilution of the specific primary antibody (*see* **Subheading 2.** and **Table 2**) and incubate for 60 min with gentle shaking.
7. Wash the membrane three times in TBS-T for 5 min with rapid shaking (*see* **Note 4**).
8. Move the membrane to TBS-T + 3% NFDM containing the specific peroxidase-conjugated secondary antibody (1/10,000) and incubate for 45 min with gentle shaking.
9. Wash the membrane in TBS-T three times for 5 min with rapid shaking, to remove any unbound secondary antibody.
10. Use the ECL kit to reveal the bound secondary antibody, following the manufacturer's recommendations.
11. Cover the blot with a transparent Saran Wrap® membrane and expose it immediately to the Kodak film. A recommended first exposure time is 1–3 min. Multiple exposures and various exposure times may be necessary to obtain optimum results. Develop the film using appropriate developing solutions and apparatus.

3.2. Functional Integrity—Binding Assay with In Vitro-Translated E1A

We have proposed a binding assay with the in-vitro-translated Adenovirus E1A (IVT-E1A) as a rapid screening method to detect functional alterations of the retinoblastoma protein in cell lines, primary cultures, or frozen leukocyte samples. This method, which evaluates the functional interaction between E1A and pRb, is based on the well-known property of the adenoviral E1A protein to bind to the pocket region of the wild-type retinoblastoma protein ***(6)***. In-vitro-translated ^{35}S labeled Adenovirus E1A oncoprotein is challenged with a tumor cell lysate, and its binding to pRb is evaluated by means of a coimmunoprecipitation assay, using an anti-pRb monoclonal antibody. The amount of coimmunoprecipitated radioactive E1A, measured by SDS-PAGE/autoradiography or by liquid scintillation, is taken as a quantitative and qualitative measure of the retinoblastoma protein in the specific sample. The same coimmunoprecipitation assay has been demonstrated to be effective in estimating the function of p107 and pRb2/p130 ***(48,53,54)***.

Figure 2 shows the results obtained using this assay to evaluate the functionality of pRb from different tumor cell lines, which possess either functional (T98G and MCF-7; ***[55,56]***) or defective (SAOS2 and H209; ***[49,57]***) RB genes.

3.2.1. *In Vitro E1A Translation*

3.2.1.1. Preparation of DNA Template

The Adenovirus E1A mRNA is synthesized in vitro using a plasmid containing the Adenovirus 5 12 S E1A coding sequence cloned in the pSP64 vector. The plasmid is linearized with EcoRI and used for in-vitro transcription system preparation (Promega).

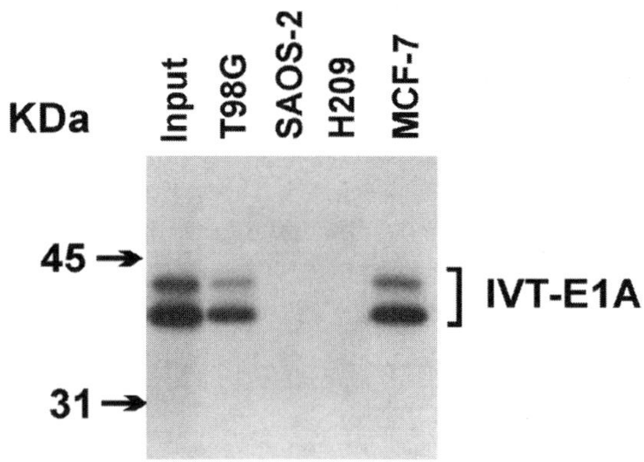

Fig. 2. Autoradiography of the analysis by SDS-PAGE (10%) of coimmunoprecipitation of IVT-E1A bound to pRb from different cell lines. [^{35}S]-labeled IVT-E1A was preincubated with unlabelled cell lysates prepared from T98G, SAOS2, H209, MCF-7 cell lines and coimmunoprecipitated using the pRb-specific C36 monoclonal antibody. Input, an equivalent amount of [^{35}S]-labeled IVT-E1A was immunoprecipitated using the E1A specific M76 monoclonal antibody. Left ordinate, molecular-weight marker migration distances.

1. In an Eppendorf tube, digest 25 µg of plasmid using 75 units of EcoRI in a final volume of 300 µL (at least 3 units for 1 µg of DNA) at 37°C, using the buffer provided by the enzyme producer.
2. After 1 h of incubation, examine an aliquot of 6 µL of the digestion mix by agarose gel electrophoresis, in order to check the completeness of the digestion. In case of incomplete digestion, add 75 more units of enzyme and continue the incubation at 37°C for one more hour.
3. When the DNA is completely digested, add 300 µL of TE-saturated phenol to the tube, mix by vortexing, then add 300 µL of chloroform:isoamyl alcohol mixture, mix by vortexing at least 1 min, and separate the aqueous from the organic phase by centrifuging in a microfuge for 5 min.
4. Save the aqueous phase and repeat the phenol–chloroform–isoamylalcohol extraction.
5. Transfer the final aqueous phase to a fresh tube, add 30 µL of DEPC-treated 3 *M* Na-acetate and 1 mL of cold 96% ethanol, mix thoroughly by inverting the tube.
6. Recover the DNA precipitate by centrifuging at the maximum speed in a microfuge for 5 min, and drain the pellet.
7. Add 500 µL of cold 70% ethanol, wash the pellet by vortexing, and spin again for 1 min at the maximun speed. Drain the pellet and wash it again using 500 µL of cold 96% ethanol.
8. Carefully drain the final pellet, eliminating all the residual ethanol, and air-dry it with the tube open on the bench for 5 min.
9. Dissolve the pellet in 50 µL of DEPC-treated H_2O.
10. Check the DNA concentration by measuring the absorbance at 260 nm. The final concentration should be about 0.5 µg/µL.

3.2.1.2. DNA-Driven RNA Transcription

1. In an Eppendorf tube on ice, set up the following mixture: 10 µL Transcription buffer 5×; 50 U RNAsin; 5 µL DTT 100 mM; 10 µL rNTP mix; 6 µL (about 3 µg) EcoRI digested E1A plasmid; 1 µL (about 40 U) SP6 RNA polymerase (40 U/µL); up to 50 µL DEPC-treated H_2O. Mix thoroughly and incubate at 37°C for 45 min, then add 40 units more of SP6 RNA polymerase and continue the incubation for a further 45 min.

2. Add 3 U of RQ1 DNase (1 U/μg of template DNA) and incubate at 37°C for 15 min.
3. Add 50 μL of DEPC-treated water and 100 μL of water-saturated phenol, mix by vortexing, add 100 μL of chloroform/isoamyl alcohol mixture, mix by vortexing for 1 min, then separate the aqueous phase by centrifuging 5 min at the maximum speed in a microfuge.
4. Transfer the aqueous phase in a fresh tube, add 200 μL of chloroform/isoamyl alchool mixture, mix by vortexing, then centrifuge to separate the aqueous from the organic phase.
5. In a fresh Eppendorf tube, add to the aqueous phase 10 μL of DEPC-treated 3 M Na-acetate and 300 μL of 96% cold ethanol, mix thoroughly, and store at –20°C.
6. Recover the precipitated RNA by centrifuging at full speed for 10 min, drain the pellet and wash twice with cold 70% ethanol, by vortexing and centrifuging.
7. Carefully drain and air-dry the RNA pellet, then dissolve it in 100 μL of DEPC-treated H2O.
8. Measure the RNA concentration by reading the absorbance at 260 nm with a spectrophotometer. The yield should be about 15 μg of transcribed RNA.
9. Check the size and the quality of the RNA on a non-denaturing 1.5% agarose gel in TBE 1×.
10. Store the RNA at –80°C, aliquoted, until the utilization.

3.2.1.3. In Vitro Translation

1. In an Eppendorf tube on ice set up the following mixture:
   35 μL Nuclease treated reticulocyte lysate; 1 μL RNasin ribonuclease inhibitor; 1 μL Amino acid mixture (1 m*M* each) (minus methionine);1.5 μg RNA template; 4 μL (40 μCi) [^{35}S]-methionine, in vitro translation grade; up to 50 μL DEPC-treated H_2O. Always set up two parallel reactions, one including and the other one omitting the E1A RNA template.
2. Incubate the reactions for 60 min at 30°C.
3. Store the translation product at –80°C. The amount of IVT-E1A produced is sufficient for 5–6 coimmunoprecipitation assays.

3.2.2. Physical Interaction and Immunoprecipitation

Coimmunoprecipitations are always done in a buffer containing no anionic detergents, such as SDS or DOC.

The rationale behind this assay requires that the E1A-pRb interaction should be analyzed with the following recommendation: a large excess of IVT-E1A, antibody, and immobilized protein A/G with respect to the amount of the RB family protein present in the sample, should always be used. The following protocol refers to the functional assay for pRb. If p107 or pRb2/p130 is analyzed, antibodies (*see* **Table 2**) and immobilized protein A or G should be chosen accordingly.

1. Thaw out the pellet of frozen cells in which pRb function should be tested (1 × 10^6 cells for each sample).
2. Add IPPT lysis buffer (0.2–0.5 mL/pellet), and incubate for 30 min on ice.
3. Spin down in a microfuge at maximum speed for 10 min.
4. Transfer the supernatant to a fresh tube, add 8 μL of ^{35}S-labeled IVT-E1A (≃ 8 × 10^4 cpm of M73-precipitable radioactivity) and 5 μg/pellet of purified mouse IgG and incubate 30 min in ice.
5. Using a cut micropipet tip, add 20 μL of immobilized protein G for each tube, and incubate for 30 min at 4°C.
6. Spin down for 10 s. Transfer carefully the supernatant (precleared supernatant) without taking any protein G particle (spin down again and repeat, if necessary).
7. Add to the precleared lysate 1 μg of C36 anti-pRb monoclonal antibody (Pharmingen or Pierce), and incubate for 1 h at 4°C using a rocking device.

8. Using a cut micropipet tip, add 10 µL of immobilized protein G suspension, and rock for 1 h at 4°C.
9. Spin down 5–10 s and resuspend in 1 mL of lysis buffer, vortex briefly.
10. Repeat **step 9** twice.
11. Spin down 5–10 s, remove the supernatant, and add 30–40 µL of Laemmli's sample buffer (*see* **Note 5**). Heat samples at 95°C for 5 min, add 0.5 µL of bromophenol blue saturated solution and spin them down to remove immobilized protein G. The samples are now ready for SDS-PAGE.

3.2.3. SDS-PAGE and Detection

To resolve the IVT 12 E1A, a minigel at a concentration of 10% acrylamide is adequate. Minigels are prepared as described before (*see* **Subheading 3.1.5.** for details), changing the volumes according to the reduced dimensions of the gels.

3.2.3.1. Preparation of Resolving Gel Mixture

For two 8 × 10-cm, 0.75-cm-thick minigels: in a 50-mL disposable tube add 6.7 mL acrylamide 30% stock solution, 7.9 mL H_2O, 5 mL 1.5 *M* Tris-HCl pH 8.8, 0.2 mL 10% SDS, 0.2 mL of 10% ammonium persulfate, and 8 µL of TEMED. Pour the resolving gel mix into the assembled gel plates, leaving a 1.5-cm space at the top for the stacking gel. Gently overlay the gel surface with water-saturated 2-butanol, and allow the gel to polymerize in the appropriate stand for at least 30 min.

3.2.3.2. Preparation of Stacking Gel Mixture

For two 8 × 10-cm, 0.75-cm-thick minigels: in a 10-mL tube, add 0.83 mL of acrylamide 30% stock solution, 3.40 mL of H_2O, 0.63 mL of 0.5 *M* Tris-HCl pH 6.8, 0.05 mL of 10% SDS, 0.05 mL of ammonium persulfate, and 5 µL of TEMED. Mix gently, pour the stacking gel solution up the top of the plates, and insert the combs immediately.

After the stacking gel has polymerized, remove the comb and rinse the wells with water gently to remove the unpolymerized acrylamide. The distance from the top of the resolving gel and the bottom of the wells should be about 1 cm.

Mount the gels in the gel running apparatus, fill it with the running buffer, and load the heat-denatured samples. As migration standards, you can use prestained gel markers, as the Life-Technology Benchmark Prestained Protein Ladder (cat. no. 10748-010). IVT 12S E1A migrates with an apparent molecular mass of about 38–44 kDa. As a reference, add a lane in which E1A has been immunoprecipitated using the M73 monoclonal antibody, following a parallel procedure.

Run the gels at 200 V for 45 min, or until the bromophenol blue stain reaches the bottom of the gel.

At the end of the run, dismount the gel from the running apparatus, open the plates carefully, and soak the gel for 10 min in the fixing solution, rinse with water, then soak for 10 min in 20% glycerol. Put it on a moistened sheet of Whatman 3MM paper and then in a gel-drying apparatus.

Finally, expose the dried gel to a phosphorimager screen (or autoradiographic Kodak film), to visualize and quantify the amount of E1A protein bound to the sample pRb.

4. Notes

1. This reagent is hazardous if employed without specific precautions ***(58)***.
2. If the apparent molecular mass of the protein of interest, as evaluated by SDS-PAGE, is close enough to that of the denatured heavy and light Ig chains (about 50 and 25 kDa, respectively) it is advisable, only in this case, not to treat the samples at 95°C, but to treat them, instead, at room temperature for 15 min. The overall yield will be reduced, but the interference of the Ig chains will be considerably lower when detecting proteins in Western blot employing a secondary anti-Ig peroxidase-conjugated antibody. In the case of the retinoblastoma family proteins, their apparent molecular mass is sufficiently different from the denatured Ig chains. Anyway, it is always advisable to remove the lower part of the PVDF, where the Ig are located, before processing the membrane.
3. To parallel samples, 100 m*M* of sodium phosphate (final concentration), pH 6.0, can be added to confirm that the observed effects are caused by dephosphorylation.
4. When it is necessary to reduce the background signal, use for the first wash TBS-T "strong wash." TBS-T "strong wash" is prepared by adding 21 g NaCl and 5 mL of Triton X-100 to 1 L of TBS-T.
5. Alternatively, instead of treating the samples with Laemmli's sample buffer and analyzing them on SDS-PAGE, they can be counted directly in a scintillation apparatus in 1 mL of water-compatible scintillation fluid. In this case, in order to have an idea of the efficiency of the coimmunoprecipitation, it is mandatory to add a reference tube in which IVT-E1A has been immunoprecipitated using the M73 monoclonal antibody.

Acknowledgments

The authors thank Mr. T. Battista and Drs. F. Bonetto (Regina Elena Institute, Rome, Italy), C. Cenciarelli, A. Magenta, and F. De Santa (CNR, Rome, Italy) for providing unpublished data and for helpful discussion.

References

1. Weinberg, R. A. (1991) Tumor suppressor genes. *Science* **254**, 1138–1146.
2. Hinds, P. W. and Weinberg, R. A. (1994) Tumor suppressor genes. *Curr. Opin. Genet. Dev.* **4**, 135–141.
3. Paggi, M. G., Baldi, A., Bonetto, F., and Giordano, A. (1996) Retinoblastoma protein family in cell cycle and cancer: a review. *J. Cell. Biochem.* **62**, 418–430.
4. Mulligan, G. and Jacks, T. (1998) The retinoblastoma gene family: cousins with overlapping interests. *Trends Genet.* **14**, 223–229.
5. Harlow, E., Whyte, P., Franza, B. R., Jr., and Schley, C. (1986) Association of adenovirus early-region 1A proteins with cellular polypeptides. *Mol. Cell. Biol.* **6**, 1579–1589.
6. Whyte, P., Buchkovich, K. J., Horowitz, J. M., et al. (1988) Association between an oncogene and an anti-oncogene: the adenovirus E1A proteins bind to the retinoblastoma gene product. *Nature* **334**, 124–129.
7. Ewen, M. E., Xing, Y. G., Lawrence, J. B., and Livingston, D. M. (1991) Molecular cloning, chromosomal mapping, and expression of the cDNA for p107, a retinoblastoma gene product-related protein. *Cell* **66**, 1155–1164.
8. Zhu, L., van den Heuvel, S., Helin, K., et al. (1993) Inhibition of cell proliferation by p107, a relative of the retinoblastoma protein. *Genes Dev.* **7**, 1111–1125.
9. Mayol, X., Graña, X., Baldi, A., Sang, N., Hu, Q., and Giordano, A. (1993) Cloning of a new member of the retinoblastoma gene family (pRb2) which binds to the E1A transforming domain. *Oncogene* **8**, 2561–2566.

10. Li, Y., Graham, C., Lacy, S., Duncan, A. M., and Whyte, P. (1993) The adenovirus E1A-associated 130-kD protein is encoded by a member of the retinoblastoma gene family and physically interacts with cyclins A and E. *Genes Dev.* **7**, 2366–2377.
11. Hannon, G. J., Demetrick, D., and Beach, D. (1993) Isolation of the Rb-related p130 through its interaction with CDK2 and cyclins. *Genes Dev.* **7**, 2378–2391.
12. Ewen, M. E. (1994) The cell cycle and the retinoblastoma protein family. [Review]. *Cancer Metastasis Rev.* **13**, 45–66.
13. Weinberg, R. A. (1995) The retinoblastoma protein and cell cycle control. *Cell* **81**, 323–330.
14. Harbour, J. W. and Dean, D. C. (2000) The Rb/E2F pathway: expanding roles and emerging paradigms. *Genes Dev.* **14**, 2393–2409.
15. DeCaprio, J. A., Ludlow, J. W., Figge, J., et al. (1988) SV40 large tumor antigen forms a specific complex with the product of the retinoblastoma susceptibility gene. *Cell* **54**, 275–283.
16. Dyson, N., Buchkovich, K., Whyte, P., and Harlow, E. (1989) The cellular 107K protein that binds to adenovirus E1A also associates with the large T antigens of SV40 and JC virus. *Cell* **58**, 249–255.
17. Dyson, N., Howley, P. M., Munger, K., and Harlow, E. (1989) The human papilloma virus-16 E7 oncoprotein is able to bind to the retinoblastoma gene product. *Science* **243**, 934–937.
18. Marsilio, E., Cheng, S. H., Schaffhausen, B., Paucha, E., and Livingston, D. M. (1991) The T/t common region of simian virus 40 large T antigen contains a distinct transformation-governing sequence. *J. Virol.* **65**, 5647–5652.
19. Dyson, N., Guida, P., Munger, K., and Harlow, E. (1992) Homologous sequences in adenovirus E1A and human papillomavirus E7 proteins mediate interaction with the same set of cellular proteins. *J. Virol.* **66**, 6893–6902.
20. Ludlow, J. W. and Skuse, G. R. (1995) Viral oncoprotein binding to pRB, p107, p130, and p300. *Virus Res.* **35**, 113–121.
21. Goodrich, D. W., Wang, N. P., Qian, Y. W., Lee, E. Y., and Lee, W. H. (1991) The retinoblastoma gene product regulates progression through the G1 phase of the cell cycle. *Cell* **67**, 293–302.
22. Starostik, P., Chow, K. N. B., and Dean, D. C. (1996) Transcriptional repression and growth suppression by the p107 pocket protein. *Mol. Cell. Biol.* **16**, 3606–3614.
23. Claudio, P. P., Howard, C. M., Baldi, A. et al. (1994) p130/pRb2 has growth suppressive properties similar to yet distinctive from those of retinoblastoma family members pRb and p107. *Cancer Res.* **54**, 5556–5560.
24. Lee, W. H., Shew, J. Y., Hong, F. D., et al. (1987) The retinoblastoma susceptibility gene encodes a nuclear phosphoprotein associated with DNA binding activity. *Nature* **329**, 642–645.
25. Buchkovich, K., Duffy, L. A., and Harlow, E. (1989) The retinoblastoma protein is phosphorylated during specific phases of the cell cycle. *Cell* **58**, 1097–1105.
26. Mittnacht, S. and Weinberg, R. A. (1991) G1/S phosphorylation of the retinoblastoma protein is associated with an altered affinity for the nuclear compartment. *Cell* **65** , 381–393.
27. Chen, P. L., Scully, P., Shew, J. Y., Wang, J. Y., and Lee, W. H. (1989) Phosphorylation of the retinoblastoma gene product is modulated during the cell cycle and cellular differentiation. *Cell* **58**, 1193–1198.
28. DeCaprio, J. A., Ludlow, J. W., Lynch, D., et al. (1989) The product of the retinoblastoma susceptibility gene has properties of a cell cycle regulatory element. *Cell* **58**, 1085–1095.
29. Mihara, K., Cao, X. R., Yen, A., et al. (1989) Cell cycle-dependent regulation of phosphorylation of the human retinoblastoma gene product. *Science* **246**, 1300–1303.
30. MacLachlan, T. K., Sang, N., and Giordano, A. (1995) Cyclins, cyclin-dependent kinases and Cdk inhibitors: implications in cell cycle control and cancer. *Eukariotic Gene Expr.* **5**, 127–156.

31. Beijersbergen, R. L., Carlée, L., Kerkhoven, R. M., and Bernards, R. (1995) Regulation of the retinoblastoma protein-related p107 by G_1 cyclin complexes. *Genes Dev.* **9**, 1340–1353.
32. Faha, B., Ewen, M. E., Tsai, L. H., Livingston, D. M., and Harlow, E. (1992) Interaction between human cyclin A and adenovirus E1A-associated p107 protein. *Science* **255**, 87–90.
33. Lees, E., Faha, B., Dulic, V., Reed, S. I., and Harlow, E. (1992) Cyclin E/cdk2 and cyclin A/cdk2 kinases associate with p107 and E2F in a temporally distinct manner. *Genes Dev.* **6**, 1874–1885.
34. Baldi, A., De Luca, A., Claudio, P. P., et al. (1995) The Rb2/p130 gene product is a nuclear protein whose phosphorylation is cell cycle regulated. *J. Cell Biochem.* **59**, 402–408.
35. Cobrinik, D., Whyte, P., Peeper, D. S., Jacks, T., and Weinberg, R. A. (1993) Cell cycle-specific association of E2F with the p130 E1A-binding protein. *Genes Dev.* **7**, 2392–2404.
36. Claudio, P. P., De Luca, A., Howard, C. M., et al. (1996) Functional analysis of pRb2/p130 interaction with cyclins. *Cancer Res.* **56**, 2003–2008.
37. Riley, D. J., Lee, E. Y. H. P., and Lee, W.-H. (1994) The retinoblastoma protein: more than a tumor suppressor. *Annu. Rev. Cell Biol.* **10**, 1–29.
38. Sidle, A., Palaty, C., Dirks, P., et al. (1996) Activity of the retinoblastoma family proteins, pRB, p107, and p130, during cellular proliferation and differentiation. *Crit. Rev. Biochem. Mol. Biol.* **31**, 237–271.
39. Herwig, S. and Strauss, M. (1997) The retinoblastoma protein: a master regulator of cell cycle, differentiation and apoptosis. *Eur. J. Biochem.* **246**, 581–601.
40. Stiegler, P., Kasten, M., and Giordano, A. (1998) The RB family of cell cycle regulatory factors. *J. Cell Biochem.* 30–31, 30–36.
41. Sherr, C. J. (2000) The Pezcoller lecture: cancer cell cycles revisited. *Cancer Res.* **60**, 3689–3695.
42. Yeung, R. S., Bell, D. W., Testa, J. R., et al. (1993) The retinoblastoma-related gene, RB2, maps to human chromosome 16q12 and rat chromosome 19. *Oncogene* **8**, 3465–3468.
43. Yandell, D. W., Campbell, T. A., Dayton, S. H., et al. (1989) Oncogenic point mutations in the human retinoblastoma gene: their application to genetic counseling. *N. Engl. J. Med.* **321**, 1689–1695.
44. Zacksenhaus, E., Bremner, R., Phillips, R. A., and Gallie, B. L. (1993) A bipartite nuclear localization signal in the retinoblastoma gene product and its importance for biological activity. *Mol. Cell Biol.* **13**, 4588–4599.
45. Cinti, C., Claudio, P. P., Howard, C. M., et al. (2000) Genetic alterations disrupting the nuclear localization of the retinoblastoma-related gene *RB2/p130* in human tumor cell lines and primary tumors. *Cancer Res.* **60**, 383–389.
46. Zerler, B., Moran, B., Maruyama, K., Moomaw, J., Grodzicker, T., and Ruley, H. E. (1986) Adenovirus E1A coding sequences that enable ras and pmt oncogenes to transform cultured primary cells. *Mol. Cell Biol.* **6**, 887–899.
47. Beijersbergen, R. L. and Bernards, R. (1996) Cell cycle regulation by the retinoblastoma family of growth inhibitory proteins. *Biochim. Biophys. Acta Rev. Cancer* **1287**, 103–120.
48. Paggi, M. G., Martelli, F., Fanciulli, M., et al. (1994) Defective human retinoblastoma protein identified by lack of interaction with the E1A oncoprotein. *Cancer Res.* **54**, 1098– 1104.
49. Bignon, Y. J., Shew, J. Y., Rappolee, D., et al. (1990) A single Cys706 to Phe substitution in the retinoblastoma protein causes the loss of binding to SV40 T antigen. *Cell Growth Differ.* **1**, 647–651.
50. Ting, A. and Morris, P. J. (1971) A technique for lymphocyte preparation from stored heparinized blood. *Vox Sang.* **20**, 561–563.
51. Mancini, M. A., Shan, B., Nickerson, J. A., Penman, S., and Lee, W. H. (1994) The retinoblastoma gene product is a cell cycle-dependent, nuclear matrix-associated protein. *Proc. Natl. Acad. Sci. USA* **91**, 418–422.

52. Laemmli, U. K. (1970) Cleavage of structural proteins during the assembly of the head of the bacteriophage T4. *Nature* **227**, 680–685.
53. Paggi, M. G., De Fabritiis, P., Bonetto, F., et al. (1995) The retinoblastoma gene product in acute myeloid leukemia: A possible involvement in promyelocytic leukemia. *Cancer Res.* **55**, 4552–4556.
54. Valente, P., Melchiori, A., Paggi, M. G., et al. (1996) RB1 oncosuppressor gene over-expression inhibits tumor progression and induces melanogenesis in metastatic melanoma cells. *Oncogene* **13**, 1169–1178.
55. Olopade, O. I., Jenkins, R. B., Ransom, D. T., et al. (1992) Molecular analysis of deletions of the short arm of chromosome 9 in human gliomas. *Cancer Res.* **52**, 2523–2529.
56. Lukas, J., Parry, D., Aagaard, L., et al. (1995) Retinoblastoma-protein-dependent cell-cycle inhibition by the tumour suppressor p16. *Nature* **375**, 503–506.
57. Shew, J. Y., Lin, B. T., Chen, P. L., Tseng, B. Y., Yang Feng, T. L., and Lee, W. H. (1990) C-terminal truncation of the retinoblastoma gene product leads to functional inactivation. *Proc. Natl. Acad. Sci. USA* **87**, 6–10.
58. A. K. Furr, ed. *CRC Handbook of Laboratory Safety*, (1995). CRC Press, Boca Raton, FL.
59. Baldi, A., Esposito, V., De Luca, A., et al. (1997) Differential expression of Rb2/p130 and p107 in normal human tissues and in primary lung cancer. *Clin. Cancer Res.* **3**, 1691–1697.
60. Helin, K., Holm, K., Niebuhr, A., et al. (1997) Loss of the retinoblastoma protein-related p130 protein in small cell lung carcinoma. *Proc. Natl. Acad. Sci. USA.* **94**, 6933–6938.
61. Susini, T., Baldi, F., Howard, C. M., et al. (1998) Expression of the retinoblastoma-related gene Rb2/p130 correlates with clinical outcome in endometrial cancer. *J. Clin. Oncol.* **16**, 1085–1093.
62. Massaro-Giordano, M., Baldi, G., De Luca, A., Baldi, A., and Giordano, A. (1999) Differential expression of the retinoblastoma gene family members in choroidal melanoma: prognostic significance. *Clin. Cancer Res.* **5**, 1455–1458.
63. Leoncini, L., Bellan, C., Cossu, A., et al. (1999) Retinoblastoma-related p107 and pRb2/p130 proteins in malignant lymphomas: distinct mechanisms of cell growth control. *Clin. Cancer Res.* **5**, 4065–4072.
64. Zamparelli, A., Masciullo, V., Bovicelli, A., et al. (2001) Expression of cell-cycle-associated proteins pRb2/p130 and $p27^{kip1}$ in vulvar squamous cell carcinomas. *Hum. Pathol.* **32**, 4–9.
65. Mutti, L., De Luca, A., Claudio, P. P., Convertino, G., Carbone, M., and Giordano, A. (1998) Simian virus 40-like DNA sequences and large-T antigen-retinoblastoma family protein pRb2/p130 interaction in human mesothelioma. *Dev. Biol. Stand.* **94**, 47–53.
66. Claudio, P. P., Howard, C. M., Pacilio, C., et al. (2000) Mutations in the retinoblastoma-related gene *RB2/p130* in lung tumors and suppression of tumor growth *in vivo* by retrovirus-mediated gene transfer. *Cancer Res.* **60**, 372–382.
67. Claudio, P. P., Howard, C. M., Fu, Y., et al. (2000) Mutations in the retinoblastoma-related gene *RB2/p130* in primary nasopharyngeal carcinoma. *Cancer Res.* **60**, 8–12.
68. Ichimura, K., Hanafusa, H., Takimoto1 H, Ohgama, Y., Akagi, T., and Shimizu, K. (2000) Structure of the human retinoblastoma-related p107 gene and its intragenic deletion in a B-cell lymphoma cell line. *Gene* **251**, 37–43.

2

The APC Tumor Suppressor Pathway

Patrice J. Morin and Ashani T. Weeraratna

1. Introduction

1.1. Colon Cancer

Colon cancer is the second most common cause of cancer mortality among adults in the United States today ***(1)***. Colon cancer arises from a pathological transformation of the normal colonic epithelium to an adenomatous polyp, which can then progress to an invasive tumor. This progression is brought about by a number of genetic changes, which include the inactivation of tumor suppressor genes and activation of protooncogenes, the net result of which is to confer a proliferative advantage to the cancerous cell ***(2,3)***. These changes are best described in the colon cancer adenoma-carcinoma sequence model as outlined in **Fig. 1** ***(4)***.

This model of the genetic changes involved in the progression of colon cancer begins with the development of benign polyps. Inactivation of both copies (alleles) of the adenomatous polyposis coli (*APC*) gene located on chromosome 5 constitutes an early, and possibly the very first event in colorectal tumorigenesis ***(5)***. Consistent with this hypothesis, germline mutation of the *APC* gene causes the inherited familial adenomatous polyposis syndrome (FAP) (see below). Another early event in early colon carcinogenesis is the activation of the K-ras protooncogene through mutation of codon 12 or 13 ***(6–8)***. Interestingly, while *APC* and K-ras alterations are considered two of the earliest events in colon tumorigenesis, the order appears crucial. Indeed, while *APC* mutations were associated with early dysplasia, lesions exhibiting K-Ras mutations without *APC* alterations appeared mostly nondysplastic lesions with limited potential to progress to carcinoma ***(7)***.

Next in the progression from adenoma to carcinoma is the loss of heterozygosity of the long arm of 18q, near the *SMAD4* (*DPC4*) locus. *DCC* and *SMAD2* are also candidate tumor suppressor genes in this region. Interestingly, germline mutation of *SMAD4* was found to cause juvenile polyposis, implicating *SMAD4* further in colorectal cancer ***(9)***. Mutation of p53 on chromosome 17p appears to be a late-stage event. This mutation is thought to allow the growing tumor with multiple genetic alterations to evade cell cycle arrest and apoptosis, or programmed cell death ***(4)***. However, it is almost certain that there are additional, undiscovered genetic events that occur along the way as indi-

From: *Methods in Molecular Biology, Vol. 222: Tumor Suppressor Genes: Pathways and Isolation Strategies*
Edited by: Wafik S. El-Deiry © Humana Press Inc., Totowa, NJ

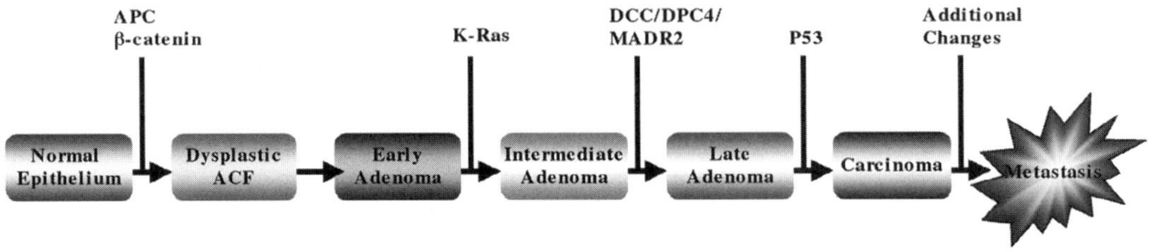

Fig. 1. A genetic model for colorectal tumorigenesis. Adapted from Fearon and Vogelstein ***(4)***.

cated by evidence of allelic loss on chromosomes 1q, 4p, 6p, 8p, 9q, and 22q in 25–50% of colorectal cancers ***(10)***.

Finally, genes involved in DNA repair have been implicated in colorectal cancer tumorigenesis. These genes include *hMSH2*, *hMLH1*, *hPMS1*, or *hPMS2* and are involved in DNA mismatch repair ***(11)***. The mutation of certain of these genes causes a colorectal cancer syndrome known as hereditary nonpolyposis colorectal cancer (HNPCC). Alteration of these genes is also prevalent in sporadic colon cancer, and may favor mutations of other crucial genes for colon cancer development such as *APC* and K-ras or may represent an alternative to the *APC* pathway for colon cancer development ***(12–14)***.

1.2. Familial Adenomatous Polyposis (FAP)

The identification of an interstitial chromosome 5q deletion in a patient with Gardner syndrome led to the approximate localization of the gene responsible for FAP ***(15)***, a rare, autosomal, dominantly inherited syndrome, characterized by the development of hundreds to thousands of colorectal adenomas ***(16)***. Some years later, several groups identified this gene as *APC* by performing linkage analysis of families with the syndrome, as well as positional cloning of the gene, localizing the gene to chromosome 5q21 ***(17–19)***. In addition to being a predominant germline mutation in many FAP patients, it is also mutated in approximately 80% of all sporadic colon cancer patients ***(5)***.

The age of onset of FAP is usually very young (20 years of age), and the disease affects approximately 1 in 10,000 individuals. Because of the large numbers of tumors as well as the fact that a subset of these adenomas tend to become malignant, the most common treatment for the disease is prophylactic resection of the colon. However, these adenomas appear outside the colon as well, and can be found in organs such as the stomach, pancreas, thyroid, periampullary region, and the central nervous system. In addition, the tumors can appear not just as adenomas, but also as adenocarcinomas, osteomas, desmoids, epidermal cysts, congenital hypertrophy of the retinal pigmented epithelium (CHRPE), and also as dental abnormalities. The pathological association of desmoid tumors and osteomas with FAP is clinically termed Gardner syndrome, and the association of CNS tumors, in particular medulloblastoma, is known as Turcot syndrome **(20).**

This diversity in the clinical profile of FAP patients is interesting, considering that virtually all these patients have mutations in the *APC* gene that result in C-terminally truncated proteins. This may be partially accounted for by the exact type and location of mutations in the *APC* gene. In fact, more than 1600 mutations, both germline and somatic, have been described for *APC*. A database is accessible online at http://perso.curie.fr/Thierry.Soussi/APC.html.

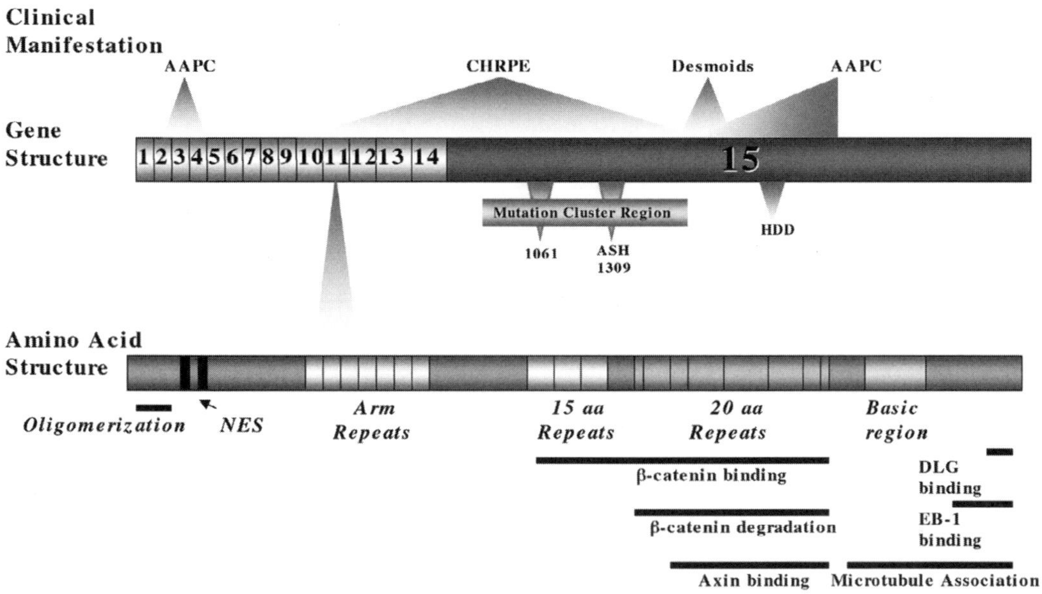

Fig. 2. The *APC* gene structure, amino acid structure, and disease profile.

2. APC Gene Structure

2.1. Gene Structure and APC Homologs

The structure of the *APC* gene is outlined in **Fig. 2**. The gene contains 21 exons which are within a 98-kb locus, and 16 of which are translated ***(21)***. The largest exon is exon 15, with a remarkable uninterrupted open reading frame of 6579 bp, which comprises more than 75% of the coding sequence for this gene. The open reading frame consists of 8538 bp, and codes for a protein of 2861 amino acids. The most abundantly occurring transcript of this gene lacks exon 10A (the smallest exon) and has a mass of 310 kDa, comprised of 2843 amino acids ***(22,23)***. Alternatively expressed exons of the gene encode alternate isoforms of the protein, ranging in molecular weight from 90 to 310 kDa. All these isoforms are expressed in different tissue-specific patterns, but their function is not as yet known, although the encoded proteins differ most dramatically in their N-termini and dimerization capacities ***(23–25)***. Another member of the *APC* family, *APC-2 /APC-L*, is located within a 50-kB genomic fragment on chromosome 19p13.3 ***(26)***. It is ubiquitously expressed during brain development, and closely resembles *APC* in its overall structure ***(27,28)***. However it does not contain the 15 amino acid repeats or the hDLG region and only has four 20 amino acid repeats, which will be described in the next section. There are several splice variants of this protein of which the function is unclear. It appears that *APC* can compensate for APC-2 loss but not vice versa, implying that there are overlapping functions of the two proteins. Overall, *APC* appears to have a wider range of function ***(29)***.

2.2. Amino Acid Sequence and Interacting Proteins

The N-terminal third of the APC protein contains seven Arm repeats (aa 453 to 767), named for an amino acid motif repeated in Armadillo, the *Drosophila* homolog of the protein β-catenin ***(30)*** (**Fig. 2**). This domain binds to APC-stimulated exchange factor (ASEF), which allows for interaction with raf, and consequently with rac and rho, which mediate cell adhesion and motility via cytoskeletal changes ***(31)***. The region is referred to as the ARD domain, or armadillo repeat domain. Arm repeats are present in several proteins that are involved in protein–protein interactions, such as plakoglobin, plakophilin, p120cas, among others. Arm repeats in β-catenin are essential for APC binding, as well as E-cadherin association, and form a superhelix, with a positively charged groove, that is responsible for structural interactions with partnering proteins ***(32,33)***. The ARD is the most conserved part of APC and is the most homologous ARD to that of β-catenin, perhaps indicating that APC and β-catenin may interact with some of the same proteins. Recently, APC, via its ARD, has been found to bind to protein phosphatase 2A ***(34)***. This enzyme can also bind to Axin, another protein involved in the APC/β-catenin signal transduction pathway, implicating PP2A as a potential antagonist of the kinase GSK3β ***(35)***. These observations have yet to be confirmed, but the ARD does indeed seem to be important for survival, as it is retained in most mutations of the *APC* gene, and deletion of this region in mice causes early lethality ***(36)***.

The central third of APC contains three 15-aa repeats between amino acid residues 1020 and 1169 and seven 20-aa repeats, between residues 1262 and 2033. These repeats are comprised of the amino acids TPXXFSXXXSL ***(19)***. They are known as the β-catenin-binding repeats, but also associate with plakoglobin ***(37)***. These repeats are highly conserved among the species, and have great similarity to one another. The 20-aa repeats require phosphorylation by the protein GSK3β, before binding β-catenin ***(38)***. Interestingly, while overexpression of the 15-aa region has no effect, the 20-aa region downregulates β-catenin levels in vivo when expressed in *APC*-deficient cells ***(39)***. It is in the 20-aa portion of APC that the binding site for Axin can be found, a protein that is also part of the β-catenin/APC/GSK3β complex and that will be discussed in detail later in this chapter. The binding site for Axin appears to be crucial for the tumor suppressor role of APC and is comprised of three SAMP (Ser-Ala-Met-Pro) repeats ***(40,41)***.

The C-terminal end of APC is rich in basic amino acids (**Fig. 2**). This region is essential for microtubule binding, particularly residues 2219–2580, which bind nonassembled tubulin and co-localize with microtubules in vivo ***(42,43)***. EB1, a member of the EB1/RP family of tubulin-binding proteins first identified by the two-hybrid system as an APC-associated protein ***(44)***, binds this region as well. EB1 also associates with microtubules and tubulin in vitro and in vivo. Consistent with these findings, APC has been found in the leading edge of migrating epithelial cells where there is an increased concentration of microtubules, further implicating the protein in cell adhesion and migration ***(45)***.

Two other proteins have important interactions with the C terminus of the APC protein: the human homolog of the *Drosophila* tumor suppressor gene, discs large (DLG), which binds the C-terminal 72 amino acids ***(46)***, and the protein phosphatase PTP-BL ***(47)***. Both proteins contain a PDZ domain that binds to C-terminal XS/TXV domains. APC contains specifically the sequence VTSV, which matches this motif, and the interaction of these two proteins was identified through yeast-two-cell hybrid screening, and by co-localization studies. DLG and APC co-localize at areas of cell–cell contact both in epithelial cells, as well as neuronal junctions along synaptic processes.

Finally, APC interacts with itself. It is the tertiary structure of APC that may mediate its dimerization and oligemerization capacities. The first 170 amino acids are sufficient for APC dimerization in vitro ***(48,49)***. The N-terminal third of the protein contains a coiled-coil motif, marked by a series of heptad repeats of hydrophobic residues. The first of these heptad repeats involves the first 45 amino acids, which are crucial for APC homodimerization ***(49)***. Although not previously demonstrated, these regions may also mediate oligomerization between wild-type and mutant proteins, resulting in a dominant negative effect ***(50,51)***.

2.3. APC Mutations in Colon and Other Cancers

As previously mentioned, more than 90% of *APC* mutations result in a premature stop codon, and thus in the truncation of the C terminus ***(52)***. A 5-bp deletion at codons 1309 and 1061 are found in 18% and 12%, respectively, of all germline mutations (*see* **Fig. 2**). These deletions, which occur in short direct repeats within *APC*, have also been identified in patients without any family history, suggesting that they may also arise sporadically. Germline mutations can predict the phenotype of the disease they are likely to cause, by their location within the *APC* gene. A phenotype of multiple tumors (over 5000) is predicted when the mutation arises between codons 1249 and 1330, whereas mutations that are upstream or downstream of this region result in a phenotype of fewer than 1000 tumors ***(53)***. Attenuated adenomatous polyposis coli (AAPC) is caused by mutations at codons 78 to 163 (5' end), and this disease is characterized by the development of very few tumors (less than 100) at a later age. Similarly, germline 3' mutations in exon 15 are also associated with an attenuated phenotype ***(54)***. Germline mutations in codons 457–1444, as well as in exon 9, can code specifically for CHRPE ***(55)***, and germline mutations between codons 1395 and 1560 are associated with desmoid tumors and mandibular osteomas ***(56)***.

Unlike germline mutations, which are found all along the gene, sporadic mutations are clustered (about 805 of all sporadic mutations) at the 5' end of exon 15, between codons 1280 and 1500 ***(57)***. This region is known as the mutation cluster region (MCR). It overlaps with the 15-amino acid domain between residues 1020 and 1169, and the 20-aa repeats between residues 1324 and 2075 that are responsible for β-catenin interactions, as well as the SAMP repeats located within the 20-aa repeats that interact with Axin. In addition, three somatic mutational hotspots can be found at codons 1309, 1450, and 1554, and account for 7%, 8%, and 5% of all somatic mutations, respectively. In addition, the promoter regions of *APC* termed 1A and 1B may be targets of an epigenetic mechanism of silencing—hypermethylation. Although there is no evidence of 1B methylation, 1A is heavily methylated in colorectal cancers, and this is thought to be the second hit in colon tumorigenesis ***(58)***. Interestingly, no evidence for the hypermethylation of 1A in normal colonic mucosa can be found ***(59)***.

In Ashkenazi Jews, about 6% of the population expresses a germline variation of APC known as the I1307K variant ***(60)***. This polymorphism leads to a hypermutable A_8 tract that can undergo slippage to produce frameshift mutations. Accordingly, I1307K allele carriers are 50–70% more likely to develop colon cancer ***(61)***. Another variant is the E1317Q allele, which is associated with multiple colorectal adenomas ***(62)***. It has been suggested that E1317Q affects APC function by subtle effects on β-catenin sequestration or degradation. In fact, most germline and somatic mutations result in the loss or alteration of β-catenin binding, as most of these point mutations occur at CpG dinucleotides, generating stop codons by a C-to-T transition in a CGA sequence.

3. APC Function

APC plays important roles in cell cycle, motility, adhesion, and signaling. It is often referred to as a multitasking protein, and the disruptions of its interactions may lead to the inability of APC to perform these functions, thereby contributing to tumor formation. The main functions attributed to APC are downregulation of the Wnt pathway (through β-catenin), modulation of cell adhesion/migration, and maintenance of chromosomal stability.

3.1. Downregulation of Wnt Signaling

The Wnt family of proteins is comprised of several members, all of which are cysteine-rich secreted proteins of about 38–45 kDa in size. Wnts were first identified in *Drosophila*, are highly conserved throughout the species, and are important in mediating cell–cell interactions during embryogenesis. In normal development, Wnts are expressed both tissue specifically and temporally ***(63)***. Deletions of Wnt in mice result in very specific phenotypes, and will be discussed in the section concerning transgenic animals. In general, the deletion of specific Wnts results in the lack of development of specific organs, stressing the importance of Wnts in development due to their signal transduction cascades during proliferation and differentiation.

3.1.1. β-Catenin Regulation

One of the main features of Wnt signaling is the regulation of β-catenin stabilization. Although β-catenin stability can be regulated by Wnt/APC-independent mechanisms, such as via integrin-linked kinase, GBP, and the c-met receptor tyrosine kinase ***(64)***, APC appears to be a key player. In the absence of Wnt signals, cytoplasmic β-catenin is phosphorylated at multiple residues in its N terminus by glycogen synthase kinase 3β (GSK3β) and then targeted for degradation (**Fig. 3**). The phosphorylation of β-catenin depends on a multiprotein complex that consists of APC, Axin/Conductin, and GSK3β ***(40,65)***. Phosphorylated β-catenin is recognized by the ubiquitin ligase β-TrCP/Slimb, ubiquitinated ***(66–68)*** and degraded by the proteasome pathway ***(69–71)***. Axin and APC are also phosphorylated by GSK3β, resulting in enhanced β-catenin binding to the complex and thus enhanced β-catenin degradation ***(72,73)***.

Wnt signaling leads to the stabilization of β-catenin and its accumulation in the cytoplasm. The free β-catenin in the cytoplasm can translocate into the nucleus and participate in the transcription of target genes through its association with T-cell factors (TCFs), which provide the DNA-binding moiety ***(74)***. Because β-catenin does not contain a nuclear localization sequence, it was originally suggested that β-catenin enters the nucleus through its association with TCF. More recent evidence suggests that β-catenin enters the nucleus through direct interactions with the nuclear pore complex in a manner similar to importin β ***(75,76)***.

The APC protein can be identified both in the cytoplasm and the nucleus and contains NLS ***(77–79)***. Sublocalization of APC can vary dramatically in epithelial cells of

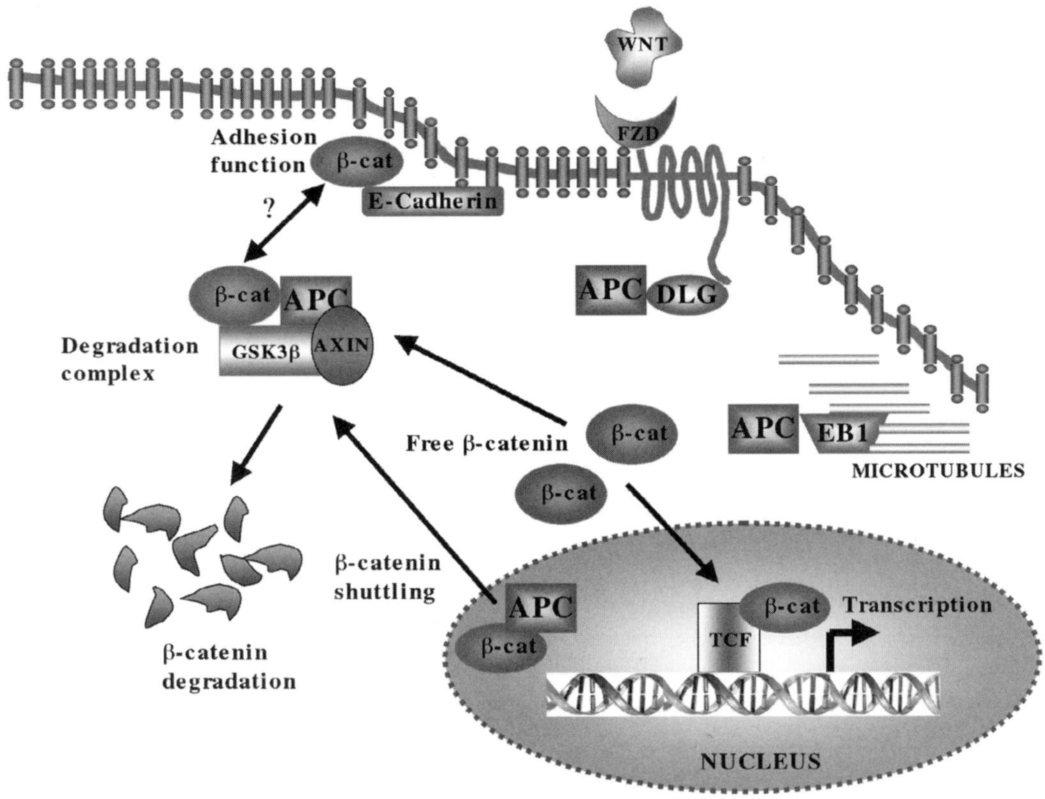

Fig. 3. The APC/β-catenin signaling pathway. See text for details.

the same lineage; for example, enterocytes at the base of the intestinal crypt are negative for APC, but the upper third of the crypt expresses some APC. The cells at the luminal surface of the crypt express even more APC, which is localized at their apical surface ***(77)***. At its N terminus, APC also contains two nuclear export sequences at amino acids 68–77 and 165–174, as well as a putative one at 1472–1481 ***(80)***. APC could thus downregulate the Wnt pathway by acting as a shuttle for β-catenin from the nucleus to the degradation complex in the cytoplasm. Mutant APC proteins lose the ability to exit from the nucleus leading to an excess nuclear β-catenin and possibly the activation of TCF/β-catenin transcriptional targets.

3.1.2. TCF/β-Catenin Transcriptional Activity and Targets of the Pathway

TCF/Lef transcription factors were originally identified as lymphoid-specific DNA-binding proteins that recognize a specific sequence- 5′CTTTGWW3′, where W = A or T ***(81–83)***. TCF/Lef bind to DNA via their high-mobility group (HMG), which induces a sharp bend in the DNA helix, but have no transactivation domains ***(74,84)***. As such, TCF/Lef are considered "silent" transcription factors, as they can activate gene transcription upon the binding of β-catenin, which contains the transactivation domains that TCF lacks, and these are contained within the C- and N-terminal ARD ***(85-87)***.

In the absence of Wnt signaling, TCFs can actively mediate the repression of Wnt-regulated genes via binding of the co-repressors TLE/Groucho ***(88–91)***. Groucho, for example, can repress transcription via the recruitment of repressive chromatin through interactions with histone H3 ***(88)***. CBP was reported to antagonize Wnt signaling by interacting directly with TCF, and acetylating a lysine residue in the β-catenin region of dTCF ***(92)***. On the other hand, CBP was also found to interact directly with β-catenin and act as a transcriptional coactivator to increase expression from TCF sites ***(93–95)***. The reason for these two apparently opposite functions of CBP on the Wnt pathway are currently unknown. TCF transcriptional activity can also be modified by phosphorylation. For example, a MAP kinase-related pathway involving a TGF-β-activated kinase (TAK1) and NEMO-like kinase can antagonize β-catenin/TCF signaling by phosphorylating TCF ***(96)***.

Many putative targets of the TCF/β-catenin transcription factors have been identified in tumorigenesis. The first two genes to be identified as targets of TCF/β-catenin, c-Myc, and Cyclin D1, provided a molecular basis for the oncogenic property of the pathway ***(97–99)***. The transcription of c-Myc can be repressed by overexpressing APC ***(97)***, and transfecting cells with a dominant negative TCF can result in the arrest of colon cancer cells in the G1 phase of the cell cycle, by interfering with the production of cyclin D1 (*98*). Other targets of this pathway include PPARδ ***(100)***, the matrix metalloproteinase matrilysin ***(91,101)***, TCF1 ***(102)***, and AP-1, another transcription factor, as well as c-jun and fra1 ***(103)***. Inappropriate activation of one or some of these genes may be responsible for the oncogenic properties of the pathway, although this has not been convincingly proven ***(104)***.

3.1.3. Mutations of the Wnt Pathway in Colon and Other Cancers

APC mutations account for 80% of sporadic colorectal cancers and have been found at low frequencies in a few other cancers ***(105–108)***. The general importance of the Wnt pathway in human cancer has been emphasized by the identification of mutations of various members of the pathway in a large number of human malignancies. The case of β-catenin is particularly interesting. As described above, APC is crucial for the downregualtion of β-catenin-mediated transcriptional activity. Interestingly, activating mutations that make β-catenin resistant to ubiquitination and proteasome degradation have been identified in about half the colon cancers with wild-type APC ***(109,110)***. These cancers presumably have no selective pressure for APC mutations, since they already have the pathway activated through a β-catenin-activating mutation. These mutations typically arise at the N-terminal GSK3β phosphorylation site, essential for β-catenin degradation ***(64,73)***. Mutations in β-catenin have now been identified in many malignancies, including melanoma ***(111)***, ovarian cancer ***(112)***, prostate cancer ***(113)***, skin cancer ***(114)***, medulloblastoma ***(115)***, liver cancer ***(116,117)***, and endometrial cancer ***(118)***. Finally, AXIN1 mutations have also been identified in hepatocellular carcinomas ***(119)***. These findings emphasize the importance of this pathway in human cancer and its relevance to the function of APC.

3.2. Adhesion and Migration

One of the interesting observations of tumors in the intestinal epithelium was their abnormal tissue architecture, and concomitant defects in cell migration and adhesion. As these tumors arise from the loss of functional APC, the role of APC in the change of the morphology of these cells was examined. In addition to its well-known function as a β-catenin regulator, and thus the regulation of β-catenin/ E-cadherin interactions in cell–cell adhesion, APC also has strong links with other molecules involved in cell migration. As mentioned previously, APC accumulates at the crypt/villus boundary, and this location is critical to the enterocyte migration from the crypt ***(120)***. When a C-terminus-truncated form of APC is introduced into mice, heterozygotes show an increase in their β-catenin levels in the enterocytes, which accumulate at the crypt/villus boundary ***(121)***. In addition, forcibly expressing the full-length gene also results in a deregulation of cell migration ***(122)*** and expression of an oncogenic β-catenin in the mouse intestine lead to abnormal architecture of the villi ***(123)***. APC-rich membranes are actively involved in cell migration in response to wounded epithelial monolayers, upon addition of scatter factor /HGF ***(124)***.

In addition, as mentioned earlier, APC binds to a rac-specific guanine nucleotide exchange factor termed ASEF ***(31)***. The rac/rho pathway has been shown to be involved in the regulation of the actin cytoskeletal network and is responsible for phenomena such as membrane ruffling, cell flattening, and motility. Microtubule reorganization may also be a target of APC signaling. These dynamic polymers of α and β subunits are involved in cell migration and polarized cellular morphogenesis, and their functions may be regulated by APC. APC has been found to be localized at the end of microtubules, regions known as puncta, where microtubules have reassembled to form protrusions, an important part of migration ***(125)***. APC has in-vivo microtubule-binding capacities and can nucleate microtubules in vitro.

3.3. Role of APC in Chromosomal Stability/Mitotic Checkpoint

Both the microtubule-binding domain and the EB1-binding region of APC implicate it in the regulation of chromosomal instability ***(44,126–128)***. Both microtubules and EB1 are associated with mitosis and cell division and spindle formation in dividing cells ***(129)***. EB1 is also associated with centrosomes and required for a microtubule-dependent checkpoint in yeast ***(130)***. These suggestive data have been supported recently by a study that shows that mouse embryonic stem cells that are homozygous for the min or APC1638T (truncated APC) alleles display extensive chromosome and spindle aberrations, implicating APC as an important player in the mitotic checkpoint ***(131,132)***. During these processes, APC associates with the checkpoint proteins Bub1 and Bub3 and can be found accumulated at the kinetochore. These results provide evidence that the C-terminal end of APC is crucial for chromosomal stability and may contribute to the observed chromosomal instability in colon cancer. Interestingly, the functions of APC as a regulator of β-catenin and as a cell-cycle checkpoint protein are likely independent, since ES cells from APC1638T mice also exhibit chromosomal instability. APC1638T homozygous mice contain a C-terminally deleted APC protein capable of degrading β-catenin but deficient in its ability to interact with EB1.

3.4. Consequences of APC Loss of Function

The exact physiologic role of the APC tumor suppressor pathway in tumorigenesis remains to be elucidated. Overexpression of APC has been reported to increase apoptosis in an APC-deficient colon cell line, suggesting that β-catenin may act as a survival factor ***(133)***. Interestingly, mice lacking *TCF4,* a *TCF* family member implicated in colon cancer ***(134)***, are characterized by a lack of intestinal stem cells ***(135)***. These findings suggest that the β-catenin/TCF pathway may be important for the maintenance of stem cell characteristics including longevity. Thus, the first step in colon tumorigenesis may be the activation of the APC/β-catenin pathway in order to inhibit differentiation and preserve stem cell characteristics. *APC* expression has also been reported to cause a G1/S block in NIH 3T3 cells ***(136)***, and this block is associated with the interaction of APC with hDLG ***(137)***. Consistent with this finding expression of a dominant negative TCF protein in a colon cancer line containing an activated β-catenin was shown to cause a G1 block ***(98)*** and overexpression of an activated β-catenin promotes G1/S transition ***(138)***. The effect of β-catenin activation on cell proliferation has been controversial, with some studies demonstrating increased proliferation of cells expressing activated β-catenin ***(138,139)*** and other studies finding an effect on tumor size but not on proliferation ***(140)***. β-Catenin expressing cells were also found to exhibit reduced contact inhibition and an increased ability to grow in soft agar ***(138)***.

As is the case for other oncogenes, β-catenin activation can in some cases have negative effects on cell proliferation. Indeed, it has been shown that APC inactivation in *Drosophila* leads to retinal defects similar to those occasionally observed in FAP patients ***(141)***. The *Drosophila* retinal defects are due to increased levels of apoptosis in retinal neurons, and this phenotype appears to be dependent on Armadillo upregulation. These findings demonstrate that activation of the APC/β-catenin pathway may have very different consequences in different tissue types and further implicates this pathway in tissue homeostasis.

4. Animal Models

Animal models have contributed significantly to our understanding of the APC tumor suppressor pathway. The previous sections have already mentioned several of these models. Here, we will describe in more details these models with deficiencies or overexpression of the various proteins important in the pathway.

4.1. Apc

The first Apc-deficient mice were generated by using ethylnitrosurea to create a point mutation in a mouse germline. One line of mutagenized mice was termed the min mouse, which has been briefly described earlier in this chapter. These mice display multiple intestinal neoplasia and have an anemic phenotype caused by multiple bleeding polyps in the intestine, very similar to the phenotype seen in patients with FAP. The Min mice were found to carry a nonsense mutation in mouse *Apc* ***(142)***. The homozygous mice are not viable *in utero,* and the heterozygous mice rarely lived past 5 mo. The tumor profile of these animals differs slightly from FAP patients in that tumors develop in the small rather than large intestine and about 10% of animals develop mammary tumors. There is recent evidence, based on a highly sensitive technique, that inactivating mutations of APC may be involved in breast cancer ***(108)***, but this finding requires confirmation. Interestingly, the crossing of these animals with other strains revealed the interaction of *APC* with modifier loci. When min mice were crossed with C57BL/6 mice (their parent strain), they developed high numbers of polyps, but when they were crossed with other strains (AKR, for example), they appeared resistant to tumor development. The gene controlling this tumor profile was identified as *Mom1* (modifier of Min 1) and was mapped to the distal end of mouse chromosome 4 ***(143)***. The human homolog to this gene has been identified as a secretory type II phospholipase A2 (*sPLA2*) and is located at 1p35 a chromosomal region frequently lost in human cancer. However, this gene was not found mutated in human colorectal cancer and is not believed to play a major role in this disease ***(144)***. The extent of environmental influences and epigenetic effects on FAP patients remains unknown.

Apc-deficient mice have also been developed using gene targeting. The Apc Δ716 knockout mice are heterozygotes, which develop intestinal adenomas at a high rate ***(145)***. These mice develop tumors also along the small intestine, characterized by an abnormal bulging or pocketing of the intestinal epithelium at the crypt/villus boundary. However, there was no increase on the rate of proliferation of polyps, which are polyclonal in origin. In both the mice strains described above, animals homozygous for the deletion die *in utero*. To circumvent this, a conditional gene targeting system was used. The *Apc* gene was inactivated by the insertion of *loxP* sites into the introns on either side of exon 14 of the *Apc* gene, and then introducing this gene into the mouse germline. These mice bearing the *Apc-loxp* gene developed normally. By then using an adenovirus encoding the *cre*-recombinase gene to infect the colorectal region, the *Apc* allele could be specifically inactivated, once *Cre* recombined with *loxP*-deleting exon 14. Mice developed adenomas within 4 wk, and these adenomas showed deletion of exon 14 of *Apc* ***(146)***. The loss of APC in these animals provides us with many insights into how FAP may develop, and what kind of role APC plays in development, as the loss of APC leads to malformation of the intestinal crypts, due to defects in many process ranging from inappropriate activation of TCF to defects in adhesion, migration, and even apoptosis and growth arrest. Finally, Apc1638T mice have contributed significantly to our understanding of APC function ***(36)***. These mice are viable as homozygotes, and the adults do not develop tumors. The Apc1638T allele can downregulate β-catenin (albeit with a lower efficiency that wild-type APC), but appears to be deficient in its ability to bind microtubules (*see* **Subheading 3.3.**).

4.2. β-Catenin

Transgenic mice engineered to express intestinal β-catenin with a truncated N terminus develop polyps due to outpocketing of the intestinal epithelium ***(123)***. This epithelium remains intact even after villi development and fusion of adjacent villi, and the crypt compartment expands abnormally, perhaps due to the β-catenin/TCF4 signaling, as the crypt cells continue to proliferate without differentiating. When oncogenic β-catenin is expressed in the skin, however, there is increased morphogenesis at the hair follicle, resulting in mice with abundant fur ***(114)***. In addition, several tumors known as pilomatricomas, tumors of the hair matrix, form, and this has been supported in humans by the observation of β-catenin mutations in pilomatricomas. Nuclear overexpression of LEF1 was found to accompany these tumors in transgenic mice, indicating that these effects were due to APC/β-catenin deregulation.

4.3. TCFs and Other Wnt Pathway Proteins

Animals engineered to have defects in the other members of the APC/β-catenin signaling pathway also provide useful information as to the role of these proteins in disease and development. TCF1 knockouts do not develop T cells, but do develop adenomas in the gut and mammary glands, and the addition of mutant APC increases these tumors, and this implicates *TCF1* as a repressor of target genes in the absence of activated β-catenin, perhaps in cohort with *APC* ***(102)***. TCF4 knockouts lead to abnormalities of the small intestine, where the stem cell compartment of the crypts is missing, implicating Wnt signaling in the proliferation of these crypt stem cells ***(135)***. Lef1 knockouts exhibit defects in the formation of teeth, hair, and mammary glands, and this is thought to be mediated via disruptions of the signaling of these epithelial cells with their underlying mesenchyme ***(147)***. Mice that are deficient for both Lef1 and Tcf1 do not form paraxial mesoderm and develop an excess of neural tubes.

Mutations of the Wnt genes can sometimes mimic these findings, indicating an overlap of functions for the different members of this pathway. Wnt3a knockouts for example, mimic the Lef1/Tcf1 knockouts ***(148)***. Most Wnt knockouts, however, have significant effects on development. The ablation of Wnt-1 results in the abolition of a part of the mid-brain ***(149)***, Wnt4 and Wnt7a ablation result in the malformation of the kidney and limbs ***(150,151)***, and Wnt 5/Frizzled 5 disruption can have effects on placental angiogenesis ***(152)***. Disheveled gene abolition appears to have no effect on development, but some data suggest that they may have effects on behavior in the adult mice. Other transgenic animals containing mutations in targets of the β-catenin, such as c-myc, which develop hepatocellular carcinomas, demonstrate concomitant mutations in the GSK3β phosphorylation site of β-catenin ***(117)***, perhaps suggesting that targets other than c-Myc are critical for tumorigenesis through the APC pathway.

5. Conclusions

APC is a large protein that wears many hats. It is involved not only in cell signaling and control of gene transcription, but also in cell cycle regulation, cell proliferation, cell migration, cell adhesion, cytoskeletal reorganization, and chromosomal instability. The loss or deregulation of APC can thus have many consequences for a cell, including tumorigenesis. Which of these functions are interrelated and which are independent is a matter of intense investigation. Another important question is whether all of these functions are important for the tumor suppressive function of *APC*. Overall based on several independent lines of evidence, it appears that the downregulation of β-catenin is the major pathway by which APC suppresses tumorigenesis. This is supported by the following evidence: 1) mutations in *APC* and β-catenin tend to be mutually exclusive in colon cancer, 2) mice containing mutant alleles of *APC* that can downregulate β-catenin do not develop tumors, and 3) β-catenin mutations are found in many cancer types. It is possible and even likely that other functions of *APC* are important in tumor development. Consistent with this hypothesis is the fact that tumors containing β-catenin appear less likely to progress to malignancy than APC deficient tumors ***(153)***. This is also consistent with an important function of APC in chromosomal stability. In any event, the rapid advances of the past 10 years have allowed a deep understanding of the most important pathway deregulated in colon tumorigenesis. There is no doubt that future studies will clarify the many remaining questions and will provide a clear view of the molecular mechanisms important in this devastating disease.

References

1. Greenlee, R. T., Hill-Harmon, M. B., Murray, T., and Thun, M. (2001) Cancer statistics, 2001. *CA Cancer J. Clin.* **51**, 15–36.

2. Vogelstein, B. and Kinzler, K. W. (1993) The multistep nature of cancer. *Trends Genet.* **9**, 138–141.
3. Vogelstein, B., Fearon, E. R., Hamilton, S. R., et al. (1988) Genetic alterations during colorectal-tumor development. *N. Engl. J. Med.* **319**, 525–532.
4. Fearon, E. R. and Vogelstein, B. (1990) A genetic model for colorectal tumorigenesis. *Cell* **61**, 759–767.
5. Nagase, H. and Nakamura, Y. (1993) Mutations of the APC (adenomatous polyposis coli) gene. *Hum. Mutat.* **2**, 425–434.
6. Forrester, K., Almoguera, C., Han, K., Grizzle, W. E., and Perucho, M. (1987) Detection of high incidence of K-ras oncogenes during human colon tumorigenesis. *Nature* **327**, 298–303.
7. Jen, J., Powell, S. M., Papadopoulos, N., et al. (1994) Molecular determinants of dysplasia in colorectal lesions. *Cancer Res.* **54**, 5523–5526.
8. Zauber, N. P., Sabbath-Solitare, M., Marotta, S. P., and Bishop, D. T. (1999) K-ras mutation and loss of heterozygosity of the adenomatous polyposis coli gene in patients with colorectal adenomas with in situ carcinoma. *Cancer* **86**, 31–36.
9. Howe, J. R., Roth, S., Ringold, J. C., et al. (1998) Mutations in the SMAD4/DPC4 gene in juvenile polyposis. *Science* **280**, 1086–1088.
10. Gryfe, R., Swallow, C., Bapat, B., Redston, M., Gallinger, S., and Couture, J. (1997) Molecular biology of colorectal cancer. *Curr. Probl. Cancer* **21**, 233–300.
11. Jiricny, J. and Nystrom-Lahti, M. (2000) Mismatch repair defects in cancer. *Curr. Opin. Genet. Dev.* **10**, 157–161.
12. Huang, J., Papadopoulos, N., McKinley, A. J., et al. (1996) APC mutations in colorectal tumors with mismatch repair deficiency. *Proc. Natl. Acad. Sci. USA* **20**, 9049–9054.
13. Herman, J. G., Umar, A., Polyak, K., et al. (1998) Incidence and functional consequences of hMLH1 promoter hypermethylation in colorectal carcinoma. *Proc. Natl. Acad. Sci. USA* **95**, 6870–6875.
14. Veigl, M. L., Kasturi, L., Olechnowicz, J., et al. (1998) Biallelic inactivation of hMLH1 by epigenetic gene silencing, a novel mechanism causing human MSI cancers. *Proc. Natl. Acad. Sci. USA* **95**, 8698–8702.
15. Herrera, L., Kakati, S., Gibas, L., Pietrzak, E., and Sandberg, A. A. (1986) Gardner syndrome in a man with an interstitial deletion of 5q. *Am. J. Med. Genet.* **25**, 473–476.
16. Lynch, H. T., Rozen, P., and Schuelke, G. S. (1985) Hereditary colon cancer: polyposis and nonpolyposis variants. *CA. Cancer J. Clin.* **35**, 95–114.
17. Joslyn, G., Carlson, M., Thliveris, A., et al. (1991) Identification of deletion mutations and three new genes at the familial polyposis locus. *Cell* **66**, 601–613.
18. Kinzler, K. W., Nilbert, M. C., Su, L. K., et al. (1991) Identification of FAP locus genes from chromosome 5q21. *Science* **253**, 661–665.
19. Groden, J., Thliveris, A., Samowitz, W., et al. (1991) Identification and characterization of the familial adenomatous polyposis coli gene. *Cell* **66**, 589–600.
20. Goss, K. H. and Groden, J. (2000) Biology of the adenomatous polyposis coli tumor suppressor. *J. Clin. Oncol.* **18**, 1967–1979.
21. Thliveris, A., Albertsen, H., Tuohy, T., et al. (1996) Long-range physical map and deletion characterization of the 1100-kb NotI restriction fragment harboring the APC gene. *Genomics* **34**, 268–270.
22. Sulekova, Z., Reina-Sanchez, J., and Ballhausen, W. G. (1995) Multiple APC messenger RNA isoforms encoding exon 15 short open reading frames are expressed in the context of a novel exon 10A-derived sequence. *Int. J. Cancer* **63**, 435–441.
23. Sieber, O. M., Tomlinson, I. P., and Lamlum, H. (2000) The adenomatous polyposis coli (APC) tumour suppressor—genetics, function and disease. *Mol. Med. Today* **6**, 462–469.

24. Santoro, I. M. and Groden, J. (1997) Alternative splicing of the APC gene and its association with terminal differentiation. *Cancer Res.* **57**, 488–494.
25. Horii, A., Nakatsuru, S., Ichii, S., Nagase, H., and Nakamura, Y. (1993) Multiple forms of the APC gene transcripts and their tissue-specific expression. *Hum. Mol. Genet.* **2**, 283–287.
26. van Es, J. H., Kirkpatrick, C., van de Wetering, M., et al. (1999) Identification of APC2, a homologue of the adenomatous polyposis coli tumour suppressor. Curr. Biol. 9, 105–108.
27. Nakamura, T., Hamada, F., Ishidate, T., et al. (1998) Axin, an inhibitor of the Wnt signalling pathway, interacts with beta-catenin, GSK-3beta and APC and reduces the beta-catenin level. *Genes Cells* **3**, 395–403.
28. Nakagawa, H., Koyama, K., Murata, Y., Morito, M., Akiyama, T., and Nakamura, Y. (2000) APCL, a central nervous system-specific homologue of adenomatous polyposis coli tumor suppressor, binds to p53-binding protein 2 and translocates it to the perinucleus. *Cancer Res.* **60**, 101–105.
29. van Es, J. H., Giles, R. H., and Clevers, H. C. (2001) The many faces of the tumor suppressor gene apc. *Exp. Cell Res.* **264**, 126–134.
30. Peifer, M., Berg, S., and Reynolds, A. B. (1994) A repeating amino acid motif shared by proteins with diverse cellular roles. *Cell* **76**, 789–791.
31. Kawasaki, Y., Senda, T., Ishidate, T., et al. (2000) Asef, a link between the tumor suppressor APC and G-protein signaling. *Science* **289**, 1194–1197.
32. Orsulic, S. and Peifer, M. (1996) An in vivo structure-function study of Armadillo, the β-catenin homolog, reveals both separate and overlapping regions of the protein required for cell adhesion and for cell signaling. *J. Cell Biol.* **134**, 1283–1300.
33. Huber, A. H., Nelson, W. J., and Weis, W. I. (1997) Three-dimensional structure of the armadillo repeat region of beta-catenin. *Cell* **90**, 871–882.
34. Seeling, J. M., Miller, J. R., Gil, R., Moon, R. T., White, R., and Virshup, D. M. (1999) Regulation of beta-catenin signaling by the B56 subunit of protein phosphatase 2A. *Science* **283**, 2089–2091.
35. Hsu, W., Zeng, L., and Costantini, F. (1999) Identification of a domain of axin that binds to the serine/threonine protein phosphatase 2A and a self-binding domain. *J. Biol. Chem.* **274**, 3439–3445.
36. Smits, R., Kielman, M. F., Breukel, C., et al. (1999) Apc1638T: a mouse model delineating critical domains of the adenomatous polyposis coli protein involved in tumorigenesis and development. *Genes Dev.* **13**, 1309–1321.
37. Rubinfeld, B., Souza, B., Albert, I., Munemitsu, S., and Polakis, P. (1995) The APC protein and E-cadherin form similar but independent complexes with alpha-catenin, beta-catenin, and plakoglobin. *J. Biol. Chem.* **270**, 5549–5555.
38. Rubinfeld, B., Albert, I., Porfiri, E., Fiol, C., Munemitsu, S., and Polakis, P. (1996) Binding of GSK3-beta to the APC-beta-catenin complex and regulation of complex assembly. *Science* **272**, 1023–1025.
39. Rubinfeld, B., Albert, I., Porfiri, E., Munemitsu, S., and Polakis, P. (1997) Loss of beta-catenin regulation by the APC tumor suppressor protein correlates with loss of structure due to common somatic mutations of the gene. *Cancer Res.* **57**, 4624–4630.
40. Behrens, J., Jerchow, B., Wurtele, M., et al. (1998) Functional interaction of an axin homolog, conductin, with beta-catenin, APC, and GSK3 beta. *Science* **280**, 596–599.
41. von Kries, J. P., Winbeck, G., Asbrand, C., et al. (2000) Hot spots in beta-catenin for interactions with LEF-1, conductin and APC. *Nat. Struct. Biol.* **7**, 800–807.
42. Smith, K. J., Levy, D. B., Maupin, P., Pollard, T. D., Vogelstein, B., and Kinzler, K. W. (1994) Wild-type but not mutant APC associates with the microtubule cytoskeleton. *Cancer Res.* **54**, 3672–3675.

43. Munemitsu, S., Souza, B., Muller, O., Albert, I., Rubinfeld, B., and Polakis, P. (1994) The APC gene product associates with microtubules in vivo and promotes their assembly in vitro. *Cancer Res.* **54**, 3676–3681.
44. Su, L. K., Burrell, M., Hill, D. E., et al. (1995) APC binds to the novel protein EB1. *Cancer Res.* **55**, 2972–2977.
45. Mimori-Kiyosue, Y., Shiina, N., and Tsukita, S. (2000) Adenomatous polyposis coli (APC) protein moves along microtubules and concentrates at their growing ends in epithelial cells. *J. Cell Biol.* **148**, 505–518.
46. Matsumine, A., Ogai, A., Senda, T., et al. (1996) Binding of APC to the human homolog of the *Drosophila* discs large tumor suppressor protein. *Science* **272**, 1020–1023.
47. Erdmann, K. S., Kuhlmann, J., Lessmann, V., et al. (2000) The Adenomatous Polyposis Coli-protein (APC) interacts with the protein tyrosine phosphatase PTP-BL via an alternatively spliced PDZ domain. *Oncogene* **19**, 3894–3901.
48. Su, L. K., Johnson, K. A., Smith, K. J., Hill, D. E., Vogelstein, B., and Kinzler, K. W. (1993) Association between wild type and mutant APC gene products. *Cancer Res.* **53**, 2728–2731.
49. Joslyn, G., Richardson, D. S., White, R., and Alber, T. (1993) Dimer formation by an N-terminal coiled coil in the APC protein. *Proc. Natl. Acad. Sci. USA* **90**, 11109–11113.
50. Bodmer, W., Bishop, T., and Karran, P. (1994) Genetic steps in colorectal cancer. *Nat. Genet.* **6**, 217–219.
51. Dihlmann, S., Gebert, J., Siermann, A., Herfarth, C., and Doeberitz, M. V. (1999) Dominant negative effect of the APC(1309) mutation: a possible explanation for genotype-phenotype correlations in familial adenomatous polyposis. *Cancer Res.* **59**, 1857–1860.
52. Laurent-Puig, P., Beroud, C., and Soussi, T. (1998) APC gene: database of germline and somatic mutations in human tumors and cell lines. *Nucleic Acids Res.* **26**, 269–270.
53. Spirio, L. N., Samowitz, W., Robertson, J., et al. (1998) Alleles of APC modulate the frequency and classes of mutations that read to colon polyps. *Nat. Genet.* **20**, 385–388.
54. Brensinger, J. D., Laken, S. J., Luce, M. C., et al. (1998) Variable phenotype of familial adenomatous polyposis in pedigrees with 3' mutation in the APC gene. *Gut* **43**, 548–552.
55. Olschwang, S., Tiret, A., Laurent-Puig, P., Muleris, M., Parc, R., and Thomas, G. (1993) Restriction of ocular fundus lesions to a specific subgroup of APC mutations in adenomatous polyposis coli patients. *Cell* **75**, 959–968.
56. Caspari, R., Olschwang, S., Friedl, W., et al. (1995) Familial adenomatous polyposis: desmoid tumours and lack of ophthalmic lesions (CHRPE) associated with APC mutations beyond codon 1444. *Hum. Mol. Genet.* **4**, 337–340.
57. Miyoshi, Y., Nagase, H., Ando, H., et al. (1992) Somatic mutations of the APC gene in colorectal tumors: mutation cluster region in the APC gene. *Hum. Mol. Genet.* **1**, 229–233.
58. Lambertz, S. and Ballhausen, W. G. (1993) Identification of an alternative 5' untranslated region of the adenomatous polyposis coli gene. *Hum. Genet.* **90**, 650–652.
59. Hiltunen, M., Alhonen, L., Koistinaho, J., et al. (1997) Hypermethylation of the APC (adenomatous polyposis coli) gene promoter region in human colorectal carcinoma. *Int. J. Cancer* **70**, 644–648.
60. Laken, S. J., Petersen, G. M., Gruber, S. B., et al. (1997) Familial colorectal cancer in Ashkenazim due to a hypermutable tract in APC. *Nat. Genet.* **17**, 79–83.
61. Gryfe, R., Di Nicola, N., Lal, G., Gallinger, S., and Redston, M. (1999) Inherited colorectal polyposis and cancer risk of the APC I1307K polymorphism. *Am. J. Hum. Genet.* **64**, 378–384.
62. Popat, S., Stone, J., Coleman, G., et al. (2000) Prevalence of the APC E1317Q variant in colorectal cancer patients. *Cancer Lett.* **149**, 203–206.
63. Cadigan, K. M. and Nusse, R. (1997) Wnt signaling: a common theme in animal development. *Genes Dev.* **11**, 3286–3305.
64. Morin, P. J. (1999) beta-Catenin signaling and cancer. *Bioessays* **21**, 1021–1030.

65. Ikeda, S., Kishida, S., Yamamoto, H., Murai, H., Koyama, S., and Kikuchi, A. (1998) Axin, a negative regulator of the wnt signaling pathway, forms a complex with GSK-3beta and beta-catenin and promotes GSK-3beta-dependent phosphorylation of beta-catenin. *EMBO J.* **17**, 1371–1384.
66. Jiang, J. and Struhl, G. (1998) Regulation of the Hedgehog and Wingless signalling pathways by the F-box/WD40-repeat protein Slimb. *Nature* **391**, 493–496.
67. Winston, J. T., Strack, P., Beer-Romero, P., Chu, C. Y., Elledge, S. J., and Harper, J. W. (1999) The SCF beta-TRCP-ubiquitin ligase complex associates specifically with phosphorylated destruction motifs in I kappa B alpha and beta-catenin and stimulates I kappa B alpha ubiquitination in vitro. *Genes Dev.* **13**, 270–283.
68. Kitagawa, M., Hatakeyama, S., Shirane, M., et al. (1999) An F-box protein, FWD1, mediates ubiquitin-dependent proteolysis of beta-catenin. *EMBO J.* **18**, 2401–2410.
69. Aberle, H., Bauer, A., Stappert, J., Kispert, A., and Kemler, R. (1997) beta-catenin is a target for the ubiquitin-proteasome pathway. *EMBO J.* **16**, 3797–3804.
70. Salomon, D., Sacco, P. A., Roy, S. G., et al. (1997) Regulation of beta-catenin levels and localization by overexpression of plakoglobin and inhibition of the ubiquitin-proteasome system. *J. Cell Biol.* **139**, 1325–1335.
71. Orford, K., Crockett, C., Jensen, J. P., Weissman, A. M., and Byers, S. W. (1997) Serine phosphorylation-regulated ubiquitination and degradation of beta-catenin. *J. Biol. Chem.* **272**, 24735–24738.
72. Willert, K., Shibamoto S., and Nusse, R. (1999) Wnt-induced dephosphorylation of Axin releases beta-catenin from the Axin complex. *Genes Dev.* **13**, 1768–1773.
73. Polakis, P. (1999) The oncogenic activation of beta-catenin. *Curr. Opin. Genet. Dev.* **9**, 15–21.
74. Clevers, H. and van de Wetering, M. (1997) TCF/LEF factors earn their wings. *Trends Genet.* **13**, 485–489.
75. Fagotto, F., Gluck, U., and Gumbiner, B. M. (1998) Nuclear localization signal-independent and importin/karyopherin-independent nuclear import of beta-catenin. *Curr. Biol.* **8**, 181–190.
76. Yokoya, F., Imamoto, N., Tachibana, T., and Yoneda, Y. (1999) beta-catenin can be transported into the nucleus in a Ran-unassisted manner. *Mol. Biol. Cell* **10**, 1119–1131.
77. Smith, K. J., Johnson, K. A., Bryan, T. M., et al. (1993) The APC gene product in normal and tumor cells. *Proc. Natl. Acad. Sci. USA* **90**, 2846–2850.
78. Neufeld, K. L. and White, R. L. (1997) Nuclear and cytoplasmic localizations of the adenomatous polyposis coli protein. *Proc. Natl. Acad. Sci. USA* **94**, 3034–3039.
79. Zhang, F., White, R. L., and Neufeld, K. L. (2000) Phosphorylation near nuclear localization signal regulates nuclear import of adenomatous polyposis coli protein. *Proc. Natl. Acad. Sci. USA* **97**, 12577–12582.
80. Rosin-Arbesfeld, R., Townsley, F., and Bienz, M. (2000) The APC tumour suppressor has a nuclear export function. *Nature* **406**, 1009–1012.
81. van de Wetering, M., Oosterwegel, M., Dooijes, D., and Clevers, H. (1991) Identification and cloning of TCF-1, a T lymphocyte-specific transcription factor containing a sequence-specific HMG box. *EMBO J.* **10**, 123–132.
82. Travis, A., Amsterdam, A., Belanger, C., and Grosschedl, R. (1991) LEF-1, a gene encoding a lymphoid-specific protein with an HMG domain, regulates T-cell receptor alpha enhancer function [corrected]. *Genes Dev.* **5**, 880–894.
83. Giese, K., Amsterdam, A., and Grosschedl, R. (1991) DNA-binding properties of the HMG domain of the lymphoid-specific transcriptional regulator LEF-1. *Genes Dev.* **5**, 2567–2578.
84. Giese, K., Cox, J., and Grosschedl, R. (1992) The HMG domain of lymphoid enhancer factor 1 bends DNA and facilitates assembly of functional nucleoprotein structures. *Cell* **69**, 185–195.

85. Hsu, S. C., Galceran, J., and Grosschedl, R. (1998) Modulation of transcriptional regulation by LEF-1 in response to wnt-1 signaling and association with beta-catenin. *Mol. Cell. Biol.* **18**, 4807–4818.
86. van de Wetering, M., Cavallo, R., et al. (1997) Armadillo coactivates transcription driven by the product of the *Drosophila* segment polarity gene dTCF. *Cell* **88**, 789–799.
87. Hecht, A., Litterst, C. M., Huber, O., and Kemler, R. (1999) Functional characterization of multiple transactivating elements in beta-catenin, some of which interact with the TATA-binding protein in vitro. *J. Biol. Chem.* **274**, 18017–18025.
88. Cavallo, R. A., Cox, R. T., Moline, M. M., et al. (1998) *Drosophila* Tcf and Groucho interact to repress Wingless signalling activity. *Nature* **395**, 604–608.
89. Levanon, D., Goldstein, R. E., Bernstein, Y., et al. (1998) Transcriptional repression by AML1 and LEF-1 is mediated by the TLE/Groucho corepressors. *Proc. Natl. Acad. Sci. USA* **95**, 11590–11595.
90. Roose, J., Molenaar, M., Peterson, J., et al. (1998) The Xenopus Wnt effector XTcf-3 interacts with Groucho-related transcriptional repressors. *Nature* **395**, 608–612.
91. Crawford, H. C., Fingleton, B. M., Rudolph-Owen, L. A., et al. (1999) The metalloproteinase matrilysin is a target of beta-catenin transactivation in intestinal tumors. *Oncogene* **18**, 2883–2891.
92. Waltzer, L. and Bienz, M. (1998) *Drosophila* CBP represses the transcription factor TCF to antagonize Wingless signalling. *Nature* **395**, 521–525.
93. Takemaru, K. I. and Moon, R. T. (2000) The transcriptional coactivator CBP interacts with beta-catenin to activate gene expression. *J. Cell Biol.* **149**, 249–254.
94. Hecht, A., Vleminckx, K., Stemmler, M. P., van Roy, F., and Kemler, R. (2000) The p300/CBP acetyltransferases function as transcriptional coactivators of beta-catenin in vertebrates. *EMBO J.* **19**, 1839–1850.
95. Miyagishi, M., Fujii, R., Hatta, M., et al. (2000) Regulation of Lef-mediated transcription and p53-dependent pathway by associating beta-catenin with CBP/p300. *J. Biol. Chem.* **275**, 35170–35175.
96. Ishitani, T., Ninomiya-Tsuji, J., Nagai, S., et al. (1999) The TAK1-NLK-MAPK-related pathway antagonizes signalling between beta- catenin and transcription factor TCF. *Nature* **399**, 798–802.
97. He, T. C., Sparks, A. B., Rago, C., et al. (1998) Identification of c-MYC as a target of the APC pathway. *Science* **281**, 1509–1512.
98. Tetsu, O. and McCormick, F. (1999) Beta-catenin regulates expression of cyclin D1 in colon carcinoma cells. *Nature* **398**, 422–426.
99. Shtutman, M., Zhurinsky, J., Simcha, I., et al. (1999) The cyclin D1 gene is a target of the beta-catenin/LEF-1 pathway. *Proc. Natl. Acad. Sci. USA* **96**, 5522–5527.
100. He, T. C., Chan, T. A., Vogelstein, B., and Kinzler, K. W. (1999) PPAR delta is an APC-regulated target of nonsteroidal anti-inflammatory drugs. *Cell* **99**, 335–345.
101. Brabletz, T., Jung, A., Dag, S., Hlubek, F., and Kirchner, T. (1999) beta-catenin regulates the expression of the matrix metalloproteinase-7 in human colorectal cancer. *Am. J. Pathol.* **155**, 1033–1038.
102. Roose, J., Huls, G., van Beest, M., et al. (1999) Synergy between tumor suppressor APC and the beta-catenin-Tcf4 target Tcf1. *Science* **285**, 1923–1926.
103. Mann, B., Gelos, M., Siedow, A., et al. (1999) Target genes of beta-catenin-T cell-factor lymphoid-enhancer-factor signaling in human colorectal carcinomas. *Proc. Natl. Acad. Sci. USA* **96**, 1603–1608.
104. Kolligs, F. T., Hu, G., Dang, C. V., and Fearon, E. R. (1999) Neoplastic transformation of RK3E by mutant beta-catenin requires deregulation of Tcf/Lef transcription but not activation of c-myc expression. *Mol. Cell Biol.* **19**, 5696–5706.

105. Kinzler, K. W. and Vogelstein, B. (1996) Lessons from hereditary colorectal cancer. *Cell* **87**, 159–170.
106. Horii, A., Nakatsuru, S., Miyoshi, Y., et al. (1992) The APC gene, responsible for familial adenomatous polyposis, is mutated in human gastric cancer. *Cancer Res.* **52**, 3231–3233.
107. Oda, H., Imai, Y., Nakatsuru, Y., Hata, J., and Ishikawa, T. (1996) Somatic mutations of the APC gene in sporadic hepatoblastomas. *Cancer Res.* **56**, 3320–3323.
108. Furuuchi, K., Tada, M., Yamada, H., et al. (2000) Somatic mutations of the APC gene in primary breast cancers. *Am. J. Pathol.* **156**, 1997–2005.
109. Morin, P. J., Sparks, A. B., Korinek, V., et al. (1997) Activation of beta-catenin-Tcf signaling in colon cancer by mutations in beta-catenin or APC. *Science* **275**, 1787–1790.
110. Sparks, A. B., Morin, P. J., Vogelstein, B., and Kinzler, K. W. (1998) Mutational analysis of the APC/beta-catenin/Tcf pathway in colorectal cancer. *Cancer Res.* **58**, 1130–1134.
111. Rubinfeld, B., Robbins, P., El-Gamil, M., Albert, I., Porfiri, E., and Polakis, P. (1997) Stabilization of b-catenin by genetic defects in melanoma cell lines. *Science* **275**, 1790–1792.
112. Palacios, J. and Gamallo, C. (1998) Mutations in the beta-catenin gene (CTNNB1) in endometrioid ovarian carcinomas. *Cancer Res.* **58**, 1344–1347.
113. Voeller, H. J., Truica, C. I., and Gelmann, E. P. (1998) beta-catenin mutations in human prostate cancer. *Cancer Res.* **58**, 2520–2523.
114. Chan, E. F., Gat, U., McNiff, J. M., and Fuchs, E. (1999) A common human skin tumour is caused by activating mutations in beta-catenin. *Nat. Genet.* **21**, 410–413.
115. Zurawel, R. H., Chiappa, S. A., Allen, C., and Raffel, C. (1998) Sporadic medulloblastomas contain oncogenic beta-catenin mutations. *Cancer Res.* **58**, 896–899.
116. Miyoshi, Y., Iwao, K., Nagasawa, Y., et al. (1998) Activation of the beta-catenin gene in primary hepatocellular carcinomas by somatic alterations involving exon 3. *Cancer Res.* **58**, 2524–2527.
117. De La Coste, A., Romagnolo, B., Billuart, P., et al. (1998) Somatic mutations of the beta-catenin gene are frequent in mouse and human hepatocellular carcinomas. *Proc. Natl. Acad. Sci. USA* **95**, 8847–8851.
118. Fukuchi, T., Sakamoto, M., Tsuda, H., Maruyama, K., Nozawa, S., and Hirohashi, S. (1998) Beta-catenin mutation in carcinoma of the uterine endometrium. *Cancer Res.* **58**, 3526–3528.
119. Satoh, S., Daigo, Y., Furukawa, Y., et al. (2000) AXIN1 mutations in hepatocellular carcinomas, and growth suppression in cancer cells by virus-mediated transfer of AXIN1. *Nat. Genet.* **24**, 245–250.
120. Nathke, I. S., Adams, C. L., Polakis, P., Sellin, J. H. and Nelson, W. J. (1996) The adenomatous polyposis coli (APC) tumor suppressor protein localizes to plasma membrane sites involved in active cell migration. *J. Cell Biol.* **134**, 165–180.
121. Mahmoud, N. N., Boolbol, S. K., Bilinski, R. T., Martucci, C., Chadburn, A., and Bertagnolli, M. M. (1997) Apc gene mutation is associated with a dominant-negative effect upon intestinal cell migration. *Cancer Res.* **57**, 5045–5050.
122. Wong, M. H., Hermiston, M. L., Syder, A. J., and Gordon, J. I. (1996) Forced expression of the tumor suppressor adenomatosis polyposis coli protein induces disordered cell migration in the intestinal epithelium. *Proc. Natl. Acad. Sci. USA* **93**, 9588–9593.
123. Wong, M. H., Rubinfeld, B. and Gordon, J. I. (1998) Effects of forced expression of an NH_2-terminal truncated beta-catenin on mouse intestinal epithelial homeostasis. *J. Cell Biol.* **141**, 765–777.
124. Pollack, A. L., Barth, A. I. M., Altschuler, Y., Nelson, W. J., and Mostov, K. E. (1997) Dynamics of beta-catenin interactions with APC protein regulate epithelial tubulogenesis. *J. Cell Biol.* **137**, 1651–1662.

125. Zumbrunn, J., Kinoshita, K., Hyman, A. A., and Nathke, I. S. (2001) Binding of the adenomatous polyposis coli protein to microtubules increases microtubule stability and is regulated by GSK3 beta phosphorylation. *Curr. Biol.* **11**, 44–49.
126. Su, L. and Qi, Y. (2001) Characterization of human mapre genes and their proteins. *Genomics* **71**, 142–149.
127. Berrueta, L., Kraeft, S. K., Tirnauer, J. S., et al. (1998) The adenomatous polyposis coli-binding protein EB1 is associated with cytoplasmic and spindle microtubules. *Proc. Natl. Acad. Sci. USA* **95**, 10596–10601.
128. Askham, J. M., Moncur, P., Markham, A. F., and Morrison, E. E. (2000) Regulation and function of the interaction between the APC tumour suppressor protein and EB1. *Oncogene* **19**, 1950–1958.
129. Tirnauer, J. S. and Bierer, B. E. (2000) EB1 proteins regulate microtubule dynamics, cell polarity, and chromosome stability. *J. Cell Biol.* **149**, 761–766.
130. Muhua, L., Adames, N. R., Murphy, M. D., Shields, C. R., and Cooper, J. A. (1998) A cytokinesis checkpoint requiring the yeast homologue of an APC-binding protein. *Nature* **393**, 487–491.
131. Kaplan, K. B., Burds, A. A., Swedlow, J. R., Bekir, S. S., Sorger, P. K., and Nathke, I. S. (2001) A role for the Adenomatous Polyposis Coli protein in chromosome segregation. *Nat. Cell Biol.* **3**, 429–432.
132. Fodde, R., Kuipers, J., Rosenberg, C., et al. (2001) Mutations in the APC tumour suppressor gene cause chromosomal instability. *Nat. Cell Biol.* **3**, 433–438.
133. Morin, P. J., Vogelstein, B., and Kinzler, K. W. (1996) Apoptosis and APC in colorectal tumorigenesis. *Proc. Natl. Acad. Sci. USA* **93**, 7950–7954.
134. Korinek, V., Barker, N., Morin, P. J., et al. (1997) Constitutive transcriptional activation by a beta-catenin-Tcf complex in APC-/- colon carcinoma. *Science* **275**, 1784–1787.
135. Korinek, V., Barker, N., Moerer, P., et al. (1998) Depletion of epithelial stem-cell compartments in the small intestine of mice lacking Tcf-4. *Nat. Genet.* **19**, 379–383.
136. Baeg, G. H., Matsumine, A., Kuroda, T., et al. (1995) The tumour suppressor gene product APC blocks cell cycle progression from G0/G1 to S phase. *EMBO J.* **14**, 5618–5625.
137. Ishidate, T., Matsumine, A., Toyoshima, K., and Akiyama, T. (2000) The APC-hDLG complex negatively regulates cell cycle progression from the G0/G1 to S phase. *Oncogene* **19**, 365–372.
138. Orford, K., Orford, C. C., and Byers, S. W. (1999) Exogenous expression of beta-catenin regulates contact inhibition, anchorage-independent growth, anoikis, and radiation-induced cell cycle arrest. *J. Cell Biol.* **146**, 855–867.
139. Kuhnen, C., Herter, P., Muller, O., et al. (2000) Beta-catenin in soft tissue sarcomas: expression is related to proliferative activity in high-grade sarcomas. *Mod. Pathol.* **13**, 1005–1013.
140. Brabletz, T., Herrmann, K., Jung, A., Faller, G., and Kirchner, T. (2000) Expression of nuclear beta-catenin and c-myc is correlated with tumor size but not with proliferative activity of colorectal adenomas. *Am. J. Pathol.* **156**, 865–870.
141. Ahmed, Y., Hayashi, S., Levine, A., and Wieschaus, E. (1998) Regulation of armadillo by a Drosophila APC inhibits neuronal apoptosis during retinal development. *Cell* **93**, 1171–1182.
142. Su, L. K., Kinzler, K. W., Vogelstein, B., et al. (1992) Multiple intestinal neoplasia caused by a mutation in the murine homolog of the APC gene. *Science* **256**, 668–670.
143. Dietrich, W. F., Lander, E. S., Smith, J. S., et al. (1993) Genetic identification of Mom-1, a major modifier locus affecting Min- induced intestinal neoplasia in the mouse. *Cell* **75**, 631–639.
144. Riggins, G. J., Markowitz, S., Wilson, J. K., Vogelstein, B., and Kinzler, K. W. (1995) Absence of secretory phospholipase A2 gene alterations in human colorectal cancer. *Cancer Res.* **55**, 5184–5186.

145. Oshima, M., Oshima, H., Kitagawa, K., Kobayashi, M., Itakura, C., and Taketo, M. (1995) Loss of Apc heterozygosity and abnormal tissue building in nascent intestinal polyps in mice carrying a truncated Apc gene. *Proc. Natl. Acad. Sci. USA* **92**, 4482–4486.
146. Shibata, H., Toyama, K., Shioya, H., et al. (1997) Rapid colorectal adenoma formation initiated by conditional targeting of the apc gene. *Science* **78**, 120–123.
147. van Genderen, C., Okamura, R. M., Farinas, I., et al. (1994) Development of several organs that require inductive epithelial- mesenchymal interactions is impaired in LEF-1-deficient mice. *Genes Dev.* **8**, 2691–2703.
148. Liu, P., Wakamiya, M., Shea, M. J., Albrecht, U., Behringer, R. R., and Bradley, A. (1999) Requirement for Wnt3 in vertebrate axis formation. *Nat. Genet.* **22**, 361–365.
149. McMahon, B. (1990) The Wnt-1 (int-1) proto-oncogene is required for development of a large region of the mouse brain. *Cell* **62**, 1073.
150. Stark, K., Vainio, S., Vassileva, G., and McMahon, A. P. (1994) Epithelial transformation of metanephric mesenchyme in the developing kidney regulated by Wnt-4. *Nature* **372**, 679–684.
151. Parr, B. A., and McMahon, A. P. (1995) Dorsalizing signal Wnt-7a required for normal polarity of D-V and A-P axes of mouse limb. *Nature* **374**, 350–353.
152. Ishikawa, T., Tamai, Y., Zorn, A. M., et al. (2001) Mouse Wnt receptor gene Fzd5 is essential for yolk sac and placental angiogenesis. *Development* **128**, 25–33.
153. Samowitz, W. S., Powers, M. D., Spirio, L. N., Nollet, F., van Roy, F., and Slattery, M. L. (1999) Beta-catenin mutations are more frequent in small colorectal adenomas than in larger adenomas and invasive carcinomas. *Cancer Res.* **59**, 1442–1444.

3

Hereditary Breast and Ovarian Cancer Genes

Ralph Scully and Nadine Puget

1. Introduction

Breast cancer affects one out of every 10 women in industrialized countries, and is a leading cause of cancer morbidity and mortality in women. Ovarian cancer, although less common than breast cancer, is very difficult to treat effectively, in part due to difficulty in early diagnosis of the disease. Most cases of breast or ovarian cancer appear to occur without a clear family history of the disease. These sporadic cases account for approximately 95% of all breast cancer. Possibly, low-penetrance genes contribute as risk factors to this group of cancers. However, about 5% of breast cancers occur clustered within families. Importantly, cancer in these familial syndromes characteristically manifests at a younger age than sporadic cancer. Therefore, in terms of both the number of women affected and the effect of disease on individual families, the familial breast and/or ovarian cancer syndromes have a large impact upon society. Over the past few years, the genetic basis of familial breast and ovarian cancer has become clear. It is caused by germline mutations affecting one of two autosomal tumor suppressor genes, *BRCA1* and *BRCA2*. Surprisingly, these two genes have been found to participate in the control of homologous recombination, suggesting that they may function as tumor suppressors by regulating genome integrity maintenance functions.

2. Identification of the BRCA1 and BRCA2 Genes

The French physician, Paul Broca, was the first to formally describe the existence of a familial breast cancer syndrome *(1)* (**Fig. 1**). His description of cancer cases in a family pointed strongly to a familial element to disease risk. This link later came to be widely accepted, but the genetics of inheritance was unclear for many decades. The search for the genes responsible focused on families in which both breast and ovarian cancer was common, and this culminated in the discovery by King's group in 1990 of a clear linkage between disease risk and a locus on chromosome 17q21 *(2,3)*. Transmission of the 17q-linked syndrome was found to be dominant and highly penetrant. This implied that a single gene—termed *BRCA1*—was likely responsible for this increased

From: *Methods in Molecular Biology, Vol. 222: Tumor Suppressor Genes: Pathways and Isolation Strategies*
Edited by: Wafik S. El-Deiry © Humana Press inc., Totowa, NJ

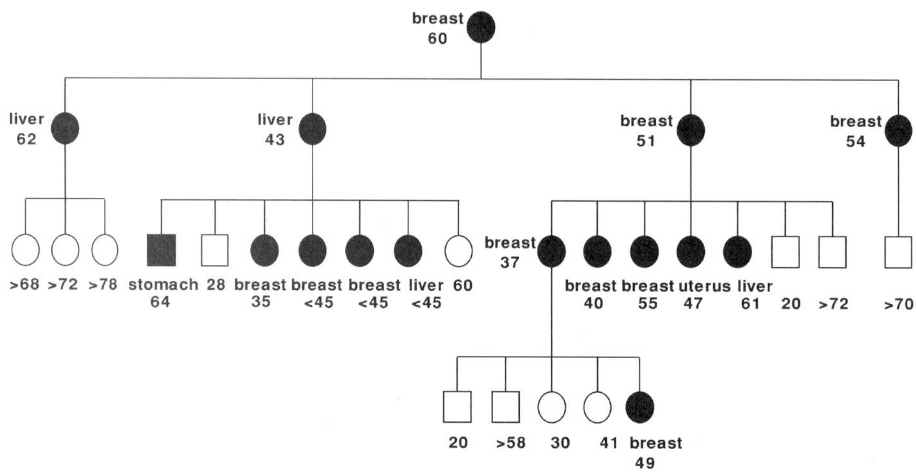

Fig. 1. Reconstruction of the breast cancer-prone family pedigree described by Paul Broca. Broca's original description *(1)* is shown here reconstructed in the form of a family tree. Ovals indicate females; squares indicate males. Filled shapes indicate cancer cases, and the age of death is given. The symbol ">" shows that the individual was still alive at the time of Broca's 1866 publication. Note that a diagnosis of "liver" cancer at that time did not distinguish the site of primary cancer.

risk. Breast and ovarian tumors arising in members of such families showed loss of heterozygosity (LOH) at the 17q locus, with retention of the disease-predisposing allele ***(4–6)***. This pattern suggested that *BRCA1* is a tumor suppressor gene, since it appeared to conform to Knudson's original "two-hit" model of tumor suppressor gene inactivation ***(7)***. This model predicts that disease-predisposing alleles of *BRCA1* should be loss-of-function alleles.

After an intense search of the *BRCA1* locus, the *BRCA1* gene was identified in 1994 by Wiseman, Skolnick, and their co-workers ***(8,9)***. Most (>90%) disease-predisposing *BRCA1* alleles are predicted to encode a truncated gene product, presumably resulting in the synthesis of a nonfunctional or unstable protein ***(10–13)***. In addition, a minority of mutant alleles encode missense mutations, for which a single amino acid change disturbs the function of the gene product. Knudson's model, as well as previous experience with the retinoblastoma gene, *p53,* and many other tumor suppressors, predicted that biallelic somatic inactivation of *BRCA1* would be a common feature of sporadic breast/ovarian cancer. Surprisingly, in fact, examination of sporadic breast or ovarian cancers—even those revealing LOH at the *BRCA1* locus—failed to provide evidence of mutation of the remaining *BRCA1* allele ***(9)***.

The second major hereditary breast–ovarian cancer susceptibility gene, *BRCA2*, was mapped in 1994 to a locus on chromosome 13q12-13 and identified one year later by Stratton's group ***(14,15)***. *BRCA2*, like *BRCA1*, has properties of a tumor suppressor gene, but conforms to Knudson's model only in familial and not in sporadic disease. As was noted for *BRCA1*, although LOH is sometimes seen at the *BRCA2* locus in sporadic breast cancer, the retained *BRCA2* allele is almost always wild-type ***(16)***. Why *BRCA1* and *BRCA2* dysfunction should be restricted to the hereditary syndrome is unclear at present. At a minimum, it suggests that some mechanisms that promote *BRCA* gene-linked breast cancer are distinct from those giving rise to sporadic breast cancer.

3. Characterization of *BRCA*-Linked Tumors

Analysis of tumor histology and other phenotypic markers has revealed some differences between *BRCA*-linked tumors and sporadic cases ***(17–19)***. However, many "hits" on the pathway to cancer are in fact shared between *BRCA*-linked cancer and sporadic cancer, and the differences that emerge are relative rather than absolute. Lynch's group found that *BRCA*-linked tumors exhibit a higher degree of aneuploidy and markedly increased mitotic indices, compared to age-matched sporadic cases ***(17)***. Despite this, the recurrence rate of *BRCA1*-linked tumors was lower than that of nonhereditary breast cancer cases. A consensus has not yet been reached regarding the prognosis of *BRCA*-linked disease relative to prognosis of sporadic disease ***(17,20–22)***. *BRCA1*-linked tumors exhibit a much higher rate of medullary breast cancer, and are very frequently estrogen receptor negative ***(23)***. This bias is not seen in age-matched *BRCA2*-linked cases. However, both *BRCA1* and *BRCA2*-linked tumors reveal greatly elevated frequencies of *p53* mutation compared to sporadic breast cancer. Gene expression profiles in breast cancer reveal differences between *BRCA1*-linked tumors, *BRCA2*-linked tumors, and sporadic cases ***(24)***. The mechanistic basis of these differences, and their actual relationship to tumor suppressor activity, are unknown.

Taken together, the data seem to suggest that *BRCA* gene mutation distorts or biases the clinical picture, but that overall phenotypic and prognostic markers are to a certain extent preserved. As noted above, many sporadic breast cancers reveal LOH at the *BRCA1* and/or *BRCA2* loci, raising the possibility that some sporadic cancers are haploinsufficient for *BRCA* gene function. Epigenetic effects may also serve to reduce the expression of *BRCA* genes in some cases of sporadic breast cancer ***(25–28)***. However, clear evidence of a haploinsufficiency syndrome for either *BRCA1* or *BRCA2* has not emerged from examination of *BRCA*+/− mice or from analysis of human tumors. Therefore, the functional significance of these effects in tumorigenesis is unclear at present.

4. Gene Structure and Protein Motifs

The *BRCA1* and *BRCA2* genes are both very large, spanning more than 80 kb and comprising 23 and 27 coding exons, respectively ***(29)***. cDNA cloning showed that alternative splicing of *BRCA1* transcripts could, in principle, give rise to several distinct gene products ***(8)***. In both *BRCA1* and *BRCA2*, a very large central exon (exon 11) contributes to a significant fraction of the open reading frame. In the case of *BRCA1*, this exon is subject to alternative splicing ***(30,31)***. However, many disease-predisposing mutations occur within this exon, implying that the full-length polypeptide is required for tumor suppression. The function of the alternatively spliced products, and their potential role as modifiers of *BRCA1* function, is a subject of continuing investigation.

Both *BRCA1* and *BRCA2* genes are unusually dense in repetitive elements. For example, Alu repeats make up a remarkable 41% of the *BRCA1* gene sequence. About 15% of inherited disease-predisposing mutations of *BRCA1* have been shown to arise by germline rearrangements ***(32–34)***. Some of these arose by recombination between Alu repeats, implying that mis-targeted homologous recombination in the germline has caused this rearrangement. It has been suggested that the high density of such repeats within the *BRCA1* and *BRCA2* genes might predispose *somatic* cells to large interstitial deletion (and hence LOH) at these loci, by mis-targeted homologous recombination ***(35)***. However, the breakpoints of large somatic interstitial deletions are difficult to identify, and often lie outside the gene. Sequence data on these breakpoints in the *BRCA1* or *BRCA2* gene is not yet available, making it difficult to test this interesting hypothesis.

Although initial reports suggested that *BRCA1* might be unique to mammals, it is now clear that homologs exist in other vertebrates, such as the chicken and frog ***(36,37)***. A possible *BRCA1* homolog has been identified in *Arabidopsis thalania* ***(37)***. However, yeast and flies appear to lack *BRCA1* or *BRCA2* homologs. *BRCA1* and *BRCA2* homologs across species show a rather low degree of sequence similarity throughout much of the open reading frame. For example, the human and mouse *BRCA1* and *BRCA2* genes are 58% identical and 59% identical, respectively. In contrast, the human and mouse gene orthologs of many other tumor suppressors—*p53*, *pRB*, *VHL*, *APC*, *WT1*, *NF1,* and *NF2*—are 90–98% identical. Within the human population, it has been estimated that simple sequence variation/polymorphism in the *BRCA2* open reading frame occurs at a frequency of 1 in 194 base pairs ***(38)***. This suggests that *BRCA1* and *BRCA2* are relatively tolerant of evolutionary "drift," at least in many parts of their open reading frames. However, many of these sequence variations occur in gene regions that are nonetheless indispensable for tumor suppression. The reasons for this may become clearer once we know the structure of the protein domains encoded by these more "plastic" regions of *BRCA1* and *BRCA2*.

The open reading frames of *BRCA1* and *BRCA2* provided few initial clues as to the function of the genes. Full-length BRCA1 is a nuclear protein of 1863 amino acids (**Fig. 2**). The polypeptide has an N-terminal RING domain, a globular domain that has recently been implicated in ubiquitin conjugation functions ***(39)***. The BRCA1 RING domain region interacts tightly, perhaps stochiometrically, with the RING domain region of BARD1 (*B*RCA1 *A*ssociated *R*ING *D*omain protein 1) ***(40–42)***. At the C terminus of both BRCA1 and BARD1 are two globular repeats, termed BRCT (*BR*CA1 *C-T*erminal domains) ***(43)***. BRCT domains have subsequently been identified in numerous proteins involved in DNA damage signaling and DNA repair ***(44,45)***. The RING and BRCT coding regions of *BRCA1* show uncharacteristically high degrees of similarity across species, implying that much more rigid constraints exist for these structures than for other parts of the gene, and they are relative "hot spots" of missense mutation. The structures of both the RING domain and the BRCT domain have been solved in recent years, as has the specific structure of the BRCA1–BARD1 RING–RING interaction ***(42,46,47)***. The latter reveals a tight interaction between α-helical bundles on either side of the RING domains, the two RING domains forming a mutually supporting structure that is predicted to interact with ubiquitin-conjugating enzymes ***(42)***.

The structure of the large central part of BRCA1 is as yet uncharacterized, although database comparisons between mammalian and nonmammalian BRCA1 orthologs indicate regions within this central part that are relatively well conserved across species ***(36)***. Numerous protein interaction surfaces have been identified within this central region of BRCA1, some—such as the Rad51 and BRCA2 interaction domains—overlapping these more conserved regions. Moreover, some residues within the central region of BRCA1 are targeted for phosphorylation in response to cell-cycle cues or DNA damage response signals ***(48,49)***. A region rich in SQ motifs, which are phosphorylation targets of the DNA damage response, is also relatively well conserved across species.

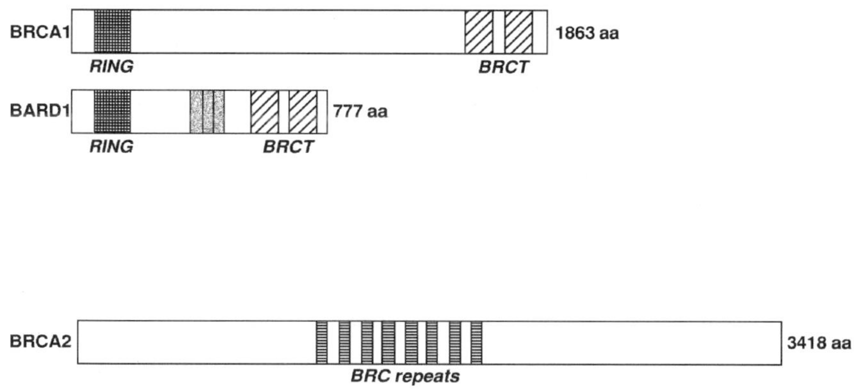

Fig. 2. Structural motifs in the BRCA1, BARD1, and BRCA2 polypeptides (not to scale). BRCA1 and its heterodimeric partner, BARD1, share similar features, including amino-terminal RING domains (stippled) and carboxyl-terminal tandem repeated BRCT domains (cross-hatched). BARD1 ankyrin repeats are shown in light gray. Eight BRC repeats of BRCA2 are shown (horizontal shading).

BRCA2 encodes a very large, nuclear polypeptide of 3418 amino acids in length (**Fig. 2**). Few structural domains have been identified in BRCA2. Most prominent are eight "BRC" repeats within the central part of BRCA2. The structure of these repeats is not yet known; they are known to function as interaction surfaces for the recombination protein Rad51, as discussed in the following section. A second Rad51 interaction surface lies within the C terminus of BRCA2—a region of the protein that is relatively well conserved between human and mouse (77% identical, as compared to 59% overall identity).

5. Functional Analysis of BRCA1 and BRCA2

A large body of evidence suggests that BRCA1 and BRCA2 function together on a pathway regulating homologous recombination (reviewed in **ref. *50***). This may be a major pathway of *BRCA* gene-mediated tumor suppression. However, several other functions have been deduced for both BRCA1 and BRCA2, including other genome integrity maintenance functions and transcriptional regulatory functions.

5.1. Recombination Functions of BRCA1 and BRCA2

Functional analysis of *BRCA1* was initially limited by the fact that *BRCA1–/–* mice die *in utero* around the time of gastrulation (~E7.5) ***(51,52)***. No cells could be cultivated from *BRCA1–/–* embryos, and there were no human *BRCA1–/–* cancer cell lines available at the time the gene was cloned, making traditional functional analysis of mutant cell lines impossible. Hakem et al. described a marked proliferative defect and greatly elevated mRNA levels of the *p53*-regulated cell-cycle checkpoint gene, *p21*, in the days preceding death of *BRCA1–/–* embryos ***(51)***. This suggested that p53 is activated in the dying *BRCA1–/–* embryos. Consistent with this, biallelic germline inactivation of either *p53* or *p21* rescues development of *BRCA1–/–* embryos by a matter of 1 or 2 d ***(53)***. For the gastrulating embryo, even this small extension of lifespan represents a huge difference in development. Thus, *p53* function appears to play a role in the death of *BRCA1–/–* embryos. Almost identical observations were made with respect to *BRCA2*, although the *BRCA2* nullizygotes die on average 1 d later (~E8.5) than *BRCA1*−/− embryos ***(52,54,55)***. Significantly, *BRCA1*−/−*BRCA2*−/− mice die at ~E7.5, suggesting that the effects of the two gene disruptions are not additive, and raising the possibility that *BRCA1* mutation is epistatic over *BRCA2* mutation ***(52)***.

The mechanisms underlying these observations were suggested by the finding that both BRCA1 and BRCA2 have roles in DNA recombination. BRCA1 localizes to characteristic nuclear foci of unknown function during DNA synthesis (the S phase) and the G2 phase of the somatic cell cycle ***(56–58)***. These foci were found to contain Rad51, a protein with a central role in homologous recombination ***(57)***. BRCA1–Rad51 complexes were detected in cell extracts, indicating that the two proteins interact physically. BRCA2 also interacts and co-localizes with Rad51, making multiple possibly stochiometric interactions with Rad51, involving both the C terminus and the BRC repeats of BRCA2 ***(54,59–63)***. BRCA1 interacts with BRCA2 through a region N terminal to the BRCT domains ***(63)***. *BRCA1*, *BRCA2,* and *Rad51* are coexpressed in development, consistent with their function on a common pathway ***(64,65)***. Notably, like *BRCA*–/– embryos, *Rad51* nullizygotes show *p53*-mediated growth arrest and early embryonic lethal phenotypes ***(66)***.

Human *Rad51* is a homolog of *S. cerevisiae RAD51*, a gene required for homologous recombination ***(67,68)***. *S. cerevisiae rad51* mutants show defects in double-strand break repair and a failure of meiotic recombination. Eukaryotic *Rad51* species are homologs of the prokaryotic recombination gene, *RecA*. All eukaryotic Rad51 species, as well as their ancestor, RecA, can bind to single-stranded (ss) DNA to form a characteristic loosely wound nucleoprotein filament, which is able to invade a homologous double-stranded (ds) DNA strand ***(67,69,70)***. This strand exchange function of Rad51/RecA is thought to be the basic catalytic function that underlies its role in recombination. Further, BRCA1, BRCA2, and Rad51 also localize to the axial element of the developing synaptonemal complex in meiotic cells, suggesting a possible role for this protein complex in meiotic recombination ***(57,63)***. The specific stage at which association is seen in meiotic prophase I corresponds to a step preceding synapsis of homologous chromosomes. Possibly, the axial element is a site of interaction between identical sister chromatids, preceding the interhomolog interactions that culminate in meiotic crossing-over ***(71)***.

These observations, together with the finding of an interaction between BRCA1 and the Rad50/MRE11/NBS1 complex ***(72–74)***, suggested that BRCA1 and BRCA2 function in homologous recombination, and in the maintenance of genomic integrity. This hypothesis was later strengthened with the availability of human and mouse cells that lacked wild-type *BRCA1* or *BRCA2*, but could nonetheless be passaged in tissue culture. *BRCA* mutant cells were found to be highly sensitive to ionizing radiation (IR), a genotoxic agent known to induce double-strand DNA breaks (DSB) ***(54,75–77)***. Expression of wild-type *BRCA1* in a *BRCA1*−/− human breast cancer cell line reverses IR hypersensitivity and restores efficient DSBR to the cells ***(77)***. In contrast, missense mutant *BRCA1* alleles, affecting any one of three distinct regions of BRCA1—the RING domain, the BRCT domains, and a central region—fail to reverse IR hypersensitivity or to restore efficient DSBR ***(77)***. This suggests that the DSBR function of *BRCA1* is part of its tumor suppressor function, and that multiple domains of BRCA1 collaborate to enforce efficient DSBR. BRCA1 interaction with a novel DEAH-type helicase, BACH 1, is required for efficient DSBR ***(78)***. However, since these experiments were performed on bulk genomic DNA, they do not give evidence of the mechanisms of DSBR used during repair.

Several distinct biochemical pathways can contribute to DSBR. These include homologous recombination, nonhomologous end joining, and single-strand annealing ***(79–81)***. With the exception of homologous recombination, each of these repair mechanisms is mutagenic, resulting in insertions or deletions of DNA sequence. Experiments designed to discriminate between these mechanisms have thus far pointed to defects in homologous recombination in *BRCA1* or *BRCA2* mutant cells ***(82,83)***, and expression of wild-type *BRCA1* in a *BRCA1−/−* embryonic stem cell clone reverses the recombination defect ***(84)***. It is not yet clear whether this recombination function of *BRCA1* can discriminate between wild-type and mutant alleles of *BRCA1*. Conditional mutation of mouse *BRCA2* reduces the efficiency of homologous recombination in favor of single-strand annealing, possibly reflecting a greater tendency toward mutagenic repair of a DSB in the absence of functional *BRCA2* ***(85)***. Gene-targeting defects in *BRCA1−/−* and *BRCA2−/−* cells are reversed by expression of the relevant wild-type *BRCA* allele, suggesting that *BRCA* genes contribute generally to homologous recombination ***(84,86)***. To date, defects in nonhomologous end joining have not been detected in either *BRCA1−/−* cells or in *BRCA2−/−* cells ***(82,83,87,88)***.

The mechanisms of action of BRCA1 and BRCA2 in recombination are poorly understood at present. Recently, a central region of the BRCA1 polypeptide was shown to have DNA binding activity in vitro, binding without sequence specificity to double-stranded DNA and preferentially to branched DNA structures ***(89)***. How this relates to BRCA1 function in the context of the cell is unknown, but it invites the suggestion that BRCA1 interacts with recombination intermediates such as the Holliday junction—an idea reinforced by the observed interaction of BRCA1 with a candidate Holliday junction-interacting protein, the Bloom syndrome helicase ***(74)***. Notably, deletion of a part of the DNA-binding region of BRCA1 by gene targeting abrogates the genome integrity maintenance function of *BRCA1* ***(75)***, and missense mutation affecting this part of the open reading frame is known to disrupt *BRCA1* DSBR function ***(77)***. BRCA1 RING domain function is required for IR resistance, and clinically described missense mutations affecting the RING domain lose both IR resistance and ubiquitination functions, suggesting that the ubiquitination functions of the BRCA1–BARD1 heterodimer are required for genome integrity maintenance and for tumor suppression ***(90)***. One possible target of BRCA1-mediated ubiquitination is the Fanconi anemia protein, FANCD2 ***(91)***. Presumably, other BRCA1-dependent ubiquitination targets remain to be identified.

BRCA2 interacts with a novel protein, BRAF35, that also has affinity for branched DNA structures in vitro ***(92)***. Recently, purified fragments of BRCA2 BRC repeats were found to inhibit Rad51 binding to single-stranded (ss) DNA in vitro ***(93)***. Possibly, BRCA2 regulates loading of Rad51 onto the nucleoprotein filament.

Double-strand breaks in mammalian cells are accompanied by extensive modifications in local chromatin structure, as exemplified by the phosphorylation of histone H2AX ***(94,95)***. Phosphorylated H2AX colocalizes with BRCA1 and other recombination proteins in the "locale" of DSBs, raising the possibility that a chromatin-remodeling function of BRCA1 contributes to its role in recombination ***(96)***. Given the interactions of BRCA1 and BRCA2 with chromatin-remodeling protein complexes, it is possible that both proteins play a part in regulating chromatin structure around recombinogenic DNA lesions ***(97–99)***.

5.1.1. DNA Polymerase Stalling and DNA Damage Signaling During S Phase

Clues as to the role of BRCA1 and BRCA2 in the somatic cell cycle came initially from analysis of the DNA damage response. S-phase cells exposed to certain DNA-damaging agents or to replication-blocking agents such as hydroxyurea (HU) reveal rapid relocalization of BRCA1, BARD1, Rad51, and BRCA2 to sites of replication, suggesting an interaction with damaged replicating DNA ***(58,63)***. In addition, BRCA1 undergoes hyperphosphorylation in synchrony with this relocalization, indicating that BRCA1 is the target of a DNA-damage signaling pathway ***(58,100)***. One possible explanation for this is that BRCA protein-containing complexes control recombination between sister chromatids—a process that has long been hypothesized to occur at the stalled replication fork ***(50,101)***. Indeed, recent work in model organisms implies that sister chromatid recombination plays an essential role in restoring stalled or degraded replication forks, and that it is required to reinitiate replication at the stalled fork ***(102,103)*** (and accompanying articles).

Some of the pathways connecting DNA damage signaling with BRCA1 have been identified over recent years. Two key DNA damage signaling kinases, ATM and ATR, each participate in phosphorylation of BRCA1 ***(48,49)***. These large nuclear kinases have orthologs in yeast. In general, yeast orthologs of ATR but not ATM have been found to be essential in mediating signals from stalled replication events to target proteins ***(104)***. Possibly, the extensive tracts of ssDNA ("daughter-strand gaps") generated by replication across DNA polymerase-stalling lesions, or by treatment with agents such as HU, might act as the nucleus for an ATR-dependent DNA damage signal and as a coincident substrate of recombinational repair ***(49,50,101,105)*** (**Fig. 3**).

Functional support for these observations came from analysis of *BRCA* gene hypomorphs, which die later in development and in some cases give rise to viable offspring. Mouse embryonic fibroblasts (MEF) explanted from these animals reveal spontaneous chromosome breakage with characteristic "chromatid-type" errors—lesions that have been interpreted as arising from a failure of recombination control during S phase ***(75,87)***. Thus, gene-targeting experiments reinforced the notion that the recombination functions of *BRCA1* and *BRCA2* are at least in part dedicated to recombinational repair of DNA damage generated during DNA replication.

5.1.2. BRCA Gene Loss and Checkpoint Dysfunction in Tumorigenesis

The suggestion that *BRCA1* and *BRCA2* regulate homologous recombination and thereby maintain genomic integrity suggested a way to interpret the lethal phenotype of the *BRCA*−/− mice, and also introduced a framework for understanding how the *BRCA* genes might function in suppressing cancer ***(57)***. Loss of *BRCA* gene function promotes spontaneous chromosome breakage, which could serve as a cue for the activation of DNA damage-dependent checkpoint pathways, such as p53 (**Fig. 4**). This is a probable cause of the growth arrest seen in *BRCA*−/− embryos and in primary cells explanted from *BRCA* mutant mice. The same fate could occur to a breast epithelial cell that loses *BRCA* gene function. However, if the cell in question had already suffered inactivation of key checkpoint functions, the chromosome breakage syndrome caused by *BRCA* gene dysfunction might be tolerated, and might allow the cell to progress rapidly toward cancer by virtue of the cell's severe genomic instability. By this argument, mutagenesis would be a key cancer-promoting event caused by *BRCA* gene dysfunction.

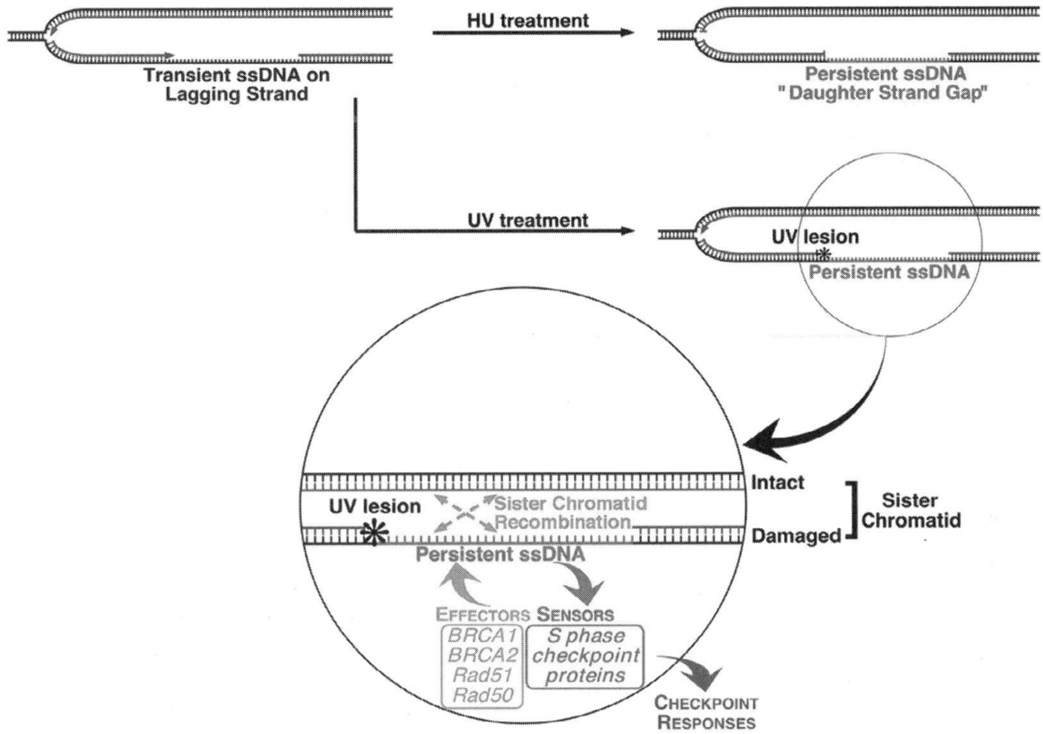

Fig. 3. Proposed role for the BRCA proteins in sister chromatid recombination. DNA polymerase stalling activates S-phase checkpoint signaling and recruits BRCA1, BRCA2, Rad51, and associated proteins to sites of arrested DNA synthesis. We propose that this recruitment arises from the existence of tracts of persistent ssDNA ("daughter-strand gaps") close to a replication fork. Such lesions, or their derivatives, may activate S-phase checkpoint signaling and act as a substrate for BRCA protein-mediated recombinational repair. (Copyright Nature Publishing Group, reproduced with permission.)

This model of *BRCA*-linked tumorigenesis is supported by work on animal models and by analysis of the human disease (reviewed in **ref.** ***106***). As noted above, p53 activation is seen in growth-arrested *BRCA−/−* embryos prior to death, and a similar cell-cycle arrest phenotype has been noted in mouse embryonic fibroblasts (MEF), where *p53* function again plays an important role ***(75,107)***. Although the full picture of checkpoint activation in *BRCA−/−* cells has yet to be clarified, one study pointed to a role for the spindle checkpoint in arrest of *BRCA2*-mutant MEFs ***(108)***. Further, tissue-specific deletion of *BRCA1* or *BRCA2* in the breast promotes breast cancer and, where examined, this phenotype is much accelerated if the mice are hemizygous for *p53* ***(109–111)***. A similar interaction with *p53* mutation has been noted for tumorigenesis in *BRCA1* hypomorphs ***(112)***. Thus, in accordance with this model, *p53* mutation cooperates with *BRCA* gene mutation in tumorigenesis. In further support of this, as noted above, *p53* mutation is very common in *BRCA*-linked human cancer. Interestingly, some rare *p53* mutations occur more frequently in *BRCA−/−* tumors, raising the possibility that p53 functions in addition to the traditional checkpoint functions may be relevant to *BRCA*-linked tumorigenesis ***(113)***. Mouse models promise to facilitate the genetic study of *BRCA*-linked tumor pathology ***(114)***.

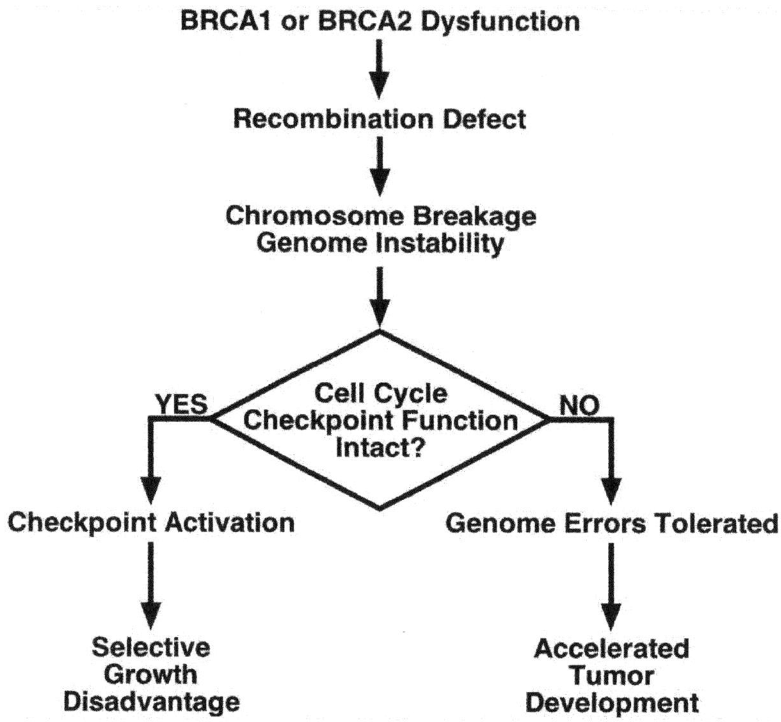

Fig. 4. Checkpoint inactivation and *BRCA* gene-mediated tumorigenesis. Inactivation of DNA damage-responsive cell-cycle checkpoints may contribute to the cellular response to *BRCA* gene dysfunction. Loss of *BRCA* function and associated chromosome breakage trigger a checkpoint that deselects it. If such checkpoints have already been disabled (as in precancerous cell), the chromosome breakage syndrome may be tolerated and lead to accelerated neoplastic progression. (Copyright Nature Publishing Group, reproduced with permission.)

5.2 Other Functions of BRCA1 and BRCA2

Both BRCA1 and BRCA2 have been proposed to regulate transcription as coactivators/corepressors of specific target genes (reviewed in **refs.** ***115*** and ***116)***. BRCA1 in particular is capable of interacting with a large number of different proteins (reviewed in **refs.** ***117*** and ***118***) that fall into three principal categories: sequence-specific transcription factors such as p53 ***(119)***, c-myc ***(120)***, and the estrogen receptor ***(121)***; general transcription factors/chromatin remodeling factors ***(98,99,122,123)***; and DNA-repair protein complexes, as discussed above.

Overexpression of *BRCA1* in human cells provokes activation of p53, with induction of the p53 target genes, *p21* and *GADD45*, and also stabilizes p53 and promotes apoptosis ***(119,124–127)***. Overexpression of BARD1 also promotes apoptosis ***(128)***. How these observations relate to the above-noted p53 activation in *BRCA1*−/− cells is not yet understood. They raise the possibility that BRCA1 acts early in oncogenesis. For example, if the predisposed *BRCA1*+/− individual were haploinsufficient for p53 coactivation, this might serve as a stimulus to cancer initiation, perhaps accelerating the loss of *p53* and of the second *BRCA1* allele. BRCA2 has also been implicated in the p53 pathway ***(129)***.

BRCA1 also has properties of a transcriptional repressor. The multifunctional BRCT domains of BRCA1 interact with CtIP, a known transcriptional corepressor ***(123)***, and BRCA1 also interacts with a novel transcriptional repressor, ZBRK1 ***(130)***. Other functions of BRCA1 and BRCA2 have been inferred from analysis of interacting proteins. A recent partial purification of BRCA1 raised several interesting connections with mismatch repair factors, replication factors, as well as with recombination proteins ***(74)***.

Other abnormalities noted in *BRCA1*−/− and *BRCA2*−/− cells include centrosome amplification and severe aneuploidy ***(131,132)***. Whether these are direct effects, or the consequence of recombination dysfunction, is unknown. Notably, however, BRCA1 has been found localized to the centrosome in some cell lines ***(133)***. Abnormalities of components of the G2/M checkpoint have also been found in some *BRCA1* mutant cells ***(132,134)***. In some *BRCA1* or *BRCA2* mutant cells, defects in transcription-associated repair of oxidative DNA damage have also been noted ***(135,136)***. Novel functions for BARD1 in polyadenylation potentially implicate BRCA1 in this process ***(137,138)***. As more functional assays for BRCA1 and BRCA2 are developed, we can expect to see these novel functions subjected to rigorous investigation as potential tumor suppressor functions.

6. Summary and Future Prospects

Although it seems likely that *BRCA1* and *BRCA2* function as tumor suppressors by regulating homologous recombination, how this process connects to the tissue-specific suppression of breast and ovarian cancer is not understood at present. In some model organisms, homologous pairing can regulate gene expression directly, as the phenomenon of transvection shows ***(139)***. Future experiments will provide a clearer picture of potential *BRCA* gene functions and their relationship to tumor suppression. As mouse models of *BRCA*-linked cancer become more sophisticated, we can expect to understand better how *BRCA* gene mutation cooperates with *p53* mutation and with "hits" in other cancer pathways to cause early-onset breast or ovarian cancer. In time, it may become possible to exploit unique aspects of the biology of *BRCA*−/− cells in the treatment of human *BRCA*-linked cancers.

References

1. Broca, P. (1866) *Influence héréditaire*, Traité des tumeurs (Asselin, P., ed.), Tome premier: Des tumeurs en général, Paris.
2. Hall, J. M., et al. (1990) Linkage of early-onset familial breast cancer to chromosome 17q21. *Science* **250,** 1684–1689.
3. Narod, S. A., et al. (1991) Familial breast-ovarian cancer locus on chromosome 17q12-q23. *Lancet* **338,** 82–83.
4. Smith, S. A., Easton, D. F., Evans, D. G., and Ponder, B. A. (1992) Allele losses in the region 17q12-21 in familial breast and ovarian cancer involve the wild-type chromosome. *Nat. Genet.* **2,** 128–131.
5. Neuhausen, S. L. and Marshall, C. J. (1994) Loss of heterozygosity in familial tumors from three BRCA1-linked kindreds. *Cancer Res.* **54,** 6069–6072.

6. Cornelis, R. S., et al. (1995) High allele loss rates at 17q12-q21 in breast and ovarian tumors from BRCAl-linked families. The Breast Cancer Linkage Consortium. *Genes Chromosomes Cancer* **13,** 203–210.
7. Knudson, A. G., Jr. (1971) Mutation and cancer: statistical study of retinoblastoma. *Proc. Natl. Acad. Sci. USA* **68,** 820–823.
8. Miki, Y., et al. (1994) A strong candidate for the breast and ovarian cancer susceptibility gene BRCA1. *Science* **266,** 66–71.
9. Futreal, P. A., et al. (1994) BRCA1 mutations in primary breast and ovarian carcinomas. *Science* **266,** 120–122.
10. Castilla, L. H., et al. (1994) Mutations in the BRCA1 gene in families with early-onset breast and ovarian cancer. *Nat. Genet.* **8,** 387–391.
11. Simard, J., et al. (1994) Common origins of BRCA1 mutations in Canadian breast and ovarian cancer families. *Nat. Genet.* **8,** 392–398.
12. Friedman, L. S., et al. (1994) Confirmation of BRCA1 by analysis of germline mutations linked to breast and ovarian cancer in ten families. *Nat. Genet.* **8,** 399–404.
13. BIC. (2001) Breast Cancer Information Core. *http://www.nhgri.nih.gov/Intramural_research/Lab_transfer/Bic/.*
14. Wooster, R., et al. (1994) Localization of a breast cancer susceptibility gene, BRCA2, to chromosome 13q12-13. *Science* **265,** 2088–2090.
15. Wooster, R., et al. (1995) Identification of the breast cancer susceptibility gene BRCA2. *Nature* **378,** 789–792.
16. Lancaster, J. M., et al. (1996) BRCA2 mutations in primary breast and ovarian cancers. *Nat. Genet.* **13,** 238–240.
17. Marcus, J. N., et al. (1996) Hereditary breast cancer: pathobiology, prognosis, and BRCA1 and BRCA2 gene linkage. *Cancer* **77,** 697–709.
18. Lakhani, S. R., et al. (1998) Multifactorial analysis of differences between sporadic breast cancers and cancers involving BRCA1 and BRCA2 mutations. *J. Natl. Cancer Inst.* **90,** 1138–1145.
19. Lakhani, S. R. (1999) The pathology of hereditary breast cancer. *Dis. Markers* **15,** 113–114.
20. Foulkes, W. D., Wong, N., Brunet, J. S., and Narod, S. A. (1998) BRCA mutations and survival in breast cancer. *J. Clin. Oncol.* **16,** 3206–3208.
21. Gaffney, D. K., et al. (1998) Response to radiation therapy and prognosis in breast cancer patients with BRCA1 and BRCA2 mutations. *Radiother. Oncol.* **47,** 129–136.
22. Wagner, T. M., et al. (1998) BRCA1-related breast cancer in Austrian breast and ovarian cancer families: specific BRCA1 mutations and pathological characteristics. *Int. J. Cancer* **77,** 354–360.
23. Osin, P. P. and Lakhani, S. R. (1999) The pathology of familial breast cancer: Immunohistochemistry and molecular analysis. *Breast Cancer Res.* **1,** 36–40.
24. Hedenfalk, I., et al. (2001) Gene-expression profiles in hereditary breast cancer. *N. Engl. J. Med.* **344,** 539–548.
25. Wilson, C. A., et al. (1999) Localization of human BRCA1 and its loss in high-grade, non-inherited breast carcinomas. *Nat. Genet.* **21,** 236–240.
26. Dobrovic, A. and Simpfendorfer, D. (1997) Methylation of the BRCA1 gene in sporadic breast cancer. *Cancer Res.* **57,** 3347–3350.
27. Magdinier, F., Ribieras, S., Lenoir, G. M., Frappart, L., and Dante, R. (1998) Down-regulation of BRCA1 in human sporadic breast cancer; analysis of DNA methylation patterns of the putative promoter region. *Oncogene* **17,** 3169–3176.
28. Rice, J. C., Massey-Brown, K. S., and Futscher, B. W. (1998) Aberrant methylation of the BRCA1 CpG island promoter is associated with decreased BRCA1 mRNA in sporadic breast cancer cells. *Oncogene* **17,** 1807–1812.

29. Welcsh, P. L., Owens, K. N., and King, M. C. (2000) Insights into the functions of BRCA1 and BRCA2. *Trends Genet.* **16,** 69–74.
30. Wilson, C. A., et al. (1997) Differential subcellular localization, expression and biological toxicity of BRCA1 and the splice variant BRCA1-delta11b. *Oncogene* **14,** 1–16.
31. Huber, L. J., et al. (2001) Impaired DNA damage response in cells expressing an exon 11-deleted murine Brca1 variant that localizes to nuclear foci. *Mol. Cell. Biol.* **21,** 4005–4015.
32. Puget, N., et al. (1999) An Alu-mediated 6-kb duplication in the BRCA1 gene: a new founder mutation? *Am. J. Hum. Genet.* **64,** 300–302.
33. Unger, M. A., et al. (2000) Screening for genomic rearrangements in families with breast and ovarian cancer identifies BRCA1 mutations previously missed by conformation-sensitive gel electrophoresis or sequencing. *Am. J. Hum. Genet.* **67,** 841–850.
34. Payne, S. R., Newman, B., and King, M. C. (2000) Complex germline rearrangement of BRCA1 associated with breast and ovarian cancer. *Genes Chromosomes Cancer* **29,** 58–62.
35. Welcsh, P. L. and King, M. C. (2001) BRCA1 and BRCA2 and the genetics of breast and ovarian cancer. *Hum. Mol. Genet.* **10,** 705–713.
36. Orelli, B. J., Logsdon, J. M., Jr., and Bishop, D. K. (2001) Nine novel conserved motifs in BRCA1 identified by the chicken orthologue. *Oncogene* **20,** 4433–4438.
37. Joukov, V., Chen, J., Fox, E. A., Green, J. B., and Livingston, D. M. (2001) Functional communication between endogenous BRCA1 and its partner, BARD1, during *Xenopus laevis* development. *Proc. Natl. Acad. Sci. USA* **98,** 12078–12083.
38. Wagner, T. M., et al. (1999) Global sequence diversity of BRCA2: analysis of 71 breast cancer families and 95 control individuals of worldwide populations. *Hum. Mol. Genet.* **8,** 413–423.
39. Lorick, K. L., et al. (1999) RING fingers mediate ubiquitin-conjugating enzyme (E2)-dependent ubiquitination. *Proc. Natl. Acad. Sci. USA* **96,** 11364–11369.
40. Wu, L. C., et al. (1996) Identification of a RING protein that can interact in vivo with the BRCA1 gene product. *Nat. Genet.* **14,** 430–440.
41. Jin, Y., et al. (1997) Cell cycle-dependent colocalization of BARD1 and BRCA1 proteins in discrete nuclear domains. *Proc. Natl. Acad. Sci. USA* **94,** 12075–12080.
42. Brzovic, P. S., Rajagopal, P., Hoyt, D. W., King, M. C., and Klevit, R. E. (2001) Structure of a BRCA1-BARD1 heterodimeric RING-RING complex. *Nat. Struct. Biol.* **8,** 833–837.
43. Koonin, E. V., Altschul, S. F., and Bork, P. (1996) BRCA1 protein products . . . Functional motifs. *Nat. Genet.* **13,** 266–268.
44. Bork, P., et al. (1997) A superfamily of conserved domains in DNA damage-responsive cell cycle checkpoint proteins. *FASEB J.* **11,** 68–76.
45. Callebaut, I. and Mornon, J. P. (1997) From BRCA1 to RAP1: a widespread BRCT module closely associated with DNA repair. *FEBS Lett.* **400,** 25–30.
46. Brzovic, P. S., Meza, J. E., King, M. C., and Klevit, R. E. (2001) BRCA1 RING domain cancer-predisposing mutations: structural consequences and effects on protein–protein interactions. *J. Biol. Chem.* **28,** 28.
47. Williams, R. S., Green, R., and Glover, J. N. (2001) Crystal structure of the BRCT repeat region from the breast cancer- associated protein BRCA1. *Nat. Struct. Biol.* **8,** 838–842.
48. Cortez, D., Wang, Y., Qin, J., and Elledge, S. J. (1999) Requirement of ATM-dependent phosphorylation of brca1 in the DNA damage response to double-strand breaks. *Science* **286,** 1162–1166.
49. Tibbetts, R. S., et al. (2000) Functional interactions between BRCA1 and the checkpoint kinase ATR during genotoxic stress. *Genes Dev.* **14,** 2989–3002.
50. Scully, R. and Livingston, D. M. (2000) In search of the tumor suppressor functions of BRCA1 and BRCA2. *Nature* **408,** 429–432.
51. Hakem, R., et al. (1996) The tumor suppressor gene Brca1 is required for embryonic cellular proliferation in the mouse. *Cell* **85,** 1009–1023.

52. Ludwig, T., Chapman, D. L., Papaioannou, V. E., and Efstratiadis, A. (1997) Targeted mutations of breast cancer susceptibility gene homologs in mice: lethal phenotypes of Brca1, Brca2, Brca1/Brca2, Brca1/p53, and Brca2/p53 nullizygous embryos. *Genes Dev.* **11,** 1226–1241.
53. Hakem, R., de la Pompa, J. l., Elia, A., Potter, J., and Mak, T. W. (1997) Partial rescue of Brca1 (5-6) early embryonic lethality by p53 or p21 null mutation. *Nat. Genet.* **16,** 298–302.
54. Sharan, S. K., et al. (1997) Embryonic lethality and radiation hypersensitivity mediated by Rad51 in mice lacking Brca2. *Nature* **386,** 804–810.
55. Suzuki, A., et al. (1997) Brca2 is required for embryonic cellular proliferation in the mouse. *Genes Dev.* **11,** 1242–1252.
56. Scully, R., et al. (1996) Location of BRCA1 in human breast and ovarian cancer cells. *Science* **272,** 123–126.
57. Scully, R., et al. (1997) Association of BRCA1 with Rad51 in mitotic and meiotic cells. *Cell* **88,** 265–275.
58. Scully, R., et al. (1997) Dynamic changes of BRCA1 subnuclear location and phosphorylation state are initiated by DNA damage. *Cell* **90,** 425–435.
59. Mizuta, R., et al. (1997) RAB22 and RAB163/mouse BRCA2: proteins that specifically interact with the RAD51 protein. *Proc. Natl. Acad. Sci. USA* **94,** 6927–6932.
60. Wong, A. K., Pero, R., Ormonde, P. A., Tavtigian, S. V., and Bartel, P. L. (1997) RAD51 interacts with the evolutionarily conserved BRC motifs in the human breast cancer susceptibility gene brca2. *J. Biol. Chem.* **272,** 31941–31944.
61. Chen, P. L., et al. (1998) The BRC repeats in BRCA2 are critical for RAD51 binding and resistance to methyl methanesulfonate treatment. *Proc. Natl. Acad. Sci. USA* **95,** 5287–5292.
62. Sarkisian, C. J., Master, S. R., Huber, L. J., Ha, S. I., and Chodosh, L. A. (2001) Analysis of murine brca2 reveals conservation of protein-protein interactions but differences in nuclear localization signals. *J. Biol. Chem.* **276,** 37640–37648.
63. Chen, J., et al. (1998) Stable interaction between the products of the BRCA1 and BRCA2 tumor suppressor genes in mitotic and meiotic cells. *Mol. Cell* **2,** 317–328.
64. Rajan, J. V., Wang, M., Marquis, S. T., and Chodosh, L. A. (1996) Brca2 is coordinately regulated with Brca1 during proliferation and differentiation in mammary epithelial cells. *Proc. Natl. Acad. Sci. USA* **93,** 13078–13083.
65. Rajan, J. V., Marquis, S. T., Gardner, H. P., and Chodosh, L. A. (1997) Developmental expression of Brca2 colocalizes with Brca1 and is associated with proliferation and differentiation in multiple tissues. *Dev. Biol.* **184,** 385–401.
66. Lim, D. S. and Hasty, P. (1996) A mutation in mouse rad51 results in an early embryonic lethal that is suppressed by a mutation in p53. *Mol. Cell Biol.* **16,** 7133–7143.
67. Baumann, P., Benson, F. E., and West, S. C. (1996) Human Rad51 protein promotes ATP-dependent homologous pairing and strand transfer reactions in vitro. *Cell* **87,** 757–766.
68. Shinohara, A., Ogawa, H., and Ogawa, T. (1992) Rad51 protein involved in repair and recombination in S. cerevisiae is a RecA-like protein. *Cell* **69,** 457–470.
69. West, S. C., Cassuto, E., and Howard-Flanders, P. (1981) Homologous pairing can occur before DNA strand separation in general genetic recombination. *Nature* **290,** 29–33.
70. Sung, P. (1994) Catalysis of ATP-dependent homologous DNA pairing and strand exchange by yeast RAD51 protein. *Science* **265,** 1241–1243.
71. Roeder, G. S. (1997) Meiotic chromosomes: it takes two to tango. *Genes Dev.* **11,** 2600–2621.
72. Zhong, Q., et al. (1999) Association of BRCA1 with the hRad50-hMre11-p95 complex and the DNA damage response. *Science* **285,** 747–750.
73. Wu, X., et al. (2000) Independence of R/M/N focus formation and the presence of intact BRCA1. *Science* **289,** 11a.
74. Wang, Y., et al. (2000) BASC, a super complex of BRCA1-associated proteins involved in the recognition and repair of aberrant DNA structures. *Genes Dev.* **14,** 927–939.

75. Shen, S. X., et al. (1998) A targeted disruption of the murine Brca1 gene causes gamma-irradiation hypersensitivity and genetic instability. *Oncogene* **17,** 3115–3124.
76. Foray, N., et al. (1999) Gamma-rays-induced death of human cells carrying mutations of BRCA1 or BRCA2. *Oncogene* **18,** 7334–7342.
77. Scully, R., et al. (1999) Genetic analysis of BRCA1 function in a defined tumor cell line. *Mol. Cell* **4,** 1093–1099.
78. Cantor, S. B., et al. (2001) BACH1, a novel helicase-like protein, interacts directly with BRCA1 and contributes to its DNA repair function. *Cell* **105,** 149–160.
79. Kanaar, R., Hoeijmakers, J. H., and van Gent, D. C. (1998) Molecular mechanisms of DNA double strand break repair. *Trends Cell Biol.* **8,** 483–489.
80. Paques, F. and Haber, J. E. (1999) Multiple pathways of recombination induced by double-strand breaks in *Saccharomyces cerevisiae. Microbiol. Mol. Biol. Rev.* **63,** 349–404.
81. Karran, P. (2000) DNA double strand break repair in mammalian cells. *Curr. Opin. Genet. Dev.* **10,** 144–150.
82. Moynahan, M. E., Chiu, J. W., Koller, B. H., and Jasin, M. (1999) Brca1 controls homology-directed DNA repair. *Mol. Cell* **4,** 511–518.
83. Moynahan, M. E., Pierce, A. J., and Jasin, M. (2001) BRCA2 is required for homology-directed repair of chromosomal breaks. *Mol. Cell* **7,** 263–272.
84. Moynahan, M. E., Cui, T. Y., and Jasin, M. (2001) Homology-directed dna repair, mitomycin-c resistance, and chromosome stability is restored with correction of a Brca1 mutation. *Cancer Res.* **61,** 4842–4850.
85. Tutt, A., et al. (2001) Mutation in Brca2 stimulates error-prone homology-directed repair of DNA double-strand breaks occurring between repeated sequences. *EMBO J.* **20,** 4704–4716.
86. Xia, F., et al. (2001) Deficiency of human BRCA2 leads to impaired homologous recombination but maintains normal nonhomologous end joining. *Proc. Natl. Acad. Sci. USA* **98,** 8644–8649.
87. Patel, K. J., et al. (1998) Involvement of Brca2 in DNA repair. *Mol. Cell* **1,** 347–357.
88. Mak, T. W., et al. (2000) Brcal required for T cell lineage development but not TCR loci rearrangement. *Nat. Immunol.* **1,** 77–82.
89. Paull, T. T., Cortez, D., Bowers, B., Elledge, S. J., and Gellert, M. (2001) Direct DNA binding by Brca1. *Proc. Natl. Acad. Sci. USA* **98,** 6086–6091.
90. Ruffner, H., Joazeiro, C. A., Hemmati, D., Hunter, T., and Verma, I. M. (2001) Cancer-predisposing mutations within the RING domain of BRCA1: loss of ubiquitin protein ligase activity and protection from radiation hypersensitivity. *Proc. Natl. Acad. Sci. USA* **98,** 5134–5139.
91. Garcia-Higuera, I., et al. (2001) Interaction of the Fanconi anemia proteins and BRCA1 in a common pathway. *Mol. Cell* **7,** 249–262.
92. Marmorstein, L. Y., et al. (2001) A human BRCA2 complex containing a structural DNA binding component influences cell cycle progression. *Cell* **104,** 247–257.
93. Davies, A. A., et al. (2001) Role of BRCA2 in control of the RAD51 recombination and DNA repair protein. *Mol. Cell* **7,** 273–282.
94. Rogakou, E. P., Pilch, D. R., Orr, A. H., Ivanova, V. S., and Bonner, W. M. (1998) DNA double-stranded breaks induce histone H2AX phosphorylation on serine 139. *J. Biol. Chem.* **273,** 5858–5868.
95. Rogakou, E. P., Boon, C., Redon, C., and Bonner, W. M. (1999) Megabase chromatin domains involved in DNA double-strand breaks in vivo. *J. Cell Biol.* **146,** 905–916.
96. Paull, T. P., et al. (2000) A critical role for histone H2AX in recruitment of repair factors to nuclear foci after DNA damage. *Curr. Biol.* **10,** 886–895.
97. Fuks, F., Milner, J., and Kouzarides, T. (1998) BRCA2 associates with acetyltransferase activity when bound to P/CAF. *Oncogene* **17,** 2531–2534.
98. Yarden, R. I. and Brody, L. C. (1999) BRCA1 interacts with components of the histone deacetylase complex. *Proc. Natl. Acad. Sci. USA* **96,** 4983–4988.

99. Bochar, D. A., et al. (2000) BRCA1 is associated with a human SWI/SNF-related complex: linking chromatin remodeling to breast cancer. *Cell* **102,** 257–265.
100. Thomas, J. E., Smith, M., Tonkinson, J. L., Rubinfeld, B., and Polakis, P. (1997) Induction of phosphorylation on BRCA1 during the cell cycle and after DNA damage. *Cell Growth Differ.* **8,** 801–809.
101. Scully, R., Puget, N., and Vlasakova, K. (2000) DNA polymerase stalling, sister chromatid recombination and the *BRCA* genes. *Oncogene* **19,** 6176–6183.
102. Cox, M. M., et al. (2000) The importance of repairing stalled replication forks. *Nature* **404,** 37–41.
103. Kowalczykowski, S. C. (2000) Initiation of genetic recombination and recombination-dependent replication. *Trends Biochem. Sci.* **25,** 156–165.
104. Zhou, B.-B. S., and Elledge, S. J. (2000) The DNA damage response: putting checkpoints in perspective. *Nature* **408,** 433–439.
105. Cordeiro-Stone, M., Makhov, A. M., Zaritskaya, L. S., and Griffith, J. D. (1999) Analysis of DNA replication forks encountering a pyrimidine dimer in the template to the leading strand. *J. Mol. Biol.* **289,** 1207–1218.
106. Brodie, S. G. and Deng, C. X. (2001) BRCA1-associated tumorigenesis: what have we learned from knockout mice? *Trends Genet.* **17,** S18– S22.
107. Connor, F., et al. (1997) Tumorigenesis and a DNA repair defect in mice with a truncating Brca2 mutation. *Nat. Genet.* **17,** 423–430.
108. Lee, H., et al. (1999) Mitotic checkpoint inactivation fosters transformation in cells lacking the breast cancer susceptibility gene, Brca2. *Mol. Cell* **4,** 1–10.
109. Xu, X., et al. (1999) Conditional mutation of Brca1 in mammary epithelial cells results in blunted ductal morphogenesis and tumour formation. *Nat Genet.* **22,** 37–43.
110. Ludwig, T., Fisher, P., Ganesan, S., and Efstratiadis, A. (2001) Tumorigenesis in mice carrying a truncating Brca1 mutation. *Genes Dev.* **15,** 1188–1193.
111. Ludwig, T., Fisher, P., Murty, V., and Efstratiadis, A. (2001) Development of mammary adenocarcinomas by tissue-specific knockout of Brca2 in mice. *Oncogene* **20,** 3937–3948.
112. Xu, X., et al. (2001) Genetic interactions between tumor suppressors Brca1 and p53 in apoptosis, cell cycle and tumorigenesis. *Nat. Genet.* **28,** 266–271.
113. Smith, P. D., et al. (1999) Novel p53 mutants selected in BRCA-associated tumours which dissociate transformation suppression from other wild-type p53 functions. *Oncogene* **18,** 2451–2459.
114. Brodie, S. G., et al. (2001) Multiple genetic changes are associated with mammary tumorigenesis in Brca1 conditional knockout mice. *Oncogene* **20,** 7514–7523.
115. Monteiro, A. N. (2000) BRCA1: exploring the links to transcription. *Trends Biochem. Sci.* **25,** 469–474.
116. Parvin, J. D. (2001) BRCA1 at a branch point. *Proc. Natl. Acad. Sci. USA* **98,** 5952–5954.
117. Irminger-Finger, I., Siegel, B. D., and Leung, W. C. (1999) The functions of breast cancer susceptibility gene 1 (BRCA1) product and its associated proteins. *Biol. Chem.* **380,** 117–128.
118. Deng, C. X. and Brodie, S. G. (2000) Roles of BRCA1 and its interacting proteins. *Bioessays* **22,** 728–737.
119. Somasundaram, K., et al. (1997) Arrest of the cell cycle by the tumour-suppressor BRCA1 requires the CDK-inhibitor p21WAF1/CiP1. *Nature* **389,** 187–190.
120. Wang, Q., Zhang, H., Kajino, K., and Greene, M. I. (1998) BRCA1 binds c-Myc and inhibits its transcriptional and transforming activity in cells. *Oncogene* **17,** 1939–1948.
121. Fan, S., et al. (1999) BRCA1 inhibition of estrogen receptor signaling in transfected cells. *Science* **284,** 1354–1356.
122. Scully, R., et al. (1997) BRCA1 is a component of the RNA polymerase II holoenzyme. *Proc. Natl. Acad. Sci. USA* **94,** 5605–5610.

123. Yu, X., Wu, L. C., Bowcock, A. M., Aronheim, A., and Baer, R. (1998) The C-terminal (BRCT) domains of BRCA1 interact in vivo with CtIP, a protein implicated in the CtBP pathway of transcriptional repression. *J. Biol. Chem.* **273,** 25388–25392.
124. Zhang, H., et al. (1998) BRCA1 physically associates with p53 and stimulates its transcriptional activity. *Oncogene* **16,** 1713–1721.
125. Somasundaram, K., et al. (1999) BRCA1 signals ARF-dependent stabilization and coactivation of p53. *Oncogene* **18,** 6605–6614.
126. Harkin, D. P., et al. (1999) Induction of GADD45 and JNK/SAPK-dependent apoptosis following inducible expression of BRCA1. *Cell* **97,** 575–586.
127. MacLachlan, T. K., et al. (2000) BRCA1 effects on the cell cycle and the DNA damage response are linked to altered gene expression. *J. Biol. Chem.* **275,** 2777–2785.
128. Irminger-Finger, I., et al. (2001) Identification of BARD1 as mediator between proapoptotic stress and p53- dependent apoptosis. *Mol. Cell* **8,** 1255–1266.
129. Marmorstein, L. Y., Ouchi, T., and Aaronson, S. A. (1998) The BRCA2 gene product functionally interacts with p53 and RAD51. *Proc. Natl. Acad. Sci. USA* **95,** 13869–13874.
130. Zheng, L., et al. (2000) Sequence-specific transcriptional corepressor function for BRCA1 through a novel zinc finger protein, ZBRK1. *Mol. Cell* **6,** 757–768.
131. Tutt, A., et al. (1999) Absence of Brca2 causes genome instability by chromosome breakage and loss associated with centrosome amplification. *Curr. Biol.* **9,** 1107–1110.
132. Xu, X., et al. (1999) Centrosome amplification and a defective G2-M cell cycle checkpoint induce genetic instability in BRCA1 exon 11 isoform-deficient cells. *Mol. Cell* **3,** 389–395.
133. Hsu, L. C. and White, R. L. (1998) BRCA1 is associated with the centrosome during mitosis. *Proc. Natl. Acad. Sci. USA* **95,** 12983–12988.
134. Deming, P. B., et al. (2001) The human decatenation checkpoint. *Proc. Natl. Acad. Sci. USA* **98,** 12044–12049.
135. Gowen, L. C., Avrutskaya, A. V., Latour, A. M., Koller, B. H., and Leadon, S. A. (1998) BRCA1 required for transcription-coupled repair of oxidative DNA damage. *Science* **281,** 1009–1012.
136. Le Page, F., et al. (2000) BRCA1 and BRCA2 are necessary for the transcription-coupled repair of the oxidative 8-oxoguanine lesion in human cells [In Process Citation]. *Cancer Res.* **60,** 5548–5552.
137. Kleiman, F. E. and Manley, J. L. (1999) Functional interaction of BRCA1-associated BARD1 with polyadenylation factor CstF-50. *Science* **285,** 1576–1579.
138. Kleiman, F. E. and Manley, J. L. (2001) The BARD1-CstF-50 interaction links mRNA 3' end formation to DNA damage and tumor suppression. *Cell* **104,** 743–753.
139. Henikoff, S. (1997) Nuclear organization and gene expression: homologous pairing and long-range interactions. *Curr. Opin. Cell. Biol.* **9,** 388–395.

4

Hereditary Colon Cancer Genes

William M. Grady and Sanford D. Markowitz

1. Introduction

Colorectal cancer develops as the result of the progressive accumulation of genetic and epigenetic alterations that lead to the transformation of normal colonic epithelium to colon adenocarcinoma. The fact that colon cancer develops over 10–15 years and progresses through parallel histologic and molecular changes has permitted the study of its molecular pathology in more detail than other cancer types. Consequently, the specific nature of many of these cancer-associated genetic alterations has been determined over the last 15 years. The subsequent effect of these alterations on the cell and molecular biology of the cancer cells in which they occur has also begun to be revealed over the last decade. From the analysis of the molecular genesis of colon cancer, three key themes concerning the molecular pathogenesis of cancer have been established. The first is that cancer emerges via a multistep progression at both the molecular and the morphologic levels ***(1)***. The second is that loss of genomic stability is a key molecular and pathophysiologic step in cancer formation ***(2)***. The third is that hereditary cancer syndromes frequently correspond to germline forms of key genetic defects whose somatic occurrences drive the emergence of sporadic colon cancers ***(3)***.

Consistent with these themes, colon cancer is most commonly initiated by alterations in elements in the Wnt signaling pathway and then progresses as the result of the accumulation of sequential events that either activate oncogenes or inactivate tumor suppressor genes. (**Fig. 1**) Some of the alterations that have been convincingly shown to promote colon carcinogenesis include *K-RAS*, *TP53,* and elements of the TGF-β signaling pathway, such as *TGFBR2*, the gene for TGF-β receptor type II, and *SMAD4*. The pathogenetic significance of many of these alterations in colon cancer formation has been clearly demonstrated by their role in the hereditary colon cancer syndromes.

2. Adenoma–Carcinoma Sequence

The evolution of normal epithelial cells to adenocarcinoma usually follows a predictable progression of histologic changes and concurrent genetic and epigenetic changes. These alterations provide a growth advantage and lead to clonal expansion of

From: *Methods in Molecular Biology, Vol. 222: Tumor Suppressor Genes: Pathways and Isolation Strategies*
Edited by: Wafik S. El-Deiry © Humana Press Inc., Totowa, NJ

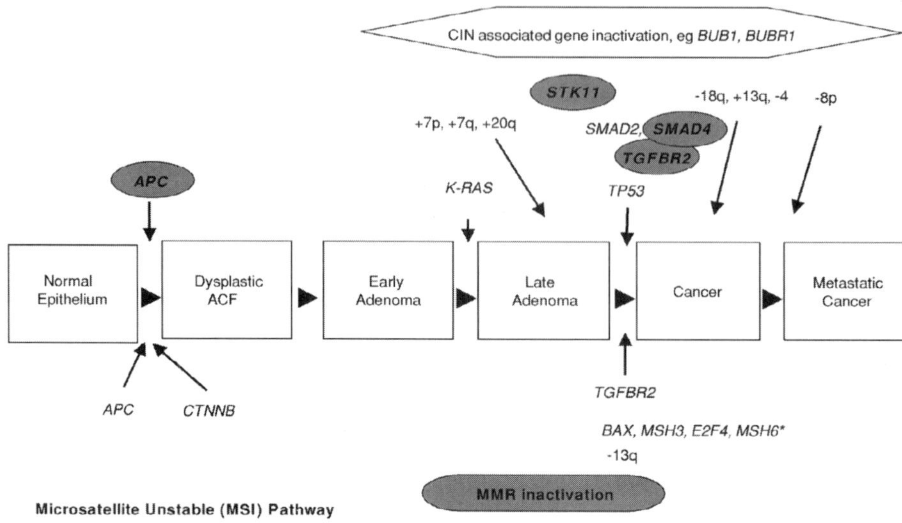

Fig. 1. Schematic representation of genetic alterations occurring in the colon adenoma–adenocarcinoma sequence. Germline mutations in the highlighted genes have been shown to cause hereditary colon cancer. Different patterns of genetic alterations can be identified in colon cancers, depending on whether they display microsatellite instability or chromosome instability. Microsatellite instability occurs when the mutation mismatch repair (MMR) system is inactivated either through mutations or epigenetic alterations. *These genes have been shown to be mutated in approximately 30–90% of MSI colon cancers.

the altered cells. Subsequent alterations with waves of clonal expansion then occur as a consequence of progressive events that provide other growth advantages to the cells, such as loss of cell contact inhibition.

The earliest identifiable lesion in colon cancer formation is the aberrant crypt focus (ACF). The true neoplastic potential of this lesion is still open to debate, but it does appear that some of these lesions can progress to frank adenocarcinoma and that some of them harbor mutations in *K-RAS* or *APC*. In particular, dysplastic aberrant crypt foci frequently carry mutations in *APC* and appear to have the highest potential for progressing to colon cancer. Thus, alterations in *APC*, which results in overactivation of the Wingless/Wnt signaling pathway (discussed below), appear to initiate tumor formation in the colon. Subsequent alterations in other genes then play a role in tumor growth and the eventual acquisition of other malignant characteristics such as tissue invasiveness and the ability to metastasize. In addition, some of these alterations do not directly affect the cell biology of the tumor but instead result in loss of genomic stability, which contributes to the accumulation of mutations in tumor suppressor genes and oncogenes. The timing of the loss of genomic stability, either chromosomal instability (CIN) or microsatellite instability (MSI), relative to adenoma initiation is controversial, but is certainly before progression to frank malignancy. In fact, both CIN and MSI appear to occur relatively early in tumor evolution ***(4–9)***. Ried et al. detected a stepwise increase in the average number of copy alterations using comparative genomic hybridization (CGH) as adenomas progressed from low grade to high grade and then finally to carcinoma ***(8)***. Furthermore, Shih et al. demonstrated allelic imbalance of chromosomes 1p, 8p, 15q, and 18q in 10%, 19%, 28%, and 28% of small adenomas (1–3 mm in size), respectively ***(10)***. The early detection of MSI and CIN in tumor progression provides evidence for the concept that genomic instability creates a permissive state allowing for the accumulation of genetic alterations that lead to tumor development and progression. Some of the specific alterations that have been identified in hereditary colon cancer syndromes will be discussed below. One key concept that has emerged is that these genetic alterations, which occur in a setting of genomic instability, have been shown to involve specific molecular pathways in colon cancer formation and that disruption of these pathways presumably results in specific biologic effects that promote carcinogenesis.

3. Hereditary Colon Cancer Genes

3.1. Familial Adenomatous Polyposis (FAP)

3.1.1. APC

The role of genetic alterations in colon cancer formation was initially suggested by the colon cancer family syndromes, familial adenomatous polyposis (FAP) and its variant, Gardner syndrome. FAP and Gardner syndromes are hereditary colon cancer predisposition syndromes that are characterized by the development of hundreds of intestinal adenomatous polyps and sometimes other extraintestinal features such as congenital hypertrophy of the pigmented retinal epithelium (CHRPE) and desmoid tumors. The clinical features of this syndrome and their autosomal dominant inheritance patterns suggested that a single gene was responsible for the development of the numerous adenomatous polyps in these individuals. The gene responsible for this syndrome, adenomatous polyposis coli (*APC*) was identified as the result of the discovery of an interstitial deletion on chromosome 5q in a patient affected with Gardner syndrome and from classical linkage analysis of families affected by familial adenomatous polyposis (FAP) ***(11–13)***.

The *APC* gene has 15 exons and encodes a large protein (310 kDa, 2843 amino acids) that possesses multiple functional domains that mediate oligomerization as well as binding to a variety of intracellular proteins including β-catenin, γ-catenin, glycogen synthase kinase (GSK)-3β, axin, tubulin, EB1, and hDLG ***(3)*** (**Fig. 2**). Germline mutations in *APC* result in FAP or one of its variants, Gardner syndrome, attenuated FAP, Turcott syndrome, or the flat adenoma syndrome ***(14–17)***.

The positional cloning of *APC* quickly led to an examination of the role of *APC* as a tumor suppressor gene in sporadic colon cancer. Loss of heterozygosity (LOH) studies of 5q21 had already demonstrated that this region was subject to allelic loss in at least 30% of colonic adenomas and adenocarcinomas, and it was presumed that *APC* was the target of these events ***(18)***. The finding of frequent *APC* locus LOH was not only consistent with the model of biallelic inactivation of tumor suppressor genes originally proposed by Knudsen, but also demonstrated that this inactivation commonly occurs through the chromosomal deletion of one of the alleles. Subsequent studies have shown that *APC* is mutated in up to 70% of all sporadic colon adenocarcinomas, which is a high *APC* mutation frequency that is unique to colorectal cancers ***(19,20)***. These mutations are present beginning in the earliest stages of colon cancer formation and precede the other alterations observed during colon cancer formation ***(18,21,22)***. In fact, dysplastic aberrant crypt foci, a presumptive precursor lesion to colon cancer, have been found by some investigators to harbor *APC* mutations ***(23,24)***. The mutations observed

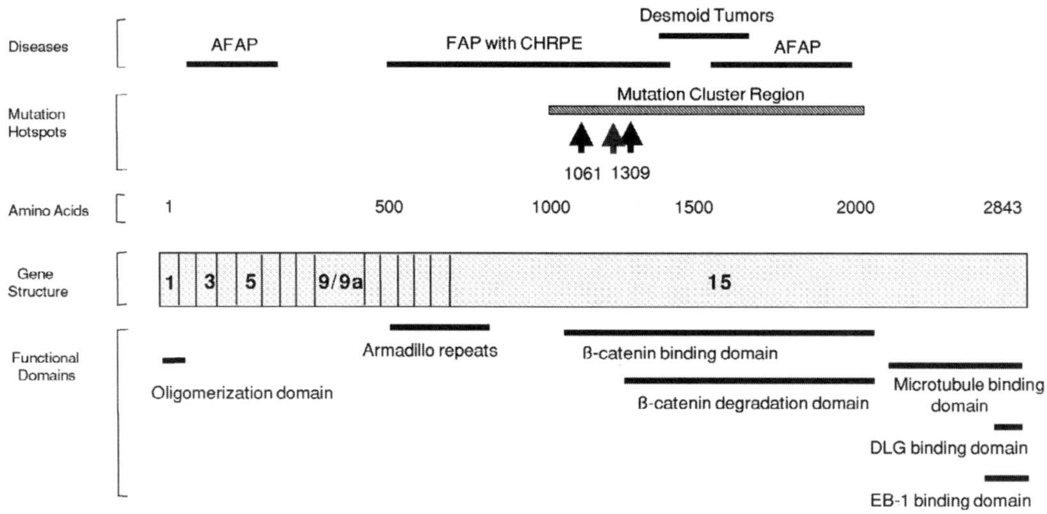

Fig. 2. Schematic diagram of the gene structure of APC. The disease phenotypes associated with mutations in different regions of APC are shown above the diagram. The functional domains of the protein encoded by the gene are indicated below the diagram. The exons in the gene are indicated by the numbers in the gene diagram. The black arrows in the mutation hot spots section indicate two of the most commonly observed mutations, 5-bp deletions, and the patterned arrow indicates the location of a nucleotide polymorphism, I1307K, in the Ashkenazi Jewish population that is associated with colon cancer. AFAP = attenuated familial adenomatous polyposis, FAP with CHRPE = familial adenomatous polyposis with congenital hypertrophy of the pigmented epithelium.

in sporadic colon cancer are found most frequently in the 5′ end of exon 15 between amino acid residues 1280 and 1500 ***(19)***. Mutations in this region can affect the domains between amino acid residues 1020–1169 and 1324–2075, which have been implicated in β-catenin interactions. These mutations can also affect the SAMP (Ser-Ala-Met-Pro) domains located between amino acids 1324 and 2075 and thus disrupt APC's interaction with axin ***(25–27)***. The vast majority of *APC* mutations (>90%) result in premature stop codons and truncated gene products ***(28)*** (**Fig. 2**). As mentioned above, these mutations are often accompanied by chromosomal deletion of the residual wild-type allele, but biallelic inactivation of *APC* can also occur through other routes. For instance, Spirio et al. demonstrated the loss of the mutant germline *APC* allele accompanied by new somatic mutations in the remaining allele in some families affected by attenuated FAP ***(29)***.

One of the central tumor promoting effects of these mutations is to lead to overactivation of the Wingless/Wnt signaling pathway, with the subsequent expression of genes that favor cell growth. The Wnt signaling pathway is overactivated by APC mutations because the *APC* mutations disrupt the ability of APC to bind β-catenin, which leads to increased β-catenin:Tcf-mediated transcription ***(20)***. Normally, GSK-3β forms a complex with APC, β-catenin, and axin, and phosphorylates these proteins. The phosphorylation of β-catenin targets it for ubiquitin-mediated proteasomal degradation. Truncating APC mutations prevent this process from happening and cause an increase in the amount of cytoplasmic β-catenin, which can then translocate to the nucleus and interact with other transcription factors such as T-cell factor/lymphoid enhancing factor (TCF/Lef). TCF-4 is the predominant TCF family member expressed in colonic epithelium. Relevant targets upregulated by TCF-4-mediated transcription that have been identified to date include *CYCLIN D1*, *C-MYC*, *MATRILYSIN*, *C-JUN*, *FRA-1*, urokinase-type plasminogen activator receptor, and the peroxisome proliferator activator receptor ∂ *(PPAR ∂)* ***(30–34)***. Other genes regulated by the Wnt signaling pathway and that have been shown to be upregulated in colon cancer include the *WISP* genes *WISP-1* and *WISP-3*. The *WISP* genes are members of the CCN family of growth factors which includes connective tissue growth factor (*CTGF*), *CYR61*, and *NOV* ***(35)***. Their role in the pathogenesis of colon cancer remains to be defined. Consistent with the concept that overexpression of β-catenin is a central tumor-promoting effect of *APC* mutation, oncogenic mutations in the β-catenin gene (*CTNNB1*) have been observed in some colorectal cancers and are discussed in more detail in the following section ***(36,37)***. In addition, transgenic mice that express activated β-catenin in the gastrointestinal tract display the same phenotype as the MIN mice, which carry heterozygous germline mutations in *APC* ***(38)***.

The clinical effects of *APC* mutations are best understood in the context of FAP. There is a clear correlation with specific clinical features and the location of the mutation in the *APC* gene. *APC* germline mutations between codons 463 and 1387 are associated with CHRPE, and mutations between 1445 and 1578 are associated with an increased risk of desmoid tumors ***(39,40)***. In addition, mutations in the 5' end of the gene proximal to codon 1517 or at the 3' end of the gene distal to codon 1900 are associated with attenuated FAP ***(41,42)***. Attenuated FAP is characterized by the occurrence of fewer colonic adenomas than seen in classic FAP. The polyps occur predominantly in the proximal colon and have a delayed age of onset, with colon cancer occurring in the sixth decade of life. The correlation of the location of the germline *APC* mutations with these features is not uniform, however, and identical mutations can be associated with differing phenotypes ***(13)***. Polymorphisms in the *APC* gene have also been identified. One of these polymorphisms, I1307K, occurs exclusively in people of Ashkenazi Jewish descent and results in a twofold increased risk of colonic adenomas and adenocarcinomas compared to the general population ***(43,44)***. The I1307K polymorphism results from a transition from T to A at nucleotide 3920 in the *APC* gene and appears to create a region of hypermutability. Interestingly, mutations occurring on this allele tend to occur within 36 bp of an 8-bp polyadenine tract created by the polymorphism. The mechanism responsible for this increased mutation susceptibility is unknown but speculated to be increased slippage of the mutation mismatch repair complex as a consequence of the microsatellite repeat generated by the polymorphism ***(43)*** (**Fig. 2**).

3.1.2. β-Catenin (CTNNB1)

β-Catenin is a member of the APC/β-catenin/T-cell factor-lymphoid enhancer factor (tcf/lef) pathway that has been recently shown to play an important role in the formation of certain tumors such as colon cancer, melanoma, and gastric cancer. Indeed, as mentioned above, transgenic mice that carry activating mutations in *CTNNB1* or inactivating mutations in *APC*, both develop multiple intestinal adenomas. However, no germline *CTNNB1* mutations in FAP families have been identified to date. β-Catenin is a homolog of armadillo, and its expression is increased by activation of the Wnt signaling pathway ***(45–47)***. APC interacts with β-Catenin and forms a macromolecular complex with it and glycogen synthase kinase 3β (GSK-3β). β-catenin is consequently directed toward degradation as a result of phosphorylation by GSK-3β ***(48–50)***. Mutations of *CTNNB1* or *APC* often render β-catenin insensitive to APC/β-catenin/GSK-3β-mediated degradation ***(51,52)***. One of the functions of β-catenin is to bind members of the Tcf family of transcription factors and activate gene transcription. Accordingly, cancers with *APC* or *CTNNB1* mutations have increased β-catenin/Tcf-mediated transcription, which leads to the overexpression of genes such as *CYCLIN D1* and *c-MYC* ***(30,53)***. The majority of these mutations are in a portion of exon 3 encoding for the GSK-3β phosphorylation consensus region of β-catenin. These mutations are often missense mutations in the highly conserved aspartic acid 32 and presumably impair the ability of GSK-3β to phosphorylate β-catenin ***(54)***. Caca et al. found *CTNNB1* mutations in the NH_2-terminal phosphorylation sites of β-catenin and found increased Tcf/Lef transcriptional activity in association with this mutation ***(55)***. Mutations that abolish β-catenin binding with E-cadherin have also been identified and have been shown to impair cell adhesion ***(56,57)***. Like *APC* mutations, *CTNNB1* mutations have an essential role in early colon tumor formation. Of interest, though, the incidence of *CTNNB1* mutations decreases from 12.5% in benign adenomas to 1.4% in invasive cancers, suggesting that *CTNNB1* mutations do not favor the progression of adenomas to adenocarcinomas ***(58)***. Of interest, frameshift mutations in a polyadenine tract in *TCF-4* have also been identified in microsatellite unstable tumors although their functional significance is unknown ***(59)***.

The pathways that interact with and counterregulate the Wnt signaling pathway have undergone preliminary characterization. The mitogen-activated protein kinase-related enzymes TAK1 and NLK are homologs of MOM-4 and LIT-1 and have been shown in *Caenorhabditis elegans* to inhibit β-catenin-mediated increases in a TCF-response element through phosphorylation of TCF-4 by NLK ***(60)***. Furthermore, protein phosphatase 2A (PP2A) also has been demonstrated to regulate β-catenin signaling, presumably through dephosphorylation of GSK-3β ***(61)***. Mutations in the *PPP2B1B* gene have been observed in 15% of colon cancers, and these mutations may be promoting colon carcinogenesis through upregulation of Wnt signaling ***(62)***.

3.2. Hereditary Nonpolyposis Colon Cancer Syndrome (HNPCC)

3.2.1. DNA Mismatch Repair System Mutation

The DNA mismatch repair system (also known as the mutation mismatch repair [MMR] system) consists of a complex of proteins that recognize and repair base-pair mismatches that occur during DNA replication. Microsatellite instability (MSI) occurs as the consequence of inactivation of the mutation mismatch repair system and is recognized by frameshift mutations in microsatellite repeats located throughout the genome. Inactivation of the MMR system due to germline gene defects accounts for the colon cancer family syndrome, hereditary nonpolyposis colon cancer syndrome (HNPCC). Somatic inactivation of the mismatch repair system additionally gives rise to approximately 15% of sporadic colon cancers. In either instance, the resulting colon cancers display the phenotype of microsatellite instability. The demonstration of microsatellite instability in cancers can most readily be performed by assaying for alterations at microsatellite loci that are particularly frequently mutated in the setting of MMR inactivation (**Fig. 3**). Since many colon cancers demonstrate frameshift mutations at a small percentage of microsatellite repeats, the designation of a colon adenocarcinoma as showing microsatellite instability depends on the detection of at least two un-

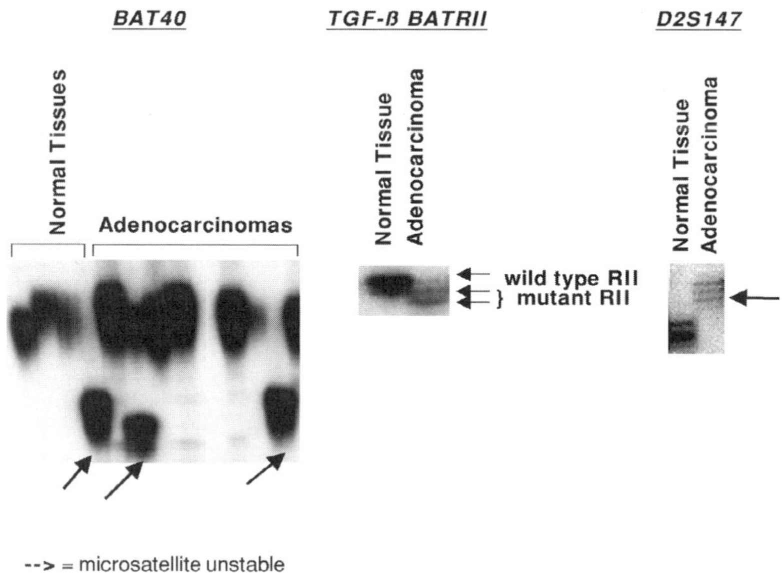

Fig. 3. Representative example of microsatellite instability analysis at *BAT40*, *TGF-βATRII*, and *D2S147*. Microsatellite instability is most commonly demonstrated by a PCR-based method that generates labeled PCR products amplified from different microsatellite loci, e.g., *BAT40*. The labeled PCR products are subjected to polyacrylamide gel electrophoresis and then visualized. The unstable loci are identified because of their size difference from the normal loci. In the example here, the PCR product has been end-labeled with ^{32}P-γATP and visualized with autoradiography. The microsatellite unstable loci are recognized by their size difference from the normal tissue loci and are marked by arrows.

stable loci out of five from a panel of loci that were selected at a recent National Cancer Institute consensus conference *(63)*.

Alterations in at least seven of the genes that encode proteins involved in the MMR system have been identified in either HNPCC or sporadic colon cancer. These genes include *hMSH2, hMSH3, hMSH6, hMLH1, hPMS1, hPMS2,* and *EXO1* ***(64,65)***. HNPCC-related colon cancers account for 3–6% of all colon cancers, and germline mutations in *hMSH2* and *hMLH1* have been found in 45–70% of families that meet strict clinical criteria for HNPCC ***(65–67)***. Since inactivation of both alleles of *hMSH2* or *hMLH1* is required to generate MSI, the cancers that arise in HNPCC kindreds frequently show loss of heterozygosity at the loci of these genes ***(68)***, or alternatively show somatic mutation of the sole wild-type MMR allele. The germline mutations that occur in *hMSH2* and *hMLH1* are widely distributed throughout either gene. *hMSH2* possesses 16 exons and spans 73 kb, and *hMLH1* has 19 exons and spans 58 kb. The mutations that occur in either gene tend to be point mutations that are single-base-pair substitutions, deletions, or insertions. These mutations result in frameshifts (60% of *hMSH2* mutations and 40% of *hMLH1* mutations), premature truncations (23% of *hMSH2* mutations), or missense mutations (31% of *hMLH1* mutations) ***(69)***. The lack of a mutation hot spot has hampered the development of an inexpensive clinical assay to detect germline mutations in the genes known to cause HNPCC. Furthermore, because one allele is sufficient to maintain MMR activity, functional assays to detect MMR gene mutation carriers have not been developed for clinical use to date. However, recently proof-of-principle studies have demonstrated that it may be possible to develop such an assay by forcing a cell to a hemizygous state, in which case a mutant MMR allele could be detected with a functional screening assay ***(70)***.

Studies of the 15% of sporadic colon cancers that demonstrate MSI have shown these arose due to somatic inactivation of MMR genes, and not due to germline MMR gene mutations with low penetrance. Although occasional somatic mutations of *hMSH2* and *hMLH1* have been detected ***(71)***, the predominant mechanism for inactivating MMR unexpectedly proved to be the epigenetic silencing of the *hMLH1* promoter due to aberrant promoter methylation ***(72,73)***. The role of epigenetic alterations in colon carcinogenesis will be discussed in further detail below.

Study of the biochemistry of the MMR proteins has revealed that recognition of the base–base mismatches and insertion/deletion loops is performed by a heterodimer of either hMSH2 and hMSH6 or hMSH2 and hMSH3. Of interest, the hMSH2-hMSH3 heterodimer preferentially recognizes insertion/deletion loops and thus cannot compensate for loss of hMSH6. Consequently, cancers arising with a loss of hMSH6 function display microsatellite instability only in mononucleotide repeats ***(74)***. The hMLH1, hPMS2, and hPMS1 proteins appear to operate primarily in performing the repair of the base–base mismatches and insertion/deletion loops. A heterodimer of hMLH1-hPMS2 operates as a "molecular matchmaker" and is involved in executing the repair of the mismatches in conjunction with DNA-polymerase ∂ and the replication factors proliferating cell nuclear antigen (PCNA), RPA, and RFC as well as the $5' \rightarrow 3'$ exo/endonucleases EXO1 and FEN1 and other unidentified $3' \rightarrow 5'$ exonucleases and helicases ***(74,75)***.

The microsatellite instability that results from loss of MMR activity affects mono-, di-, and tri-nucleotide tracts predominantly. However, cell lines from these tumors also show up to a 1000-fold increased mutation rate at expressed gene sequences, and in particular show instability of short sequence repeats with expressed sequences ***(76)***. Genes that possess such "microsatellite-like" repeats in their coding regions appear to be the targets relevant to carcinogenesis. Indeed, frequently, many genes that possess microsatellite repeats are observed to be mutated in MSI colon cancers. The cancers that demonstrate the presence of multiple genes with insertion/deletion mutations in microsatellite repeats have been termed to possess a microsatellite mutator phenotype. This pathway to tumor formation appears to be distinct from that seen in colon cancers that are microsatellite stable ***(77)***. The most frequently targeted gene for mutation in this pathway is the TGF-β receptor type II tumor suppressor gene, which as discussed in greater detail below, and is ubiquitously mutated in MSI colon cancers. Other, less frequently targeted genes include the *IGF2 receptor*; *BAX* and *CASPASE 5,* proteins which regulate apoptosis; *E2F4* and *TCF4*, transcription factors; *MSH3* and *MSH6*, DNA mismatch repair proteins; *RIZ*, the retinoblastoma protein-interacting zinc finger gene; and *CDX2*, an intestinal homeobox factor ***(77–82)***. Importantly, MSI and the subsequent target gene mutations appear to occur throughout the adenoma-to-carcinoma progression. The timing of many of these events during tumor formation remains to be mapped, but preliminary studies have shown they occur at distinct phases of tumor progression. For example, *TGFBR2* mutations have been mapped to the adenoma–carcinoma transition, suggesting that loss of wild-type *TGFBR2* is important for the acquisition of the malignant phenotype ***(5)***. Thus, MSI creates a favorable state for accumulating mutations in vulnerable genes that control cell growth and death, and these alterations lead ultimately to the generation of colon cancers.

The relationship between the microsatellite mutator pathway and other genetic alterations frequently found in colon cancer is only partially understood. Alteration of the Wnt/Wingless pathway can be observed in tumors regardless of MSI ***(83)***. Mutations in *APC* and *CTNNB1* can be found in 21% and 43% of MSI tumors, respectively ***(84,85)***. In addition, the incidence of *K-RAS* mutations appears to be as high as 22–31%, which is similar to the incidence observed in MSS colon cancers ***(39,86)***. Mutations in *TP53* do appear to be less frequent in MSI cancers than in MSS cancers. The *TP53* mutation incidence in MSI colon cancers has been demonstrated to range between 0 and 40%, whereas the incidence in MSS tumors is between 31 and 67% ***(84,86–88)***. Of interest, monoallelic and biallelic *BAX* mutations are found frequently in MSI colon cancers and may serve to replace the role of mutant *TP53* in colon carcinogenesis. Thus, the microsatellite mutator pathway appears to be initiated through changes in the Wnt/Wingless pathway and share some alterations with the MSS colon cancer pathway. However, other events, such as *TP53* and *TGFBR2* mutations, occur at different frequencies in the MSI versus the MSS pathway.

An appreciation of the genotype:phenotype correlation in HNPCC has predictably lagged behind an understanding of the molecular biology of MMR inactivation. There are presently 228 presumptive disease causing mutations that have been identified and 47 presumptive polymorphisms in the seven MMR genes associated with HNPCC ***(89)***. The number of germline mutations is anticipated to continue to grow until it numbers several hundred. Founder mutations have been identified in HNPCC, with a prominent example being the genomic deletion of exon 16 of *hMLH1*, which has been termed the "Finland 1" mutation, observed in 40 ostensibly unrelated families in Finland and Sweden ***(89)***. There are a number of other recurrent mutations that have been identified in HNPCC families, however, there is no clear mutational hot spot that has lent itself to the development of a screening assay.

In HNPCC, in general there is improved survival for individuals with colon cancer even when corrected for stage ***(90)***. In regards to the effects of the particular MMR germline mutations, mutations of *hMLH1* are more commonly associated with gastric cancer ***(91)***. Furthermore, germline mutations in *MSH6* are associated with an atypical form of HNPCC that does not meet Amsterdam criteria because of a later age of onset and tumors that display low-level microsatellite instability ***(92,93)***. A large catalog of mutations observed in HNPCC families has been deposited on the ICG-Hereditary Nonpolyposis Colon Cancer website (http://nfdht.nl).

3.2.2. Epigenetic Alterations: Aberrant CpG Island Methylation of the MLH1 Promoter

The finding of aberrant *hMLH1* promoter methylation in sporadic MSI colon cancers dramatically illustrated the role of epigenetic changes as potential pathogenetic alterations in cancer ***(72,73,94)***. Aberrant methylation of the cancer genome, and associated silencing of the genes whose promoters demonstrated such methylation, has been well described at multiple genetic loci ***(95,96)***. Reversion of the methylation using demethylating agents such as 5-deoxy-azacytidine frequently restores expression of these, demonstrating methylation, in fact, induces gene silencing. As inactivation of hMLH1 presumptively plays an initiating role in the pathogenesis of MSI colon cancers, the finding of aberrant methylation of *hMLH1* in sporadic MSI colon cancers, and the restoration of hMLH1 expression by demethylating the *hMLH1* promoter in cell lines derived from such cancers, strongly suggested that such aberrant methylation could be a cause rather than a consequence of colon carcinogenesis ***(72,73,94)***. Fine-structure analysis of the methylation status of specific CpG's in the *hMLH1* promoter has shown that the methylation status of small clusters of CpG's in the 5′ region of the *hMLH1* promoter appears to dictate the transcriptional status of the gene ***(97)***. Additional genetic support for a primary role for aberrant methylation in gastrointestinal carcinogenesis was provided by the findings of Grady et al. that loss of expression of E-cadherin in association with CpG methylation of the wild-type *CDH1* allele in tumors occurring in the setting of the cancer family syndrome Hereditary Diffuse Gastric Cancer ***(98)***. It also appears that the epigenetic and genetic changes cooperate to promote cancer formation ***(95)***. Moreover, it is likely that the aberrant hypermethylation of 5' CpG dinucleotides that has been demonstrated to silence a variety of tumor suppressor genes including *CDH1*, *CKN2A/p16*, *thrombospondin-1* (*TSP1*), *hMLH1*, and *GSTP1* may be similarly pathogenetic in the tumors in which these changes have been identified ***(73,94,95,99–101)***. In particular, though mutation of *CDKN2A/p16* has not been described in colon cancer, methylation of *CDKN2A/p16* is detected in 40% of colon cancers and has been found not only in colon cancer but also in colon adenomas ***(100)***. This observation demonstrates that aberrant promoter methylation is occurring early in the adenoma sequence, although it does not confirm that the aberrant *CDKN2A/p16* methylation is a primary rather than a secondary event in the tumorigenesis process. More broadly, early work has suggested that colon cancers that hypermethylated *hMLH1* and/or *CDKN2A/p16* may belong to a distinct subclass of colon cancers that demonstrate genome-wide aberrant methylation of gene promoters and that may arise by a distinct and unique mechanisms ***(100,101)***. Also worthy of note is recent progress that has elucidated how CpG island methylation can mediate transcriptional silencing by recruiting methyl-binding proteins, MeCP2, MBD2, and MBD3, that recognize methylated sequence and recruit histone deacetylases (HDACs). The HDACs then induce changes in chromatin structure that impede the access of transcription factors to the promoter ***(95)***.

3.2.3. TGFBR2 (TGF-β-Receptor Type II)

TGFBR2 encodes the type II transforming growth factor β-receptor type II protein that is an essential component of the heteromeric transforming growth factor-β (TGF-β) receptor. As mentioned above, it appears to play an important role in HNPCC colon cancers since it is biallelically inactivated in essentially all of these tumors. Furthermore, at least one atypical HNPCC family has been shown to carry an inactivating germline mutation in *TGFBR2* (T315M) ***(102)***.

Transforming growth factor-β (TGF-β) is a multifunctional cytokine that can induce growth inhibition, apoptosis, and differentiation in intestinal epithelial cells ***(103,104)***. TGF-β mediates its effects on cells through a heteromeric receptor complex that consists of type I (RI) and type II (RII) components. RI and RII are serine–threonine kinases that phosphorylate downstream signaling proteins upon activation ***(105)***. The receptor complex is activated by TGF-β binding to the RII component of the receptor complex, causing formation of the heteromeric RI–RII receptor complex. The activated RII component then phosphorylates the RI component in the GS box of RI, a glycine–serine-rich region of the receptor. RI then propagates the signal from the receptor through the phosphorylation of downstream proteins such as the smad proteins, Smad2 and Smad3 ***(103)***. Smad2 and Smad3 then form a hetero-oligomeric complex, which can also include Smad4, and translocate to the nucleus ***(105,106)***. In the nucleus they modulate transcription of specific genes through cis-regulatory Smad-binding sequences and through binding with other transcription factors such as p300/CBP, TFE3, Ski, and c-jun ***(20,107,108)***.

The downstream transcriptional targets of the TGF-β signaling pathway are involved in the regulation of cell proliferation, extracellular matrix production, and immune surveillance. These functions not only are an integral part of tissue homeostasis but also are logical targets for dysregulation in colon carcinogenesis. Elements involved in growth regulation that have been clearly shown to be controlled in part by TGF-β include the cyclin-associated proteins cyclin D1, cdk4, p21, p27, p15, and Rb ***(109–113)***. C-myc is also a downstream target of TGF-β and has been shown to be transcriptionally repressed in MvLu1 cells after treatment with TGF-β1 ***(112,114)***. In addition to the cyclin-associated proteins, the extracellular matrix proteins and regulators of extracellular matrix proteins, fibronectin, tenascin, and plasminogen activator inhibitor 1, also appear to be regulated by TGF-β ***(115,116)***. The disruption of the normal extracellular matrix production may play a role in tumor invasion and metastasis. In support of this concept, in a study of a group of MSI colon adenomas that encompassed the spectrum of progression from tubular adenoma to adenocarcinoma, mapping of *TGFBR2* mutations demonstrated that they are temporally associated with the progression of adenomas to adenocarcinomas ***(5)***.

Considerable evidence has accumulated over the last decade demonstrating that the TGF-β signaling pathway can serve as a tumor suppressor in a variety of human tumor types, including colon, gastric, and pancreatic cancer ***(103,117)***. Evidence of TGF-β's role in colon cancer formation came first from studies that demonstrated colon cancer cell lines were resistant to the normal growth inhibitory effects of TGF-β ***(118)***. It is now known that a common mechanism through which colon cancers acquire TGF-β resistance is through genetic alterations of the *TGFBR2* gene. In fact, functionally significant alterations of *TGFBR2* have been identified in up to 30% of colon cancers and are the most common mechanism identified to date for inactivating the TGF-β signaling pathway ***(78,119)***. No alterations in *TGFBRI* or the type III TGF-β receptor (*TGFBR3*) have been observed in studies of TGF-β-resistant colon cancer cell lines, suggesting that mutational inactivation of *TGFBR2* is a particularly favorable event that leads to tumor formation (He and Markowitz, unpublished data). In the study that first elucidated the role of *TGFBR2* mutations in colon cancer formation, Markowitz et al. demonstrated that mutational inactivation of *TGFBR2* is an extremely common event in cancers that display microsatellite instability. *TGFBR2* has a microsatellite-like region in exon 3 that consists of a 10-bp polyadenine tract, making it particularly susceptible to mutation in the setting of MSI ***(78,120,121)***. The mutations in this region, which has been named *BAT-RII* (*B*ig *A*denine *T*ract in TGF-β *R*eceptor type *II*), are frameshift mutations that result in the insertion or deletion of one or two adenines between nucleotides 709 and 718, introducing nonsense mutations that encode a truncated TGF-β RII protein (**Fig. 4**). This truncated protein is only 129–161 amino acids in length, compared to the wild-type protein (565 amino acids), and lacks the receptor's transmembrane domain and intracellular kinase domain ***(78)***. In a series of 110 MSI colon cancers, 100 were found to carry *BAT-RII* mutations and in almost all of these cases the mutations were biallelic,

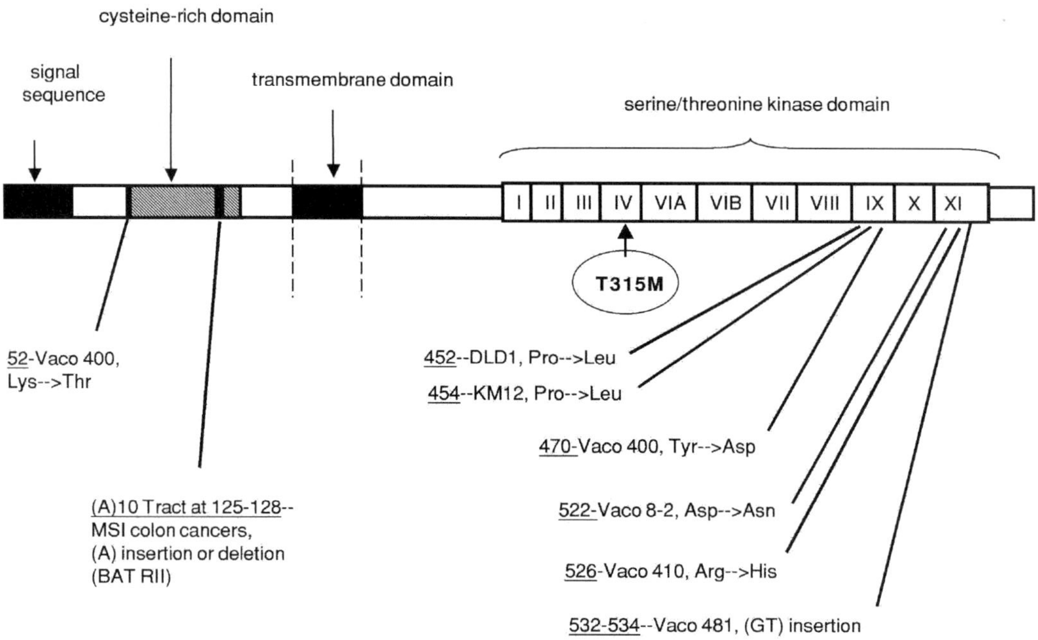

Fig. 4. Schematic diagram of the protein stucture of TGF-βRII, showing the locations of mutations identified in colon cancers identified to date. The domains of the protein are indicated above the protein figure. The codon location of the mutations are shown below the figure. The codon is listed first, followed by the colon cancer cell line in which the mutation was identified and the effect of the mutation on the coded amino acid sequence. The T315M mutation (circled) has been observed in the germline DNA of one family with atypical HNPCC.

consistent with the tumor suppressor function of TGF-β RII ***(120)***. The tumors or cell lines that did not possess biallelic *BAT-RII* mutations were found to have missense mutations in the residual *TGFBR2* allele (e.g., codon 452, Pro→Leu; codon 454, Pro→Leu) ***(120)***. *TGFBR2*'s role as a tumor suppressor gene in colon cancer has been further elucidated by studies showing reconstitution of wild-type *TGFBR2* in HCT116, a MSI colon cancer cell line with a mutant *TGFBR2* gene, suppresses the tumor phenotype of the cell line ***(122)***. This study demonstrated that the introduction of wild-type *TGFBR2* into HCT116 caused a significant growth delay in two clones of HCT116 stably transfected with *TGFBR2*. This effect is likely an underestimate of the tumor suppressor activity of *TGFBR2*, since the clones expressing *TGFBR2* have been selected to tolerate TGF-β RII expression. Independent genetic evidence demonstrating *TGFBR2's* role in colon cancer formation was provided by the identification of a germline mutation in *TGFBR2* in a colon cancer family. This family was characterized by having an HNPCC-like presentation but with an older age of onset of the tumors. Importantly, the tumors arising in this family were shown to have inactivated the residual wild-type allele ***(102)***.

Further support for *TGFBR2's* role as a tumor suppressor gene in colon cancer in general was provided by the demonstration of *TGFBR2* mutations in colon cancer cell lines that are microsatellite stable (MSS). *TGFBR2* mutations have been found in 15% (n = 3/14) of TGF-β-resistant MSS colon cancer cell lines. These mutations are not frameshift mutations in *BAT-RII* but are missense mutations in the kinase domain or putative binding domain of *TGFBR2*. The mutations in these cell lines have been clearly demonstrated to inactivate TGF-β-mediated growth inhibition and Smad-dependent transcription ***(119)*** (**Fig. 4**). However, missense mutations in *TGFBR2* may vary in their effect on the TGF-β receptor's function, which is different from the BAT-RII mutations, which are believed to completely abrogate TGF-β RII's function. For example, a mutation resulting in a substitution of methionine for threonine at codon 315 in the kinase subdomain IV of *TGFBR2*, which inactivates the growth-inhibitory effects of the gene product but does not affect the transcription of plasminogen activator inhibitor type I (*PAI 1*), has been identified ***(123)***. In aggregate, the overall incidence of *TGFBR2* mutation in both microsatellite stable (MSS) and MSI colon cancers appears to be 30% ***(119)***. Interestingly, in a study of colon cancer cell lines, the incidence of TGF-β resistance was found to be 55% despite frequently having wild-type *TGFBR1* and *TGFBR2* ***(119)***. These cancers have presumably inactivated the TGF-β signaling pathway through genetic or epigenetic alterations in postreceptor signaling elements, further underscoring the significance of the TGF-β signaling pathway in colon cancer formation.

The clinical correlations of *TGFBR2* mutations are currently being defined. Colon cancers with MSI and *TGFBR2* mutations display unique clinicopathologic features including an increased incidence in the proximal colon, presentation at an early stage, and better prognosis than microsatellite stable (MSS) colon cancer ***(124)***. In addition, Watanabe et al. have found that Stage III MSI colon cancers that carry *BAT-RII* mutations are associated with a favorable outcome after adjuvant chemotherapy with fluorouracil-based regimens (relative risk of death 2.9, 95% confidence interval 1.14–7.35) ***(125)***. The specific mechanisms through which mutational inactivation of *TGFBR2* mediates these clinicopathologic features remain to be determined.

3.3. Familial Juvenile Polyposis (FJP)

3.3.1. MADH4(SMAD4)

Familial juvenile polyposis (JP; OMIM 174900) is an autosomal dominant gastrointestinal hamartomatous polyposis syndrome that is associated with an increase risk of gastrointestinal cancer. Affected individuals have numerous hamartomatous polyps throughout their gastrointestinal tract and have a several-fold increased risk of colon cancer ***(126)***. Studies performed by Howe et al. in 1998 demonstrated that a locus for FJP mapped to 18q21.1 and that germline mutations in *MADH4* associated with the affected individuals in these families (**Fig. 5**). Of interest, this locus has been intensively studied in sporadic colon cancer because loss of heterozygosity (LOH) of chromosome 18q occurs in approximately 70% of colon adenocarcinomas. The region of deletion on 18q that is shared among colon cancers that demonstrate allelic loss involving a contiguous segment of 18q is the locus of a number of tumor suppressor genes implicated in colon cancer formation including *DCC*, *SMAD2*, and *SMAD4*. All of these genes have been shown to be mutated in sporadic colorectal cancers ***(127–129)***.

The Smad proteins are a family of proteins that serve as intracellular mediators to regulate TGF-β superfamily signaling. The Smad proteins compose an evolutionarily conserved signaling pathway that has been demonstrated in *C. elegans*, *Drosophila melanogaster*, *Xenopus*, as well as humans. These proteins are characterized by two regions that are homologous to the *Drosophila* ortholog, Mad, and that are located at the N- and C-termini of the protein. These regions are termed the Mad-homology domains MH1 and MH2, respectively, and are connected by a less well conserved, proline-rich

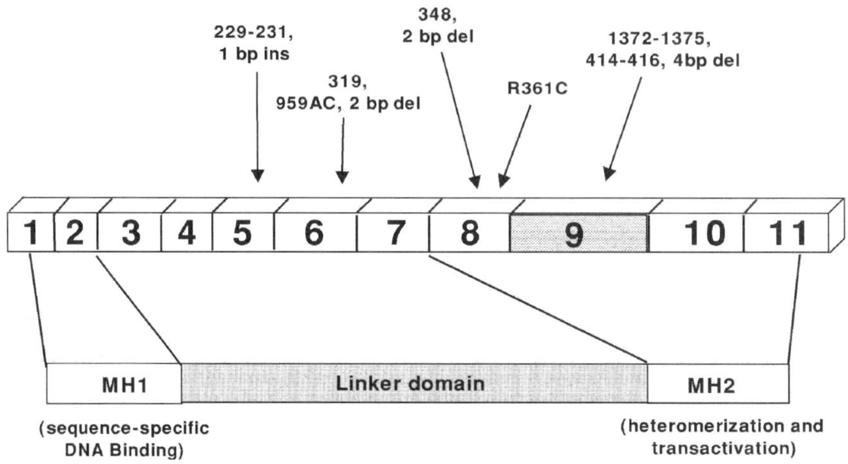

Fig 5. Schematic diagram of *SMAD4* gene structure and encoded protein. The gene has 11 exons that encode a protein that had 3 domains, the MH1 domain, the Linker domain, and the MH2 domain. The location of the germline mutations observed in familial juvenile polyposis families are shown above the gene. The most common mutation is the 1372–1375 deletion that is in exon 9.

linker domain (**Fig. 5**). Numerous studies have identified three major classes of Smad proteins: (a) the receptor-regulated Smads (R-Smads), which are either direct targets of the TGF-β receptor family type I kinases and include Smads1, 2, 3, and 5; (b) the common Smads (Co-Smads: Smad4), which form heteromeric complexes with the R-Smads and propagate the TGF-β-mediated signal; and (c) the inhibitory Smads (I-Smads: Smad6 and Smad7), which antagonize TGF-β signaling through the Smad pathway. Ligand binding to the TGF-β receptor complex results in TGF-β receptor type I mediated phosphorylation of Smad2 and Smad3 on two serine residues in a conserved –SS(M/V)S motif located at the C-terminus of the R-Smads ***(130,131)***. Phosphorylation of these serine residues is required for downstream signaling pathway activation ***(132,133)***. The specificity of Smad2 and Smad3 for the TGF-β receptor is dictated by a matching set of two residues within the C-terminal MH2 domain of these proteins ***(134)***. In addition, Smad2 and Smad3 are recruited to the TGF-β receptor type I (TGF-β RI) through an interaction with a membrane-associated, lipid-binding FYVE domain protein, Smad-Anchor for Receptor Activation (SARA) ***(135)***. Activation of Smad2 and Smad3 induces their partnering with Smad4. This complex then translocates to the nucleus and activates transcriptional responses ***(105,106)***. Smad4 acts as a convergent node in the Smad pathways that are downstream of the TGF-β superfamily receptors and can interact with all the R-Smad proteins. Smad4 appears to regulate the transcriptional activity of the Smad complexes by controlling nuclear trafficking ***(136–138)***. Although initially felt to be an essential component of the TFG-β signaling pathway, Smad4 has since been shown to be dispensible for some TGF-β-mediated responses, such as fibronectin induction and TGF-β-induced cell cycle arrest ***(137,139–142)***. The transcriptional targets of the Smad signaling pathway that may play a role in carcinogenesis include Smad7, c-Jun, Jun-B, p21, and TGF-β1 ***(143–151)***.

In light of the known tumor suppressor effects of the TGF-β signaling pathway and the role the Smad proteins play in propagating this signal, it is not surprising that alterations of some of the SMAD genes have been found in sporadic colon cancer as well as in FJP. Mutational inactivation of *SMAD2* and *SMAD4* has been observed in a high percentage of pancreatic cancers and in 5–10% of colon cancers ***(128,129,152,153)***. *SMAD4* alterations have been found in up to 16% of colon cancers ***(128)***. The effect of these mutations on colon carcinogenesis is being investigated in a number of different animal models. One murine model, a compound heterozygote *Smad4*$^{-/+}$/*APC*$^{\Delta 716}$, develops colon cancer unlike the APC$^{\Delta 716}$ mouse, which only develops small intestinal adenomas ***(154)***. This model suggests that *SMAD4* inactivation may play a role in the progression of colon cancers as opposed to their initiation. However, in some contexts *SMAD4* mutations also appear to initiate tumor formation and appear to contribute to tumor initiation while in a state of haploid insufficiency. The *Smad4*$^{-/+}$ mouse develops gastric and intestinal juvenile polyps and invasive gastric cancer after several months, however, it does not appear to develop colon cancer ***(155,156)***. The concept that haploinsufficiency could predispose to cancer formation has been supported by the identification of familial juvenile polyposis (FJP) kindreds that carry germline mutations in *MADH4* ***(157,158)***. Importantly, though, the polyps observed in FJP and the invasive cancers in the *Smad4*$^{-/+}$ mouse have been shown to have allelic loss of *SMAD4*, supporting the idea that biallelic inactivation of *SMAD4* is needed for cancer formation ***(156,159)***.

3.3.2. BMPR1A

The identification of *MADH4* mutations in FJP families has implicated inactivation of TGF-β superfamily signaling in this syndrome. This superfamily includes not only TGF-β1, TGF-β2, and TGF-β3, but also the BMPs (Bone Morphogenetic Proteins), activin, and inhibin. The recent identification of germline mutations in *BMPR1A* in kindreds with FJP has suggested that FJP results from disruption of signaling from a specific subclass of TGF-β ligands, the BMPs. Howe et al. found nonsense and missense germline mutations in *BMPR1A* in four FJP families, 44-47delTGTT, 715C>T, 812G>A, and 961delC affecting exons 1, 7, 7, and 8, respectively ***(160)***.

The BMPs are disulfide-linked dimeric proteins that number at least 15 in total and include BMP-2, BMP-4, and BMP-7 (OP-1). They have a wide range of biologic activities including the regulation of morphogenesis of various tissues and organs during development as well as the regulation of growth, differentiation, chemotaxis, and apoptosis in monocytes, epithelial cells, mesenchymal cells, and neuronal cells ***(161)***. The BMPs transduce their signals through a heteromeric receptor that consists of a type I and a type II receptor. BMPR1A is one of two different type I BMP receptors (BMPR1A and BMPR1B). It serves predominantly to bind BMP-4 and BMP-2 as well as other BMPs and transduces their signals when partnered with a BMP type II receptor. As with the TGF-β receptor, the best understood post-BMP receptor pathway is the Smad pathway. The R-Smads, Smads 1 and 5, partner with Smad4 (Co-Smad) to transduce BMP-mediated signals from the BMP receptors ***(161)***. Thus, the identification of both *BMPR1A* and *MADH4* germline mutations in FJP families strongly implicates BMP signaling disruption in the pathogenesis of this syndrome. Interestingly, *BMPR1A* homozygous null mice die at E9.5 due to a defect in mesoderm formation during gastrulation, demonstrating that BMPR1A plays an important role in mesoderm formation ***(162)***. This effect of BMP signaling on mesoderm may explain the histology of FJP polyps, which are characterized by an excessive proliferation of mesenchymal components. Of particular

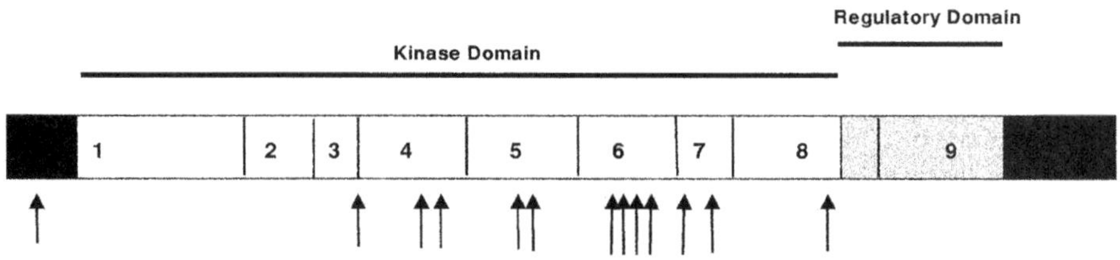

Fig. 6. Schematic diagram of *STK11* gene structure and encoded protein. The gene has 9 exons that encode a protein that has at least 2 domains, a kinase domain and a regulatory domain. The locations of the germline mutations observed in Peutz-Jeghers syndrome families are shown below the gene.

note, the identification of *BMPR1A* germline mutations in FJP is the first piece of significant genetic evidence implicating BMP signaling in colon carcinogenesis.

3.4. Peutz-Jeghers Syndrome

3.4.1. STK11

STK11 was recently cloned from the locus linked to Peutz-Jeghers syndrome. This syndrome is characterized most prominently by gastrointestinal hamartomatous polyps and melanocytic macules of the lips, buccal mucosa, and digits. In addition, it is associated with an increased risk of various neoplasms including sex cord tumors, pancreatic cancer, melanoma, and gastric cancer ***(163–166)***. Germline mutations in *STK11*, also known as *LKB1*, appear to be responsible for most if not all cases of Peutz-Jeghers syndrome ***(167–171)***. *STK11* maps to 19p13.3 and encodes a serine–threonine kinase (**Fig. 6**). The function of the gene product is presently unknown, but the mouse homolog appears to be a nuclear protein ***(172)***. Inactivating mutations of *STK11* have been found in a subset of sporadic tumors including melanomas, pancreatic cancers, and gastric cancers. Park et al. screened a collection of 28 sporadic gastric cancers from Korea for *STK11* mutations using SSCP and DNA sequencing and found 1 (3%) missense mutation and 2 silent mutations. The role of *STK11* mutations in sporadic gastric cancer in low-risk areas remains to be determined ***(173)***. Thus, its role in cancer formation is best appreciated in the context of Peutz-Jeghers syndrome.

4. Conclusion

Colon cancer genetics has yielded an extraordinary cornucopia of new insights and paradigms that have broadly informed the studies of most solid tumors. Key insights that have been contributed include the multistep nature of carcinogenesis, the central role of tumor suppressor pathways, the key role in cancer of mutational inactivation of *TP53*, the role of DNA repair genes and genomic stability in cancer prevention, and the role of TGF-β signaling in tumor suppression. Nonetheless, many challenges remain. The molecular genesis of the metastatic phenotype that accounts directly for cancer lethality still remains hidden. A mechanistic understanding of the basis of chromosomal instability, aneuploidy, and aberrant methylation of the cancer genome has yet to be achieved. Furthermore, there is not yet an understanding of the genetic basis within the general population of individual susceptibility to colon or other cancers. Lastly, the translation of molecular genetics to new diagnostic, prognostic, and therapeutic modalities remains a challenge. The promise for the future is that this field of inquiry, which has been so fruitful to date, will continue to yield the important answers to these and other key questions.

References

1. Fearon, E. and Vogelstein, B. (1990) A genetic model for colorectal tumorigenesis. *Cell* **61**, 759–767.
2. Lengauer, C., Kinzler, K., and Vogelstein, B. (1998) Genetic instabilities in human cancers. *Nature* **396**, 643–649.
3. Kinzler, K. and Vogelstein, B. (1996) Lessons from hereditary colorectal cancer. *Cell* **87**, 159–170.
4. Aaltonen, L., Peltomaki, P., Mecklin, J.-P., et al. (1994) Replication errors in benign and malignant tumors from hereditary nonpolyposis colorectal cancer patients. *Cancer Res.* **54**, 1645–1648.
5. Grady, W., Rajput, A., Myeroff, L., et al. (1998) Mutation of the type II transforming growth factor-β receptor is coincident with the transformation of human colon adenomas to malignant carcinomas. *Cancer Res.* **58**, 3101–3104.
6. Jacoby, R., Marshall, D., Kailas, S., Schlack, S., Harms, B., and Love, R. (1995) Genetic instability associated with adenoma to carcinoma progression in hereditary nonpolyposis colon cancer. *Gastroenterology* **109**, 73–82.
7. Bomme, L., Bardi, G., Pandis, N., Fenger, C., Kronborg, O., and Heim, S. (1998) Cytogenetic analysis of colorectal adenomas: karyotypic comparisons of synchronous tumors. *Cancer Genet. Cytogene.* **106**, 66–71.
8. Ried, T., Heselmeyer-Haddad, K., Blegen, H., Schrock, E., and Auer, G. (1999) Genomic changes defining the genesis, progression, and malignancy potential in solid human tumors: a phenotype/genotype correlation. *Genes Chromosomes Cancer* **25**, 195–204.
9. Rooney, P., Murray, G., Stevenson, D., Haites, N., Cassidy, J., and McLeod, H. (1999) Comparative genomic hybridization and chromosomal instability in solid tumors. *Br. J. Cancer* **80**, 862–873.
10. Shih, I. M., Zhou, W., Goodman, S. N., Lengauer, C., Kinzler, K. W., and Vogelstein, B. (2001) Evidence that genetic instability occurs at an early stage of colorectal tumorigenesis. *Cancer Res.* **61**, 818–822.
11. Herrera, L., Kakati, S., Gibas, L., Pietrzak, E., and Sandberg, A. A. Gardner (1986) syndrome in a man with an interstitial deletion of 5q. *Am. J. Med. Genet.* **25**, 473–476.
12. Groden, J., Thliveris, A., Samowitz, W., et al. (1991) Identification and characterization of the familial adenomatous polyposis coli gene. *Cell* **66**, 589–600.
13. Nishisho, I., Nakamura, Y., Miyoshi, Y., et al. (1991) Mutations of chromosome 5q21 genes in FAP and colorectal cancer patients. *Science* **253**, 665–669.
14. Mori, T., Nagase, H., Horii, A., et al. (1994) Germ-line and somatic mutations of the APC gene in patients with Turcot syndrome and analysis of APC mutations in brain tumors. *Genes Chromo. Cancer* **9**, 168–172.
15. Spirio, L., Otterud, B., Stauffer, D., et al. (1992) Linkage of a variant or attenuated form of adenomatous polyposis coli to the adenomatous polyposis coli (APC) locus. *Am. J. Hum. Genet.* **51**, 92–100.
16. Soravia, C., Berk, T., Madlensky, L., et al. (1998) Genotype-phenotype correlations in attenuated adenomatous polyposis coli. *Am. J. Hum. Genet.* **62**, 1290–1301.
17. Foulkes, W. D. (1995) A tale of four syndromes: familial adenomatous polyposis, Gardner syndrome, attenuated APC and Turcot syndrome. *Q. J. Med.* **88**, 853–863.
18. Vogelstein, B., Fearon, E. R., Hamilton, S. R., et al. (1988) Genetic alterations during colorectal-tumor development. *N. Engl. J. Med.* **319**, 525–532.

19. Miyaki, M., Konishi, M., Kikuchi-Yanoshita, R., et al. (1994) Characteristics of somatic mutation of the adenomatous polyposis coli gene in colorectal tumors. *Cancer Res.* **54**, 3011–3020.
20. Chung, D. (2000) The genetic basis of colorectal cancer:insights into critical pathways of tumorigenesis. *Gastroenterology* **119**, 854–865.
21. Powell, S. M., Zilz, N., Beazer-Barclay, Y., et al. (1992) APC mutations occur early during colorectal tumorigenesis. *Nature* **359**, 235–237.
22. Miyoshi, Y., Nagase, H., Ando, H., et al. (1992) Somatic mutations of the APC gene in colorectal tumors: mutation cluster region in the APC gene. *Hum. Mol. Genet.* **1**, 229–233.
23. Jen, J., Powell, S. M., Papadopoulos, N., et al. (1994) Molecular determinants of dysplasia in colorectal lesions. *Cancer Res.* **54**, 5523–5526.
24. Smith, A. J., Stern, H. S., Penner, M., et al. (1994) Somatic APC and K-ras codon 12 mutations in aberrant crypt foci from human colons. *Cancer Res.* **54**, 5527–5530.
25. Su, L. K., Vogelstein, B., and Kinzler, K. W. (1993) Association of the APC tumor suppressor protein with catenins. *Science* **262**, 1734–1737.
26. Rubinfeld, B., Souza, B., Albert, I., et al. (1993) Association of the APC gene product with beta-catenin. *Science* **262**, 1731–1734.
27. Behrens, J., Jerchow, B. A., Wurtele, M., et al. (1998) Functional interaction of an axin homolog, conductin, with beta- catenin, APC, and GSK3beta. *Science* **280**, 596–599.
28. Powell, S. M., Petersen, G. M., Krush, A. J., et al. (1993) Molecular diagnosis of familial adenomatous polyposis [see comments]. *N. Engl. J. Med.* **329**, 1982–1987.
29. Spirio, L. N., Samowitz, W., Robertson, J., et al. (1998) Alleles of APC modulate the frequency and classes of mutations that lead to colon polyps. *Nat. Genet.* **20**, 385–388.
30. He, T. C., Sparks, A. B., Rago, C., et al. (1998) Identification of c-MYC as a target of the APC pathway [see comments]. *Science* **281**, 1509–1512.
31. Tetsu, O. and McCormick, F. (1999) Beta-catenin regulates expression of cyclin D1 in colon carcinoma cells. *Nature* **398,** 422–426.
32. Crawford, H. C., Fingleton, B. M., Rudolph-Owen, L. A., et al. (1999) The metalloproteinase matrilysin is a target of beta-catenin transactivation in intestinal tumors. *Oncogene* **18**, 2883–2891.
33. He, T. C., Chan, T. A., Vogelstein, B., and Kinzler, K. W. (1999) PPARdelta is an APC-regulated target of nonsteroidal anti-inflammatory drugs. *Cell* **99**, 335–345.
34. Mann, B., Gelos, M., Siedow, A., et al. (1999) Target genes of beta-catenin-T cell-factor/lymphoid-enhancer-factor signaling in human colorectal carcinomas. *Proc. Natl. Acad. Sci. U S A* **96**, 1603–1608.
35. Pennica, D., Swanson, et al. (1998) WISP genes are members of the connective tissue growth factor family that are up-regulated in wnt-1-transformed cells and aberrantly expressed in human colon tumors. *Proc. Natl. Acad. Sci. U S A* **95**, 14717–14722.
36. Sparks, A. B., Morin, P. J., Vogelstein, B., and Kinzler, K. W. (1998) Mutational analysis of the APC/beta-catenin/Tcf pathway in colorectal cancer. *Cancer Res.* **58**, 1130–1134.
37. Kitaeva, M., Grogan, L., Williams, J., et al. (1997) Mutations in β-catenin are uncommon in colorectal cancer occurring in occasional replication error-positive tumors. *Cancer Res.* **57**, 4478–4481.
38. Harada, N., Tamai, Y., Ishikawa, T., et al. (1999) Intestinal polyposis in mice with a dominant stable mutation of the beta-catenin gene. *EMBO J.* **18**, 5931–5942.
39. Olschwang, S., Tiret, A., Laurent-Puig, P., Muleris, M., Parc, R., and Thomas, G. (1993) Restriction of ocular fundus lesions to a specific subgroup of APC mutations in adenomatous polyposis coli patients. *Cell* **75**, 959–968.
40. Caspari, R., Olschwang, S., Friedl, W., et al. (1995) Familial adenomatous polyposis: desmoid tumours and lack of ophthalmic lesions (CHRPE) associated with APC mutations beyond codon 1444. *Hum. Mol. Genet.* **4**, 337–340.

41. Spirio, L., Olschwang, S., Groden, J., et al. (1993) Alleles of the APC gene: an attenuated form of familial polyposis. *Cell* **75**, 951–957.
42. Gardner, R. J., Kool, D., Edkins, E., et al. (1997) The clinical correlates of a 3' truncating mutation (codons 1982-1983) in the adenomatous polyposis coli gene. *Gastroenterology* **113**, 326–331.
43. Laken, S. J., Petersen, G. M., Gruber, S. B., et al. (1997) Familial colorectal cancer in Ashkenazim due to a hypermutable tract in APC. *Nat. Genet.* **17**, 79–83.
44. Lothe, R. A., Hektoen, M., Johnsen, H., et al. (1998) The APC gene I1307K variant is rare in Norwegian patients with familial and sporadic colorectal or breast cancer. *Cancer Res.* **58**, 2923–2924.
45. Hulsken, J., Birchmeier, W., and Behrens, J. (1994) E-cadherin and APC compete for the interaction with beta-catenin and the cytoskeleton. *J. Cell Biol.* **127**, 2061–2069.
46. Aberle, H., Butz, S., Stappert, J., Weissig, H., Kemler, R., and Hoschuetzky, H. (1994) Assembly of the cadherin-catenin complex in vitro with recombinant proteins. *J. Cell Sci.* **107**, 3655–3663.
47. Moon, R. T., Brown, J. D., Yang-Snyder, J. A., and Miller, J. R. (1997) Structurally related receptors and antagonists compete for secreted Wnt ligands. *Cell* **88**, 725–728.
48. Rubinfeld, B., Albert, I., Porfiri, E., Munemitsu, S., and Polakis, P. (1997) Loss of beta-catenin regulation by the APC tumor suppressor protein correlates with loss of structure due to common somatic mutations of the gene. *Cancer Res.* **57**, 4624–4630.
49. Munemitsu, S., Albert, I., Souza, B., Rubinfeld, B., and Polakis, P. (1995) Regulation of intracellular beta-catenin levels by the adenomatous polyposis coli (APC) tumor-suppressor protein. *Proc. Natl. Acad. Sci. USA* **92**, 3046–3050.
50. Munemitsu, S., Albert, I., Rubinfeld, B., and Polakis, P. (1996) Deletion of an amino-terminal sequence beta-catenin in vivo and promotes hyperphosporylation of the adenomatous polyposis coli tumor suppressor protein. *Mol. Cell Biol.* **16**, 4088–4094.
51. Morin, P. J., Sparks, A. B., Korinek, V., et al. (1997) Activation of beta-catenin-Tcf signaling in colon cancer by mutations in beta-catenin or APC [see comments]. *Science* **275**, 1787–1790.
52. Rubinfeld, B., Robbins, P., El-Gamil, M., Albert, I., Porfiri, E., and Polakis, P. (1997) Stabilization of beta-catenin by genetic defects in melanoma cell lines [see comments]. *Science* **275**, 1790–1792.
53. Shtutman, M., Zhurinsky, J., Simcha, I., et al. (1999) The cyclin D1 gene is a target of the beta-catenin/LEF-1 pathway. *Proc. Natl. Acad. Sci. USA* **96**, 5522–5527.
54. Park, W. S., Oh, R. R., Park, J. Y., et al. (1999) Frequent somatic mutations of the beta-catenin gene in intestinal-type gastric cancer. *Cancer Res.* **59**, 4257–4260.
55. Caca, K., Kolligs, F. T., Ji, X., et al. (1999) Beta- and gamma-catenin mutations, but not E-cadherin inactivation, underlie T-cell factor/lymphoid enhancer factor transcriptional deregulation in gastric and pancreatic cancer. *Cell Growth Differ.* **10**, 369–376.
56. Kawanishi, J., Kato, J., Sasaki, K., Fujii, S., Watanabe, N., and Niitsu, Y. (1995) Loss of E-cadherin-dependent cell-cell adhesion due to mutation of the beta-catenin gene in a human cancer cell line, HSC-39. *Mol. Cell Biol.* **15**, 1175–1181.
57. Luber, B., Candidus, S., Handschuh, G., et al. (2000) Tumor-derived mutated E-cadherin influences beta-catenin localization and increases susceptibility to actin cytoskeletal changes induced by pervanadate [In Process Citation]. *Cell Adhes. Commun.* **7**, 391–408.
58. Samowitz, W. S., Powers, M. D., Spirio, L. N., Nollet, F., van Roy, F., and Slattery, M. L. (1999) Beta-catenin mutations are more frequent in small colorectal adenomas than in larger adenomas and invasive carcinomas. *Cancer Res.* **59**, 1442–1444.
59. Duval, A., Gayet, J., Zhou, X. P., Iacopetta, B., Thomas, G., and Hamelin, R. (1999) Frequent frameshift mutations of the TCF-4 gene in colorectal cancers with microsatellite instability. *Cancer Res.* **59**, 4213–4215.

60. Ishitani, T., Ninomiya-Tsuji, J., Nagai, S., et al. (1999) The TAK1-NLK-MAPK-related pathway antagonizes signalling between beta- catenin and transcription factor TCF. *Nature* **399**, 798–802.
61. Seeling, J. M., Miller, J. R., Gil, R., Moon, R. T., White, R., and Virshup, D. M. (1999) Regulation of beta-catenin signaling by the B56 subunit of protein phosphatase 2A. *Science* **283**, 2089–2091.
62. Wang, S. S., Esplin, E. D., Li, J. L., et al. (1998) Alterations of the PPP2R1B gene in human lung and colon cancer. *Science* **282**, 284–287.
63. Boland, C., Thibodeau, S., Hamilton, S., et al. (1998) National Cancer Institute workshop on microsatellite instability for cancer detection and familial predispostion:development of international criteria for the determination of microsatellite instability in colorectal cancer. *Cancer Res.* **58**, 5248–5257.
64. Wu, Y., Berends, M. J., Post, J. G., et al. (2001) Germline mutations of exo1 gene in patients with hereditary nonpolyposis colorectal cancer (hnpcc) and atypical hnpcc forms. *Gastroenterology* **120**, 1580–1587.
65. Marra, G. and Boland, C. (1995) Hereditary nonpolyposis colorectal cancer:the syndrome, the genes, and historical perspective, *J. Natl. Cancer Inst.* **87**, 1114–1125.
66. Wijnen, J. T., Vasen, H. F., Khan, P. M., et al. (1998) Clinical findings with implications for genetic testing in families with clustering of colorectal cancer. *N. Engl. J. Med.* **339**, 511–518.
67. Liu, B., Parsons, R., Papadopoulos, N., et al. (1996) Analysis of mismatch repair genes in hereditary non-polyposis colorectal cancer patients [see comments]. *Nat. Med.* **2**, 169–174.
68. Hemminki, A., Peltomaki, P., Mecklin, J. P., et al. (1994) Loss of the wild type MLH1 gene is a feature of hereditary nonpolyposis colorectal cancer. *Nat. Genet.* **8**, 405–410.
69. Peltomaki, P. and Vasen, H. F. (1997) Mutations predisposing to hereditary nonpolyposis colorectal cancer: database and results of a collaborative study. The International Collaborative Group on Hereditary Nonpolyposis Colorectal Cancer. *Gastroenterology* **113**, 1146–1158.
70. Yan, H., Papadopoulos, N., Marra, G., et al. (2000) Conversion of diploidy to haploidy. *Nature* **403**, 723–724.
71. Liu, B., Nicolaides, N., Markowitz, S., et al. (1995) Mismatch repair gene defects in sporadic colorectal cancers with microsatellite instability. *Nature Genet.* **9**, 48–53.
72. Kane, M., Loda, M., Gaida, G., et al. (1997) Methylation of the hMLH1 promoter correlates with lack of expression of hMLH1 in sporadic colon tumors and mismatch repair-defective human tumor cell lines. *Cancer Res.* **57**, 808–811.
73. Veigl, M., Kasturi, L., Olechnowicz, J., et al. (1998) Biallelic inactivation of hMLH1 by epigenetic gene silencing, a novel mechanism causing human MSI cancers. *Proc. Natl. Acad. Sci.* **95**, 8698–8702.
74. Jiricny, J. (1998) Replication errors: cha(lle)nging the genome. *EMBO J.* **17**, 6427–6436.
75. Kolodner, R. D. and Marsischky, G. T. (1999) Eukaryotic DNA mismatch repair. *Curr. Opin. Genet. Dev.* **9**, 89–96.
76. Eshleman, J., Lang, E., Bowerfind, G., et al. (1995) Increased mutation rate at the hprt locus accompanies microsatellite instability in colon cancer. *Oncogene* **10**, 33–37.
77. Yamamoto, H., Sawai, H., Weber, T., Rodriguez-Bigas, M., and Perucho, M. (1998) Somatic frameshift mutations in DNA mismatch repair and proapoptosis genes in hereditary nonpolyposis colorectal cancer. *Cancer Res.* **58**, 997–1003.
78. Markowitz, S., Wang, J., Myeroff, L., et al. (1995) Inactivation of the type II TGF-β receptor in colon cancer cells with microsatellite instability. *Science* **268**, 1336–1338.
79. Schwartz, S., Yamamoto, H., Navarro, M., Maestro, M., Reventos, J., and Perucho, M. (1999) Frameshift mutations at mononucleotide repeats in *caspase-5* and other target genes in endometrial and gastrointestinal cancer of the microsatellite mutator phenotype. *Cancer Res.* **59**, 2995–3002.

80. Ikeda, M., Orimo, H., Moriyama, H., et al. (1998) Close correlation between mutations of E2F4 and hMSH3 genes in colorectal cancers with microsatellite instability. *Cancer Res.* **58**, 594–598.
81. Piao, Z., Fang, W., Malkhosyan, S., et al. (2000) Frequent frameshift mutations of RIZ in sporadic gastrointestinal and endometrial carcinomas with microsatellite instability [In Process Citation]. *Cancer Res.* **60**, 4701–4704.
82. Wicking, C., Simms, L. A., Evans, T., et al. (1998) CDX2, a human homologue of *Drosophila* caudal, is mutated in both alleles in a replication error positive colorectal cancer. *Oncogene* **17**, 657–659.
83. Huang, J., Papadopoulos, N., McKinley, A., et al. (1996) APC mutations in colorectal tumors with mismatch repair deficiency. *Proc. Natl. Acad. Sci. USA* **93**, 9049–9054.
84. Konishi, M., Kikuchi-Yanoshita, R., Tanaka, K., et al. (1996) Molecular nature of colon tumors in hereditary nonpolyposis colon cancer, familial polyposis, and sporadic colon cancer. *Gastroenterology* **111**, 307–317.
85. Miyaki, M., Iijima, T., Kimura, J., et al. (1999) Frequent mutation of beta-catenin and APC genes in primary colorectal tumors from patients with hereditary nonpolyposis colorectal cancer. *Cancer Res.* **59**, 4506–4509.
86. Fujiwara, T., Stolker, J. M., Watanabe, T., et al. (1998) Accumulated clonal genetic alterations in familial and sporadic colorectal carcinomas with widespread instability in microsatellite sequences. *Am. J. Pathol.* **153,** 1063–1078.
87. Eshleman, J., Casey, G., Kochera, M., et al. (1998) Chromosome number and structure both are markedly stable in RER colorectal cancers and are not destabilized by mutation of p53. *Oncogene* **17**, 719–725.
88. Olschwang, S., Hamelin, R., Laurent-Puig, P., et al. (1997) Alternative genetic pathways in colorectal carcinogenesis. *Proc. Natl. Acad. Sci. USA* **94**, 12122–12127.
89. Lynch, H. T. and de la Chapelle, A. (1999) Genetic susceptibility to non-polyposis colorectal cancer. *J. Med. Genet.* **36**, 801–818.
90. Watson, P., Lin, K. M., Rodriguez-Bigas, M. A., et al. (1998) Colorectal carcinoma survival among hereditary nonpolyposis colorectal carcinoma family members [see comments]. *Cancer* **83**, 259–266.
91. Aarnio, M., Mecklin, J. P., Aaltonen, L. A., Nystrom-Lahti, M., and Jarvinen, H. J. (1995) Life-time risk of different cancers in hereditary non-polyposis colorectal cancer (HNPCC) syndrome. *Int. J. Cancer* **64**, 430–433.
92. Miyaki, M., Konishi, M., Tanaka, K., et al. (1997) Germline mutation of MSH6 as the cause of hereditary nonpolyposis colorectal cancer [letter]. *Nat. Genet.* **17**, 271–272.
93. Akiyama, Y., Sato, H., Yamada, T., et al. (1997) Germ-line mutation of the hMSH6/GTBP gene in an atypical hereditary nonpolyposis colorectal cancer kindred. *Cancer Res.* **57**, 3920–3923.
94. Herman, J., Umar, A., Polyak, K., et al. (1998) Incidence and functional consequences of hMLH1 promoter hypermethylation in colorectal carcinoma. *Proc. Natl. Acad. Sci. USA* **95,** 6870–6875.
95. Baylin, S. B. and Herman, J. G. (2000) DNA hypermethylation in tumorigenesis: epigenetics joins genetics. *Trends Genet.* **16**, 168–174.
96. Jones, P. and Laird, P. (1999) Cancer epigenetics comes of age. *Nature Genet.* **21**, 163– 167.
97. Deng, G., Chen, A., Hong, J., Chae, H., and Kim, Y. (1999) Methylation of CpG in a small region of the hMLH1 promoter invariably correlates with the absence of gene expression. *Cancer Res.* **59**, 2029–2033.
98. Grady, W. M., Willis, J., Guilford, P. J., et al. (2000) Methylation of the CDH1 promoter as the second genetic hit in hereditary diffuse gastric cancer [In Process Citation] *Nat. Genet.* **26**, 16–17.

99. Herman, J. G., Merlo, A., Mao, L., et al. (1995) Inactivation of the CDKN2/p16/MTS1 gene is frequently associated with aberrant DNA methylation in all common human cancers. *Cancer Res.* **55**, 4525–4530.
100. Toyota, M., Ho, C., Ahuja, N., et al. (1999) Identification of differentially methylated sequences in colorectal cancer by methylated CpG island amplification. *Cancer Res.* **59**, 2307–2312.
101. Toyota, M., Ahuja, N., Ohe-Toyota, M., Herman, J. G., Baylin, S. B., and Issa, J. P. (1999) CpG island methylator phenotype in colorectal cancer. *Proc. Natl. Acad. Sci. USA* **96**, 8681–8686.
102. Lu, S. L., Kawabata, M., Imamura, T., et al. (1998) HNPCC associated with germline mutation in the TGF-beta type II receptor gene [letter]. *Nat. Genet.* **19**, 17–18.
103. Markowitz, S. and Roberts, A. (1996) Tumor supressor activity of the TGF-β pathway in human cancers. *Cytokine Growth Factor Rev.* **7**, 93–102.
104. Fynan, T. M. and Reiss, M. (1993) Resistance to inhibition of cell growth by transforming growth factor- beta and its role in oncogenesis. *Crit. Rev. Oncog.* **4**, 493–540.
105. Massague, J. (1996) TGF-β signaling: receptors, transducers, and mad proteins. *Cell* **85**, 947–950.
106. Wrana, J. and Pawson, T. (1997) Signal transduction. Mad about SMADs [news; comment]. *Nature* **388**, 28–29.
107. Luo, K., Stroschein, S. L., Wang, W., et al. (1999) The Ski oncoprotein interacts with the Smad proteins to repress TGFbeta signaling. *Genes Dev.* **13**, 2196–2206.
108. Hua, X., Liu, X., Ansari, D. O., and Lodish, H. F. (1998) Synergistic cooperation of TFE3 and smad proteins in TGF-beta-induced transcription of the plasminogen activator inhibitor-1 gene. *Genes Dev.* **12**, 3084–3095.
109. Geng, Y. and Weinberg, R. A. (1993) Transforming growth factor beta effects on expression of G1 cyclins and cyclin-dependent protein kinases. *Proc. Natl. Acad. Sci. USA* **90**, 10315–10319.
110. Howe, P. H., Draetta, G., and Leof, E. B. (1991) Transforming growth factor beta 1 inhibition of p34cdc2 phosphorylation and histone H1 kinase activity is associated with G1/S-phase growth arrest. *Mol. Cell Biol.* **11**, 1185–1194.
111. Ewen, M. E., Sluss, H. K., Whitehouse, L. L., and Livingston, D. M. (1993) TGF beta inhibition of Cdk4 synthesis is linked to cell cycle arrest. *Cell* **74**, 1009–1020.
112. Alexandrow, M. and Moses, H. (1995) Transforming growth factor β and cell cycle regulation. *Cancer Res.* **55**, 1452–1457.
113. Hannon, G. and Beach, D. (1994) $p15^{INK4B}$ is a potential effector of TGF-β-induced cell cycle arrest. *Nature* **371**, 257–261.
114. Moses, H., Yang, E., and Pietonpol, J. (1990) TGF-β stimulation and inhibition of cell proliferation: new mechanistic insights. *Cell* **63,** 245–247.
115. Keeton, M. R., Curriden, S. A., van Zonneveld, A. J., and Loskutoff, D. J. (1991) Identification of regulatory sequences in the type 1 plasminogen activator inhibitor gene responsive to transforming growth factor beta. *J. Biol. Chem.* **266**, 23048–23052.
116. Zhao, Y. (1999) Transforming growth factor-beta (TGF-beta) type I and type II receptors are both required for TGF-beta-mediated extracellular matrix production in lung fibroblasts. *Mol. Cell Endocrinol.* **150**, 91–97.
117. Kim, S. J., Im, Y. H., Markowitz, S. D., and Bang, Y. J. (2000) Molecular mechanisms of inactivation of TGF-beta receptors during carcinogenesis. *Cytokine Growth Factor Rev.* **11,** 159–168.
118. Hoosein, N., McKnight, M., Levine, A., et al. (1989) Differential sensitivity of subclasses of human colon carcinoma cell lines to the growth inhibitory effects of transforming growth factor-β1. *Exp. Cell Res.* **181**, 442–453.

119. Grady, W., Myeroff, L., Swinler, S., et al. (1999) Mutational inactivation of transforming growth factor β receptor type II in microsatellite stable colon cancers. *Cancer Res.* **59**, 320–324.
120. Parsons, R., Myeroff, L., Liu, B., et al. (1995) Microsatellite instability and mutations of the transforming growth factor β type II receptor gene in colorectal cancer. *Cancer Res.* **55**, 5548–5550.
121. Myeroff, L., Parsons, R., Kim, S.-J., et al. (1995) A transforming growth factor β receptor type II gene mutation common in colon and gastric but rare in endometrial cancers with microsatellite instability. *Cancer Res.* **55**, 5545–5547.
122. Wang, J., Sun, L., Myeroff, L., et al. (1995) Demonstration that mutation of the type II transforming growth factor β receptor inactivates its tumor supressor activity in replication error-positive colon carcinoma cells. *J. Biol. Chem.* **270**, 22044–22049.
123. Lu, S. L., Kawabata, M., Imamura, T., Miyazono, K., and Yuasa, Y. (1999) Two divergent signaling pathways for TGF-beta separated by a mutation of its type II receptor gene. *Biochem. Biophys. Res. Commun.* **259**, 385–390.
124. Lynch, H. and Smyrk, T. (1996) Hereditary nonpolyposis colorectal cancer (Lynch syndrome): an updated review. *Cancer* **78**, 1149–1167.
125. Watanabe, T., Wu, T. T., Catalano, P. J., et al. (2001) Molecular predictors of survival after adjuvant chemotherapy for colon cancer. *N. Engl. J. Med.* **344**, 1196–1206.
126. Jarvinen, H. and Franssila, K. O. (1984) Familial juvenile polyposis coli; increased risk of colorectal cancer. *Gut.* **25**, 792–800.
127. Fearon, E. R., Cho, K. R., Nigro, J. M., et al. (1990) Identification of a chromosome 18q gene that is altered in colorectal cancers. *Science* **247**, 49–56.
128. Takagi, Y., Kohmura, H., Futamura, M., et al. (1996) Somatic alterations of the DPC4 gene in human colorectal cancers in vivo. *Gastroenterology* **111**, 1369–1372.
129. Eppert, K., Scherer, S., Ozcelik, H., et al. (1996) MADR2 maps to 18q21 and encodes a TGFβ-regulated MAD-related protein that is functionally mutated in colorectal cancer. *Cell* **86**, 543–552.
130. Kretzschmar, M., Liu, F., Hata, A., Doody, J., and Massague, J. (1997) The TGF-beta family mediator Smad1 is phosphorylated directly and activated functionally by the BMP receptor kinase. *Genes Dev.* **11**, 984–995.
131. Zhang, Y., Feng, X.-H., Wu, R.-Y., and Derynck, R. (1996) Receptor-associated Mad homologues synergize as effectors of the TGF-β response. *Nature* **383**, 168–172.
132. Souchelnytskyi, S., Tamaki, K., Engstrom, U., Wernstedt, C., ten Dijke, P., and Heldin, C. H. (1997) Phosphorylation of Ser465 and Ser467 in the C terminus of Smad2 mediates interaction with Smad4 and is required for transforming growth factor-beta signaling. *J. Biol. Chem.* **272**, 28107–28115.
133. Abdollah, S., Macias-Silva, M., Tsukazaki, T., Hayashi, H., Attisano, L., and Wrana, J. L. (1997) TbetaRI phosphorylation of Smad2 on Ser465 and Ser467 is required for Smad2-Smad4 complex formation and signaling. *J. Biol. Chem.* **272**, 27678–27685.
134. Chen, Y. G., Hata, A., Lo, R. S., Wotton, D., Shi, Y., Pavletich, N., and Massague, J. (1998) Determinants of specificity in TGF-beta signal transduction. *Genes Dev.* **12**, 2144– 2152.
135. Tsukazaki, T., Chiang, T. A., Davison, A. F., Attisano, L., and Wrana, J. L. (1998) SARA, a FYVE domain protein that recruits Smad2 to the TGFbeta receptor. *Cell* **95**, 779–791.
136. Hata, A., Lo, R. S., Wotton, D., Lagna, G., and Massague, J. (1997) Mutations increasing autoinhibition inactivate tumour suppressors Smad2 and Smad4 [see comments]. *Nature* **388**, 82–87.
137. Liu, F., Pouponnot, C., and Massague, J. (1997) Dual role of the Smad4/DPC4 tumor suppressor in TGFbeta-inducible transcriptional complexes. *Genes Dev.* **11**, 3157–3167.

138. de Caestecker, M., Piek, E., and Roberts, A. The role of TGF-β signaling in cancer. *J. Natl. Cancer Instit.*, in press.
139. de Caestecker, M. P., Hemmati, P., Larisch-Bloch, S., Ajmera, R., Roberts, A. B., and Lechleider, R. J. (1997) Characterization of functional domains within Smad4/DPC4. *J. Biol. Chem.* **272**, 13690–13696.
140. Hocevar, B. A., Brown, T. L., and Howe, P. H. (1999) TGF-beta induces fibronectin synthesis through a c-Jun N-terminal kinase-dependent, Smad4-independent pathway. *EMBO J.* **18**, 1345–1356.
141. Dai, J. L., Schutte, M., Bansal, R. K., Wilentz, R. E., Sugar, A. Y., and Kern, S. E. (1999) Transforming growth factor-beta responsiveness in DPC4/SMAD4-null cancer cells. *Mol. Carcinog.* **26**, 37–43.
142. Lagna, G., Hata, A., Hemmati-Brivanlou, A., and Massague, J. (1996) Partnership between DPC4 and SMAD proteins in TGF-beta signalling pathways. *Nature* **383**, 832–836.
143. Kim, S. J., Angel, P., Lafyatis, R., et al. (1990) Autoinduction of transforming growth factor beta 1 is mediated by the AP-1 complex. *Mol. Cell Biol.* **10**, 1492–1497.
144. Ashcroft, G. S., Yang, X., Glick, A. B., et al. (1999) Mice lacking Smad3 show accelerated wound healing and an impaired local inflammatory response [see comments]. *Nat. Cell Biol.* **1**, 260–266.
145. Nagarajan, R. P., Zhang, J., Li, W., and Chen, Y. (1999) Regulation of Smad7 promoter by direct association with Smad3 and Smad4. *J. Biol. Chem.* **274**, 33412–33418.
146. von Gersdorff, G., Susztak, K., Rezvani, F., Bitzer, M., Liang, D., and Bottinger, E. P. (2000) Smad3 and Smad4 mediate transcriptional activation of the human Smad7 promoter by transforming growth factor beta. *J. Biol. Chem.* **275**, 11320–11326.
147. Wong, C., Rougier-Chapman, E. M., Frederick, J. P., et al. (1999) Smad3-Smad4 and AP-1 complexes synergize in transcriptional activation of the c-Jun promoter by transforming growth factor beta. *Mol. Cell Biol.* **19**, 1821–1830.
148. Jonk, L. J., Itoh, S., Heldin, C. H., ten Dijke, P., and Kruijer, W. (1998) Identification and functional characterization of a Smad binding element (SBE) in the JunB promoter that acts as a transforming growth factor-beta, activin, and bone morphogenetic protein-inducible enhancer. *J. Biol. Chem.* **273**, 21145–21152.
149. Grau, A. M., Zhang, L., Wang, W., et al. (1997) Induction of p21waf1 expression and growth inhibition by transforming growth factor beta involve the tumor suppressor gene DPC4 in human pancreatic adenocarcinoma cells. *Cancer Res.* **57**, 3929–3934.
150. Datto, M. B., Hu, P. P., Kowalik, T. F., Yingling, J., and Wang, X. F. (1997) The viral oncoprotein E1A blocks transforming growth factor beta-mediated induction of p21/WAF1/Cip1 and p15/INK4B. *Mol. Cell Biol.* **17**, 2030–2037.
151. Li, J. M., Nichols, M. A., Chandrasekharan, S., Xiong, Y., and Wang, X. F. (1995) Transforming growth factor beta activates the promoter of cyclin-dependent kinase inhibitor p15INK4B through an Sp1 consensus site. *J. Biol. Chem.* **270**, 26750–26753.
152. Hahn, S., Schutte, M., Shamsul Hoque, A., et al. (1996) *DPC4*, a candidate tumor supressor gene at human chromosome 18q21.1. *Science* **271**, 350–353.
153. Riggins, G., Thiagalingam, S., Rozenblum, E., et al. (1996) Mad-related genes in the human. *Nat. Genet.* **13**, 347–349.
154. Takaku, K., Oshima, M., Miyoshi, H., Matsui, M., Seldin, M., and Taketo, M. (1998) Intestinal tumorigenesis in compound mutant mice of both *Dpc4* (*Smad4*) and *Apc* genes. *Cell* **92**, 645–656.
155. Takaku, K., Miyoshi, H., Matsunaga, A., Oshima, M., Sasaki, N., and Taketo, M. M. (1999) Gastric and duodenal polyps in Smad4 (Dpc4) knockout mice. *Cancer Res.* **59**, 6113–6117.
156. Xu, X., Brodie, S. G., Yang, X., et al. (2000) Haploid loss of the tumor suppressor Smad4/Dpc4 initiates gastric polyposis and cancer in mice. *Oncogene* **19**, 1868–1874.

157. Friedl, W., Kruse, R., Uhlhaas, S., et al. (1999) Frequent 4-bp deletion in exon 9 of the SMAD4/MADH4 gene in familial juvenile polyposis patients. *Genes Chromosomes Cancer* **25**, 403–406.
158. Roth, S., Sistonen, P., Salovaara, R., et al. (1999) A. SMAD genes in juvenile polyposis. *Genes Chromosomes Cancer* **26**, 54–61.
159. Woodford-Richens, K., Williamson, J., Bevan, S., et al. (2000) Allelic loss at SMAD4 in polyps from juvenile polyposis patients and use of fluorescence in situ hybridization to demonstrate clonal origin of the epithelium. *Cancer Res.* **60**, 2477–2482.
160. Howe, J. R., Bair, J. L., Sayed, M. G., et al. (2001) Germline mutations of the gene encoding bone morphogenetic protein receptor 1A in juvenile polyposis. *Nat. Genet.* **28**, 184–187.
161. Kawabata, M., Imamura, T., and Miyazono, K. (1998) Signal transduction by bone morphogenetic proteins. *Cytokine Growth Factor Rev.* **9**, 49–61.
162. Mishina, Y., Suzuki, A., Ueno, N., and Behringer, R. R. (1995) Bmpr encodes a type I bone morphogenetic protein receptor that is essential for gastrulation during mouse embryogenesis. *Genes Dev.* **9**, 3027–3037.
163. Rowan, A., Bataille, V., MacKie, R., et al. (1999) Somatic mutations in the Peutz-Jeghers (LKB1/STKII) gene in sporadic malignant melanomas. *J. Invest. Dermatol.* **112**, 509–511.
164. Su, G. H., Hruban, R. H., Bansal, R. K., et al. (1999) Germline and somatic mutations of the STK11/LKB1 Peutz-Jeghers gene in pancreatic and biliary cancers. *Am. J. Pathol.* **154**, 1835–1840.
165. Wang, Z. J., Churchman, M., Campbell, I. G., et al. (1999) Allele loss and mutation screen at the Peutz-Jeghers (LKB1) locus (19p13.3) in sporadic ovarian tumours. *Br. J. Cancer* **80**, 70–72.
166. Itzkowitz, S. and Kim, Y. (1998) Colonic polyps and polyposis syndromes, in: M. Feldman, B. Scharschmidt, and M. Sleisenger, eds., *Sleisenger and Fordtran's Gastrointestinal and Liver Disease*, 6th ed., Saunders, Philadelphia, vol. 2, p. 1891.
167. Jenne, D. E., Reimann, H., Nezu, J., et al. (1998) Peutz-Jeghers syndrome is caused by mutations in a novel serine threonine kinase. *Nat. Genet.* **18**, 38–43.
168. Hemminki, A., Markie, D., Tomlinson, I., et al. (1998) A serine/threonine kinase gene defective in Peutz-Jeghers syndrome. *Nature* **391**, 184–187.
169. Ylikorkala, A., Avizienyte, E., Tomlinson, I. P., et al. (1999) Mutations and impaired function of LKB1 in familial and non-familial Peutz-Jeghers syndrome and a sporadic testicular cancer. *Hum. Mol. Genet.* **8**, 45–51.
170. Yoon, K. A., Ku, J. L., Choi, H. S., et al. (2000) Germline mutations of the STK11 gene in Korean Peutz-Jeghers syndrome patients. *Br. J. Cancer* **82**, 1403–1406.
171. Westerman, A. M., Entius, M. M., Boor, P. P., et al. (1999) Novel mutations in the LKB1/STK11 gene in Dutch Peutz-Jeghers families. *Hum. Mutat.* **13**, 476–481.
172. Smith, D. P., Spicer, J., Smith, A., Swift, S., and Ashworth, A. (1999) The mouse Peutz-Jeghers syndrome gene Lkb1 encodes a nuclear protein kinase. *Hum. Mol. Genet.* **8**, 1479–1485.
173. Park, W. S., Moon, Y. W., Yang, Y. M., et al. (1998) Mutations of the STK11 gene in sporadic gastric carcinoma. *Int. J. Oncol.* **13**, 601–604.

5

Patched, Hedgehog, and Skin Cancer

Anthony G. Quinn and Ervin Epstein, Jr.

1. Introduction

The Hedgehog signaling pathway plays a key role in directing growth and patterning during embryonic development and is required in vertebrates for the normal development of many structures, including the axial skeleton, neural tube, lungs, hair, and teeth. Recent evidence has implicated a pathologic role for the Hh signaling pathway in the development of the common skin cancer, basal cell carcinoma (BCC). This chapter provides an overview of the biology and molecular genetics of BCCs and reviews the mechanisms by which activation of Hh signaling may lead to the development of this common cancer.

2. Basal Cell Carcinoma—Clinical and Pathologic Aspects

Basal cell carcinoma (BCC) is the most common human malignancy in Caucasian populations worldwide, with an estimated 3–4 million occurrences each year in the United States alone. Although the case fatality rate of BCCs is low, the high incidence of these tumors and the frequent occurrence of multiple primary tumors in affected individuals means that BCC is an important cause of morbidity in the population. Over the last 40 years there has been a dramatic increase in the incidence of this cancer, and it is estimated that nearly 30% of the population of European ancestry in areas of high ambient sun exposure will develop a BCC of the skin ***(1)***.

BCCs appear most frequently as translucent skin-colored to whitish papules and typically are recognized at 1–4 mm in diameter. Untreated tumors can grow to a large size and involve large areas of the nose, face, or scalp. Although BCCs only rarely metastasize, they may cause considerable disfiguring and local tissue destruction, and they have the potential to invade and damage critical structures including the orbit of the eye, the ears, and even the brain. The three accepted environmental insults predisposing one to BCCs are ultraviolet radiation, ionizing radiation, and arsenic.

Histologically, BCCs are comprised of nests of cells with large nuclei and relatively scant cytoplasm that are palisaded along the stroma–tumor interface; i.e., they resemble cells of the normal basal layer of the epidermis, hence their name. By contrast, the cells

From: *Methods in Molecular Biology, Vol. 222: Tumor Suppressor Genes: Pathways and Isolation Strategies*
Edited by: Wafik S. El-Deiry © Humana Press Inc., Totowa, NJ

of squamous cell carcinomas (SCCs), the other common cutaneous nonmelanoma skin cancer, resemble cells of the upper, suprabasal epidermis. Unlike BCCs, SCCs show histologic evidence of terminal differentiation analagous to the process seen in the outermost epidermal layer of normal human skin. SCCs are also distinct from BCCs in that the main serious risk to the patient is dissemination to regional lymphatics leading to metastasis rather than damage through local invasion.

A range of distinct histologic subtypes of BCCs are recognized, most of which show characteristic clinical features. Superfical BCCs, characterized by small islands of basaloid cells budding off the epidermis, are frequently found on the trunk and present as red scaly patches that are often multiple. Other variants include morpheic or sclerosing BCCs, which appear as scarlike lesions with small, firm papules that microscopically show relatively few basaloid cells but have an exuberant surrounding new stroma. Less common tumors with histologic resemblance to BCCs include trichoepitheliomas, trichoblastomas, and basaloid follicular hamartomas. These differ from BCCs in having relatively subtle histologic characteristics as well as in their generally less aggressive local growth.

3. The Genetics of Inherited Susceptibility to BCC Development

Interindividual differences in the susceptibility to BCC development have been recognized for many years. Epidemiologic studies have identified phenotypic features such as fair skin and freckling tendency that are associated with an increased susceptibility to BCCs, and there is evidence that a subgroup of patients with BCCs are predisposed to the development of multiple primary tumors. Genetic factors that determine inherited susceptibility to BCC development can be divided broadly into two main types. The first type is a small number of genes associated with rare, highly penetrant cancer predisposition syndromes and includes conditions such as the basal cell nevus syndrome (BCNS) (naevoid basal cell carcinoma syndrome, Gorlin syndrome, MIM #109400) and Bazex syndrome (MIM #301845). The second type are multiple, low-penetrant genetic loci that may contribute to BCC susceptibility in the general population. Evidence for the importance of these genes has come from quantitative trait loci mapping of other cancers in murine models. Although mapping of similar loci in humans is difficult, association studies provide some evidence that supports a role for high-frequency, low-penetrant traits such as DNA damage repair capacity and xenobiotic metabolism in BCC susceptibility ***(2,3)***.

3.1. Familial Cancer Syndromes Predisposing to BCC Development

BCCs, like many other common human cancers, occur in both sporadic and familial forms. As has been the case for other cancers such as those of the breast and colon, understanding of the molecular basis of the common, sporadic form has been critically dependent on the availability of rare families with highly penetrant autosomal dominant susceptibility syndromes. Several clinical syndromes predispose affected individuals to BCC development. The best-characterized and most common is the BCNS (also known as NBCCS or Gorlin syndrome). Other, rarer syndromes that point to the existence of other genes integral to BCC development include Bazex syndrome and Rombo syndrome. Bazex syndrome is a rare genodermatosis that predisposes affected individuals to the development of multiple BCCs at an early age ***(4)***. Additional clinical features that allow distinction from BCNS include follicular atrophoderma, hypotrichosis, and hypohidrosis. The absence of male-to-male transmission suggests X-linked inheritance, and linkage analysis has mapped the gene to Xq24-q27 ***(5)***. Rombo syndrome is a very rare autosomal dominant syndrome first described in 1981 ***(6)***, which is frequently complicated by the development of BCCs (OMIM #180730). Although there are some similarities with Bazex syndrome (follicular atrophoderma and milia), there are a number of distinctive features including cyanotic discoloration of the hands and lips in childhood and telangiectasia. The genetic locus for Rombo syndrome has not yet been mapped.

3.2. Basal Cell Nevus Syndrome

BCNS is a highly penetrant autosomal dominant familial cancer syndrome in which affected individuals develop a variable combination of a large number of phenotypic abnormalities ***(7)***. In particular, they are predisposed to the development of multiple (tens to hundreds or even thousands) BCCs starting at an early age—usually in the midteens but sometimes even before puberty. Like that of sporadic BCCs, the growth of BCCs in BCNS patients ranges from indolent to aggressive. BCNS patients also develop, with frequencies lower than their development of BCCs but higher than that in non-BCNS patients, other tumors—cysts of the jaw and skin, meningiomas, ovarian fibromas (which often are calcified and hence readily detected radiographically), rhabdomyosarcomas, cardiac fibromas, and, in particular, medulloblastomas. Approximately 3% of BCNS patients develop medulloblastomas, and about 3% of patients with medulloblastomas have BCNS ***(8)***. The peak incidence of medulloblastomas in BCNS patients, like that of BCCs, is younger than in sporadic cases. In addition to a propensity to develop tumors postnatally, these patients develop other diagnostically useful phenotypic abnormalities, including a highly characteristic facies (with large forehead), bifid or otherwise misshapen ribs, vertebral and other skeletal anomalies, pits of the skin of the palms and soles, dysgenesis of the corpus callosum, calcification of the falx cerebri (at an earlier age than is seen in non-BCNS individuals), and macrocephaly. Although BCNS differs from other autosomal dominant cancer syndromes in that many of the associated features appear to be developmental abnormalities, the presence in BCNS of multiple BCCs starting at an age earlier than that in sporadic cases is consistent with the "two-hit model" for inherited cancers first proposed by Knudson, and this is supported by observations from loss-of-heterozygosity studies of DNA from both sporadic and familial BCCs. The presence of BCCs in smaller numbers occurring at a later age in sporadic cases and in higher numbers at an earlier age in the autosomal dominant BCNS suggested that the susceptibility gene for BCCs might follow the Rb paradigm, with inactivating germline mutations in BCNS patients and somatic mutations in sporadic BCCs.

4. Molecular Genetics of BCNS and of Sporadic BCCs

4.1. Identification of the PATCHED Gene as the Culprit

As with other cancers, two complementary approaches were used to map BCC/BCNS tumor suppressor genes. The first approach was to study the pattern of chromosome loss in sporadic BCCs to identify regions that were lost frequently during BCC formation.

Such comparison of DNA from tumors and nontumorous tissue identified a large region on chromosome 9q that is lost in approximately 50% of BCCs, from sporadic or BCNS patients ***(9,10)***. This pattern of losses mostly confined to this one region is in keeping with earlier studies showing a predominantly diploid DNA content in BCCs, and contrasts with the extensive complex pattern of chromosome loss seen in most epithelial malignancies including squamous cell carcinomas of the skin ***(11)***. Once this region was identified through comparison of DNA from tumors and control tissue, it was possible to use the second approach—linkage analysis—in a straightforward manner to demonstrate that the gene whose mutations underlie the hereditary BCNS also mapped to the same region of chromosome 9q ***(9)***.

The mapping of the BCNS/BCC locus to chromosome 9q initiated a hunt at that site for the susceptibility gene, an exercise which ended in 1996 with the identification of mutations in the *PATCHED (PTCH1)* gene—heritable mutations in patients with BCNS and somatic mutations in sporadic BCCs ***(12,13)***. Underlying mutations have been identified at all regions of the *PTCH1* gene—there appear not to be mutational hot spots—and frequently the mutations identified are predicted to code for prematurely truncated proteins, which suggests that most of the mutations produce functionally null rather than dominant negative alleles. Furthermore, there is no obvious correlation between the site/type of the mutation and the severity of the clinical phenotype. For BCNS, this is not surprising as it has been recognized clinically for many years that the variability in phenotypic severity of affected patients is as great within families as between families. Although mutations in *PTCH1* have not been identified in all patients with BCNS, it seems likely that this is owing to technical problems rather than to locus heterogeneity, since no family in which the *PTCH1* site has been excluded by linkage analysis has been reported.

This identification of *PTCH1* mutations (in the germline of patients with BCNS and in tumor cells of sporadic BCCs) pointed to a key role for the Hedgehog (Hh) signaling pathway in the pathogenesis of BCCs. *PTCH1* plays an important regulatory role in this pathway, was first elucidated by elegant genetic studies of embryonic segmentation and imaginal disk specification in *Drosophila*, and is highly conserved from insects to vertebrates. Vertebrates differ from *Drosophila* in that there are multiple homologs of a number of components of the pathway (**Table 1**). Although this adds to the complexity, the basic principles of the signaling mechanisms appear to be maintained, and there is evidence that a number of the mammalian homologs can substitute for specific activities when expressed in *Drosophila* ***(14,15)***. The earlier characterization of Hh signaling in *Drosophila* meant that it was possible following the cloning of the BCNS gene to make some predictions about the downstream consequences of inactivation of PTCH1. Although most of these predictions have been confirmed experimentally, it has become increasingly clear that we are a long way away from fully understanding the regulatory mechanisms that control the activity of this pathway in humans and the function of this pathway in normal and diseased adult human skin.

5. Hedgehog Signal Transduction

Ptch1 protein is the receptor for all three forms of vertebrate hedgehog (Sonic, Indian, and Desert) and normally acts as a negative regulator of the pathway by specifically inhibiting constitutive signaling from another transmembrane protein, Smoothened

Table 1
Major Components of Hh Signalling

Components	*Drosophila*	Vertebrates (homologs)
Hedgehog	Hh	Sonic hedgehog (SHH) Indian hedgehog (IHH) Desert hedgehog (DHH)
Patched	Ptch	Patched 1 (Ptch1) Patched 2 (Ptch2)
Smoothened	Smo	Smo
Fused	Fu	Fu
Suppressor of fused	Su(Fu)	Su(Fu)
Costal-2		
Cubitus interruptus	CI	Gli1 Gli2 Gli3

(SMO). In mammalian cells Ptch1 localizes predominantly to intracellular vesicles and co-localizes with Caveolin-1, which binds and transports cholesterol from the Golgi to the plasma membrane ***(16)***. Ptch1 shows a high degree of homology in its transmembrane domain to the sterol-sensing domain of a number of proteins, including HMG CoA reductase and the putative proton-driven lipid translocator mutated in Niemann-Pick C1 disease ***(17)***.

A number of lines of evidence indicate that Hh activates signaling by binding to PTCH1 protein, but the precise mechanism by which PTCH inhibits Smo signaling remains in dispute. Smo is predicted to be a seven-pass membrane protein and has recently been shown in a frog melanophore assay to act as a G-protein-coupled receptor ***(18)***. Although initial studies supported a model in which Smo and Ptc bind to form a receptor complex to which binding of hedgehog protein releases SMO from PTCH-mediated inhibition ***(19,20)***, more recent studies support an alternative model in which the regulation of SMO by PTCH may be indirect, with a role for PTCH in destabilizing SMO protein and removing it from the cell surface ***(21)***. Binding of Hh causes internalization of both PTCH and Hedgehog proteins, and the two proteins have been reported to co-localize in endocytic vesicles. This is accompanied by temporally related posttranscriptional changes in SMO protein—increased phosphorylation, increased stabilization, and increased localization to the cell surface ***(22,23)***. The exact mechanism by which Hh induces these changes in SMO protein remains unclear. Although the increased expression of SMO at the cell surface may reflect increased protein levels, an intriguing possible mechanism proposed on the basis of immunohistochemical analysis of wild-type and transgenic *Drosophila* embryos is that SMO protein is present, but that in the presence of PTCH it is in a conformation that renders it nonfunctional and interferes with the binding of some antibodies ***(24)***. Further support for the hypothesis that endogenous PTCH may be acting on SMO by a catalytic mechanism and not simply by direct interaction comes from studies expressing SMO in both *Drosophila* and mammalian cells, which have shown that SMO activity is still Hh-dependent even when expressed at high level. Mutations that render SMO resistant to PTCH1 inhibition have been identified ***(25,26)***, and it is interesting that one of the mutations identified is predicted to alter the conformation of the carboxy-terminal tail, which is the region that shows differential binding to an anti-Smo antibody in *Drosophila* in the presence of Hh signaling ***(24)***.

The changes downstream of SMO following activation of Hh signaling are also subject to a number of complex regulatory mechanisms. Several evolutionarily conserved genes have been identified that link SMO to the transcriptional changes associated with Hh signaling. These include the serine/threonine protein kinase Fused (Fu), which activates signaling, and Suppressor of fused [Su(fu)], which contains a PEST sequence in *Drosophila* but not in mammals ***(27)***, the kinesin-like protein Costal-2 (Cos2; vertebrate homolog not identified), and protein kinase A (PKA), all of which inhibit signaling. In the cytosol, Hh signaling regulates a multiprotein complex that includes Cos-2, Fu, Su(Fu), and Cubitus interruptus (ci), the major transcriptional effector of the Hh signaling pathway in *Drosophila*. In the absence of Hh signaling this complex is anchored to the microtubules, and ci is proteolyzed to a smaller amino-terminal fragment that localizes to the nucleus, where it binds to the same consensus sequence as does the activated, full-length protein but acts as a transcriptional repressor ***(28)***. Activation of SMO releases this complex from the microtubules and inhibits conversion of full-length Ci to its repressor form ***(29)***. As is the case for Hh, PTC, and SMO, there are broad similarities in the mechanisms between *Drosophila* and vertebrate cells, but there are a number of important known differences between phyla, particularly with respect to the main final transcriptional effectors.

In vertebrates, ci function is subserved by the Gli family of transcription factors, Gli1, Gli2, and Gli3. These proteins share a DNA-binding motif that recognizes the same consensus DNA sequence recognized by Ci. *Gli1* was originally identified as a gene highly amplified in a glioblastoma multiforme, and it subsequently has been shown to be located on chromosome 12q in a region of DNA that is amplified in some childhood sarcomas, gliomas, and in B-cell lymphomas, although the evidence in toto suggests that the relevant amplifications in these tumors may be of other, adjacent genes ***(30–37)***. *Gli2* and *Gli3* were identified based on their homology within a tandem C2–H2 zinc finger domain, and functional analyses have shown that Gli1 acts as a transcriptional activator, Gli3 acts as a repressor whose repressing activity is enhanced by proteolytic cleavage, and Gli2 can act as either an activator or repressor. The distinct activities of the different *Gli* genes and the spatial and temporal differences observed during embryonic development suggest that they are not functionally equivalent ***(38)***. Gli1 and Ptch1 mRNA expression are increased adjacent to Shh-expressing tissues. Although Gli1 can induce expression of Shh targets, the normal development of mice homozygous for a deletion of the Gli1 DNA-binding domain suggests that in embryos there is redundancy and that the other *Gli* genes may be able to substitute for deficiency of Gli1 ***(39)***.

In addition to the transcriptional regulation of Gli1 and Gli2 and the complex regulation of the activity of Gli2 and Gli3 by alternative splicing or proteolytic processing in response to Hh signaling, there also is evidence that the subcellular localization of the Gli proteins may be subject to regulation. Immunostaining of BCCs for Gli1 has shown that the protein is predominately cytoplasmic, and this has been confirmed in transfected cell lines. Although the localization to the cytoplasm was initially thought to be due to cytoplasmic retention, the predominant nuclear localization of Gli1 in cells treated with a nuclear protein export inhibitor (leptomycin B) suggests that a significant amount of Gli1 may be shuttling between the nucleus and cytoplasm ***(40,41)***.

6. Hedgehog Signaling in Normal Skin and BCCs

6.1. BCCs, Hair Follicles, and Stem Cells

While genetic studies have established that *Ptch1* is a tumor suppressor gene, the mechanisms by which inactivation of *Ptch1* lead to BCC development are still not clear. Much of the information to date on human BCCs is restricted to studies of gene and protein expression, as it has not been possible to isolate cells in culture from BCCs for functional analysis. Loss of Ptch1 activity leads to activation of Smo, and this is supported by the observation of increased Gli1 and Ptch1 mRNA levels in BCCs. The consistent upregulation of these hedgehog target genes, even in those BCCs in which PTCH1 mutation cannot be identified, indicates that dysregulated hedgehog signaling is pivotal to the development of all BCCs.

Functional studies in mice have implicated hedgehog signaling in normal skin in hair follicle development and in the normal cyclical growth of hair follicles ***(42)***. In fetal skin Shh is expressed in the epidermis overlying the mesenchymal condensation of early hair follicles, and PTCH1, Smo, Gli1, and Gli2 are expressed in both the epidermal and dermal components. Studies in Shh-knockout mice indicate that Shh signaling is not required for hair follicle induction but is essential for the normal growth, downward elongation, and differentiation during the later aspects of hair follicle morphogenesis. Conversely, expression of Shh in basal keratinocytes is associated with expansive downgrowths of basaloid keratinocytes, which do not express suprabasal differentiation markers and which show an immunohistologic phenotype similar to that of BCCs in humans. These observations suggest that inactivation of PTCH1 may favor BCC development by stimulating proliferation and inhibiting terminal differentiation of basal keratinocytes, which in turn allows expansion and downgrowth of populations of relatively undifferentiated keratinocytes. A similar model has been proposed to explain medulloblastoma development. This tumor, which also occurs in patients with BCNS, is characterized by expansion of neural progenitor cells, which are thought to originate from the external germinal layer (EGL) of the cerebellum. EGL cell differentiation in vitro is inhibited by Shh, which allows expansion of this population by maintaining their high intrinsic proliferation rate ***(43)***.

Inhibition of terminal differentiation is also a requirement for stem cell maintenance in vivo. There is evidence in *Drosophila* that supports a potential role for Hh in stem cell maintenance. In the *Drosophila* ovary, Hh has been shown to act as a somatic stem cell factor, and loss of Hh signaling is associated with loss of stem cells in the somatic epithelial compartment ***(43)***. Although there is no direct evidence implicating Shh in stem cell maintenance in mammals, it is interesting that Gli1 protein is expressed in some human hair follicles in the vicinity of the bulge region of the hair follicle, which site is thought to be the location of putative epidermal stem cells in skin. This observation raises the possibility that one of the roles for Shh signaling in human skin, in addition to its role in hair follicle cycling, is in the maintenance of a subpopulation of keratinocytes in a state that permits unlimited proliferative capacity ***(45)***.

6.2. Cell–Cell Interactions and Hedgehog Signaling

BCC development and hair follicle formation are both critically dependent on orchestrated interactions between the epidermis and the mesenchyme. Although the signaling mechanisms responsible for these interactions are still not fully characterized, information from transgenic and knockout mice generated to investigate Shh signaling in the skin has provided new insights into these interactions and their contribution to BCC development. The observation that the initial downgrowth of the epidermis in hair follicle formation occurs in Shh−/−mice points to the existence of additional signaling mechanisms that initiate hair follicle formation. Given the similarities between hair follicle formation and BCC development, it is likely that similar signaling mechanisms may be necessary for initiating BCC development. In Shh and mut-Smo transgenic mice in which these genes are expressed using promoters that drive expression in basal keratinocytes, a number of the changes in gene and protein expression are restricted to the epidermal downgrowths and are not found uniformly in the Shh- and/or mut-Smo-expressing keratinocytes ***(25,46)***. The enhanced accumulation of PTCH1 RNA within the downgrowths in mut-Smo transgenic mice suggests that additional factors may contribute to the high-level expression of PTCH1 in areas of activated Shh signaling in these models. While the identity of these factors is at present unknown, their existence is supported by studies in talpid3 mice, which have shown that the ta3 gene product is required for normal expression of some but not all hedgehog target genes during embryonic development ***(47)***. Elucidation of these additional signaling mechanisms may shed light on some of the genetic factors that contribute to the phenotypic variability within families with BCNS.

6.3. BCCs, Trichoepitheliomas, and Other Skin Tumors

In patients with BCNS who have germline mutations in the *PTCH1* gene, BCCs are by far the most common type of skin tumor. Increased Ptch1 mRNA expression and Ptch1 mutations also have been identified in trichoepitheliomas, as are hair follicle-derived tumors that show some histologic similarities to BCCs ***(48)***. Although trichoepitheliomas are unusual in BCNS patients, BCCs can occur in patients with the multiple familial trichoepithelioma syndrome (OMIM 601606). In mice the spectrum and clinical behavior of tumors induced by genetic manipulation of Shh signaling in keratinocytes are different from that seen in BCNS patients. The epidermal downgrowths seen in mice expressing Shh ***(46)*** or mutant Smo ***(25)*** in basal keratinocytes differ from human BCCs in that they do not progressively enlarge but instead show evidence of hair follicle differentiation. Trichoepitheliomas and BCCs have been described in mice expressing Gli1 in basal keratinocytes ***(49)***. In addition, these mice develop another skin appendage tumor—cylindromas. The coexistence of BCCs, trichoepitheliomas, and cylindromas in mice is interesting, as although both BCCs and cylindromas can occur in association with trichoepitheliomas, there is at present no direct link between BCCs and cylindromas, and no obvious clues from the structure of the recently identified gene for familial cylindromatosis (OMIM 605018) ***(50)*** as to how these tumors may be linked. Heterozygous Ptch1 knockout mice develop spontaneous BCC, trichoblastoma, and other BCC-like tumors, which can be induced at a higher frequency by exposure to UVR and ionizing radiation ***(51)***. Ptch1+/− mice, in contrast to BCNS patients, also show an increased susceptiblity to SCC induction by UVR, and this murine bias in favor of SCC induction may be one of the reasons why it has been difficult to use wild-type mice to study BCC formation in response to UVR and chemical carcinogens. The basis for the increased incidence of SCCs in Ptch1+/− mice is not clear, and it is interesting that, in contrast to BCCs, the SCCs do not show evidence of full activation of the Hh signaling pathway, as evidenced by their lack of increased Ptch1 expression. Perhaps of relevance to this finding is the identification of PTCH1 mutations in a subset of SCCs that arise in patients with multiple BCCs *(52)*. These findings suggest that factors in addition to dysregulated Shh signaling may be important in determining tumor histogenesis, and it will be vital, in future studies, to determine the basis for these species-specific differences in the types of tumors induced by activation of Hh signaling.

References

1. Miller, D. L. and Weinstock, M. A. (1994) Non-melanoma skin cancer in the United States: incidence. *J. Am. Acad. Dermatol.* **30**, 774–778.
2. Bastiaens, M. T., Huurne, J. A., Kielich, C., et al. (2001) Melanocortin-1 receptor gene variants determine the risk of nonmelanoma skin cancer independently of fair skin and red hair. *Am. J. Hum. Genet.* **68**, 884–894.
3. Ramachandran, S., Fryer, A. A., Smith, A. G., et al. (2001) Basal cell carcinomas: association of allelic variants with a high-risk subgroup of patients with the multiple presentation phenotype. *Pharmacogenetics* **11**, 247–254.
4. Plosila, M., Kiistala, R., and Niemi, K. M. (1981) The Bazex syndrome: follicular altrophoderma with multiple basal cell carcinomas, hypotrichosis and hypohidrosis. *Clin. Exp. Dermatol.* **6**, 31–41.
5. Vabres, P., Lacombe, D., Rabinowitz, L. G., et al. (1995) The gene for Bazex-Dupre-Christol syndrome maps to chromosome Xq. *J. Invest. Dermatol.* **105**, 87–91.
6. Michaelsson, G., Olsson, E., and Westermark, P. (1981) The Rombo syndrome: a familiar disorder with verniculate atrophoderma, milia, hypotrichosis, trichoepitheliomas, basal cell carcinomas and peripheral vasodilation with cyanosis. *Acta. Dermatol. Venereol.* **61**, 497–503.
7. Gorlin, R. J. (1987) Nevoid basal-cell carcinoma syndrome. *Medicine* **66**, 98–113.
8. Evans, D. G. R., Farndon, P. A., Burnell, L. D., Gattamaneni, H. R., and Birch, J. M. (1991) The incidence of Gorlin syndrome in 173 consecutive cases of medulloblastoma. *Br. J. Cancer* **64**, 959–961.
9. Gailani, M. R., Bale, S. J., Leffell, D. J., et al. (1992) Developmental defects in Gorlin syndrome related to a putative tumor suppressor gene on chromosome 9. *Cell* **69**, 111–117.
10. Quinn, A. G., Campbell, C., Healy, E., and Rees, J. L. (1994) Chromosome 9 allele loss occurs in both basal and squamous cell carcinomas of the skin. *J. Invest. Dermatol.* **102**, 300–303.
11. Quinn, A. G., Sikkink, S., and Rees, J. L. (1994) Basal cell carcinomas and squamous cell carcinomas of human skin show distinct patterns of chromosome loss. *Cancer Res.* **54**, 4756–4759.
12. Hahn, H., Wicking, C., Zaphiropoulos, P. G., et al. (1996) Mutations of the human homologue of *Drosophila* patched in the nevoid basal cell carcinoma syndrome. *Cell* **85**, 841–851.
13. Johnson, R. L., Rothman, A. L., Xie, J., et al. (1996) Human homolog of patched, a candidate gene for the basal cell nevus syndrome. *Science* **272**, 1668–1671.
14. Mering, C. V. and Basler, K. (1999) Distinct and regulated activities of human Gli proteins in *Drosophila*. *Curr. Biol.* **9**, 1319–1322.
15. Aza-Blanc, P., Lin, H. Y., and Kornberg, T. B. (2000) Expression of the vertebrate Gli proteins in *Drosophila* reveals a distribution of activator and repressor activities. *Development* **127**, 4293–4301.

16. Karpen, H. E., Bukowski, J. T., Hughes, T., Gratton, J. P., Sessa, W. C., and Gailani, M. R. (2001) The sonic hedgehog receptor patched associates with caveolin-1 in cholesterol-rich microdomains of the plasma membraine. *J. Biol. Chem.* **276**, 19503–19511.
17. Carstea, E. D., Morris, J. A., Coleman, K. G., et al. (1997) Niemann-Pick C1 disease gene: homology to mediators of cholesterol homeostasis. *Science* **277**, 228–231.
18. DeCamp, D. L., Thompson, T. M., deSauvage, F. J., and Lerner, M. R. (2000) Smoothened activates Galphai-mediated signaling in frog melanophores. *J. Biol. Chem.* **275**, 26322–26327.
19. Stone, D., Hynes, M., Armanini, M., et al. (1996) The tumour-suppressor gene patched encodes a candidate receptor for Sonic hedgehog. *Nature* **384**, 129–134.
20. Marigo, V., Davey, R. A., Zuo, Y., Cunningham, J. M., and Tabin, C. (1996) Biochemical evidence that Patched is the Hedgehog receptor. *Nature* **14**, 176–179.
21. Taipale, J., Cooper, M. K., Maiti, T., and Beachy, P. A. (2002) Patched acts catalytically to suppress the activity of Smoothened. *Nature* **418**, 892–897.
22. Alcedo, J., Y. Zou, and Noll, M. (2000) Posttranscriptional regulation of smoothened is part of a self-correcting mechanism in the hedgehog signaling system. *Mol. Cell* **6**, 457–465.
23. Denef, N., Neubuser, D., Perez, L., and Cohen, S. M. (2000) Hedgehog induces opposite changes in turnover and subcellular localization of patched and smoothened. *Cell* **102**, 521–531.
24. Ingham, P. W., Nystedt, S., Nakano, Y., et al. (2000) Patched represses the Hedgehog signaling pathway by promoting modification of the Smoothened protein. *Curr. Biol.* **10**, 1315–1318.
25. Xie, J., Murone, M., Luoh, S., et al. (1998) Activating Smoothened mutations in sporadic basal-cell carcinoma. *Nature* **391**, 90–92.
26. Reifenberger, J., Wolter, M., Weber, R. G., et al. (1998) Missense mutations in SMOH in sporadic basal cell carcinomas of the skin and primitive neuroectodermal tumors of the central nervous system. *Cancer Res.* **58**, 1798–1803.
27. Delattre, M., Briand, S., Paces-Fessy, M., and Blanchet-Tournier, M. F. (1999) The Suppressor of fused gene, involved in Hedgehog signal transduction in Drosophila, is conserved in mammals. *Dev. Genes Evol.* **209**, 294–300.
28. Aza-Blanc, P., Ramirez-Weber, F. A., Laget, M. P., Schwartz, C., and Kornberg, T. B. (1997) Proteolysis that is inhibited by hedgehog targets Cubitus interruptus protein to the nucleus and converts it to a repressor. *Cell* **89**, 1043–1053.
29. Methot, N. and Basler, K. (1999) Hedgehog controls limb development by regulating the activities of distinct transcriptional activator and repressor forms of Cubitus interruptus. *Cell* **96**, 819–831.
30. Kinzler, K. W., Bigner, S. H., Bigner, D. D., et al. (1987) Identification of an amplified, highly expressed gene in a human glioma. *Science* **236**, 70–73.
31. Roberts, W. M., Douglas, E. C., Peiper, S. C., Houghton, P. J., and Look, A. T. (1989) Amplification of the gli gene in childhood sarcomas. *Cancer Res.* **49**, 5407–5413.
32. Smith, S. H., Weiss, S. W., Jankowski, S. A., Coccia, M. A., and Meltzer, P. S. (1992) SAS amplification in soft tissue sarcomas. *Cancer Res.* **52**, 3746–3749.
33. Reifenberger, G., Reifenberger, J., Ichimura, K., Meltzer, P. S., and Collins, V. P. (1994) Amplification of multiple genes from chromosomal region 12q13-14 in human malignant gliomas: preliminary mapping of the amplicons shows preferential involvement of CDK4, SAS, and MDM2. *Cancer Res.* **54**, 4299–4303.
34. Collins, V. P. (1995) Gene amplification in human gliomas. *Glia* **15**, 289–296.
35. Reifenberger, G., Ichimura, K., Reifenberger, J., Elkabloun, A. G., Meltzer, P. S., and Collins, V. P. (1996) Refined mapping of 12q13-q15 amplicons in human malignant gliomas suggests CDK4/SAS and MDM2 as independent amplification targets. *Cancer Res.* **56**, 5141–5145.

36. Werner, C. A., Dohner, H., Joos, S., et al. (1997) High-level DNA amplifications are common genetic aberrations in B-cell neoplasms. *Am. J. Pathol.* **151**, 335–42.
37. Rao, P. H., Houldsworth, J., Dyomina, K., et al. (1998) Chromosomal and gene amplification in diffuse large B-cell lymphoma. *Blood* **92**, 234–240.
38. Matise, M. P. and Joyner, A. L. (1999) Gli genes in development and cancer. *Oncogene* **18**, 7852–7859.
39. Matise, M. P., Epstein, D. J., Park, H. L., Platt, K. A., and Joyner, A. L. (1998) Gli2 is required for induction of floor plate and adjacent cells, but not most ventral neurons in the mouse central nervous system. *Development* **125**, 2759–2770.
40. Chen, C.-H., Kessler, D. V., Park, W., Wang, B., Ma, Y., and Beachy, P. A. (1999) Nuclear trafficking of cubitus interruptus in the transcriptional regulation of hedgehog target gene expression. *Cell* **98**, 305–316.
41. Kogerman, P., Grimm, T., Kogerman, L., et al. (1999) Mammalian suppressor-of-fused modulates nuclear-cytoplasmic shuttling of Gli-1. *Natl. Cell. Biol.* **1**, 312–319.
42. Chiang, C., Swan, R. Z., Grachtchouk, M., et al. (1999) Essential role for sonic hedgehog during hair follicle morphogenesis. *Dev. Bio.* **205**, 1–9.
43. Wechsler-Reya, R. J. and Scott, M. P. (1999) Control of neuronal precursor proliferation in the cerebellum by Sonic Hedgehog. *Neuron* **22**, 103–114.
44. Zhang, Y. and Kalderon, D. (2001) Hedgehog acts as a somatic stem cell factor in the *Drosophila* ovary. *Nature* **410**, 599–604.
45. Bonifas, J. M., Pennypacker, S., Chuang, P. T., et al. (2001) Activation of expression of hedgehog target genes in basal cell carcinomas. *J. Invest. Dermatol.* **116**, 739–742.
46. Oro, A. E., Higgins, K. M., Hu, Z., Bonifas, J. M., Epstein, E. H., Jr., and Scott, M. (1997) Basal cell carcinomas in mice overexpressing sonic hedgehog. *Science* **276**, 817–821.
47. Lewis, K. E., Drossopoulous, G., Paton, I. R., et al. (1999) Expression of ptc and gli genes in talpid3 suggests bifurcation in Shh pathway. *Development* **126**, 2397–2407.
48. Vorechovsky, I., Unden, A. B., Sandstedt, B., Toftgard, R., and Stahle-Backdahl, M. (1997) Trichoepitheliomas contain somatic mutations in the overexpressed PTCH gene: support for a gatekeeper mechanism in skin tumorigenesis. *Cancer* **57**, 4677–4681.
49. Nilsson, M., Unden, A. B., Krause, D., et al. (2000) Induction of basal cell carcinomas and trichoepitheliomas in mice overexpressing GLI-1. *Proc. Natl. Acad. Sci. USA* **97**, 3438–3443.
50. Bignell, G.R., Warren, W., Seal, S., et al. (2000) Identification of thc familial cylindromatosis tumor-suppressor gene. *Nat. Genet.* **25**, 160–165.
51. Aszterbaum, M., Epstein, J., Oro, A., et al. (1999) Ultraviolet and ionizing radiation enhance the growth of BCCs and trichoblastomas in patched heterozygous knockout mice. *Nature Med.* **5**, 1285–1291.
52. Ping, X. L., Ratner, D., Zhang, H., et al. (2001) PTCH mutations in squamous cell carcinoma of the skin. *J. Invest. Dermatol.* **116**, 614–616.

6

Tumor Suppressor Genes in Lung Cancer

Arvind K. Virmani and Adi F. Gazdar

1. Introduction

1.1. Lung Cancer Incidence and Subtypes

Lung cancer is the number-one cancer killer of both men and women in the United States, killing more than 150,000 people every year. Lung cancer consists of two broad groups: small-cell lung cancers (SCLC), which account for about 25% of bronchogenic carcinomas, and the remaining tumors, non-small-cell lung cancers (NSCLC). SCLC cells possess neuroendrocrine-like properties, whereas most NSCLC lack these functions and can be divided into three major subtypes, adenocarcinoma, squamous cell carcinoma (SCC), and large cell carcinoma (LC). SCLC is one of the most virulent forms of human cancer, characterized by early dissemination and aggressive clinical evolution as compared to NSCLC. The initial responsiveness of SCLC to anticancer drugs and radiation treatment contrasts with the relative resistance of NSCLC to the same types of interventions. The reasons for the differences between SCLC and NSCLC are not fully understood, and one possible explanation may be resistance to killing via programmed cell death or apoptosis.

1.2. The Origins of Lung Cancers

Lung cancers, like other epithelial malignancies, arise after a series of morphological and molecular changes, which take many years to develop. The changes associated with squamous cell carcinomas are well documented, while the antecedents of adenocarcinomas and SCLC are more poorly understood ***(1)***. Molecular changes commence in histologically normal epithelium, and thus precede the onset of pathologically recognizable lesions ***(2,3)***. Many other molecular changes characteristic of lung cancers can be detected in preneoplastic lesions, including deregulation of the *p53, p16, FHIT*, and *telomerase* genes, and neoangiogenesis ***(4–10)***. These changes are not random, but follow a defined pattern ***(8)***.

1.3. How Cancers Evolve

The pathology and molecular biology of lung cancer demonstrate that these tumors evolve through a series of mutations, molecular abnormalities, and concomitant mor-

From: *Methods in Molecular Biology, Vol. 222: Tumor Suppressor Genes: Pathways and Isolation Strategies*
Edited by: Wafik S. El-Deiry © Humana Press inc., Totowa, NJ

phologic changes. Because many occupational and environmental lung cancers are caused by multiple etiologic agents ***(11)*** the integration of histology with cellular, biochemical, and molecular biomarker analysis may provide new approaches toward understanding this disease process.

Hanahan and Weinberg ***(12)*** have suggested that the vast catalog of cancer cell genotypes is a manifestation of six essential alterations in cell physiology that collectively dictate malignant growth: (a) self-sufficiency in growth signals, (b) insensitivity to growth-inhibitory (antigrowth) signals, (c) evasion of programmed cell death (apoptosis), (d) limitless replicative potential, (e) sustained angiogenesis and (f) tissue invasion and metastasis. Each of these physiologic changes involving novel capabilities acquired during tumor development represents the successful breaching of an anticancer defense mechanism hard-wired into cells and tissues. Hanahan and Weinberg proposed that these six capabilities are shared in common by most and perhaps all types of human tumors. This multiplicity of defenses may explain why cancer is relatively rare during an average human's lifetime.

1.4. The causes of lung cancers

Although other factors, including ionizing radiation and exposure to industrial carcinogens including asbestos, may contribute to its pathogenesis, tobacco use is the cause of the vast majority of lung cancers ***(13,14)***. Lung cancer is related to the extent of smoke exposure, and people who stop smoking, even well into middle age, avoid most of their subsequent risk of lung cancer ***(15)***. The particulate phase, and, to a lesser extent, the vapor phase of tobacco smoke contain numerous carcinogens, at least 20 of which induce lung cancer ***(16)***. Of these, polycyclic aromatic hydrocarbons and tobacco-specific nitrosamines are likely to play major roles. Tobacco smoke carcinogens interact with DNA, form adducts, and cause genetic changes, including mutations in oncogenes and tumor suppressor genes.

There are major differences in the molecular pathogenesis of lung cancers arising in ever-smokers and never-smokers. While *p53* mutations are found in lung cancers arising in both smokers and nonsmokers, the mutational spectrums of the gene in these two groups are very different ***(17–19)***. In addition, point mutations of the *ras* oncogene are much more frequent in smokers ***(20)***. Although there is no direct evidence that carcinogens in tobacco smoke induces DNA methylation and silencing of tumor suppressor genes, aberrant methylation of the *p16* gene is more frequent in smokers than in never-smokers ***(21)***. Cloutier et al. reported that carcinogens present in tobacco smoke result in selective methylation of DNA regions ***(22)***. The mutational spectrum of the *p53* gene reveals certain hotspots and types of mutations in lung cancers arising in smokers. A disproportionally high number of mutations in *p53* (and other genes) are found at methylated CpG dinucleotides. These sequences are particularly prone to mutagenesis involving both endogenous events as well as modification by exogenous carcinogens ***(23,24)***. These data provide a rational explanation of the differences in the molecular pathogenesis of lung cancers in smokers and never-smokers.

2. Oncogenes and Tumor Suppressor Genes (TSGs)

Tumor growth is the result of the deregulation of intrinsic cell proliferation (cell division). Genes known as protooncogenes code for proteins that stimulate cell division. Mutated forms, called oncogenes, can cause the stimulatory proteins to be overactive, with the result that cells proliferate excessively. Tumor suppressor genes, on the other hand, code for proteins that inhibit cell division.

The activation of oncogenes is well known as one of the possible mechanisms to transform normal cells. Now, it has become clear that the inactivation of the various tumor suppressor genes (TSGs), which can be thought of as brakes of cell growth, also are crucial to the development and progression of lung cancer.

Tumor suppressor genes (TSGs), like dominant oncogenes, play a vital role in normal cell growth control. They can also be involved in DNA repair and the response to DNA damage. Overall they inhibit the process of tumorigenesis. Kinzler and Vogelstein ***(25)*** have recently proposed a new system to categorize TSGs based on their roles as "gatekeepers" and "caretakers." Gatekeepers are TSGs that are involved directly in controlling proliferation by regulating cell cycle checkpoints (e.g., the Rb and INK families of cdk inhibitors). Mutations of these genes results in high penetrance. In contrast, the caretakers are of rather low penetrance and have an indirect effect on cell growth. They are responsible for genomic integrity, and changes in such genes lead to genome instability (e.g., the MSH2 DNA repair gene, which is involved in hereditary nonpolyposis colon cancer or HNPCC). It is likely that more categories will follow, for instance, "landscaper" for genes involved in epithelial-stromal and epithelial interactions (E-cadherin and alpha catenin).

2.1. Mechanism of TSG Inactivation

TSGs are considered to act mostly in a recessive fashion and according to Knudson's two-hit hypothesis, loss of function of a TSG requires that both alleles be inactivated. One allele is inactivated by mutation, methylation, deacetylation, or other changes that target the individual TSG. The other allele is usually inactivated as part of a chromosomal loss involving the gene of interest as well as many genetic markers in the vicinity. This large-scale loss can occur by chromosomal deletion, nonreciprocal translocation, or mitotic recombination. Occasionally homozygous deletion may occur, especially in tumor cell lines, and are very useful for localization and cloning of the putative TSG within the deleted region. The loss is referred to as allele loss or loss of heterozygosity (LOH). Chromosomal loss is mostly analyzed by karyotypic studies or loss of heterozygosity (LOH), and mutations are mostly frequently studied by sequencing of the gene of interest or by single-strand conformational polymorphism analysis (SSCP). In many cases mutations can result in truncated protein products that are relatively simple to detect. However, recently it has been shown that functional inactivation of TSG can be caused by many other mechanisms besides mutations, including aberrant promoter methylation, increased degradation, or relocalization of the gene product.

2.2 Chromosomal Localization of TSGs

Consistent LOH for genetic markers at a given locus in tumors from multiple patients is strong evidence for the presence of an underlying TSG in that region. Allelotyping studies for LOH indicate that multiple other TSGs associated with lung cancer may be present at novel chromosomal regions including 1p, 3p, 4p, 4q, 5q, 6p, 6q, 9p, 8p, 11p, 13q.17p, 18q, and 22q ***(26,27)***. Presumably, many of these regions are the locations of TSGs that play an important role in lung cancer pathogenesis, although only a few of the relevant genes have been identified. Some of these genes, discussed in greater detail below, are retinoic acid receptor. β *(RARβ)* on 3p24, *RASSF1* on 3p21-22, *p16*^{INK4a} on 9p21, *PTEN* on 10q22, *RB* on 13q14, and *p53* on 17p13.

3. Tumor Suppressor Genes on Chromosome 3p

Chromosome 3p allelic loss has been shown to be the most frequent molecular alteration in lung cancer and commences early during the progression of lung cancers ***(28,29)***. This region is also frequently deleted in many other cancers, including breast, head and neck, cervix, and renal cell. Allelic loss of one copy of 3p is found in 90% of SCLC and 70% of NSCLC. Multiple distinct 3p chromosomal regions, 3p25, 3p24, 3p21.3, 3p14.2, and 3p12, undergo allele loss and are suspected to harbor TSGs. Several independent homozgyous deletions at 3p24 *(RARα)* 3p21.3 *(RASSF1)*, 3p14.2 *(FHIT)*, and 3p12 identified in lung cancer are strong indicators for the presence of TSGs. The Von Hippel-Lindau (VHL) gene at 3p25, which is inactivated in a sizable fraction of renal cell cancers is rarely found to be mutated in lung tumors. Other candidate 3p TSGs include the BRCA1-binding protein BAP1 (3p21), the base excision repair gene *hOGG1* (3p25), the DNA mismatch repair gene *hMLH1* (3p21), and a series of several genes such as the semaphorins, which were found in the 3p21.3 homozygous deletion region. An intensive search is ongoing to identify and characterize the putative 3p TSGs, and several promising candidate genes have been identified in a 630-kb lung cancer homozygous deletion region harboring one or more tumor suppressor genes on chromosome 3p21.3 ***(30)***. This location was identified through somatic genetic mapping in tumors, cancer cell lines, and premalignant lesions of the lung and breast, including the discovery of several homozygous deletions. The combination of molecular and manual methods and computational predictions permitted the detection, isolation, and characterization of 19 of genes within the deleted region.

4. The Importance of Methylation in Gene Silencing

In the past few years, it has become apparent that methylation is the major mechanism of gene silencing in human tumors ***(31,32)***. One allele is silenced by aberrant promoter methylation, while the other is deleted. Methylation of DNA can only occur in vertebrates at cytosine residues of CpG dinucleotides. It is an epigenetic change that does not modify the nucleotide structure, it is reversible, it is passed on to daughter cells after cell division, but it cannot be transmitted directly via the germ line. CpG sites are not randomly distributed in the human genome, and the promoter regions of about half of the genes are rich in these sites ("CpG islands"). Normally, the islands are nonmethylated. Promoter methylation is associated with gene silencing. To maintain methylation during cell division, cytosine residues in newly synthesized DNA are methylated by DNA cytosine methyltransferase (DNMT) enzymes, which transfer a methyl group from the methyl donor S-adenosylmethionine using the hemimethylated DNA template. Methylation plays a crucial role in the normal organism, and its roles include gene imprinting, X-chromosome inactivation, normal development, and repression of gene transcription. The mechanism of gene silencing is slowly being elucidated. Methylation may prevent binding of sequence specific transcription factors or recruit transcriptional corepressors. Methylation's relationship to chromatin structure, in particular, is crucial. A family of proteins bind to methyl CpG sites and frequently associate with histone deacetylases ***(33)***. Deacetylation of lysine groups in histones results in a compact nucleosome structure that inhibits transcription.

Aberrant CpG-island methylation has nonrandom and tumor type-specific patterns. In a global analysis of the methylation status of over 1000 human tumors, Costello et. al. ***(34)*** estimated that an average of 600 CpG islands (range 0–4500) were aberrantly methylated in individual tumors. They identified patterns of CpG-island methylation that were shared within each tumor type, together with patterns and targets that displayed distinct tumor-type specificity. Lung tumors have a characteristic pattern of genes methylated and inactivated at frequencies greater than 20% **(35–37)**. However, the number of genes whose methylation status has been studied in lung cancers is modest (less than 40), and many more, perhaps hundreds, remain to be discovered. Methylation commences during preneoplasia and can be detected in the bronchial epithelium of smokers ***(10)***. Treatment with demethylating agents such as 5-aza-2'-deoxycytidine restores gene expression. The effect of demethylating agents may be enhanced by combination with inhibitors of histone deacetylation such as trichostatin A ***(32)***.

5. Retinoic Acid Receptor β *(RARβ)*

The retinoic acid receptor. β *(RARβ)* gene has been intensively studied in lung cancer and was found to have defective function making it a candidate TSG. This is particularly important given the interest in using retinoids as chemoprevention agents for lung cancer ***(38)***. Studies show that lung cancer cells are resistant to retinoids in vitro. RARβ functions in vivo as a heterodimer with RXR receptor to stimulate the transcription of many genes after binding a retinoid ligand. One of the genes that is upregulated by this mechanism is *RARβ* itself ***(39)***. Diminished or absent RARβ protein expression is present in most NSCLC tumors and the bronchial epithelium of smokers ***(40,41)***. One of the mechanisms for this loss of expression is methylation of the RARβ promoter region, as has been reported in other cancers. Because receptor isoforms RARβ2 and RARβ4 are repressed in human lung cancers, we investigated whether methylation of their promoter, P2, might lead to silencing of the *RARβ* gene in human lung tumors and cell lines ***(42)***. The P2 promoter was methylated in 72% (63 of 87) of SCLC and in 41% (52 of 127) of NSCLC tumors and cell lines, and the difference was statistically significant. By contrast, in 57 of 58 control samples, we observed only the unmethylated form of the gene. Four tumor lines with methylated promoter regions lacked expression of these isoforms, but demethylation by exposure to 5-aza-2′-deoxycytidine restored their expression. Promoter methylation of the RARβ is one mechanism that silences RARβ2 and RARβ4 expression in many lung cancers, particularly SCLC.

6. *RASSF1* Gene

One of the key TSG candidates at chromosome 3p21.3 is *RASSF1*. Using a positional cloning approach from a series of nested homozygous deletions in SCLC, a 630-kb region was recently narrowed down to a 120-kb subregion by a breast cancer homozygous deletion. Nine genes border this region. One of them is *RASSF1* (7.6 kb) and has a predicted RAS association domain. The presence of a Ras association domain in both RASSF1 isoforms (RASSF1A and RASSF1C) suggests that these proteins may function as effectors of Ras signaling in normal cells. If so, the observation that RASSF1 can function as a TSG implies that RASSF1 acts in opposition to the Ras effector pathways that stimulate cell proliferation. Ras mutations rarely occur in SCLC and are found in only 30% of NSCLC (mostly in adenocaricnomas). Thus the finding of RASSF1 methylation in 90% of SCLC and many of 30% NSCLC without Ras mutation suggests that inactivation of RASSF1 expression may be a mechanism distinct from Ras mutations leading to activation of Ras signaling in tumors. Thus methylation serves as an alternative to the genetic loss of a TSG function by deletion or mutation ***(43)***.

Both RASSF1 isoforms also possess a putative ATM kinase phosphorylation site in their common exon based on in-vitro phosphorylation studies (RASSF1 is referred to as PTS by Kim et. al.). Dammann et. al. ***(44)*** reported they have isolated RASSF1 transcript in a yeast 2 hybrid assay with DNA repair protein XPA as bait. These findings suggest the hypothesis that *RASSF1* gene products may participate in DNA damage response or DNA damage-induced regulation of other cellular signaling events. The putative diacylglycerol-binding domain in RASSF1A, but not in RASSF1C, suggests studies to test the role of tumor promoters interacting with the RASSF1A isoform, while the loss of RASSF1A expression in tumors predicts that if such interaction exists, tumor promoter may actually help control growth when RASSF1A is expressed.

7. The *FHIT* Gene

The fragile histidine triad (*FHIT*) gene has been identified as a candidate TSG that spans a fragile site at 3p14.2 that spans FRA3B, the most common of the aphidicolin-inducible fragile sites ***(45)***. The protein product of the *FHIT* gene is involved in the metabolism of diadenoside tetraphosphate to adenosine triphosphate (ATP) and adenosine monophosphate (AMP). Molecular abnormalities of the *FHIT* gene and the FRA3B region are frequent in lung cancer ***(45)***. Analyses of microsatellite markers within the *FHIT* gene have shown that LOH is found in 70% of both major types of lung cancer ***(46)***. LOH at the *FHIT* gene locus was more frequent in smokers than in nonsmokers, suggesting that FHIT is the molecular target of tobacco smoke carcinogen. Abnormal transcripts were observed in 80% of SCLCs, 40% of NSCLCs, and several other tumors ***(46–50)***. These alterations include deletions of one or more exons and insertion of various-sized intronic sequences at the RNA level. Absent *FHIT,* protein expression occurs in 50% of all lung cancers. Aberrant methylation of this gene has been found in 64% of NSCLC and SCLC cell lines and in 37% of primary NSCLC tumors. *FHIT* gene promoter methylation was significantly correlated with loss of mRNA expression, and methylation was found in 17% of bronchial brushes from heavy cigarette smokers. FHIT gene LOH commences very early in the lung cancer pathogenesis, and it remains to be determined whether methylation represents a marker for risk assesement (51). Transfection of a wild-type copy of the *FHIT* gene into lung cancer cells can reverse the malignant phenotype and induce tumor cell apoptosis. It appears that FHIT is one of the TSG in the chromosome 3p14.2 LOH region in lung cancer. Replacement gene therapy may thus be a logical step for clinical trials.

8. Adenomatous Polyposis Coli *(APC)*

The adenomatous polyposis coli *(APC)* gene encodes a large multidomain protein that plays an integral role in the wnt signaling pathway and intracellular adhesion. There is preliminary evidence that the region of the *APC* gene that may be involved in the regulation of apoptosis coincides with the region that mediates β-catenin binding and degradation. Germline mutations in the *APC* gene are responsible for the autosomal dominant inherited disease familial adenomatous polyposis coli (FAP), while somatic mutations in *APC* occur in 80% of sporadic colorectal tumors. *APC* mutation almost always results in a truncated protein product with abnormal function. Recent advances in research have concentrated on the interdependence of *APC* mutations in colorectal tumorigenesis, the contribution of *APC* missense variants to inherited risk of colorectal cancer and biological interaction of the APC protein and its partner.

The *APC* gene has two promoter regions, 1A and 1B. Promoter 1A is most commonly active. Aberrant methylation of the APC promoter region has been postulated as a possible second hit mechanism in colorectal tumors where *APC* mutation is present. More recent research, however, indicates that hypermethylation of APC promoter occurs in both colorectal adenomas and carcinomas but not in adjacent normal colon mucosa. These finding suggest that APC promoter methylation may provide an alternative mechanism for APC inactivation in the early stages of colorectal tumorigenesis. Nevertheless, the preponderance of APC mutation and LOH at the *APC* locus in colorectal tumors precludes hypermethylation from being a major event in this process.

There are very few studies describing LOH and mutation analysis at the *APC* locus in lung cancer patients despite very low mutation rates and high allelic loss (70–85%) in SCLC and NSCLC at the 5q21-22 region. The *APC* gene has a relatively minor role in lung cancer pathogenesis. We have recently found a high incidence of APC gene promoter 1A methylation (53%) in NSCLC primary tumors and cell lines and a modest methylation (26%) in SCLC cell lines despite a very high rate of LOH in SCLC. APC functions by binding to β-catenin, and this usually take place in the cytoplasm. It has been recognized for some time that APC is found within the nucleus as well in cytoplasm in most tumors and cell lines, but mostly only in the nucleus of the methylated cell lines. The loss of exon 1A transcript due to methylation of promoter 1A may play a vital role in APC cellular trafficking (unpublished observation). An improved understanding of both the genetics and biology of *APC* gene alterations may in time culminate in preventive or therapeutic strategies targeted specifically at reducing the burden of lung cancer. Whether APC inactivation contributes to lung cancer development cannot be determined at the present time.

9. The *p16* Gene

A recessive oncogene region relevant to lung cancer is located in the interferon gene cluster region on chromosome 9p21-22 **(52,53)**. A tumor suppressor gene on 9p has been identified (54, 55). The gene, *CDKN2* (also known as *MTS1* or $p16^{INK4}$) encodes a previously identified inhibitor (*p16*) of cyclin-dependent kinase 4. A second related cyclin/cdk inhibitor, p15, was also found on chromosome 9p adjacent to the gene coding for $p16^{INK4}$ ***(56)***, and homozygous loss in lung cancer at chromosome 9p frequently encompasses both *p15* and *p16* ***(57,58)***. A recent study demonstrated that there are two distinct regions of deletion on 9p21 and 9p21-22 ***(59)***. LOH at 9p21-22 locus has been reported in 33% in NSCLC, with a much higher frequency in squamous cell carcinomas (55–59%) than in adenocarcinomas (21–36%)***(60,61)***. Several methods of inactivation (based on Knudson principles) have been described in lung cancer for the *p16* gene, including point mutations (20–30%) or aberrant methylation (30%) (both combined with monoallelic loss), or homozygous deletions (30%).

In SCLC the *RB* gene on chromosome 13q is usually inactivated whereas in NSCLC it is the *CDKN2* gene on 9p21 that is most frequently deleted or mutated. Inactivation of either of these TSGs appears to be mutually exclusive. The reason for this became apparent when it was discovered that these TSGs encode proteins that participate in the same G1-S growth regulatory pathway and that only one protein needs to be inactivated for loss of functional growth regulation ***(62)***.

The $p16^{INK4}$ locus encodes a second alternative reading frame protein, $p19^{ARF}$, which also is believed to be important in growth regulation. By binding to the MDM2-p53 complex it leads to p53 inactivation. A recent study found that $p19^{ARF}$ was more frequently lost in lung tumors with neuroendocrine features. Alterations in the 9p21 locus may inactivate both $p16^{INK4}$ and $p19^{ARF}$ leading to dysfunction of the p16/RB and the p53 pathways.

10. Retinoblastoma *(Rb)* Gene

The *RB* gene is located on chromosome 13q14, and loss of this region is important in pathogenesis of lung cancer. The *RB* gene encodes a nuclear protein that was initially determined to be abnormal in patients with retinoblastoma. Knudson predicted the TSG nature of this gene by studying inheritance patterns of familial retinoblastoma. The 106-kDa protein encoded by the *RB* gene is phosphorylated in a cell cycle-dependent manner, interacts with the transcription factor E2F, and is important in regulating the cell cycle during G0/G1 phase. The cell cycle is a complex set of events regulated by multiple proteins. The phosphorylation status of RB is one of the most critical determinants of progression through the various phases of the cell cycle.

When cells leave the late G1 phase to transit into the S phase, the RB protein becomes phosphorylated. At the end of mitosis the RB protein is dephosphorylated and is complexed to E2F, a transcription factor important in activating genes that regulate the cell cycle and suppress the transition from the G1 to the S phase. A crucial event in the progression of the cell cycle is the dissociation of the RB–E2F complex when RB becomes phosphorylated by cyclin-dependent kinase in late G1 phase and leads to its inactivation. Once E2F is no longer complexed to RB, it activates the transcription of various genes needed in the cell cycle such as dihydrofolate reductase, thymidine kinase, thymidylate synthase, and DNA polymerase. The binding of RB to E2F has been shown to be regulated by p53, $p21^{WAF1}$, cyclins, and $p16^{INK4}$, among other cellcycle regulatory proteins.

Besides retinoblastoma, SCLC was a tumor initially identified to have abnormalities of RB ***(63)***. In SCLC, there is a rearrangement of chromosome 13, and up to 70% of SCLC cell lines show structural alterations in the *RB* gene or abnormal expression of the protein ***(64,65)***. Abnormal expression of the RB protein in over 90% of SCLC cell lines is due either to lack of expression of Rb protein or hypo-phosphorylation of RB protein or to deletion or mutation of the pocket domain. In NSCLC such abnormalities are present in a much smaller fraction of tumors and cell lines. Thus, the pathogenesis of lung cancer could occur by either mutational disruption of the RB protein (in SCLC) or by absence of $p16^{INK4}$ inhibitor (in NSCLC) that functions to keep RB phosphorylated and thereby disrupts the cell-cycle regulatory mechanism. Usually, inactivation of RB and $p16^{INK4}$ are mutually exclusive. Abnormalities of RB expression were present in 20% of Stage I and II and in 60% of Stage III and IV NSCLC tumors.

11. *PTEN/MMAC1*

Recently, a new candidate tumor suppressor gene, *PTEN* (Phosphatase and Tensin homolog deleted on chromosome TEN), also called *MMAC1* (Mutated in Multiple Advanced Cancers) was identified and localized to chromosome region 10q23.3 ***(66,67)***. Simultaneously, Li et al. cloned a novel tyrosine phosphatase, TEP1, which has been shown to be identical to PTEN/MMAC1 ***(68)***. PTEN/MMAC1 appears to be frequently mutated in gliomas, prostate cancers, melanomas, and endometrial cancers, with less frequent mutations in breast cancers ***(66,67,69–72)***. Importantly, germline mutations were found in the *PTEN/MMAC1* gene for individuals affected with Cowden disease, a hereditary, autosomal dominant cancer syndrome associated with a predisposition to thyroid, breast and skin cancers, and in Bannayan-Zonana syndrome ***(73–75)***. Recently the PTEN phosphatase has been described as having tumor suppressor activity by negatively regulating cell interactions with the extracellular matrix ***(76)***. Overexpression of PTEN inhibited cell migration, whereas antisense PTEN enhanced migration. Integrin mediated cell spreading and the formation of focal adhesions was downregulated by wild-type PTEN but not by PTEN with an inactive phosphatase domain. PTEN interacts with the focal adhesion kinase (FAK) and reduces its tyrosine phosphorylation.

Allelotyping studies utilizing microsatellite markers in close proximity to the *PTEN/MMAC1* gene (D10S541, D10S185, and D10S597) have demonstrated a high incidence of loss of heterozygosity (LOH) in lung cancers, with 91% LOH in small-cell lung cancer (SCLC) and 41% LOH in non-small-cell lung cancer (NSCLC) ***(26)***. In addition, comparative genomic hybridization data show frequent deletions on chromosome 10q in SCLC, both in primary tumors and in their cell lines ***(77–79)***. These observations suggest that genetic changes in one or more tumor suppressor genes (TSGs) located in the 10q region may be involved in the development or progression of lung cancer, particularly in the highly metastatic SCLC. We have recently shown that the *PTEN/MMAC1* gene is mutated in a small subset of lung cancers ***(80)***. We have identified homozygous deletions in 5 of 127 (4%) (5/66 SCLC, 8% and 0/61 NSCLCs), and point mutations in 6 of 53 (11%) (3/35 SCLC and 3/18 NSCLC) lung cancer cell lines analyzed. However, we only found two silent mutations, and two apparent homozygous deletions in 22 uncultured primary or metastatic SCLC tumors. The frequency of homozygous deletions (2/22, 9% in uncultured SCLC specimens) compared well with that found in the SCLC cell lines (5/66, 8%). Taken together, these results indicate a *PTEN/MMAC1* mutation rate of about 5–15% in lung cancers. The alterations in the *PTEN/MMAC1* gene may play a role in the pathogenesis of a relatively small subset of lung cancers, since they demonstrate a high a very high incidence of loss at 10q23-25, the MXIL region (unpublished data).

12. *TSG101*

TSG101, a recently discovered candidate tumor suppressor gene, was cloned based on a novel strategy that uses regulated antisense RNA initiated within a "retrovirus-based" gene search vector to identify previously unknown autosomal genes whose inactivation is associated with a defined phenotype. In this case a functional knockout in mouse fibroblast cells leads to spontaneous lung metastases in a nude mice model ***(81)***. While the normal function of *TSG101* is unknown, several structural motifs have been identified, including a proline-rich region and a leucine heptad repeat (coiled-coil) domain, which is nearly identical to that of its murine homolog. The murine coiled-coil domain has been shown to interact with a regulator of cell growth and differentiation called stathmin ***(81,82)***. Elevated stathmin levels have been reported in acute leukemias/aggressive lymphomas ***(83)*** and neuroblastomas ***(84)***. Furthermore, a structural analysis of the amino-terminal domain ofTSG101 indicates that it mimics a group of ubiquitin-conjugating enzymes, implying that TSG101 may function in ubiquitin-mediated proteolysis and cell cycle progression ***(85,86)***. Li, L. et al. ***(87)*** have shown that the ubiquitin domain of TSG101 interferes with ubiquitination of MDM2, that TSG101 inhibits MDM2 decay and elevates its steady-state level, and that these events are associated with downregulation of the p53 protein.

Oh et al. ***(88)*** found that at least one truncated TSG101 transcript was expressed simultaneously with wild-type TSG101 transcript in almost all SCLC cell lines as determined by reverse transcription polymerase chain reaction (RT-PCR) and Northern blot analysis. In contrast, normal lung tissue, normal colonic mucosa, as well as primary non-small-cell lung cancer (NSCLC) specimens express only a wild-type transcript. Sequence analysis of the wild-type transcript did not reveal any mutations, while the truncated transcript common to all SCLC cell lines exhibited an identical deletion in exons 2–4 and parts of exons 1 and 5. The results demonstrate that both the normal and truncated transcripts were expressed in ~90% of SCLC cell lines, while in normal tissue or NSCLC samples tested only the normal transcript was seen. This abnormal transcript represents the most frequent TSG101 abnormality found in human cancers so far. A recent study ***(89)*** has shown that a normal healthy person's lymphocytes can have more than one aberrant transcript of *FHIT, TSG101*, and *PTEN/MMAC1* genes. The size and the number of the transcripts vary and the diversity is unconstrained. It is not dependent on the time, condition of the reaction, or the isolation method. They have suggested that these aberrant transcripts may be the imperfect products of splicesomes that occur in one of every thousand, ten thousand, or more cases and imply no direct association with tumorigenesis.

13. *H-cadherin*

Cadherins are members of a large family of transmembrane glycoproteins that mediate calcium-dependent homophilic, and to a lesser extent, heterophilic cell–cell adhesion, and play an important role in the maintenance of normal tissue architecture ***(90–92)***. It is now clear that cadherin dysfunction is implicated in tumor development ***(93,94)***. Cell–cell association is often disorganized in human tumors and is thought to be a cause of the unregulated behavior of tumor cells, including invasion and metastasis. When cadherins are functional, epithelial cells remain tightly connected to each other. However, when cadherin products are inactivated or lost, cells lose their association to neighboring cells and migrate into surrounding tissues. The prototype family member is e-cadherin (CDH1), whose expression is dysregulated in many NSCLC and SCLC tumors ***(95,96)***.

Recent study clearly revealed that the human CDH13 (H-cadherin) is the human homolog of chicken T-cadherin, a unique glycosyl phosphatidyl inositol (GPI)-linked membrane ***(97)***. This candidate gene, which was strongly expressed in heart, was termed *H-cadherin* and appeared to be identical to *CDH13*. The *CDH13* gene was mapped on 16q24, at which loss of heterozygosity in patients with sporadic breast, liver, lung, and prostrate cancer were reported, suggesting that a candidate tumor suppressor gene or genes localized in this regions could play an important role in cancer development ***(98–101)***. Sato et al. ***(102)*** reported that the *CDH13* gene is inactivated in a considerable number of human lung cancer specimens. It has been reported that the aberrant methylation related to inactivation of H-cadherin molecule was found in 43% of primary NSCLC tumor and 20% of SCLC cell lines ***(102,103)***. Zhong et al. ***(104)*** have also reported loss of expression in 43% of NSCLC resected tumor specimens. They have further demonstrated that loss of H- and/or E-cadherin is associated with enhanced tumorigenicity of human NSCLC tumors in nude mice. Lee et al. ***(105)*** further demonstrated that H-cadherin overexpression after transfection, into breast cancer cell lines resulted in significant growth inhibition in vitro. The H-cadherin is a unique member of the cadherin family, because it lacks the cytoplasmic region, which is well conserved in the other cadherins. This cytoplasmic region is necessary for the other cadherins to bind cytoskeletal protein for e.g. β-catenin. Studies by Takeuchi et al. ***(106)*** suggest that H-cadherin, besides cell adhesion, has a unique role in the regulation of surfactant protein (SP-D) gene expression in human bronchioloaveolar type II cells.

14. The *P53* Gene

The *p53* gene is located in chromosome region 17p13.1 and encodes a 53-kDa nuclear protein that acts as a transcription factor to turn on the expression of a DNA damage-response genetic program. It blocks the progression of cells through the cell cycle in the late G1 phase and triggers apoptosis. DNA damage is a major signal for p53 activation by phosphorylation by the ataxia telangiectasia gene *(ATM)* encoded kinase. Phosphorylated p53 acts as a specific DNA-binding transcription factor for several genes, including p21/WAF1/CIP1, MDM2, Growth arrest DNA damage protein (GADD45), BAX, and cyclin G. Activation of these genes results in apoptotic cell cycle arrest and DNA repair.

Knockout *p53* mice are structurally normal. However, homozygously deleted mice develop tumors at an early age, and heterozygous mice that contain one wild-type *p53* allele develop tumors at a reduced frequency and at a slower rate. The most common genetic change associated with cancer involves mutation of the *p53* gene ***(107)***. The prevalent type of point mutation is a GC-to-TA transversion, causing missense mutations, and is believed to be related to benzo(a) pyrene-induced damage from cigarette smoking. The *p53* missense mutation usually leads to an increase in protein half-life, with subsequent higher levels of 53 protein that are easily detected by immunohistochemistry. Abnormal p53 expression by immuno-histochemistry was found in 40% to 70% of SCLC and in 40–60% of NSCLC.

There are a number of reports investigating the association of p53 abnormalities with the prognosis of NSCLC patients. Although several studies observed that p53 alterations are poor prognostic indicators in NSCLC patients, other studies did not confirm these results. A recent meta-analysis studying 3-yr and 5-yr survival rates concluded that p53 alterations (overexpression of the protein or DNA sequence change) are poor prognostic markers, especially for patients with adenocarcinomas ***(108)***. Another study demonstrated that null mutations are a significant indicator of poor outcome in early-stage NSCLC patients, suggesting that the type of mutation is important for prediction of the outcome in this group of patients ***(109)***. As a novel therapy for lung cancer, p53 has been introduced in clinical trials with retroviral and adenoviral-delivered gene therapy, with initially promising antitumor responses. In addition, vaccine trials using mutant p53 peptides are ongoing ***(110,111)***.

15. Neurofibromatosis Type 2 (NF2)

Neurofibromatosis type 2 (NF2) is an autosomal dominantly inherited disorder predisposing affected individuals to tumors of multiple cell types in the central nervous system, including meningiomas. A candidate tumor suppressor gene for this disorder has recently been cloned; the protein product of this gene has a predicted role in linking integral membrane proteins with the cytoskeleton ***(112)***. Mutations were detected in multiple tumor types related to the NF2 disorder and also in NF2-unrelated tumors such as melanoma and breast carcinoma. Sekido et al., ***(113)*** found that *NF2* gene mutation is involved in 41% of mesothelioma tumors but in none of 75 lung cancer cell lines. These results suggest that the *NF2* gene participates in the pathogenesis in a subset of mesotheliomas but not in lung cancers.

16. Potential Clinical Applications of Molecular Markers in Lung Cancer

Abnormalities of protooncogenes, genetic and epigenetic changes of tumor suppressor genes, the role of angiogenesis in the multistage development of lung cancer, as well as detection of molecular abnormalities in preinvasive respiratory lesions have recently come into focus. Efforts are ongoing to translate these findings into new clinical strategies for risk assessment, chemoprevention, early diagnosis, treatment selection, prognosis, and to provide new targets and methods of treatment for lung cancer patients. These strategies should aid in reducing the number of newly diagnosed lung cancer cases and in increasing the survival and quality of life of patients with lung cancer. Understanding the molecular basis of lung cancer should allow translation of the research into early diagnosis and therapy. A comprehensive review by Mulshine ***(114)*** has catergorized several potential diagnostic targets in bronchial fluids and or sputum, including differentiation markers (glycolipid), specific tumor products, e.g., mucin, matrix protein, surfactant, etc., oncogenes or oncogene products, cytogenetic abnormalities, specific chromosomal deletion or rearrangements, and growth factors. Anti-p53 antibodies have been detected in sputum samples from patients with bronchial dyplasia as well as squamous cell carcinomas by immunohistochemistry techniques. In addition, anti-p53 antibodies have been detected in a small number of patients with occult lung cancer, which suggests use of this method for screening early lung cancer. With rapid technological advances in molecular or cell biology such as detection of the target cells in low abundance, by polymerase chain reaction ***(115)***, it is feasible to detect abnormalities such as *p53* and *RAS* mutations ***(116,117)***, methylation, loss of heterozygosity (LOH), or microsatellite alterations in sputum ***(2,118)*** bronchial ***(119,120)***, cytological specimens ***(121)***, blood ***(122)*** or plasma ***(123,124)***, and microdissected cells from various stages of cancer as well as histologically normal-appearing cells ***(125)*** by various applications of PCR. Thus, the entire spectrum of clinical specimens is available for genetic analysis, risk assessment, and early detection.

Taken together, these data clearly suggest that there are both common and distinct genetic pathways in the pathogenesis of the major types of lung cancer that are consistent with their markedly different biologic and clinical features. In order to make an impact in the race for a cure of this devastating cancer, we must continue to understand the basic underlying molecular abnormalities that are involved in lung cancer and study phenotypic and genotypic differences between the major forms of the tumor.

References

1. Wistuba, I. I. and Gazdar, A. F. (1999) Molecular abnormalities in the sequential development of lung carcinoma,in *Molecular Pathology of Early Cancer,* Srivastava, S., Henson, D. E., and Gazdar, A. F., eds.), IOS Press, Amsterdam, pp. 265–276.
2. Mao, L., Hruban, R. H., Boyle, J. O., Tockman, M., and Sidransky, D. (1994). Detection of oncogene mutations in sputum precedes diagnosis of lung cancer. *Cancer Res.* **54**, 1634–1637.
3. Wistuba, I. I., Barry, J., Behrens, C., et al. (2000) Molecular changes in the bronchial epithelium of patients with small cell lung cancer. *Clin. Cancer Res.* **6**, 2604–2610.
4. Sozzi, G., Sard, L., De Gregorio, L., et al. (1997) Association between cigarette smoking and FHIT gene alterations in lung cancer. *Cancer Res.* **57**, 2121–2123.
5. Franklin, W. A., Gazdar, A. F., Haney, J., et al. (1997) Widely dispersed *p53* mutation in respiratory epithelium. *J. Clin. Invest.* **100**, 2133-2137.
6. Orfanidou, D., Kalomenidis, J., Rasidakis, A., et al. (1998) Immunohistochemical detection of p53 protein in neoplastic, preneoplastic and normal bronchial mucosa specimens obtained during diagnostic bronchoscopy, *Oncol. Rep.* **5**, 763–769.
7. Keith, R. L., Miller, Y. E., Gemmill, R. M., et al. (2000) Angiogenic squamous dysplasia in bronchi of individuals at high risk for lung cancer. *Clin. Cancer Res.* **6**, 1616–1625.
8. Hirsch, F. R., Franklin, W. A., Gazdar, A. F., and Bunn, P. A., Jr. (2001) Early detection of lung cancer: clinical perspectives of recent advances in biology and radiology. *Clin Cancer Res.* **7**, 5–22.
9. Yashima, K., Litzky, L. A., Kaiser, L., et al. (1997) Telomerase expression in respiratory epithelium during the multistage pathogenesis of lung carcinomas. *Cancer Res.* **57**, 2373–2377.
10. Palmisano, W. A., Divine, K. K., Saccomanno, G., et al. (2000) Predicting lung cancer by detecting aberrant promoter methylation in sputum. *Cancer Res.* **60**, 5954–5958.
11. Ginsberg, R. J., Kris, M. G., and Armstrong, J. G. (1993) Cancer of the lung, in *Cancer: Principles and Practice of Oncology*, 3rd ed. DeVita, V. T. J. Hellman, S., and Rosenberg, S. A., eds.), Lippincott, Philadelphia , pp. 673–758.

12. Hanahan, D. and Weinberg, R. A. (2000) The hallmarks of cancer. *Cell* **100**, 57–70.
13. Parkin, D. M., Pisani, P., Lopez, A. D., and Masuyer, E. (1994) At least one in seven cases of cancer is caused by smoking. Global estimates for 1985. *Int. J. Cancer* **59**, 494–504.
14. Peto, R., Lopez, A. D., Boreham, J., Thun, M., Heath, C., Jr., and Doll, R. (1996) Mortality from smoking worldwide. *Br. Med. Bull.* **52**, 12–21.
15. Peto, R., Darby, S., Deo, H., Silcocks, P., Whitley, E., and Doll, R. Smoking, smoking cessation, and lung cancer in the UK since 1950: combination of national statistics with two case-control studies [see comments]. *Br. Med. J* **321**, 323–329.
16. Hecht, S. S. (1999) Tobacco smoke carcinogens and lung cancer. *J. Natl. Cancer Inst.* **91**, 1194–1210.
17. Suzuki, H., Takahashi, T., Kuroishi, T., et al. (1992) p53 Mutations in non-small cell lung cancer in Japan: association between mutations and smoking. *Cancer Res.* **52**, 734–736.
18. Hernandez-Boussard, T. M. and Hainaut, P. (1998) A specific spectrum of p53 mutations in lung cancer from smokers: review of mutations compiled in the IARC p53 database. *Environ Health Perspect.* **106**, 385–391.
19. Gealy, R., Zhang, L., Siegfried, J. M., Luketich, J. D., and Keohavong, P. (1999) Comparison of mutations in the p53 and K-ras genes in lung carcinomas from smoking and nonsmoking women. *Cancer Epidemiol. Biomarkers Prev.* **8**, 297–302.
20. Slebos, R. J., Hruban, R. H., Dalesio, O., Mooi, W. J., Offerhaus, G. J., and Rodenhuis, S. (1991) Relationship between K-ras oncogene activation and smoking in adenocarcinoma of the human lung. *J. Natl. Cancer Inst.* **83**, 1024–1027.
21. Kim, D. H., Nelson, H. H., Wiencke, J. K., et al. (2001) p16(INK4a) and histology-specific methylation of CpG islands by exposure to tobacco smoke in non-small cell lung cancer. *Cancer Res.* **61**, 3419–3424.
22. Cloutier, J. F., Drouin, R., and Castonguay, A. (1999) Treatment of human cells with N-nitroso(acetoxymethyl)methylamine: distribution patterns of piperidine-sensitive DNA damage at the nucleotide level of resolution are related to the sequence context. *Chem. Res. Toxicol.* **12**, 840–849.
23. Pfeifer, G. P. (2000) p53 mutational spectra and the role of methylated CpG sequences. *Mutat. Res.* **450**, 155–166.
24. Pfeifer, G. P., Tang, M., and Denissenko, M. F. (2000) Mutation hotspots and DNA methylation. *Curr. Top. Microbiol. Immunol.* **249**, 1–19.
25. Kinzler, K. W. and Vogelstein, B. (1998) Landscaping the cancer terrain. *Science* **280**, 1036–1037..
26. Virmani, A. K., Fong, K. M., Kodagoda, D., et al. (1998) Allelotyping demonstrates common and distinct patterns of chromosomal loss in human lung cancer types. *Genes Chromosomes Cancer*, **21**, 308–319.
27. Girard, L., Zochbauer-Muller, S., Virmani, A. K., Gazdar, A. F., and Minna, J. D. (2000) Genome-wide allelotyping of lung cancer identifies new regions of allelic loss, differences between small cell lung cancer and non-small cell lung cancer, and loci clustering. *Cancer Res.* **60**, 4894–4906.
28. Wistuba, II, Montellano, F. D., Milchgrub, S., et al. (1997) Deletions Of Chromosome 3p are frequent and early events in the pathogenesis of uterine cervical carcinoma. *Cancer Res.* **57**, 3154–3158.
29. Wistuba, II, Behrens, C., Virmani, A. K., et al. (2000) High resolution chromosome 3p allelotyping of human lung cancer and preneoplastic/preinvasive bronchial epithelium reveals multiple, discontinuous sites of 3p allele loss and three regions of frequent breakpoints. *Cancer Res.* **60**, 1949–1960.
30. Lerman, M. I. and Minna, J. D. (2000) The 630-kb lung cancer homozygous deletion region on human chromosome 3p21.3: identification and evaluation of the resident candidate tumor

suppressor genes. The International Lung Cancer Chromosome 3p21.3 Tumor Suppressor Gene Consortium. *Cancer Res.* **60**, 6116–6133.

31. Baylin, S. B., Herman, J. G., Graff, J. R., Vertino, P. M., and Issa, J. P. (1998) Alterations in DNA methylation: a fundamental aspect of neoplasia. *Adv. Cancer Res.* **72**, 141–196.
32. Baylin, S. B., Esteller, M., Rountree, M. R., Bachman, K. E., Schuebel, K., and Herman, J. G. (2001) Aberrant patterns of DNA methylation, chromatin formation and gene expression in cancer. *Hum. Mol. Genet.* **10**, 687–692.
33. Ng, H. H. and Bird, A. (2000) Histone deacetylases: silencers for hire. *Trends Biochem. Sci.* **25**, 121–126.
34. Costello, J. F., Fruhwald, M. C., Smiraglia, D. J., Rush, et al. (2000) Aberrant CpG-island methylation has non-random and tumour-type-specific patterns. *Nat. Genet.* **24**, 132–138
35. Esteller, M., Sanchez-Cespedes, M., Rosell, R., Sidransky, D., Baylin, S. B., and Herman, J. G. (1999) Detection of aberrant promoter hypermethylation of tumor suppressor genes in serum DNA from non-small cell lung cancer patients. *Cancer Res.* **59**, 67–70.
36. Zochbauer-Muller, S., Fong, K. M., Virmani, A. K., Geradts, J., Gazdar, A. F., and Minna, J. D. (2001) Aberrant promoter methylation of multiple genes in non-small cell lung cancers. *Cancer Res.* **61**, 249–255.
37. Esteller, M., Corn, P. G., Baylin, S. B., and Herman, J. G. (2001) A gene hypermethylation profile of human cancer. *Cancer Res.* **61**, 3225–3229.
38. Lippman, S. M., Clayman, G. L., Huber, M. H., Benner, S. E., and Hong, W. K. (1995) Biology and reversal of aerodigestive tract carcinogenesis. *Cancer Treat. Res.* **74**, 89–115.
39. Sun, S. Y., Wan, H., Yue, P., Hong, W. K., and Lotan, R. (2000) Evidence that retinoic acid receptor beta induction by retinoids is important for tumor cell growth inhibition. *J. Biol. Chem.* **275**, 17149–17153.
40. Lotan, R. (1999) Aberrant expression of retinoid receptors and lung carcinogenesis [editorial]. *J. Natl. Cancer Inst.* **91**, 989–991.
41. Xu, X. C., Lee, J. S., Lee, J. J., Morice, et al. (1999) Nuclear Retinoid Acid Receptor Beta in Bronchial Epithelium of Smokers Before and During Chemoprevention. *J. Natl. Cancer Inst.* **91**, 1317–1321.
42. Virmani, A. K., Rathi, A., Zochbauer-Muller, S., et al. (2000) Promoter methylation and silencing of the retinoic acid receptor-beta gene in lung carcinomas. *J. Natl. Cancer Inst.* **92**, 1303–1307.
43. Burbee, D. G., Forgacs, E., Zochbauer-Muller, S., et al. (2001) Epigenetic inactivation of RASSF1A in lung and breast cancers and malignant phenotype suppression. *J. Natl. Cancer Inst.* **93**, 691–699.
44. Dammann, R., Li, C., Yoon, J. H., Chin, P. L., Bates, S., and Pfeifer, G. P. (2000) Epigenetic inactivation of a RAS association domain family protein from the lung tumour suppressor locus 3p21.3. *Nat. Genet.* **25**, 315–319.
45. Ohta, M., Inoue, H., Cotticelli, M. G., et al. (1996) The FHIT gene, spanning the chromosome 3p14.2 fragile site and renal carcinoma-associated t(3;8) breakpoint, is abnormal in digestive tract cancers. *Cell* **84**, 587–597.
46. Fong, K. M., Eric J. Biesterveld, Arvind K. Virmani, et al. (1997) FHIT and FRA3B 3p14.2 allele loss are common in lung cancer and preneoplastic bronchial lesions and are associated with cancer-related FHIT cDNA splicing abberrations. *Cancer Res.* **57**, 2256–2267.
47. Hayashi, S., Tanimoto, K., Hajiro-Nakanishi, K., et al. (1997) Abnormal FHIT transcripts in human breast carcinomas: a clinicopathological and epidemiological analysis of 61 Japanese cases. *Cancer Res.* **57**, 1981–1985.
48. Lin, P. M., Liu, T. C., Chang, J. G., Chen, T. P., and Lin, S. F. (1997) Aberrant FHIT transcripts in acute myeloid leukaemia. *Br. J. Haematol.* **99**, 612–617.

49. Negrini, M., Monaco, C., Vorechovsky, I., et al. (1996) The *FHIT* gene at 3p14.2 is abnormal in breast carcinomas. *Cancer Res.* **56**, 3173–3179.
50. Yoshino, K., Enomoto, T., Nakamura, T., et al. (1998) Aberrant FHIT transcripts in squamous cell carcinoma of the uterine cervix. *Int. J. Cancer* **76**, 176–181.
51. Zochbauer-Muller, S., Fong, K. M., Maitra, A., et al. (2001) 5' CpG island methylation of the FHIT gene is correlated with loss of gene expression in lung and breast cancer. *Cancer Res.* **61**, 3581–3585.
52. Olopade, O. I., Buchhagen, D. L., Minna, J. D., et al. (1991) Deletions of the short arm of chromosome 9 that include the alpha and beta interferon genes are associated with lung cancers (meeting abstract). *Proc. Annu. Meet. Am. Assoc. Cancer Res.* **32**, A1814.
53. Olopade, O. I., Buchhagen, D. L., Malik, K., et al. (1993) Homozygous loss of the interferon genes defines the critical region on 9p that is deleted in lung cancers. *Cancer Res.* **53**, 2410–2415.
54. Kamb, A., Gruis, N. A., Weaver-Feldhaus, J., et al. (1994) A cell cycle regulator potentially involved in genesis of many tumor types [see comments]. *Science* **264**, 436–440.
55. Hayashi, N., Sugimoto, Y., Tsuchiya, E., Ogawa, M., and Nakamura, Y. (1994) Somatic mutations of the MTS (multiple tumor suppressor)1/CDK41 (cyclin-dependent kinase-4 inhibitor) gene in human primary non-small cell lung carcinomas. *Biochem. Biophy. Res. Commun.* **202**, 1426–1430.
56. Hannon, G. J. and Beach, D. (1994) p15INK4B is a potential effector of TGF-beta-induced cell cycle arrest. *Nature* **371**, 257–261.
57. Washimi, O., Nagatake, M., Osada, H., et al. (1995) In vivo occurrence of p16 (MTS1) and p15 (MTS2) alterations preferentially in non-small cell lung cancers. *Cancer Res.* **55**, 514–517.
58. Xiao, S., Li, D., Corson, J. M., Vijg, J., and Fletcher, J. A. (1995) Codeletion of p15 and p16 genes in primary non-small cell lung carcinoma. *Cancer Res.* **55**, 2968–2971.
59. Hamada, K., Kohno, T., Takahashi, M., et al. (2000) Two regions of homozygous deletion clusters at chromosome band 9p21 in human lung cancer. *Genes Chromosomes Cancer* **27**, 308–318.
60. Sato, S., Nakamura, Y., and Tsuchiya, E. Difference of allelotype between squamous cell carcinoma and adenocarcinoma of the lung. *Cancer Res.* **54**, 5652–5655.
61. Kishimoto, Y., Sugio, K., Mitsudomi, T., et al. (1995) Frequent loss of the short arm of chromosome 9 in resected non-small-cell lung cancers from Japanese patients and its association with squamous cell carcinoma. *J. Cancer Res. Clin. Oncol.* **121**, 291–296.
62. Otterson, G. A., Kratzke, R. A., Coxon, A., Kim, Y. W., and Kaye, F. J. (1994) Absence of p16INK4 protein is restricted to the subset of lung cancer lines that retains wildtype RB. *Oncogene* **9**, 3375–3378.
63. Harbour, J. W., Sali, S. L., Whang-Peng, J., Gazdar, A. F., Minna, J. D., and Kaye, F. J. (1988) Abnormalities in structure and expression of the human retinoblastoma gene in SCLC. *Science* **241**, 353–357.
64. Shimizu, E., Coxon, A., Otterson, G. A., et al. (1994) RB protein status and clinical correlation from 172 cell lines representing lung cancer, extrapulmonary small cell carcinoma and mesothelioma. *Oncogene* **9**, 2441–2448.
65. Otterson, G. A., Kratzke, R. A., Coxon, A., Kim, Y. W., and Kaye, F. J. (1994) Absence of p16INK4 protein is restricted to the subset of lung cancer lines that retains wildtype RB. *Oncogene* **9**, 3375–3378.
66. Steck, P. A., Pershouse, M. A., Jasser, S. A., et al. (1997) Identification of a candidate tumour suppressor gene, MMAC1, at chromosome 10q23.3 that is mutated in multiple advanced cancers. *Nat. Genet.* **15**, 356–362.
67. Li, J., Yen, C., Liaw, D., et al. (1997) Pten, a putative protein tyrosine phosphatase gene mutated in human brain, breast, and prostate cancer. *Science* **275**, 1943–1947.

68. Li, D. M. and Sun, H. (1997) Tep1, encoded by a candidate tumor suppressor locus, is a novel protein tyrosine phosphatase regulated by transforming growth factor beta. *Cancer Res.* **57**, 2124–2129.
69. Ried, T., Petersen, I., Holtgreve-Grez, H., et al. (1994) Mapping of multiple DNA gains and losses in primary small cell lung carcinomas by comparative genomic hybridization. *Cancer Res.* **54**, 1801–1806.
70. Guldberg, P., Straten, P. T., Birck, A., Ahrenkiel, V., Kirkin, A. F., and Zeuthen, J. (1997) Disruption of the Mmac1/Pten gene by deletion or mutation is a frequent event in malignant melanoma. *Cancer Res.* **57**, 3660–3663.
71. Rhei, E., Kang, L., Bogomolniy, F., Federici, M. G., Borgen, P. I., and Boyd, J. (1997) Mutation analysis of the putative tumor suppressor gene Pten/Mmac1 in primary breast carcinomas. *Cancer Res.* **57**, 3657–3659.
72. Tashiro, H., Blazes, M. S., Wu, R., et al. (1997) Mutations in Pten are frequent in endometrial carcinoma but rare in other common gynecological malignancies. *Cancer Res.* **57**, 3935–3940.
73. Arch, E. M., Goodman, B. K., Vanwesep, R. A.,et al. (1997) Deletion of PTENin a patient with Bannayan-Riley-Ruvalcaba-Syndrome suggests allelism with Cowden-Disease. *Am. J. Med. Genet.* **71**, 489–493.
74. Liaw, D., Marsh, D. J., Li, J., et al. (1997) Germline mutations of the Pten gene in Cowden-Disease, an inherited breast and thyroid cancer syndrome. *Nat. Genet.* **16**, 64–67.
75. Marsh, D. J., Dahia, P. L. M., Zheng, Z. M., et al. (1997) Germline mutations in Pten are present in Bannayan-Zonana Syndrome. *Nat. Genet.* **16**, 333–334.
76. Tamura, M., Gu, J. G., Matsumoto, K., Aota, S., Parsons, R., and Yamada, K. M. (1998) Inhibition of cell migration, spreading, and focal adhesions by tumor suppressor PTEN. *Science* **280**, 1614–1617.
77. Levin, N. A., Brzoska, P., Gupta, N., Minna, J. D., Gray, J. W., and Christman, M. F. (1994) Identification of frequent novel genetic alterations in small cell lung carcinoma. *Cancer Res.* **54**, 5086–5091.
78. Petersen, I., Langreck, H., Wolf, G., et al. (1997) Small-cell lung cancer is characterized by a high incidence of deletions on chromosomes 3p, 4q, 5q, 10q, 13q and 17p. *Br. J. Cancer* **75**, 79–86.
79. Schwendel, A., Langreck, H., Reichel, M., et al. (1997) Primary small-cell lung carcinomas and their metastases are characterized by a recurrent pattern of genetic alterations. *Int. J. Cancer* **74**, 86–93.
80. Forgacs, E., Biesterveld, E. J., Sekido, Y., et al. (1998) Mutation analysis of the *PTEN/MMAC1* gene in lung cancer. *Oncogene* **17**, 1557–1565.
81. Li, L. M. and Cohen, S. N. (1996) Tsg101—a novel tumor susceptibility gene isolated by controlled homozygous functional knockout of allelic loci in mammalian cells. *Cell* **85**, 319–329.
82. Marklund, U., Brattsand, G., Shingler, V., and Gullberg, M. (1993) Serine 25 of oncoprotein 18 is a major cytosolic target for the mitogen-activated protein kinase. *J. Biol. Chem.* **268**, 15039–15047.
83. Roos, G., Brattsand, G., Landberg, G., Marklund, U., and Gullberg, M. (1993) Expression of oncoprotein 18 in human leukemias and lymphomas. *Leukemia* **7**, 1538–1546.
84. Hailat, N., Strahler, J., Melhem, R., et al. (1990) N-myc gene amplification in neuroblastoma is associated with altered phosphorylation of a proliferation related polypeptide (Op18). *Oncogene* **5**, 1615–1618.
85. Koonin, E. V. and Abagyan, R. A. (1997) TSG101 may be the prototype of a class of dominant negative ubiquitin regulators [letter]. *Nat. Genet.* **16**, 330–331.
86. Ponting, C. P., Cai, Y. D., and Bork, P. (1997) The breast cancer gene product Tsg101—a regulator of ubiquitination. *J. Mol. Med.* **75**, 467–469.

87. Li, L., Liao, J., Ruland, J., Mak, T. W., and Cohen, S. N. A (2001) TSG101/MDM2 regulatory loop modulates MDM2 degradation and MDM2/p53 feedback control. *Proc. Natl. Acad. Sci. USA* **98**, 1619–1624.
88. Oh, Y., Proctor, M. L., Fan, Y. H., et al. (1998) TSG101 is not mutated in lung cancer but shortened transcript is frequently expressed in small cell lung cancer. *Oncogene* **16**,.
89. Wang, N. M., Chang, J. G., Liu, T. C.,et al. (2000) Aberrant transcripts of FHIT, TSG101 and PTEN/MMAC1 genes in normal peripheral mononuclear cells. *Int. J. Oncol.* **16**, 75–80.
90. Takeichi, M. Cadherins in cancer: implications for invasion and metastasis, Curr Opin Cell Biol. *5:* 806-11., 1993.
91. Takeichi, M., Watabe, M., Shibamoto, S., and Ito, F. (1994) Cadherin-dependent organization and disorganization of epithelial architecture. *Princess Takamatsu Symp.* **24**, 28–37.
92. Takeichi, M. (1995) Morphogenetic roles of classic cadherins. *Curr. Opin. Cell Biol.* **7**, 619–627.
93. Behrens, J. (1993) The role of cell adhesion molecules in cancer invasion and metastasis. *Breast Cancer Res. Treat.* **24**, 175–184.
94. Lee, S. W. (1996) H-cadherin, a novel cadherin with growth inhibitory functions and diminished expression in human breast cancer. *Nat. Med.* **2**, 776–782.
95. Kase, S., Sugio, K., Yamazaki, K., Okamoto, T., Yano, T., and Sugimachi, K. (2000) Expression of E-cadherin and beta-catenin in human non-small cell lung cancer and the clinical significance. *Clin. Cancer Res.* **6**, 4789–4796.
96. Clavel, C. E., Nollet, F., Berx, G., et al. (2001) Expression of the E-cadherin-catenin complex in lung neuroendocrine tumours. *J. Pathol.* **194**, 20–6.
97. Ranscht, B. and Dours-Zimmermann, M. T. (1991) T-cadherin, a novel cadherin cell adhesion molecule in the nervous system lacks the conserved cytoplasmic region. *Neuron* **7**, 391–402.
98. Tsuda, H., Callen, D. F., Fukutomi, T., Nakamura, Y., and Hirohashi, S. (1994) Allele loss on chromosome 16q24.2-qter occurs frequently in breast cancers irrespectively of differences in phenotype and extent of spread. *Cancer Res.* **54**, 513–517.
99. Tsuda, H., Zhang, W. D., Shimosato, Y., et al. (1990) Allele loss on chromosome 16 associated with progression of human hepatocellular carcinoma. *Proc. Natl. Acad. Sci. USA* **87**, 6791–6794.
100. Sato, M., Mori, Y., Sakurada, A., et al. (1998) Identification of a 910-kb region of common allelic loss in chromosome bands 16q24.1-q24.2 in human lung cancer. *Genes Chromosomes Cancer* **22**, 1–8.
101. Carter, B. S., Ewing, C. M., Ward, W. S., et al. (1990) Allelic loss of chromosomes 16q and 10q in human prostate cancer. *Proc. Natl. Acad. Sci. USA* **87**, 8751–8755.
102. Sato, M., Mori, Y., Sakurada, A., Fujimura, S., and Horii, A. (1998) The H-cadherin (CDH13) gene is inactivated in human lung cancer. *Hum. Genet.* **103**, 96–101.
103. Toyooka, K. O., Toyooka, S., Virmani, A. K.,et al. (2001) Loss of expression and aberrant methylation of the CDH13 (H-cadherin) gene in breast and lung carcinomas. *Cancer Res.* **61**, 4556–4560.
104. Zhong, Y., Delgado, Y., Gomez, J., Lee, S. W., and Perez-Soler, R. (2001) Loss of H-cadherin protein expression in human non-small cell lung cancer is associated with tumorigenicity. *Clin. Cancer Res.* **7**, 1683–1687.
105. Lee, S. W., Reimer, C. L., Campbell, D. B., Cheresh, P., Duda, R. B., and Kocher, O. (1998) H-cadherin expression inhibits in vitro invasiveness and tumor formation in vivo. *Carcinogenesis* **19**, 1157–1159.
106. Takeuchi, T., Misaki, A., Fujita, J., Sonobe, H., and Ohtsuki, Y. (2001) T-cadherin (CDH13, H-cadherin) expression downregulated surfactant protein D in bronchioloalveolar cells. *Virchows Arch.* **438**, 370–375.

107. Harris, C. C. (1996) p53: At the crossroads of molecular carcinogenesis and molecular epidemiology. *J. Investig. Dermatol. Symp. Proc.* **1**, 115–118.
108. Mitsudomi, T., Hamajima, N., Ogawa, M., and Takahashi, T. (2000) Prognostic significance of p53 alterations in patients with non-small cell lung cancer: a meta-analysis. *Clin. Cancer Res.* **6**, 4055–4063.
109. Huncharek, M., Kupelnick, B., Geschwind, J. F., and Caubet, J. F. (2000) Prognostic significance of p53 mutations in non-small cell lung cancer: a meta-analysis of 829 cases from eight published studies. *Cancer Lett.* **153**, 219–226.
110. Fujiwara, T., Cai, D. W., Georges, R. N., Mukhopadhyay, T., Grimm, E. A., and Roth, J. A. (1994) Therapeutic effect of a retroviral wild-type p53 expression vector in an orthotopic lung cancer model [see comments]. *J. Natl. Cancer Inst.* **86**, 1458–1462.
111. Minna, J. D. and Gazdar, A. F. (1996) Translational research comes of age [comment]. *Nat. Med.* **2**, 974–975.
112. Hovens, C. M. and Kaye, A. H. (2001) The tumour suppressor protein NF2/merlin: the puzzle continues. *J. Clin. Neurosci.* **8**, 4–7.
113. Sekido, Y., Pass, H. I., Bader, S., et al. (1995) Neurofibromatosis type 2 (NF2) gene is somatically mutated in mesothelioma but not in lung cancer. *Cancer Res.* **55**, 1227–1231.
114. Mulshine, J. L., Scott, F., Zhou, J., Avis, I., Vos, M., and Treston, A. M. (1997) Recent molecular advances in the approach to early lung cancer detection and intervention. *Environ. Health Perspect.* **105 Suppl 4**, 935–939.
115. Ronai, Z. and Yakubovskaya, M. (1995) PCR in clinical diagnosis. *J. Clin. Lab. Anal.* **9**, 269–283.
116. Ronai, Z., Yabubovskaya, M. S., Zhang, E., and Belitsky, G. A. (1996) K-ras mutation in sputum of patients with or without lung cancer. *J. Cell Biochem. Suppl.* **25**, 172–176.
117. Yakubovskaya, M. S., Spiegelman, V., Luo, F. C., et al. (1995) High frequency of K-ras mutations in normal appearing lung tissues and sputum of patients with lung cancer. *Int. J. Cancer* **63**, 810–814.
118. Mao, L., Lee, D. J., Tockman, M. S., Erozan, Y. S., Askin, F., and Sidransky, D. (1994) Microsatellite alterations as clonal markers for the detection of human cancer. *Proc. Natl. Acad. Sci. USA* **91**, 9871–9875.
119. Miozzo, M., Sozzi, G., Musso, K., et al. (1996) Microsatellite alterations in bronchial and sputum specimens of lung cancer patients. *Cancer Res.* **56**, 2285–2288.
120. Sidransky, D., Von Eschenbach, A., Tsai, Y. C., et al. (1991) Identification of p53 gene mutations in bladder cancers and urine samples. *Science* **252**, 706–709.
121. Yamashita, K., Kuba, T., Shinoda, H., Takahashi, E., and Okayasu, I. (1998) Detection of K-Ras point mutations in the supernatants of peritoneal and pleural effusions for diagnosis complementary to cytologic examination. *Am. J. Clin. Pathol.* **109**, 704–711.
122. Shibata, K., Mori, M., Kitano, S., and Akiyoshi, T. (1998) Detection of Ras gene mutations in peripheral blood of carcinoma patients using Cd45 immunomagnetic separation and nested mutant allele specific amplification. *Int. J. Oncol.* **12**, 1333–1338.
123. Sorenson, G. D., Pribish, D. M., Valone, F. H., Memoli, V. A., Bzik, D. J., and Yao, S. L. (1994) Soluble normal and mutated DNA sequences from single-copy genes in human blood. *Cancer Epidemiol. Biomarkers Prev.* **3**, 67–71.
124. Chen, X. Q., Stroun, M., Magnenat, J.-L., et al. (1996) Microsatellite alterations in plasma DNA of small cell lung cancer patients. *Nat. Med.* **2**, 1033–1035.
125. Wistuba, II, Lam, S., Behrens, C., et al. (1997) Molecular damage in the bronchial epithelium of current and former smokers *J. Natl. Cancer Inst.* **89**, 1366–1373.

7

TP53, hChk2, and the Li-Fraumeni Syndrome

Jenny Varley

1. Introduction

In 1969 Li and Fraumeni ***(1,2)*** reported a systematic epidemiologic study of the incidence of cancers in relatives of children with rhabdomyosarcoma (RMS). They noted five families of 648 probands in which a sib or cousin was also affected by sarcoma, and furthermore prospective studies of four of these families revealed a strikingly increased frequency of several types of cancer. Twenty-four families were subsequently studied in detail by the same groups ***(3)***, and a definition of what has now been termed Li-Fraumeni syndrome (LFS) was derived. Classic LFS is defined by a proband with a sarcoma aged under 45 years with a first degree relative with any cancer under 45 and an additional first- or second-degree relative in the same lineage with any cancer under 45, or a sarcoma at any age. This definition is still widely used, but other groups have introduced more relaxed definitions taking into account a more extensive knowledge of the range of tumors identified in families, and the ages of onset. One of the most widely used was defined by Birch et al. ***(4)*** as Li-Fraumeni-like (LFL) and is a proband with any childhood tumor, or sarcoma, brain tumor, or adrenocortical tumor under 45 years, plus a first- or second-degree relative with a typical LFS tumor at any age and another first- or second-degree relative with any cancer under the age of 60. In all the discussions that follow, the above definitions of LFS and LFL will be used. Typical LFS tumors in this context include sarcomas, brain and breast tumors, and childhood adrenocortical carcinoma.

In 1990 Malkin et al. ***(5)*** identified germline mutations to the tumor suppressor gene *TP53* as the causative defects in the Li-Fraumeni cancer predisposition syndrome. Initially it appeared that not only were germline *TP53* mutations responsible for cancer predisposition in most, if not all LFS families, but also that the mutations were clustered in a small region of exon 7 within the highly conserved region IV of the gene. However, as more extensive studies were reported, it was clear that this was not the general situation, and that mutations could occur throughout the gene ***(4,6–9)***. Furthermore, it is now clear that germline *TP53* mutations do not account for the cancer predisposition in all LFS or LFL families. The most comprehensive reports to date indicate that only approximately 80% of LFS and 40% of LFL families have detectable *TP53* mutations

From: *Methods in Molecular Biology, Vol. 222: Tumor Suppressor Genes: Pathways and Isolation Strategies*
Edited by: Wafik S. El-Deiry © Humana Press Inc., Totowa, NJ

(8,9). There are currently over 200 reports of germline *TP53* mutations in the literature, although fewer than 50% of the mutations that have been reported occur in LFS or LFL families. The remainder of the mutations have been described in a variety of cohorts of patients or families, including patients with sarcomas ***(10–12)***, brain tumors ***(13–15)***, or breast cancer ***(16–18)***, and children with adrenocortical carcinoma ***(19–21)***. Individuals with multiple primary tumors, particularly those typical of LFS, have also been described with germline *TP53* mutations ***(22–30)***.

2. The Spectrum of *TP53* Mutations in the Germline

As a generalization, the position of mutations in *TP53* detected in the germline reflects that found in sporadic tumors. Mutations in codons 248, 273, 245, 175, and 282 are the most frequent in the germline as they are somatically in tumors (see http://www.iarc.fr/P53/index.html). However, some *TP53* mutations that are frequent in the germline are far less common in tumors. The most striking of these are in codons 235 (2.9% of germline and 0.26% of somatic), 133 (1.4% vs 0.2%), and 181 (1.4% vs 0.43%). The reasons for this are currently unclear, although there appear to be no indication of founder germline mutations, as the families originate from distinct populations and indeed some of the mutations are different from one another. Mutations at two other sites, codons 152 and 337, are extremely common in the germline, but this is biased by the finding that these mutations are associated with the development of adrenocortical carcinomas in children unselected for a positive family history ***(19,21)***. This will be discussed at greater length below. However, in summary, contrary to initial hopes, there are no hot spots for mutation in the germline, indicating all types of mutation can be tolerated and complicating the search for germline mutations in LF families.

Germline *TP53* mutations can be of several kinds, including missense, nonsense, splicing, and deletion or insertion mutations. The majority of mutations within the core DNA-binding domain encoded by exons 4–8 are missense mutations, and this is almost exclusively the case for mutations within conserved domains II–V ***(31)***. An exception is that over half of the mutations in exon 6 in the germline are either nonsense mutations or are deletion/insertion events. It is not clear why this should be the case. While many studies have analyzed only the central part of the gene (exons 5–8), encoding the DNA-binding domain, a few have systematically examined the entire gene. The most thorough studies have originated from our own group, where the "gold standard" of direct sequencing of all exons (coding and noncoding), including the splice sites, promoter, and 3′ untranslated region, has been undertaken ***(9,32–34)***. These studies indicate that not all LFS and LFL families contain a germline *TP53* mutation, but also that it is essential to analyze the entire gene when looking for constitutional mutations. Of 45 mutations found by our group to date, 13 (29%) fall outside the regions of exons 5–8 commonly examined by other groups. This figure is consistent with that deduced by other groups carrying out systematic studies of somatic mutations in series of specific tumors, where approximately a quarter of somatic mutations fall outside the central part of the gene ***(35,36)***. We have found seven splicing mutations in a total of 45 patients or families with germline *TP53* defects, and have analyzed the consequences at the RNA level ***(34)***. Until recently it appeared that splicing mutations were relatively rare, but our own study together with that of Verselis et al. ***(37)*** now suggest that they are more common (around

8% of all reported germline *TP53* mutations) than previously thought. The apparent underreporting of germline splicing mutations may again reflect the bias introduced by analysis by some groups of only exons 5–8.

As yet there are no reports in the literature of large-scale germline deletions removing the entire *TP53* gene, of restricted expression from just one allele in the absence of a detectable germline mutation (indicating, for example, a promoter mutation), or of a proven promoter mutation. We have examined our cohort of *TP53* mutation-negative LFS and LFL families for such defects and have found none with the exception of the following. We have detected an identical single nucleotide deletion in the promoter in probands from two separate families, but we have been unable to show that expression is affected. Although the sequence variant is significantly more frequent in Li-Fraumeni families in which no other germline *TP53* mutation has been detected than in a control population ($p = 0.006$, ***38***), in the absence of any expression differences we must conclude that it is a rare polymorphism. It should be noted, however, that in-vivo expression studies were carried out in lymphocytes, a cell type not normally associated with malignant transformation in germline *TP53* mutation carriers.

3. Genotype–Phenotype Correlations

Careful and critical ascertainment of Li-Fraumeni families, and stringent verification of tumor diagnoses, ages of onset of disease, and pedigree information on both affected and unaffected family members allows a detailed study of the consequences of different types of *TP53* mutations. We have carried out such a study on a cohort of 34 LFS and LFL families in which we have both extensive pedigree data and defined germline mutation status. It is not possible to incorporate into this analysis information from other published data, because the information provided in most reports is insufficiently precise or unverified. Because of the relatively small numbers of families available for this study, plus the wide spectrum of types and positions of *TP53* mutations, we were only able to compare the consequences of three classes of mutation. We classified the mutations as follows: germline missense mutations within the core DNA-binding domain (type A); germline mutations leading to nonfunctional p53 protein products or null mutations (type B); and those families in which no *TP53* germline mutation could be found. These results have been reported previously ***(33)*** and are consistent with an enhanced oncogenic potential for mutations within the DNA-binding domain. In families with type A mutations cancers occur more frequently and at younger ages than in families with type B mutations. In particular, the incidence of breast and CNS tumors is higher in carriers of type A mutations. These observations are compatible with a plethora of functional studies of mutant p53 proteins in which certain mutations within the central DNA-binding domain show either gain of function or dominant-negative properties. Further support for this comes from analysis of the patterns of allele loss in tumors in carriers of germline *TP53* mutations. Loss of the wild-type allele is more frequent in carriers of type B mutations, which are loss of function or null mutants. The wild-type allele is commonly retained in tumors from type A mutation carriers, where the mutation is dominant and hence there is no strong selective pressure for loss of the wild-type allele (unpublished data and **refs.** ***33,34,*** and ***39***).

Clearly, larger systematic studies need to be undertaken to refine these types of analysis, and for these it will be essential to have precise clinical and epidemiologic data on the families, as well as comprehensive mutation screening. Such studies will be valuable to clinicians managing and counseling Li-Fraumeni families.

4. The Consequences of a Germline *TP53* Mutation

As p53 plays a fundamental role in maintaining genomic integrity by acting as "the guardian of the genome" ***(40)*** and is ubiquitously expressed, it might be anticipated that germline mutations would result in an increased predisposition to all malignancies. A number of studies have attempted to determine the frequencies of different tumor types in carriers of *TP53* mutations ***(41–43)***. However, taking information from literature reports as these studies have cannot produce an accurate analysis. Many literature studies contain verified pedigree, clinical, and genetic information, but a significant proportion do not, and so there is an inherent level of inaccuracy in carrying out any analysis incorporating such studies. All three reports analyzing published data ***(41–43)*** have also included proband cancers, which introduces a huge bias into the analysis. Finally, no account has been taken of variations of cancer incidences both ethnically and with time between different studies.

We have analyzed a cohort of 28 families with germline *TP53* mutations ascertained according to the strict criteria for LFS and LFL. Full details of this study are given by Birch et al. ***(32)*** but in summary, within the cohort all the cancers have been verified by examining medical records, death certificates, cancer registries, and/or by histopathologic examination. Detailed information was available on all family members (affected and unaffected). Unique to this study was that the incidence of cancers in the 28 families could be compared with national cancer statistics. This analysis could therefore directly compare the incidence of cancers in the families with national data for the same time period within which those cancers were diagnosed. In this way an accurate picture of the component tumors associated with a germline *TP53* mutation could be determined. Proband and defined support cancers were excluded from the analysis because by definition they are of specific types, and would therefore introduce a bias.

The distribution of cancers within the cohort of 28 families is highly significantly different from the expected cancer distribution for the population. Tumors strongly associated with a germline *TP53* mutation are breast and adrenocortical carcinoma, brain and spinal cord tumors, bone and soft tissue sarcomas, all of which have been highlighted in many previous studies. Leukemia, which was initially identified by Li et al. ***(3)***, was not observed as a major component tumor in our cohort, but Wilms' and malignant phyllodes tumors were ***(32)***. There was a very clear trend for an increasing risk with a decreasing age at onset. Apart from breast cancer, the only other common adult epithelial cancer to show any increased association with a germline *TP53* mutation was pancreas, and there was no association with common cancers such as lung, bowel, cervix, bladder, ovary, or prostate ***(32)***. This study provides the first formal demonstration that the inheritance of a mutant *TP53* allele does not lead to an increased risk of cancer in general, but does have to a tissue/cell-type specific effect. The basis for this is still not understood.

5. The Frequency of Germline *TP53* Mutations in Component Tumors

A number of studies have attempted to determine the frequency of germline *TP53* mutations in cohorts of patients with tumors typical of those seen in LFS/LFL families (for review, *see* **ref.** ***8***). In general, the greater the level of selection, the higher is the frequency of germline *TP53* mutations. So, for example, if patients with sarcomas are selected for study, higher frequencies of mutations are found in younger patients, those with a positive family history of cancer (particularly those cancers typical of LFS), and those with either multifocal disease or particular subtypes of sarcoma. Thus only 3% of children with osteosarcoma were found to have a germline *TP53* mutation ***(12)***, but the figure is much higher in patients with a family history or multiple primary tumors ***(11)***.

The tumor type arising with the most strikingly high incidence of germline *TP53* mutations is childhood adrenocortical carcinoma (ACC). An increased incidence of childhood ACC was noted in the initial studies from Li and Fraumeni ***(3,44)***, and two small studies analyzed families or children with ACC for germline *TP53* mutations ***(20,45)***. In our cohort of LFS/LFL families, all those in which there is a case of ACC are associated with a germline *TP53* mutation. We therefore investigated the frequency of such mutations in a series of children with ACC who were unselected for a family history of cancer. In our initial studies we identified 14 patients with ACC, and were able to identify germline mutations in over 80% ***(19)***. In total we have now analyzed 24 families in which there is a case of ACC, or individuals with ACC unselected for a family history, and 20 of these have germline *TP53* mutations (83%).

The spectrum of mutations seen in both families in which there is a case of ACC and individuals with ACC unselected for a family history is interesting. Taking all the literature reports into account ***(20,21,30,46–60)*** together with data from our own studies (unpublished data and **refs.** ***4,9,19,*** and ***34***), only approx 10% of the *TP53* germline mutations in individuals of families with a case of ACC affect one of the four most commonly mutated codons (175, 245, 248, and 273): this contrasts to a figure closer to 30% in all reports of germline mutations.

In our study of children with ACC unselected for family history, the majority of mutations occurred at codon 152 or 158 ***(19)***. Although the patients were derived from a population-based series there was no evidence for a founder effect. Of particular interest from this study was the observation that a number of unaffected relatives were confirmed or obligate carriers of the germline mutations. Previous estimates for lifetime penetrance of germline *TP53* mutations have been derived from studies of LFS/LFL families, which by definition carry high-penetrance alleles ***(61)***. The data from our study indicate that there may be low-penetrance *TP53* alleles. This hypothesis is supported by a recent study by Ribeiro et al. ***(21)***, who found an identical codon 337 mutation (Arg-His) in 35/36 patients with childhood ACC from southern Brazil. These authors excluded a founder effect by haplotype analysis, and propose that this mutation represents a low-penetrance *TP53* allele predisposing to childhood ACC. Although the mutation is different to that seen in our cohort of patients, it has been previously reported in the germline ***(57)***, interestingly in a child with ACC and an unaffected father. The differences in the spectra of mutations found in the two studies from our group ***(19)*** and that of Ribeiro ***(21)*** may reflect exposure of the cohort of children studied by the latter group to a carcinogenic insult. It has been recognized that there is an elevated incidence of childhood ACC in some areas of southern Brazil ***(62,63)***, and it is from this population that Ribeiro et al. took their study group. It is not clear why there is an increased incidence of ACC among these children, but it has been postulated that it could be due to exposure to environmental pollutants such as pesticides ***(63)***.

There have been a number of studies determining the incidence of germline *TP53* mutations in either patients with sporadic breast cancer, or in breast cancer families in which there is no germline *BRCA1* or *BRCA2* mutation ***(16–18,64–67)***. Germline *TP53* mutations have been detected only infrequently. Two studies have analyzed consecutive series of unselected breast cancer patients, and have detected mutations in only 2 patients out of a total of 303 screened ***(16,17)***. Sidransky et al. ***(18)*** examined 126 consecutive breast cancer patients aged under 40 years, and detected one germline mutation in codon 181. It is of interest that this is the same codon, although a different mutation, as reported in the study by Borresen et al. ***(16)***. The latter study additionally screened 40 female patients with breast cancer aged less than 35 years, and identified one patient with a germline mutation. These studies together show that germline *TP53* mutations are not responsible for a significant proportion of breast cancer cases, but there is an indication that with a younger age of onset this frequency may increase. Several studies have analyzed breast cancer families for germline *TP53* mutations, and again the frequency is low. However, one clear feature to emerge from these studies is that in families in which there is a clustering of cases of breast cancer in the absence of other tumor types, germline *TP53* mutations are seldom found, and in fact the significance of some of the mutations is questionable (for example, **ref.** ***67***). In contrast, however, in breast cancer families in which there are also cases of tumors typical of LFS (for example, sarcomas and/or brain tumors), the frequency of germline *TP53* mutations increases ***(65,66)***.

We have analyzed a cohort of 99 women with breast cancer aged less than 30 years for germline *TP53* mutations. In addition, the same patients have been screened for *BRCA1* and *BRCA2* mutations. The patients fall into groups who have a family history of breast and/or ovarian cancer (n = 31), a family history conforming to LFS/LFL (n = 5), and women with no family history (n = 63). Two of five women in LFL/LFL families but none of those with a family history of only breast or ovarian cancer had germline *TP53* mutations. Within the latter group, 14 women had *BRCA1* or *BRCA2* mutations. Only one of 63 women with no family history had a mutation in either of the latter two genes, but two women had germline *TP53* mutations ***(68)***. Therefore, in this cohort 4% of women had a germline *TP53* mutation, and although the numbers are low, such mutations were more common in patients with sporadic breast cancers than were mutations in either *BRCA1* or *BRCA2*.

6. LF Families with No Detectable Germline *TP53* Mutation

As described briefly under **Subheading 1**, not all LFS or LFL families carry germline *TP53* mutations. It is always difficult to prove a negative result in these circumstances, as a failure to find a mutation could be the result of lack of sensitivity of the mutation detection methods used even when the entire gene is analyzed. With respect to the *TP53* gene, this could also be in part due to the failure of some groups to screen the entire gene, concentrating instead on the central exons encoding the DNA-binding domains of the protein. We have analyzed the *TP53* gene in our cohort of LFS/LFL families by direct sequencing of all exons (coding and noncoding), plus all splice junctions, the promoter and the 3′ untranslated region. In addition we have confirmed the expression of both alleles wherever possible (ruling out any mutations that would affect expression), and determined whether there are cryptic intronic splice mutations by examination of the transcripts. We are therefore confident that we have correctly assigned these families as negative for a germline *TP53* mutation.

There could be a number of explanations for the cancer incidence observed in the families that are negative for a germline *TP53* mutation, including a chance aggregation of early-onset cancers. Certainly this could be the case for some LFL families, but is unlikely for the highly penetrant LFS families. In LFL families it is possible that germline mutations in one of the breast cancer susceptibility genes could be responsible. However, there is still a significant proportion of families in which we do not understand the mechanism(s) of cancer predisposition. A number of studies have analyzed genes that could be candidates based on the involvement of their gene products in p53 pathways, but until recently these have all proved negative ***(69–72)***.

7. *hChk2* and Li-Fraumeni Syndrome

In 1999 Bell et al. ***(73)*** examined a number of LFS, LFL, and variant LF families (all negative for germline *TP53* mutations) for mutations in two genes encoding kinases with key functions in cell cycle checkpoints, *hChk1* and *hChk2*. No germline mutations were detected in *hChk1*, but two were identified in *hChk2*. One was in a classic LFS family in which the mutation (1100delC, leading to premature termination at codon 381) segregated with the disease, and in which loss of the wild-type allele was subsequently demonstrated ***(74)***. The second was in a patient with multiple primary tumors, and was a missense mutation within the forkhead (FHA) domain of Chk2 (Ile157Thr). A third reported germline mutation ***(73)*** was subsequently shown to be an artifact derived from an additional genomic copy of the 3' part of *hChk2* ***(75)***.

Subsequent studies have extended the numbers of families with germline mutations, including seven breast cancer families (four of which also are LFL), with identical mutations at codon 157 (Ile157Thr, *see* **ref.** ***76***), and an LFS family with the same mutation ***(74)***. The consequences of this mutation are still somewhat uncertain. The Finnish group ***(76)*** have identified the Ile157Thr variant in a similar proportion of control samples to familial cases, whereas this variant has not been detected in any control samples from a control series in the Boston area ***(73,74)***. A limited number of functional studies have been carried out on this mutant protein, and in some assays it behaves as wild-type Chk2 ***(74,77)***. However, recently it has been demonstrated to be defective in binding p53 ***(78)*** and shows a reduced ability to bind to and phosphorylate Cdc25A ***(79)***. These studies provide support that the Ile157Thr variant is a pathogenic mutation, but further studies will be necessary to determine whether the results from the Finnish population do indeed reflect a high proportion of individuals with a compromised Chk2 function.

In addition to the family initially reported with a deletion at position 1100delC ***(73)***, two other families have now been reported with the same mutation ***(80)***. Neither of the latter families conforms to the definition of LFS/LFL, but both show an unusual clustering of adult tumors suggestive of LFS. The Chk2 protein encoded by the mutant gene is truncated at position 381 and is kinase dead, showing neither baseline nor γ-irradiated activity ***(74,77)***.

One other mutant was originally identified in a colorectal cell line (HCT15, *see* **ref.** *73*), but has subsequently also been described in a variant LF family ***(74)***. A number of groups have studied the mutant protein (Arg145Trp) for functional alterations. The mutant Chk2 protein shows reduced kinase activity using Cdc25C as a substrate ***(74,77)***, reduced ability to autophosphorylate at Thr383 and Thr387 ***(77,81)***, and reduced levels of phosphorylation at Thr68 by the ATM protein ***(77)***. It is interesting to note that it does not function as a dominant-negative mutant in these assays, but rather loses function. This observation ties in with the finding that there is loss of the wild-type allele in a tumor from a patient carrying the defective germline *hChk2* allele ***(74)***. The Arg145Trp mutant protein appears to be unstable, and has a reduced half-life. There is evidence that the mutant protein is targeted for degradation through the proteosome pathway ***(74)***.

One further germline *hChk2* variant has been identified (Arg3Trp, *see* **ref.** *74*), but no functional alterations to the protein have been described. There was no loss of the wild-type *hChk2* allele in a tumor from a patient carrying this germline variant, but there was a homozygous somatic *TP53* mutation. The same report describes two independent *TP53* somatic mutations in a tumor from a patient with a germline Ile157Thr *hChk2* mutation ***(74)***, also with no LOH at *hChk2*. As p53 and Chk2 are two components of a linear pathway activated by DNA damage, these observations could be interpreted as an indication that the germline *hChk2* variant is of uncertain consequence. Simplistically, as p53 is a substrate for DNA damage- and ATM-dependent phosphorylation by Chk2 (for recent reviews, *see* **refs**. ***82*** and ***83***), mutations in the two components of the pathway could be functionally redundant. Recently, a direct physical interaction has been identified between Chk2 and p53, with the interaction requiring an intact FHA domain of the former, and tetramerization of the latter ***(78)***. Ionizing radiation (IR)-induced DNA damage activates a pathway in which the ATM protein phosphorylates Chk2 on Thr68 ***(84)***, and this protein in turn phosphorylates p53 at a number of sites including Ser20 ***(85)***. Phosphorylation of p53 at Ser20 prevents the binding of HDM2 to p53, thus stabilizing p53 and ultimately activating a G_1 cell cycle checkpoint. In addition, phosphorylation and activation of Chk2 also results in Chk2 phosphorylating Cdc25C on Ser216, suggesting a role in the G_2/M checkpoint. However, it is not clear whether Chk2 is required in mammalian cells for the regulation of G_2/M transition (for review, *see* **ref.** ***82***). Recently, Chk2 has been implicated in IR-induced S-phase arrest ***(79)***. This checkpoint also involves ATM, but the downstream target of the activated Chk2 kinase in S-phase arrest is Cdc25A rather than Cdc25C. A defective S-phase checkpoint can result in radioresistant DNA synthesis, which has been shown to be independent of p53 status ***(86)***. Clearly Chk2 is a key component in the network of checkpoints at different stages of the cell cycle, including G_1, S, and G_2/M, through the phosphorylation of key targets such as Cdc25C, Cdc25A, p53. The identification of the involvement of Chk2 in a p53-independent cell cycle checkpoint provides evidence that inactivation of the two proteins does not have identical consequences. This may support the classification of some germline *hChk2* sequence variants as genuine pathogenic mutations, even though tumors from patients carrying these mutations show somatic inactivation of p53.

The colorectal cell line HCT15 has both copies of *hChk2* inactivated by mutation ***(74)***, and Falck et al. ***(78)*** have recently confirmed that it also has functional loss of p53. These authors show that concomitant loss of both these proteins does confer a selective advantage, adding credence to the suggestion that the germline missense *hChk2* variants are pathogenic, as discussed above. However, HCT15 has additional mutations inactivating both copies of the *hMsh6* gene ***(87)***. HCT15 cells have been reported to have a defective G_2 checkpoint when treated with methylating agents, but this defect is restored in cells transfected with a construct expressing wild-type Msh6 ***(88)***. Intriguingly, a family has been reported in which there is both a germline *hChk2* and an *hMsh6* mutation ***(80)***, although no functional analyses have been carried out in cells with the two mutations.

There is an increasing body of evidence for both functional and direct interactions between components of the DNA damage recognition and repair machinery, and cell cycle checkpoint controls ***(83,89,90)***. Studies of germline mutations in many of these genes may provide valuable support for these interactions, and confirmatory evidence that many of these genes are tumor suppressors.

8. Summary

Germline *TP53* mutations are responsible for the large majority of classic LFS families, and a smaller proportion of LFL families. In some of the families shown to have no germline *TP53* mutation, germline *hChk2* mutations have been described. In some cases the functional consequences of the latter have been demonstrated, although there are still relatively few reports of such mutations. Due to the paucity of families currently described with *hChk2* mutations, it is not possible to reach any conclusions concerning the phenotypic/clinical differences between the two types of germline mutation. At least one family with a germline *hChk2* mutation is a classic LFS family ***(73,74)***, whereas others are LFL ***(76)***, variant-LFS, or phenotypically suggestive of LFS ***(73,74,80)***. However, there is still a significant number of LFS/LFL families for which no underlying genetic determinant has been identified. It will be fascinating to see what genetic defects are responsible, and whether they involve additional components of DNA damage recognition, repair, or cell cycle checkpoint pathways.

References

1. Li, F. P. and Fraumeni, J. F. (1969) Rhabdomyosarcoma in children; epidemiologic study and identification of a cancer family syndrome. *J. Natl. Cancer. Inst.* **43**, 1365–1373.
2. Li, F. P. and Fraumeni, J. F. (1969) Soft-tissue sarcomas, breast cancer and other neoplasms: a familial syndrome? *Ann. Intern. Med.* **71**, 747–752.
3. Li, F. P., Fraumeni, J. F., Mulvihill, J. J., et al. (1988) A cancer family syndrome in twenty-four kindreds. *Cancer Res.* **48**, 5358–5362.
4. Birch, J. M., Hartley, A. L., Tricker, K. J., et al. (1994) Prevalence and diversity of constitutional mutations in the *p53* gene among 21 Li-Fraumeni families. *Cancer Res.* **54**, 1298–1304.
5. Malkin, D., Li, F. P., Strong, L. C., et al. (1990) Germ line p53 mutations in a familial syndrome of breast cancer, sarcomas, and other neoplasms. *Science* **250**, 1233–1238.
6. Santibanez-Koref, M. F., Birch, J. M., Hartley, A. L., et al. (1991) p53 germline mutations in Li-Fraumeni syndrome. *Lancet* **338**, 1490–1491.
7. Frebourg, T., Barbier, N., Yan, Y., et al. (1995) Germ-line p53 mutations in 15 families with Li-Fraumeni syndrome. *Am. J. Hum. Genet.* **56**, 608–615.
8. Varley, J. M., Evans, D. G. R., and Birch, J. M. (1997) Li-Fraumeni syndrome—a molecular and clinical review. *Br. J. Cancer* **76**, 1–14.
9. Varley, J. M., McGown, G., Thorncroft, M., et al. (1997) Germ-line mutations of *TP53* in Li-Fraumeni families: an extended study of 39 families. *Cancer Res.* **57**, 3245–3252.

10. Diller, L., Sexsmith, E., Gottlieb, A., Li, F. P., and Malkin, D. (1995) Germline p53 mutations are frequently detected in young children with rhabdomyosarcoma. *J. Clin. Invest.* **95**, 1606–1611.
11. Toguchida, J., Yamaguchi, T., Dayton, S. H., et al. (1992) Prevalence and spectrum of germline mutations of the p53 gene among patients with sarcoma. *N. Engl. J. Med.* **326**, 1301–1308.
12. McIntyre, J. F., Smith-Sorensen, B., Friend, S. H., et al. (1994) Germline mutations of the p53 tumor suppressor gene in children with osteosarcoma. *J. Clin. Oncol.* **12**, 925–930.
13. Li, Y.-J., Sanson, M., Hoang-Xuan, K., et al. (1995) Incidence of germ-line *p53* mutations in patients with gliomas. *Int. J. Cancer* **64**, 383–387.
14. Kyritsis, A. P., Bondy, M. L., Xiao, M., et al. (1994) Germline p53 mutations in subsets of glioma patients. *J. Natl. Cancer Inst.* **86**, 344–349.
15. Chen, P., Iavarone, A., Fick, J., Edwards, M., Prados, M., and Israel, M. A. (1995) Constitutional p53 mutations associated with brain tumors in young adults. *Cancer Genet. Cytogenet.* **82**, 106–115.
16. Børresen, A.-L., Andersen, T. I., Garber, J., et al. (1992) Screening for germ line *TP53* mutations in breast cancer patients. *Cancer Res.* **52**, 3234–3236.
17. Prosser, J., Porter, D., Coles, C., et al. (1992) Constitutional p53 mutation in a non-Li-Fraumeni cancer family. *Br. J. Cancer* **65**, 527–528.
18. Sidransky, D., Tokino, T., Helzlsouer, K., et al. (1992) Inherited *p53* gene mutations in breast cancer. *Cancer Res.* **52**, 2984–2986.
19. Varley, J. M., McGown, G., Thorncroft, M., et al. (1999) Are there low penetrance *TP53* alleles? Evidence from childhood adrenocortical tumors. *Am. J. Hum. Genet.* **65**, 995–1006.
20. Wagner, J., Portwine, C., Rabin, K., Leclerc, J.-M., Narod, S. A., and Malkin, D. (1994) High frequency of germline p53 mutations in childhood adrenocortical cancer. *J. Natl. Cancer Inst.* **86**, 1707–1710.
21. Ribeiro, R. C., Sandrini, F., Figueiredo, B., et al. (2001) An inherited *p53* mutation that contributes in a tissue-specific manner to pediatric adrenal cortical carcinoma. *Proc. Natl. Acad. Sci. USA* **98**, 9330–9335.
22. Malkin, D., Jolly, K. W., Barbier, N., et al. (1992) Germline mutations of the p53 tumor-suppressor gene in children and young adults with second malignant neoplasms. *N. Engl. J. Med.* **326**, 1309–1315.
23. Mazoyer, S., Lalle, P., Moyret-Lalle, C., et al. (1994) Two germ-line mutations affecting the same nucleotide at codon 257 of p53 gene, a rare site for mutations. *Oncogene* **9**, 1237–1239.
24. Shiseki, M., Nishikawa, R., Yamamoto, H., et al. (1993) Germ-line *p53* mutation is uncommon in patients with triple primary cancers. *Cancer Lett.* **73**, 51–57.
25. Iavarone, A., Matthay, K. K., Steinkirchner, T. M., and Israel, M.A. (1992) Germ-line and somatic p53 mutations in multifocal osteogenic sarcoma. *Proc. Natl. Acad. Sci. USA* **89**, 4207–4209.
26. Quesnel, S., Verselis, S., Portwine, C., et al. (1999) p53 compound heterozygosity in a severely affected child with Li-Fraumeni Syndrome. *Oncogene* **18**, 3970–3978.
27. Gallo, O., Sardi, I., Pepe, G., et al. (1999) Multiple primary tumors of the upper aerodigestive tract: is there a role for constitutional mutations in the *p53* gene. *Int. J. Cancer* **82**, 180–186.
28. Bot, F. J., Sleddens, H. F., and Dinjens, W. N. (1998) Molecular assessment of clonality leads to the identification of a new germ line TP53 mutation associated with malignant cystosarcoma phylloides and soft tissue sarcoma. *Diagn. Mol. Pathol.* **7**, 295–301.
29. Speiser, P., Gharehbaghi-Schnell, E., Eder, S., et al. (1996) A constitutional *de novo* mutation in exon 8 of the p53 gene in a patient with multiple primary malignancies. *Br. J. Cancer* **74**, 269–273.

30. Gutierrez, M. I., Bhatia, K. G., Barreiro, C., et al. (1994) A *de novo* p53 germline mutation affecting codon 151 in a six year old child with multiple tumors. *Hum. Mol. Genet.* **3**, 2247–2248.
31. Soussi, T., Caron de Fromentel, C., and May, P. (1990) Structural aspects of the p53 protein in relation to gene evolution. *Oncogene* **5**, 945–952.
32. Birch, J. M., Alston, R. D., McNally, R. J. Q., et al. (2001) Relative frequency and morphology of cancers in carriers of germline TP53 mutations. *Oncogene* **20**, 4621–4628.
33. Birch, J. M., Blair, V., Kelsey, A. M., et al. (1998) Cancer phenotype correlates with constitutional *TP53* genotype in families with the Li-Fraumeni Syndrome. *Oncogene* **17**, 1061–1068.
34. Varley, J. M., Attwooll, C., White, G., et al. (2001) Characterisation of germline *TP53* splicing mutations and their genetic and functional analysis. *Oncogene* **20**, 2647–2654.
35. Casey, G., Lopez, M. E., Ramos, J. C., et al. (1996) DNA sequence analysis of exons 2 though 11 and immunohistochemical staining are required to detect all known p53 alterations in human malignancies. *Oncogene* **13**, 1971–1981.
36. Bergh, J., Norberg, T., Sjogren, S., Lindgren, A., and Holmberg, L. (1995) Complete sequencing of the *p53* gene provides prognostic information in breast cancer patients, particularly in relation to adjuvant systemic therapy and radiotherapy. *Nature Med.* **1**, 1029–1034.
37. Verselis, S. J., Rheinwald, J. G., Fraumeni, J. F., and Li, F. P. (2000) Novel p53 splice site mutations in three families with Li-Fraumeni syndrome. *Oncogene* **19**, 4230–4235.
38. Attwooll, C. L., McGown, G., Thorncroft, M., Stewart, F. J., Birch, J. M., and Varley, J. M. (2002) Identification of a rare polymorphism in the human *TP53* promoter. *Cancer Genet. Cytogenet.* **135**, 165–172.
39. Varley, J. M., Thorncroft, M., McGown, G., et al. (1997) A detailed study of loss of heterozygosity on chromosome 17 in tumours from Li-Fraumeni patients carrying a mutation to the *TP53* gene. *Oncogene* **14**, 865–871.
40. Lane, D. P. (1992) p53, guardian of the genome. *Nature* **358**, 15–16.
41. Nichols, K. E., Malkin, D., Garber, J. E., Fraumeni, J. F., and Li, F. P. (2001) Germ-line *p53* mutations predispose to a wide spectrum of early-onset cancers. *Cancer Epidemiol. Biomarkers Prev.* **10**, 83–91.
42. Kleihues, P., Schäuble, B., zur Hausen, A., Estève, J., and Ohgaki, H. (1997) Tumors associated with *p53* germline mutations. A synopsis of 91 families. *Am. J. Pathol.* **150**, 1–13.
43. Wang, Q., Lasset, C., Sobol, H., and Ozturk, M. (1996) Evidence of a hereditary *p53* syndrome. *Int. J. Cancer* **65**, 554–557.
44. Garber, J. E., Goldstein, A. M., Kantor, A. F., Dreyfus, M. G., Fraumeni, J. F., and Li, F. P. (1991) Follow-up study of twenty-four families with Li-Fraumeni syndrome. *Cancer Res.* **51**, 6094–6097.
45. Sameshima, Y., Tsunematsu, Y., Watanabe, S., et al. (1992) Detection of novel germ-line p53 mutations in diverse-cancer-prone families identified by screening patients with childhood adrenocortical carcinoma. *J. Natl. Cancer Inst.* **84**, 703–707.
46. Stolzenberg, M.-C., Brugières, L., Gardes, M., et al. (1994) Germ-line exclusion of a single p53 allele by premature termination of translation in a Li-Fraumeni family. *Oncogene* **9**, 2799–2804.
47. Felix, C. A., Strauss, E. A., D'Amico, D., et al. (1993) A novel germline p53 splicing mutation in a pediatric patient with a second malignant neoplasm. *Oncogene* **8**, 1203–1210.
48. Warneford, S. G., Witton, L. J., Townsend, M. L., et al. (1992) Germ-line splicing mutation of the *p53* gene in a cancer-prone family. *Cell Growth Differ.* **3**, 839–846.
49. Grayson, G. H., Moore, S., Schneider, B. G., Saldivar, V., and Hensel, C. H. (1994) Novel germline mutation of the p53 tumor suppressor gene in a child with incidentally discovered adrenal cortical carcinoma. *Am. J. Pediatr. Hematol. Oncol.* **16**, 341–347.

50. Lubbe, J., von Ammon, K., Watanabe, K., Hegi, M. E., and Kleihues, P. (1995) Familial brain tumour syndrome associated with a *p53* germline deletion at codon 236. *Brain Pathol.* **5**, 15–23.
51. Strauss, E. A., Hosler, M. R., Herzog, P., Salhany, K., Louie, R., and Felix, C. A. (1995) Complex replication error causes *p53* mutation in a Li-Fraumeni family. *Cancer Res.* **55**, 3237–3241.
52. Bardeesy, N., Falkoff, D., Petruzzi, M.-J., et al. (1994) Anaplastic Wilms' tumour, a subtype displaying poor prognosis, harbours p53 gene mutations. *Nature Genet.* **7**, 91–97.
53. Giunta, C., Youil, R., Venter, D., et al. (1996) Rapid diagnosis of germline p53 mutation using the enzyme mismatch cleavage method. *Diagn. Mol. Pathol.* **5**, 265–270.
54. Moutou, C., Le Bihan, C., Chompret, A., et al. (1996) Genetic transmission of susceptibility to cancer in families of children with soft tissue sarcomas. *Cancer* **78**, 1483–1491.
55. Sedlacek, Z., Kodet, R., Kriz, V., et al. (1998) Two Li-Fraumeni syndrome families with novel germline p53 mutations: loss of the wild type p53 allele in only 50% of tumours. *Br. J. Cancer* **77**, 1034–1039.
56. Pivnick, E. K., Furman, W. L., Velagaleti, G. V. N., Jenkins, J. J., Chase, N. A., and Ribiero, R. C. (1998) Simultaneous adrenocortical carcinoma and ganglioneuroblastoma in a child with Turner syndrome and germline p53 mutation. *J. Med. Genet.* **35**, 328–332.
57. Chompret, A., Brugieres, L., Ronsin, M., et al. (2000) p53 germline mutations in childhood cancers and cancer risk for carrier individuals. *Br. J. Cancer* **82**, 1932–1937.
58. Vital, A., Bringuier, P.-P., Huang, H., et al. (1998) Astrocytomas and choroid plexus tumors in two families with identical *p53* germline mutations. *J. Neuropathol. Exp. Neurol.* **57**, 1061–1069.
59. Yonemoto, T., Tatezaki, S.-I., Ishii, T., Satoh, T., and Inoue, M. (1999) Two cases of osteosarcoma occurring as second malignancy of childhood cancer. *Anticancer Res.* **19**, 5563–5566.
60. Bougeard, G., Limacher, J.-M., Martin, C., et al. (2001) Detection of 11 germline inactivating *TP53* mutations and absence of *TP63* and *hCHK2* mutations in 17 French families with Li-Fraumeni or Li-Fraumeni-like syndrome. *J. Med. Genet.* **38**, 253–256.
61. LeBihan, C., Moutou, C., Brugières, L., Feunteun, J., and Bonaïti-Pellié, C. (1995) ARCAD: a method for estimating age-dependent disease risk associated with mutation carrier status from family data. *Genet. Epidem.* **12**, 13–25.
62. Marigo, C., Muller, H., and Davies, J. N. P. (1968) Survey of cancer in children admitted to a Brazilian charity hospital. *J. Natl. Cancer Inst.* **43**, 1231–1240.
63. Sandrini, R., Ribeiro, R. C., and DeLacerda, L. (1997) Childhood adrenocortical tumors. *J. Clin. Endocrinol. Metab.* **82**, 2027–2031.
64. Prosser, J., Elder, P. A., Condie, A., MacFadyen, I., Steel, C. M., and Evans, H. J. (1991) Mutations in p53 do not account for heritable breast cancer: A study in five affected families. *Br. J. Cancer* **63**, 181–184.
65. Huusko, P., Castren, K., Launonen, V., et al. (1999) Germ-line TP53 mutations in Finnish cancer families exhibiting features of the Li-Fraumeni syndrome and negative for BRCA1 and BRCA2. *Cancer Genet. Cytogenet.* **112**, 9–14.
66. Rapakko, K., Allinen, M., Syrjakoski, K., et al. (2001) Germline *TP53* alterations in Finnish breast cancer families are rare and occur at conserved mutation-prone sites. *Br. J. Cancer* **84**, 116–119.
67. Sun, X.-F., Johannsson, O., Hakansson, S., et al. (1996) A novel *p53* germline alteration identified in a late onset breast cancer kindred. *Oncogene* **13**, 407–411.
68. Lalloo, F., Varley, J., Ellis, D., et al. (2002) Family history is predictive of pathogenic, high penetrance mutations in *BRCA1*, *BRCA2* and *TP53* in early-onset breast cancer. *Lancet* in press
69. Burt, E. C., McGown, G., Thorncroft, M., James, L. A., Birch, J. M., and Varley, J. M. (1999) Exclusion of the genes *CDKN2* and *PTEN* as causative gene defects in Li-Fraumeni Syndrome. *Br. J. Cancer* **80**, 9–10.

70. Brown, L. T., Sexsmith, E., and Malkin, D. (2000) Identification of a novel PTEN intronic deletion in Li-Fraumeni Syndrome and its effect on RNA processing. *Cancer Genet. Cytogenet.* **123**, 65–68.
71. Portwine, C., Lees, J., Verselis, S., Li, F. P., and Malkin, D. (2000) Absence of germline p16(INK4a) alterations in p53 wild type Li-Fraumeni syndrome families. *J. Med. Genet.* **37**, E11.
72. Stone, J. G., Eeles, R. A., Sodha, N., Murday, V., Sheriden, E., and Houlston, R. S. (1999) Analysis of Li-Fraumeni Syndrome and Li-Fraumeni-like families for germline mutations in *Bcl10*. *Cancer Lett.* **147**, 181–185.
73. Bell, D. W., Varley, J. M., Szydlo, T. E., et al. (1999) Heterozygous germ line *hCHK2* mutations in Li-Fraumeni Syndrome. *Science* **286**, 2528–2531.
74. Lee, S. B., Kim, S. H., Bell, D. W., et al. (2001) Destabilization of CHK2 by a missense mutation associated with Li-Fraumeni Syndrome. *Cancer Res.* **61**, 8062–8067.
75. Sodha, N., Williams, R., Mangion, J., et al. (2000) Screening *hCHK2* for mutations. *Science* **289**, 359a.
76. Allinen, M., Huusko, P., Mantyniemi, S., Launonen, V., and Winqvist, R. (2001) Mutation analysis of the *CHK2* gene in families with hereditary breast cancer. *Br. J. Cancer* **85**, 209–212.
77. Wu, X., Webster, S. R., and Chen, J. (2001) Characterization of tumor-associated Chk2 mutations. *J. Biol. Chem.* **276**, 2971–2974.
78. Falck, J., Lukas, C., Protopopova, M., Lukas, J., Selivanova, G., and Bartek, J. (2001) Functional impact of concomitant *versus* alternative defects in the Chk2-p53 tumour suppressor pathway. *Oncogene* **20**, 5503–5510.
79. Falck, J., Mailand, N., Syljuasen, R. G., Bartek, J., and Lukas, J. (2001) The ATM-Chk2-Cdc25A checkpoint pathway guards against radioresistant DNA synthesis. *Nature* **410**, 842–847.
80. Vahteristo, P., Tamminen, A., Karvinen, P., et al. (2001) *p53*, *CHK2*, and *CHK1* genes in Finnish families with Li-Fraumeni syndrome: further evidence of *CHK2* in inherited cancer predisposition. *Cancer Res.* **61**, 5718–5722.
81. Lee, C.-H. and Chung, J. H. (2001) The hCds1 (Chk2)-FHA domain is essential for a chain of phosphorylation events on hCds1 that is induced by ionizing radiation. *J. Biol. Chem.* **276**, 30537–30541.
82. Rhind, N. and Russell, P. (2000) Chk1 and Cds1: linchpins of the DNA damage and replication checkpoint pathways. *J. Cell Sci.* **113**, 3889–3896.
83. Zhou, B.-B. S. and Elledge, S. J. (2000) The DNA damage response: putting checkpoints into perspective. *Nature* **408**, 433–439.
84. Matsuoka, S., Rotman, G., Ogawa, A., Shiloh, Y., Tamai, K., and Elledge, S. J. (2000) Ataxia telangiectasia-mutated phosphorylates Chk2 *in vivo* and *in vitro*. *Proc. Natl. Acad. Sci. USA* **97**, 10389–10394.
85. Chehab, N. H., Malikzay, A., Appel, M., and Halazonetis, T. D. (2000) Chk2/hCds1 functions as a DNA damage checkpoint in G_1 by stabilizing p53. *Genes Dev.* **14**, 278–288.
86. Xie, G., Habbersett, R. C., Jia, Y., et al. (1998) Requirements for the p53 and ATM gene product in the regulation of G1/S and S phase checkpoints. *Oncogene* **16**, 721–736.
87. Papadopoulos, N., Nicolaides, N. C., Liu, B., et al. (1995) Mutations of *GTBP* in genetically unstable cells. *Science* **268**, 1915–1917.
88. Lettieri, T., Marra, G., Aquilina, G., et al. (1999) Effect of *hMSH6* cDNA expression on the phenotype of mismatch repair-deficient colon cancer cell line HCT15. *Carcinogenesis* **20**, 373–382.
89. Lee, J.-S., Collins, K. M., Brown, A. L., Lee, C.-H., and Chung, J. H. (2000) hCds1-mediated phosphorylation of BRCA1 regulates the DNA damage response. *Nature* **404**, 201–204.
90. Wang, Y., Cortez, D., Yazdi, P., Neff, N., Elledge, S. J., and Qin, J. (2000) BASC, a super complex of BRCA1-associated proteins involved in the recognition and repair of aberrant DNA structures. *Genes Dev.* **14**, 927–939.

8

Genetic Alterations in Esophageal Cancer

Jun-ichi Okano, Lorraine Snyder, and Anil K. Rustgi

1. Epidemiology and Etiologic Factors

In the United States, esophageal cancers are of the squamous cell type (ESCC) or adenocarcinoma (EADC). Other esophageal malignant neoplasms are rare. Smoking and alcohol, particularly consumption of hard liquor, are strongly associated with development of ESCC in the United States and Europe. Dietary factors such as vitamin deficiencies, trace-metal deficiencies, fungal contamination of food, and high nitrite and nitrosamine content are contributing factors in China and Africa. Other predisposing influences include gastrointestinal disorders such as achalasia, Plummer-Vinson syndrome, long-standing celiac disease, and tylosis. Males and African Americans are more likely to develop ESCC in the United States, whereas EADC is more common in males and Caucasians. The incidence of EADC has increased markedly in the past two decades, both in the United States and abroad. In many countries, the incidence of EADC is increasing at a rate that exceeds that of any other malignancy. It is well accepted that esophageal adenocarcinomas arise from Barrett's esophagus, a premalignant condition of the esophagus. The risk of EADC appears to be at least 30 times greater in those with Barrett's esophagus than in those without. Barrett's esophagus represents conversion of the normal squamous epithelium to a columnar intestinal metaplasia, in response (in part) to chronic injury from chronic gastroesophageal reflux. Similarly to other types of gastrointestinal cancers, an accumulation of genetic alterations follows the dysplasia–adenocarcinoma sequence in esophageal adenocarcinoma. Although certain genetic alterations appear more often or earlier than others, it does not appear that there is a uniform sequence of mutations leading from normal esophageal mucosa to invasive esophageal adenocarcinoma, although some mutations tend to occur early and others late.

Genetic changes associated with the development of esophageal cancers include mutations of the *p53* gene, disruption of cell-cycle control in G1 phase by inactivation of *p16*, amplification of *cyclin D1*, and activation of oncogenes such as *epidermal growth factor receptor (EGFR)* and *c-myc* ***(1,2)***, whereas *ras* activation is rare ***(3)***. Recognizing that ESCC and EADC are different entities, investigation of genetic alterations have often grouped these cancers, and therefore, discussion of particular genetic changes will distinguish between ESCC and EADC whenever possible.

From: *Methods in Molecular Biology, Vol. 222: Tumor Suppressor Genes: Pathways and Isolation Strategies*
Edited by: Wafik S. El-Deiry © Humana Press inc., Totowa, NJ

Historically, one of the earliest genetic changes identified in Barrett's esophagus and EADC is related to aneuploidy. Cells with abnormal DNA content appear early in the progression from normal esophagus to esophageal adenocarcinoma. The presence of aneuploidy may further be a prognostic factor for malignant change in dysplastic esophagus. Some studies have reported that aneuploidy is associated with lymph node involvement, metastasis, and decreased survival.

2. Oncogene Activation

2.1. EGFR and Related Growth Factor Receptors

The epidermal growth factor receptor (EGFR) family comprises EGFR, ErbB2, ErbB3, and ErbB4, all of which are tyrosine kinase receptors. When ligands, such as EGF and transforming growth factor (TGF)-α, bind to an extracellular domain of the EGFR, EGFR family members are homo- or heterodimerized in cytoplasmic domains, and tyrosine kinase activity is stimulated. The phosphorylated cytoplasmic domains then serve as docking sites for downstream signaling molecules ***(4, 5)*** such as the mitogen-activated protein kinase (MAPK) and phosphatidylinositol 3-kinase (PI3K) ***(6)***.

Overexpression of EGFR has been reported in ESCC tissues ***(7)*** and cell lines derived from ESCC patients ***(8)***, in which *EGFR* gene amplification is one of the mechanisms ***(9,10)*** of activation. Deregulation of TGF-α and EGFR overexpression have been implicated to contribute to autonomous cell growth in esophageal cancers ***(11,12)***. Although overexpression of ErbB2 has been noted in 6.7–26% of human esophageal cancers ***(13–15)***, it does not appear to be involved in esophageal carcinogenesis ***(16)***, which also applies for ErbB3 and ErbB4 ***(17)***.

Selective inhibitors for EGFR tyrosine kinase induced apoptosis in colon cancer cells ***(18)*** and EGFR was essential for skin tumor development ***(19)***, representing a proneoplastic role for EGFR in carcinogenesis. By contrast, activation of EGFR promoted late terminal differentiation of keratinocytes ***(20)***. In terms of esophageal cancers, an anti-EGFR antibody was cytotoxic to ESCC-derived cell lines ***(21)*** and overexpression of EGFR in ESCC patients was correlated with poor prognosis ***(22–26)***, supporting a role for EGFR in esophageal carcinogenesis. By contrast, there was no correlation of EGFR overexpression with patients' prognosis in another study ***(27,28)***. It is clear that in cells overexpressing EGFR there is activation of AKT (protein kinase B, PKB), an anti-apoptotic serine/threonine kinase, but the degree of EGFR overexpression appears to regulate different Akt isoforms in yet to be defined mechanisms ***(8)***.

Both esophageal adenocarcinoma and Barrett's esophagus have high expression levels of epidermal growth factor (EGF) and EGFR. Whereas overexpression of EGF does not correlate with the degree of dysplasia, overexpression of the EGFR does. Expression of the structurally related growth factor TGF-α also increases with degree of dysplasia progressing to esophageal adenocarcinoma. The c-erbB2 (HER2/neu) oncogene encodes a truncated form of EGFR that contains a continuously active tyrosine kinase. Cells expressing this oncogene behave as though they are constantly being stimulated by growth factors. Amplification of the c-erbB2 gene and/or overexpression of its product are found to occur in esophageal adenocarcinoma, but do not characterize earlier stages of dysplasia. The acidic form of fibroblast growth factor (aFGF) increases in expression from the beginning with Barrett's metaplasia through the adenocarcinoma stage.

Transforming growth factor-β (TGF-β) generally inhibits cell proliferation, but also induces cell differentiation. Whereas TGF-β is expressed in Barrett's metaplasia as well as esophageal adenocarcinoma, these lesions have also been found to have decreased expression of TGF-β type I and II receptors. Finally, Src expression and activity are increased in Barrett's metaplasia and increase further with successive degrees of dysplasia. Src is a tyrosine kinase that can associate noncovalently with different cell surface receptors to initiate signal transduction. Like erbB2, the Src oncogene product contains a constitutively active tyrosine kinase.

2.2. Cyclin D1

Cyclins are proteins that interact with cyclin-dependent kinases (CDKs) to influence cell cycle progression. Cyclin–CDK binding, along with phosphorylation of a threonine side chain and dephosphorylation of a tyrosine side chain, activates the CDK to send a positive signal. Activation of the CDK4/6 pathway by cyclin D1 leads to phosphorylation of the retinoblastoma protein (pRb), which triggers release of E2F transcription factors. These transcription factors activate genes necessary for entry into the S-phase, in which DNA replication takes place. Amplification or overexpression of the cyclin D1 gene leads to constitutive activation of the cyclin D1–CDK4/6 pathway.

Gene amplification of the *cyclin D1* gene ***(29–32)***, often in squamous cell type ***(33)***, and overexpression of the cyclin D1 protein ***(34)*** have been reported in esophageal cancer patients and in several rat esophageal cancer models ***(35–37)***. Since nuclear accumulation of the cyclin D1 protein has been observed in precancerous dysplastic lesions ***(38)***, cyclin D1 overexpression is believed to be an early event in the esophageal squamous cell carcinogenesis ***(35,36,39)***. In the metaplasia–dysplasia–neoplasia sequence of esophageal adenocarcinoma, increased nuclear expression of cyclin D1 begins in Barrett's metaplasia. It is also found in esophageal adenocarcinoma, although not necessarily to a greater degree than in less dysplastic lesions.

The positive rates for cyclin D1 gene amplification and overexpression are 16–36.4% ***(40–44)***, and 29–75% ***(32,34,45–52)***, respectively. Increased expression of cyclin D1 was observed in both adenocarcinomas and squamous carcinomas of the esophagus ***(53)***. A study by Roncalli ***(33)***, which investigated cyclin D1 gene amplification and product accumulation and Rb allelic loss and protein expression in a series of 75 esophageal cancer cases, found that no cases of adenocarcinoma showed cyclin D1 gene amplification (41% for squamous cell cases), whereas 22% showed cyclin D1 product accumulation (35% for squamous cell cases). Patients in this study were followed up to 5 yr and showed a trend toward increased lymph node metatasis and reduced overall survival rate with increased cyclin D1 accumulation.

There is some direct evidence that cyclin D1 overexpression contributes to abnormal growth and tumorigenicity of esophageal cancer cells ***(54)***. Introduction of an antisense cyclin D1 cDNA construct into cyclin D1-overexpressing cells reversed the transformed phenotype of human esophageal cancer cells ***(55)***. The targeting of the *cyclin D1* oncogene by an Epstein-Barr virus promoter in transgenic mice caused dysplasia in aero-upper digestive squamous epithelial tissues including the esophagus ***(56)***.

The cyclin D1 status appears to be an appropriate marker for predicting patients' prognosis, although there are some conflicting reports ***(46,47)***. Cyclin D1 overexpression could be a useful biomarker in identifying Barrett's esophagus patients at high risk of esophageal adenocarcinoma ***(57)***. Cyclin D1 overexpression indicated a poor prognosis in advanced, but not in superficial, esophageal cancer patients ***(34)***. Gene amplification, but not overexpression, of cyclin D1 in ESCC-derived cell lines was significantly correlated with poor prognosis in patients from whom the cell lines were established ***(10)***. Gene amplification ***(33,40,43,58)*** and overexpression ***(41,48,50,59,60)*** of cyclin D1 in esophageal cancers correlated with patients' prognosis. Expression of cyclin D1 in surgically treated esophageal cancers correlated negatively with tumor differentiation, but positively with mitotic activity and nodal status ***(61)***. Cyclin D1 overexpression may be a useful predictor of sensitivity to concurrent chemoradiotherapy for ESCC ***(62,63)***. Three of four cases with lymph node spread were cyclin D1-positive, indicating that cyclin D1-positive tumors may have a greater propensity for spread ***(64)***.

2.3. COX-2

Cyclooxygenase (COX, PGH synthase) catalyzes the rate-limiting step in prostaglandin synthesis. Whereas COX-1 is constitutively expressed in most mammalian tissues, and is thought to carry out housekeeping functions such as cytoprotection of the gastric mucosa and control of platelet aggregation, COX-2 levels are normally undetectable but inducible by proinflammatory and mitogenic stimuli. Chronic esophagitis is associated with increased production of prostaglandin E2, a prostaglandin with increased expression in both colon and gastric cancers. PGE2 is expressed in both Barrett's metaplasia and increasingly dysplastic tissues. In experimental studies using esophageal cancer cell lines, inhibition of COX-2 induced apoptosis and reduced proliferation, and indomethacin had antitumor activity. In a study evaluating COX-2 expression in 172 squamous cell cancers and 27 adenocarcinomas of the esophagus, Zimmerman et al. ***(117)*** found that normal esophageal squamous epithelium stained weakly positive for COX-2 by immunohistochemistry, whereas metaplastic Barrett's mucosa showed no immunoreactivity. In the small number of samples studied, 77.8% of the adenocarcinomas showed COX-2 expression, but this expression did not correlate with tumor size, node positivity, or tumor differentiation. In contrast, a study by Wilson et al. ***(65)*** evaluating paired Barrett's and gastric mucosa control samples from 21 patients, and paired samples from 6 patients, found that 80% of the Barrett's cases and all of the adenocarcinoma cases showed increased COX-2 mRNA levels compared with gastric mucosa.

3. Tumor Suppressor Gene Inactivation

3.1. p53

The *p53* gene encodes a transcription factor involved in facilitating cell cycle arrest, DNA repair and apoptosis. The p53 protein is a homotetramer. A mutation in one of the four monomers will not only cause the entire complex to function abnormally, but often will be destroyed less readily than normal protein. Over time, the concentration of mutant protein increases, with the result of increasing inhibition of normal protein through a dominant negative effect.

The normal p53 gene product binds to DNA to induce transcription of another gene, whose product inhibits the activity of cyclin-CDK2 complex and thereby stops the cell cycle at the G1 checkpoint. P53 also downregulates bcl-2 and transcriptionally upregulates genes such as $p21^{WAF1/CIP1}$, *bax*, and *PIG3*, among other genes. Mutations of *p53* are frequently associated with esophageal cancers. Since *p53* mutations are also found in esophageal dysplasia, they are considered as an early event in esophageal carcinogenesis, unlike in the colon ***(66–77)***. A pivotal role of *p53* modulation in esophageal carcinogenesis was substantiated in a mouse model, in which *p53*-knockout mice with jejunoesophageal reflux developed esophageal adenocarcinoma ***(78)***. In studies of human esophageal adenocarcinoma tissues, p53 accumulation increased with increasing degree of dysplasia, beginning with low-grade dysplasia, with marked p53 overexpression in up to 70% of esophageal adenocarcinomas. Familial inheritance of p53 mutations is rare (Li-Fraumeni syndrome), and p53 mutations in esophageal adenocarcinoma may occur through allelic loss of chromosome 17p.

The methods for detecting mutated *p53* genes include immunohistochemistry, polymerase chain reaction (PCR)–single-strand conformation polymorphism (SSCP), PCR–restriction fragment length polymorphism (RFLP) ***(79,80)***, and direct sequencing of PCR products in exons 5–8, which cover 90% of *p53* mutations ***(81,82)***. Although immunohistochemical analysis is a widely used technique, its interpretation must be tempered because positive p53 staining does not necessarily imply the existence of *p53* mutations, and absence of p53 staining may occur with gene deletion, failure of transcription, or a nonstabilizing mutation ***(83)***. For example, 86% of ESCC that had p53 accumulation did contain *p53* gene mutations when assessed by PCR–SSCP analysis ***(76)***, and 59.5% of esophageal cancers were p53 immunopositive but only 40.5% were proven to contain *p53* gene mutations when analyzed by direct DNA sequencing of PCR products ***(82)***. When immunohistochemical and PCR–SSCP analyses were compared, there were 38% discordant cases ***(84)***. Problems of direct DNA sequencing include the tedious nature and expense of this technique, and the masking of mutant *p53* allele by the wild-type *p53* contaminating from normal tissue ***(85)***. Therefore, a combinatory analysis of immunohistochemistry and PCR–SSCP would be recommended for the determination of *p53* mutations.

Most recently, circulating anti-p53 antibodies (p53-Abs) in the serum have been utilized as a potential indicator for *p53* mutations in various cancers as well as esophageal cancers. Serum p53-Abs were present in 25–60% of EDAC patients ***(86,87)***, and there was a close correlation between positivity for *p53* mutations in tissues and serum p53-Abs ***(81,87,88)***. Since patients with Barrett's esophagus developed serum p53-Abs predating the clinical diagnosis of malignancy ***(89)***, serum p53-Abs might be useful for following these patients.

The positive rates for mutated or accumulated p53 in esophageal cancers are ranging from 33.3% to 84% as summarized in **Table 1**. Epidemiologically, high p53 accumulation was correlated with habitual drinking and smoking ***(90)***, and frequent consumption of hot beverages. Histologically, undifferentiated esophageal cancer patients had a high positive rate for p53 expression ***(91,92)***, and p53 protein expression was significantly related to the coexistence and spreading of intraepithelial carcinoma ***(93)***. There was a positive correlation between the presence of *p53* mutations and EGFR overexpression ***(94)***.

Intensive efforts have been undertaken to predict the radio- and chemosensitivity of esophageal cancer patients by utilizing p53 status.When human esophageal cancer cells bearing mutated *p53* gene were retrovirally transduced with the wild-type *p53* gene, sensitivity to radiation was significantly improved and cells became susceptible to cisplatin ***(95)***. Esophageal cancer cells expressing either mutated *p53* or no p53 protein were more resistant to the growth-inhibitory effects of chemotherapeutic drugs ***(96)***. In vivo, ESCC patients with positive p53 staining in pretreatment biopsy samples were resistant to cis-

Table 1
***p53* Mutations in Esophageal Cancer**

Histology	Methods	*p53* Mutations (%)	Correlation to prognosis	References
ESCC	IH	56.7	+	119
ESCC	IH	71.7	+	47
ESCC	IH	59	+	120
ESCC	PCR–SSCP	33.3	+	121
EA	IH	45	+	122
EA	IH	54	+	123
EC	IH	56.5	+	124
EC	IH	84	+	98
EC	IH	54	+	125
EC	IH	53	+	126
EC	PCR–SSCP	34	+	127
ESCC	IH	41	–	128
ESCC	IH	48.1	–	129
ESCC	IH	42	–	45
ESCC	IH	64	–	130
ESCC	IH	56.8	–	74
ESCC	IH	64	–	131
ESCC	IH	67.2	–	63
ESCC	PCR–SSCP	55.5	–	132
EA	PCR–SSCP	36	–	85
EA	IH	79	–	131
EC	IH	48.4	–	46
EC	PCR–SSCP	38	–	133
ESCC	IH	70	N/D	76
ESCC	IH	51.4	N/D	93
ESCC	IH	40.9	N/D	75
ESCC	IH	67	N/D	134
ESCC	PCR–SSCP	47	N/D	135
EC	IH	57.1	N/D	136
EC	IH	48.4	N/D	137

ESCC: esophageal squamous cell carcinoma
EADC: esophageal adenocarcima
EC: combined results of ESCC and EA
IH: immunohistochemistry
PCR–SSCP: polymerase chain reaction–single-strand conformation polymorphism
ND: not determined

platin ***(97)*** or chemoradiation ***(98)*** in retrospective studies, and overexpression of p53 protein was associated with unfavorable treatment response ***(99,100)***. Overexpression of p53 protein was associated with decreased responsiveness to chemoradiotherapy in esophageal adenocarcinoma, but not in ESCC ***(101)***, implying a possible differential role of mutated *p53* gene depending upon histologic types. By contrast, p53 overexpression had no significant relationship with responsiveness to chemotherapy ***(102)***. Further studies must be employed to determine a subgroup in which therapeutic responses can be predictable by the p53 status.

3.2. P16

The *p16* gene encodes a protein that complexes with CDK4 and CDK6, inhibiting their ability to phosphorylate the retinoblastoma protein. Inactivation of the *p16* gene, which may occur through mutation, deletion, or methylation of the promoter, may lead to uncontrolled cell cycling and growth. The gene is located at 9p21, a region of frequent allelic loss in esophageal adenocarcinoma. Studies have shown p16 inactivation via p16 promoter methylation in nondysplastic Barrett's esophagus and more dysplastic lesions. In contrast to squamous cell carcinomas of the esophagus, point mutations in the *p16* gene are not common in esophageal adenocarcinoma.

3.3. Other Growth-Related Genes

The expression of the *KAI1/CD82* metastasis suppressor gene, which may be a cellular target of p53 ***(103)***, correlated with lymph node metastasis of ESCC patients ***(104)***.

The protein p73, which is structurally related to p53 and is possibly a tumor suppressor, showed increased expression in ESCC patients, including that by loss of imprinting (LOI), suggesting a partial compensatory mechanism for defective p53 ***(105)***, although this finding is controversial ***(106)***.

Loss of $p27^{KIP1}$ expression, which is a cdk inhibitor regulating the progression from G1 to S phase, has been reported in ESCC patients, demonstrating a positive correlation ***(107,108)*** or no correlation with patients' prognosis ***(48,109,110)***. Although increased expression of p27 is found in high-grade Barrett's dysplasia, expression levels are lower in esophageal adenocarcinomas. Loss of p27 correlates with increasing grade of dysplasia, as well as invasiveness, lymph node metastasis, and decreased survival.

GML, whose expression is regulated in a p53-dependent fashion, sensitized esophageal cancer cells to ionizing radiation ***(111)***. Loss of E-cadherin expression in ESCC was correlated with poor prognosis ***(22,112)***. The *Tylosis oesophageal cancer (TOC)* gene locus, which might contain a tumor suppressor gene, was frequently deleted in sporadic esophageal cancers ***(113)***. Clarification of the precise roles of these growth suppressor genes awaits further elucidation.

In one study, 64% of 22 esophageal adenocarcinoma tumors studied showed loss of heterozygosity in the 3p region. Candidates for genes in the region retaining one (mutated) copy include the von Hippel Lindau gene and the peroxidase proliferator activated receptor-γ gene. Studies report loss of heterozygosity at 5q in 45–72% of esophageal adenocarcinoma tumors. Candidate tumor suppressor genes there include MSH3, APC, MCC, and IRF-1. Although loss of heterozygosity at the mutated in colon cancer (MCC) locus is frequent, mutation of the retained MCC allele is not. Similarly, although loss of heterozygosity at the APC locus on 5q in high-grade Barrett's dysplasia and esophageal adenocarcinoma is present, a low rate of APC mutations has been found. Allelic loss of 18q is common in esophageal adenocarcinoma. Target genes in this area include the DPC4 or SMAD4 gene; however, mutations in this gene have not been identified in esophageal adenocarcinoma tumors.

Fas/APO-1 encodes a transmembrane protein involved in apoptosis expressed on cell surfaces (CD95). Expression of Fas has in some studies been found to be reduced or absent despite high levels of mRNA in the tumors. The failure of expression of Fas may thus prevent apoptosis in esophageal adenocarcinoma cells.

4. Mismatch repair genes

Defects in the mismatch repair system result in point mutation accumulation as well as variation in microsatellite length. Although microsatellite instability (MSI) and germline mutations in the *MSH2* and *MLH1* genes have been found at high frequency in hereditary nonpolyposis colon cancer (HNPCC), there are only a few reports investigating alterations of the mismatch repair genes in esophageal cancers. Based on the limited information, mutations of mismatch repair genes do no appear to play a major role in development of esophageal squamous cell carcinoma ***(114, 115)***; however, MSI has been reported in a subset of Barrett's metaplasia and dysplasia ***(116)***.

5. Summary and Future Directions

Esophageal squamous cell carcinoma and esophageal adenocarcinoma have different etiologic bases and their epidemiology vary in the United States and worldwide. Nevertheless, one can view the genetic basis of these two common forms of esophageal neoplasms in the context of oncogene activation, tumor suppressor gene inactivation and alteration of other genes involved in cellular proliferation and differentation. Cell culture and animal models have been emerging, and these provide the foundation for the discovery of new genes. Ultimately, current and future knowledge about the genetic basis for esophageal cancer will translate into innovative diagnostic and therapeutic approaches.

Acknowledgments

This work was supported by the Abramson Family Cancer Research Institute and P01-DE012467.

References

1. Mandard, A. M., Hainaut, P., and Hollstein, M. (2000) Genetic steps in the development of squamous cell carcinoma of the esophagus. *Mutat. Res.* **462**, 335–342.
2. Montesano, R., Hollstein, M., and Hainaut, P. (1996) Genetic alterations in esophageal cancer and their relevance to etiology and pathogenesis: a review. *Int. J. Cancer* **9**, 225–235.
3. Hollstein, M. C., Smits, A. M., Galiana, C., et al. (1988) Amplification of epidermal growth factor receptor gene but no evidence of ras mutations in primary human esophageal cancers. *Cancer Res.* **48**, 5119–5123.
4. Olayioye, M. A., Neve, R. M., Lane, H. A., and Hynes N. E. (2000) The ErbB signaling network: receptor heterodimerization in development and cancer. *EMBO J.* **19**, 3159–3167.
5. Alroy, I. and Yarden, Y. (1997) The ErbB signaling network in embryogenesis and oncogenesis: signal diversification through combinatorial ligand-receptor interactions. *FEBS Lett.* **410**, 83–86.
6. Moghal, N. and Sternberg, P. W. (1999) Multiple positive and negative regulators of signaling by the EGF-receptor. *Curr. Opin. Cell Biol.* **11**, 190–196.
7. Itakura, Y., Sasano, H., Shiga C., et al. (1994) Epidermal growth factor receptor overexpression in esophageal carcinoma. An immunohistochemical study correlated with clinicopathologic findings and DNA amplification. *Cancer* **74**, 795–804.
8. Okano, J., Gaslightwala, I., Birnbaum, M. J., Rustgi, A. K., and Nakagawa, H. (2000) Akt/Protein kinase B isoforms are differentially regulated by epidermal growth factor stimulation. *J. Biol. Chem.* **275**, 30934–30942.

9. Kitagawa, Y., Ueda, M., Ando, N., Ozawa, S., Shimizu, N., and Kitajima, M. (1996) Further evidence for prognostic significance of epidermal growth factor receptor gene amplification in patients with esophageal squamous cell carcinoma. *Clin. Cancer Res.* **2**, 909–914.
10. Kanda, Y., Nishiyama, Y., Shimada, Y., et al. (1994) Analysis of gene amplification and overexpression in human esophageal-carcinoma cell lines. *Int. J. Cancer* **58**, 291–297.
11. Yoshida, K., Kuniyasu, H., Yasui, W., Kitadai, Y., Toge, T., and Tahara, E. (1993) Expression of growth factors and their receptors in human esophageal carcinomas: regulation of expression by epidermal growth factor and transforming growth factor alpha. *J. Cancer Res. Clin. Oncol.* **119**, 401–407.
12. Wang, Q. S., Sabourin, C. L., Bijur. G. N., Robertson, F. M., and Stoner, G. D. (1996) Alterations in transforming growth factor-alpha and epidermal growth factor receptor expression during rat esophageal tumorigenesis. *Mol. Carcinog.* **15,** 144–153.
13. Lam, K. Y., Tin, L., and Ma, L. (1998) C-erbB-2 protein expression in oesophageal squamous epithelium from oesophageal squamous cell carcinomas, with special reference to histological grade of carcinoma and pre-invasive lesions. *Eur. J. Surg. Oncol.* **24**, 431–435.
14. Hardwick, R. H., Barham, C. .P., Ozua, P., et al. (1997) Immunohistochemical detection of p53 and c-erbB-2 in oesophageal carcinoma; no correlation with prognosis. *Eur. J. Surg. Oncol.* **23**, 30–35.
15. Shiga, K., Shiga, C., Sasano, H., et al. (1993) Expression of c-erbB-2 in human esophageal carcinoma cells: overexpression correlated with gene amplification or with GATA-3 transcription factor expression. *Anticancer Res.* **13**, 1293–1301.
16. Suo, Z., Su, W., Holm, R., and Nesland, J. M. (1995) Lack of expression of c-erbB-2 oncoprotein in human esophageal squamous cell carcinomas. *Anticancer Res.* **15**, 2797–2798.
17. Friess, H., Fukuda, A., Tang, W. H., et al. (1999) Concomitant analysis of the epidermal growth factor receptor family in esophageal cancer: overexpression of epidermal growth factor receptor mRNA but not of c-erbB-2 and c-erbB-3. *World J. Surg.* **23**, 1010–1018.
18. Karnes, W. E. Jr., Weller, S. G., Adjei, P. N., et al. (1998) Inhibition of epidermal growth factor receptor kinase induces protease-dependent apoptosis in human colon cancer cells.*Gastroenterology* **114**, 930–939.
19. Sibilia, M., Fleischmann, A., Behrens, A., et al. (2000) The EGF receptor provides an essential survival signal for SOS-dependent skin tumor development. *Cell* **102**, 211–220.
20. Wakita, H., and Takigawa, M. (1999) Activation of epidermal growth factor receptor promotes late terminal differentiation of cell-matrix interaction-disrupted keratinocytes. *J. Biol. Chem.* **274**, 37285–37291.
21. Suwa, T., Ueda, M., Jinno, H., et al. (1999) Epidermal growth factor receptor-dependent cytotoxic effect of anti-EGFR antibody-ribonuclease conjugate on human cancer cells. *Anticancer Res.* **19**, 4161–4165.
22. Inada, S., Koto, T., Futami, K., Arima, S., and Iwashita, A. (1999) Evaluation of malignancy and the prognosis of esophageal cancer based on an immunohistochemical study (p53, E-cadherin, epidermal growth factor receptor). *Surg. Today* **29**, 493–503.
23. Iihara, K., Shiozaki, H., Tahara, H., et al. (1993) Prognostic significance of transforming growth factor-alpha in human esophageal carcinoma. Implication for the autocrine proliferation. *Cancer* **71**, 2902–2909.
24. Mukaida, H., Toi, M., Hirai, T., Yamashita, Y., and Toge, T. (1991) Clinical significance of the expression of epidermal growth factor and its receptor in esophageal cancer. *Cancer* **68**, 142–148.
25. Ozawa, S., Ueda, M., Ando, N., Shimizu, N., and Abe, O. (1989) Prognostic significance of epidermal growth factor receptor in esophageal squamous cell carcinomas. *Cancer* **63**, 2169–2173.
26. Hickey, K., Grehan, D., Reid, I. M., O'Briain, S., Walsh, T. N., and Hennessy, T.P. Expression of epidermal growth factor receptor and proliferating cell nuclear antigen predicts response of esophageal squamous cell carcinoma to chemoradiotherapy. *Cancer* **74**, 1693–1698.

27. Torzewski, M., Sarbia, M., Verreet, P., et al. (1997) The prognostic significance of epidermal growth factor receptor expression in squamous cell carcinomas of the oesophagus. *Anticancer Res.* **17**, 3915–3919.
28. Wang, L. S., Chow, K. C., Chi, K. H., et al. (1999) Prognosis of esophageal squamous cell carcinoma: analysis of clinicopathological and biological factors. *Am. J. Gastroenterol.* **94**, 1933–1940.
29. Jiang, W., Kahn, S. M., Tomita, N., Zhang, Y. J., Lu, S. H., and Weinstein, I. B. (1992) Amplification and expression of the human cyclin D gene in esophageal cancer. *Cancer Res.* **52**, 2980–2983.
30. Jiang, W., Zhang, Y. J., Kahn, S. M., et al. (1993) Altered expression of the cyclin D1 and retinoblastoma genes in human esophageal cancer. *Proc. Natl. Acad. Sci. USA* **90**, 9026–9030.
31. Tsuruta, H., Sakamoto, H., Onda, M., and Terada, M. (1993) Amplification and overexpression of EXP1 and EXP2/Cyclin D1 genes in human esophageal carcinomas. *Biochem. Biophys. Res. Commun.* **196**, 1529–1536.
32. Nakagawa, H., Zukerberg, L., Togawa, K., Meltzer, S. J., Nishihara, T., and Rustgi, A. K. (1995) Human cyclin D1 oncogene and esophageal squamous cell carcinoma. *Cancer* **76**, 541–549.
33. Roncalli, M., Bosari, S., Marchetti, A., et al. (1998) Cell cycle-related gene abnormalities and product expression in esophageal carcinoma. *Lab. Invest.* **78**, 1049–1057.
34. Toyoda, H., Nakamura, T., Shinoda, M., et al. (2000) Cyclin D1 expression is useful as a prognostic indicator for advanced esophageal carcinomas, but not for superficial tumors. *Dig. Dis. Sci.* **45**, 864–869.
35. Wang, Q. S., Sabourin, C. L., Wang, H., and Stoner, G. D. (1996) Overexpression of cyclin D1 and cyclin E in N-nitrosomethylbenzylamine-induced rat esophageal tumorigenesis. *Carcinogenesis* **17**, 1583–1588.
36. Youssef, E. M., Hasuma, T., Morishima, Y., et al. (1997) Overexpression of cyclin D1 in rat esophageal carcinogenesis model. *Jpn. J. Cancer Res.* **88**, 18–25.
37. Fong, L. Y., Nguyen, V. T., Farber, J. L., Huebner, K., and Magee, P. N. (2000) Early deregulation of the the p16ink4a-cyclin D1/cyclin-dependent kinase 4-retinoblastoma pathway in cell proliferation-driven esophageal tumorigenesis in zinc-deficient rats. *Cancer Res.* **60**, 4589–4595.
38. Shamma, A., Doki, Y., Shiozaki, H., et al. (2000) Cyclin D1 overexpression in esophageal dysplasia: a possible biomarker for carcinogenesis of esophageal squamous cell carcinoma. *Int. J. Oncol.* **16**, 261–266.
39. Arber, N., Lightdale, C., Rotterdam, H., et al. (1996) Increased expression of the cyclin D1 gene in Barrett's esophagus. *Cancer Epidemiol. Biomarkers Prev.* **5**, 457–459.
40. Gramlich, T. L., Fritsch, C. R., Maurer, D., Eberle, M., and Gansler, T. S. (1994) Differential polymerase chain reaction assay of cyclin D1 gene amplification in esophageal carcinoma. *Diagn. Mol. Pathol.* **3**, 255–259.
41. Inomata, M., Uchino, S., Tanimura, H., Shiraishi, N., Adachi, Y., and Kitano, S. (1998) Amplification and overexpression of cyclin D1 in aggressive human esophageal cancer. *Oncol. Rep.* **5**(1), 71–176.
42. Morgan, R. J., Newcomb, P. V., Hardwick, R. H., and Alderson, D. (1999) Amplification of cyclin D1 and MDM-2 in oesophageal carcinoma. *Eur. J. Surg. Oncol.* **25**, 364–367.
43. Shinozaki, H., Ozawa, S., Ando, N., et al. (1996) Cyclin D1 amplification as a new predictive classification for squamous cell carcinoma of the esophagus, adding gene information. *Clin. Cancer Res.* **2**, 1155–1161.
44. Tsuruta, H., Sakamoto, H., Onda, M., and Terada, M. (1993) Amplification and overexpression of EXP1 and EXP2/Cyclin D1 genes in human esophageal carcinomas. *Biochem. Biophys. Res. Commun.* **196**, 1529–1536.

45. Chetty, R. and Chetty, S. (1997) Cyclin D1 and retinoblastoma protein expression in oesophageal squamous cell carcinoma. *Mol. Pathol.* **50**, 257–260.
46. Hirai, T., Kuwahara, M., Yoshida, K., Osaki, A., and Toge, T. (1999) The prognostic significance of p53, p21 (Waf1/Cip1), and cyclin D1 protein expression in esophageal cancer patients. *Anticancer Res.* **19**, 4587–4591.
47. Ikeda, G., Isaji, S., Chandra, B., Watanabe, M., and Kawarada, Y. (1999) Prognostic significance of biologic factors in squamous cell carcinoma of the esophagus. *Cancer* **86**, 1396–1405.
48. Itami, A., Shimada, Y., Watanabe, G., and Imamura, M. (1999) Prognostic value of p27(Kip1) and CyclinD1 expression in esophageal cancer. *Oncology* **57**, 311–317.
49. Kuwahara, M., Hirai, T., Yoshida, K., et al. (1999) p53, p21(Waf1/Cip1) and cyclin D1 protein expression and prognosis in esophageal cancer. *Dis. Esophagus* **12**, 116–119.
50. Matsumoto, M., Furihata, M., Ishikawa, T., Ohtsuki, Y., and Ogoshi, S.(1999) Comparison of deregulated expression of cyclin D1 and cyclin E with that of cyclin-dependent kinase 4 (CDK4) and CDK2 in human oesophageal squamous cell carcinoma. *Br. J. Cancer* **80**, 256–261.
51. Shamma, A., Doki, Y., Shiozaki, H., et al. (1998) Effect of cyclin D1 and associated proteins on proliferation of esophageal squamous cell carcinoma. *Int. J. Oncol.* **13**, 455–460.
52. Sheyn, I., Noffsinger, A. E., Heffelfinger, S., Davis, B., Miller, M. A., and Fenoglio-Preiser, C. M. (1997) Amplification and expression of the cyclin D1 gene in anal and esophageal squamous cell carcinomas. *Hum. Pathol.* **28**, 270–276.
53. Arber, N., Gammon, M. D., Hibshoosh, H., et al. (1999) Overexpression of cyclin D1 occurs in both squamous carcinomas and adenocarcinomas of the esophagus and in adenocarcinomas of the stomach. *Hum. Pathol.* **30**, 1087–1092.
54. Watanabe, M., Kuwano, H., Tanaka, S., Toh, Y., Masuda, H., and Sugimachi, K. (1999) A significant morphological transformation is recognized in human esophageal cancer cells with an amplification/overexpression of the cyclin D1 gene. *Int. J. Oncol.* **15**, 1103–1108.
55. Zhou, P., Jiang, W., Zhang, Y. J., et al. (1995) Antisense to cyclin D1 inhibits growth and reverses the transformed phenotype of human esophageal cancer cells. *Oncogene* **11**, 571–580.
56. Nakagawa, H., Wang, T. C., Zukerberg, L., et al. (1997) The targeting of the cyclin D1 oncogene by an Epstein-Barr virus promoter in transgenic mice causes dysplasia in the tongue, esophagus and forestomach. *Oncogene* **14**, 1185–1190.
57. Bani-Hani, K., Martin, I. G., Hardie, L. J., et al. (2000) Prospective study of cyclin D1 overexpression in Barrett's esophagus: association with increased risk of adenocarcinoma. *J. Natl. Cancer Inst.* **92**, 1316–1321.
58. Naitoh, H., Shibata, J., Kawaguchi, A., Kodama, M., and Hattori, T. (1995) Overexpression and localization of cyclin D1 mRNA and antigen in esophageal cancer. *Am. J. Pathol.* **146**, 1161–1169.
59. Sarbia, M., Porschen, R., Borchard, F., Horstmann, O., Willers, R., and Gabbert, H. E. (1994) p53 protein expression and prognosis in squamous cell carcinoma of the esophagus. *Cancer* **74**, 2218–2223.
60. Takeuchi, H., Ozawa, S., Ando, N., et al. (1997) Altered p16/MTS1/CDKN2 and cyclin D1/PRAD-1 gene expression is associated with the prognosis of squamous cell carcinoma of the esophagus. *Clin. Cancer Res.* **3**, 2229–2236.
61. Sarbia, M., and Gabbert, H. E., (2000) Modern pathology: prognostic parameters in squamous cell carcinoma of the esophagus. *Recent Results Cancer Res.* **155**, 15–27.
62. Samejima, R., Kitajima, Y., Yunotani, S., and Miyazaki, K. (1999) Cyclin D1 is a possible predictor of sensitivity to chemoradiotherapy for esophageal squamous cell carcinoma. *Anticancer Res.* **19**, 5515–5521.

63. Sarbia, M., Stahl, M., Fink, U., et al. (1999) Prognostic significance of cyclin D1 in esophageal squamous cell carcinoma patients treated with surgery alone or combined therapy modalities. *Int. J. Cancer* **84**, 86–91.
64. Chetty, R. and Simelane, S. (1999) p53 and cyclin A protein expression in squamous carcinoma of the oesophagus. *Pathol. Oncol. Res.* **5**, 193–196.
65. Wilson, K. T., Fu, S., Ramanujam, K. S., and Meltzer, S. J. (1998) Increased expression of inducible nitric oxide synthase and cyclooxyengase-2 expression in human esophageal carcinoma. *Cancer Res* **58**, 2929–2934.
66. Jaskiewicz, K. and De Groot, K. M. (1994) p53 gene mutants expression, cellular proliferation and differentiation in oesophageal carcinoma and non-cancerous epithelium. *Anticancer Res.* **14**, 137–140.
67. Casson, A. G., Manolopoulos, B., Troster, M,, et al. (1994) Clinical implications of p53 gene mutation in the progression of Barrett's epithelium to invasive esophageal cancer. *Am. J. Surg.* **167**, 52–57.
68. Symmans, P. J., Linehan, J. M., Brito, M. J., and Filipe, M. I. (1994) p53 expression in Barrett's oesophagus, dysplasia, and adenocarcinoma using antibody DO-7. *J. Pathol.* **173**, 221–226.
69. Wang, L. D., Hong, J. Y., Qiu, S. L., Gao, H., and Yang, C. S. (1993) Accumulation of p53 protein in human esophageal precancerous lesions: a possible early biomarker for carcinogenesis. *Cancer Res.* **53**, 1783–1787.
70. Wang, L. D., Shi, S. T., Zhou, Q., et al. (1994) Changes in p53 and cyclin D1 protein levels and cell proliferation in different stages of human esophageal and gastric-cardia carcinogenesis. *Int. J. Cancer* **59**, 514–519.
71. Volant, A., Nousbaum, J. B., Giroux, M. A., et al. (1995) p53 protein accumulation in oesophageal squamous cell carcinomas and precancerous lesions. *J. Clin. Pathol.* **48**, 531–534.
72. Parenti, A. R., Rugge, M., Frizzera, E., et al. (1995) p53 overexpression in the multistep process of esophageal carcinogenesis. *Am. J. Surg. Pathol.* **19**, 1418–1422.
73. Ohashi, Y., Sasano, H., Yamaki, H., et al. (1999) Cell cycle inhibitory protein p27 in esophageal squamous cell carcinoma. *Anticancer Res.* **19**, 1843–1848.
74. Chaves, P., Pereira, A. D., Pinto, A., et al. (1997) p53 protein immunoexpression in esophageal squamous cell carcinoma and adjacent epithelium. *J. Surg. Oncol.* **65**, 3–9.
75. Ikeguchi, M., Saito, H., Katano, K., Tsujitani, S., Maeta, M., and Kaibara, N. (1997) Clinicopathologic significance of the expression of mutated p53 protein and the proliferative activity of cancer cells in patients with esophageal squamous cell carcinoma. *J. Am. Coll. Surg.* **185**, 398–403.
76. Shi ST, Yang GY, Wang LD, et al. (1999) Role of p53 gene mutations in human esophageal carcinogenesis: results from immunohistochemical and mutation analyses of carcinomas and nearby non-cancerous lesions. *Carcinogenesis* **20**, 591–597.
77. Tian, D., Feng, Z., Hanley, N. M., Setzer, R. W., Mumford, J. L., and DeMarini, D. M. (1998) Multifocal accumulation of p53 protein in esophageal carcinoma: evidence for field cancerization. *Int. J. Cancer* **78**, 568–575.
78. Fein, M., Peters, J. H., Baril, N., et al. (1999) Loss of function of Trp53, but not Apc, leads to the development of esophageal adenocarcinoma in mice with jejunoesophageal reflux. *J. Surg. Res.* **83**, 48–55.
79. Whetsell, L. H., Ringer, D. P., and Schaefer, F. V. (1994) Molecular approach to rapid assessment of p53 tumor suppressor mutations in esophageal tumors from stained histological slides. *Diagn. Mol. Pathol.* **3**, 132–141.
80. Kajiyama, Y., Kanno, H., Ueno, M., et al. (1998) p53 gene mutation in 150 dissected lymph nodes in a patient with esophageal cancer. *Dis. Esophagus.* **11**, 279–283.
81. Ralhan, R., Arora, S., Chattopadhyay, T. K., Shukla, N. K., and Mathur, M. (2000) Circulating p53 antibodies, p53 gene mutational profile and product accumulation in esophageal squamous-cell carcinoma in India. *Int. J. Cancer.* **85**, 791–795.

82. Ribeiro, U. Jr., Finkelstein, S. D., Safatle-Ribeiro, A. V., et al. (1998) p53 sequence analysis predicts treatment response and outcome of patients with esophageal carcinoma. *Cancer* **83**, 7–18.
83. Wynford-Thomas, D. (1992) P53 in tumour pathology: can we trust immunocytochemistry? *J. Pathol.* **166**, 329–330.
84. Coggi, G., Bosari, S., Roncalli, M., et al. (1997) p53 protein accumulation and p53 gene mutation in esophageal carcinoma. A molecular and immunohistochemical study with clinicopathologic correlations. *Cancer* **79**, 425–432.
85. Soontrapornchai, P., Elsaleh, H., Joseph, D., Hamdorf, J. M., House, A. K., and Iacopetta, B. (1999) TP53 gene mutation status in pretreatment biopsies of oesophageal adenocarcinoma has no prognostic value. *Eur. J. Cancer* **35**, 1683–1687.
86. Sobti, R. C. and Parashar, K. (1998) A study on p53 protein and anti-p53 antibodies in the sera of patients with oesophageal cancer. *Mutat. Res.* **422**, 271–277.
87. von Brevern, M. C., Hollstein, M. C., Cawley, H. M., et al. (1996) Circulating anti-p53 antibodies in esophageal cancer patients are found predominantly in individuals with p53 core domain mutations in their tumors. *Cancer Res.* **56**, 4917–4921.
88. Hagiwara, N., Onda, M., Miyashita, M., and Sasajima, K. (2000) Detection of circulating anti-p53 antibodies in esophageal cancer patients. *J. Nippon. Med. Sch.* **67**, 110–117.
89. Cawley, H. M., Meltzer, S. J., De Benedetti, V. M., et al. (1998) Anti-p53 antibodies in patients with Barrett's esophagus or esophageal carcinoma can predate cancer diagnosis. *Gastroenterology* **115**, 19–27.
90. Saeki, H., Ohno, S., Araki, K., et al. (2000) Alcohol consumption and cigarette smoking in relation to high frequency of p53 protein accumulation in oesophageal squamous cell carcinoma in the Japanese. Br. J. Cancer 82, 1892–1894.
91. Matsumoto, M., Natsugoe, S., Nakashima, S., et al. (2000) Biological evaluation of undifferentiated carcinoma of the esophagus. *Ann. Surg. Oncol.* **7**, 204–209.
92. Casson, A. G., Tammemagi, M., Eskandarian, S., Redston, M., McLaughlin, J., and Ozcelik, H. (1998) p53 alterations in oesophageal cancer: association with clinicopathological features, risk factors, and survival. *Mol. Pathol.* **51**, 71–79.
93. Nozoe, T., Kuwano, H., Toh, Y., Watanabe, M., Kitamura, M., and Sugimachi, K. (1998) Significance of p53 protein expression in growth pattern of esophageal squamous cell carcinoma. *Oncol. Rep.* **5**, 1119–1123.
94. Esteve, A., Lehman, T., Jiang, W., et al. (1993) Correlation of p53 mutations with epidermal growth factor receptor overexpression and absence of mdm2 amplification in human esophageal carcinomas. *Mol. Carcinog.* **8**, 306–311.
95. Matsubara, H., Kimura, M., Sugaya, M., et al. (1999) Expression of wild-type p53 gene confers increased sensitivity to radiation and chemotherapeutic agents in human esophageal carcinoma cells. *Int. J. Oncol.* **14**, 1081–1085.
96. Nabeya, Y., Loganzo, F. Jr., Maslak, P., et al. (1995) The mutational status of p53 protein in gastric and esophageal adenocarcinoma cell lines predicts sensitivity to chemotherapeutic agents. *Int. J. Cancer* **64**, 37–46.
97. Shimada, Y., Watanabe, G., Yamasaki, S., et al. (2000) Histological response of cisplatin predicts patients' survival in oesophageal cancer and p53 protein accumulation in pretreatment biopsy is associated with cisplatin sensitivity. *Eur. J. Cancer* **36**, 987–993.
98. Krasna, M. J., Mao, Y. S., Sonett, J. R., et al. (1999) P53 gene protein overexpression predicts results of trimodality therapy in esophageal cancer patients. *Ann. Thorac. Surg.* **68**, 2021–2024.
99. Monges, G. M., Seitz, J. F., Giovannini, M. F., Gouvernet, J. M., Torrente, M. A., and Hassoun, J. A. (1996) Prognostic value of p53 protein expression in squamous cell carcinoma of the esophagus. *Cancer Detect. Prev.* **20**, 63–67.
100. Moreira, L. F., Naomoto, Y., Hamada, M., Kamikawa, Y., and Orita, K. (1995) Assessment of apoptosis in oesophageal carcinoma preoperatively treated by chemotherapy and radiotherapy. *Anticancer Res.* **15**, 639–644.

101. Yang, B., Rice, T. W., Adelstein, D. J., Rybicki, L. A., and Goldblum, J. R. (1999) Overexpression of p53 protein associates decreased response to chemoradiotherapy in patients with esophageal carcinoma. *Mod. Pathol.* **12**, 251–256.
102. Lam, K. Y., Law, S., Ma, L. T., Ong, S. K., and Wong, J. (1997) Pre-operative chemotherapy for squamous cell carcinoma of the oesophagus: do histological assessment and p53 overexpression predict chemo-responsiveness? *Eur. J. Cancer* **33**, 1221–1225.
103. Jackson, P. and Puisieux, A. (2000) Is the KAI1 metastasis suppressor gene a cellular target of p53? A review of current evidence. *Biochem. Biophys. Res. Commun.* **278**, 499–502.
104. Miyazaki, T., Kato, H., Shitara, Y., et al. (2000) Mutation and expression of the metastasis suppressor gene KAI1 in esophageal squamous cell carcinoma. *Cancer* **89**, 955–962.
105. Cai, Y. C., Yang, G. Y., Nie, Y., et al. (2000) Molecular alterations of p73 in human esophageal squamous cell carcinomas: loss of heterozygosity occurs frequently; loss of imprinting and elevation of p73 expression may be related to defective p53. *Carcinogenesis* **21**, 683–689.
106. Nimura, Y., Mihara, M., Ichimiya, S., et al. (1998) p73, a gene related to p53, is not mutated in esophageal carcinomas. *Int. J. Cancer* **78**, 437–440.
107. Anayama, T., Furihata, M., Ishikawa, T., Ohtsuki, Y., and Ogoshi, S. (1998) Positive correlation between p27Kip1 expression and progression of human esophageal squamous cell carcinoma. *Int. J. Cancer* **79**, 439–443.
108. Shamma, A., Doki, Y., Tsujinaka, T., et al. (2000) Loss of p27(KIP1) expression predicts poor prognosis in patients with esophageal squamous cell carcinoma. *Oncology* **58**, 152–158.
109. Nita, M. E., Nagawa, H., Tominaga, O., et al. (1999) p21Waf1/Cip1 expression is a prognostic marker in curatively resected esophageal squamous cell carcinoma, but not p27Kip1, p53, or Rb. Ann. *Surg. Oncol.* **6**, 481–488.
110. Ohashi, K., Nemoto, T., Eishi, Y., Matsuno, A., Nakamura, K., and Hirokawa, K. (1997) Proliferative activity and p53 protein accumulation correlate with early invasive trend, and apoptosis correlates with differentiation grade in oesophageal squamous cell carcinomas. *Virchows Arch.* **430**, 107–115.
111. Kagawa, K., Inoue, T., Tokino, T., Nakamura, Y., and Akiyama, T. (1997) Overexpression of GML promotes radiation-induced cell cycle arrest and apoptosis. Biochem. Biophys. Res. Commun. 241, 481–485.
112. Shimada, Y., Imamura, M., Watanabe, G., et al. (1999) Prognostic factors of oesophageal squamous cell carcinoma from the perspective of molecular biology. *Br. J. Cancer* **80**, 1281–1288.
113. Iwaya, T., Maesawa, C., Ogasawara, S., Tamura, G. (1998) Tylosis esophageal cancer locus on chromosome 17q25.1 is commonly deleted in sporadic human esophageal cancer. *Gastroenterology* **114**, 1206–1210.
114. Muzeau, F., Flejou, J. F., Belghiti, J., Thomas, G., and Hamelin, R. (1997) Infrequent microsatellite instability in oesophageal cancers. *Br. J. Cancer* **75**, 1336–1339.
117. Naidoo, R. and Chetty, R. (1999) DNA repair gene status in oesophageal cancer. *Mol. Pathol.* **52**, 125–130.
116. Meltzer, S. J., Yin, J., Manin, B., et al. (1994) Microsatellite instability occurs frequently and in both diploid and aneuploid cell populations of Barrett's-associated esophageal adenocarcinomas. *Cancer Res.* **54**, 3379–3382.
117. Zimmermann, K.C., Sarbia, M., Weber, A. A., Borchard, F., Gabbert, H. E., and Schrör, K. (1999) Cyclooxygenase-2 Expression in Human Esophageal Carcinoma. *Cancer Res* **59**, 198–204.
118. Nakashima, S., Natsugoe, S., Matsumoto, M., et al. (2000) Expression of p53 and p21 is useful for the prediction of preoperative chemotherapeutic effects in esophageal carcinoma. Anticancer Res. 20, 1933–1937

119. Goukon, Y., Sasano, H., Nishihira, T., Nagura, H., and Mori, S. (1994) p53 overexpression in human esophageal carcinoma: a correlation with tumor DNA ploidy and two parameter flow cytometric study. *Anticancer Res. 14*, 1305–1312.
120. Kobayashi, S., Koide, Y., Endo, M., Isono, K., and Ochiai, T. (1999) The p53 gene mutation is of prognostic value in esophageal squamous cell carcinoma patients in unified stages of curability. *Am. J. Surg.* **177**, 497–502.
121. Moskaluk, C. A., Heitmiller, R., Zahurak, M., Schwab, D., Sidransky, D., Hamilton, S. R. (1996) p53 and p21(WAF1/CIP1/SDI1) gene products in Barrett esophagus and adenocarcinoma of the esophagus and esophagogastric junction. *Hum. Pathol.* **27**, 1211– 1220.
122. Casson, A. G., Kerkvliet, N., and O'Malley, F. (1995) Prognostic value of p53 protein in esophageal adenocarcinoma. J. Surg. Oncol. 60, 5–11.
123. Kihara, C., Seki, T., Furukawa, Y., et al. (2000) Mutations in zinc-binding domains of p53 as a prognostic marker of esophageal-cancer patients. *Jpn. J. Cancer Res.* **91,** 190–198.
124. Pomp, J., Davelaar, J., Blom, J., et al. (1998) Radiotherapy for oesophagus carcinoma: the impact of p53 on treatment outcome. *Radiother. Oncol.* **46**, 179–184.
125. Shimaya, K., Shiozaki, H., Inoue, M., et al. (1993) Significance of p53 expression as a prognostic factor in oesophageal squamous cell carcinoma. *Virchows Arch. A. Pathol. Anat. Histopathol.* **422**, 271–276.
126. Uchino, S., Saito, T., Inomata, M., et al. (1996) Prognostic significance of the p53 mutation in esophageal cancer. *Jpn. J. Clin. Oncol.* **26**, 287–292.
127. Ikeguchi, M., Saito, H., Katano, K., Tsujitani, S., Maeta, M., and Kaibara, N. (1997) Clinicopathologic significance of the expression of mutated p53 protein and the proliferative activity of cancer cells in patients with esophageal squamous cell carcinoma. J. Am. Coll. Surg. 185, 398–403.
128. Kanamoto A, Kato H, Tachimori Y, et al. (1999) No prognostic significance of p53 expression in esophageal squamous cell carcinoma. *J. Surg. Oncol.* **72**, 94–98.
129. Lam, K. Y., Law, S., Tin, L, Tung, P. H., and Wong, J. (1999) The clinicopathological significance of p21 and p53 expression in esophageal squamous cell carcinoma: an analysis of 153 patients. *Am. J. Gastroenterol.* **94**, 2060–2068.
130. Vijeyasingam, R., Darnton, S. J., Jenner, K., Allen, C. A., Billingham, C., and Matthews, H. R. (1994) Expression of p53 protein in oesophageal carcinoma: clinicopathological correlation and prognostic significance. *Br. J. Surg.* **81**, 1623–1626.
131. Wang, L. S., Chow, K. C., Liu, C. .C., and Chiu, J. H. (1998) p53 gene alternation in squamous cell carcinoma of the esophagus detected by PCR-cold SSCP analysis. *Proc. Natl. Sci. Counc. Repub. China B.* **22**, 114–121.
132. Tamura, G., Maesawa, C., Suzuki, Y., Satodate, R., Ishida, K., and Saito, K. (1992) p53 gene mutations in esophageal cancer detected by polymerase chain reaction single-strand conformation polymorphism analysis. *Jpn. J. Cancer Res.* **83**, 559–562.
133. Baron, P. L., Gates, C. E., Reed, C. E., et al. (1997) p53 overexpression in squamous cell carcinoma of the esophagus. *Ann. Surg. Oncol.* **4**, 37–45.
134. Wagata, T., Shibagaki, I., Imamura, M., et al. (1993) Loss of 17p, mutation of the p53 gene, and overexpression of p53 protein in esophageal squamous cell carcinomas. *Cancer Res.* **53**, 846–850.
135. Kato, H., Yoshikawa, M., Fukai, Y., et al. (2000) An immunohistochemical study of p16, pRb, p21 and p53 proteins in human esophageal cancers. *Anticancer Res.* **20**, 345–349.
136. Kuwahara, S., Ajioka, Y., Watanabe, H., Hitomi, J., Nishikura, K., and Hatakeyama, K. Heterogeneity of p53 mutational status in esophageal squamous cell carcinoma. *Jpn. J. Cancer Res.* **89**, 405–410.

9

PTEN and Cancer

Ramon Parsons and Laura Simpson

1. A Tumor Suppressor on Chromosome 10q23

Genetic evidence strongly suggested that a tumor suppressor was located on chromosome 10. During the development of glioblastoma, one copy of chromosome 10 was typically lost ***(1)***. Cytogenetic and molecular analysis revealed partial or complete loss of chromosome 10 in bladder, endometrial, and prostate cancer ***(2,3)***. These studies implicated 10q23 as a chromosomal region likely to contain one or more important tumor suppressor genes. When wild-type chromosome 10 was reintroduced into glioblastoma cell lines, it reduced the ability of these cell lines to form tumors in nude mice, in part due to inhibition of angiogensis ***(4)***. During the same year (1996), a linkage analysis report of a cancer predisposition syndrome, Cowden disease (CD), determined that a CD locus was present on chromosome10q23 ***(5)***. These findings greatly bolstered the idea that chromosome 10q23 contained a novel tumor suppressor gene whose loss was key in the formation of several types of cancer.

In 1997, Li et al. ***(6)*** identified a putative tumor suppressor on 10q23 through the use of representational difference analysis (RDA). RDA was performed on 12 primary breast tumors and a probe derived from one tumor mapped to chromosome 10q23. Six homozygous deletions were identified in cancer xenografts by walking away from the RDA marker on the STS-based YAC map. By trapping exons from a bacterial artificial chromosome (BAC) that spanned the smallest homozygous deletion, two exons were identified that matched expressed sequence tags (EST) present in the dBEST database. Sequence analysis of the 403 amino acid open reading frame revealed a protein tyrosine phosphatase domain and a large region of homology to chicken tensin and bovine auxilin. The gene was, therefore, designated PTEN; for *P*hosphatase and *Ten*sin homolog deleted on chromosome *Ten*.

Steck et al. ***(7)*** independently reported the identification of the same candidate tumor suppressor gene on 10q23, which was termed MMAC1 for *M*utated in *M*ultiple *A*dvanced *C*ancers. They performed a high-density scan between microsatellite markers D10S191 and D10S221 in 21 glioma cell lines and primary cultures to narrow the critical region of deletion on 10q. Homozygous deletions were detected in four cell lines, with the critical region of loss bordered by D10S541 and AFM280. Using fragments

From: *Methods in Molecular Biology, Vol. 222: Tumor Suppressor Genes: Pathways and Isolation Strategies*
Edited by: Wafik S. El-Deiry © Humana Press Inc., Totowa, NJ

from a BAC containing a portion of the region that was homozygously deleted, exon trapping was performed to identify exons of the tumor suppressor gene. The gene discovered proved to be identical to PTEN.

Finally, in a search for novel human protein tyrosine phosphatases, a gene termed TEP1 (*T*GF-β-regulated and *e*pithelial cell-enriched *p*hosphatase) was identified ***(8)***. This search was conducted using an entirely different method than previously mentioned. PCR using degenerate primers pairs that correspond to the phosphatase catalytic domain was used to screen human cDNA libraries for clones. Also, conserved phosphatase sequence motifs were used to search the GenBank cDNA database. Through a combination of these two approaches TEP1 was identified. TEP1 also proved to be identical to PTEN. The purified TEP1 protein was capable of dephosphorylating phosphotyrosyl RCML, an in-vitro substrate for many tyrosine phosphatases. When the essential cysteine residue in the tyrosine phosphatase signature motif was mutated to serine (C124S), the phosphatase activity was abolished, showing that PTEN was indeed a phosphatase.

2. PTEN in Sporadic Cancers

Glioblastoma mutiforme is the most aggressive form of glioma, and patients diagnosed with the disease usually survive less than 2 yr. Loss of chromosome 10q occurs in the vast majority of glioblastomas. The initial reports identifying PTEN as a candidate tumor suppressor also examined several cell lines and primary tumors for mutations in PTEN ***(6,7)***. These reports demonstrated that glioblastoma cell lines and primary tumors had a high frequency of PTEN mutations accompanied by loss of heterozygosity (LOH). Subsequently, other reports that examined glioblastomas for mutations in PTEN determined the mutation rate of PTEN to be high (20–44%) ***(9–13)***. In lower-grade glioma and glioneuronal tumors, however, PTEN mutations are rare ***(11)***. One study examining the level of expression of PTEN in glioblastomas vs lower grades of gliomas by reverse transcription polymerase chain reaction found a significant difference between the two groups ***(14)***. Furthermore, immunostaining of glioblastomas revealed little or no PTEN expression in about two-thirds of the tumors examined ***(14)***. The reduction or loss of PTEN appears to be very important in the progression of gliomas to glioblastoma mutiforme, the most aggressive form of glioma.

LOH studies reveal that chromosome 10q is frequently lost in prostate cancer ***(15–18)***. The most common region of loss (~50% of tumors studied) spans region 10q23. One study examining clinically localized prostate carcinoma determined that inactivation of PTEN by homozygous deletion in 10–15% of the cases. In a study that examined both clinically localized and metastatic prostate carcinomas, researchers found PTEN mutations in approximately 10% of the cases ***(19)***. They also found that mutations occurred more frequently in metastatic disease. Together, these reports suggest that the inactivation of PTEN by mutation occurred predominately in advanced prostate cancer. In addition to examining the mutational status of PTEN in prostate cancer, one group analyzed the expression of PTEN in xenografts derived from metastatic prostate cancer ***(20)***. Expression of PTEN at the mRNA and protein level was reduced in at least 50% of the cases. In a study of 109 cases of prostate cancer, PTEN expression was assessed by immunohistochemistry and the loss of PTEN protein was correlated with pathologic markers of poor prognosis ***(21)***. Inactivation or loss of expression of PTEN appears to be important in the development of advanced prostate cancer.

Another cancer to exhibit a high frequency of loss of 10q is melanoma. Sporadic tumors have alterations in the region of 10q22-10qter and LOH studies show frequent early loss of 10q ***(22–26)***. Although the loss or mutation of PTEN is high in malignant melanoma cell lines (~40%) ***(27)***, the frequency of PTEN aberrations in primary tumors appears to be much lower with the majority of the mutations occurring in metastatic disease ***(28,29)***. As with several other tumors, PTEN mutations appear to be associated with late-stage disease. Examination of 4 primary melanomas and 30 metastases found little to no expression of PTEN in 65% of the cases ***(30)***. Reduction of expression, therefore, may also play a role in the development of metastatic melanoma. One study examined the mutational status of both PTEN and N-RAS in 53 cutaneous melanoma cell lines. They found 16 cell lines (30%) with alterations in PTEN and 11 cell lines (21%) with activating NRAS mutations, with only one cell line having mutations in both ***(31)***. This suggests that PTEN and N-RAS may function on the same pathway and that perturbation of this pathway is necessary for the development of advanced melanomas.

Breast cancer has a high frequency of LOH at chromosome 10q. PTEN, however, is mutated only in a small fraction of breast cancer cases (~5%) ***(32–35)***. Immunohistochemical analysis of sporadic primary breast carcinomas had either no or decreased expression in 33% of these tumors. Loss of PTEN may therefore play a more important role in the development of sporadic breast cancer than previously thought ***(34,36)***. PTEN mutations have also been found to a lesser extent in cancers of the bladder, lung, and lymphatic systems ***(37–41)***.

Another tumor type that has a high frequency of mutations in PTEN is endometrial carcinomas. Studies examining endometrial carcinomas reported significant LOH on chromosome 10, with two commonly deleted regions, 10q22-24 and 10q25-26 ***(42,43)***. The frequency of PTEN mutations in endometrial carcinomas of the endometrioid type is approximately 50% ***(44,45)***. In addition, endometrioid endometrial carcinomas showed complete loss of PTEN protein expression in 61% of cases and reduced expression in 97% ***(46)***. Unlike other tumor types, inactivation of PTEN happens early in tumor development and may represent the initial transforming event of tumorigenesis. Examination of other gynecologic malignancies revealed that endometrioid ovarian tumors had frequent mutations in PTEN (21%), while in serous and mucinous epithelial ovarian tumors mutations of PTEN were absent ***(45,47)***. PTEN mutations in cervical cancer were also determined to be rare ***(45,48)***.

Overall, PTEN is altered in a wide variety of tumors. Alteration happens early or late in tumor development, depending on the affected tissue. Alterations may occur through genetic inactivation or the regulation of expression.

3. PTEN Mutation Is Detected in Hamartoma Syndromes

Cowden disease (CD) is a rare, autosomal dominant familial cancer syndrome associated with a high risk for the development of breast cancer. The hallmark for this disease is the presence of harmatomas of the skin, breast, thyroid, oral mucosa, and intestinal epithelium. In 1996, the gene responsible for CD was localized to chromosome 10q22–23 by linkage analysis ***(5)***. When PTEN was discovered less than a year later, researchers were eager to see if PTEN was the gene involved in this cancer syndrome. The first study involving 5 families with CD identified mutations in PTEN in 4 of the families ***(49)***. They found missense and nonsense mutations that disrupted the phosphatase domain. One of the mutations resided in the catalytic pocket changing a glycine at position 129 to a glutamic acid. A follow-up study involving 37 CD families found PTEN in 81% of the families. These included missense, nonsense, insertions, deletions and splice-site mutations ***(50)***. Mutations in the phosphatase domain comprised 43% of all mutations. The remaining mutations were scattered over the entire length of PTEN. In a study of 10 families with CD in which 8 had germline mutations in PTEN, the importance of PTEN in the development of the characteristic hamartomas was investigated ***(51)***. Twenty hamartomas were examined for LOH of markers flanking and within PTEN. LOH within the CD interval and including PTEN was identified in two fibroadenomas of the breast, a thyroid adenoma, and a pulmonary hamartoma. The wild allele was lost in these harmartomas. Reduced levels of PTEN RNA were found in hamartomas from different tissues in a CD patient. From the above evidence, it appears that PTEN functions as a tumor suppressor in CD.

Bannayan-Zonana syndrome (BZS) is another autosomal dominant hamartomous disease. Some of the characteristic features of the disease are polyposis, macrocephaly, cutaneous lipomas, and high birthweight, and, in males, a speckled penis. Unlike CD, BZS is not associated with an increased risk of malignancy. The clinical overlap between CD and BZS led researchers to examine BZS for possible germline mutations in PTEN. One study examined 2 families for germline mutations in PTEN and found mutations in both families ***(52)***. These mutations segregated with affected family members but were absent in unaffected members and 100 control alleles. Interestingly, one mutation was identical to a mutation found in an unrelated family with CD. Another study of 7 BZS families discovered germline mutations in 57% of the cases ***(50)***. They also observed a mutation that was common to 2 unrelated CD families and 1 BZS family. CD and BZS may represent variable penetrance of the same disorder. It is also possible to speculate that there are modifying genetic or epigenetic factors that account for the varying manifestations of the two syndromes. Whatever the case, it is clear that PTEN plays a significant role in the development of both diseases and fulfills the criteria for a tumor suppressor.

4. PTEN Induces Cell Cycle Arrest and Apoptosis in Several Systems

To examine PTEN's affect on cellular processes such as cell cycle and apoptosis, several groups expressed exogenous PTEN in different cell lines. Expression of PTEN in PTEN-null glioma cell lines caused growth suppression ***(53,54)***. Apoptotic and cell cycle analysis showed that PTEN-mediated growth suppression was due to G1 arrest. Phosphatase mutants of PTEN did not suppress cellular growth or induce cell cycle arrest. One group found that expression of PTEN made glioma cell lines more susceptible to anoikis, apoptosis initiated by disruption of cell interaction with the extracellular matrix ***(55)***. Expression of PTEN in a variety of breast cancer cell lines caused growth suppression, but in this case the mechanism was apoptosis ***(56)***. Another study using tetracycline-inducible system to express PTEN in a PTEN-positive breast cancer cell line found that PTEN induced a G1 arrest followed by apoptosis ***(57)***. Similar to the studies in glioblastoma cell lines, a catalytically inactive mutant of PTEN did not induce G1 arrest or apoptosis. Clearly, PTEN's phosphatase domain is necessary for its role in cell cycle arrest and apoptosis when overexpressed. Introduction of PTEN by retrovirus into endometrial cancer lines lacking PTEN resulted in nearly physiologic levels of PTEN expression. Such cells had a reduced level of proliferation and an increase of cells in G1 ***(58)***. Another study using the same retroviral approach in glioblastoma cell lines determined that PTEN did not affect proliferation or the cell cycle in tissue culture ***(59)***. However, when lines with reintroduced PTEN were grown as xenografts, their rate of growth was restricted relative to the parental line or clones expressing a catalytically inactive PTEN protein. Such cells appeared to be unable to stimulate angiogenesis as potently as the parental line. This is consistent with observation that PTEN regulates the expression of thrombospondin and VEGF in glioblastoma ***(59,60)***.

5. PTEN and Animal Model Systems

Mice that are heterozygously deficient for PTEN were created to examine the role of PTEN in development and tumor suppression. PTEN was shown to be essential in embryonic development. PTEN−/− mice did not survive beyond d 9.5 of embryonic development, and embryos appeared disorganized ***(61–63)***. One group found abnormal patterning in the PTEN−/− embryo and overproliferation of the cephalic and caudal regions with no alterations in apoptosis ***(63)***. PTEN-heterozygous mice developed tumors in multiple tissues such as endometrium, prostate, thyroid, adrenal medulla, and colon ***(61–63)***. These tumors were associated with an increase in proliferation with no change in apoptosis. Some, but not all, of the tumors had loss of the wild-type allele. Another interesting phenotype of PTEN+/− mice is the development of nonneoplastic hyperplasia of the lymph nodes. An inherited defect in apoptosis in B-cells and macrophages was observed in the hyperplastic lymph nodes by Podsypanina et al. ***(61)***. The lymphoadenopathy of PTEN+/− mice was associated with autoimmune glomerulonephropathy ***(64)***. Fas-mediated apoptosis was impaired in these mice and T-lymphocytes showed reduced activation-induced cell death and increased proliferation upon activation. Immortalized mouse embryonic fibroblasts (PTEN+/−, and PTEN−/−) were generated to analyze PTEN's role in cellular processes such as growth arrest and apoptosis ***(63)***. PTEN−/−cells did not have an increased rate of proliferation when compared to PTEN+/−cells. However, PTEN−/− cells exhibited a decrease in sensitivity to a number of apoptotic stimuli, and reintroduction of PTEN into these cells restored the cells' sensitivity to apoptosis. These mouse studies revealed the importance of PTEN in the regulation of both proliferation and apoptosis, which varied depending on the organ.

A few examples of the ability of PTEN to synergize with other signaling pathways have been examined in mice. PTEN mutation hastens the morbidity of mammary tumorigenesis initiated by wnt-1, indicating synergy between PTEN and the wnt/β-catenin/APC pathway ***(65)***. Synergy was also observed between Pten and p27 in prostate tumorigenesis ***(66)***.

The *Drosophila* counterpart to mammalian PTEN has been recently cloned and characterized ***(67–70)***. The amino-terminal half of the dPTEN protein is highly conserved, sharing about 65% identity with the human PTEN in this region. As seen in the mouse, *Drosophila* dPTEN mutants die during early development. To study the function of PTEN, somatic dPTEN−/− cells in both the eye and the wing were created ***(68–70)***.

Analysis in both the eye and the wing revealed that dPTEN−/− clones were larger in size than their wild-type counterparts. This phenotype is cell autonomous, since the mutant cells had no affect on wild-type neighbors. Lack of dPTEN also appeared to affect the regulation of S phase. Cells lacking dPTEN in the imaginal disc also had an increase in proliferation. Overexpression of dPTEN during eye development was shown to inhibit cell proliferation ***(68)***. This overexpression of PTEN caused a cell cycle arrest in the G2 or G2/M phase, in contrast to human tissue culture experiments, which show that PTEN expression arrests cells in the G1 phase. This difference could be due to stage of development, cell type, and/or species-specific factors. Interestingly, overexpression of dPTEN in differentiating cells during eye development induced apoptosis, showing that PTEN may act in a different fashion depending on the context of its expression.

6. PTEN's Phosphatase Domain Is Essential for Its Activity as a Tumor Suppressor

The mutation spectra in patients and experiments in tissue culture strongly implicated the role of the phosphatase domain in the suppression of tumorigenesis. Efforts were therefore made to determine the substrate of PTEN's phosphatase domain. Recombinant PTEN was shown to dephosphorylate protein and peptide substrates phosphorylated on serine, threonine, and tyrosine residues, which indicated that PTEN is a dual-specificity phosphatase ***(71)***. PTEN also exhibited a high degree of substrate specificity for acidic substrates in vitro. The finding that a PTEN mutant, G129E, found in two CD kindreds, still exhibited wild-type protein phosphatase activity toward poly(Glu_4Tyr_1), however, suggested that the actual substrate in vivo had not yet been identified ***(71)***. To address this question, Maehama et al. ***(72)*** investigated the possibility that the second messenger phospholipid phosphatidylinositol-3,4,5-trisphosphate was a substrate for PTEN. They found that wild-type PTEN catalyzed the dephosphorylation of PtdIns-3,4,5-P3 on the D3 position of the inositol ring, whereas a mutant PTEN could not. Overexpression of PTEN in 293 cells reduced insulin-induced PtdIns-3,4,5-P3 levels without effecting PI3K activity. On the other hand, expression of a catalytically inactive mutant caused PtdIns-3,4,5-P3 accumulation. This study revealed that PTEN is a lipid phosphatase in vitro and in vivo. Studies examining the PTEN-G129E mutant revealed the importance of the lipid phosphatase activity of PTEN for its function as a tumor suppressor. This mutation in PTEN specifically ablated the lipid phosphatase activity of PTEN while retaining its protein phosphatase activity ***(73,74)***. Expression of the PTEN-G129E mutant in a PTEN-null glioblastoma cell line did not induce the growth suppression seen when expressing wild-type PTEN, proving that the lipid phosphatase activity is necessary for this effect ***(54)***.

7. PTEN Regulates the PI3K Pathway

The formation of PtdIns-3,4,5-P3 by PI3K is essential for the activation of Akt. Akt is a serine/threonine kinase that is activated by a variety of growth factors including insulin and IGF-1 ***(75)***. PI3K activation through addition of growth factors results in an increase in PtdIns-3,4,5-P3. This leads to the translocation of Akt to the membrane and a conformational change that allows for PDK1 and PDK1 to phosphorylate Ser-473 and Thr-308 activating Akt ***(75)***. If PTEN desphosphorylates PtdIns-3,4,5-P3 in vivo, then PTEN should act as a negative regulator of the PI3K/Akt pathway.

Genetic studies involving the *Caenorhabditis elegans* model system have been instrumental in the formation of our current understanding of the PI3K/Akt pathway and its role in insulin signaling. DAF-2 is the *C. elegans* homolog of the mammalian insulin/insulin-growth factor receptor ***(76)***. daf-2 mutant animals constitutively form dauers, a diapausing larval stage in which their metabolism is shifted toward energy storage ***(76–79)***. DAF-2 has been shown to reside upstream of AGE-1, a homolog of the mammalian p110 catalytic subunit of PI3K, since daf-2 and age-1 mutant animals display similar phenotypes, which includes increased lifespan ***(80)***. AKT-1 and AKT-2, two *C. elegans* homologs of Akt, act downstream of AGE-1 in transducing signals from DAF-2 ***(81)***. Researchers studying AGE-1 signaling in *C. elegans* discovered DAF-18, a PTEN homolog ***(82–84)***. Studies of DAF-18 have established that it functions as a regulator of insulin signaling. Animals with mutations inactivating daf-18 fail to form dauers under appropriate conditions, and mutations of daf-18 suppress the age-1 constitutive dauer phenotype and suppress a similar phenotype in daf-2 mutants ***(82,83)***. Inactivation of daf-18 using RNA interference also suppresses the dauer phenotype of daf-2 and age-1, whereas introduction of a wild-type daf-18 transgene rescues the dauer defect ***(82,84)***. Inhibition of akt-1 and akt-2 by RNA interference causes dauer formation in both wild-type and daf-18 mutants, showing that daf-18 is upstream of akt-1 and akt-2 ***(82)***.

Studies with *Drosophila* also place dPTEN on the insulin signaling pathway. Homozygous DPTEN mutant cells have similar growth phenotypes to cells expressing a constitutively active form of Dp110, a PI3K homolog ***(85)***. Overexpression of DPTEN completely suppresses the growth-promoting activity of overexpressed Dp110 ***(70)***. Conversely, the small-eye phenotype of dPTEN overexpression is suppressed by Dp110 and enhanced by a dominant negative Dp110 ***(68)***. As in mammals, DPTEN appears to act as an antagonist to the PI3K homolog. Further confirmation that DPTEN is on the insulin signaling pathway in *Drosophila* is seen in studies involving chico, the homolog of IRS-1-4. Eye clones mutant for chico usually display a reduced growth phenotype, but in cells mutant for both chico and dPTEN, this phenotype is masked by the overgrowth phenotype associated with loss of dPTEN function ***(70)***. Additionally, removal of one copy of chico enhances the small-eye phenotype of dPTEN overexpression ***(68)***. This strongly suggests that dPTEN is downstream of chico in a negative regulatory role in insulin signaling. Finally, one study reported that Dakt1, an Akt homolog, could suppress the dPTEN overexpression phenotype ***(69)***. Studies with *C. elegans* and *Drosophila* revealed that PTEN is evolutionarily conserved and have placed it on the insulin signaling pathway. Since PTEN is highly conserved, these model systems provide researchers with an excellent way in which to study the function of PTEN.

The effects of PTEN on Akt in mammals has also been studied. Expression of PTEN in the cell leads to the downregulation of Akt activity, as evidenced by decreased levels of phospho-Akt ***(53,55–57,86,87)***. Additionally, constitutively active Akt, but not wild-type Akt, can rescue cells from PTEN-mediated G1 arrest and apoptosis ***(56,86,87)***. PI3K, the upstream activator of Akt cannot rescue cells from apoptosis, showing that PTEN acts below PI3K but above Akt on this signaling pathway. In agreement with these findings, tumor cell lines that are PTEN-deficient showed elevated levels of phospho-Akt as compared to PTEN-positive cell lines ***(86,88)***. Studies with PTEN−/− mouse fibroblasts have yielded similar results ***(63)***. The level of Akt kinase activity was elevated in PTEN−/−cells when compared to PTEN+/− cells. Pretreatment of serum starved cells with PI3K inhibitors eliminated basal as well as PDGF-dependent increase in Akt phosphorylation in both PTEN−/− and +/− cells. Also, PTEN−/− cells have elevated levels of PtdIns-3,4,5-P3 when compared to PTEN+/− cells. Finally, all spontaneous epithelial tumors that arise in PTEN+/− mice have an elevation in the level of phospho-AKT, which is associated in most cases with a reduction in the expression or mutation of wild-type PTEN ***(89)***.The role of PTEN in the regulation of p70/S6K is a rapidly evolving area of inquiry. The activation of S6K contributes to the cell's ability to grow, both at the level of cell size and proliferation ***(90)***. Under starved conditions, S6K is inactive. Activation of PI-3 kinase stimulates glucose uptake and in parallel leads to the phosphorylation of several amino acids on S6K. This alone, however, is not sufficient to activate S6K. Amino acids in the media are also required to activate mTOR kinase, which in turn enhances the steady-state phosphorylation of S6K through the regulation of phosphatase and kinase activities in the cell. Therefore, inhibition of either PI-3 kinase or mTOR leads to the inactivation of S6K. Genetic studies of homologs of S6K and mTOR in flies has strengthened the link between PTEN and these enzymes ***(91,92)***. DTOR mutation is lethal and leads to small cells that fail to proliferate. This mutation is epistatic to dPTEN mutation, since double-mutant cells remain small and hypoproliferative. DS6K−/− flies are viable but small, due to small cell size, and overexpression of dS6K is able to rescue hypomorphic alleles of dTor from lethality. In sum, these data suggest that PTEN may affect the regulation of S6K in mammalian cells.

Recently, studies have focused on the role of PTEN as a regulator of the insulin signaling pathway in mammals. Expression of PTEN in 3T3-L1 adipocytes inhibited insulin-induced 2-deoxy-glucose, GLUT4 translocation, and membrane ruffling ***(93)***. Both Akt and p70S6 kinase were significantly inhibited by PTEN overexpression. Perhaps more convincing evidence that PTEN regulates S6K is the finding that PTEN−/− tumor cells have elevated levels of phosphorylated S6K ***(94)***. Expression of PTEN in a PTEN-null breast cancer cell line has been shown to cause upregulation of insulin receptor substrate-2 (IRS-2), on the RNA, protein, and posttranslational phosphorylation levels ***(95)***. The IRS-2 produced by the expression of PTEN is tyrosine-phosphorylated and associated with PI3K. In addition, PTEN alters the posttranslational modification of IRS-1, which inhibits the interaction with grb2 ***(96)***. IRS-1 and IRS-2 are docking proteins, which are recruited by receptor tyrosine kinases such as the insulin receptor. In turn, IRS-1 and IRS-2 recruit PI3K and grb-2/Sos to signal through the PI-3 kinase and MAPK pathways. These data suggest that PTEN can both positively and negatively regulate insulin signaling.

PI3K/Akt pathway has an established role in regulating translation and apoptosis within the cell. Overexpression of a constitutively active, membrane-targeted form of Akt induces phosphorylation of 4E-BP1, a repressor of mRNA translation, inactivating the protein ***(97,98)***. When PTEN is transfected into cells, levels of active, phosphorylated Akt are diminished and 4E-BP1 phosphorylation is reduced ***(86)***. Akt has also been shown to phosphorylate Bad, a Bcl-2 family member. Dephosphorylated Bad inactivates pro-survival factors such as Bcl-X_L precipitating apoptosis. Expression of PTEN into PTEN-null cells resulted in decreased levels of phosphorylated Akt and phosphorylated BAD ***(55)***. Reciprocally, PTEN-null ES cells showed increased levels of phosphorylated Akt and phosphorylated BAD ***(99)***. In *C. elegans*, DAF-16 has been shown to be antagonized by the insulin signaling pathway and encode a Fork head transcription factor proving that insulin signaling ultimately alters transcription ***(77–79,81,100–103)***. The mammalian counterparts of DAF-16, AFX, FKHR, and FKHRL1, have been shown to be involved in transcriptional control and to be negatively regulated through phosphorylation by Akt ***(104–108)***. Overexpression of Forkhead transcription factors induces cell cycle arrest and apoptosis ***(104–106)***. AFX has been shown to mediate cell cycle arrest by transcriptionally activating $p27^{kip1}$ ***(109)***, and similarly, FKHRL1 induces transcriptional upregulation of Fas ligand ***(104)***. Therefore, PTEN may mediate cell cycle arrest and apoptosis through these transcription factors. Indeed, a recent report shows that expression of PTEN in PTEN-deficient cells induces the nuclear localization and transcriptional activity of FKHR ***(110)***. The results of these studies suggest that PTEN can control translation, thereby regulating the cell's entry into G1 and its progression through the cell cycle. In addition, these studies reveal the mechanisms by which PTEN may induce apoptosis.

PTEN's emerging role as an antagonist of the PI3K pathway led to studies examining PTEN's role in angiogenesis, since hypoxia, insulin, and IGF-1 induce angiogenic gene expression that may be regulated through the PI3K/Akt pathway ***(111–113)***. Reintroduction of wild-type PTEN into PTEN-null glioblastoma cell lines substantially reduced hypoxia, and IGF-1 induction of HIF-1 regulated genes such as vascular endothelial growth factor (VEGF) ***(60)***. PTEN suppresses the stabilization of hypoxia-mediated HIF-1α, which when stabilized through the PI3K/Akt pathway upregulates VEGF expression. Similar results were seen in epidermal growth factor (EGF)-stimulated prostate cancer cell lines ***(114)***. Stimulation of cells with EGF leads to increased levels of HIF-1α, HIF-1 transcriptional activity, and VEGF protein expression, which is blocked by expression of PTEN. This suggests a possible role for PTEN in angiogenesis.

Although PTEN's ability to inhibit the PI3K pathway has been examined mainly in glial and epithelial cell lines, it is possible PTEN may play a similar role in other cell types. When PTEN is exogenously expressed in Jurkat-T cells, it induces apoptosis ***(115)***. Apoptosis could be prevented by coexpression of a constitutively active, membrane bound form of Akt. PTEN also decreased T-cell receptor (TCR)-induced activation of mitogen-activated protein kinase, extracellular signal-related kinase (ERK2) that is dependent on PI3K activity. A recent report found that Jurkat T-cells exhibit constitutive phosphorylation of Akt, indicating high basal levels of PtdIns-3,4,5-P3 ***(116)***. In addition, there was constitutive membrane association of inducible T-cell kinase (Itk), a pleckstrin homology (PH)-containing tyrosine kinase downstream of PI3K. This kinase facilitates TCR-dependent calcium influxes and increases in extracellular-regulated kinase activity. Expression of PTEN or treatment of the cells with PI3K inhibitors blocked constitutive phophorylation of Akt and redistributed Itk to the cytosol. The PTEN-deficient Jurkat cells were hyperresponsive to TCR stimulation as measured by Itk activity, tyrosine phosphorylation of phospholipase C-γ 1, and activation of Erk. These studies provide evidence that PTEN may have an important role in TCR signaling in T-cells. The above-mentioned studies clearly provide evidence that PTEN acts antagonistically to PI3K by dephosphorylation of PtdIns-3,4,5-P3, and by doing so may affect a variety of cellular processes.

8. Domain Structure of PTEN

The initial study of PTEN showed that it was a 403-amino acid protein that contained an amino-terminal region with homology to auxilin and tensin and a protein phosphatase domain ***(6,7)*** (**Fig. 1**). Solving the crystal structure of PTEN revealed a phosphatase domain that was similar to protein phosphatases but had an enlarged active site, necessary

Fig. 1. Protein domains of PTEN. PTEN is a 403-amino acid protein that contains a phosphatase domain (residues 14–185) in the N-terminal region with the phosphatase motif (HCXXGXXR; residues 123–130) essential for its tumor suppressor activity. PTEN contains a C2 domain (residues 186–351) that allows for the binding of PTEN to phospholipids, perhaps for the effective positioning of PTEN at the membrane. PEST sequences (degradation motifs) are located between residues 350–375 and 379–396 within the tail domain. The tail region contains CK2 phosphorylation sites important for the stability and activity of PTEN. There is also a PDZ-binding domain that allows PTEN to bind MAGI proteins, which may enhance the efficiency of PTEN signaling through the formation of PTEN/MAGI complexes at the membrane.

for the accommodation of a phosphoinositide substrate that accounted for its ability to dephosphorylate lipid substrates ***(117)***. PTEN's phosphatase domain has been the focus of numerous studies, and its function is essential in PTEN's tumor suppressor function.

More recently, other domains of PTEN have been discovered and examined to determine their function. The crystal structure of PTEN revealed the existence a C2 domain that had affinity for phospholipid membranes in vitro ***(117)***. Mutation of the solvent-exposed residues in the CBR3 and Cα2 elements of the C2 domain had reduced affinity for membranes in vitro when compared to wild-type PTEN. These mutations reduced the growth-suppression activity of PTEN in a glioma cell line. Cells expressing these mutants also proliferated better than those expressing wild-type PTEN and demonstrated anchorage-independent growth. In addition, the structure revealed that the phosphatase domain and C2 domains associate across an extensive surface, which suggests that the C2 domain may have a role in positioning the catalytic domain on the membrane. Georgescu et al., in a previous study, had found that the expression of various C-terminal mutants in glioma cell lines allowed for anchorage-dependent growth that is usually suppressed by wild-type PTEN ***(118)***. Various mutations created were in predicted β-strands (now known to be in the C2 domain), and the level of phosphatase activity of these mutants depended on the degree to which the mutation disrupted these structural elements. The mutants were unstable and subject to rapid degradation.

PTEN also contains two PEST sequences and a PDZ motif in the carboxy-terminal (C-terminal region), and these elements were found to be dispensable for tumor suppressor function ***(118)***. The function of PEST sequences is to target proteins with short intracellular half-life for protein degradation. Deletion of these sequences in PTEN, however, led to decreased protein expression, perhaps by impaired protein folding. From the results of these studies, the C-terminal region appears to be important in PTEN stability and enzymatic activity. In another study focusing on the C-terminal region, Vazquez et al. ***(119)*** examined PTEN in which the 50-amino acid tail was removed, and found that it was necessary for maintaining protein stability. Surprisingly, the protein had increased activity as measured by the ability of the mutant to induce G1 arrest and FKHR transcriptional activity. The tail was also shown to be phosphorylated on serine 370 and one or more sites of a serine/threonine cluster (ser380, thr382, thr383, and ser385). Mutation of ser380, thr382, and thr383 each reduced both the steady-state protein levels and the protein half-life. However, these mutants were more active in inducing G1 arrest and FKHR transcriptional activation. Phosphorylation of the tail, therefore, may mediate the regulation of PTEN function.

To determine what kinase phosphorylates the tail of PTEN, researchers examined the amino acid sequence of the tail region and found that casein kinase-2 phosphorylation sites were located within a C-terminal cluster of serine/threonine residues (ser370, ser380, thr383, and ser385) ***(120)***. Casein kinase-2 is a serine/threonine kinase that is ubiquitously expressed and phosphorphorylates a variety of substrates involved in cell cycle and cell growth ***(121–123)***. PTEN was shown to be constitutively phosphorylated by CK2 in vivo, and PTEN phosphorylation-defective mutants showed decreased stability in comparison to wild-type PTEN. In addition, these mutants showed accelerated proteasome-mediated degradation. Treatment with a CK2 inhibitor also resulted in a reduction in PTEN protein levels, but only in the absence of a proteasome inhibitor. This data suggest that levels of PTEN in the cell could be regulated by its phosphorylation status.

Since PTEN contains a PDZ domain, this raised the possibility that PTEN may interact with other proteins that may be essential in PTEN's function. PDZ domains are protein–protein interaction domains that bind to consensus motifs (S/TXV) in the C-terminus of partner proteins, although other, more amino-terminal residues may be involved in the specificity of PDZ domain binding. Using yeast two-hybrid systems, PTEN was found to associate with members of the membrane-associated guanylate kinase family protein with multiple PDZ domains called MAGI for (membrane-associated guanylate kinase inverted***) (124,125)***. There are three distinct MAGI proteins (MAGI-1, -2, -3) with a high degree of conservation of the functional domains, including the PTEN-binding PDZ domains. These proteins localize to epithelial cell tight junctions. PTEN was found to associate with both MAGI2 and MAGI3 in two separate studies. The ability of low levels of PTEN to inhibit Akt activity was enhanced by the expression of either MAGI2 or MAGI3 as compared to low levels of PTEN alone.

Removal of the PDZ domain from PTEN greatly reduced its ability to inhibit Akt activity, which points to the potential importance of the interaction of PTEN and MAGI proteins in facilitating PTEN's function as a phosphatase. A catalytically inactive mutant of PTEN (C124S) that contains amino acids 1–377 was subject to rapid degradation, and the addition of the PDZ domain back to this protein conferred a greater degree of stability to the protein ***(124)***. This suggests that the interaction of PTEN and MAGI may enhance the stability of PTEN, though this interaction is probably not the only stabilizing factor needed by PTEN. Given these data, it is possible that the interaction between PTEN and MAGI proteins improves the efficiency of PTEN signaling by the formation of a complex at the cell membrane.

9. PTEN May Be More than a Lipid Phosphatase

Although PTEN clearly acts as a lipid phosphatase, it has been shown to dephosphorylate protein substrates as well and may have other roles in the cell besides antagonizing PI3K. Tamura et al. ***(126)*** overexpressed PTEN in both fibroblasts and glioblastoma cell lines and observed that PTEN inhibited cell migration. Integrin-mediated cell spreading and the formation of focal adhesions were downregulated by wild-type but not by PTEN with an inactive phosphatase domain. In addition, PTEN interacted with focal adhesion kinase (FAK) and reduced its tyrosine phosphorylation. A mutant of PTEN, G129E, which had lost its lipid phosphatase activity but retained its tyrosine phosphatase activity, still inhibited cell spreading and dephosphorylated FAK. The tyrosine phosphatase activity of PTEN appears to be essential in regulating cell interactions with the cellular matrix. Further examination of the role of PTEN in focal contact formation and cell spreading implicated PTEN in the dephosphorylation of Shc ***(127)***. The dephosphorylation of Shc by PTEN inhibited the recruitment of Grb2, which led to downregulation of the MAP kinase pathway. The ability of PTEN to dephosphorylate FAK is not universal, however. There appears to be no difference in the phosphoryation status of FAK in PTEN−/− versus PTEN+/+ ES cells ***(99)***. Another group studying the role of PTEN in invasion found that introduction of wild-type PTEN or phosphatase-deficient (C124S) PTEN both inhibited invasion in glioma cell lines ***(128)***. Reduction of FAK phosphorylation was not observed in conjunction with inhibition of invasion. This raises the possibility that domains other than the phosphatase domain may play a role in inhibiting invasion. Regarding PTEN's ability to regulate the MAP kinase pathway, it is possible that PTEN can inhibit it without direct dephosphorylation of Shc. Expression of PTEN can inhibit the movement of Gab1 to the plasma membrane by dephosphorylation of PtdIns-3,4,5-P3, since the pleckstrin homology domain of Gab1binds this phospholipid in its translocation to the membrane ***(129)***. This inhibition would effectively inhibit signaling through the MAP kinase pathway. The *Drosophila* model also gives us a tantalizing clue that the role of PTEN may be broader than just the dephosphorylation of PtdIns-3,4,5-P3. Researchers have noted that the bristle, hair, and rhabdomere phenotypes observed in DPTEN mutants have not been reported in Dp110 or insulin signaling mutants, which raises the possibility that these effects are not caused simply by an increase in PtdIns-3,4,5-P3 levels ***(70)***. In DPTEN mutants, the subcellular regulation of the assembly of actin microfilaments appears to be abnormal, which may account for the structural defects observed. Goberdhan et al. ***(70)*** postulated that the loss of DPTEN function may alter communication between the peripheral actin cytoskeleton, the plasma membrane, and the extracellular matrix. The study of how PTEN may function in capacities other than that of a lipid phosphatase still remains in its infancy.

10. Future Directions

Since the discovery of PTEN in 1997, its importance as a tumor suppressor has been validated by its mutation and/or loss of expression in a variety of sporadic cancers and its association with Cowden disease, an autosomal dominant cancer syndrome. Reintroduction of PTEN into cells causes G1 cell cycle arrest and/or apoptosis, depending on the cell type, revealing how PTEN may function as a tumor suppressor. Studies examining PTEN's role as a lipid phosphatase have shown that it dephosphorylates the D3 position of PtdIns-3,4,5-P3, acting in opposition to PI3K, which allows it to affect a number of cellular processes. Through the use of mouse models, PTEN has been shown to be important in embryonic development, tumor formation, and apoptosis. PTEN homologs have been identified in both *C. elegans* and *Drosophila*. Studies with these models have placed PTEN on the insulin signaling pathway as a negative regulator of a PI3K homolog in concordance with studies in mammals. Many vital questions about PTEN's role in cellular processes still remain to be answered. During human development, high levels of PTEN expression were observed in the skin, thyroid, central nervous system, autonomic nervous system, and upper gastrointestinal tract ***(130)***. PTEN's involvement in the development of these tissues is currently unknown. Regulation of PTEN levels through transcriptional, translational, and posttranslational mechanisms is another area that remains uncharted. Phosphorylation of residues in the PTEN tail has been shown to affect protein stability and activity ***(119)***. A recent report identified CK2 as the kinase that constitutively phosphorylates the tail region of PTEN ***(131)***. This phosphorylation confers stability to PTEN, and phosphorylation-defective mutants are subject to rapid turnover via the proteasome. Identification of a phosphatase that can dephosphorylate PTEN still remains to be discovered. The nature of this phosphatase will provide clues as to when PTEN can be dephosphorylated and under what conditions this takes place. It is possible that the selective dephosphorylation of PTEN may be crucial in regulating its specific activity under particular circumstances in the cell. PTEN affects transcription, translation, the cell cycle, and apoptosis. The continuing investigation into the many aspects of PTEN's function will give us a better understanding of how the loss of PTEN affects tumor development and progression.

References

1. Bigner, S. H., Mark, J., Mahaley, M. S., and Bigner, D. D. (1984) Patterns of the early, gross chromosomal changes in malignant human gliomas. *Hereditas* **101**, 103–113.
2. Lundgren, R., Kristoffersson, U., Heim, S., Mandahl, N., and Mitelman, F. (1988) Multiple structural chromosome rearrangements, including del(7q) and del(10q), in an adenocarcinoma of the prostate. *Cancer Genet. Cytogenet.* **35**, 103–108.
3. Gibas, Z., Prout, G. R., Jr., Connolly, J. G., Pontes, J. E., and Sandberg, A. A. (1984) Nonrandom chromosomal changes in transitional cell carcinoma of the bladder. *Cancer Res.* **44**, 1257–1264.
4. Hsu, S. C., Volpert, O. V., Steck, P. A., et al. (1996) Inhibition of angiogenesis in human glioblastomas by chromosome 10 induction of thrombospondin-1. *Cancer Res.* **56**, 5684–5691.
5. Nelen, M. R., Padberg, G. W., Peeters, E. A., et al. (1996) Localization of the gene for Cowden disease to chromosome 10q22-23. *Nat. Genet.* **13**, 114–116.
6. Li, J., Yen, C., Liaw, D., et al. (1997) R. PTEN, a putative protein tyrosine phosphatase gene mutated in human brain, breast, and prostate cancer [see comments]. *Science* **275**, 1943–1947.
7. Steck, P. A., Pershouse, M. A., Jasser, S. A., et al. (1997) Identification of a candidate tumour suppressor gene, MMAC1, at chromosome 10q23.3 that is mutated in multiple advanced cancers. *Nat. Genet.* **15**, 356–362.
8. Li, D. M. and Sun, H. (9197) TEP1, encoded by a candidate tumor suppressor locus, is a novel protein tyrosine phosphatase regulated by transforming growth factor beta. *Cancer Res.* **57**, 2124–2129.
9. Wang, S. I., Puc, J., Li, J., et al. (1997) Somatic mutations of PTEN in glioblastoma multiforme. *Cancer Res.* **57**, 4183–4186.
10. Liu, W., James, C. D., Frederick, L., Alderete, B. E., and Jenkins, R. B. (1997) PTEN/MMAC1 mutations and EGFR amplification in glioblastomas. *Cancer Res.* **57**, 5254–5257.
11. Duerr, E. M., Rollbrocker, B., Hayashi, Y., et al. (1998) PTEN mutations in gliomas and glioneuronal tumors. *Oncogene* **16**, 2259–2264.
12. Teng, D. H., Hu, R., Lin, H., et al. (1997) MMAC1/PTEN mutations in primary tumor specimens and tumor cell lines. *Cancer Res.* **57**, 5221–5225.
13. Chiariello, E., Roz, L., Albarosa, R., Magnani, I., and Finocchiaro, G. (1998) PTEN/MMAC1 mutations in primary glioblastomas and short-term cultures of malignant gliomas. *Oncogene* **16**, 541–545.
14. Sano, T., Lin, H., Chen, X., et al. (1999) Differential expression of MMAC/PTEN in glioblastoma multiforme: relationship to localization and prognosis. *Cancer Res.* **59**, 1820–1824.

15. Ittmann, M. (1996) Allelic loss on chromosome 10 in prostate adenocarcinoma. *Cancer Res.* **56**, 2143–2147.
16. Gray, I. C., Phillips, S. M., Lee, S. J., Neoptolemos, J. P., Weissenbach, J., and Spurr, N. K. (1995) Loss of the chromosomal region 10q23-25 in prostate cancer. *Cancer Res.* **55**, 4800–4803.
17. Komiya, A., Suzuki, H., Ueda, T., et al. (1996) Allelic losses at loci on chromosome 10 are associated with metastasis and progression of human prostate cancer. *Genes Chromosomes Cancer* **17**, 245–253.
18. Trybus, T. M., Burgess, A. C., Wojno, K. J., Glover, T. W., and Macoska, J. A. (1996) Distinct areas of allelic loss on chromosomal regions 10p and 10q in human prostate cancer. *Cancer Res.* **56**, 2263–2267.
19. Cairns, P., Okami, K., Halachmi, S., et al. (1997) Frequent inactivation of PTEN/MMAC1 in primary prostate cancer. *Cancer Res.* **57**, 4997–5000.
20. Whang, Y. E., Wu, X., Suzuki, H., et al. (1998) Inactivation of the tumor suppressor PTEN/MMAC1 in advanced human prostate cancer through loss of expression. *Proc. Natl. Acad. Sci. USA* **95**, 5246–5250.
21. McMenamin, M. E., Soung, P., Perera, S., Kaplan, I., Loda, M., and Sellers, W. R. (1999) Loss of PTEN expression in paraffin-embedded primary prostate cancer correlates with high Gleason score and advanced stage. *Cancer Res.* **59**, 4291–4296.
22. Parmiter, A. H., Balaban, G., Clark, W. H., Jr., and Nowell, P. C. (1988) Possible involvement of the chromosome region 10q24–q26 in early stages of melanocytic neoplasia. *Cancer Genet. Cytogenet.* **30**, 313–317.
23. Richmond, A., Fine, R., Murray, D., Lawson, D. H., and Priest, J. H. (1986) Growth factor and cytogenetic abnormalities in cultured nevi and malignant melanomas. *J. Invest. Dermatol.* **86**, 295–302.
24. Healy, E., Belgaid, C. E., Takata, M., et al. (1996) Allelotypes of primary cutaneous melanoma and benign melanocytic nevi. *Cancer Res.* **56**, 589–593.
25. Healy, E., Rehman, I., Angus, B., and Rees, J. L. (1995) Loss of heterozygosity in sporadic primary cutaneous melanoma. *Genes Chromosomes Cancer* **12**, 152–156.
26. Herbst, R. A., Weiss, J., Ehnis, A., Cavenee, W. K., and Arden, K. (1994) C. Loss of heterozygosity for 10q22-10qter in malignant melanoma progression. *Cancer Res.* **54**, 3111–3114.
27. Guldberg, P., thor Straten, P., Birck, A., Ahrenkiel, V., Kirkin, A. F., and Zeuthen, J. (1997) Disruption of the MMAC1/PTEN gene by deletion or mutation is a frequent event in malignant melanoma. *Cancer Res.* **57**, 3660–3663.
28. Tsao, H., Zhang, X., Benoit, E., and Haluska, F. G. (1998) Identification of PTEN/MMAC1 alterations in uncultured melanomas and melanoma cell lines. *Oncogene* **16**, 3397–3402.
29. Reifenberger, J., Wolter, M., Bostrom, J., et al. (2000) Allelic losses on chromosome arm 10q and mutation of the PTEN (MMAC1) tumour suppressor gene in primary and metastatic malignant melanomas. *Virchows Arch.* **436**, 487–493.
30. Zhou, X. P., Gimm, O., Hampel, H., Niemann, T., Walker, M. J., and Eng, C. (2000) Epigenetic PTEN silencing in malignant melanomas without PTEN mutation [In Process Citation]. *Am. J. Pathol.* **157**, 1123–1128.
31. Tsao, H., Zhang, X., Fowlkes, K., and Haluska, F. G. (2000) Relative reciprocity of NRAS and PTEN/MMAC1 alterations in cutaneous melanoma cell lines. *Cancer Res.* **60**, 1800–1804.
32. Ueda, K., Nishijima, M., Inui, H., et al. (1998) Infrequent mutations in the PTEN/MMAC1 gene among primary breast cancers. *Jpn. J. Cancer Res.* **89**, 17–21.
33. Rhei, E., Kang, L., Bogomolniy, F., Federici, M. G., Borgen, P. I., and Boyd, J. (1997) Mutation analysis of the putative tumor suppressor gene PTEN/MMAC1 in primary breast carcinomas. *Cancer Res.* **57**, 3657–3659.

34. Bose, S., Wang, S. I., Terry, M. B., Hibshoosh, H., and Parsons, R. (1998) Allelic loss of chromosome 10q23 is associated with tumor progression in breast carcinomas. *Oncogene* **17**, 123–127.
35. Feilotter, H. E., Coulon, V., McVeigh, J. L., et al. (1999) Analysis of the 10q23 chromosomal region and the PTEN gene in human sporadic breast carcinoma. *Br. J. Cancer* **79**, 718–723.
36. Perren, A., Weng, L. P., Boag, A. H., et al. (1999) Immunohistochemical evidence of loss of PTEN expression in primary ductal adenocarcinomas of the breast. *Am. J. Pathol.* **155**, 1253–1260.
37. Yokomizo, A., Tindall, D. J., Drabkin, H., et al. (1998) PTEN/MMAC1 mutations identified in small cell, but not in non-small cell lung cancers. *Oncogene* **17**, 475–479
38. Kohno, T., Takahashi, M., Manda, R., and Yokota, J. (1998) Inactivation of the PTEN/MMAC1/TEP1 gene in human lung cancers. *Genes Chromosomes Cancer* **22**, 152–156.
39. Cairns, P., Evron, E., Okami, K., et al. (1998) Point mutation and homozygous deletion of PTEN/MMAC1 in primary bladder cancers. *Oncogene* **16**, 3215–3218.
40. Gronbaek, K., Zeuthen, J., Guldberg, P., Ralfkiaer, E., and Hou-Jensen, K. (1998) Alterations of the MMAC1/PTEN gene in lymphoid malignancies [letter]. *Blood* **91**, 4388–4390.
41. Kim, S. K., Su, L. K., Oh, Y., Kemp, B. L., Hong, W. K., and Mao, L. (1998) Alterations of PTEN/MMAC1, a candidate tumor suppressor gene, and its homologue, PTH2, in small cell lung cancer cell lines. *Oncogene* **16**, 89–93.
42. Peiffer, S. L., Herzog, T. J., Tribune, D. J., Mutch, D. G., Gersell, D. J., and Goodfellow, P. J. (1995) Allelic loss of sequences from the long arm of chromosome 10 and replication errors in endometrial cancers. *Cancer Res.* **55**, 1922–1926.
43. Nagase, S., Sato, S., Tezuka, F., Wada, Y., Yajima, A., and Horii, A. (1996) Deletion mapping on chromosome 10q25-q26 in human endometrial cancer. *Br. J. Cancer* **74**, 1979–1983.
44. Risinger, J. I., Hayes, A. K., Berchuck, A., and Barrett, J. C. (1997) PTEN/MMAC1 mutations in endometrial cancers. *Cancer Res.* **57**, 4736–4738.
45. Tashiro, H., Blazes, M. S., Wu, R., et al. (1997) Mutations in PTEN are frequent in endometrial carcinoma but rare in other common gynecological malignancies. *Cancer Res.* **57**, 3935–3940.
46. Mutter, G. L., Lin, M. C., Fitzgerald, J. T., et al. (2000) Altered PTEN expression as a diagnostic marker for the earliest endometrial precancers [see comments]. *J. Natl. Cancer Inst.* **92**, 924–930.
47. Obata, K., Morland, S. J., Watson, R. H., et al. (1998) Frequent PTEN/MMAC mutations in endometrioid but not serous or mucinous epithelial ovarian tumors. *Cancer Res.* **58**, 2095–2097.
48. Su, T. H., Chang, J. G., Perng, L. I., et al. (2000) Mutation analysis of the putative tumor suppressor gene PTEN/MMAC1 in cervical cancer. *Gynecol. Oncol.* **76**, 193–199.
49. Liaw, D., Marsh, D. J., Li, J., et al. (1997) Germline mutations of the PTEN gene in Cowden disease, an inherited breast and thyroid cancer syndrome. *Nat. Genet.* **16**, 64-67.
50. Marsh, D. J., Coulon, V., Lunetta, K. L., et al. (1998) Mutation spectrum and genotype-phenotype analyses in Cowden disease and Bannayan-Zonana syndrome, two hamartoma syndromes with germline PTEN mutation. *Hum. Mol. Genet.* **7**, 507–515.
51. Marsh, D. J., Dahia, P. L., Coulon, V., et al. (1998) Allelic imbalance, including deletion of PTEN/MMACI, at the Cowden disease locus on 10q22-23, in hamartomas from patients with Cowden syndrome and germline PTEN mutation. *Genes Chromosomes Cancer* **21**, 61–69.
52. Marsh, D. J., Dahia, P. L., Zheng, Z., et al. (1997) Germline mutations in PTEN are present in Bannayan-Zonana syndrome [letter]. *Nat. Genet.* **16**, 333–334.

53. Li, D. M. and Sun, H. (1998) PTEN/MMAC1/TEP1 suppresses the tumorigenicity and induces G1 cell cycle arrest in human glioblastoma cells. *Proc. Nat. Acad. Sci. USA* **95**, 15406–15411.
54. Furnari, F. B., Huang, H. J., and Cavenee, W. K. (1998) The phosphoinositol phosphatase activity of PTEN mediates a serum-sensitive G1 growth arrest in glioma cells. *Cancer Res.* **58**, 5002–5008.
55. Davies, M. A., Lu, Y., Sano, T., et al. (1998) Adenoviral transgene expression of MMAC/PTEN in human glioma cells inhibits Akt activation and induces anoikis. *Cancer Res.* **58**, 5285–5290.
56. Li, J., Simpson, L., Takahashi, M., et al. (1998) The PTEN/MMAC1 tumor suppressor induces cell death that is rescued by the AKT/protein kinase B oncogene. *Cancer Res.* **58**, 5667–5672.
57. Weng, L. P., Smith, W. M., Dahia, P. L., et al. (1999) PTEN suppresses breast cancer cell growth by phosphatase activity-dependent G1 arrest followed by cell death. *Cancer Res.* **59**, 5808–5814.
58. Zhu, X., Kwon, C. H., Schlosshauer, P. W., Ellenson, L. H., and Baker, S. J. (2001) PTEN induces G(1) cell cycle arrest and decreases cyclin D3 levels in endometrial carcinoma cells. *Cancer Res.* **61**, 4569–4575.
59. Wen, S., Stolarov, J., Myers, M. P., et al. (2001) PTEN controls tumor-induced angiogenesis. *Proc. Natl. Acad. Sci. USA.* **98**, 4622–4627.
60. Zundel, W., Schindler, C., Haas-Kogan, D., et al. (2000) Loss of PTEN facilitates HIF-1-mediated gene expression. *Genes Dev.* **14**, 391–396.
61. Podsypanina, K., Ellenson, L. H., Nemes, A., et al. (1999) Mutation of Pten/Mmac1 in mice causes neoplasia in multiple organ systems. *Proc. Natl. Acad. Sci. USA* **96**, 1563–1568.
62. Di Cristofano, A., Pesce, B., Cordon-Cardo, C., and Pandolfi, P. P. (1998) Pten is essential for embryonic development and tumour suppression. *Nat. Genet.* **19**, 348–355.
63. Stambolic, V., Suzuki, A., de la Pompa, J. L., et al. (1998) Negative regulation of PKB/Akt-dependent cell survival by the tumor suppressor PTEN. *Cell* **95**, 29–39.
64. Di Cristofano, A., Kotsi, P., Peng, Y. F., Cordon-Cardo, C., Elkon, K. B., and Pandolfi, P. P. (1999) Impaired Fas response and autoimmunity in Pten+/− mice. *Science* **285**, 2122–2125.
65. Li, Y., Podsypanina, K., Liu, X., et al. (2001) Deficiency of Pten accelerates mammary oncogenesis in MMTV-Wnt-1 transgenic mice. *BMC Mol. Biol.* **2**, 2.
66. Di Cristofano, A., De Acetis, M., Koff, A., Cordon-Cardo, C., and Pandolfi, P. P. (2001) Pten and p27KIP1 cooperate in prostate cancer tumor suppression in the mouse. *Nat. Genet.* **27**, 222–224.
67. Scanga, S. E., Ruel, L., Binari, R. C., et al. (2000) The conserved PI3′K/PTEN/Akt signaling pathway regulates both cell size and survival in *Drosophila*. *Oncogene* **19**, 3971–3977.
68. Huang, H., Potter, C. J., Tao, W., et al. (1999) PTEN affects cell size, cell proliferation and apoptosis during *Drosophila* eye development. *Development* **126**, 5365–5372.
69. Gao, X., Neufeld, T. P., and Pan, D. *Drosophila* (2000) PTEN regulates cell growth and proliferation through PI3K-dependent and -independent pathways. *Dev. Biol.* **221**, 404–418.
70. Goberdhan, D. C., Paricio, N., Goodman, E. C., Mlodzik, M., and Wilson, C. (1999) *Drosophila* tumor suppressor PTEN controls cell size and number by antagonizing the Chico/PI3-kinase signaling pathway. *Genes Dev.* **13**, 3244–3258.
71. Myers, M. P., Stolarov, J. P., Eng, C., et al. (1997) P-TEN, the tumor suppressor from human chromosome 10q23, is a dual-specificity phosphatase. *Proc. Natl. Acad. Sci. USA.* **94**, 9052–9057.
72. Maehama, T. and Dixon, J. E. (1998) The tumor suppressor, PTEN/MMAC1, dephosphorylates the lipid second messenger, phosphatidylinositol 3,4,5-trisphosphate. *J. Biol. Chem.* **273**, 13375–13378.
73. Myers, M. P., Pass, I., Batty, I. H., et al. (1998) The lipid phosphatase activity of PTEN is critical for its tumor supressor function. *Proc. Natl. Acad. Sci. USA* **95**, 13513–13518.

74. Furnari, F. B., Huang, H. J., and Cavenee, W. K. (1998) The phosphoinositol phosphatase activity of PTEN mediates a serum-sensitive G1 growth arrest in glioma cells. *Cancer Res.* **58**, 5002–5008.
75. Datta, S. R., Brunet, A., and Greenberg, M. E. (1999) Cellular survival: a play in three Akts. *Genes Dev.* **13**, 2905–2927.
76. Kimura, K. D., Tissenbaum, H. A., Liu, Y., and Ruvkun, G. (1997) daf-2, an insulin receptor-like gene that regulates longevity and diapause in Caenorhabditis elegans [see comments]. *Science* **277**, 942–946.
77. Vowels, J. J. and Thomas, J. H. (1992) Genetic analysis of chemosensory control of dauer formation in *Caenorhabditis elegans. Genetics* **130**, 105–123.
78. Gottlieb, S. and Ruvkun, G. (1994) daf-2, daf-16 and daf-23: genetically interacting genes controlling Dauer formation in *Caenorhabditis elegans*. *Genetics* **137**, 107–120.
79. Larsen, P. L., Albert, P. S., and Riddle, D. L. (1995) Genes that regulate both development and longevity in *Caenorhabditis elegans. Genetics* **139**, 1567–1583.
80. Morris, J. Z., Tissenbaum, H. A., and Ruvkun, G. (1996) A phosphatidylinositol-3-OH kinase family member regulating longevity and diapause in *Caenorhabditis elegans. Nature* **382**, 536–539.
81. Paradis, S. and Ruvkun, G. (1998) Caenorhabditis elegans Akt/PKB transduces insulin receptor-like signals from AGE-1 PI3 kinase to the DAF-16 transcription factor. *Genes Dev.* **12**, 2488–2498.
82. Ogg, S. and Ruvkun, G. (1998) The C. elegans PTEN homolog, DAF-18, acts in the insulin receptor-like metabolic signaling pathway [In Process Citation]. *Mol. Cell* **2**, 887–893.
83. Gil, E. B., Malone Link, E., Liu, L. X., Johnson, C. D., and Lees, J. A. (1999) Regulation of the insulin-like developmental pathway of *Caenorhabditis elegans* by a homolog of the PTEN tumor suppressor gene. *Proc. Natl. Acad. Sci. USA* **96**, 2925–2930.
84. Rouault, J. P., Kuwabara, P. E., Sinilnikova, O. M., Duret, L., Thierry-Mieg, D., and Billaud, M. (1999) Regulation of dauer larva development in Caenorhabditis elegans by daf-18, a homologue of the tumour suppressor PTEN. Curr. Biol. 9, 329–332.
85. Leevers, S. J., Weinkove, D., MacDougall, L. K., Hafen, E., and Waterfield, M. D. (1996) The *Drosophila* phosphoinositide 3-kinase Dp110 promotes cell growth. *EMBO J.* **15**, 6584–6594.
86. Wu, X., Senechal, K., Neshat, M. S., Whang, Y. E., and Sawyers, C. L. (1998) The PTEN/MMAC1 tumor suppressor phosphatase functions as a negative regulator of the phosphoinositide 3-kinase/Akt pathway. *Proc. Natl. Acad. Sci. USA* **95**, 15587–15591.
87. Ramaswamy, S., Nakamura, N., Vazquez, F., et al. (1999) Regulation of G1 progression by the PTEN tumor suppressor protein is linked to inhibition of the phosphatidylinositol 3-kinase/Akt pathway [In Process Citation]. *Proc. Natl. Acad. Sci. USA* **96**, 2110–2115.
88. Dahia, P. L., Aguiar, R. C., Alberta, J., et al. (1999) PTEN is inversely correlated with the cell survival factor Akt/PKB and is inactivated via multiple mechanismsin haematological malignancies. *Hum. Mol. Genet.* **8**, 185–193.
89. Stambolic, V., Tsao, M. S., Macpherson, D., Suzuki, A., Chapman, W. B., and Mak, T. W. (2000) High incidence of breast and endometrial neoplasia resembling human Cowden syndrome in pten+/− mice. *Cancer Res.* **60**, 3605–3611.
90. Dufner, A. and Thomas, G. (1999) Ribosomal S6 kinase signaling and the control of translation. *Exp. Cell Res.* **253**, 100–109.
91. Zhang, H., Stallock, J. P., Ng, J. C., Reinhard, C., and Neufeld, T. P. (2000) Regulation of cellular growth by the *Drosophila* target of rapamycin dTOR. *Genes Dev.* **14**, 2712–2724.
92. Oldham, S., Montagne, J., Radimerski, T., Thomas, G., and Hafen, E. (2000) Genetic and biochemical characterization of dTOR, the *Drosophila* homolog of the target of rapamycin. *Genes Dev.* **14**, 2689–2694.

93. Nakashima, N., Sharma, P. M., Imamura, T., Bookstein, R., and Olefsky, J. M. (2000) The tumor suppressor PTEN negatively regulates insulin signaling in 3T3-L1 adipocytes. *J. Biol. Chem.* **275**, 12889–12895.
94. Lu, Y., Lin, Y. Z., LaPushin, R., et al. (1999) The PTEN/MMAC1/TEP tumor suppressor gene decreases cell growth and induces apoptosis and anoikis in breast cancer cells. *Oncogene* **18**, 7034–7045.
95. Simpson, L., Li, J., Liaw, D., et al. (2001) PTEN expression causes feedback upregulation of insulin receptor substrate 2. *Mol. Cell Biol.* **21**, 3947–3958.
96. Weng, L. P., Smith, W. M., Brown, J. L., and Eng, C. (2001) PTEN inhibits insulin-stimulated MEK/MAPK activation and cell growth by blocking IRS-1 phosphorylation and IRS-1/Grb-2/Sos complex formation in a breast cancer model. *Hum. Mol. Genet.* **10**, 605–616.
97. Gingras, A. C., Kennedy, S. G., O'Leary, M. A., Sonenberg, N., and Hay, N. (1998) 4E-BP1, a repressor of mRNA translation, is phosphorylated and inactivated by the Akt(PKB) signaling pathway. *Genes Dev.* **12**, 502–513.
98. Kohn, A. D., Barthel, A., Kovacina, K. S., et al. (1998) Construction and characterization of a conditionally active version of the serine/threonine kinase Akt. *J. Biol. Chem.* **273**, 11937–11943.
99. Sun, H., Lesche, R., Li, D. M., et al. (1999) PTEN modulates cell cycle progression and cell survival by regulating phosphatidylinositol 3,4,5,-trisphosphate and Akt/protein kinase B signaling pathway. *Proc. Natl. Acad. Sci.* USA **96**, 6199–6204.
100. Kenyon, C., Chang, J., Gensch, E., Rudner, A., and Tabtiang, R. (1993) A *C. elegans* mutant that lives twice as long as wild type [see comments]. *Nature* **366**, 461–464.
101. Dorman, J. B., Albinder, B., Shroyer, T., and Kenyon, C. (1995) The age-1 and daf-2 genes function in a common pathway to control the lifespan of *Caenorhabditis elegans*. *Genetics* **141**, 1399–1406.
102. Lin, K., Dorman, J. B., Rodan, A., and Kenyon, C. (1997) daf-16: An HNF-3/forkhead family member that can function to double the life-span of *Caenorhabditis elegans* [see comments]. *Science* **278**, 1319–1322.
103. Ogg, S., Paradis, S., Gottlieb, S., et al. (1997) The Fork head transcription factor DAF-16 transduces insulin-like metabolic and longevity signals in *C. elegans*. *Nature* **389**, 994–999.
104. Brunet, A., Bonni, A., Zigmond, M. J., et al. (1999) Akt promotes cell survival by phosphorylating and inhibiting a Forkhead transcription factor. *Cell* **96**, 857–868.
105. Kops, G. J., de Ruiter, N. D., De Vries-Smits, A. M., Powell, D. R., Bos, J. L., and Burgering, B. M. (1999) Direct control of the Forkhead transcription factor AFX by protein kinase B. *Nature* **398**, 630–634.
106. Tang, E. D., Nunez, G., Barr, F. G., and Guan, K. L. (1999) Negative regulation of the forkhead transcription factor FKHR by Akt. *J. Biol. Chem.* **274**, 16741–16746.
107. Guo, S., Rena, G., Cichy, S., He, X., Cohen, P., and Unterman, T. (1999) Phosphorylation of serine 256 by protein kinase B disrupts transactivation by FKHR and mediates effects of insulin on insulin-like growth factor-binding protein-1 promoter activity through a conserved insulin response sequence. *J. Biol. Chem.* **274**, 17184–17192.
108. Nakae, J., Park, B. C., and Accili, D. (1999) Insulin stimulates phosphorylation of the forkhead transcription factor FKHR on serine 253 through a Wortmannin-sensitive pathway. *J. Biol. Chem.* **274**, 15982–15985.
109. Medema, R. H., Kops, G. J., Bos, J. L., and Burgering, B. M. (2000) AFX-like Forkhead transcription factors mediate cell-cycle regulation by Ras and PKB through p27kip1. *Nature* **404**, 782–787.
110. Nakamura, N., Ramaswamy, S., Vazquez, F., Signoretti, S., Loda, M., and Sellers, W. R. (2000) Forkhead transcription factors are critical effectors of cell death and cell cycle arrest downstream of PTEN. *Mol. Cell. Biol.* **20**, 8969–8982.

111. Warren, R. S., Yuan, H., Matli, M. R., Ferrara, N., and Donner, D. B. (1996) Induction of vascular endothelial growth factor by insulin-like growth factor 1 in colorectal carcinoma. *J. Biol. Chem.* **271**, 29483–29488.
112. Mazure, N. M., Chen, E. Y., Laderoute, K. R., and Giaccia, A. J. (1997) Induction of vascular endothelial growth factor by hypoxia is modulated by a phosphatidylinositol 3-kinase/Akt signaling pathway in Ha-ras-transformed cells through a hypoxia inducible factor-1 transcriptional element. *Blood* **90**, 3322–3331.
113. Zelzer, E., Levy, Y., Kahana, C., Shilo, B. Z., Rubinstein, M., and Cohen, B. (1998) Insulin induces transcription of target genes through the hypoxia-inducible factor HIF-1alpha/ARNT. *EMBO J.* **17**, 5085–5094.
114. Zhong, H., Chiles, K., Feldser, D., et al. (2000) Modulation of hypoxia-inducible factor 1alpha expression by the epidermal growth factor/phosphatidylinositol 3-kinase/PTEN/AKT/FRAP pathway in human prostate cancer cells: implications for tumor angiogenesis and therapeutics. *Cancer Res.* **60**, 1541–1545.
115. Wang, X., Gjorloff-Wingren, A., Saxena, M., Pathan, N., Reed, J. C., and Mustelin, T. (2000) The tumor suppressor PTEN regulates T cell survival and antigen receptor signaling by acting as a phosphatidylinositol 3-phosphatase. *J. Immunol.* **164**, 1934–1939.
116. Shan, X., Czar, M. J., Bunnell, S. C., et al. (2000) Deficiency of PTEN in Jurkat T cells causes constitutive localization of Itk to the plasma membrane and hyperresponsiveness to CD3 stimulation. *Mol. Cell. Biol.* **20**, 6945–6957.
117. Lee, J. O., Yang, H., Georgescu, M. M., et al. (1999) Crystal structure of the PTEN tumor suppressor: implications for its phosphoinositide phosphatase activity and membrane association. *Cell* **99**, 323–334.
118. Georgescu, M. M., Kirsch, K. H., Akagi, T., Shishido, T., and Hanafusa, H. (1999) The tumor-suppressor activity of PTEN is regulated by its carboxyl-terminal region. *Proc. Natl. Acad. Sci. USA* **96**, 10182–10187.
119. Vazquez, F., Ramaswamy, S., Nakamura, N., and Sellers, W. R. (2000) Phosphorylation of the PTEN tail regulates protein stability and function [In Process Citation]. *Mol. Cell. Biol.* **20**, 5010–5018.
120. Torres, J. and Pulido, R. (2001) The tumor suppressor PTEN is phosphorylated by the protein kinase CK2 at its C-terminus: implications for PTEN stability to proteasome-mediated degradation. *J. Biol. Chem.* **12**, 993–998.
121. Pinna, L. A. (1990) Casein kinase 2: an "eminence grise" in cellular regulation? *Biochim. Biophys. Acta* **1054**, 267–284.
122. Meisner, H. and Czech, M. P. (1991) Phosphorylation of transcriptional factors and cell-cycle-dependent proteins by casein kinase II. *Curr. Opin. Cell Biol.* **3**, 474–483.
123. Allende, J. E. and Allende, C. C. (1995) Protein kinases. 4. Protein kinase CK2: an enzyme with multiple substrates and a puzzling regulation. *FASEB J.* **9**, 313–323.
124. Wu, X., Hepner, K., Castelino-Prabhu, S., et al. (2000) Evidence for regulation of the PTEN tumor suppressor by a membrane-localized multi-PDZ domain containing scaffold protein MAGI-2. *Proc. Natl. Acad. Sci. USA* **97**, 4233–4238.
125. Wu, Y., Dowbenko, D., Spencer, S., et al. (2000) Interaction of the tumor suppressor PTEN/MMAC with a PDZ domain of MAGI3, a novel membrane-associated guanylate kinase. *J. Biol. Chem.* **275**, 21477–21485.
126. Tamura, M., Gu, J., Matsumoto, K., Aota, S., Parsons, R., and Yamada, K. M. (1998) Inhibition of cell migration, spreading, and focal adhesions by tumor suppressor PTEN. *Science* **280**, 1614–1617.
127. Gu, J., Tamura, M., Pankov, R., et al. (1999) Shc and FAK differentially regulate cell motility and directionality modulated by PTEN. *J. Cell Biol.* **146**, 389–403.
128. Maier, D., Jones, G., Li, X., et al. (1999) The PTEN lipid phosphatase domain is not required to inhibit invasion of glioma cells. *Cancer Res.* **59**, 5479–5482.

129. Rodrigues, G. A., Falasca, M., Zhang, Z., Ong, S. H., and Schlessinger, J. (2000) A novel positive feedback loop mediated by the docking protein Gab1 and phosphatidylinositol 3-kinase in epidermal growth factor receptor signaling. *Mol. Cell. Biol.* **20**, 1448–1459.
130. Gimm, O., Attie-Bitach, T., Lees, J. A., Vekemans, M., and Eng, C. (2000) Expression of the PTEN tumour suppressor protein during human development. *Hum. Mol. Genet.* **9**, 1633–1639.
131. Torres, J. and Pulido, R. (2001) The tumor suppressor PTEN is phosphorylated by the protein kinase CK2 at its C terminus. Implications for PTEN stability to proteasome-mediated degradation. *J. Biol. Chem.* **276**, 993–998.

10

VHL and Kidney Cancer

Michael Ohh and William G. Kaelin, Jr.

1. Introduction

Kidney cancer affects approximately 30,000 individuals in the United States and is responsible for more than 12,000 deaths each year ***(1,2)***. It has been estimated that 3% of adult malignancies are kidney cancers ***(1)***. Like many cancers, such as breast and colon cancers, kidney cancer occurs in both sporadic and familial forms. The vast majority of kidney cancer—up to 96%—is sporadic or nonhereditary, while 4% is thought to have a hereditary basis ***(2)***. There are at least four types of familial kidney cancer that have been categorized: renal carcinoma associated with Birt-Hogg-Dube syndrome, hereditary papillary renal carcinoma, hereditary clear cell renal carcinoma, and renal carcinoma associated with the von Hippel-Lindau (VHL) disease ***(3)***. This chapter will focus on the recent molecular understanding of VHL disease.

2. VHL Disease

VHL disease is a hereditary cancer syndrome characterized by a predisposition to developing tumors in a number of organs, including kidney, eye, cerebellum, spinal cord, adrenal gland, pancreas, inner ear, and epididymis ***(4-7)***. VHL-associated renal carcinomas are characterized by earlier onset than their sporadic counterparts and are frequently multifocal ***(7)***. They are invariably of clear cell type and are often associated with renal cysts ***(7–9)***. Renal clear cell carcinoma is also the predominant form of kidney cancer in the general population and has been reported to metastasize in up to 40% of affected individuals ***(8,9)***. Currently, renal clear cell carcinoma is the major cause of morbidity and mortality in VHL patients ***(10)***.

VHL disease afflicts 1 in 40,000 individuals ***(7)***. Clinically, it displays an autosomal dominant pattern of inheritance due to its variable but virtually complete penetrance ***(7,11)***. Molecularly, however, VHL disease is a recessive disorder ***(7,12,13)***. Specifically, affected individuals are heterozygous at the *VHL* locus, harboring a wild-type *VHL* allele and a mutated *VHL* allele. Tumor development occurs when the remaining wild-type *VHL* allele is inactivated or lost. The likelihood that at least one susceptible cell will undergo this event during the lifetime of a *VHL* germline heterozygote is approxi-

From: *Methods in Molecular Biology, Vol. 222: Tumor Suppressor Genes: Pathways and Isolation Strategies*
Edited by: Wafik S. El-Deiry © Humana Press Inc., Totowa, NJ

mately 90%, thus accounting for the seemingly autosomal dominant pattern of inheritance ***(7)***. The requirement of biallelic inactivation of the *VHL* locus for the development of VHL-associated neoplasms conforms to the Knudson's two-hit model of carcinogenesis by tumor suppressor genes ***(7,14)***. In keeping with this model, functional inactivation of both *VHL* alleles has been demonstrated in the majority of sporadic renal clear cell carcinoma (75%) ***(15–19)*** and cerebellar haemangioblastoma (50–60%) ***(20–23)***. Unexpectedly, however, *VHL* mutations are rare in sporadic pheochromocytoma, even though VHL patients are at increased risk for this tumor ***(24,25)***. The explanation for this exception to the Knudson two-hit model is not known.

Inactivation of the remaining wild-type *VHL* allele, manifest as loss of heterozygosity at the *VHL* locus, has been documented in premalignant renal cysts in VHL patients ***(26)***. These cysts arise from the epithelial lining of the proximal renal tubule ***(26)***. Thus it appears that biallelic inactivation of *VHL* in the kidney is sufficient for renal cyst formation and it is presumed that additional mutations at non-*VHL* genetic loci are required for conversion to a frank renal carcinoma. In summary, inactivation of the *VHL* gene product (pVHL) is an early step in the pathogenesis of renal clear cell carcinoma. The *VHL* gene appears to play the role of a "gatekeeper" for kidney cancer, much as the APC (adenomatous polyposis coli) gene plays the role of gatekeeper in colorectal cancer ***(27)***.

3. The VHL Gene

Two lines of investigation pointed to the possible existence of a renal carcinoma suppressor gene on the short arm of chromosome 3. Firstly, it was noted that 3p deletions were common in a variety of solid tumors, including, notably, renal carcinomas ***(28,29)***. Second, several studies identified kindreds with early-onset, bilateral, multifocal renal clear cell carcinoma and documented in such families germline translocations in which chromosome 3p was fused to chromosome 6, 8, or 11 ***(30–32)***. Moreover, family members who did not display this genetic abnormality remained kidney cancer-free, whereas all affected family members inherited the translocation ***(30–32)***.

In an effort to isolate the gene on chromosome 3p responsible for renal carcinoma, scientists explored the hereditary renal carcinoma associated with VHL disease with the supposition the *VHL* gene may be involved in the nonhereditary form of kidney cancer. Toward this end, Seizinger and co-workers, in 1988, using genetic linkage analysis, mapped the *VHL* gene to a 6- to 8-centimorgan (cM) region of chromosome 3p25-26 ***(33)***. The putative site of *VHL* locus was further narrowed by multipoint linkage analysis to a 4-cM interval on 3p26 between RAF1 and the anonymous marker D3S18 ***(34)***. Yao et al. and Richards et al. used pulse-field gel electrophoresis to demarcate the minimal genomic region that was commonly deleted among unrelated VHL families and identified a collection of cDNAs that mapped to this region ***(34,35)***. Ultimately, this information enabled Latif and co-workers to isolate and clone the *VHL* gene in 1993 ***(36)***.

The authenticity of the *VHL* gene was confirmed by the detection of intragenic mutations that segregated with the disease in VHL kindreds ***(36,37)***. Currently, mutations in the *VHL* gene can be detected in virtually all patients displaying phenotypic features of the VHL disease ***(38)***. The *VHL* gene is expressed ubiquitously in fetal and adult tissues ***(39,40)***. Hence, *VHL* expression is not restricted to the organs affected by VHL disease. Nonetheless, the expression pattern in the fetal kidney suggest a role in normal renal tubular development and differentiation ***(39,40)***. The *VHL* gene contains 3 exons ***(36)***. Moreover, a second *VHL* mRNA isoform, in which exon 1 is spliced directly to exon 3, has been observed in some VHL patients, suggesting that this isoform is insufficient to suppress tumor development ***(15)***.

Tumor-derived *VHL* gene mutations have been identified throughout the *VHL* open reading frame with the exception of the sequences encoding the first 54 amino acid residues ***(41,42)***. However, striking correlations between certain germline mutations and specific clinical manifestations of the disease are emerging. Such genotype–phenotype correlations have led to the classification of VHL disease into subcategories depending on their propensity to develop pheochromocytoma: type I families have a low risk and type II families have a high risk of developing pheochromocytoma. Type II is further subdivided into IIa (low risk of renal carcinoma), IIb (high risk of renal carcinoma), and IIc (develop exclusively pheochromocytoma without the other stigmata of the disease) ***(7)***. Whereas germline mutations associated with type I disease are regularly deletions, microinsertions, and nonsense mutations, germline mutations associated with type II disease are almost invariably (>90%) missense mutations ***(37,42)***.

4. The VHL Proteins

The human *VHL* gene encodes two proteins with apparent molecular weights of approximately 30 ($pVHL^{30}$) and 19 kDa ($pVHL^{19}$) ***(43–45)***. The former contains 213 amino acids and the latter contains 160 amino acids as a result of translation from the second methionine at residue 54. The N-terminus of $pVHL^{30}$ contains a region homologous to an acidic pentameric repeat found in *Trypanosoma brucei* ***(36)***.

To date, the expression of $pVHL^{30}$ and $pVHL^{19}$ are not mutually exclusive. Furthermore, reconstitution of either $pVHL^{30}$ or $pVHL^{19}$ into *VHL−/−* renal carcinoma cells suppresses their ability to form tumors in nude mice ***(43,45–47)***. Intriguingly, *VHL* tumor-derived mutations invariably map outside the first 54 amino acid residues ***(41,42)***. This likely is due to the fact that mutations in this region would still, in principle, be compatible with the production of biologically active $pVHL^{19}$.

There are subtle differences between $pVHL^{30}$ and $pVHL^{19}$ in terms of subcellular localization. Biochemical fractionation and immunohistochemical studies have shown that although $pVHL^{30}$ is predominantly cytoplasmic, significant amounts can be detected in the nucleus and associated with the endoplasmic reticulum ***(46,48–51)***. $pVHL^{19}$ is found equally distributed between the nucleus and the cytoplasm, but is excluded from the membrane fraction ***(44)***. This raises the possibility that the NH_2-terminal acid pentameric repeat is required for membrane association.

Complicating the interpretation of pVHL subcellular localization further is the fact that the $pVHL^{30}$ shuttles between the nucleus and cytoplasm ***(52,53)***. Whether this is also true for $pVHL^{19}$ is not known. It appears that the steady-state distribution of pVHL can be influenced by cell density ***(52)***. Furthermore, treatment of cells with RNA polymerase II inhibitors causes pVHL to accumulate in the nucleus ***(53)***. Unraveling the all physiologic and biochemical signals that influence pVHL shuttling is an area of intensive investigation. The importance of shuttling, however, is underscored by the observation that pVHL mutants that are unable to shuttle by virtue of fusion to strong heterologous nuclear import or export signals are functionally defective ***(53)***.

Whereas $pVHL^{30}$ is a phosphoprotein, $pVHL^{19}$ is not. The specific phosphorylated residues and the physiological relevant kinase (s) have not been determined. The functional significance of $pVHL^{30}$ phosphorylation is unknown.

5. The pVHL–Multiprotein Complex

When the VHL gene was isolated, it was clear that neither the nucleotide nor the amino acid sequence of the pVHL had any significant homology to proteins with a known function. Thus, to discern the function (s) of pVHL, several groups sought to identify cellular proteins that bound to pVHL, with the supposition that pVHL-associated proteins might have identifiable functions.

5.1. Elongin B and Elongin C

The first pVHL-associated proteins to be isolated were elongins B and C ***(54,55)***. Elongins B and C had earlier been found as a binary complex that enhances the transcriptional elongation activity of a protein called elongin A ***(56)***. Comparison of elongin A and pVHL revealed a short stretch of residues that were highly conserved between the two proteins. This stretch corresponding to the pVHL amino acid residues 157–172 was determined to be the minimal region necessary and sufficient for binding elongins B and C ***(54,56)***. Not so surprisingly, the homologous region on elongin A likewise associates with elongins B and C ***(57)***, thereby establishing these residues on pVHL and elongin A as an elongin B/C-binding motif.

In a cell-free system, pVHL can negatively regulate the elongin A transcriptional elongation activity by competing away elongins B and C ***(55)***. In other words, the association of elongin B/C complex to elongin A and pVHL is mutually exclusive. Moreover, residues defining the elongin B/C-binding motif on pVHL are frequently mutated in VHL kindreds, underscoring the significance of elongin B/C binding to tumor suppressor function of pVHL ***(54,58)***. These observations led to the hypothesis that the failure of pVHL to regulate elongin/SIII activity negatively in vivo leads to tumorigenesis. However, it should be noted that, to date, there is no evidence that elongin/SIII regulates the transcription of specific genes in vivo. Perhaps more important, elongins B and C are in vast molar excess of elongin A and pVHL ***(59)***. Furthermore, other cellular proteins with the elongin B/C-binding motif have since been identified ***(59)***. Although the above arguments do not formally disprove the elongation theory, the isolation of the next pVHL-associated protein strongly suggested an alternative model of tumor suppression by pVHL.

5.2. Cullin 2

In 1996, Kipreos et al. identified a novel gene family called cullins in *Caenorhabditis elegans* during a screen for mutants that displayed excess postembryonic cell divisions ***(60)***. There are at least 3 cullins in budding yeast, five in nematodes, and six in humans. The product of the human *Cul2* gene, an approximately ~80-kDa protein, was found to bind stoichiometrically to elongin B, elongin C, and pVHL ***(51,61)***.

Cul2 bears remarkable similarity to the yeast cullin, Cdc53 ***(51,60,61)***. Cdc53 is part of a multiprotein ubiquitin–ligase complex known as SCF that includes Skp1 (S-phase kinase-associated protein 1, suppressor of Cdk inhibitor proteolysis and of kinetochore

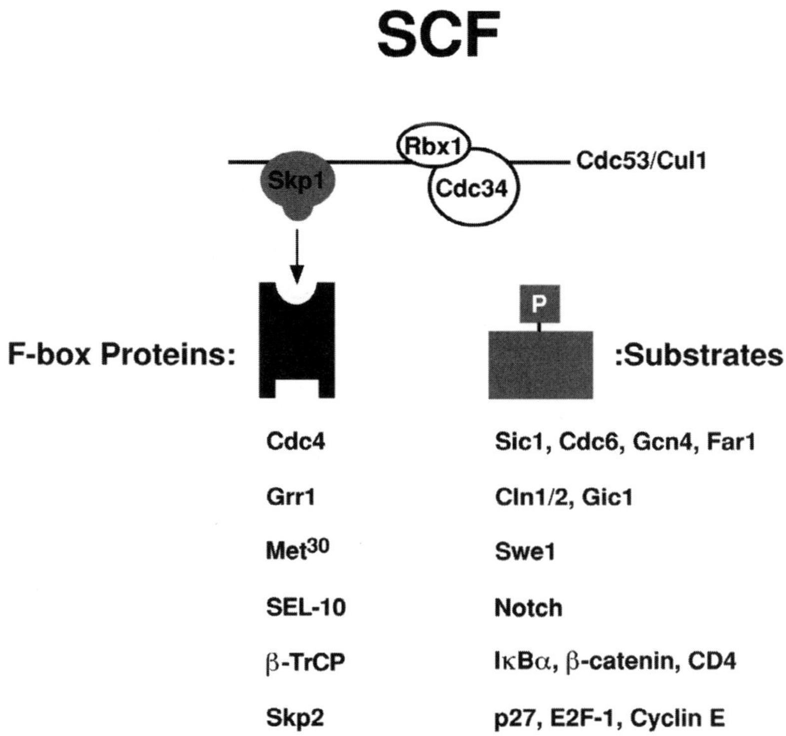

Fig. 1. Modular SCF multiprotein complex. Skp1 of the SCF complex acts as an adapter protein bridging multiple F-box proteins that confer substrate specificity to the rest of the E3 ligase complex. The nucleated SCF complex, in concert with a specific E2 ubiquitin-conjugating enzyme such as Cdc34, targets the substrate for ubiquitylation.

protein) and a substrate-conferring F-box protein (so-called because of a collinear motif first identified in cyclin F) (**Fig. 1**) ***(62–66)***. SCF complexes, with the cooperation of ubiquitin-conjugating enzyme, such as Cdc34, target specific proteins for ubiquitylation ***(64,65)***. For example, SCF^{Cdc4} recognizes G1 cyclin-dependent kinase (cdk)-phosphorylated Sic1 through Cdc4 for ubiquitylation (**Fig. 2**) ***(67,68)***. This polyubiquitin-tagged Sic1 is then selectively recognized by the 26S proteasome for destruction, thus allowing G1/S transition in yeast ***(67,68)***. Human Cul1 has also been demonstrated to be part of the mammalian SCF complexes ***(66)***.

Intriguingly, elongin C bears significant homology to Skp1 and, using sophisticated sequence comparison programs, one can identify a collinear sequence of pVHL that loosely resembles an F-box ***(62,69)***. Furthermore, these similarities based on sequence comparisons are borne out in the pVHL/elongin B/elongin C structure ***(69)***. Specifically, Pavletich and co-workers showed that pVHL has two domains: an α-domain and a β-domain ***(69)***. The α-domain is essential for directly binding elongin C to nucleate the pVHL complex VEC (pVHL-elongin B/C-Cul2). This domain contains a region loosely resembling the F-box, which corresponds to the elongin B/C-binding motif. The β-domain is predicted to function as a putative substrate-docking site. Moreover, disease-associated mutations found in VHL kindreds frequently map to the surface residues of α- and β-domains, suggesting the importance of both domains in the tumor suppressor function of pVHL ***(69)***.

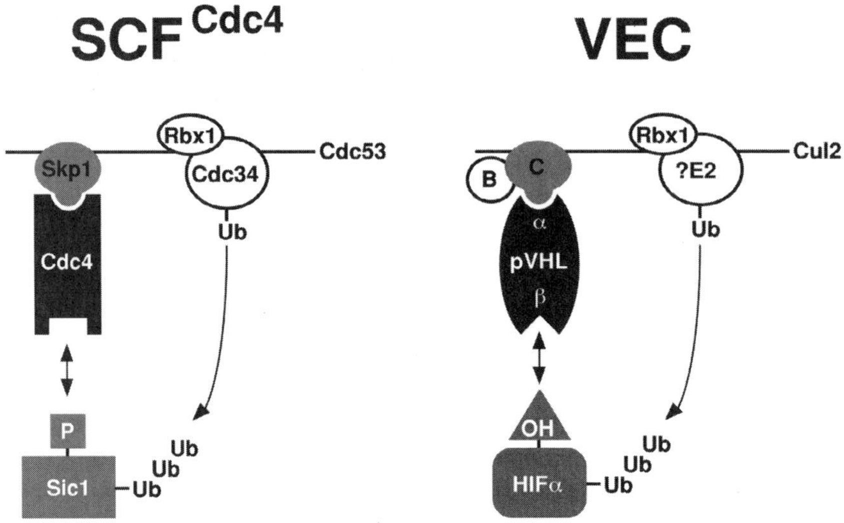

Fig. 2. Comparison of SCF and VEC complexes. There are significant structural similarities between SCF complexes and the VEC complex. In this figure SCF^{cdc4} is chosen as a representative SCF complex. Both complexes contain the Rbx1 (also known as ROC1/Hrt1) ring-H2 finger protein. Cdc53 and Cul2 are orthologs. Elongin C (denoted as C; elongin B is denoted as B) is highly conserved with Skp1, and similar to the role Skp1 plays in SCF complexes, elongin C serves as an adapter or bridge between pVHL and Cul2 in the VEC complex. The pVHL α-domain loosely resembles an F-box and the β-domain binds to substrates. The F-box protein Cdc4 recognizes phosphorylated Sic1, whereupon Sic1 is targeted for ubiquitylation by the SCF^{Cdc4} E3 complex in collaboration with an E2 Cdc34 enzyme. The polyubiquitylated Sic1 is subsequently degraded by the 26S proteasome, allowing G1/S transition in yeast. In regards to VEC complex, pVHL, via the β-domain, specifically recognizes the prolyl hydroxylated HIFα subunits, which are then ubiquitylated by the VEC complex, presumably in collaboration with a cognate E2 enzyme.

Thus, the remarkable structural homology between VEC and SCF complexes led to the notion that pVHL may have a role in targeting certain protein(s) for ubiquitin-mediated proteolysis (**Fig. 2**). In support of this idea, Lisztwan et al. and Iwai et al. concurrently showed that pVHL immunoprecipitates with a ubiquitin ligase activity when supplemented with a ubiquitin-conjugating enzyme ***(70,71)***. Importantly, tumor-derived pVHL mutants failed to immunoprecipitate with this activity ***(70,71)***.

5.3. Rbx1/ROC1/Hrt1

Several groups independently identified a cullin partner protein called Rbx1 (also called ROC1 or Hrt1) with an associated ubiquitin ligase activity ***(72–75)***. Rbx1 is thought to aid in the recruitment of an appropriate ubiquitin-conjugating enzyme and thereby to enhance ubiquitin ligase activity ***(73–75)***. Rbx1 is a common component of the VEC and SCF complexes, further strengthening the notion that pVHL is an active component of a ubiquitin ligase complex (**Fig. 2**) ***(66,76)***.

6. pVHL Function

6.1. Targeting HIF for Ubiquitylation

VHL-associated tumors such as retinal and CNS hemangioblastomas and renal clear cell carcinoma are known to be hypervascular ***(7)***. This has been attributed to the inability of such tumors to suppress the expression of angiogenic peptides such as vascular endothelial growth factor (VEGF) that are normally induced by hypoxia ***(7,11,12)***. Indeed, several groups showed that pVHL-defective cells routinely overproduce hypoxia-inducible mRNAs even in the presence of oxygen ***(47,77–80)***. Furthermore, this defect can be corrected by reconstitution of pVHL function using gene transfer methodologies ***(47,61,78,81,82)***.

HIF is a heterodimeric transcriptional activator composed of α- and β-subunits that binds DNA in a sequence-specific manner and regulates the transcription of hypoxia-inducible genes such as the aforementioned VEGF ***(83–86)***. Whereas the α-subunit is labile under normal oxygen tension, the abundance of the β-subunit (also known as ARNT, aryl hydrocarbon receptor nuclear translocator) is not influenced by changes in oxygen availability ***(83,84)***. Furthermore, the destruction of the α-subunits is linked to the polyubiquitylation of a specific region of the HIFα polypeptide called the oxygen-dependent degradation (ODD) domain ***(87,88)***. Hence, the regulation of HIF occurs through the control of the unstable α-subunit. Recently, Maxwell et al. observed that the α subunit is stabilized in cells devoid of wild-type pVHL ***(80)***. More important, they found that HIFα subunits are physically associated with pVHL ***(80)***. These observations strongly suggested that the α-subunits of HIF are the physiologic targets of pVHL.

In proof of this notion, several groups independently demonstrated a direct physical interaction of HIF1 and 2α subunits via the predicted β-domain of pVHL ***(89–91)***. Intriguingly, the ODD domain of HIFα subunit is both necessary and sufficient for binding to pVHL ***(89–91)***. Furthermore, pVHL is specifically required for the ubiquitylation of HIFα subunits in vitro via the ODD domain ***(89–92)***. pVHL mutations associated with the classical stigmata of VHL disease almost always compromise the integrity of the α-domain, which is required for elongin C binding, or the β-domain, which is required for binding to HIFα subunits ***(89)***. Hypoxia mimetics attenuate the ubiquitin ligase activity of pVHL, reconfirming the notion that VEC selectively targets HIFα subunits in the presence of oxygen ***(89)***. Failure to degrade HIFα subunit provides a mechanistic explanation for the aforementioned overproduction of hypoxia-inducible mRNAs in cells lacking wild-type pVHL. These observations established pVHL as a vital component of a bona-fide ubiquitin ligase that marks HIFα subunits with a polyubiquitin chain for destructive targeting by the 26S proteasome.

6.2. Prolyl Hydroxylation of HIFα Subunit: Marked for Death

How pVHL recognized HIFα subunits selectively under normoxia was no longer a question limited to the VHL tumor suppressor field but an outstanding mystery in the general area of biology. Recently, however, Ivan et al. and Jaakkola et al. independently showed that the interaction of pVHL and HIFα is governed by an oxygen-dependent posttranslational modification (**Fig. 3**) ***(93,94)***. In the presence of oxygen and iron, HIFα is hydroxylated at a highly conserved proline residue at position 564 ***(93,94)***. Hydroxylation at this proline residue by a yet unidentified prolyl hydroxylase enzyme is necessary for binding to pVHL ***(93,94)***. Thus, under high oxygen tension or normoxia, pVHL, as part of a ubiquitin ligase complex, proceeds to ubiquitylate the prolyl hydroxylated HIFα. Thus, studies of pVHL have provided new insights into how cells sense and respond to changes in oxygen availability.

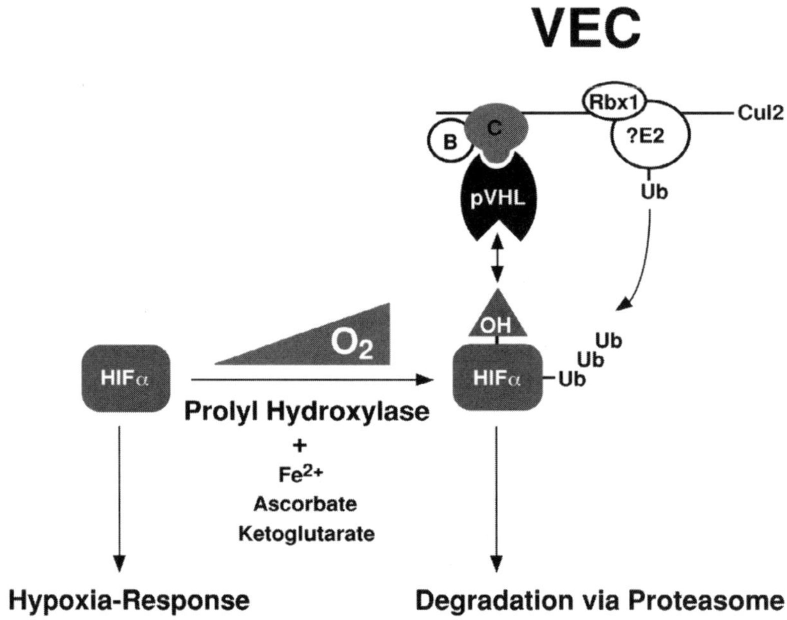

Fig. 3. Oxygen sensing and regulation of hypoxia-inducible proteins by the VEC complex. Under low oxygen tension or hypoxia, HIFα subunit is unmodified and escapes recognition by pVHL. Thus, the untargeted HIFα subunit binds the constitutively stable HIFβ subunit, forming a heterodimeric transcriptional activator. The HIF transcription factor upregulates the expression of numerous hypoxia-inducible genes including *VEGF* and *GLUT1*. However, under normal oxygen tension (or normoxia) and in the presence of the cofactors iron, ascorbate, and ketoglutarate, the HIFα subunit is post-translationally modified by prolyl hydroxylase at a conserved proline residue 564. The prolyl hydroxylated HIFα subunit is then specifically recognized by the β-domain of pVHL and is subsequently ubiquitylated by the VEC complex. The polyubiquitylated HIFα subunit is then selectively targeted for destruction by the 26S proteasome, thus suppressing the hypoxic response.

7. Other pVHL-Associated Functions

7.1. Assembly of Extracellular Fibronectin Matrix

Dysregulated extracellular matrix (ECM) formation, as well as abnormal cell–matrix interactions, are hallmarks of solid tumors ***(95)***. Alterations in the fibronectin component of the ECM has been correlated with cellular transformation ***(95)***. Conversely, multimeric forms of fibronectin or the overproduction of fibronectin receptor, such as α5β1 integrin, have been shown to promote differentiation and suppress the proliferative and metastatic potentials of transformed cells in various model systems ***(95–97)***.

The pVHL binds directly to fibronectin ***(50,98)***. All disease-associated pVHL mutants examined to date exhibit decreased or undetectable binding to fibronectin ***(50,98)***. Moreover, the physical interaction of fibronectin with the wild-type pVHL has been linked to the proper assembly of an extracellular fibronectin matrix ***(50,98)***. The mechanism by which pVHL promotes fibronectin matrix assembly is unknown. However, in light of pVHL involvement in the VEC ubiquitin ligase, one model predicts that VEC targets malprocessed or malfolded fibronectin, which would otherwise disrupt or interfere with the formation of matrix in the extracellular space, for ubiquitin-mediated proteolysis. Specifically, defective fibronectin might undergo retrograde transport to the cytoplasmic surface of the endoplasmic reticulum, whereupon it would be targeted for ubiquitylation by the VEC complex for subsequent recognition and destruction by the 26S proteasome. In should be noted that $pVHL^{30}$ binds fibronectin in the endoplasmic reticulum ***(50)***, and $pVHL^{19}$, which is excluded from the endoplasmic reticulum, fails to associate fibronectin in cells ***(44)***. Hence, to date, this is the first and only variance detected between $pVHL^{30}$ and $pVHL^{19}$ attributed to differential subcellular localization. The significance of this contrariety is unknown.

Several other lines of investigation have also pointed to an important role of pVHL in the control of extracellular matrix formation and cell-matrix interactions. Kerbel and co-workers demonstrated that renal carcinoma cells lacking pVHL grow as tightly packed amorphous spheroids in a three-dimensional growth assay, indicative of an undifferentiated phenotype ***(99)***. In contrast, renal carcinoma cells reconstituted with pVHL form loose aggregates, which, on microscopic and ultrastructural examination, exhibit evidence of epithelial differentiation, such as trabecular and tubular structures ***(99)***. Furthermore, reintroduction of pVHL suppresses the growth of multicellular tumor spheroids ***(99)***. These observations correlate with differences in fibronectin matrix deposition, with pVHL-expressing cells producing appreciable fibronectin arrays in the extracellular space, suggesting that fibronectin is actively involved in growth and differentiation of cells through the interaction with pVHL ***(99)***.

Furthermore, Burk and colleagues demonstrated that pVHL-expressing renal carcinoma cells grown on ECM differentiate into organized epithelial monolayers, whereas pVHL-deficient cells are branched, disorganized, and fail to arrest under high cell density ***(100)***. The reintroduction of pVHL in renal carcinoma cells markedly increases the expression of hepatocyte nuclear factor 1α, a transcription factor and global activator of proximal tubule-specific genes in the nephron ***(100)***. These observations strengthen the notion that pVHL plays a causal role in renal cell differentiation and mediates growth arrest through the integration of cell–ECM signals. Vande Woude and colleagues showed that renal carcinoma cells lacking a functional pVHL produce increased levels of matrix metalloproteinases (MMPs) and attenuated levels of tissue inhibitors of metalloproteinases (TIMPs), and demonstrate increased invasiveness on growth factor-reduced matrigel (an artificial ECM) in response to hepatocyte growth factor/scatter factor ***(101)***. Thus, pVHL is required for regulating the changes that control tissue invasiveness in the context of ECM.

7.2. Regulation of Extracellular pH

Lerman and co-workers, using differential display, observed that *VHL−/−* cells overproduce carbonic anhydrases 9 (CA9) and 12 (CA12) ***(102)***. CA9 and CA12 belong to the larger family of CAs and promote the extracellular hydration of CO_2 and the formation of unstable carbonic acid ***(103)***. This activity affects the function of several ionic channels and leads ultimately to the acidification of the microenvironment ***(103,104)***. Interestingly, CA9 and CA12 are frequently overproduced in renal carcinomas and have been proposed as diagnostic markers ***(102,105)***. Moreover, the extracellular pH of tumors is generally more acidic than normal cells and, at least in vitro, acidic pH favors the growth, spreading, and invasive properties of tumor cells ***(104)***. The mechanism by which pVHL negatively regulates CA9 and CA12 is unknown.

7.3. Link to Cell Cycle

Perturbation of the cell cycle is a common underlying mechanism to achieving cellular transformation. Cells lacking functional pVHL fail to exit the cell cycle following serum starvation ***(106,107)***. This failure has been correlated with the inability to upregulate the cdk (cyclin-dependent kinase) inhibitor p27, and overexpression of pVHL leads to the accumulation of p27 ***(106,107)***. Although these observations suggest a potential role of pVHL in cell cycle, a direct causal link between pVHL and p27 has not been established.

Transforming growth factor (TGF) α is an epithelial mitogen that is upregulated in renal carcinoma cells lacking pVHL ***(108,109)***. TGFα can also stimulate the proliferation of pVHL containing renal carcinoma cells as well as primary renal proximal tubule epithelial cells, the suspected origin of renal clear cell ***(108,109)***. Importantly, a drug that specifically blocks the epidermal growth factor receptor (EGFR), the receptor through which TGFα stimulates proliferation, suppresses the growth of cells lacking functional pVHL ***(109)***. These observations suggest a potential TGFα–EGFR autocrine loop in renal proximal tubule epithelial cells lacking nonfunctional pVHL, possibly accounting for the failure of such cells to exit the cell-cycle following growth factor withdrawal.

8. VHL-Knockout Mice

To examine the role of pVHL in development and tumor suppression, Gnarra and coworkers generated mice with a homozygous deletion of the *VHL* loci ***(110)***. *VHL*-nullizygous mice died *in utero* between d 10.5 and d 12.5 of gestation. The cause of death appeared to be a lack of embryonic placental vasculogenesis. Paradoxically, the trophoblast exhibited reduced level of VEGF, which potentially explains the attenuated vascularization of the placenta. In addition, *VHL*−/− embryos exhibit defects in fibronectin matrix assembly, in keeping with the considerations above ***(50)***. *VHL*-heterozygous mice, however, developed normally and remained tumor-free throughout life ***(110)***. Thus, the autosomal dominant pattern of the VHL disease in humans was not recapitulated in mice.

Since *VHL*−/− mice are not viable, thus precluding the study of pVHL inactivation in adults, Haase et al. generated conditional *VHL*-knockout mice using the Cre-loxP-mediated recombination in which *VHL* can be inactivated in a spatially and temporally controlled manner ***(111)***. Targeted disruption of *VHL* in the liver resulted in severe steatosis, numerous blood-filled vascular cavities, and foci of increased vascularization within the hepatic parenchyma ***(111)***. Hepatic steatosis was caused by the accumulation of neutral fat in hepatocytes ***(111)***. Interestingly, similar lipid accumulation has been observed in the stromal component of CNS hemangioblastomas associated with VHL disease ***(21,112,113)***. Furthermore, hypoxia-inducible mRNAs were markedly increased ***(111)***. Thus, the conditional mouse model for VHL disease underscores the significance of pVHL in the regulation of hypoxia-inducible genes. Unexpectedly, Haase et al. also observed similar vascular tumors in the livers of *VHL*+/− mice ***(111)***. Why tumors developed in the *VHL*+/− mice generated by Haase et al. and not those of Gnarra et al. is not known. Experiments are ongoing to address the consequences of specifically inactivated *VHL* in the kidney.

9. Summary and Perspectives

VHL disease is a rare familial cancer syndrome characterized by a development of hypervascular tumors affecting a number of organs including the retina, brain, spinal cord, and the kidney ***(7,11)***. Biallelic inactivation of the *VHL* gene has also been demonstrated in the majority of sporadic renal clear cell carcinomas and cerebellar hemangioblastomas ***(7,11)***. The vascular phenotype of the various VHL-associated tumors has been attributed to the overexpression of HIF target genes such as the *VEGF* gene ***(7)***. Under normoxia, HIFα subunits are prolyl-hydroxylated at a conserved residue at position 564 ***(93,94)***. The hydroxylated HIFα subunits are selectively recruited via the β-domain of pVHL and are subsequently ubiquitylated by the action of VEC ubiquitin ligase complex ***(89,93,94)***. The disease-associated pVHL mutants are either defective in recognizing the hydroxylated HIFα or is incapable of nucleating the full compliment of the ubiquitin ligase complex. The former is associated with the β-domain mutations and the latter is associated with the α-domain mutations.

Biallelic inactivation of *VHL* in the kidney causes premalignant renal cysts ***(7)***. It is presumed that additional mutations are required for conversion to frank renal carcinomas. Cystic diseases of the kidney are often characterized by abnormal renal epithelial cell proliferation and abnormal cell–matrix interactions ***(7,8,114,115)***. pVHL controls the production of a known epithelial mitogen TGFα ***(108,109)***, cell–matrix interactions through its ability to regulate fibronectin matrix assembly ***(50,98,99)***, the production of TIMPs and MMPs ***(101)***, as well as the regulation of the extracellular pH ***(102)***.

A number of questions remain. First, it is unclear why pVHL inactivation is intimately linked to the pathogenesis of kidney cancer and not other common epithelial cancers. Second, the identity of the HIF prolyl hydroxylase remains unknown. Answers to questions such as these, and further delineation of the pathways governing the cellular response to hypoxia, will undoubtedly aid in designing novel therapeutic approaches to combat diseases characterized by deregulated angiogenesis such as cancer.

References

1. Wingo, P. A., Tong, T., and Bolden, S. (1995) Cancer statistics, 1995. *CA Cancer J. Clin.* **45**, 8–30.
2. Linehan, W. M., Lerman, M. I., and Zbar, B. (1995) Identification of the von Hippel-Lindau (VHL) gene. *J. Am. Med. Assoc.* **273**, 564–570.
3. Toro, J. R., Glenn, G., Duray, P., et al. (1999) Birt-Hogg-Dube syndrome: a novel marker of kidney neoplasia. *Arch. Dermatol.* **135**, 1195–1202.
4. Iliopoulos, O. and Kaelin, W. G. (2000) von Hippel-Lindau disease, in *Hereditary Cancer*. (Fisher, D. E., ed.), Humana Press, Totowa, NJ.
5. Maher, E. R., Yates, J. R. W., Harries, R., et al. (1990) Clinical features and natural history of von Hippel-Lindau disease. *Q. J. Med.* **77**, 1151–1163.
6. Maher, E. R., Webster, A. R., and Moore, A. T. (1995) Clinical features and molecular genetics of von Hippel-Lindau disease. *Ophthal. Genet.* **16**, 79–84.
7. Maher, E. and Kaelin, W. G. (1997) von Hippel-Lindau disease. *Medicine* **76**, 381–391.
8. Poston, C. D., Jaffe, G. S., Lubensky, I. A., et al. (1995) Characterization of the renal pathology of a familial form of renal cell carcinoma associated with von Hippel-Lindau disease: clinical and molecular genetic implications. *J. Urol.* **153**, 22–26.
9. Walther, M., Lubensky, I., Venzon, D., Zbar, B., and Linehan, W. (1995) Prevalence of microscopic lesions in grossly normal renal parenchyma from patients with von Hippel-Lindau disease, sporadic renal cell carcinoma and no renal disease: clinical implications. *J. Urol.* **154**, 2010–2014.
10. Richards, F. M., Webster, A. R., McMahon, R., Woodward, E. R., Rose, S., and Maher, E. R. (1998) Molecular genetic analysis of von Hippel-Lindau disease. *J. Intern. Med.* **243**, 527–533.

11. Ohh, M. and Kaelin, W. G. J. (1999) The von Hippel-Lindau tumour suppressor protein: new perspectives. *Mol. Med. Today* **5**, 257–263.
12. Ivan, M. and Kaelin, W. G. (2001) The von Hippel-Lindau tumor suppressor protein. *Curr. Opin. Genet. Dev.* **11**, 27–34.
13. Kondo, K. and Kaelin, W. G., Jr. (2001) The von Hippel-Lindau tumor suppressor gene. *Exp. Cell Res.* **264**, 117–125.
14. Knudson, A. G., Jr. (1971) Mutation and cancer: statistical study of retinoblastoma. *Proc. Natl. Acad. Sci. USA* **68**, 820–823.
15. Gnarra, J. R., Tory, K., Weng, Y., et al. (1994) Mutations of the VHL tumor suppressor gene in renal carcinoma. *Nat. Genet.* **7**, 85–90.
16. Foster, K., Prowse, A., van den Berg, A., et al. (1994) Somatic mutations of the von Hippel-Lindau disease tumor suppressor gene in non-familial clear cell renal carcinoma. *Hum. Mol. Gen.* **3**, 2169–2173.
17. Shuin, T., Kondo, K., Torigoe, S., et al. (1994) Frequent somatic mutations and loss of heterozygosity of the von Hippel-Lindau tumor suppressor gene in primary human renal cell carcinomas. *Cancer Res.* **54**, 2852–2855.
18. Shuin, T., Kondo, K., Ashida, S., et al. (1999) Germline and somatic mutations in von Hippel-Lindau disease gene and its significance in the development of kidney cancer. *Contrib. Nephrol.* **128**, 1–10.
19. Whaley, J. M., Naglich, J., Gelbert, L., et al. (1994) Germ-line mutations in the von Hippel-Lindau tumor suppressor gene are similar to somatic von Hippel-Lindau abberations in sporadic renal cell carcinoma. *Am. J. Hum. Genet.* **55**, 1092–1102.
20. Kanno, H., Kondo, K., Ito, S., et al. (1994) Somatic mutations of the von Hippel-Lindau tumor supressor gene in sporadic central nervous systems hemangioblastomas. *Cancer Res.* **54**, 4845–4847.
21. Lee, J.-Y., Dong, S.-M., Park, W.-S., et al. (1998) Loss of heterozygosity and somatic mutations of the VHL tumor suppressor gene in sporadic cerebellar hemangioblastomas. *Cancer Res.* **58**, 504–508.
22. Oberstrass, J., Reifenberger, G., Reifenberger, J., Wechsler, W., and Collins, V. (1996) Mutation of the Von Hippel-Lindau tumour suppressor gene in capillary haemangioblastomas of the central nervous system. *J. Pathol.* **179**, 151–156.
23. Tse, J., Wong, J., Lo, K.-W., Poon, W.-S., Huang, D. and Ng, H.-K. (1997) Molecular genetic analysis of the von Hippel-Lindau disease tumor suppressor gene in familial and sporadic cerebellar hemangioblastomas. *Am. J. Clin. Pathol.* **107**, 459–466.
24. Eng, C., Crossey, P., Mulligan, L., et al. (1995) Mutations in the RET proto-oncogene and the von Hippel-Lindau disease tumor suppressor gene in sporadic and syndromic phaeochromocytomas. *J. Med. Genet.* **32**, 934–937.
25. Hofstra, R. M. W., Stelwagen, T., Stulp, R. P., et al. (1996) Extensive mutation scanning of RET in sporadic medullary thyroid carcinoma and of RET and VHL in sporadic pheochromocytoma reveals involvement of these genes in only a minority of cases. *J. Clin. Endocrinol. Metab.* **81**, 2881–2884.
26. Lubensky, I. A., Gnarra, J. R., Bertheau, P., Walther, M. M., Linehan, W. M., and Zhuang, Z. (1996) Allelic deletions of the VHL gene detected in multiple microscopic clear cell renal lesions in von Hippel-Lindau disease patients. *Am. J. Pathol.* **149**, 2089–2094.
27. Kinzler, K. and Vogelstein, B. (1996) Lessons from hereditary colorectal cancer. *Cell* **87**, 159–170.
28. Zbar, B., Brauch, H., Talmadge, C, and Linehan, M. (1987) Loss of alleles of loci on the short arm of chromosome 3 in renal cell carcinoma. *Nature* **327**, 721–724.
29. Kovacs, G., Erlandsson, R., Boldog, F., et al. (1988) Consistent chromosome 3p deletion and loss of heterozygosity in renal cell carcinoma. *Proc. Natl. Acad. Sci. USA* **85**, 1571–1575.

30. Kovacs, G., Brusa, P., and De Riese, W. (1989) Tissue-specific expression of a constitutional 3;6 translocation: development of multiple bilateral renal-cell carcinomas. *Int. J. Cancer* **43**, 422–427.
31. Cohen, A. J., Li, F. P., Berg, S., et al. (1979) Hereditary renal-cell carcinoma associated with a chromosomal translocation. *N. Engl. J. Med.* **301**, 592–595.
32. Pathak, S., Strong, L. C., Ferrell, R. E., and Trindade, A. (1982) Familial renal cell carcinoma with a 3;11 chromosome translocation limited to tumor cells. *Science* **217**, 939–941.
33. Seizinger, B. R., Rouleau, G. A., Ozelius, L. J., et al. (1988) Von-hippel lindau disease maps to the region of chromosome 3 associated with renal cell carcinoma. *Nature* **332**, 268–269.
34. Richards, F. M., Maher, E. R., Latif, F., et al. (1993) Detailed genetic mapping of the von Hippel-Lindau disease tumour suppressor gene. *J. Med. Genet.* **30**, 104–107.
35. Yao, M., Latif, F., Orcutt, M., et al. (1993) Von Hippel-Lindau disease: identification of deletion mutations by pulsed field gel electrophoresis. *Hum. Genet.* **92**, 605–614.
36. Latif, F., Tory, K., Gnarra, J., et al. (1993) Identification of the von Hippel-Lindau disease tumor suppressor gene. *Science* **260**, 1317–1320.
37. Chen, F., Kishida, T., Yao, M., et al. (1995) Germline mutations in the von Hippel-Lindau disease tumor suppressor gene:correlations with phenotype. *Hum. Mutat.* **5**, 66–75.
38. Stolle, C., Glenn, G., Zbar, B., et al. (1998) Improved detection of germline mutations in the von Hippel-Lindau disease tumor suppressor gene. *Hum. Mutat.* **12**, 417–423.
39. Kessler, P., Vasavada, S., Rackley, R., et al. (1995) Expression of the von Hippel-Lindau tumor-suppressor gene, VHL, in human fetal kidney and during mouse embryogenesis. *Mol. Med.* **1**, 457–466.
40. Richards, F., Schofield, P., Fleming, S., and Maher, E. (1996) Expression of the von Hippel-Lindau disease tumour suppressor gene during human embryogenesis. *Hum. Mol. Genet.* **5**, 639–644.
41. Crossey, P. A., Richards, F. M., Foster, K., et al. (1994) Identification of intragenic mutations in the von Hippel-Lindau disease tumor suppressor gene and correlation with disease phenotype. *Hum. Mol. Gen.* **3**, 1303–1308.
42. Zbar, B., Kishida, T., Chen, F., et al. (1996) Germline mutations in the von Hippel-Lindau (VHL) gene in families from North America, Europe, and Japan. *Hum. Mutat.* **8**, 348–357.
43. Schoenfeld, A., Davidowitz, E. J., and Burk, R. D. (1998) A second major native von Hippel-Lindau gene product, initiated from an internal translation start site, functions as a tumor suppressor. *Proc. Natl. Acad. Sci. USA* **95**, 8817–8822.
44. Iliopoulos, O., Ohh, M., and Kaelin, W. (1998) pVHL19 is a biologically active product of the von Hippel-Lindau gene arising from internal translation initiation. *Proc. Natl. Acad. Sci. USA* **95**, 11661–11666.
45. Blankenship, C., Naglich, J., Whaley, J., Seizinger, B., and Kley, N. (1999) Alternate choice of initiation codon produces a biologically active product of the von Hippel Lindau gene with tumor suppressor activity. *Oncogene* **18**, 1529–1535.
46. Iliopoulos, O., Kibel, A., Gray, S., and Kaelin, W. G. (1995) Tumor suppression by the human von Hippel-Lindau gene product. *Nat. Med.* **1**, 822–826.
47. Gnarra, J. R., Zhou, S., Merrill, M. J., et al. (1996) Post-transcriptional regulation of vascular endothelial growth factor mRNA by the VHL tumor suppressor gene product. *Proc. Natl. Acad. Sci. USA* **93**, 10589–10594.
48. Corless, C. L., Kibel, A., Iliopoulos, O., and Kaelin, W. G. J. (1997) Immunostaining of the von Hippel-Lindau gene product (pVHL) in normal and neoplastic human tissues. *Hum. Pathol.* **28**, 459–464.
49. Los, M., Jansen, G. H., Kaelin, W. G., Lips, C. J. M., Blijham, G. H., and Voest, E. E. (1996) Expression pattern of the von Hippel-Lindau protein in human tissues. *Lab. Invest.* **75**, 231–238.

50. Ohh, M., Yauch, R. L., Lonergan, K. M., et al. (1998) The von Hippel-Lindau tumor suppressor protein is required for proper assembly of an extracellular fibronectin matrix. *Mol. Cell* **1**, 959–968.
51. Pause, A., Lee, S., Worrell, R. A., et al. (1997) The von Hippel-Lindau tumor-suppressor gene product forms a stable complex with human CUL-2, a member of the Cdc53 family of proteins. *Proc. Natl. Acad. Sci. USA* **94**, 2156–2161.
52. Lee, S., Chen, D. Y. T., Humphrey, J. S., Gnarra, J. R., Linehan, W. M., and Klausner, R. D. (1996) Nuclear/cytoplasmic localization of the von Hippel-Lindau tumor suppressor gene product is determined by cell density. *Proc. Natl. Acad. Sci. USA* **93**, 1770–1775.
53. Lee, S., Neumann, M., Stearman, R., et al. (1999) Transcription-dependent nuclear-cytoplasmic trafficking is required for the function of the von Hippel-Lindau tumor suppressor protein. *Mol. Cell. Biol.* **19**, 1486–1497.
54. Kibel, A., Iliopoulos, O., DeCaprio, J. D., and Kaelin, W. G. (1995) Binding of the von Hippel-Lindau tumor suppressor protein to elongin B and C. *Science* **269**, 1444–1446.
55. Duan, D. R., Pause, A., Burgress, W., et al. (1995) Inhibition of transcriptional elongation by the VHL tumor suppressor protein. *Science* **269**, 1402–1406.
56. Aso, T., Lane, W. S., Conaway, J. W., and Conaway, R. C. (1995) Elongin (SIII): a multisubunit regulator of elongation by RNA polymerase II. *Science* **269**, 1439–1443.
57. Aso, T., Haque, D., Barstead, R., Conaway, R., and Conaway, J. (1996) The inducible elongin A elongation activation domain: structure, function and interaction with elongin BC complex. *EMBO J.* **15**, 101–110.
58. Ohh, M., Takagi, Y., Aso, T., et al. (1999) Synthetic peptides define critical contacts between elongin C, elongin B, and the von Hippel-Lindau protein. *J. Clin. Invest.* **104**, 1583–1591.
59. Kamura, T., Sato, S., Haque, D., et al. (1998) The elongin BC complex interacts with the conserved SOCS-box motif present in members of the SOCS, ras, WD-40 repeat, and ankyrin repeat families. *Genes Dev.* **12**, 3872–3881.
60. Kipreos, E. T., Lander, L. E., Wing, J. P., He, W. W., and Hedgecock, E. M. (1996) cul-1 is required for cell cycle exit in *C. elegans* and identifies a novel gene family. *Cell* **85**, 829–839.
61. Lonergan, K.M., Iliopoulos, O., Ohh, M., et al. (1998) Regulation of hypoxia-inducible mRNAs by the von Hippel-LIndau protein requires binding to complexes containing elongins B/C and Cul2. *Mol. Cell. Biol.* **18**, 732–741.
62. Bai, C., Sen, P., Hofmann, K., et al. (1996) SKP1 connects cell cycle regulators to the ubiquitin proteolysis machinery through a novel motif, the F-box. *Cell* **86**, 263–274.
63. Skowyra, D., Craig, K. L., Tyers, M., Elledge, S. J., and Harper, J. W. (1997) F-box proteins are receptors that recruit phosphorylated substrates to the SCF ubiquitin-ligase complex. *Cell* **91**, 209–219.
64. Patton, E. E., Willems, A. R., and Tyers, M. (1998) Combinatorial control in ubiquitin-dependent proteolysis: don't Skp the F-box hypothesis. *Trends Genet.* **14**, 236–243.
65. Peters, J. M. (1998) SCF and APC: the Yin and Yang of cell cycle regulated proteolysis. *Cur. Opin. Cell. Biol.* **10**, 759–768.
66. Deshaies, R. (1999) SCF and Cullin/Ring H2-based ubiquitin ligases. *Annu. Rev. Cell. Dev. Biol.* **15**, 435–467.
67. Verma, R., Annan, R. S., Huddleston, M. J., Carr, S. A., Reynard, G., and Deshaies, R. J. (1997) Phosphorylation of Sic1p by G1 Cdk required for its degradation and entry into S phase. *Science* **278**, 455–460.
68. Feldman, R. M., Correll, C. C., Kaplan, K. B., and Deshaies, R. J. (1997) A complex of Cdc4p, Skp1p, and Cdc53p/cullin catalyzes ubiquitination of the phosphorylated CDK inhibitor Sic1p. *Cell* **91**, 221–230.
69. Stebbins, C. E., Kaelin, W. G., and Pavletich, N. P. (1999) Structure of the VHL-elonginC-elonginB complex: implications for VHL tumor suppressor function. *Science* **284**, 455–461.

70. Lisztwan, J., Imbert, G., Wirbelauer, C., Gstaiger, M., and Krek, W. (1999) The von Hippel-Lindau tumor suppressor protein is a component of an E3 ubiquitin-protein ligase activity. *Genes Dev.* **13**, 1822–1833.
71. Iwai, K., Yamanaka, K., Kamura, T., et al. (1999) Identification of the von Hippel-Lindau tumor-suppressor protein as part of an active E3 ubiquitin ligase complex. *Proc. Natl. Acad. Sci. USA* **96**, 12436–12441.
72. Kamura, T., Koepp, D. M., Conrad, M. N., et al. (1999) Rbx1, a component of the VHL tumor suppressor complex and SCF ubiquitin ligase. *Science* **284**, 657–661.
73. Ohta, T., Michel, J. J., Schottelius, A. J., and Xiaong, Y. (1999) ROC1, a homolog of APC11, represents a family of cullin partners with an associated ubiquitin ligase activity. *Mol. Cell* **3**, 535–541.
74. Tan, P., Fuchs, S. Y., Chen, A., et al. (1999) Recruitment of a ROC1-CUL1 ubiquitin ligase by Skp1 and HOS to catalyze the ubiquitination of IκBα. *Mol. Cell* **3**, 527–533.
75. Skowyra, D., Koepp, D. M., Kamura, T., et al. (1999) Reconstitution of G1 cyclin ubiquitination with complexes containing SCFGrr1 and Rbx1. *Science* **284**, 662–665.
76. Ciechanover, A., Orian, A., and Schwartz, A. L. (2000) Ubiquitin-mediated proteolysis: biological regulation via destruction. *Bioessays* **22**, 442–451.
77. Wizigmann-Voos, S., Breier, G., Risau, W., and Plate, K. (1995) Up-regulation of vascular endothelial growth factor and its receptors in von Hippel-Lindau disease-associated and sporadic hemangioblastomas. *Cancer Res.* **55**, 1358–1364.
78. Iliopoulos, O., Jiang, C., Levy, A. P., Kaelin, W. G., and Goldberg, M. A. (1996) Negative regulation of hypoxia-inducible genes by the von Hippel-Lindau protein. *Proc Natl Acad Sci USA* **93**, 10595–10599.
79. Krieg, M., Marti, H., and Plate, K. H. (1998) Coexpression of erythropoietin and vascular endothelial growth factor in nervous system tumors associated with von Hippel-Lindau tumor suppressor gene loss of function. *Blood* **92**, 3388–3393.
80. Maxwell, P., Weisner, M., Chang, G.-W., et al. (1999) The von Hippel-Lindau gene product is necessary for oxgyen-dependent proteolysis of hypoxia-inducible factor α subunits. *Nature* **399**, 271–275.
81. Siemeister, G., Weindel, K., Mohrs, K., Barleon, B., Martiny-Baron, G., and Marme, D. (1996) Reversion of deregulated expression of vascular endothelial growth factor in human renal carcinoma cells by von Hippel-Lindau tumor suppressor protein. *Cancer Res.* **56**, 2299–2301.
82. Stratmann, R., Krieg, M., Haas, R., and Plate, K. (1997) Putative control of angiogenesis in hemangioblastomas by the von Hippel-Lindau tumor suppressor gene. *J. Neuropathol. Exp. Neurol.* **56**, 1242–1252.
83. Semenza, G. (1999) Regulation of mammalian O_2 homeostasis by hypoxia-inducible factor 1. *Annu. Rev. Cell Dev. Biol.* **15**, 551–578.
84. Semenza, G. (1999) Perspectives on oxygen sensing. *Cell* **98**, 281–284.
85. Wenger, R. (2000) Mammalian oxygen sensing, signalling and gene regulation. *J. Exp. Biol.* **203**, 1253–1263.
86. Zhu, H. and Bunn, F. (1999) Oxygen sensing and signaling: impact on the regulation of physiologically important genes. *Resp. Phys.* **115**, 239–247.
87. Huang, L. E., Gu, J., Schau, M., and Bunn, H. F. (1998) Regulation of hypoxia-inducible factor 1alpha is mediated by an O_2-dependent degradation domain via the ubiquitin-proteasome pathway. *Proc. Natl. Acad. Sci. USA* **95**, 7987–7992.
88. Srinivas, V., Zhang, L., Zhu, X., and Caro, J. (1999) Characterization of an oxygen/redox-dependent degradation domain of hypoxia-inducible factor alpha (HIF-alpha) proteins. *Biochem. Biophys. Res. Commun.* **260**, 557–561.
89. Ohh, M., Park, C.W., Ivan, M., et al. (2000) Ubiquitination of HIF requires direct binding to the von Hippel-Lindau protein beta domain. *Nat. Cell Biol.* **2**, 423–427.

90. Cockman, M., Masson, N., Mole, D., et al. (2000) Hypoxia inducible factor-alpha binding and ubiquitylation by the von Hippel-Lindau tumor suppressor protein. *J. Biol. Chem.* **275**, 25733–25741.
91. Tanimoto, K., Makino, Y., Pereira, T., and Poellinger, L. (2000) Mechanism of regulation of the hypoxia-inducible factor-1alpha by the von Hippel-Lindau tumor suppressor protein. *EMBO J.* **19**, 4298–4309.
92. Kamura, T., Sato, S., Iwain, K., Czyzyk-Krzeska, M., Conaway, R. C., and Conaway, J. W. (2000) Activation of HIF1a Ubiquitination by a reconstituted von Hippel-Lindau tumor suppressor complex. *Proc. Natl. Acad. Sci. USA* **97**, 10430–10435.
93. Ivan, M., Kondo, K., Yang, H., et al. (2001) HIFalpha targeted for VHL-mediated destruction by proline hydroxylation: implications for O_2 sensing. *Science* **292**, 464–468.
94. Jaakkola, P., Mole, D. R., Tian, Y. M., et al. (2001) Targeting of HIF-alpha to the von Hippel-Lindau ubiquitylation complex by O_2-regulated prolyl hydroxylation. *Science* **292**, 468–472.
95. Hynes, R. (1992) Integrins: versatility, modulation, and signalling in cell adhesion. *Cell* **69**, 11–25.
96. Pasqualini, R., Bourdoulous, S., Koivunen, E., Woods, V., Jr., and Ruoslahti, E. (1996) A polymeric form of fibronectin has antimetastatic effects against multiple tumor types. *Nat. Med.* **2**, 1197–1203.
97. Ruoslahti, E. (1992) Control of the cell motility and tumour invasion by extracellular matrix interactions. *Br. J. Cancer* **66**, 239–242.
98. Hoffman, M. A., Ohh, M., Yang, H., Klco, J. M., Ivan, M., and Kaelin, W. G., Jr. (2001) von Hippel-Lindau protein mutants linked to type 2C VHL disease preserve the ability to downregulate HIF. *Hum. Mol. Genet.* **10**, 1019–1027.
99. Lieubeau-Teillet, B., Rak, J., Jothy, S., Iliopoulos, O., Kaelin, W., and Kerbel, R. (1998) von Hippel-Lindau gene-mediated growth suppression and induction of differentiation in renal cell carcinoma cells grown as multicellular tumor spheroids. *Cancer Res.* **58**, 4957–4962.
100. Davidowitz, E. J., Schoenfeld, A. R., and Burk, R. D. (2001) VHL induces renal cell differentiation and growth arrest through integration of cell-cell and cell-extracellular matrix signaling. *Mol. Cell. Biol.* **21**, 865–874.
101. Koochekpour, S., Jeffers, M., Wang, P., et al. (1999) The von Hippel-Lindau tumor suppressor gene inhibits hepatocyte growth factor/scatter factor-induced invasion and branching morphogenesis in renal carcinoma cells. *Mol. Cell. Biol. 1999*, **19**, 5902–5912.
102. Ivanov, S., Kuzmin, I., Wei, M.-H., et al. (1998) Down-regulation of transmembrance carbonic anhydrases in renal cell carcinoma cell lines by wild-type von Hippel-Lindau transgenes. *Proc. Natl. Acad. Sci. USA* **95**, 12596–12601.
103. Sly, W. S. and Hu, P. Y. (1995) Human carbonic anhydrases and carbonic anhydrase deficiencies. *Annu. Rev. Biochem.* **64**, 375–401.
104. Martinez-Zaguilan, R., Seftor, E. A., Seftor, R. E., Chu, Y. W., Gillies, R. J., and Hendrix, M. J. (1996) Acidic pH enhances the invasive behavior of human melanoma cells. *Clin. Exp. Metastasis* **14**, 176–186.
105. Ivanov, S., Liao, S.Y., Ivanova, A., et al. (2001) Expression of hypoxia-inducible cell-surface transmembrane carbonic anhydrases in human cancer. *Am. J. Pathol.* **158**, 905–919.
106. Pause, A., Lee, S., Lonergan, K. M., and Klausner, R. D. (1998) The von Hippel-Lindau tumor suppressor gene is required for cell cycle exit upon serum withdrawal. *Proc. Natl. Acad. Sci. USA* **95**, 993–998.
107. Kim, M., Katayose, Y., Li, Q., et al. (1998) Recombinant adenovirus expressing Von Hippel-Lindau-mediated cell cycle arrest is associated with the induction of cyclin-dependent kinase inhibitor p27Kip1. *Biochem. Biophy. Res. Commun.* **253**, 672–677.

108. Knebelmann, B., Ananth, S., Cohen, H., and Sukhatme, V. (1998) Transforming growth factor alpha is a target for the von Hippel-Lindau tumor suppressor. *Cancer Res.* **58**, 226–231.
109. de Paulsen, N., Brychzy, A., Fournier, M.C., et al. (2001) Role of transforming growth factor-alpha in von Hippel-Lindau (VHL) (−/−) clear cell renal carcinoma cell proliferation: a possible mechanism coupling VHL tumor suppressor inactivation and tumorigenesis. *Proc. Natl. Acad. Sci. USA* **98**, 1387–1392.
110. Gnarra, J., Ward, J., Porter, F., et al. (1997) Defective placental vasculogenesis causes embryonic lethality in VHL-deficient mice. *Proc. Natl. Acad. Sci. USA* **94**, 9102–9107.
111. Haase, V. H., Glickman, J. N., Socolovsky, M. and Jaenisch, R. (2001) Vascular tumors in livers with targeted inactivation of the von Hippel-Lindau tumor suppressor. *Proc. Natl. Acad. Sci. USA* **98**, 1583–1588.
112. Vortmeyer, A., Gnarra, J., Emmert-Buck, M., et al. (1997) von Hippel-Lindau gene deletion detected in the stromal cell component of a cerebellar hemangioblastoma associated with von Hippel-Lindau disease. *Hum. Pathol.* **28**, 540–543.
113. Lach, B., Gregor, A., Rippstein, P., and Omulecka, A. (1999) Angiogenic histogenesis of stromal cells in hemangioblastoma: ultrastructural and immunohistochemical study. *Ultrastruct. Pathol.* **23**, 299–310.
114. Harris, P., Ward, C., Peral, B., and Hughes, J. (1995) Autosomal dominant polycystic kidney disease: molecular analysis. *Hum. Mol. Genet.* **4**, 1745–1749.
115. Carone, F., Bacallao, R., and Kanwar, Y. (1994) Biology of polycystic kidney disease. *Lab. Invest.* **70**, 437–448.

11

p16^{INK4A} and Familial Melanoma

Kapaettu Satyamoorthy and Meenhard Herlyn

1. Introduction

Cell division is controlled by a group of positive and negative regulatory molecules that act at experimentally defined phases throughout the cell-division cycle. Perturbations in these critical phases contribute to tumor development by allowing uncontrolled cell proliferation. During the cell-division cycle, cells in transit pass from a stage termed G1 to the DNA synthesis (or S) phase. At G1 phase, cells continue to synthesize RNA and proteins and irrevocably commit to DNA synthesis. This phase is controlled by several protein complexes including those consisting of cyclin-dependent kinases (CDK4 and CDK6) and cyclin D. These complexes are inhibited by the INK4 family of proteins—p16^{INK4A}, p15^{INK4B}, p18^{INK4C}, and p19^{INK4D}. Genetic alterations affecting p16^{INK4A} and cyclin D1 are so frequent in human cancers that inactivation of these proteins is believed to be necessary for tumor development. Broadly applicable anticancer therapies might be based on restoration of p16^{INK4A} CDK inhibitory function.

Extensive biochemical data are currently available on the effects of mutations in p16^{INK4A}, especially those mutations that are found in human cancers. To date, three mechanisms that lead to tumor-associated inactivation of p16^{INK4A} have been characterized. These include point mutation, homozygous deletion, and hypermethylation of the CpG island spanning the promoter region of the p16^{INK4A} gene (**Fig. 1**). More than 170 missense mutations have been identified, and it now appears that deletion and frameshift mutations in p16^{INK4A} are even more frequent. This chapter will summerize causes and consequences due to alterations in p16^{INK4A} in relation to human melanoma.

2. Melanocytes

The melanocytes in the human skin represent an intriguing and intricate group of cells that synthesize and distribute melanin to impart skin color. Melanocytes originate from the pluripotent neural crest cells, and their maturation, migration, and maintenance have been the subject of considerable research and are well understood in animal models. This has provided invaluable insight into the biologic and its associated pathologic states in humans. The melanocyte in human skin is normally embedded as a single cell in the

From: *Methods in Molecular Biology, Vol. 222: Tumor Suppressor Genes: Pathways and Isolation Strategies*
Edited by: Wafik S. El-Deiry © Humana Press Inc., Totowa, NJ

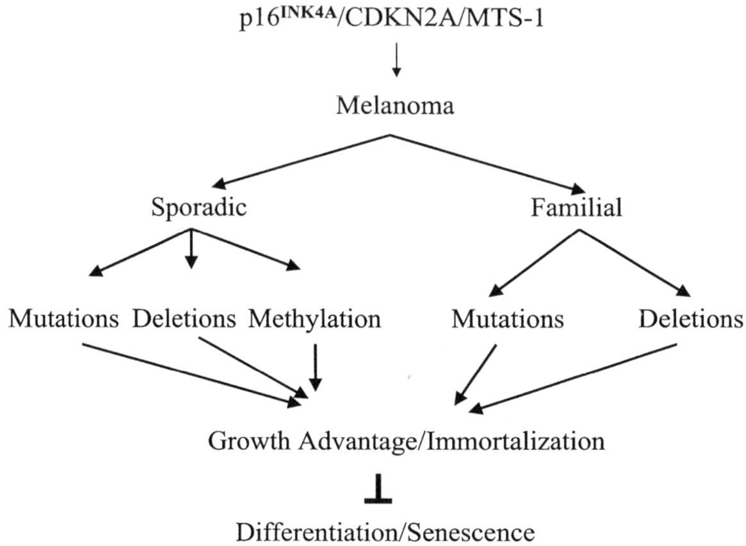

Fig. 1. Abnormalities reported in the p16^{INK4A} locus in melanoma. Functional consequence of mutations, deletions, or methylations in p16^{INK4A} locus leads to impairment of growth arrest, differentiation, or senescence in melanoma.

basal layer of keratinocytes and anchored to the basement membrane of the epidermis. Each melanocyte extends its dendrites into the upper layers of the epidermis and establishes contacts with the keratinocytes, forming the "epidermal melanin unit" that comprises 20–35 cells. Expansion of the total skin surface during growth leads to melanocyte proliferation to maintain a stable ratio with the keratinocytes in the basal layer. In adults, melanocytes are generally resting and proliferate at a slow rate, if at all. Environmental, genetic, and epigenetic factors induce abnormal growth and invasion during nevus and melanoma progression and represent many of the critical events of early melanocyte development ***(1)***.

3. Melanoma

Human melanoma originates from pigment-producing melanocytes and is found mostly in white Caucasians. Melanoma develops from a series of several architectural and phenotypically distinct stages and progressively becomes aggressive. This aggressiveness leads to tumor invasion and metastasis to distinct organs with fatal subsequences. Over the years, progress has been made in understanding the biologic, pathologic, genetic, and immunologic behavior of human melanoma. However, precise molecular determinants responsible for melanoma progression are not understood. This has considerably impeded the efforts to understand genetic abnormalities for further study. Available evidence suggests that there is no universality with respect to selected molecules or their deregulation specifically for influencing the growth of melanoma.

The worldwide incidence of melanoma is rising. In the United States, there are some 50,000 new cases reported each year, with a mortality rate of approximately 22% that accounts for 1.5% of all cancer deaths. Alarmingly, over the past three decades there has been a yearly 2–3% increase in the incidence of cutaneous melanoma in the Caucasian population. Although the etiology is multifactorial, excessive sun exposure (UVB

290–320 nm) appears to be one of the most important causative agents. Recent experiments in animals and epidemiologic observations provide strong correlation that solar radiation causes cutaneous melanoma ***(2,3)***. To date, a number of families have been identified throughout the world whose kindreds are predisposed to melanoma. Overall, approximately 10% of melanomas arise due to a genetic predisposition.

Human melanoma can develop in a sequence of steps from benign proliferative lesions. The lesions representing each step are common acquired nevus, dysplastic (atypical) nevus, radial growth-phase primary melanoma, which does not metastasize, and vertical growth phase primary melanoma which has metastatic competence, and metastatic melanoma ***(4)***. The progression of melanoma evolves from dysplastic melanocytes, which have lost their architectural integrity but do not multiply indefinitely. Further environmental and genetic insult lead to genomic instability that triggers proliferation into radial growth-phase (RGP) melanomas. These RGP cells also lack expression of key molecules that provide a tumorigenic phenotype. The phenotypic characteristics that convert RGP melanoma into vertical growth-phase melanoma (VGP) include acquired invasive ability into the dermis, aggressive growth and expression of key molecules that facilitate growth, and survival and escape of cells from the primary site. Metastasis as a whole favors the survival and growth of a few subpopulations of cells that preexist within the parent neoplasm. Metastasis can have clonal origin, or different metastasis can originate from proliferation of single cells. The outcome of metastasis depends on the interaction of metastatic cells with host factors ***(5,6)***.

4. Early Uncertainties

Multiple causative factors determine the etiology of familial melanoma. These include families with or without dysplastic nevus (DN), diagnostic inconsistencies, founder effect, and finally, multiple gene abnormalities that influence genetic model fit and candidate gene identification. Investigations from the National Cancer Institute and the University of Pennsylvania, wherein all family members with cutaneous malignant melanoma (CMM) also had dysplastic nevi on their skin, first observed a strong correlation between cutaneous malignant melanoma and dysplastic nevus ***(7)***. Therefore DN was considered as a marker for increased risk. Although a survey of primary data from melanoma-prone families rejected dominant inheritance, earlier segregation analysis suggested an autosomal dominant pattern in these families. Additional analysis identified a CMM/DN gene located on chromosome 1p36 with an estimated penetrance of 82% at the age of 72 ***(8)***. However, no candidate gene was identified, and several attempts to corroborate the gene assignment were unsuccessful. It is indeed possible that a subset of the population with etiologic and diagnostic heterogeneity may have contributed to this discrepancy. In addition to genetic heterogeneity, inheritance according to the two-hit model and confusing classification parameters may also have contributed to this effect. Presence of atypical mole syndrome was also found to be higher in melanoma patients than the normal population. Although a genetic basis for nevi is suggested by a strong inherited basis for total nevus count, genes associated with moles have not been identified ***(9)***. This may be again due to the inclusion of DN in the total mole density in the given sample contributing to pleiotropic manifestations. Taken together, it has been suggested that DN patients are at increased risk for melanoma. However, it has been found that new melanomas frequently arose from new lesions but not from preexisting DN.

4.1. Chromosome 9p21

Several studies describe the cytogenetic alterations in melanoma. Aberrations in chromosomes 1, 6, 7, 9, and 11 have been frequently noted, and, to a lesser extent, in 10 other chromosomes. Aberrations in chromosomes 1 and 9 were then identified in cases with familial predisposition to melanoma, and homozygous loss was reported in chromosome 9p21 in both cell lines and in melanoma tumor biopsies ***(10,11)***. Cytogenetic analyses have revealed chromosome 9p as an area of frequent genetic abnormality in melanoma ***(12,13)***. In 1992, Cannon-Albright et al., ***(10)*** and subsequently Australian and Dutch groups ***(14,15)***, published convincing data from Utah kindreds on the linkage in familial melanoma to around chromosome 9p21 between the markers D9S126 and interferon α (IFNα). These studies also suggested the presence of a melanoma-predisposition locus containing a putative tumor suppressor gene. The Utah group analyzed 11 CMM pedigrees without taking DN into consideration and found penetrance of 9p21 to be 53% by age 80. Genetic and environmental influences were demonstrated with the inclusion of patients with melanoma who were more sun-exposed. It is concluded that exposure to sunlight and mutations in 9p21 increased the risk for hereditary CMM. Subsequently, the Utah investigators found that 13 families from 38 melanoma-prone families were linked to 9p21. Further analysis suggested that nearly 18% of the families and 37% of the subset carry germline mutations in this gene. Linking DN to 9p21 increased the significant risk for melanoma ***(16)***. These studies also suggested the existence of more than one melanoma gene at 9p21. The CMM2 gene subsequently identified as $p16^{INK4A}$ has been variously called MTS1 and CDKN2A (**Fig. 2**). A National Cancer Institute group described germline mutation in 33 of 36 melanoma patients from 9 families. Kindreds without mutations in p16 have a risk of melanoma that is increased by a factor of 38, and those with inactivating p16 mutations have increased risk by a factor of 75. The average age at diagnosis of melanoma in DNS families is 34 years, compared to 55 years for melanoma diagnosed in the general population. Around 20–25% of melanoma-prone kindreds are linked to $p16^{INK4A}$ ***(17)***. It has been concluded that hereditary melanoma develops when cells inherit a mutant CDK2A allele and then lose the wild-type allele in a secondary somatic event.

5. $p16^{INK4A}$

The $p16^{INK4A}$ protein contains 156 amino acids and five 32-residue ankyrin-like repeats that impart high structural similarity within the INK4 family members such as $p15^{INK4B}$, $p18^{INK4C}$, and $p19^{INK4D}$. The $p16^{INK4A}$ gene contains four exons termed exon 1β, exon 1α, exons 2 and 3, which also contains a gene encoding for $p14^{ARF}$ (**Fig. 3**). The protein product of $p14^{ARF}$ is encoded by alternative exon 1β and exon 2 of the $p16^{INK4A}$ gene ***(18)***. Exon 1β lies about 12 kb upstream of exon 1α, which is under its own transcriptional control. Exon 1β is spliced to $p16^{INK4A}$ in an alternative reading frame (ARF) to $p14^{ARF}$. Although both act as cell cycle regulators to prevent G1 progression, $p16^{INK4A}$ and $p14^{ARF}$ do not share homology at the amino acid level and possess entirely different function ***(19)***. Some mutations in exon 2 would be predicted to affect amino acid composition of both proteins; however, the cell-cycle inhibitory function of $p14^{ARF}$ is encoded by exon 1β and therefore a role for such mutations on $p14^{ARF}$ is unclear, although deletions in this region are expected to affect both proteins ***(20)***.

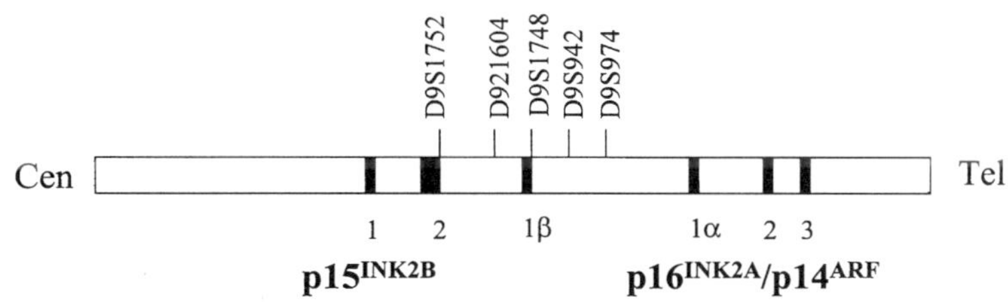

Fig. 2. Overview and arrangement of critical genes in chromosome 9p21. The $p15^{INK4B}$ is centromeric from the $p16^{INK4A}$ gene and is separated by approximately 40 kb. Some of the markers for the chromosome in this region are also shown.

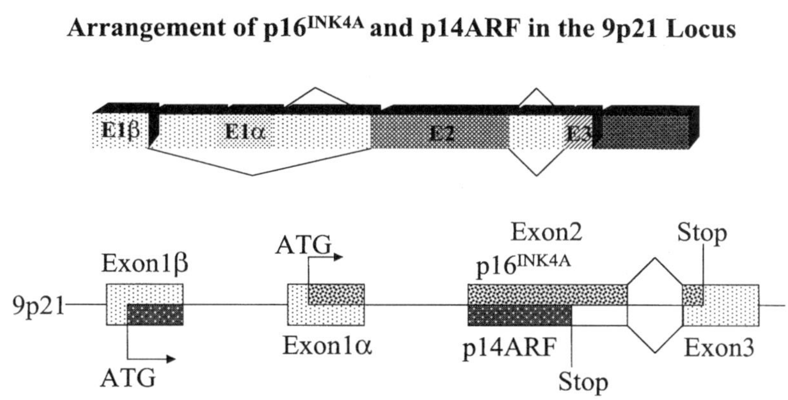

Fig. 3. Arrangement of $p16^{INK4A}$ and $p14^{ARF}$ in the $p16^{INK4A}$ locus. Three exons of the $p16^{INK4A}$ locus contain coding sequences for two proteins, $p16^{INK4A}$ and $p14^{ARF}$. Although both proteins share identical nucleotides, $p14^{ARF}$ is translated from a different reading frame of exon 1α. Both genes share a similar 3′-untranslated sequence; however, they are driven by separate promoter sequences.

When compared to other ankyrin repeat-containing proteins such as GABP, 53BP2, and myotrophin, $p16^{INK4A}$ shows similar structures with specific hydrogen-bonding patterns and hydrophobic interactions. These comparisons revealed the splayed-loop geometry that is specific to INK4 inhibitors, which results from the modified ankyrin structure in the second repeat. Deleterious mutations are found throughout the tumor suppressor protein, rather than a cluster of mutations near a single binding region ***(22)***, with the exception of melanoma. These mutations are most likely to lead to the destabilization of the three-dimensional structure, preventing them from binding to CDK4 and CDK6. It is clear, however, that most of the mutations may not participate directly in binding to the CDKs. Some of these mutations may lead to defective folding of the protein and prevent functions similar to the wild-type $p16^{INK4A}$ protein.

5.1. Implications for $p16^{INK4a}$ Mutations

It is now well established that the loss of $p16^{INK4A}$ leads to inhibition of the cyclin D–CDK4/6 complex activity or inhibition of entry into the S phase and is directly connected to the decrease/loss of CDK4/6 binding by the $p16^{INK4A}$ mutants. It includes mutations that are not only identified in tumors, but those that also reduce CDK4/6-binding activity in vitro. Mutational spectra reveal the prevalence of C:G → T:A transitions,

C:G → A:T traversions, and CC → TT tandem mutations. CC → TT mutations are unique to UV radiation effects in the formation of pyrimidine dimers. These transitional C → T UV signature mutations were in the range of 14–33% of all the changes, but they were not specific for any particular exon in the p16^{INK4A} gene *(21,22)*. Whether these mutations arose from spontaneously deaminated 5-methylcytosine remains to be established. It is evident that mutations are neither clustered within a specific region of the protein nor are there any hot spots *(23)*. The locations of the mutations, regardless of whether they are deleterious or not, show no correlation with posttranslational modifications. The amino acid composition suggests that p16^{INK4A} is largely composed of four contiguous ankyrin repeats that form helices with interposed turns *(24)*. The p16^{INK4A} mutant proteins that carried mutations at position 26, 66, 84, 92, or 124 had nuclear magnetic resonance spectra that were consistent with the retention of the wild-type secondary structure *(25)*. The tertiary structure of wild-type p16 has a putative cleft for CDK binding.

Given the difficulties of assessing the structure of p16 mutants directly, they are more frequently analyzed for functional activity. Pollock et al. *(26)* analyzed 120 point mutations in p16 from cell lines and primary tumors, but the functional importance of the majority of these mutations is not known. To date, mutations involved in the pathogenesis of familial melanoma have received the most attention. The p16 mutants in families tend to cluster in the central portions of the protein and generally result in nonconservative amino acid substitutions. When comparing the tumor-associated missense mutations across the ankyrin repeat structures of p16^{INK4A}, it was found that more than 50% of the mutations in ankyrin repeats I and IV but fewer than 30% of the mutations in ankyrin repeats II and III resulted in conservative amino acid changes. Likewise, those tumor-associated mutations in the second, third, and early fourth ankyrin repeats occur more frequently at residues that are invariant in all members of the p16^{INK4A} family that are characterized. Because highly conserved residues in protein families may identify positions critical to protein structure or function, it was hypothesized that mutations of these highly conserved residues would be more likely to alter p16^{INK4A} function *(25)*. Consequently, those highly conserved amino acids in the second, third, and early fourth ankyrin repeats may be critical for p16^{INK4A} function, and tumor-associated mutations elsewhere may not cause functional abnormalities.

Three assays are available to assess the functional ability of p16^{INK4A} proteins to (a) bind to CDKs, (b) inhibit the activity of CDKs, and (c) arrest cell proliferation. CDK binding is the most commonly used test of p16^{INK4A} function because all inactivating p16 mutations characterized to date result in mutants that lack CDK-binding activity *(27,28)*. All mutations located in the central ankyrin repeats (ankyrin repeats II and III) had detectable alterations of function. All p16 mutants with defective CDK binding were inactive in each of the known functions of p16, confirming that CDK binding is necessary for p16 function. Several mutants retained CDK binding yet had other functional defects, suggesting that CDK binding, although necessary for p16^{INK4A} activity, is insufficient for full inhibition of CDK or for growth arrest. Mutations outside the CDK-binding cleft could theoretically alter p16^{INK4A} function by decreasing protein stability or subcellular localization. Protein stability, localization, or as yet undetermined defects resulting from p16 missense mutations would not be detected by in vitro assays but could be confirmed by growth-arrest assays. Mutants that retained partial function were found in primary tumors, suggesting that incomplete loss of p16^{INK4A} function might be sufficient for carcinogenesis. However, gene deletion and promoter methylation, both of which totally eliminate expression of the p16^{INK4A} protein, appear to be the mechanisms of p16^{INK4A} gene inactivation that are most often observed in tumors.

5.2. Cell Cycle, Cyclin-Dependent Kinases, and Inhibitors

The activation and inactivation of cyclin-dependent kinases (CDKs) control progression of cells through the cell cycle. The phosphorylation status and binding of cyclin partners are common features of CDK regulation. Over the past few decades, studies by many groups have shown that passage of cells through the cell cycle depends on the activity of enzymes known as cyclin-dependent kinases (Cdks), a name indicating that they become active only when they associate with protein partners called cyclins ***(29)***. Cdks are located at the core of the cell cycle engine and drive cell proliferation forward by phosphorylating specific substrates in a cell cycle-dependent fashion. In order to become active kinases, the Cdks must associate with cyclins as well as undergo an activating phosphorylation. There are two types of primary G1-phase cyclins: D-type cyclins and the cyclin E family ***(30)***. The D family of cyclins assembles into holoenzymes with the kinase catalytic subunits Cdk4 and Cdk6, while the principal partner of cyclin E is Cdk2. The p16^{INK4A} gene product, which is an inhibitor of cyclin D–Cdk4 and cyclin D–Cdk6 complexes, is required for retinoblastoma (Rb) protein phosphorylation. The cdc25 phosphatases activate the CDKs by dephosphorylating their inhibitory tyrosine and threonine phosphorylated residues. Loss of p16^{INK4A} by 9p21 deletion or mutation allows cyclin–CDK complexes to proceed with phosphorylation of Rb. Phosphorylation of RB leads to its disengagement from the E2F family of transcription factors (**Fig. 4**). E2Fs stimulate DNA-synthesis by activating several DNA-synthesizing enzymes and releases the cells to the S phase of the cell-cycle. In malignant cells, the cell-cycle control pathway governed by the D-type cyclins is commonly mutated in tumor cells. In normal cells, the p16/CDKN2 family of Cdk-inhibitor proteins counteracts the positive effects of cyclin D complexes on cell cycle progression.

6. Sporadic Melanoma

In an attempt to identify the underlying mechanisms of neoplasia, many studies have surveyed tumors and tumor cell lines for the presence of mutations in genes encoding cell cycle-related proteins. As a result of these studies, mutations in cell cycle genes constitute the most common genetic event in tumor cells. In fact, almost all of the tumors have mutations in one of the genes involved in controlling progression through the cell cycle. Malignant cells may acquire independence from regulatory signals that are normally required for a controlled cell cycle progression. In sporadic melanoma, it might well harbor inactivated p16 locus due to any combination of abnormalities. It has been reported that virtually 100% of melanoma cell lines contain deletion in the 9p21 locus, mutations in the p16 gene, or p16 can be inactivated through methylation of its promoter. Walker et al. ***(31)*** have compiled the extent of various abnormalities in tumors and cell lines to show inactivation of this locus. Recently, it has been found that methylation of p16^{INK4A} is potentially reversible with exposure to demethylating agents, such as 5-aza-2'-deoxycytidine, which is a well-established inhibitor of DNA methylation and may open up new ways to effective tumor management ***(32)***. Since abnormalities in p16 affect Rb function and that of p14 affects p53 function, it is conceivable that aberrant

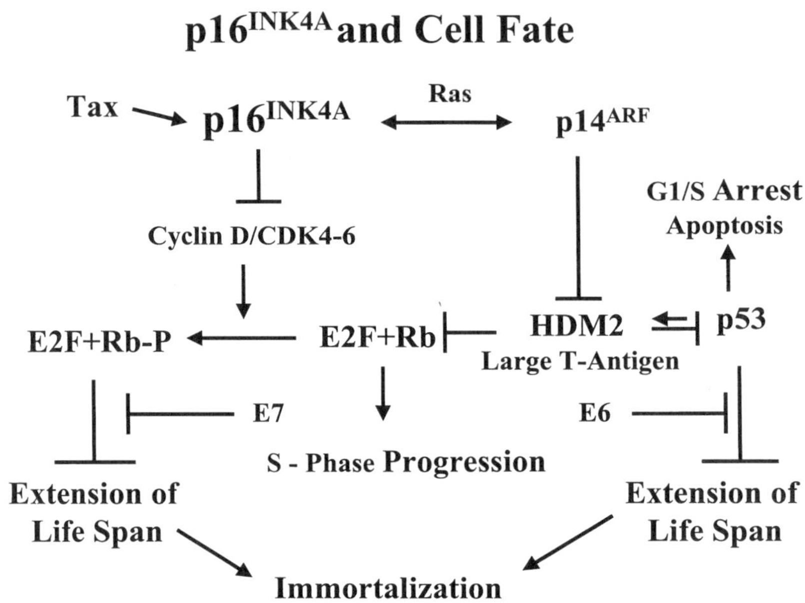

Fig. 4. Functional and biochemical consequences of abnormalities in the p16^{INK4A} and p14^{ARF} cell cycle. Inactivation of p16^{INK4A} and p14^{ARF} leads to deregulation of both Rb and p53-mediated cell cycle control. This may result in extension of lifespan and immortalization.

inactivation of this locus may lead to melanomagenesis (**Fig. 2**). Therefore, it is surprising that some of the families, which show linkage to 9p21, harbor mutations in the p16^{INK4A} gene while most of the families, which do not show linkage to this region, do not present mutations in the p16^{INK4A} gene. Even though the strong evidence that the p16^{INK4A} gene and its protein product are involved in the susceptibility to malignant melanoma, the role of this gene in sporadic melanoma is still unclear. Although initial studies in cell lines showed homozygous deletions in 60% of the cases, few mutations have been found in primary and metastasic melanoma.

7. Therapy

Introduction of wild-type p16^{INK4A} in the tumors to inhibit aberrant growth is an attractive option. Several experimental studies have shown that such an approach using a number of delivery strategies is beneficial in vitro *(33,34)*. However, targeted delivery in vivo is still a vexing question due to the disseminating nature of melanoma. Other strategies to inhibit malignant cellular proliferation involve inhibiting CDK activity or enhancing function of the CDKI with the use of chemicals. Inhibitors of CDKs showing promise in the clinic include flavopiridol and UCN-01 *(35–37)*, which show early evidence of human tolerability in clinical trials.

The p16^{INK4A} is a small molecule and its function as a G1 CDK-dependent kinase inhibitor can be mimicked by a 10-amino acid peptide that corresponds to the third ankyrin-like repeat of the full-length protein. This peptide can be synthesized together with a small carrier peptide derived from the third homeodomain of the Antennapedia protein and be directly delivered in vitro into cells with 100% efficiency, leading to arrest of cells in G1 *(38,39)*. However, at concentrations that induce G1 arrest, most cells are viable and can be rescued back into the cell cycle simply by changing the medium. G1 arrest can be maintained for weeks using human keratinocyte-derived cells and still allow cells to enter DNA replication once the peptides have been removed. This suggests that the effects of the p16^{INK4A}-mimicking peptides are reversible. Interestingly, cells treated with p16^{INK4A}-mimicking peptides show a matrix-specific inhibition of spreading on vitronectin, even though their capacity to attach to this substratum is not affected. This effect is related to dissociation of the vitronectin integrin receptor αvβ3 from the focal adhesion contacts. A similar effect is also observed using cdk4 and cdk6 inhibitory Antennapedia fusion peptides derived from the p21$^{Waf1/Cip1}$ and p18^{INK4C} proteins, suggesting that inhibition of G1-associated kinases regulates αvβ3 integrin function ***(40,41)***. This is supported by observations that transfection of glioblastoma and melanoma cells with a full-length p16 gene construct results in inhibition of spreading and migration, suggesting a novel function for p16 that is related to cell migration. Interestingly, loss of p16 correlates with upregulated expression of αvβ3 in late stages of both glioblastoma and melanoma tumor development. The αvβ3 integrin is strongly associated with melanoma ***(42,43)*** with respect to invasive cancer, and it is essential for angiogenesis ***(44)***. It is possible that inhibitors of G1 phase CDKs will have an impact on cancer treatment not only as cell cycle inhibitors, but also as inhibitors of tumor cell spreading and neovascularization.

A novel aspect of the p16^{INK4A} locus is that it encodes not just one but two separate gene products that are transcribed in alternative reading frames. Both products function as negative regulators of cell cycle progression. The p16^{INK4A} protein itself executes its effects by competitively inhibiting cyclin-dependent kinase 4 and thereby inhibiting the Rb-mediated effect on cell cycle. Inherited and acquired deletions or point mutations in the p16^{INK4A} gene increase the likelihood that potentially deleterious DNA damage will escape repair before cell division. The second product of the locus, p14^{ARF} regulates cell growth through independent effects on the p53 pathway. Although there is little evidence that p14^{ARF} by itself is involved in the pathogenesis of melanoma, deletions at the p16^{INK4A} locus disable two separate pathways that control cell growth. These recent advances open up the possibility of novel therapeutic strategies for melanoma based on gene therapy or small-molecule mimicry targeted to the correction of defects in the p16^{INK4A} and p14^{ARF} regulatory pathway.

References

1. Herlyn, M. (1993) *Molecular and Cellular Biology of Melanoma,* Landes, Austin, TX.
2. Satyamoorthy, K., Meier, F., Hsu, M. Y., Berking, C., and Herlyn, M. (1999) Human xenografts, human skin and skin reconstructs for studies in melanoma development and progression. *Cancer Metastasis Rev.* **18**, 401–405.
3. Chin, L., Merlino, G., and DePinho, R. A. (1998) Malignant melanoma: modern black plague and genetic black box. *Genes Dev.* **12**, 3467–3481.
4. Herlyn, M. and Satyamoorthy, K. (2001) Molecular biology of cutaneous melanoma, in *Cancer: Principles and Practice of Oncology,* 6th ed. DeVita, V. T., Jr., Hellman, S., and Rosenberg, S. A. (eds.), Lippincott Williams & Wilkins, Philadelphia, pp. 2003–2012.
5. Meier, F., Satyamoorthy, K., Nesbit, M., et al. (1998) Molecular events in melanoma development and progression. *Front. Biosci.* **3**, D1005–D1010.
6. Fidler, I. J. (1996) Critical determinants of melanoma metastasis. *J. Invest. Dermatol. Symp. Proc.* **1**, 203–208.

7. Greene, M. H., Clark, W. H., Jr., Tucker, M. A., Kraemer, K. H., Elder, D. E., and Fraser, M. C. (1985) High risk of malignant melanoma in melanoma-prone families with dysplastic nevi. *Ann. Intern. Med.* **102**, 458–465.
8. Tucker, M. A., Fraser, M. C., Goldstein, A. M., Elder, D. E., Guerry, D. T., and Organic, S. M. (1993) Risk of melanoma and other cancers in melanoma-prone families. *J. Invest. Dermatol.* **100**, 350S–355S.
9. Greene, M. H. (1991) Rashomon and the procrustean bed: a tale of dysplastic nevi. *J. Natl. Cancer Inst.* **83**, 1720–1724.
10. Cannon-Albright, L. A., Goldgar, D. E., Meyer, L. J., et al. (1992) Assignment of a locus for familial melanoma, MLM, to chromosome 9p13–p22. *Science* **258**, 1148–1152.
11. Skolnick, M. H., Cannon-Albright, L. A., and Kamb, A. (1994) Genetic predisposition to melanoma. *Eur. J. Cancer* **13**, 1991–1995.
12. Cowan, J. M., Halaban, R., and Francke, U. (1988) Cytogenetic analysis of melanocytes from premalignant nevi and melanomas. *J. Natl. Cancer Inst.* **80**, 1159–1164.
13. Fountain, J. W., Karayiorgou, M., Taruscio, D., et al. (1992) Genetic and physical map of the interferon region on chromosome 9p. *Genomics* **14**, 105–112.
14. Gruis, N. A., Sandkuijl, L. A., Weber, J. L.,et al. (1993) Linkage analysis in Dutch familial atypical multiple mole-melanoma (FAMMM) syndrome families. Effect of naevus count. *Melanoma Res.* **3**, 271–277.
15. Nancarrow, D. J., Mann, G. J., Holland, E. A., et al. (1993) Confirmation of chromosome 9p linkage in familial melanoma. *Am. J. Hum. Genet.* **53**, 936–942.
16. Goldstein, A. M., Dracopoli, N. C., Engelstein, M., Fraser M. C., Clark, W. H., Jr., and Tucker, M. A. (1994) Linkage of cutaneous malignant melanoma/dysplastic nevi to chromosome 9p, and evidence for genetic heterogeneity. *Am. J. Hum. Genet.* **54**, 489–496.
17. Lindor, N. M. and Greene, M. H. (1998) The concise handbook of family cancer syndromes. Mayo Familial Cancer Program. *J. Natl. Cancer. Inst.* **90**, 1039–1071.
18. Stone, S., Jiang, P., Dayananth, P., et al. (1995) Complex structure and regulation of the P16 (MTS1) locus. *Cancer Res.* **55**, 2988–2994.
19. Mao, L., Merlo, A., Bedi, G., Shapiro, G. I., Edwards, C. D., Rollins, B. J., and Sidransky, D. (1995) A novel p16INK4A transcript. *Cancer Res.* **55**, 2995–2997.
20. Roussel, M. F. (1999) The INK4 family of cell cycle inhibitors in cancer. *Oncogene* **18**, 5311–5317.
21. Piccinin, S., Doglioni, C., Maestro, R., Vukosavljevic, T., Gasparotto, D., D'Orazi, C., and Boiocchi, M. (1997) p16/CDKN2 and CDK4 gene mutations in sporadic melanoma development and progression. *Int. J. Cancer* **74**, 26–30.
22. Gruis, N. A., van der Velden, P. A., Sandkuijl, L. A., et al. (1995) Homozygotes for CDKN2 (p16) germline mutation in Dutch familial melanoma kindreds. *Nat. Genet.* **10**, 351–353.
23. Kamb, A., Gruis, N. A., Weaver-Feldhaus, J., et al. (1994) A cell cycle regulator potentially involved in genesis of many tumor types. *Science* **264**, 436–440.
24. Baumgartner, R., Fernandez-Catalan, C., Winoto, A., Huber, R., Engh, R. A., and Holak, T. A. (1998) Structure of human cyclin-dependent kinase inhibitor p19INK4d: comparison to known ankyrin-repeat-containing structures and implications for the dysfunction of tumor suppressor p16INK4a. *Structure* **6**, 1279–1290.
25. Borg, A., Sandberg, T., Nilsson, K., et al. (2000) High frequency of multiple melanomas and breast and pancreas carcinomas in CDKN2A mutation-positive melanoma families. *J. Natl. Cancer Inst.* **92**, 1260–1236.
26. Pollock, P. M., Yu, F., Qiu, L., Parsons, P. G., and Hayward, N. K. (1995) Evidence for u.v. induction of CDKN2 mutations in melanoma cell lines. *Oncogene* **11**, 663–668.
27. Zuo, L., Weger, J., Yang, Q., et al. (1996) Germline mutations in the p16INK4a binding domain of CDK4 in familial melanoma. *Nat. Genet.* **12**, 97–99.

28. Ruas, M., Brookes, S., McDonald, N. Q., and Peters, G. (1999) Functional evaluation of tumour-specific variants of p16INK4a/CDKN2A: correlation with protein structure information. *Oncogene* **18**, 5423–5434.
29. Nasmyth, K. (1996) Viewpoint: putting the cell cycle in order. *Science* **274**, 1643–1645.
30. Sherr, C. J. (1996) Cancer cell cycles. *Science* **274**, 1672–1677.
31. Walker, G. J., Flores, J. F., Glendening, J. M., Lin, A. H., Markl, I. D., and Fountain, J. W. (1998) Virtually 100% of melanoma cell lines harbor alterations at the DNA level within CDKN2A, CDKN2B, or one of their downstream targets. *Genes Chromosomes Cancer* **22**, 157–163.
32. Rocco, J. W. and Sidransky, D. (2001) p16(Mts-1/cdkn2/ink4a) in cancer progression. *Exp. Cell Res.* **264**, 42–55.
33. Schreiber, M., Muller, W. J., Singh, G., and Graham, F. L. (1999) Comparison of the effectiveness of adenovirus vectors expressing cyclin kinase inhibitors p16INK4A, p18INK4C, p19INK4D, p21(WAF1/CIP1) and p27KIP1 in inducing cell cycle arrest, apoptosis and inhibition of tumorigenicity. *Oncogene* **18**, 1663–1676.
34. Kawabe, S., Roth, J. A., Wilson, D. R., and Meyn, R. E. (2000) Adenovirus-mediated p16INK4a gene expression radiosensitizes non-small cell lung cancer cells in a p53-dependent manner. *Oncogene* **19**, 5359–5366.
35. Mani, S., Wang, C., Wu, K., Francis, R., and Pestell, R. (2000) Cyclin-dependent kinase inhibitors: novel anticancer agents. *Expert Opin. Investig. Drugs* **9**, 1849–1870.
36. Moorthamer, M., Panchal, M., Greenhalf, W., and Chaudhuri, B. (1998) The p16(INK4A) protein and flavopiridol restore yeast cell growth inhibited by Cdk4. Biochem. *Biophys. Res. Commun.* **250**, 791–797.
37. Omura-Minamisawa, M., Diccianni, M. B., Batova, A., et al. (2000) In vitro sensitivity of T-cell lymphoblastic leukemia to UCN-01 (7-hydroxystaurosporine) is dependent on p16 protein status: a Pediatric Oncology Group study. *Cancer Res.* **60**, 6573–6576.
38. Fujimoto, K., Hosotani, R., Miyamoto, Y., et al. (2000) Inhibition of pRb phosphorylation and cell cycle progression by an antennapedia-p16(INK4A) fusion peptide in pancreatic cancer cells. *Cancer Lett.* **159**, 151–158.
39. Kato, D., Miyazawa, K., Ruas, M., et al. (1998) Features of replicative senescence induced by direct addition of antennapedia-p16INK4A fusion protein to human diploid fibroblasts. *FEBS Lett.* **427**, 203–208.
40. Fahraeus, R., Lain, S., Ball, K. L., and Lane, D. P. (1998) Characterization of the cyclin-dependent kinase inhibitory domain of the INK4 family as a model for a synthetic tumour suppressor molecule. *Oncogene* **16**, 587–596.
41. Fahraeus, R. and Lane, D. P. (1999) The p16(INK4a) tumour suppressor protein inhibits alphavbeta3 integrin-mediated cell spreading on vitronectin by blocking PKC-dependent localization of alphavbeta3 to focal contacts. *EMBO J.* **18**, 2106–2118.
42. Hsu, M. Y., Shih, D. T., Meier, F. E., et al. (1998) Adenoviral gene transfer of beta3 integrin subunit induces conversion from radial to vertical growth phase in primary human melanoma. *Am. J. Pathol.* **153**, 1435–1442.
43. Van Belle, P. A., Elenitsas, R., Satyamoorthy, K., et al. (1999) Progression-related expression of beta3 integrin in melanomas and nevi. *Hum. Pathol.* **30**, 562–567.
44. Shattil, S. J. (1995) Function and regulation of the beta 3 integrins in hemostasis and vascular biology. *Thromb. Haemost.* **74**, 149–155.

12

The INK4a/ARF Locus and Human Cancer

Greg H. Enders

1. Introduction

This review focuses on the role of the INK4a/ARF locus in human cancer. Several excellent related reviews have recently been published *(1–5)*. Rather than presenting protocols, this review will describe the current state of the field. However, in keeping with the goals of these monographs, emphasis will be placed on the methods used to generate our current understanding. In particular, the methodologic challenges posed by the unusually dense structure of the locus will be highlighted.

2. Locus Structure

The INK4a/ARF locus resides on chromosome band 9p21 and directs the expression of two major tumor suppressor proteins, $p16^{INK4a}$ (p16) and $p14^{ARF}$ (ARF) (**Fig. 1**). A nearby locus encodes $p15^{INK4b}$, a p16-related cyclin-dependent kinase (Cdk) inhibitor. mRNAs for p16 and ARF are initiated at distinct sites, generating unique first exons, α for p16 and β for ARF. These exons are spliced to common second and third exons. Translation of each protein initiates in exon 1 and continues in exon 2, in alternative reading frames. Expression of proteins from overlapping reading frames is also seen in the compact genomes of plant and animal viruses but is unprecedented in the genomes of eucaryotes. p16 and ARF share no amino acid sequence homology and the reason for the unusual gene structure remains uncertain. Both genes are conserved throughout mammals, but the ARF coding region has major differences in size (mouse ARF is 19 kDa) and sequence between the human and mouse genomes. Neither p16 nor ARF is conserved in *Drosophila*, suggesting that these proteins may perform functions of unique value to large, long-lived metazoans.

3. p16 Is a Tumor Suppressor

Interest in the INK4a/ARF locus derives from abundant evidence that it plays a key role in mammalian tumor suppression. p16 was discovered as a binding partner and *in*hibitor of C*dk4* and its close relative, Cdk6 *(6)* (**Fig. 2a**). These enzymes are activated by D-type cyclins and directly inactivate proteins of the retinoblastoma (pRb) family, by

From: *Methods in Molecular Biology, Vol. 222: Tumor Suppressor Genes: Pathways and Isolation Strategies*
Edited by: Wafik S. El-Deiry © Humana Press Inc., Totowa, NJ

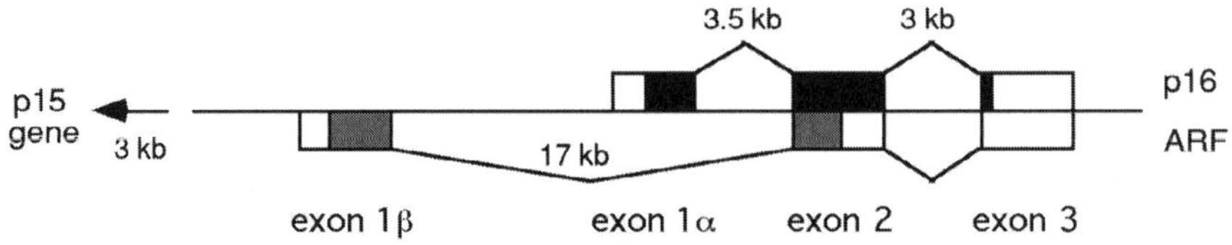

Fig. 1. Structure of the INK4a/ARF locus. Boxes above the line represent p16 exons and boxes below represent ARF exons. Shaded portions depict the approximate coding regions (black for p16, gray for ARF). Note that the human locus is depicted; thus, the ARF coding region directs synthesis of a 14-kDa protein and occupies only half of exon 2, whereas it directs synthesis of a 19-kDa protein and occupies most of exon 2 in the mouse. Introns are depicted as tented lines above and below the exons and their approximate sizes are given. The $p15^{INK4b}$ locus is 3 kb centromeric, in the direction of the arrow.

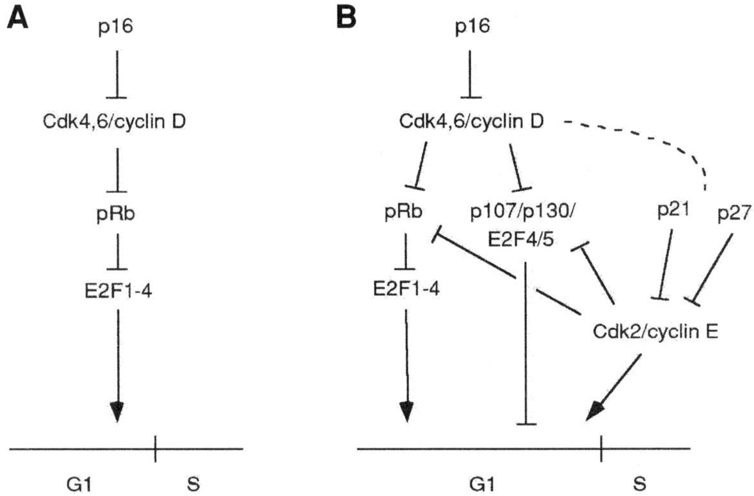

Fig. 2. The p16 response pathway. Arrows depict stimulation of function and lines with flat ends depict inhibition of function. Cell cycle position is given below. (A) Canonical p16 pathway. (B) Revised pathway, illustrating recently identified branches. The dashed line represents displacement of p21 and p27 from Cdk4/6. E2F4 and E2F5 appear to function primarily as transcriptional corepressors, rather than as activators.

phosphorylation. pRb is the prototype tumor suppressor protein ***(7)***. Mutations in pRb are responsible for familial retinoblastoma and occur in many sporadic tumors. Cdk-mediated phosphorylation of pRb family proteins relieves their repression of E2F transcription factors and induces expression of a variety of genes involved in DNA replication ***(8)***. Conversely, p16 mediates a G1 phase arrest, dependent on functional pRb family proteins ***(9–13)***. In several tumor types, inactivation of p16 or pRb is found in most derived cell lines, but simultaneous inactivation of both proteins is almost never seen, consistent with the notion that they are elements of a common and powerful tumor suppressor pathway ***(14)***. Thus, the efficacy of p16 in tumor suppression derives from its ability to activate a well-characterized tumor suppressor pathway and potently inhibit cell cycle entry.

Mutations in p16 segregate with disease in many pedigrees with familial melanoma ***(15,16)***. This observation provided some of the initial evidence that p16 was a tumor suppressor protein. The discovery of ARF complicated this interpretation. However, careful analysis suggests strongly that p16 is the key target of the mutations. Most tellingly, some mutations in familial melanoma demonstrably alter key residues in p16 without affecting the ARF coding sequence ***(9)***.

p16 is comprised of four ankyrin repeat motifs. Crystal structure data indicate that the third ankyrin repeat forms the interface with the Cdk ***(17)***. Most point mutations in the INK4a/ARF locus identified in melanoma pedigrees change amino acids that are either conversed among ankyrin repeat motifs and likely establish the basic structure of the protein or lie within the predicted Cdk interface. The functional significance of these mutations for p16 was established by studies demonstrating that the encoded proteins are typically defective in Cdk4/6 binding in vitro and in imposing cell cycle arrest ***(9,18,19)***. Some of these point mutations are in exon 1α, lie outside ARF coding sequences in exon 2, or alter "wobble bases" within the ARF coding region and are silent in the ARF reading frame ***(20)***. A few studied mutations did not demonstrably compromise p16 function, but the assays utilized may not be fully sensitive for p16 function in vivo ***(21)***, and even fewer point mutations have been shown to disrupt ARF function. Moreover, other pedigrees have been identified in which melanoma segregates with mutations in Cdk4 that disrupt p16 binding ***(22)***. Thus, there is no doubt that p16 suppresses human melanoma. The reader is referred Chapter 11 for further discussion of this issue. Genetic evidence suggests that p16 and/or ARF suppresses melanoma formation in mice ***(23)***.

Homozygous deletion of the INK4a/ARF locus is encountered commonly in human tumor cell lines, another of the original observations that pointed to the importance for the locus in tumor suppression ***(24,25)***. The coding region for $p15^{INK4b}$ is excluded from a number of these deletions, suggesting that p16, ARF, or both proteins were likely the targets. After these initial publications, it became evident that homozygous deletion of the INK4a/ARF locus is more common in tumor-derived cell lines than in primary tumors ***(26)***. In addition, many studies of primary tumors found few mutations in the p16 coding sequence. These observations raised the possibility that the INK4a/ARF locus was a "paper tiger," more relevant to growth of tumor cells in vitro than to tumor suppression in vivo. However, further investigations have substantiated p16's role in suppressing sporadic tumors. Homozygous deletion of the INK4a/ARF locus has been shown to be common in brain, head and neck, and other types of primary tumors ***(27)***. Furthermore, p16 transcription is selectively blocked epigenetically in many primary tumors of the colon and other tissues, in association with methylation of the promoter ***(27,28)***. Though infrequent on the whole, point mutations in p16 are found in a significant fraction of primary, sporadic esophageal and pancreatic carcinomas ***(5)***. These mutations preferentially alter p16 over ARF, as is found in familial melanoma ***(9,19)***. Cyclin D, the major activating subunit of Cdk4 and Cdk6, is also overexpressed in a number of malignancies, underscoring the importance of the p16/pRb pathway as a whole in tumor suppression ***(29)***. A careful analysis that takes into account all known modes of inactivation of p16, including mutation, deletion, and promoter methylation, indicates that p16 function is abrogated in nearly all pancreatic adenocarcinoma ***(30)***. This is consistent with evidence from several studies that germline mutations in p16 confer susceptibility to this tumor type ***(31–33)***, in addition to melanoma. Thus, there is no doubt that p16 plays an important role in suppression of human cancer.

The p16 gene was "knocked out" in the mouse, yielding a highly tumor-prone phenotype ***(34)***. Most mice die of sarcomas and lymphomas in the first year of life. Primary

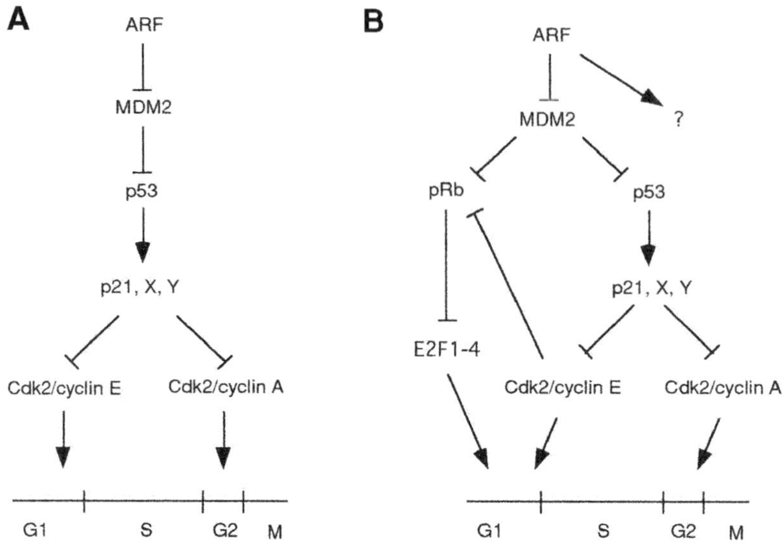

Fig. 3. The ARF response pathway, illustrated as per legend to **Fig. 2**, with the canonical pathway in (A) and recent identified branches in (B). p53 has multiple effectors, as represented by X and Y. p21 is singled out, to highlight the convergence with the p16 response pathway.

embryo fibroblasts cultured from these mice form colonies with long-term growth potential at a much higher frequency than their wild-type counterparts. However, the engineered mutation deletes INK4a/ARF exons 2 and 3 and likely eliminates function of both p16 and ARF. A mouse strain in which exon 1β was replaced by a neomycin resistance gene yielded a very similar phenotype ***(35,36)***. No deficit in p16 expression was found, although it remains possible that the mutation limits p16 expression in settings critical for tumorigenesis. This result implies that loss of ARF, rather than p16, is primarily responsible for the phenotype (see **Subheading 4**).

4. ARF Is a Tumor Suppressor

ARF was discovered through detection of an alternative transcript from the INK4a/ARF locus ***(37–40)***. Like p16, ARF has been shown to regulate a well-characterized tumor suppressor pathway. ARF expression can arrest cells in both G1 and G2 phases ***(40)*** (**Fig. 3a**). ARF activates the potent tumor suppressor p53, by binding to MDM2 ***(41,42)***. ARF binding prevents MDM2 from binding the transcriptional activation domain of p53 and targeting p53 for ubiquitin-dependent proteolyis ***(43,44)***. ARF sequesters MDM2 in the nucleolus, although the mechanism appears to be surprisingly complex. ARF has a nucleolar localization sequence, but also activates a cryptic nucleolar localization sequence in MDM2 ***(45)***. The net result is increased p53 levels and transcriptional activity ***(41,42)***.

Although point mutations of ARF are rare in human tumors, the ARF protein sequence is less conserved and may form a more open, flexible structure than that of p16, perhaps insulating ARF from inactivation by point mutation. Indeed, much of the coding region of the ARF gene is dispensable for its function in cell cycle inhibition ***(42)***. In contrast, deletion or point mutation of any of the four ankyrin repeats of p16 can disrupt its function ***(9,46,47)***. Analysis of ARF alterations in tumors has lagged behind such analysis for p16. Nonetheless, it is becoming clear that specific methylation of the ARF promoter occurs in some human tumors independent of p16 promoter methylation ***(28)***. The most compelling evidence that ARF suppresses tumorigenesis has derived from the knockout mouse studies described above ***(35,36)***. Thus far, ARF appears to be the dominant tumor suppressor in the mouse, whereas p16 appears to be the dominant tumor suppressor in humans, for unknown reasons.

5. Response Pathways: Recent Evidence for Branching

New information continues to emerge regarding the mechanisms of action of p16 and ARF. The view of the p16 response pathway as a simple, linear pathway terminating in the sequestration of E2F transcription factors by pRb (**Fig. 2a**) has been revised (**Fig. 2b**). Early evidence suggested that inhibition of Cdk4 or 6 was insufficient to arrest the cell cycle ***(48)***. Expression of a dominant negative mutant of Cdk2 can block the G1/S phase transition in many cell types, whereas similarly designed mutants of Cdk4 and Cdk6 are ineffective, alone or together ***(48,49)***. Even though p16 binds selectively to Cdk4 and Cdk6, the activity of Cdk2 is typically inhibited to a greater degree following p16 expression ***(49–53)***. The mechanism appears to primarily involve redistribution of Cip/Kip inhibitors from Cdk4 and Cdk6 to Cdk2, augmented by increased $p21^{Waf1/Cip1}$ translation in some settings.

It has been known for some time that the absence of pRb renders mouse embryo fibroblasts (MEFs) resistant to cell cycle inhibition by p16. Examination of MEFs singly and multiply deficient in different pRb family members has recently revealed that simultaneous deficiency of the other two pRb family members, p107 and p130, also renders the cells resistant ***(13)***. Likewise, MEFs deficient in both E2F4 and E2F5, the major binding partners of p107 and p130, are p16-resistant ***(54)***. pRb function does not appear to be perturbed by these deficiencies, suggesting that p107 and p130, together with E2F4 and E2F5, comprise a distinct, essential branch of the p16 response pathway. Recently, mutations in p130 have been detected at a significant frequency in primary lung cancers and Burkitt's lymphoma ***(55–57)***.

Despite abundant evidence that ARF activates p53 by antagonizing MDM2 (**Fig. 3a**), evidence has emerged for p53- and MDM2-independent functions of ARF in tumor suppression in the mouse (**Fig. 3b**). Mice defective for all three proteins are more tumor prone than those defective in p53 alone or in both p53 and MDM2 ***(58)***. The targets of ARF in these settings are unknown, and the relevance to human tumor suppression is as yet unclear. Evidence from earlier work suggests the existence of p53-independent effects of MDM2. MDM2 can stimulate E2F activity, either by binding and antagonizing the function of pRb ***(59)*** or by binding E2F itself ***(60)***. Consistent with these findings, MDM2 can drive S-phase entry in mammary epithelial cells in vivo ***(61)*** and can abrogate cell cycle arrest mediated by transforming growth factor-β in vitro ***(62)***. Both of these effects occur in the presence or absence of p53.

6. Convergence of Response Pathways

One rationale for the coevolution of p16 and ARF from a single genetic locus would be that the two pathways might cooperate mechanistically to block tumor growth. Evidence that the two pathways act synergistically to arrest neoplastic cell proliferation is lacking, but there are points at which they converge (**Figs. 2b** and **3b**). For example, ARF activation of p53 leads to transcriptional induction of p21 ***(41,42)***. Cell cycle inhibition by p21 appears to be due primarily to inhibition of Cdk2 ***(63,64)***. p16 expression leads to redistribution of p21 and p27 from Cdk4 andCdk6 to Cdk2 and, in some settings, increased translation of p21. Thus, p16 and ARF pathways converge on p21-mediated inhibition of Cdk2. Perhaps reflecting this convergence, basal expression of both p16 and ARF is required for some cells to arrest efficiently in G1 in response to gamma irradiation ***(65, 66)***. Cdk2 appears to have roles within S and G2/M phases ***(67–70)***, in addition to its role at the G1/S transition ***(48,71)***. Nonetheless, p16 expression does not arrest cells in G2 phase and does not augment the G2 arrest seen in U2-OS osteogenic sarcoma cells following γ–irradiation and p53 activation (S. Maude and G.H. Enders, unpublished). Use of antisense constructs in MEFs has provided evidence that the ARF-mediated senescence depends on both p53 and pRb ***(72)***. The pRb dependence may reflect pRb-dependent effects of p21 induced by ARF and/or activation of pRb by inhibition of MDM2.

In a different light, there is strong evidence that ARF activation serves as a "backup" to the p16/pRb pathway in tumor suppression. ARF is induced by E2F transcription factors ***(73)***. As a result, abrogation of the p16/pRb pathway activates the ARF pathway, blunting cell proliferation. Thus, there is evidence that the two pathways may cooperate via simultaneous or sequential activation.

7. Coregulation

Another rationale for the structure of the INK4a/ARF locus is that it may facilitate coregulation of expression of p16 and ARF, through shared cis-acting transcriptional regulatory elements. Evidence exists for coregulation. Transcription studies are complicated by the structural overlap of the two genes. mRNAs for p16 and ARF are distinguished by use of primers or probes derived from the unique exons, 1α and 1β. For example, a common strategy for reverse transcription polymerase chain reaction (RT-PCR) amplification of p16 mRNA employs one primer from exon1α and another from exon 2. In this way, amplification of contaminating DNA or cDNA derived from either ARF transcripts or unprocessed INK4a/ARF mRNA (which retains the intervening intron) is avoided. Distinguishing expression of the proteins poses no special problems, because there is no homology at this level. Expression of both proteins is often found to be increased in settings of abnormally sustained cell proliferation, in neoplasia or senescence of primary cells in culture. For example, both p16 and ARF are induced in response to overexpression of an activated ras allele in primary fibroblasts in vitro ***(74)***. When assayed by transient transfection in a panel of colon carcinoma cell lines, the activity of the two promoters shows both some correlation and instances of divergence ***(75)***. In mice, overexpression of the oncogene and transcriptional repressor Bmi1 inhibits expression of both p16 and ARF ***(76)***. Conversely, deletion of Bmi-1 results in proliferative defects in lymphoid cells in vivo and increased p16 and ARF expression and premature senescence in fibroblasts in vitro. This suggests that Bmi-1 may negatively regulate expression from the INK4a/ARF locus as a whole. No specific cis-acting transcriptional regulatory elements, such as enhancers or sites regulating polyadenylation, have been shown to be shared between p16 and ARF. Nonetheless, coregulation of these two tumor suppressor proteins in neoplastic states remains a leading rationale for their conjoined genomic structure.

8. Physiologic Settings of Tumor Suppression

The data summarized above have established that p16 and ARF are important tumor suppressors. A key issue facing the field now is to identify the steps in tumorigenesis at which these proteins intervene. Clues have been derived from studies of regulation of expression of the proteins in vitro and, to a lesser degree, in vivo.

p16 is typically difficult to detect in normal tissues. A recent immunohistochemical survey of p16 expression in human tissues found evidence for expression in selected cells of several organs, including proliferative cells of the breast and uterine epithelium ***(77)***. The p16 mRNA is undetectable in most normal murine tissues but becomes detectable in lung and spleen of older animals ***(78)***. ARF is more readily detected than p16 in normal tissues. However, neither protein is required for normal development of the mouse, consistent with the notion that their roles may be primarily confined to inhibition of premalignant and malignant cells ***(34)***.

p16 is expressed strongly in a number of primary human carcinomas, c.f. ***(79,80)***, but the protein may not be functional in these settings, because of mutation of pRb. Inactivation of pRb appears to augment p16 induction, but this does not offer a model of functional p16 regulation ***(81)***. p16 and ARF are strongly expressed in cultured cells approaching senescence, a setting in which pRb is active ***(82)***. Abrogation of p16-mediated arrest is, in fact, required for epithelial cells cultured under standard conditions to escape senescence ***(83)***. This requirement can be bypassed by culturing the cells on feeder layers and supplying active telomerase ***(84)***. This observation implies that p16 may be induced by one or more stresses present in cells growing under suboptimal conditions, rather than by a clock tied strictly to generation number. Consistent with this view, overexpression of an oncogenic allele of ras in early passage primary mouse and human fibroblasts induces p16 and ARF expression, and both proteins appear to contribute to the ensuing cellular senescence ***(74)***. p16 is a potent inducer of senescence in many cell types ***(81,82,85)***, including transformed human tumor cell lines ***(86–88)***. ARF appears to be the more potent inducer of senescence in the corresponding mouse cells ***(89)***. These observations suggest that p16 and ARF may respond to oncogenic signals by imposing senescence. Subsequent studies have indicated that ARF is induced by E2F, myc, ras, and raf signaling ***(1)***. The specific factors that regulate p16 and ARF expression in this setting are unknown. Whether senescence has a true in-vivo counterpart, in normal tissues or tumors, also remains unclear ***(84,90)***.

Studies of the p16 promoter have suggested a potential role for AP-1 sites and Jun B in p16 activation ***(91)***. p16 also appears to be upregulated in skin keratinocytes exposed to ultraviolet irradiation, but the responsible cis or trans-acting factors have not been identified ***(92)***. Neither p16 nor ARF is induced in fibroblasts in response to γ-irradiation, even though basal expression of each is required for efficient G1 arrest in this setting ***(65,66)***. ARF can be induced in vitro by the transcription factor DMP1 and DMP-1-deficient mice are mildly tumor-prone ***(93)***. However, the role of DMP-1 in tumor suppression is poorly understood. Likewise, although Bmi-1 can also regulate p16 and ARF expression, as mentioned earlier, the role of Bmi-1 in tumor suppression is poorly defined.

Colorectal neoplasia presents an attractive setting in which to examine p16 and ARF regulation and function during human tumorigenesis because tissue is readily available from multiple stages. In addition, the promoters become methylated independently in a substantial fraction of colon carcinomas, implying that p16 and ARF may each suppress colon tumor growth ***(28,94)***. p16 or pRb are rarely mutated in this tumor type, and p53 mutations occur late in tumorigenesis ***(27,95)***. Therefore, where they are detected, p16 and ARF are likely to be functional. p16 is found in rare cells in normal intestinal epithelial crypts, the compartments that house stem cells and their immediate, rapidly proliferating progeny ***(96)***. p16 expression is distinctly higher in aberrant crypt foci, the earliest detectable preneoplastic lesions, and in clusters of cells throughout later stages of tumorigenesis ***(96)***. The p16-expressing cells are routinely negative for markers of cell proliferation, consistent with the notion that p16 and/or coexpressed proteins may be inhibiting their proliferation ***(96)***. The heterogeneous nature of p16 expression suggests that the clonal mutations that drive neoplastic progression may not be sufficient to induce p16 and that lineage-specific factors, cellular stress, or extracellular factors may also influence p16 expression. DNA methylation patterns are generally stably inherited during mitosis ***(27)***. Therefore, differences in DNA methylation seem unlikely to account for the striking cell to cell variation in p16 staining in some colon tumors ***(96)***. ARF expression has yet to be characterized in this tumor type.

Further functional studies of p16 in vivo have been stymied by the lack of mice selectively deficient in p16. Such mice have now been generated and are undergoing phenotypic analysis (N. Sharpless and R. DePinho, personal communication).

9. Summary and Future Directions

Both products of the INK4a/ARF locus have been firmly established as tumor suppressors. Although largely distinct, there are points of convergence in their respective response pathways, suggesting opportunities for direct functional cooperation in cell cycle inhibition. Some coregulation of p16 and ARF is seen, at least in several in-vitro settings of oncogenic stimulation and/or cellular senescence, suggesting that the opportunity for direct cooperation is there. In addition, ARF is activated when the p16/pRb pathway is abrogated, an example of indirect cooperation between the two pathways. Further work will be required to elucidate the mechanisms regulating expression of the two proteins and to relate this regulation to specific settings of tumor suppression in vivo.

An understanding of p16 function will be fostered by the recent generation of mice selectively deficient in p16, with the caveat that p16 appears to play a more prominent tumor suppressor role in humans than in the mouse. The availability of mice with selective and combined deficiencies in p16 and ARF will allow direct tests of the notion that these proteins may cooperate to suppress tumorigenesis. Perhaps then we will understand better the rationale for the unusually dense structure of the INK4a/locus. Further work will, however, remain to better define the role of the locus in human cancer. Progress here may depend on methodological advances that permit the culture of a broader array of primary human cells, including stem cells and their immediate progeny. It seems plausible that the INK4a/ARF locus may have evolved to curtail the proliferation of such cells, in the face of sustained proliferative stimuli and/or suboptimal environmental conditions.

Note Added in Proof

Selective inactivation of p16 in the mouse germline has confirmed that this protein suppresses multiple tumor types, including melanoma ***(97, 98)***.

References

1. Sherr, C. J. and Weber, J. D. (2000) The ARF/p53 pathway. *Curr. Opin. Genet. Dev.* **10**, 94–99.
2. Sherr, C. J. (1998) Tumor surveillance via the ARF-p53 pathway. *Genes Dev.* **12**, 2984–2991.
3. Sharpless, N. E. and DePinho, R. A. (1999) The INK4A/ARF locus and its two gene products. *Curr. Opin. Genet. Dev.* **9**, 22–30.
4. Roussel, M. F. (1999) The INK4 family of cell cycle inhibitors in cancer. *Oncogene* **18**, 5311–5317.
5. Liggett, W. H., Jr. and Sidransky, D. (1998) Role of the p16 tumor suppressor gene in cancer. *J. Clin. Oncol.* **16**, 1197–1206.
6. Serrano, M., Hannon, G. J., and Beach, D. (1993) A new regulatory motif in cell-cycle control causing specific inhibition of cyclin D/CDK4. *Nature* **366**, 704–707.
7. Weinberg, R. A. (1995) The retinoblastoma protein and cell cycle control. *Cell* **81**, 323–330.
8. Dyson, N. (1998) The regulation of E2F by pRB-family proteins. *Genes Dev.* **12**, 2245–2262.
9. Koh, J., Enders, G. H., Dynlacht, B. D., and Harlow, E. (1995) Tumor-derived p16 alleles encoding proteins defective in cell cycle inhibition. *Nature* **375**, 506–510.
10. Medema, R. H., Herrera, R. E., Lam, F., and Weinberg, R. A. (1995) Growth suppression by p16ink4 requires functional retinoblastoma protein. *Proc. Natl. Acad. Sci. USA* **92**, 6289–6293.
11. Lukas, J., Parry, D., Aagarrd, L., et al. (1995) Retinoblastoma-protein-dependent inhibition by the tumor-suppressor p16. *Nature* **375**, 503–506.
12. Serrano, M., Gomez-Lahoz, E., DePinho, R. A., Beach, D., and Bar-Sagi, D. (1995) Inhibition of ras-induced proliferation and cellular transformation by p16INK4. *Science* **267**, 249–252.
13. Bruce, J. L., Hurford, R. K., Jr., Classon, M., Koh, J., and Dyson, N. (2000) Requirements for cell cycle arrest by p16INK4a. *Mol. Cell* **6**, 737–742.
14. Otterson, G. A., Kratzke, R. A., Coxon, A., Kim, Y. W., and Kaye, F. J. (1994) Absence of p16INK4 protein is restricted to the subset of lung cancer lines that retains wildtype RB. *Oncogene* **9**, 3375–3378.
15. Kamb, A., Gruis, N. A., Weaver-Feldhaus, J., et al. (1994) A cell cycle regulator potentially involved in genesis of many tumor types. *Science* **264**, 436–440.
16. Kamb, A., Shattuck-Eidens, D., Eeles, R., et al. (1994) Analysis of the p16 gene (cdkN2) as a candidate for the chromosome 9p melanoma susceptibility locus. *Nat. Genet.* **8**, 22–26.
17. Pavletich, N. P. (1999) Mechanisms of cyclin-dependent kinase regulation: structures of Cdks, their cyclin activators, and Cip and INK4 inhibitors. *J. Mol. Biol.* **287**, 821–828.
18. Ranade, K., Hussussian, C. J., Sikorski, R. S., et al. (1995) Mutations associated with familial melanoma impair p16INK4 function. *Nat. Genet.* **10**, 114–116.
19. Quelle, D. E., Cheng, M., Ashmun, R. A., and Sherr, C. J. (1997) Cancer-associated mutations at the INK4a locus cancel cell cycle arrest by p16INK4a but not by the alternative reading frame protein p19ARF. *Proc. Natl. Acad. Sci. USA* **94**, 669–673.
20. Haber, D. (1997) Splicing into senescence: the curious case of p16 and p19ARF. *Cell* **91**, 555–558.
21. Yarbrough, W. G., Buckmire, R. A., Bessho, M., and Liu, E. T. (1999) Biologic and biochemical analyses of p16(INK4a) mutations from primary tumors. *J. Natl. Cancer Inst.* **91**, 1569–1574.
22. Wolfel, T., Hauer, M., Schneider, J., et al (1995) A p16INK4a-insensitive CDK4 mutant targeted by cytolytic T lymphocytes in a human melanoma. *Science* **269**, 1281–1284.

23. Chin, L., Pomerantz, J., Polsky, D., et al. (1997) Cooperative effects of INK4a and ras in melanoma susceptibility in vivo. *Genes Dev.* **11**, 2822–2834.
24. Nobori, T., Miura, K., Wu, D. J., Lois, A., Takabayashi, K., and Carson, D. A. (1994) Deletions of the cyclin-dependent kinase-4 inhibitor gene in multiple human cancers. *Nature* **368**, 753–756.
25. Mori, T., Miura, K., Aoki, T., Nishihira, T., Mori, S., and Nakamura, Y. (1994) Frequent somatic mutation of the MTS1/CDK4I (multiple tumor suppressor/cycline dependent kinase 4 inhibitor) gene in esophageal squamous cell carcinoma. *Cancer Res.* **54**, 3396–3397.
26. Spruck, C. H., 3rd, Gonzalez-Zulueta, M., Shibata, A., et al. (1994) p16 gene in uncultured tumours. *Nature* **370**, 183–184.
27. Baylin, S. B., Herman, J. G., Graff, J. R., Vertino, P. M., and Issa, J.-P. (1998) Alterations in DNA methylation: a fundamental aspect of neoplasia. *Adv. Cancer Res.* **72**, 141–196.
28. Esteller, M., Tortola, S., Toyota, M., et al. (2000) Hypermethylation-associated inactivation of p14(ARF) is independent of p16(INK4a) methylation and p53 mutational status. *Cancer Res.* **60**, 129–133.
29. Jiang, W., Kahan, S., Tomita, N., Zhang, Y., Lu, S., and Weinstein, B. (1992) Amplification and expression of the human cyclin D gene in esophageal cancer. *Cancer Res.* **52**, 2980–2983.
30. Schutte, M., Hruban, R. H., Geradts, J., et al. (1997) Abrogation of the Rb/p16 tumor-suppressive pathway in virtually all pancreatic carcinomas. *Cancer Res.* **57**, 3126–3130.
31. Goldstein, A. M., Fraser, M. C., Struewig, J. P., et al. (1995) Increased risk of pancreatic cancer in melanoma-prone kindreds with p16INK4 mutations. *N. Engl. J. Med.* **333**, 970–974.
32. Borg, A., Sandberg, T., Nilsson, K., et al. (2000) High frequency of multiple melanomas and breast and pancreas carcinomas in CDKN2A mutation-positive melanoma families. *J. Natl. Cancer Inst.* **92**, 1260–1266.
33. Vasen, H. F., Gruis, N. A., Frants, R. R., van Der Velden, P. A., Hille, E. T., and Bergman, W. (2000) Risk of developing pancreatic cancer in families with familial atypical multiple mole melanoma associated with a specific 19 deletion of p16 (p16-Leiden). *Int. J. Cancer* **87**, 809–811.
34. Serrano, M., Lee, H., Chin, L., Cordon-Cardo, C., Beach, D., and DePinho, R. A. (1996) Role of the INK4a locus in tumor suppression and cell mortality. *Cell* **85**, 27–37.
35. Kamijo, T., Zindy, F., Roussel, M. F., et al. (1997) Tumor suppression at the mouse INK4a locus mediated by the alternative reading frame product p19ARF. *Cell* **91**, 649–660.
36. Kamijo, T., Bodner, S., van de Kamp, E., Randle, D. H., and Sherr, C. J. (1999) Tumor spectrum in ARF-deficient mice. *Cancer Res.* **59**, 2217–2222.
37. Stone, S., Jiang, P., Dayananth, P., et al. (1995) Complex structure and regulation of the P16 (MTS1) locus. *Cancer Res.* **55**, 2988–2994.
38. Mao, L., Merlo, A., Bedi, G., et al. (1995) A novel p16INK4A transcript. *Cancer Res.* **55**, 2995–2997.
39. Duro, D., Bernard, O., Della Valle, V., Berger, R., and Larsen, C. J. (1995) A new type of p16INK4/MTS1 gene transcript expressed in B-cell malignancies. *Oncogene* **11**, 21–29.
40. Quelle, D. E., Zindy, F., Ashmun, R. A., and Sherr, C. J. (1995) Alternative reading frames of the INK4a tumor suppressor gene encode two unrelated proteins capable of inducing cell cycle arrest. *Cell* **83**, 993–1000.
41. Pomerantz, J., Schreiber-Agus, N., Liegeois, N. J., et al. (1998) The Ink4a tumor suppressor gene product, p19Arf, interacts with MDM2 and neutralizes MDM2's inhibition of p53. *Cell* **92**, 713–723.
42. Zhang, Y., Xiong, Y., and Yarbrough, W. G. (1998) ARF promotes MDM2 degradation and stabilizes p53: ARF-INK4a locus deletion impairs both the Rb and p53 tumor suppression pathways. *Cell* **92**, 725–734.
43. Haupt, Y., Maya, R., Kazaz, A., and Oren, M. (1997) Mdm2 promotes the rapid degradation of p53. *Nature* **387**, 296–299.

44. Kubbutat, M. H., Jones, S. P., and Vousden, K. H. (1997) Regulation of p53 stability by Mdm2. *Nature* **387**, 299–303.
45. Weber, J. D., Kuo, M. L., Bothner, B., et al. (2000) Cooperative signals governing ARF-mdm2 interaction and nucleolar localization of the complex. *Mol. Cell Biol.* **20**, 2517–2528.
46. Pollack, P. M., Pearson, J. V., and Hayward, N. K. (1996) Compilation of somatic mutations of the CDKN2 gene in human cancers: non-random distribution of base substitutions. *Genes Chromosomes Cancer* **15**, 77–88.
47. Foulkes, W., D,, Flanders, T. Y., Pollock, P. M., and Hayward, N. K. (1997) The CDKN2 (p16) gene and human cancer. *Mol. Med.* **3**, 5–20.
48. van den Heuvel, S., and Harlow, E. (1993) Distinct roles for cyclin-dependent kinases in cell cycle control. *Science* **262**, 2050–2054.
49. Jiang, H., Chou, H. S., and Zhu, L. (1998) Requirement of cyclin E-Cdk2 inhibition in p16(INK4a)-mediated growth suppression. *Mol. Cell. Biol.* **18**, 5284–5290.
50. McConnell, B. B., Gregory, F. J., Stott, F. J., Hara, E., and Peters, G. (1999) Induced expression of p16(INK4a) inhibits both CDK4- and CDK2-associated kinase activity by reassortment of cyclin-CDK-inhibitor complexes [In Process Citation]. *Mol. Cell. Biol.* **19**, 1981–1989.
51. Mitra, J., Dai, C. Y., Somasundaram, K., et al. (1999) Induction of p21(WAF1/CIP1) and inhibition of Cdk2 mediated by the tumor suppressor p16(INK4a). *Mol. Cell. Biol.* **19**, 3916– 3928.
52. Cheng, M., Olivier, P., Diehl, J. A., et al. (1999) The p21(Cip1) and p27(Kip1) CDK 'inhibitors' are essential activators of cyclin D-dependent kinases in murine fibroblasts. *EMBO J.* **18**, 1571–1583.
53. Sherr, C. J., and Roberts, J. M. (1999) CDK inhibitors: positive and negative regulators of G1-phase progression. *Genes Dev.* **13**, 1501–1512.
54. Gaubatz, S., Lindeman, G. J., Ishida, S., et al. (2000) E2F4 and E2F5 play an essential role in pocket protein-mediated G1 control. *Mol. Cell.* **6**, 729–735.
55. Claudio, P. P., Howard, C. M., Pacilio, C., et al. (2000) Mutations in the retinoblastoma-related gene RB2/p130 in lung tumors and suppression of tumor growth in vivo by retrovirus-mediated gene transfer. *Cancer Res.* **60**, 372–382.
56. Cinti, C., Claudio, P. P., Howard, C. M., et al. (2000) Genetic alterations disrupting the nuclear localization of the retinoblastoma-related gene RB2/p130 in human tumor cell lines and primary tumors. *Cancer Res.* **60**, 383–389.
57. Cinti, C., Leoncini, L., Nyongo, A., et al. (2000) Genetic alterations of the retinoblastoma-related gene RB2/p130 identify different pathogenetic mechanisms in and among Burkitt's lymphoma subtypes. *Am. J. Pathol.* **156**, 751–760.
58. Weber, J. D., Jeffers, J. R., Rehg, J. E., et al. (2000) p53-independent functions of the p19(ARF) tumor suppressor. *Genes Dev.* **14**, 2358–2365.
59. Xiao, Z. X., Chen, J., Levine, A. J., et al. (1995) Interaction between the retinoblastoma protein and the oncoprotein MDM2. *Nature* **375**, 694–698.
60. Martin, K., Trouche, D., Hagemeier, C., Sorensen, T. S., La Thange, N. B., and Kouzarides, T. (1995) Stimulation of E2F1/DP1 transcriptional activity by MDM2 oncoprotein. *Nature* **375**, 691–694.
61. Lundgren, K., Montes de Oca Luna, R., McNeill, Y. B., et al. (1997) Targeted expression of MDM2 uncouples S phase from mitosis and inhibits mammary gland development independent of p53. *Genes Dev.* **11**, 714–725.
62. Sun, P., Dong, P., Dai, K., Hannon, G. J., and Beach, D. (1998) p53-independent role of MDM2 in TGF-beta1 resistance. *Science* **282**, 2270–2272.
63. Chen, J., Saha, P., Kornbluth, S., Dynlacht, B. D., and Dutta, A. (1996) Cyclin-binding motifs are essential for the function of p21CIP1. *Mol. Cell. Biol.* **16**, 4673–4682.
64. Chen, J., Jackson, P., Kirschner, M. W., and Dutta, A. (1995) Separate domains of p21 involved in the inhibition of cdk kinase and PCNA. *Nature* **374**, 386–388.

65. Shapiro, G. I., Edwards, C. D., Ewen, M. E., and Rollins, B. J. (1998) p16INK4A participates in a G1 arrest checkpoint in response to DNA damage. *Mol. Cell. Biol.* **18**, 378–387.
66. Khan, S. H., Moritsugu, J., and Wahl, G. M. (2000) Differential requirement for p19ARF in the p53-dependent arrest induced by DNA damage, microtubule disruption, and ribonucleotide depletion. *Proc. Natl. Acad. Sci. USA* **97**, 3266–3271.
67. Hu, B., Mitra, J., Heuvel, S. v. d., and Enders, G. (2001) S and G2 phase roles for Cdk2 revealed by inducible expression of a dominant negative mutant in human cells. *Mol. Cell. Biol.*, **21**, 2755–2766.
68. Lukas, C., Sorensen, C. S., Kramer, E., et al. (1999) Accumulation of cyclin B1 requires E2F and cyclin-A-dependent rearrangement of the anaphase-promoting complex. *Nature* **401**, 815–818.
69. Guadagno, T. M., and Newport, J. W. (1996) Cdk2 kinase is required for entry into mitosis as a positive regulator of Cdc2-cyclin B kinase activity. *Cell* **84**, 73–82.
70. Furuno, N., den Elzen, N., and Pines, J. (1999) Human cyclin A is required for mitosis until mid prophase. *J. Cell Biol.* **147**, 295–306.
71. Tsai, L.-H., Lees, E., Faha, B., Harlow, E., and Riabowol, K. (1993) The cdk2 kinase is required for the G1-to-S transition in mammalian cells. *Oncogene* **8**, 1593–1602.
72. Carnero, A., Hudson, J. D., Price, C. M., and Beach, D. H. (2000) p16INK4A and p19ARF act in overlapping pathways in cellular immortalization. *Nat. Cell Biol.* **2**, 148–155.
73. Bates, S., Phillips, A. C., Clark, P. A., Stott, F., Peters, G., Ludwig, R. L., and Vousden, K. H. (1998) p14ARF links the tumour suppressors RB and p53. *Nature* **395**, 124–125.
74. Serrano, M., Lin, A. W., McCurrach, M. E., Beach, D., and Lowe, S. W. (1997) Oncogenic ras provokes premature cell senescence associated with accumulation of p53 and p16INK4a. *Cell* **88**, 593–602.
75. Robertson, K., and Jones, P. (1998) The human ARF cell cycle regulatory gene promoter is a CpG island which can be silenced by DNA methylation and down-regulated by wild-type p53. *Mol. Cell. Biol.* **18**, 6457–6473.
76. Jacobs, J. J., Kieboom, K., Marino, S., DePinho, R. A., and van Lohuizen, M. (1999) The oncogene and Polycomb-group gene bmi-1 regulates cell proliferation and senescence through the ink4a locus. *Nature* **397**, 164–168.
77. Nielsen, G. P., Stemmer-Rachamimov, A. O., Shaw, J., Roy, J. E., Koh, J., and Louis, D. N. (1999) Immunohistochemical survey of p16INK4A expression in normal human adult and infant tissues. *Lab. Invest.* **79**, 1137–1143.
78. Zindy, F., Quelle, D. E., Roussel, M. F., and Sherr, C. J. (1997) Expression of the p16INK4a tumor suppressor versus other INK4 family members during mouse development and aging. *Oncogene* **15**, 203–211.
79. Geradts, J., and Wilson, P. A. (1996) High frequency of aberrant p16INK4A expression in human breast carcinoma. *Am. J. Pathol.* **149**, 15–20.
80. Geradts, J., Kratzke, R. A., Niehans, G. A., and Lincoln, C. E. (1995) Immunohistochemical detection of the cyclin-dependent kinase inhibitor 2/multiple tumor suppressor gene 1 (CDKN2/MTS1) product p16INK4A in archival human solid tumors: correlation with retinoblastoma protein expression. *Cancer Res.* **55**, 6006–6011.
81. Hara, E., Smith, R., Parry, D., Tahara, H., Stone, S., and Peters, G. (1996) Regulation of p16CDKN2 expression and its implications for cell immortalization and senescence. *Mol. Cell. Biol.* **16**, 859–867.
82. Alcorta, D. A., Xiong, Y., Phelps, D., Hannon, G., Beach, D., and Barret, J. C. (1996) Involvement of the cyclin-dependent kinase inhibitor p16(INK4a) in replicative senescence of normal human fibroblasts. *Proc. Natl. Acad. Sci. USA* **93**, 13742–13747.
83. Kiyono, T., Foster, S. A., Koop, J. I., McDougall, J. K., Galloway, D. A., and Klingelhutz, A. J. (1998) Both Rb/p16INK4a inactivation and telomerase activity are required to immortalize human epithelial cells. *Nature* **396**, 84–88.

84. Ramirez, R. D., Morales, C. P., Herbert, B. S., et al. (2001) Putative telomere-independent mechanisms of replicative aging reflect inadequate growth conditions. *Genes Dev.* **15**, 398–403.
85. Reznikoff, C. A., Yeager, T. R., Belair, C. D., Savelia, E., Puthenveetil, J. A., and Stadler, W. M. (1996) Elevated p16 at senescence and loss of p16 at immortalization in human papillomovirus 16 E6, but not E7, transformed human uroepithelial cells. *Cancer Res.* **56**, 2886–2890.
86. Dai, C. Y. and Enders, G. H. (2000) p16 INK4a can initiate an autonomous senescence program. *Oncogene* **19**, 1613–1622.
87. Uhrbom, L., Nister, M., and Westermark, B. (1997) Induction of senescence in human malignant glioma cells by p16INK4A. *Oncogene* **15**, 505–514.
88. Huschtscha, L. I. and Reddel, R. R. (1999) p16(INK4a) and the control of cellular proliferative life span. *Carcinogenesis* **20**, 921–926.
89. Kamijo, T., van de Kamp, E., Chong, M. J., et al. (1999) Loss of the ARF tumor suppressor reverses premature replicative arrest but not radiation hypersensitivity arising from disabled atm function. *Cancer Res.* **59**, 2464–2469.
90. Blackburn, E. H. (2000) Telomere states and cell fates. *Nature* **408**, 53–56.
91. Passegue, E. and Wagner, E. F. (2000) JunB suppresses cell proliferation by transcriptional activation of p16(INK4a) expression. *EMBO J.* **19**, 2969–2979.
92. Pavey, S., Conroy, S., Russell, T., and Gabrielli, B. (1999) Ultraviolet radiation induces p16CDKN2A expression in human skin. *Cancer Res.* **59**, 4185–4189.
93. Inoue, K., Wen, R., Rehg, J. E., et al. (2000) Disruption of the ARF transcriptional activator DMP1 facilitates cell immortalization, Ras transformation, and tumorigenesis. *Genes Dev.* **14**, 1797–1809.
94. Gonzalez-Zulueta, M., Bender, C. M., Yang, A. S., et al. (1995) Methylation of the 5' CpG island of the p16/CDKN2 tumor suppressor gene in normal and transformed human tissues correlates with gene silencing. *Cancer Res.* **55**, 4531–4535.
95. Kinzler, K. W. and Vogelstein, B. (1996) Lessons from hereditary colorectal cancer. *Cell* **87**, 159–170.
96. Dai, C. Y., Furth, E. E., Mick, R., et al. (2000) p16(INK4a) expression begins early in human colon neoplasia and correlates inversely with markers of cell proliferation. *Gastroenterology* **119**, 929–942.
97. Sharpless, N. E., Bardeesy, N., Lee, K. H., et al. (2001) Loss of p16Ink4a with retention of p19Arf predisposes mice to tumorigenesis. *Nature* **413**, 86–91.
98. Krimpenfort, P., Quon, K. C., Mooi, W. J., Loonstra, A., and Berns, A. (2001) Loss of p16Ink4a confers susceptibility to metastatic melanoma in mice. *Nature* **413**, 83–86.

13

Progression Model of Prostate Cancer

Teresa Acosta Almeida and Nickolas Papadopoulos

1. Introduction

Prostate cancer is the most common form of cancer diagnosed in men (other than skin cancer). In 1999, in the United States alone there were approximately 179,300 new cases and 37,000 deaths due to prostate cancer *(1)*. In 1999, the prostate cancers diagnosed accounted for 29% of cancers in men and 14.7% of all cancers (other than skin) in both sexes, making it the most diagnosed cancer in the total population. In the prostate, the formation of histologically identifiable neoplastic lesions is a very frequent event, occurring in nearly one-third of the male population over 45 years of age *(2)*. Fortunately, it has been estimated that only a small percentage of these men will actually die from prostate cancer *(3–6)*. There is evidence to support environmental (including cell–cell interactions), epigenetic, and other factors playing a role in prostate tumorigenesis *(7–17)*. However, that cancer in general is a genetic disease is currently the most widely accepted model of tumor etiology *(18,19)*. A number of genetic changes have been documented in prostate cancer. Consistently, allelic losses of specific chromosomes are observed in prostate tumors, but this is not true for mutations on specific genes *(20–22)*. Prostate cancers require many years to develop. It is believed that during these years they accumulate a number of genetic alterations. However, the molecular events that underlie the development of prostate neoplasia are not well defined *(22)*. Thus, it has been difficult to establish a genetic model for the progression of prostate cancer.

2. Paradigm for Tumor Progression

The scheme of prostate cancer progression is modeled after the one described for colon cancer. Pioneering work of Vogelstein, Kinzler, and co-workers have identified a number of genetic events required for the progression of colon cancer *(23,24)*. The basis of the model is on clonal expansion, which is driven by the orderly acquisition of specific mutation in an oncogene or tumor suppressor gene. Each mutation apparently alters a specific biochemical pathway that controls the growth of the cell and promotes progression through well-defined histopathologic stages. The construction of such a model depends on the identification of a genetic change associated with a specific stage of

From: *Methods in Molecular Biology, Vol. 222: Tumor Suppressor Genes: Pathways and Isolation Strategies*
Edited by: Wafik S. El-Deiry © Humana Press Inc., Totowa, NJ

tumor progression. Colon cancer has well-defined pathologic stages, from normal colon to different stages of adenomas and finally carcinoma. Furthermore, biologic material from such stages is relatively easy to isolate. Genetic changes that appear with the same frequency in early, intermediate, and late stages of tumor progression are considered early genetic events in the development of tumors. On the other hand, genetic changes that appear less frequently in earlier stages but more frequently in late stages are considered late events. Genetic changes observed only in early but not late tumors probably do not provide the tumor with an advantage for clonal expansion. The assumption is that prostate cancer progression follows the same rules.

3. Clinical/Pathological Progression Model of Prostate Cancer

Prostatic intraepithelial neoplasia (PIN) is considered an early premalignant lesion, which progresses to organ–confined disease *(25,26)*. Then the tumor becomes metastatic, and finally, hormone refractory. Prostate tumors are initially sensitive to androgen depletion *(27)*. Classification of prostate-confined tumor is based on size, invasion of the prostate capsule, and clinical state. The tumor progression steps of organ-confined tumors can follow the staging system for prostate cancer. Organ-confined disease includes stages T1 and T2, while organ-confined tumors with local invasion are stage T3. Invasive tumors are stage T4 *(28)*. Based on this information, a clinical and pathologic model of prostate progression is constructed (**Fig. 1**).

The task now is to associate such stages to well-defined genetic alterations.

4. Heterogeneity and Multifocality

The unambiguous identification of genetic alterations requires the availability of clinical samples that are enriched for cells of a specific stage of prostate cancer progression. This, however, has been difficult because of the heterogeneous and multifocal nature of prostate cancers. Histologic inspection of prostate cancer tissue typically shows a mixture of juxtaposed benign glands, PIN foci, and neoplastic foci of varying severity. Gleason score is a grading system used by pathologists that accounts for such heterogeneity, and the score is the average of the two most prominent histologies. The higher the score, the worse the prognosis is. In the progression model shown in **Fig. 1**, the later stages of prostate-confined disease have higher Gleason scores *(29)*. The presence of multiple independent foci of prostatic adenocarcinoma within the same gland is a common finding in men with prostate cancer *(30,31)*. Individual lesions from the same prostate tested for the presence of genetic markers were identified as genetically distinct *(32,33)*. This observation suggests that multiple tumors with distinct genetic alterations can emerge in the same prostate. The small size of the prostate gland adds to the problem that heterogeneity and multiple lesions present in the isolation of homogeneous tissue.

How can this be a problem? Imagine the following situation in colon cancer, the prototype of such progression models. It is possible that there are multiple neoplastic lesions present in the same colon, each in a different stage of progression. According to the model, mutations and other genetic alterations will be different in each lesion, although the same biologic pathway is affected. For example, APC is mutated in almost all polyps

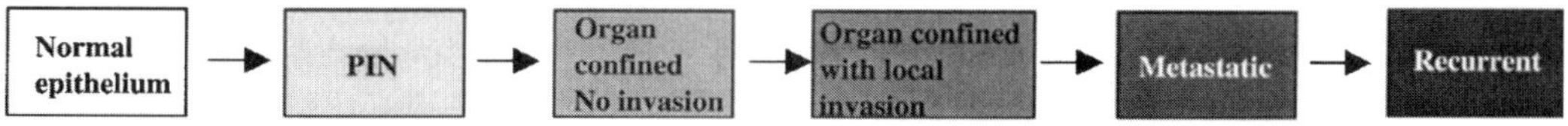

Fig. 1. Tumor progression based on clinical and pathologic parameters.

and colon cancers, but the mutation is not the same in every lesion. Furthermore, β-catenin may be mutated instead. This is the normal situation. Now imagine these polyps and carcinomas in very close proximity and in some cases overlapping each other, say, within the dimensions of a prostate gland. It would be very difficult to isolate material to study these lesions independently, and therefore difficult to identify the genetic events that gave rise to the lesions. This appears to be the case in prostate cancer.

Technologies were developed to circumvent this problem. Laser-capture microscopy (LCM) can facilitate analysis of individual neoplastic foci ***(33,34)***, while cell shorting approaches allow the relatively pure isolation of carcinoma cells ***(35,36)***. LCM has as an advantage the targeting of small lesions and provides material enriched for a specific cell type. However, the material is limited, and therefore projects involving large numbers of genetic markers are not easy to perform. Cell shorting enriches for cancerous cells. However, it cannot distinguish between cells from different lesions.

5. Cell Lines and Xenografts

The field has been also hampered by the lack of cell lines. It has been difficult for researchers to generate in-vitro cell lines ***(37,38)***. One of the potential reasons is the inherently slow growth rate of the tumors ***(39,40)***. A good alternative is the growth of xenografts. Xenografts in the context of genetic analysis of prostate tumors can be used as a means for enrichment of cancer cells. A number of groups have been able to develop such material from advanced metastatic but not early tumors. This is not surprising, because such tumors are invasive and grow well ***(41)***. Cell lines and xenografts provide a source for large amounts of pure tumor cells, which circumvents some of the problems referred to above. However, it is not immediately clear that changes identified in cell lines and xenografts represent changes that took place in the primary tumor. There are cases in which the results on cell lines, xenografts, and primary tumors are discordant. PTEN on chromosome 10q has been shown to be mutated in almost all prostate cell lines. This is not true for xenografts and primary tumors ***(42–44)***. In xenografts the PTEN expression was absent due to methylation. Epigenetic events are even more difficult to extrapolate to the primary tumors. On the other hand, the rate of mutations of PTEN in xenografts and primary tumors is similar. One thing is for sure: genetic changes that take place in the primary tumor will not be missed in cell lines and xenografts derived from this tumor. The task, then, is to short out which of the genetic changes present in cell lines and xenografts are also present in primary tumors. Given the issues outlined above, we can assume that an approach to identify all of the genetic changes in prostate cancer is to study advanced tumors in the form of xenografts. Once high-frequency genetic alterations have been identified, primary tumors can be tested for the presence of the same alterations in relatively pure material isolated by LCM or other techniques.

6. Methods for Genetic Analysis of Prostate Tumors

The methods used for identification of genetic alterations require pure cancer and paired, noncancerous DNA from the same individual. The most common techniques used are PCR-based analysis of DNA markers or genes. In addition, comparative genomic hybridization (CGH) and different variations of fluorescent in-situ hybridization (FISH) have been used. The advantage over the PCR-based techniques is their ability to identify gains of whole chromosomes or regions of specific chromosomes. On the other hand, the resolution is low and it is not easy to pinpoint alterations that involve a small region of the genome or are to be used for fine mapping of a genomic region.

7. Evidence for a Progression Model in Prostate Cancer

There is evidence for a genetic progression of prostate cancer. First, studies have shown that there is an overall accumulation of loss of heterozygy (LOH) or alterations picked up by CGH from PIN and earlier stages to later stages of prostate cancer ***(45–49)***. Second, in some cases the specific alterations have been associated with different stages of progression. Specifically, losses of chromosome 8p have been identified in large numbers of PIN specimens ***(50)*** and thus, 8p loss appears to be an early event in prostate cancer. In other cases, chromosome alterations have been controversial with respect to the stage of progression involved. For example, chromosome 7q losses have been reported as both early and late event ***(46,51,52)***. Studies of late-stage tumors have been useful for two reasons. One, they result in an association of some markers with late disease, which should be useful for management of the individuals. Two, these samples should have accumulated all of the necessary changes that are associated with progression of prostate cancer and should result in the identification of the predominant genetic alterations common to most of prostate cancers ***(53–58)***.

8. Chromosomes, Genes, and Mutations

Overall, in the past decade, a number of genetic alterations have been linked to prostate cancer development ***(21,59,60)***. Losses on chromosomes 8p and 13q are the most common ones ***(61–68)***. In addition, trisomy of chromosome 7, and losses in chromosomes 7q, 10q, 16q, 17p, and 18q, have been reported in smaller percentages ***(69–80)***. Presumably, the targets of these deletions are genes that are involved in prostate tumorigenesis and there is a focus on the isolation of the candidate genes either by positional cloning or the candidate gene approach ***(61,62,81)***. The most convincing evidence that a gene is causatively involved in the development of cancer is the identification of mutations. There are candidate genes isolated from genomic loci implicated in prostate development that did not have mutations. A good example is Nkx3.1, a good candidate gene in chromosome 8p with prostate-specific expression. Unfortunately, it is not mutated in prostate tumors ***(82–84)***. In other cases, there are more than one candidate gene for a specific chromosome locus lost in prostate cancers. For example, chromosome 10q contains PTEN, Mxi, and ANX7 ***(44,85,86)***. PTEN mutation status is unresolved, Mxi1 is infrequently mutated, and ANX1 involvement to prostate cancer is based on changes in expression of the gene determined on tissue microarrays. Other genes that have been found to be mutated in prostate cancer include pRb, p53, and p16 ***(44,87–94)***.

Androgen receptor is also mutated in prostate cancers. The androgen receptor pathway is involved in the maintenance and development of normal prostate. In addition, it is involved in the late stages of disease. The recurrent tumors are androgen-independent. However, the evidence point toward the need for the androgen pathway to remain "on" throughout the development of the tumor. The mutations associated with the androgen receptor make the receptor promiscuous to other hormones and signals besides androgen ***(92,93,95,96)***. So, it appears that androgen receptor activity is probably required in the development of prostate cancer, but alterations in the pathway itself do not give a selective advantage to the tumor cells until they are deprived of androgens.

Some genes have reduced levels of expression associated with prostate cancer progression, such as p27, bcl-2, and E-cadherin, c-CAM, and integrins. These proteins play integral roles in cell cycle, apoptosis, and cell adhesion, respectively. Processes that have been shown in rodent models and human cells are involved in prostate cancer development ***(97–104)***. Some proteins have been implicated in prostate cancer development based on biologic function. Kai1 suppresses metastasis but is not mutated ***(105,106)***. ST7, on chromosome 7q31, suppresses in-vivo proliferation of PC3 prostate cancer cell line ***(107)***.

There is no single distinct gene that has been identified as being mutated in a large numbers of prostate tumors. However, genetic alterations serve as beacons for the involvement of specific pathways in tumorigenesis. Presumably, alteration of various components of the same pathway will have the same result on the growth of the cell. This is supported by the fact that mutations of different components of the same pathway are mutually exclusive within the same tumor type, e.g., MDM2 and p53 in sarcomas ***(108)***. It has been shown that 5% of prostate tumors harbor β-catenin mutations ***(109)***. This suggested to us that the β-catenin pathway may be altered in a large number of tumors. We were able to identify mutations in other genes involved in the same pathway ***(110)***.

9. Progression Model

Putting together the information presented above, the progression model of prostate cancer is as follows. The identification of losses of chromosome 8p in a high percentage of PIN as well as late prostate tumors establishes it as an early event. Androgen receptor mutations appear to be involved in the advanced stages of the disease. The other genetic changes are harder to be associated with specific stages of progression (**Fig. 2**).

10. Familial Cancer

The identification of the genes that predispose to familial prostate cancer should help in understanding some of the early events that lead to prostate cancer (for review, see **ref.** ***111***). A number of loci have been shown to be linked to familial form of prostate cancer. However, only one candidate gene has been isolated from one of the loci, and it is not clear if indeed mutations of the gene predispose to prostate cancer ***(112–114)***. Not surprisingly, genetic changes in familial prostate cancers appear to have the same type of genetic alterations as the sporadic cancers ***(115)***.

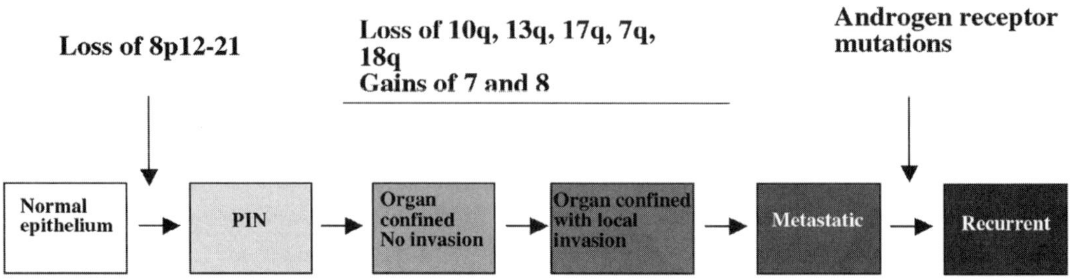

Fig. 2. Progression model of prostate cancer. The events are not well defined. Androgen receptor is probably always needed for the growth of the cancer cells. However, mutations give advantage to the cancer cells at advanced disease and during androgen depletion.

11. Conclusion

What will aid identification of the genetic alteration involved in the initiation and progression of prostate cancer? (a) The study of late tumors, because they tend to be larger and therefore it is easier to isolate material for analysis. In addition, they should have accumulated the genetic alterations responsible for neoplastic growth. (b) The accumulation of large numbers of samples enriched for cancer cells, by developing either xenografts or technologies that will allow this. (c) The availability of advanced tumors. Such material is not always available because of current patient management protocols. (d) Mutation analysis of multiple genes within pathways and not only single genes.

Gene mutations and genetic alteration can be of diagnostic or prognostic value when they are associated with a certain stage of the disease and/or other clinical parameters ***(116–120)***. Prostate cancer is readily curable by prostatectomy ***(121–123)***. The problem is that 50% of the tumors thought to be organ-confined have micrometastasis upon surgery ***(117,124)***. Molecular markers that can identify the stage of the disease are desperately needed ***(125,126)***. Construction of a genetic model for prostate cancer progression will not only help us understand the biology of the tumor, but also will provide information useful in the management of individuals who suffer from the disease. Early events will help diagnose the disease early, intermediate events should help determine the prognosis of prostate cancer, and late events will guide efforts at intervention.

References

1. Landis, S. H., Murray, T., Bolden, S., and Wingo, P. A. (1999) Cancer statistics. *CA Cancer J. Clin.* **49**, 8–31.
2. Dhom, G. (1983) Epidemiologic aspects of latent and clinically manifest carcinoma of the prostate. *J. Cancer Res. Clin. Oncol.* **106**(3), 210–218.
3. Stamey, T. (1983) Cancer of the prostate. *Monogr. Urol.* **4**, 1–21.
4. Sakr, W. A., Haas, G. P., Cassin, B., Pontes, J. E., and Crissman, J. (1993) The frequency of carcinoma and intraepithelial neoplasia of the prostate in young male patients. *J. Urol.* **150**(2 Pt 1), 379–385.
5. Franks, L. (1954) Latent carcinoma of the prostate. *J. Clin. Pathol.* **68**, 603–616.
6. Andrews, G. (1954) Latent carcinoma of the prostate. *J. Clin. Pathol.* **2**, 197–208.
7. Olumi, A. F., Grossfeld, G. D., Hayward, S. W., Carroll, P. R., Tlsty, T. D., and Cunha, G. R. (1999) Carcinoma-associated fibroblasts direct tumor progression of initiated human prostatic epithelium. *Cancer Res.* **59**(19), 5002–5011.

8. Hedlund, T. E., Duke, R. C., and Miller, G. J. (1999) Three-dimensional spheroid cultures of human prostate cancer cell lines. *Prostate* **41**(3), 154–165.
9. Miller, G. J. (1998) Vitamin D and prostate cancer: biologic interactions and clinical potentials. *Cancer Metastasis Rev.* **17**(4), 353–360.
10. Olumi, A. F., Dazin, P., and Tlsty, T. D. (1998) A novel coculture technique demonstrates that normal human prostatic fibroblasts contribute to tumor formation of LNCaP cells by retarding cell death. *Cancer Res.* **58**(20), 4525–4530.
11. Hayward, S. W., Grossfeld, G. D., Tlsty, T. D., and Cunha, G. R. (1998) Genetic and epigenetic influences in prostatic carcinogenesis (review). *Int. J. Oncol.* **13**(1), 35–47.
12. Cerhan, J. R., Torner, J. C., Lynch, C. .F., et al. (1997) Association of smoking, body mass, and physical activity with risk of prostate cancer in the Iowa 65+ Rural Health Study (United States). *Cancer Causes Control* **8**(2), 229–238.
13. Konishi, N., Hiasa, Y., Tsuzuki, T., et al. (1997) Comparison of ras activation in prostate carcinoma in Japanese and American men. *Prostate* **30**(1) 53–57.
14. Blair, A. and Fraumeni, J. F. (1978) Geographic patterns of prostate cancer in the United States. *J. Natl. Cancer Inst.* **61**(6), 1379–1384.
15. Peto, R., Doll, R., Buckley, J. D., and Sporn, M. B. (1981) Can dietary beta-carotene materially reduce human cancer rates? *Nature* **290**, 201–208.
16. Rooney, C., Beral, V., Maconochie, N., Fraser, P., and Davies, G. (1993) Case-control study of prostatic cancer in employees of the United Kingdom Atomic Energy Authority [see comments]. *Br. Med. J.* **307**, 1391–1397.
17. Gu, F. L., Xia, T. L., and Kong, X. T. (1994) Preliminary study of the frequency of benign prostatic hyperplasia and prostatic cancer in China. *Urology* **44**, 688–691.
18. Bishop, J. M. (1995) Cancer: the rise of the genetic paradigm. *Genes Dev.* **9**(11), 1309–1315.
19. Vogelstein, B. and Kinzler, K. W. (1997) Preface, in: *The Genetic Basis of Human Cancer*. Vogelstein, B. and Kinzler, K. W., eds. McGraw Hill. New York, p. xv.
20. Isaacs, W. (1996) Molecular Genetics of Prostate cancer, in *Comprehensive Textbook of Genitourinary Oncology*. Vogelzang, Scardino, Shipley, and Coffey, eds., Williams & Wilkins, Baltimore, 579–592.
21. Bova, G. S. and Isaacs, W. B. (1996) Review of allelic loss and gain in prostate cancer. *World of Urol.* **14**, 338–346.
22. Verma, R. S., Manikal, M., Conte, R. A., and Godec, C. J. (1999) Chromosomal basis of adenocarcinoma of the prostate. *Cancer Invest.* **17**(6), 441–447.
23. Fearon, E. R. and Vogelstein, B. (1990) A genetic model for colorectal tumorigenesis. *Cell* **61**, 759–767.
24. Isaacs, W. and Bova, S. (1997) Prostate cancer, in: *The Genetic Basis of Human Cancer*. Vogelstein, B., and Kingler, K. W., eds., McGraw Hill, New York, pp. 653–660.
25. Bostwick, D. G., Pcelli, A., and Lopez-Beltran, A. (1996) Molecular biology of prostatic intraepithelial neoplasia. *Prostate* **29**,117–134.
26. Bostwick, D. G. (1992) Prostatic intraepithelial neoplasia (PIN): current concepts. *J. Cell Biochem. Suppl.* **10**, 9.
27. Grayhac, J. T., Keeler, T. C., and Kozlowski, J. M. (1987) Carcinoma of the prostate. Hormonal therapy. *Cancer* **60**(3 Suppl), 589–601.
28. Montie, J. E. (1995) Staging of prostate cancer. *Cancer* **74**(1), 1–3
29. Gleason, D. F. (1992) Histological grading of prostate cancer: perspective. *Hum. Pathol.* **23**, 273–279.
30. Villers, A., McNeal, J. E., Freiha, F. S., and Stamey, T. A. (1992) Multiple cancers in the prostate. Morphologic features of clinically recognized versus incidental tumors. *Cancer* **9**, 2313–2318.
31. Byar, D. P. and Mostofi, F. K. (1972) Carcinoma of the prostate: prognostic evaluation of certain pathologic features in 208 radical prostatectomies. *Cancer* **30**, 5–13.

32. Bostwick, D. G., Shan, A., Qian, J., et al. (1998) Independent origin of multiple foci of prostatic intraepithelial neoplasia: comparison with matched foci of prostate carcinoma. *Cancer* **83**, 1995–2002.
33. Macintosh, C. A., Stower, M., Reid, N., and Maitland, N. J. (1998) Precise microdissection of human prostate cancers reveals genotypic heterogeneity. *Cancer Res.* **58**, 23–28.
34. Emmert-Buck, M. R., Bonner, R. F., Smith, P. D., et al. (1996) Laser capture microdissection. *Science* **274**, 998–1001.
35. Liu, A. Y., True, L. D., LaTray, L., et al. (1997) Cell-cell interaction in prostate gene regulation and cytodifferentiation. *Proc. Natl. Acad. Sci. USA* **94**, 10705–10710.
36. Liu, A. Y., Corey, E., Vessella, R. L., et al. (1997) Identification of differentially expressed prostate genes: increased expression of transcription factor ETS-2 in prostate cancer. *Prostate* **30**, 145–153.
37. Bright, R. K., Vocke, C. D., Emmert-Buck, M. R., et al. (1997) Generation and genetic characterization of immortal human prostate epithelial cell lines derived from primary cancer specimens. *Cancer Res.* **57**, 995–1002.
38. Navone, N. M., Logothetis, C. J., von Eschenbach, A. C., and Troncoso, P. (1999) Model systems of prostate cancer: uses and limitations. *Cancer Metastasis Rev.* **17**, 361–371.
39. Isaacs, J. T. and Coffey, D. S. (1989) Etiology and disease process of benign prostatic hyperplasia. *Prostate Suppl.* **2**, 33–50.
40. Berges, R. R., Vukanovic, J., Epstein, J. I., et al. (1995) Implication of cell kinetic changes during the progression of human prostatic cancer. *Clin. Cancer Res.* **1**, 473–480.
41. Klein, K. A., Reiter, R. E., Redula, J., et al. (1997) Progression of metastatic human prostate cancer to androgen independence in immunodeficient SCID mice. *Nat. Med.* **3**, 402–408.
42. Whang, Y. E., Wu, X., Suzuki, H., et al. (1998) Inactivation of the tumor suppressor PTEN/MMAC1 in advanced human prostate cancer through loss of expression. *Proc. Natl. Acad. Sci. USA* **95**, 5246–5250.
43. Dong, J. T., Sipe, T. W., Hyytinen, E. R., et al. (1998) PTEN/MMAC1 is infrequently mutated in pT2 and pT3 carcinomas of the prostate. *Oncogene* **17**, 1979–1982.
44. Li, J., Yen, C., Liaw, D., et al. (1997) PTEN, a putative protein tyrosine phosphatase gene mutated in human brain, breast, and prostate cancer [see comments]. *Science* **275**(5308), 1943–1947.
45. Takimoto, Y., Shimazui, T., Akaza, H., Sato, N., and Noguchi, M. (2001) Genetic heterogeneity of surgically resected prostate carcinomas and their biopsy specimens is related to their histologic differentiation. *Cancer* **91**, 362–370.
46. Alers, J. C., Rochat, J., Krijtenburg, P. J., et al. (2000) Identification of genetic markers for prostatic cancer progression. *Lab. Invest.* **80**, 931–942.
47. Fu, W., Bubendorf, L., Willi, N., et al. (2000) Genetic changes in clinically organ-confined prostate cancer by comparative genomic hybridization. *Urology* **56**, 880–885.
48. Zitzelsberger, H., Engert, D., Walch, A., et al. (2001) Chromosomal changes during development and progression of prostate adenocarcinomas. *Br. J. Cancer* **84**, 202–208.
49. Saric, T., Brkanac, Z., Troyer, D. A., et al. (1999) Genetic pattern of prostate cancer progression. *Int. J. Cancer* **81**, 219–224.
50. Emmert-Buck, M. R., Vocke, C. D., Pozzatti, R. O., et al. (1995) Allelic loss on chromosome 8p12-21 in microdissected prostatic intraepithelial neoplasia. *Cancer Res.* **55**, 2959–2962.
51. Latil, A., Cussenot, O., Fournier, G., Baron, J. C., and Lidereau, R. (1995) Loss of heterozygosity at 7q31 is a frequent and early event in prostate cancer. *Clin. Cancer Res.* **1**, 1385–1389.
52. Qian, J., Jenkins, R. B., and Bostwick, D. G. (1999) Genetic and chromosomal alterations in prostatic intraepithelial neoplasia and carcinoma detected by fluorescence in situ hybridization. *Eur. Urol.* **35**, 479–483.

53. Padalecki, S. S., Troyer, D. A., Hansen, M. F., et al. (2000) Identification of two distinct regions of allelic imbalance on chromosome 18Q in metastatic prostate cancer. *Int. J. Cancer* **85**, 654–658.
54. Strup, S. E., Pozzatti, R. O., Florence, C. D., et al. (1999) Chromosome 16 allelic loss analysis of a large set of microdissected prostate carcinomas. *J. Urol.* **162**, 590–594.
55. Sakr, W. A., Macoska, J. A., Benson, P., et al. (1994) Allelic loss in locally metastatic, multisampled prostate cancer. *Cancer Res.* **54**, 3273–3277.
56. Rubin, M. A., Gerstein, A., Reid, K., et al. (2000) 10q23.3 loss of heterozygosity is higher in lymph node-positive (pT2-3,N+) versus lymph node-negative (pT2-3,N0) prostate cancer. *Hum. Pathol.* **31**, 540–508.
57. Visakorpi, T., Kallioniemi, O. P., Heikkinen, A., Koivula, T., and Isola, J. (1992) Small subgroup of aggressive, highly proliferative prostatic carcinomas defined by p53 accumulation. *J. Natl. Cancer Inst.* **84**, 883–887.
58. Nupponen, N. N., Kakkola, L., Koivisto, P., Visakorpi, T. (1998) Genetic alterations in hormone-refractory recurrent prostate carcinomas. *Am. J. Path.* **153**, 141–148.
59. Sanberg, A. (1992) Chromosomal abnormalities and related events in prostate cancer. *Hum. Path.* **23**, 368–380.
60. Abate-Shen, C., Shen, M. M. (2000) Molecular genetics of prostate cancer. *Genes Dev.* **14**, 2410–2434.
61. Yin, Z., Spitz, M. R., Babaian, R. J., Strom, S. S., Troncoso, P., and Kagan, J. (1999) Limiting the location of a putative human prostate cancer tumor suppressor gene at chromosome 13q14.3. *Oncogene* **18**:7576–7583.
62. Afonso, A., Emmert-Buck, M. R., Duray, P. H., Bostwick, D. G., Linehan, W. M., and Vocke C. D. (1999) Loss of heterozygosity on chromosome 13 is associated with advanced stage prostate cancer [see comments]. *J Urol.* **162**(3 Pt 1), 922–926.
63. Ueda, T., Emi, M., Suzuki, H., et al. (1999) Identification of a I-cM region of common deletion on 13q14 associated with human prostate cancer. *Genes Chromosomes Cancer* **24**(3), 183–190.
64. Hyytinen, E. R., Frierson, H. F., Jr., Sipe, T. W., et al. (1999) Loss of heterozygosity and lack of mutations of the XPG/ERCC5 DNA repair gene at 13q33 in prostate cancer. *Prostate* **41**(3),190–195.
65. Wang, J. C., Radford, D. M., Holt, M. S., et al. (1999) Sequence-ready contig for the 1.4-cM ductal carcinoma in situ loss of heterozygosity region on chromosome 8p22-p23. *Genomics* **60**(1), 1–11.
66. Prasad, M. A., Trybus, T. M., Wojno, K. J., and Macoska, J. A. (1998) Homozygous and frequent deletion of proximal 8p sequences in human prostate cancers: identification of a potential tumor suppressor gene site. *Genes Chromosomes Cancer* **23**(3), 255–262.
67. Haggman, M. J., Wojno, K. J., Pearsall, C. P., and Macoska, J. A. (1997) Allelic loss of 8p sequences in prostatic intraepithelial neoplasia and carcinoma. *Urology* **50**(4), 643–647.
68. Bova, G. S., Mac Grovan, D., Levy, A., Pin, S. S., Bookstein, R., and Isaacs, W. B. (1996) Physical mapping of chromosome 8p22 markers and their homozygous deletion in a metastatic prostate cancer. *Genomics* **35**, 46–54.
69. Bandyk, M. G., Zhao, L., Troncoso, P., et al. (1994) Trisomy 7: a potential cytogenetic marker of human prostate cancer progression. *Genes Chromosomes Cancer* **9**(1), 19–27.
70. Zenklusen, J. C., Thompson, J. C., Troncoso, P., Kagan, J., and Conti, C. J. (1994) Loss of heterozygosity in human primary prostate carcinomas: a possible tumor suppressor gene at 7q31.1. *Cancer Res.* **54**(24), 6370–6373.
71. Takahashi, S., Shan, A. L., Ritland, S. R., et al. (1995) Frequent loss of heterozygosity at 7q31.1 in primary prostate cancer is associated with tumor aggressiveness and progression. *Cancer Res.* **55**(18), 4114–4119.

72. Carter, B. S., Ewing, C. M., Ward, W. S., et al. (1996) Allelic loss of chromosomes 16q and 10q in human prostate cancer. *Proc. Natl. Acad. Sci. USA* **87**(22), 8751–8755.
73. Ittmann, M. (1996) Allelic loss on chromosome 10 in prostate adenocarcinoma. *Cancer Res.* **56**(9), 2143–2147.
74. Gray, I. C., Phillips, S. M., Lee, S. J., Neoptolemos, J. P., Weissenbach, J., and Spurr, N. K. (1995) Loss of the chromosomal region 10q23-25 in prostate cancer. *Cancer Res,* **55**(21), 4800–4803.
75. Bergerheim, U. S., Kunimi, K., Collins, V. P., and Ekman, P. (1991) Deletion mapping of chromosomes 8, 10, and 16 in human prostatic carcinoma. *Genes Chromosomes Cancer* **3**(3), 215–220.
76. Li, C., Berx, G., Larsson, C., et al. (1999) Distinct deleted regions on chromosome segment 16q23-24 associated with metastases in prostate cancer. *Genes Chromosomes Cancer* **24**(3),175–182.
77. Brothman, A. R., Steele, M. R., Williams, B. J., et al. (1995) Loss of chromosome 17 loci in prostate cancer detected by polymerase chain reaction quantitation of allelic markers. *Genes Chromosomes Cancer* **13**(4), 278–284.
78. Williams, B. J., Jones, E., Zhu, X. L., et al. (1996) Evidence for a tumor suppressor gene distal to BRCA1 in prostate cancer. *J. Urol.* **155**(2), 720–725.
79. Latil, A., Baron, J. C., Cussenot, O., et al. (1994) Genetic alterations in localized prostate cancer: identification of a common region of deletion on chromosome arm 18q. *Genes Chromosomes Cancer* **11**(2), 119–125.
80. Cunningham, J. M., Shan, A., Wick, M. J., et al. (1996) Allelic imbalance and microsatellite instability in prostatic adenocarcinoma. *Cancer Res.* **56**(19), 4475–4482.
81. Yin Z., Babaian, R. J., Troncoso, P., et al. (2001) Limiting the location of putative prostate cancer tumor suppressor genes on chromosome 18q. *Oncogene* **26**, 2273–2280.
82. Bhatia-Gaur, R., Donjacour, A. A., Sciavolino, P. J., et al. (1999) Roles for Nkx3.1 in prostate development and cancer. *Genes Dev.* **13**, 966–977.
83. He, W. W., Sciavolino, P. J., Wing, J., et al. (1997) A novel human prostate-specific, androgen-regulated homeobox gene (NKX3.1) that maps to 8p21, a region frequently deleted in prostate cancer. *Genomics* **43**, 69–77.
84. Voeller, H. J., Augustus, M., Madike, V., Bova, G. S., Carter, K. C., and Gelmann, E. P. (1997) Coding region of NKX3.1, a prostate-specific homeobox gene on 8p21, is not mutated in human prostate cancers. *Cancer Res.* **57**, 4455–4459.
85. Prochownik, E. V., Eagle Grove, L., Deubler, D., et al. (1998) Commonly occurring loss and mutation of the MXI1 gene in prostate cancer. *Genes Chromosomes Cancer* **22**(4), 295–304.
86. Srivastava, M., Bubendorf, L., Srikantan, V., et al. (2001) ANX7, a candidate tumor suppressor gene for prostate cancer. *PNAS* **98**, 4575–4580.
87. Kubota, Y., Fujinami, K., Uemura, H., et al. (1995) Retinoblastoma gene mutations in primary human prostate cancer. *Prostate* **27**(6), 314–320.
88. Roy-Burman, P., Zheng, J., and Miller, G. J. (1997) Molecular heterogeneity in prostate cancer: can TP53 mutation unravel tumorigenesis? *Mol. Med. Today* **3**(11), 476–482.
89. Salem, C. E., Tomasic, N. A., Elmajian, D. A., et al. (1997) p53 protein and gene alterations in pathological stage C prostate carcinoma [see comments]. *J. Urol.* **158**(2), 510–514.
90. Pesche, S., Latil, A., Muzeau, F., et al. (1998) PTEN/MMAC1/TEP1 involvement in primary prostate cancers. *Oncogene* **16**(22), 2879–2883.
91. Cairns, P., Okami, K., Halachmi, S., et al. (1997) Frequent inactivation of PTEN/MMAC1 in primary prostate cancer. *Cancer Res.* **57**(22), 4997–5000.
92. Wallen, M. J., Linja, M., Kaartinen, K., Schleutker, J., and Visakorpi, T. (1999) Androgen receptor gene mutations in hormone-refractory prostate cancer [In Process Citation]. *J. Pathol.* **189**(4): 559–563.

93. Taplin, M. E., Bubley, G. J., Shuster, T. D., et al. (1995) Mutation of the androgen-receptor gene in metastatic androgen- independent prostate cancer [see comments]. *N. Engl. J. Med.* **332**(21), 1393–1398.
94. Gaddipati, J. P., McLeod, D. G., Sesterhenn, I. A., et al. (1997) Mutations of the p16 gene product are rare in prostate cancer. *Prostate* **30**(3), 188–194.
95. Zhao, X. Y., Malloy, P. J., Krishnan, A. V., et al. (2000) Glucocorticoids can promote androgen-independent growth of prostate cancer cells through a mutated androgen receptor. *Nat. Med.* **6**, 703–706.
96. Brinkmann, A. O., Trapman, J. (2000) Prostate cancer schemes for androgen escape. *Nat. Med.* **6**, 628–629.
97. Morrissey, C., Bennett, S., Nitsche, E., Guenette, R. S., Wong, P., and Tenniswood, M. (1999) Expression of p190A during apoptosis in the regressing rat ventral prostate. *Endocrinology* **140**(7), 3328–3333.
98. Qiu, G., Ahmed, M., Sells, S. F., Mohiuddin, M., Weinstein, M. H., Rangnekar, V. M. (1999) Mutually exclusive expression patterns of Bcl-2 and Par-4 in human prostate tumors consistent with down-regulation of Bcl-2 by Par-4. *Oncogene* **18**(3), 623–631.
99. Tenniswood, M. (1997) Apoptosis, tumour invasion and prostate cancer. *Br. J. Urol.* **79** Suppl 2, 27–34.
100. Cohen, M. B., Griebling, T. L., Ahaghotu, C. A., Rokhlin, O. W., and Ross, J. S. (1997) Cellular adhesion molecules in urologic malignancies. *Am. J. Clin. Pathol.* **107**(1), 56–63.
101. Rokhlin, O. W. and Cohen, M. B. (1995) Expression of cellular adhesion molecules on human prostate tumor cell lines. *Prostate* **26**(4), 205–212.
102. Tlsty, T. D. (1998) Cell-adhesion-dependent influences on genomic instability and carcinogenesis. *Curr. Opin. Cell Biol.* **10**(5), 647–653.
103. McDonnell, T. J., Troncoso, P., Brisbay, S. M., et al. (1992) Expression of the protooncogene bcl-2 in the prostate and its association with emergence of androgen-independent prostate cancer. *Cancer Res.* **52**(24), 6940–6944.
104. Umbas, R., Isaacs, W. B., Bringuier, P. P., et al. (1994) Decreased E-cadherin expression is associated with poor prognosis in patients with prostate cancer. *Cancer Res* **54**(14), 3929–3933.
105. Dong, J. T., Lamb, P. W., Rinker-Schaeffer, C. W., et al. (1995) KAI1, a metastasis suppressor gene for prostate cancer on human chromosome 11p11.2. *Science* **268**, 884–886.
106. Dong, J. T., Suzuki, H., Pin, S. S., et al. (1996) Down-regulation of the KAI1 metastasis suppressor gene during the progression of human prostatic cancer infrequently involves gene mutation or allelic loss. *Cancer Res.* **56**, 4387–4390.
107. Zenklusen, J. C., Conti, C. J., and Green, E. D. (2001) Mutational and functional analyses reveal that ST7 is a highly conserved tumor-suppressor gene on human chromosome 7q31. *Nat. Genet.* **27**, 392–398.
108. Leach, F. S., Tokino, T., Meltzer, P., et al. (1998) p53 Mutation and MDM2 amplification in human soft tissue sarcomas. *Cancer Res.* **53**(10 Suppl), 2231–2234.
109. Voeller, H. J., Truica, C. I., and Gelmann, E. P. (1998) Beta-catenin mutations in human prostate cancer. *Cancer Res.* **58**(12), 2520–2523.
110. Gerstein, A. V., Almeida, T. A., Zhao, G., et al. (2002) APC/CTNNB1 (β-catenin) pathway alterations in human prostate cancer. *Gene. Chromosome. Canc.* **34**, 9–10.
111. Ostrander, E. A. and Stanford, J. L. (2000) Genetics of prostate cancer: Too many loci, too few genes. *Am. J. Hum. Genet.* **67**, 1367–1375.
112. Tavtigian, S. V., Simard, J., Teng, D. H., et al. (2001) A candidate prostate cancer susceptibility gene at chromosome 17p. *Nat. Genet.* **27**, 172–180.
113. Vesprini, D., Nam, R. K., Trachtenberg, J., et al. (2001) HPC2 variants and screen-detected prostate cancer. *Am. J. Hum. Genet.* **68**, 912–917.

114. Xu, J., Zheng, S. L., Carpten, J. D., et al. (2001) Evaluation of linkage and association of HPC2/ELAC2 in patients with familial or sporadic prostate cancer. *Am. J. Hum. Genet.* **68**, 901–911.
115. Rokman, A., Koivisto, P. A., Matikainen, M. P., et al. (2001) Genetic changes in familial prostate cancer by comparative genomic hybridization. *Prostate* **46**, 233–239.
116. Hughes, J. H. and Cohen, M. B. (1998) Nuclear matrix proteins and their potential applications to diagnostic pathology. *Am. J. Clin. Pathol.* **111**: 267–274.
117. Ruijter, E. T., Werahera, P. N., van de Kaa, C. A., Stewart, J. S., Schalken, J. A., and Miller, G. J. (1998) Detection of abnormal E-cadherin expression by stimulated prostate biopsy. *J. Urology* **160**, 1368.
118. Jen, J., Kim, H., Piantadosi, S., et al. (1994) Allelic loss of chromosome 18q and prognosis in colorectal cancer [see comments]. *N. Engl. J. Med.* **331**(4), 213–221.
119. Jenkins, R., Takahashi, S., DeLacey, K., Bergstralh, E., and Lieber, M. (1998) Prognostic significance of allelic imbalance of chromosome arms 7q, 8p, 16q, and 18q in stage T3N0M0 prostate cancer. *Genes Chromosomes Cancer* **21**(2), 131–143.
120. Sato, K., Qian, J., Slezak, J. M., et al. (1999) Clinical significance of alterations of chromosome 8 in high-grade, advanced, nonmetastatic prostate carcinoma. *J. Natl. Cancer Inst.* **91**(18), 1574–1580.
121. Partin, A. W., Yoo, J., Carter, H. B., et al. (1993) The use of prostate specific antigen, clinical stage and Gleason score to predict pathological stage in men with localized prostate cancer [see comments]. *J. Urol.* **150**(1), 110–114.
122. Sgrignoli, A. R., Walsh, P. C., Steinberg, G. D., Steiner, M. S., and Epstein, J. I. (1994) Prognostic factors in men with stage D1 prostate cancer: identification of patients less likely to have prolonged survival after radical prostatectomy [see comments]. *J. Urol.* **152**(4), 1077–1081.
123. Zincke, H., Oesterling, J. E., Blute, M. L., Bergstralh, E. J., Myers, R. P., and Barrett, D. M. (1994) Long-term (15 years) results after radical prostatectomy for clinically localized (stage T2c or lower) prostate cancer [see comments]. *J. Urol.* **152**(5 Pt 2), 1850–1857.
124. Scardino, P. T., Weaver, R., and Hudson, M. A. (1992) Early detection of prostate cancer. *Hum. Pathol.* **23**(3), 211–222.
125. Isaacs, J. T. (1997) Molecular markers for prostate cancer metastasis. Developing diagnostic methods for predicting the aggressiveness of prostate cancer. *Am. J. Pathol.* **150**(5), 1511–1521.
126. Menon, M. (1997) Predicting biological aggressiveness in prostate cancer—desperately seeking a marker [editorial; comment]. *J Urol.* **157**(1), 228–229.

14

Neurofibromatosis Type 1

Margaret E. McLaughlin and Tyler Jacks

1. Introduction

Neurofibromatosis type 1 (NF1) is a complex neurocutaneous disorder also referred to as a "phakomatosis," a term derived from the Greek *phakos*, meaning lentil or birthmark. Patients with NF1 often suffer from multiple lesions of diverse type, including hyperplasias, hypoplasias, hamartomas, and neoplasms, indicating that the *NF1* gene product has dual functions in development and cell cycle control. Many NF1 lesions arise from tissues of neural crest origin, leading to the hypothesis that this syndrome results from maldevelopment of the neural crest. The term "neurocristopathy" is often used to describe NF1 and other disorders in which tissues of neural crest origin are affected ***(1)***. However, NF1 lesions also involve tissues arising from the neural tube, mesoderm, and endoderm, suggesting a widespread role for the *NF1* gene product.

This chapter summarizes what is currently known about the *NF1* gene product and its role as a tumor suppressor. Where relevant to tumorigenesis, the *NF1* gene product's role in development is also discussed. More detailed discussions of the developmental defects associated with NF1 can be found in other recent reviews ***(2,3)***.

2. Clinical Features of NF1

Exactly who reported the first case of NF1 and when is a matter of debate. Possible illustrations of NF1 sufferers with plexiform neurofibromas date back as far as the fourth century ***(4)***, but it was von Recklinghausen in 1882 who published the first comprehensive description ***(5)***. NF1 is now recognized as a common autosomal dominant disorder with an estimated prevalence of 1 in 2500 to 3500 worldwide ***(6)***. Approximately, 30–50% of cases occur sporadically, suggesting that 1 in 10,000 gametes contains a new mutation in the *NF1* gene ***(6)***. The penetrance of NF1 is 100%; however, the expression is variable, with slightly over half of patients only mildly affected ***(7,8)***. Most studies have failed to reveal clear genotype–phenotype correlations ***(8)***. In fact, clustering of specific symptoms in monozygotic twins supports the notion that much of the variablity is due to modifier loci ***(9)***. Diagnostic features of NF1 include two benign tumors (neurofibromas and optic gliomas), pigmented lesions (café-au-lait spots, freck-

From: *Methods in Molecular Biology, Vol. 222: Tumor Suppressor Genes: Pathways and Isolation Strategies*
Edited by: Wafik S. El-Deiry © Humana Press Inc., Totowa, NJ

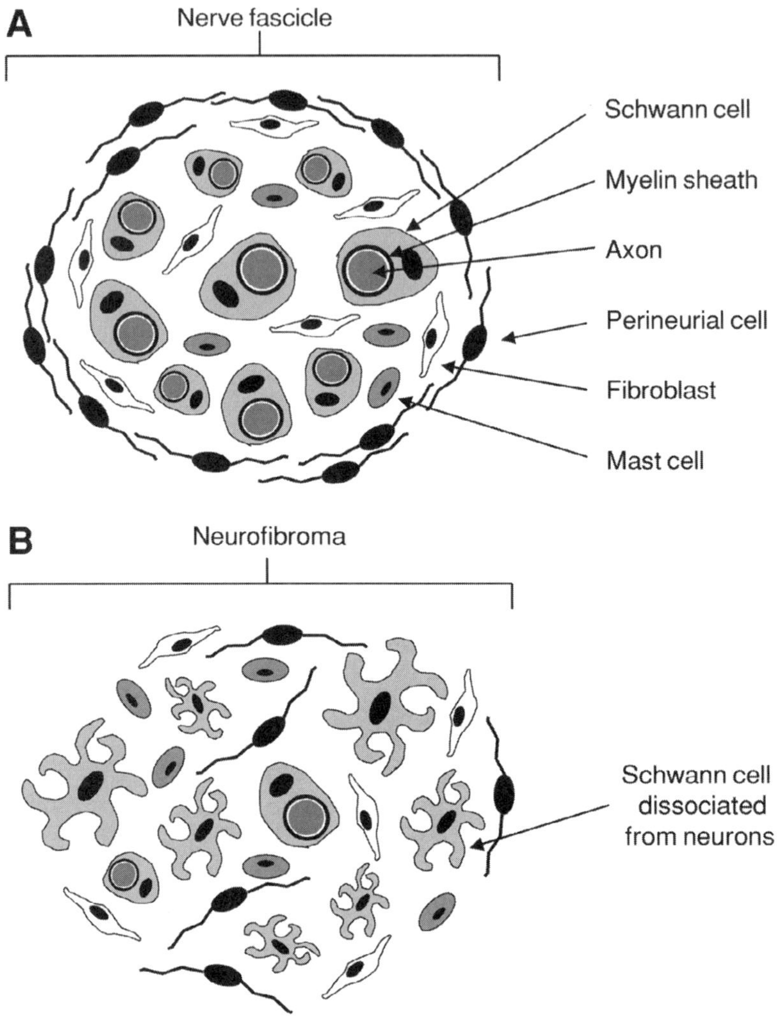

Fig. 1. Schematic drawing of a normal peripheral nerve fascicle and a neurofibroma. **(A)** Multiple nerve fascicles, bound by loose connective tissue, constitute an individual peripheral nerve. Each fascicle is composed of a variety of cell types. Nerve fibers (or axons) are encircled by either myelinating or unmyelinating Schwann cells. Between nerve fibers lies a collagenous matrix containing fibroblasts, mast cells, macrophages, and capillaries. Fascicles are surrounded by concentric layers of perineurial cells, which serve as a diffusion barrier to macromolecules. **(B)** Neurofibromas exhibit an increased number of Schwann cells, perineurial cells, fibroblasts, and mast cells. These cell types are arranged haphazardly; many Schwann cells are dissociated from axons, and the perineurial cell layer is disrupted.

ling in the axilla or groin, and iris hamartomas termed "Lisch nodules"), and skeletal malformations ***(10)***.

Neurofibromas are a frequent yet relatively poorly understood manifestation of NF1. Neurofibromas are unusual in that they are composed of a variety of cell types found in normal peripheral nerves (Schwann cells, perineurial cells, fibroblasts, and mast cells) (**Fig. 1**). Residual interspersed myelinated and unmyelinated axons may also be present. Their cellular heterogeneity led to the proposal that neurofibromas may be hyperplasias as opposed to neoplasms.

Several subtypes of neurofibroma have been described ***(11)***. While morphologically similar, their clinicopathologic features differ considerably. Localized and diffuse cutaneous neurofibromas affect the dermis and subcutis. Since these proliferations are extraneural, the nerve of origin is difficult to identify. Only a minority, approximately 10%, of cutaneous neurofibromas are associated with NF1. NF1 patients can develop hundreds, even thousands of cutaneous, neurofibromas, which can be painful and disfiguring.

Localized intraneural and plexiform neurofibromas proliferate intraneurally. While localized intraneural neurofibromas affect a segment of nerve, plexiform neurofibromas involve either a plexus of nerves or multiple fascicles within a large nerve. Like the cutaneous subtypes, the majority of localized intraneural neurofibromas are sporadic. By contrast, plexiform neurofibromas are found almost exclusively in patients with NF1. Most NF1-associated neurofibromas appear around puberty and increase in number later in life. Plexiform neurofibromas present in early childhood and are thought to be congenital ***(7)***. Importantly, about 5% of plexiform neurofibromas undergo malignant progression to malignant peripheral nerve sheath tumors (MPNSTs). Massive soft tissue neurofibromas represent the rarest subtype and are restricted to NF1 patients. Massive soft tissue neurofibromas may cause localized gigantism of an extremity. Microscopically, extraneural and plexiform components are usually present. Despite the enormous size of these lesions, malignant progression is rare.

Optic gliomas are pilocytic astrocytomas that occur within the optic nerve, a central nervous system structure. Bilateral nerve involvement, when present, is characteristic of NF1. Optic gliomas occur in about 15% of NF1 patients, but only 1.5% become symptomatic ***(12)***. The symptomatic tumors usually present in young children by causing nonparalytic squint with amblyopia or unilateral proptosis. Even at first presentation, there is typically evidence of chronicity, consistent with a congenital origin ***(13)***. Optic gliomas may remain static for many years, and some even regress ***(14)***. Tumors that involve the chiasm, located near the hypothalamus, are frequently associated with precocious puberty ***(15)***. Disinhibition of the hypothalamic–pituitary–gonadal axis is thought to be the mechanism.

Similar to neurofibromas, optic gliomas exhibit cellular heterogeneity. They contain bipolar piloid cells with long hairlike processes and elongated nuclei, and protoplasmic astrocytes with round nuclei and cobweblike processes ***(16)***. The bipolar piloid cells are strongly positive for glial fibrillary acidic protein (GFAP), while the protoplasmic astrocytes are only weakly positive for GFAP. Occasionally, cells resembling oligodendrocytes are also present. Like neurofibromas, the indolent nature and cellular heterogeneity of optic gliomas has led some to question whether they represent true neoplasms ***(13)***.

A variety of other lesions, including numerous tumors (MPNSTs, pheochromocytomas, paragangliomas, ganglioneuromas, cerebral and cerebellar astrocytomas, gastrointestinal stromal tumors, forgut carcinoid tumors, rhabdomyosarcomas, and juvenile myelomonocytic leukemia [formerly, juvenile chronic myelogenous leukemia]), are occasionally seen in association with NF1 ***(11)***.

3. Structure and Function of Neurofibromin

The gene responsible for NF1 maps to the pericentric region of chromosome 17q ***(17–19)*** and was identified by positional cloning in 1990 ***(20–22)***. It consists of 57 con-

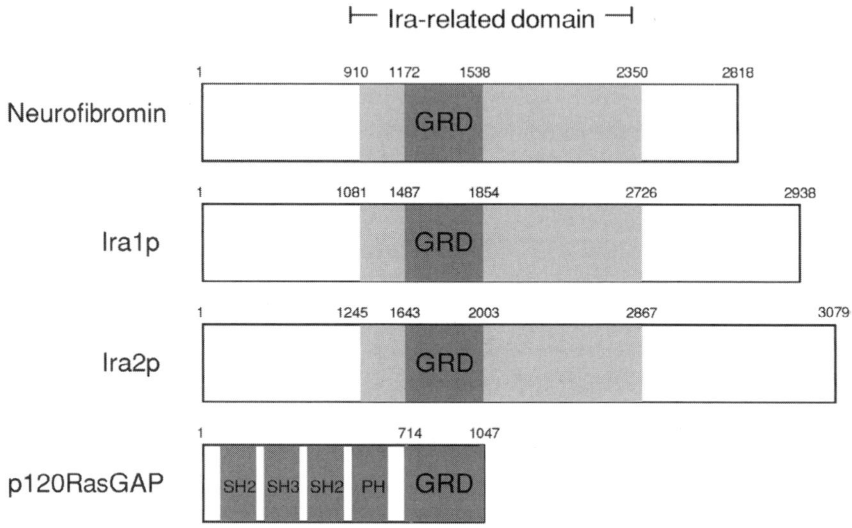

Fig. 2. Sequence similarity between neurofibromin, mammalian p120 RasGAP, and yeast RasGAP proteins. All proteins contain a 360-amino acid segment corresponding to the catalytic GAP-related domain (GRD). The similarity with the yeast RasGAP proteins, Ira1p and Ira2p, extends beyond the GRD. Src-homology 2 (SH2), Src-homology 3 (SH3), and pleckstrin homology (PH) domains are only present in p120 RasGAP.

stitutive and 3 alternatively spliced exons and spans almost 350 kb of genomic DNA. *NF1* mRNA, without any of the alternatively spliced exons, encodes a protein of 2818 amino acids. This protein, called "neurofibromin," is expressed ubiquitously but is most abundant in the adult peripheral and central nervous systems *(8)*. Neurofibromin contains a 360-amino acid segment homologous to the catalytic domains of Ras-specific GTPase activating proteins (RasGAPs), and shares broader, but lower-level, similarity with two RasGAPs, Ira1p and Ira2p, from *Saccharomyces cerevisiae* *(23,24)* (**Fig. 2**).

There is abundant evidence that neurofibromin functions as a RasGAP, accelerating the conversion of active GTP-bound Ras to inactive GDP-bound Ras (**Fig. 3A**). First, expression of an NF1 GAP-related domain protein rescues the heat-shock sensitivity of yeast *ira1* or *ira2* mutants *(25,26)*. Second, baculovirus-produced neurofibromin stimulates the GTPase activity of Ras in vitro *(25)*. Third, elevated levels of active GTP-bound Ras are present in NF1-associated tumors *(27–30)*. Fourth, single-point mutations, affecting Ras-GAP activity or Ras binding, have been detected in NF1 patients *(8)*. Finally, mutations that activate Ras or inactivate neurofibromin define separate subsets of juvenile myelomonocytic leukemia, providing genetic evidence that neurofibromin regulates Ras *(31)*.

Activation of the Ras pathway by growth factors induces cellular proliferation, and mutations and amplifications of many genes in the Ras pathway have been detected in human tumors (**Fig. 3B**). Thus, failure to inactivate Ras in the absence of neurofibromin maybe sufficient to initiate tumor formation in NF1. In this regard, therapeutic agents that inhibit Ras, such as prenylation inhibitors, hold great promise for the treatment of NF1. To date, mixed results have been observed using farnesyltransferase inhibitors (FTIs). While FTI treatment inhibited the growth of an *NF1*-deficient MPNST cell line *(32)*, an *Nf1*-deficient mouse model of myeloproliferative disease failed to respond *(33)*.

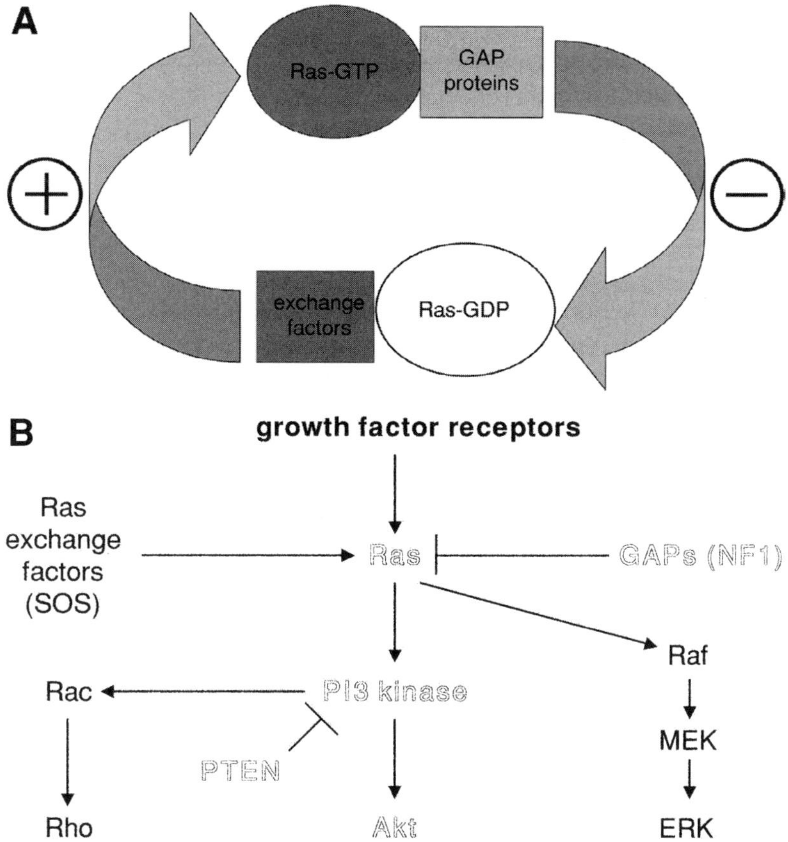

Fig. 3. Neurofibromin functions as a RasGAP. **(A)** Like other RasGAP proteins, neurofibromin inactivates Ras by catalyzing the hydrolysis of Ras-GTP to Ras-GDP. Exchange factors activate Ras by promoting the exchange of GDP for GTP. **(B)** Simplified version of the Ras pathway. Genes encoding outlined proteins have been found mutated or amplified in human cancer.

4. Is *NF1* a Tumor Suppressor?

The presence of multiple, discrete tumors in NF1 patients suggested that the *NF1* gene may behave as a tumor suppressor gene. In accordance with Knudson's two-hit model, the first *NF1* allele is inactivated by germline mutation ***(34)***. To date over 240 germline mutations have been reported in the *NF1* gene ***(2)***. The majority are deletions and insertions that either remove or inactivate the gene. With the exception of patients with large deletions, who appear more likely to exhibit a severe phenotype that includes mental impairment, numerous early-onset neurofibromas, and facial dysmorphic features ***(35,36)***, no obvious correlations between germline mutation and phenotype have been found.

Knudson's two-hit model predicts that the second *NF1* allele is inactivated at a later point in time by somatic mutation, initiating formation of a lesion. Consistent with this model, loss of the wild-type *NF1* allele has been detected in NF1 patient-derived MP-NSTs ***(37,38)***, pheochromocytomas ***(39)***, and bone marrow samples involved by juvenile myelomonocytic leukemia ***(40)***. Confirmation of a tumor suppressor role for neurofibromin has come from studies of mice carrying a targeted disruption of one *Nf1* allele. While *Nf1*+/− mice do not exhibit the classical symptoms of the human syndrome, they are predisposed to formation of a variety of tumors including two tumors associated with NF1 (pheochromocytoma and myeloid leukemia) ***(41)***. In both NF1-associated tumor types, loss of the wild-type *Nf1* allele is observed at high frequency.

The requirement for second-hit mutations in neurofibromas has been controversial. Initial studies either failed to detect somatic *NF1* mutations or found loss of heterozygosity (LOH) of the wild-type *NF1* allele in only a minority of tumors ***(42–47)***. The difficulty in detecting somatic *NF1* mutations was likely due to the heterogeneity of cell types (not all of which would be predicted to have undergone a mutational event) and the large size of the *NF1* gene. Recently, using a combination of microsatellite genotyping for LOH and cDNA single-strand conformation polymorphism (SSCP)-heteroduplex analysis for detecting point mutations, Serra and co-workers directly identified somatic *NF1* mutations from 9 out of 10 neurofibromas, providing strong evidence that second-hit mutations play a key role in tumor development ***(48)***. By selectively culturing Schwann cells and fibroblasts from neurofibromas, Kluwe, Serra, and co-workers have shown that somatic *NF1* mutations are restricted to the Schwann cell population ***(48,49)***, identifying the Schwann cell as the potential cell of origin of neurofibromas.

The question of second-hit mutations in neurofibromas has also been addressed through the creation of chimeric mice partially composed of *Nf1*−/− cells ***(50)***. Although *Nf1*+/− mice are cancer-prone, they do not develop neurofibromas. *Nf1*−/− mice die in midgestation from a cardiac defect and therefore are uninformative with respect to tumor development in vivo ***(41,51,52)***. However, nearly all chimeric mice develop numerous neurofibromas, which histologically and ultrastructurally resemble human plexiform tumors. Unlike in the human disorder, cutaneous neurofibromas are not observed, suggesting that species-specific differences in mice versus humans and/or developmental differences between cutaneous and plexiform neurofibromas exist. Although the study of *Nf1* chimeric mice clearly demonstrates that loss of both *Nf1* alleles is a required step in the development of murine plexiform neurofibromas, in-vivo proof that loss of both alleles must occur in Schwann cells will require the creation of mice in which the *Nf1* gene can be selectively inactivated in the Schwann cell lineage.

Far less information is available regarding the role of second-hit mutations in optic gliomas. Lau and co-workers have observed a loss of neurofibromin expression and elevated Ras-GTP levels in a single pilocytic astrocytoma from an NF1 patient ***(53)***. Gutmann and co-workers have gone a step further and identified LOH at the *NF1* locus in two of three NF1-associated pilocytic astrocytomas ***(54)***. However, the pilocytic astrocytomas in both studies were atypical, in that they were located in either the frontal lobe or the brainstem, not in the optic pathway. Furthermore, tissue was available for study because the patients underwent craniotomies for progressive symptoms or radiologic increase in tumor size, evidence that these tumors may belong to a more aggressive subset of pilocytic astrocytomas. Surgical resection is not recommended for optic gliomas, because blindness is a serious complication and the majority of tumors have an indolent course. As a result, the issue of second-hit mutations is unlikely to be resolved through the study of human tissue samples. Therefore, creation of a mouse optic glioma model will be critical in defining the importance of second-hit mutations and the identity of the cell of origin in this tumor type.

5. Evidence for *NF1* Haploinsufficiency

Some nonfocal features of *NF1* (short stature and intellectual impairment) are difficult to reconcile with Knudson's two-hit model. Even focal lesions, such as café-au-lait spots, do not obey Knudson's rules. Cultured melanocytes from café-au-lait spots express neurofibromin and show no evidence of LOH at the *NF1* locus ***(55)***, raising the possibility that heterozygous inactivation of *NF1* may have phenotypic consequences. Ingram and co-workers have recently examined the effects of *NF1* heterozygosity on receptor signaling in vivo ***(56)***. A hypomorphic mutation in the c-kit receptor tyrosine kinase (W^{41}) results in coat color deficiency and mast cell hypoplasia in mice. Based on previous findings implicating neurofibromin in c-kit signaling pathways in vitro, these investigators crossed *Nf1*+/− mice with W^{41} mice and found that heterozygosity at the *NF1* locus substantially rescued these defects. Since c-kit signaling is mediated by Ras, these findings suggest that *Nf1* haploinsufficiency is due to upregulation of Ras pathways.

Tumor-associated *NF1*+/− cells may contribute to tumorigenesis. In the case of neurofibromas, the tumors are thought to contain a mixture of *NF1*−/− and *NF1*+/− Schwann cells ***(48,57)***, *NF1*+/− perineurial cells, *NF1*+/− fibroblasts, and *NF1*+/− mast cells. Reciprocal signaling among these various cell types is known to occur in normal peripheral nerves ***(58,59)*** and likely occurs between cells in a neurofibroma. However, cells within a neurofibroma may respond inappropriately to these signals because of decreased neurofibromin levels. Indeed, defects in several neurofibroma-associated *NF1*+/− cell types have been observed. In cellular studies *Nf1*+/− Schwann cells are able to induce angiogenesis and are more invasive than their wild-type counterparts ***(60)***. Embryonic fibroblasts from *Nf1*+/− mice hyperproliferate, respond abnormally to wound cytokines, and secrete higher amounts of collagen in vitro ***(61)***. In vivo, in *Nf1*+/− mice, wounding incites an excessive proliferation of fibroblasts, leading to an increased amount of granulation tissue ***(61)***. *Nf1*+/− mast cells also show increased proliferation in vitro and in vivo in *Nf1*+/− mice ***(56)***. *Nf1*−/− perineurial cells derived from *Nf1*-deficient mouse embryos fail to form fascicles around nerve bundles and appear to be more abundant in an in-vitro co-culture assay ***(62)***. The phenotype of *Nf1*+/− perineurial cells has not yet been characterized. Whether tumor-associated *NF1*+/− cells are absolutely required for neurofibroma formation or simply assist in the process remains to be determined.

Tumor-associated *NF1*+/− cells may also contribute to formation of tumors in the CNS. Like *Nf1*+/− fibroblasts and mast cells, *Nf1*+/− astrocytes hyperproliferate in culture ***(63)***. Furthermore, an increased number of astrocytes has been observed in the brains of NF1 patients and *Nf1*+/− mice ***(63–65)***. The defects in *Nf1*+/− astrocytes may not be completely cell autonomous, as mice carrying a neuronal-specific targeted disruption of the *Nf1* gene also exhibit an increase in astrocytes in the brain ***(66)***. The *NF1*+/− astrocytes may serve as the cell of origin of optic gliomas and diffusely infiltrating astrocytomas and/or assist in tumor formation through intercellular signaling events.

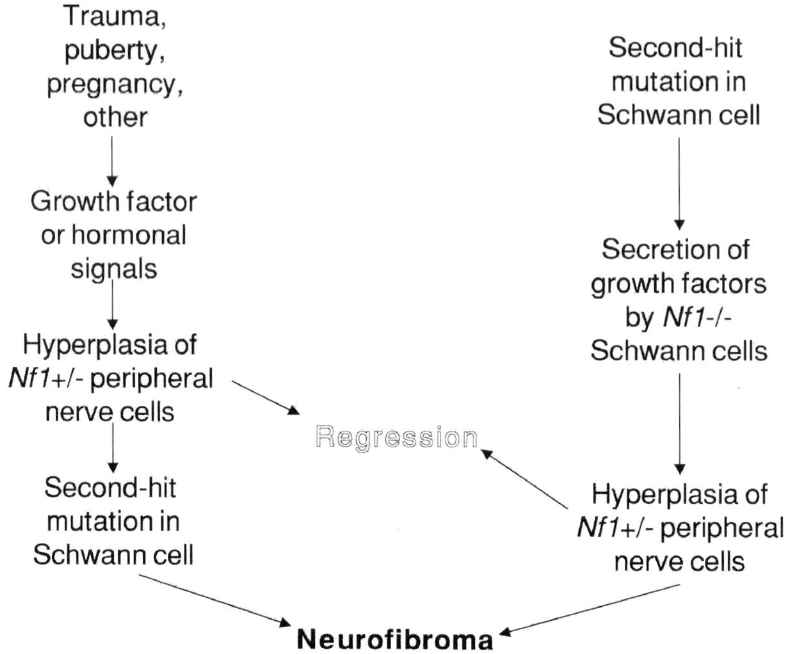

Fig. 4. Models of neurofibroma development. We propose that neurofibromas have hyperplastic and neoplastic components. Increases in growth factors or hormonal signals may induce hyperplasia of *Nf1*+/− peripheral nerve cells. Hyperplasia, in turn, may increase the chance of a second-hit mutation occurring within a Schwann cell, resulting in neurofibroma formation. Alternatively, a second-hit mutation in a Schwann cell may initiate the process. *Nf1*−/− Schwann cells may secrete abnormal amounts of growth factors, promoting hyperplasia of nearby *Nf1*+/− cells. When growth factors or hormonal signals are withdrawn, the hyperplastic component may be able to regress.

6. Hyperplasia or Neoplasia?

Is a neurofibroma a hyperplasia or a neoplasm? A hyperplasia is an increase in cell number in an organ or tissue. Unlike a neoplasm, which is a largely uncontrolled proliferation of cells, a hyperplasia remains under control of growth factors or hormonal signals. In the case of neurofibromas, the tumor-associated *NF1*+/− cells have features of hyperplastic cells. They are increased in number, and while they maybe hypersensitive to growth factors or hormonal signals due to *NF1* haploinsufficiency, they still require extracellular signals to proliferate. However, the *NF1*−/− Schwann cells that are thought to have undergone a second-hit mutation resemble neoplastic cells. They form a clonal population, and in cell culture assays are more invasive than wild-type or *Nf1*+/− Schwann cells ***(60)***. To account for all of these features, we propose that neurofibromas are hybrid lesions with hyperplastic and neoplastic components.

Which comes first the hyperplasia or the neoplasia? The answer may be "both." Trauma and hormonal changes, which have been implicated in the development of neurofibromas ***(7)***, could provide signals stimulating *NF1*+/− cells to proliferate excessively forming a hyperplasia. Hyperproliferation, in turn, may increase the chance that a second-hit mutation occurs within a Schwann cell (**Fig. 4**). In other tissues, such as the endometrium, hyperplastic lesions are known to provide fertile ground for tumor development. Alternatively, a second-hit mutation in a Schwann cell may initiate the process. *NF1*−/−

Schwann cells may secrete abnormal amounts of growth factors that promote the proliferation of nearby *NF1+/−* cells. While the former mechanism may apply to cutaneous neurofibromas that occur superficially in areas susceptible to trauma and increase in number during puberty and pregnancy, the later mechanism may apply to plexiform neurofibromas, which are thought to be congenital. The concept that neurofibromas are hybrid lesions can even explain the peculiar observation that, after increasing in size during pregnancy, neurofibromas shrink after giving birth ***(7)***. We hypothesize that, after giving birth, hormonal levels return to normal, and the now less stimulated hyperplastic component of the neurofibroma involutes. The effects of hormones may be indirect, as little to no steroid receptor-binding activity has been detected in neurofibromas ***(67,68)***.

There are striking similarities between neurofibromas and optic gliomas, suggesting that perhaps optic gliomas are also hybrid lesions. Both tumors exhibit cellular heterogeneity. In addition, like neurofibromas, an inappropriate response to injury has been speculated to play a role in optic glioma formation. *NF1* is normally expressed at relatively low levels in astrocytes. However, *Nf1* is dramatically upregulated in response to cAMP and proinflammatory cytokines in astrocytes in vitro, and in response to ischemia in vivo in a rat model system ***(69)***. These findings raise the possibility that, in an NF1 patient, CNS injury may trigger a hyperplasia of astrocytes, which increases the likelihood of a second-hit mutation and optic glioma formation. Similar to neurofibromas, decreased levels of growth factors or hormonal signals may lead to involution of the hyperplastic component of optic gliomas, resulting in the observed spontaneous regression of a subset of these tumors.

7. Progression to Malignancy

In benign NF1-associated tumors, absence of neurofibromin maybe sufficient for tumor formation. However, when neurofibromas undergo malignant change to MPNSTs, additional genetic mutations appear to be required. The p53 pathway is an important target of these additional genetic events. Chromosome 17p deletions (the chromosomal region in which *p53* resides) have been detected in MPNSTs, and in some cases point mutations in the remaining *p53* allele have been found ***(70,71)***. Homozygous deletions in the *INK4* locus, which inactivate both alleles of $p16^{INK4a}$ and $p14^{ARF}$ tumor suppressor genes, have also been identified in MPNSTs ***(72,73)***. The $\mathrm{p16^{INK4a}}$ protein is a cyclin-dependent kinase inhibitor of the cyclin D/cdk 4 complex. This complex phosphorylates the retinoblastoma (Rb) gene product, allowing cells to enter S phase. The $\mathrm{p14^{ARF}}$ protein, on the other hand, activates the p53 pathway by inhibiting Mdm-2-induced p53 degradation and transactivational silencing ***(74)***. Thus, both the $\mathrm{p16^{INK4a}}$-cyclin D/cdk4-Rb pathway and the p53 pathway may be involved in malignant progression to MPNST.

The cooperation between *NF1* and *p53* tumor suppressor genes in malignancy has also been demonstrated by studies of mice carrying heterozygous mutations of both the *Nf1* and *p53* genes ***(50,75)***. The mouse *Nf1* and *p53* genes are closely linked on chromosome 11 (they are both on chromosome 17 in humans, but on opposite arms). Therefore, mice carrying mutations on the same chromosomal arm (NP *cis*) could be generated. NP *cis* mice develop MPNSTs at a high frequency, and LOH of both the wild-type *Nf1* and *p53* alleles is observed. Mutational events involving loss of an entire chromosome have been detected in *Min* mice heterozygous for mutations in the *Apc* tumor suppressor gene ***(76)***. LOH in the NP *cis* mice likely occurs by a similar mechanism. NP *cis* mice also develop diffusely infiltrating astrocytomas which show LOH at both loci ***(77)***. Interestingly, astrocytoma development appears to be highly dependent on the genetic background of these mice, implicating a role of modifier genes in this phenotype.

In addition to the genetic mutations detected in MPNSTs, DeClue and co-workers have observed increased expression of the EGF receptor (EGF-R) in MPNST cell lines and tissue sections ***(78)***. Schwann cells from normal nerves do not express this receptor, but instead express other members of the EGF-R family, Erb-2 and -3, which bind glial growth factor (GGF). Only rare Schwann cells in neurofibromas and transformed (but not untransformed) *Nf1*-deficient mouse Schwann cells express EGF-R, supporting the hypothesis that upregulation of EGF-R is an important step in the progression to malignancy. Importantly, proliferation of MPNST cell lines and transformed *Nf1*-deficient mouse Schwann cells is inhibited by EGF-R antagonists, suggesting that EGF-R is a potential therapeutic target. The degree to which EGF-R exerts its mitogenic effects through the Ras pathway, which is already deregulated in the absence of neurofibromin, remains to be determined.

8. Neurofibromin May Have Ras-Independent Functions

Evidence of a Ras-independent function for neurofibromin is accumulating. Guo, The, and co-workers have found that flies homozygous for null mutations in an *NF1* homolog do not exhibit any of the widespread developmental defects associated with activated Ras. Instead, *NF1*-mutant flies are small, and exhibit learning defects, short-term memory loss, diminished escape response, and defects in potassium channel function at the neuromuscular junction ***(79–81)***. These phenotypes are not rescued by a constitutively active Ras transgene, but are restored by either pharmacologic manipulation of the cAMP-protein kinase A (PKA) pathway or a constitutively active PKA transgene, implicating a role for *NF1* in the regulation of the later pathway. These reports also suggest that *NF1*-dependent cAMP activity is mediated through the *rutabaga*-encoded adenylyl cyclase. Whether neurofibromin regulates the cAMP-PKA pathway in mammalian cells and whether this function of neurofibromin contributes to tumor suppression are important questions that need to be addressed.

Additional, although indirect, evidence for a Ras-independent function for neurofibromin comes from studies of human tumor cell lines. While neither melanoma nor neuroblastoma occurs at increased frequency in NF1 patients, melanoma and neuroblastoma cell lines that have lost both *NF1* alleles have been identified ***(82–84)***. Unexpectedly, these cell lines show little to no increase in Ras-GTP levels ***(82,83)***. Furthermore, although overexpression of neurofibromin inhibits proliferation of NIH3T3 cells, the level of Ras-GTP is not reduced ***(85)***. Taken together, these cell culture studies raise the possibility that neurofibromin is exerting its effects on cell proliferation through targets other than Ras.

9. Concluding Remarks

Since the cloning of the *NF1* gene in 1990, great progress has been made toward understanding the role of neurofibromin in tumor suppression. Convincing evidence has emerged that the *NF1* gene behaves as a traditional tumor suppressor gene in NF1-associated malignancies. In the majority of these tumors, upregulation of Ras pathways due to the absence of NF1-GAP activity is the primary defect that leads to uncontrolled cell proliferation. Not unexpectedly, progress has revealed new questions. What is the role of second-hit mutations and haploinsufficiency in benign NF1-associated tumors (neurofibromas and optic gliomas)? Are these benign tumors, as we suggest, hybrid lesions with hyperplastic and neoplastic components? Which genetic mutations cooperate with or modify the effects of *NF1* loss? And, does neurofibromin have Ras-independent functions? Partial answers to these questions have been discussed. However, continued investigation, making use of human tissue samples, model systems, and cell culture assays as well as newer genome-based technologies, is essential.

Note Added in Proof

Recently, Zhu and coworkers generated a conditional *Nf1* mouse model in which a floxed *Nf1* allele is deleted by a Cre transgene under control of the Schwann cell-specific promoter, Krox20 ***(86)***. All progeny with the $Nf1^{flox/-}$; Krox20-cre genotype develop plexiform neurofibromas, confirming that the Schwann cell is the cell of origin.

To determine the importance of *Nf1* haploinsufficiency in the tumor environment, Zhu and coworkers compared the size and frequency of neurofibromas occurring in $NF1^{flox/-}$; Krox20-cre mice in which all non-neoplastic tumor-associated cells are phenotypically heterozygous for *Nf1* and $Nf1^{flox/flox}$; Krox20-cre mice in which all non-neoplastic tumor-asociated cells are phenotypically wild type. In striking contrast to the widespread plexiform neurofibromas of the $Nf1^{flox/-}$; Krox20-cre mice, the $Nf1^{flox/flox}$; Krox20-cre mice only developed small, infrequent hyperplastic lesions in the cranial nerves. Although the number of mice in this study was limited, the data supports the conclusion that *Nf1* haploinsufficiency in the tumor environment promotes neurofibroma formation. What is the mechanism by which *Nf1* haploinsufficiency exerts its effect? The first step toward answering this question will be to identify the relevant cell type(s) in which *Nf1* heterozygosity is required to see frank tumor formation.

While focused on neurofibromas, this work raises the exciting possibility that the genotype of non-neoplastic cells in the tumor environment may influence the growth of other tumor types. Moreover, an increased understanding of the role of non-neoplastic tumor-associated cells may lead to new directions for cancer therapy and prevention.

References

1. Bolande, R. P. (1974) The neurocristopathies. *Hum. Pathol.* **5**, 409–429.
2. Bernards, A. and McClatchey, A. I. (2001) Neurofibromatoses, in *Tumor Suppressor Genes in Human Cancer* (Fisher, D. E., ed.), Humana Press, Totowa, NJ, pp. 253–263.
3. Cichowski, K. and Jacks, T. (2001) *NF1* tumor suppressor gene function: narrowing the GAP. *Cell* **104**, 593–604.
4. Zanca, A. (1980) Antique illustrations of neurofibromatosis. *Int. J. Dermatol.* **19**, 55–58.
5. von Recklinghausen, F. (1882) *Ueber die multiplen Fibrome der Haut und ihre Beziehung zu den multiplen Neuromen*. August Hirschwald, Berlin.
6. Huson, S. M., Compston, D. A., Clark, P., and Harper, P. S. (1989) A genetic study of von Recklinghausen neurofibromatosis in south east Wales. I. Prevalence, fitness, mutation rate, and effect of parental transmission on severity. *J. Med. Genet.* **26**, 704–711.
7. Riccardi, V. M. (1992) *Neurofibromatosis: Phenotype, Natural History, and Pathogenesis*. The Johns Hopkins University Press, Baltimore and London.

8. Upadhyaya, M. and Cooper, D. N. (1998) *Neurofibromatosis Type 1 from Genotype to Phenotype*. BIOS Scientific, Oxford, UK.
9. Easton, D. F., Ponder, M. A., Huson, S. M., and Ponder, B. A. (1993) An analysis of variation in expression of neurofibromatosis (NF) type 1 (NF1): evidence for modifying genes. *Am. J. Hum. Genet.* **53**, 305–313.
10. Gutmann, D.H., Aylsworth, A., Carey, J. C., et al. (1997) The diagnostic evaluation and multidisciplinary management of neurofibromatosis 1 and neurofibromatosis 2. *J. Am. Med. Assoc.* **278**, 51–57.
11. Scheithauer, B. W., Woodruff, J. M., and Erlandson, R. A. (1999) *Tumors of the Peripheral Nervous System*. Armed Forces Institute of Pathology, Washington, DC.
12. Huson, S. M. and Hughes, R. A. C. *The Neurofibromatoses: A Pathogenic and Clinical Overview*, Chapman & Hall Medical, London, UK, 1994.
13. Hoyt, W. F. and Baghdassarian, S. A. (1969) Optic glioma of childhood. *Br. J. Ophthal.* **53**, 793–798.
14. Leisti, E. L., Pyhtinen, J., and Poyhonen, M. (1996) Spontaneous decrease of a pilocytic astrocytoma in neurofibromatosis type 1. *Am. J. Neuroradiol.* **17**, 1691–1694.
15. Habiby, R., Silverman, B., Listernick, R., and Charrow, J. (1995) Precocious puberty in children with neurofibromatosis type 1. *J. Pediatr.* **126**, 364–367.
16. Kleihues, P. and Cavenee, W. K. (eds.) (2000)*Tumours of the Nervous System*. IARC Press, Lyon, France.
17. Barker, D., Wright, E., Nguyen, K., et al. (1987) Gene for von Recklinghausen neurofibromatosis is in the pericentromeric region of chromosome 17. *Science* **236**, 1100–1102.
18. Seizinger, B.R., Rouleau, G. A., Ozelius, L. J., et al. (1987) Genetic linkage of von Recklinghausen neurofibromatosis to the nerve growth factor receptor gene. *Cell* **49**, 589–594.
19. Goldgar, D. E., Green, P., Parry, D. M., and Mulvihill, J. J. (1989) Multipoint linkage analysis in neurofibromatosis type I: an international collaboration. *Am. J. Hum. Genet.* **44**, 6–12.
20. Viskochil, D., Buchberg, A. M., Xu, G., et al. (1990) Deletions and a translocation interrupt a cloned gene at the neurofibromatosis type 1 locus. *Cell* **62**, 187–192.
21. Cawthon, R. M., Weiss, R., Xu, G. F., et al. (1990) A major segment of the neurofibromatosis type 1 gene: cDNA sequence, genomic structure, and point mutations. *Cell* **62**, 193–201.
22. Wallace, M. R., Marchuk, D., Anderson, L. B., et al. (1990) Type 1 neurofibromatosis gene: identification of a large transcript disrupted in three NF1 patients. *Science* **249**, 181–186.
23. Buchberg, A. M., Cleveland, L. S., Jenkins, N. A., and Copeland, N. G. (1990) Sequence homology shared by neurofibromatosis type-1 gene and IRA-1 and IRA-2 negative regulators of the RAS cyclic AMP pathway. *Nature* **347**, 291–294.
24. Xu, G.F., Lin, B., Tanaka, K., et al. (1990) The catalytic domain of the neurofibromatosis type 1 gene product stimulates ras GTPase and complements *ira* mutants of *S. cerevisiae*. *Cell* **63**, 835–841.
25. Martin, G. A., Viskachil, D., Bollag, G., et al. (1990) The GAP-related domain of the neurofibromatosis type 1 gene product interacts with ras p21. *Cell* **63**, 843–849.
26. Ballester, R., Marchuk, D., Baguski, M., et al. (1990) The *NF1* locus encodes a protein functionally elated to mammalian GAP and yeast IRA proteins. *Cell* **63**, 851–859.
27. Basu, T. N., Gutmann, D. H., Fletcher, J. A., et al. (1992) Aberrant regulation of ras proteins in malignant tumour cells from type 1 neurofibromatosis patients. *Nature* **356**, 713–715.
28. DeClue, J. E., Papageorge, A., Fletcher, J. A., et al. (1992) Abnormal regulation of mammalian p21ras contributes to malignant tumor growth in von Recklinghausen (type 1) neurofibromatosis. *Cell* **69**, 265–273.
29. Guha, A., Lau, N., Huvar, I., et al. (1996) Ras-GTP levels are elevated in human NF1 peripheral nerve tumors. *Oncogene* **12**, 507–513.
30. Bollag, G., Clapp, D. W., Shih, S., et al. (1996) Loss of *NF1* results in activation of the Ras signaling pathway and leads to aberrant growth in haematopoietic cells. *Nat. Genet.* **12**, 144–148.

31. Kalra, R., Paderanga, D. C., Olson, K., and Shannon, K. M. (1994) Genetic analysis is consistent with the hypothesis that NF1 limits myeloid cell growth through p21ras. *Blood* **84**, 3435–3439.
32. Yan, N., Ricca, C., Fletcher, J., et al. (1995) Farnesyltransferase inhibitors block the neurofibromatosis type I (NF1) malignant phenotype. *Cancer Res.* **55**, 3569–3575.
33. Mahgoub, N., Taylor, B. R., Gratiot, M., et al. (1999) In vitro and in vivo effects of a farnesyltransferase inhibitor on *Nf1*-deficient hematopoietic cells. *Blood* **94**, 2469–2476.
34. Knudson, A. G. J. (1971) Mutation and cancer: statistical study of retinoblastoma. *Proc. Natl. Acad. Sci. USA* **68**, 820–823.
35. Wu, B. L., Schneider, G. H., and Korf, B. R. (1997) Deletion of the entire *NF1* gene causing distinct manifestations in a family. *Am. J. Med. Genet.* **69**, 98–101.
36. Tonsgard, J. H., Yelavarthi, K. K., Cushner, S., Short, M. P., and Lindgren, V. (1997) Do *NF1* gene deletions result in a characteristic phenotype? *Am. J. Med. Genet.* **73**, 80–86.
37. Skuse, G. R., Kosciolek, B. A., and Rowley, P. T. (1989) Molecular genetic analysis of tumors in von Recklinghausen neurofibromatosis: loss of heterozygosity for chromosome 17. *Genes Chromosomes Cancer* **1**, 36–41.
38. Legius, E., Marchuk, D. A., Collins, F. S., and Glover, T. W. (1993) Somatic deletion of the neurofibromatosis type 1 gene in a neurofibrosarcoma supports a tumour suppressor gene hypothesis. *Nat. Genet.* **3**, 122–126.
39. Xu, W., Mulligan, L. M., Ponder, M. A., et al. (1992) Loss of *NF1* alleles in phaeochromocytomas from patients with type I neurofibromatosis. *Genes Chromosomes Cancer* **4**, 337–342.
40. Side, L., Taylor, B., Cayouette, M., et al. (1997) Homozygous inactivation of the *NF1* gene in bone marrow cells from children with neurofibromatosis type 1 and malignant myeloid disorders. *N. Engl. J. Med.* **336**, 1713–1720.
41. Jacks, T., Shih, T. S., Schmitt, E. M., et al. (1994) Tumour predisposition in mice heterozygous for a targeted mutation in *Nf1*. *Nat. Genet.* **7**, 353–361.
42. Colman, S. D., Williams, C. A., and Wallace, M. R. (1995) Benign neurofibromas in type 1 neurofibromatosis (NF1) show somatic deletions of the *NF1* gene. *Nat. Genet.* **11**, 90–92.
43. Serra, E., Puig, S., Otero, D., et al. (1997) Confirmation of a double-hit model for the *NF1* gene in benign neurofibromas. *Am. J. Hum. Genet.* **61**, 512–519.
44. Daschner, K., Assum, G., Eisenbarth, I., et al. (1997) Clonal origin of tumor cells in a plexiform neurofibroma with LOH in *NF1* intron 38 and in dermal neurofibromas without LOH of the *NF1* gene. *Biochem. Biophys. Res. Commun.* **234**, 346–350.
45. Kluwe, L., Friedrich, R. E., and Mautner, V. F. (1999) Allelic loss of the *NF1* gene in NF1-associated plexiform neurofibromas. *Cancer Genet. Cytogenet.* **113**, 65–69.
46. Eisenbarth, I., Beyer, K., Krone, W., and Assum, G. (2000) Toward a survey of somatic mutation of the *NF1* gene in benign neurofibromas of patients with neurofibromatosis type 1. *Am. J. Hum. Genet.* **66**, 393–401.
47. John, A. M., Ruggieri, M., Ferner, R., and Upadhyaya, M. (2000) A search for evidence of somatic mutations in the *NF1* gene. *J. Med. Genet.* **37**, 44–49.
48. Serra, E., Rosenbaum, T., Winner, V., et al. (2000) Schwann cells harbor the somatic *NF1* mutation in neurofibromas: evidence of two different Schwann cell subpopulations. *Hum. Mol. Genet.* **9**, 3055–3064.
49. Kluwe, L., Friedrich, R., and Mautner, V. F. (1999) Loss of *NF1* allele in Schwann cells but not in fibroblasts derived from an NF1-associated neurofibroma. *Genes Chromosomes Cancer* **24**, 283–285.
50. Cichowski, K., Shih, T. S., Schmitt, E., et al. (1999) Mouse models of tumor development in neurofibromatosis type 1. *Science* **286**, 2172–2176.
51. Brannan, C. I., Perkins, A. S., Vogel, K. S., et al. (1994) Targeted disruption of the neurofibromatosis type-1 gene leads to developmental abnormalities in heart and various neural crest-derived tissues. *Genes Dev.* **8**, 1019–1029.

52. Lakkis, M. M. and Epstein, J. A. (1998) Neurofibromin modulation of ras activity is required for normal endocardial-mesenchymal transformation in the developing heart. *Development* **125**, 4359–4367.
53. Lau, N., Feldkamp, M. M., Roncari, L., et al. (2000) Loss of neurofibromin is associated with activation of RAS/MAPK and PI3- K/AKT signaling in a neurofibromatosis 1 astrocytoma. *J. Neuropathol. Exp. Neurol.* **59**, 759–767.
54. Gutmann, D. H., Donahoe, J., Brown, T., James, C. D., and Perry, A. (2000) Loss of neurofibromatosis 1 (*NF1*) gene expression in NF1-associated pilocytic astrocytomas. *Neuropathol. Appl. Neurobiol.* **26**, 361–367.
55. Eisenbarth, I., Assum, G., Kaufmann, D., and Krone, W. (1997) Evidence for the presence of the second allele of the neurofibromatosis type 1 gene in melanocytes derived from cafe-au-lait macules of NF1 patients. *Biochem. Biophys. Res. Commun.* **237**, 138–141.
56. Ingram, D. A., Yang, F. C., Travers, J. B., et al. (2000) Genetic and biochemical evidence that haploinsufficiency of the *Nf1* tumor suppressor gene modulates melanocyte and mast cell fates in vivo. *J. Exp. Med.* **191**, 181–188.
57. Sherman, L. S., Atit, R., Rosenbaum, T., Cox, A. D., and Ratner, N. (2000) Single cell Ras-GTP analysis reveals altered Ras activity in a subpopulation of neurofibroma Schwann cells but not fibroblasts. *J. Biol. Chem.* **275**, 30740–30745.
58. Jessen, K. R. and Mirsky, R. (1998) Origin and early development of Schwann cells. *Microsc. Res. Tech.* **41**, 393–402.
59. Kioussi, C. and Gruss, P. (1996) Making of a Schwann. *Trends Genet.* **12**, 84–86.
60. Kim, H. A., Ling, B., and Ratner, N. (1997) *Nf1*-deficient mouse Schwann cells are angiogenic and invasive and can be induced to hyperproliferate: reversion of some phenotypes by an inhibitor of farnesyl protein transferase. *Mol. Cell. Biol.* **17**, 862–872.
61. Atit, R. P., Crowe, M. J., Greenhalgh, D. G., Wenstrup, R. J., and Ratner, N. (1999) The *Nf1* tumor suppressor regulates mouse skin wound healing, fibroblast proliferation, and collagen deposited by fibroblasts. *J. Invest. Dermatol.* **112**, 835–842.
62. Rosenbaum, T., Boissy, Y. L., Kombrinck, K., et al. (1995) Neurofibromin-deficient fibroblasts fail to form perineurium in vitro. *Development* **121**, 3583–3592.
63. Gutmann, D. H., Loehr, A., Zhang, Y., et al. (1999) Haploinsufficiency for the neurofibromatosis 1 (*NF1*) tumor suppressor results in increased astrocyte proliferation. *Oncogene* **18**, 4450–4459.
64. Nordlund, M. L., Rizvi, T. A., Brannan, C. I., and Ratner, N. (1995) Neurofibromin expression and astrogliosis in neurofibromatosis type 1 brains. *J. Neuropathol. Exp. Neurol.* **54**, 588–600.
65. Rizvi, T. A., et al. (1999) Region-specific astrogliosis in brains of mice heterozygous for mutations in the neurofibromatosis type 1 (*Nf1*) tumor suppressor. *Brain Res.* **816**, 111–123.
66. Zhu, Y., Romero, M. I., Ghosh, P., et al. (2001) Ablation of *Nf1* function in neurons induces abnormal development of cerebral cortex and reactive gliosis in the brain. *Genes Dev.* **15**, 859–876.
67. Martuza, R. L., MacLaughlin, D. T., and Ojemann, R. G. (1981) Specific estradiol binding in schwannomas, meningiomas, and neurofibromas. *Neurosurgery* **9**, 665–671.
68. Chaudhuri, P. K., Walker, M. J., Das Gupta, T. K., and Beattie, C. W. (1982) Steroid receptors in tumors of nerve sheath origin. *J. Surg. Oncol.* **20**, 205–206.
69. Giordano, M. J., Manadeo, D., He, Y. Y., et al. (1996) Increased expression of the neurofibromatosis 1 (*NF1*) gene product, neurofibromin, in astrocytes in response to cerebral ischemia. *J. Neurosci. Res.* **43**, 246–253.
70. Menon, A. G., Anderson, K., Riccardi, V. M., et al. (1990) Chromosome 17p deletions and p53 gene mutations associated with the formation of malignant neurofibrosarcomas in von Recklinghausen neurofibromatosis. *Proc. Natl. Acad. Sci. USA* **87**, 5435–5439.

71. Lothe, R. A., Smith-Sorensen, B., Hektoen, M., et al. (2001) Biallelic inactivation of TP53 rarely contributes to the development of malignant peripheral nerve sheath tumors. *Genes Chromosomes Cancer* **30**, 202–206.
72. Kourea, H. P., Orlow, I., Scheithauer, B. W., Cordon-Cardo, C., and Woodruff, J. M. (1999) Deletions of the *INK4A* gene occur in malignant peripheral nerve sheath tumors but not in neurofibromas. *Am. J. Pathol.* **155**, 1855–1860.
73. Nielsen, G. P., Stemmer-Rachamimov, A. O., Ino, Y., et al. (1999) Malignant transformation of neurofibromas in neurofibromatosis 1 is associated with CDKN2A/p16 inactivation. *Am. J. Pathol.* **155**, 1879–1884.
74. Sherr, C. J. (1998) Tumor surveillance via the ARF-p53 pathway. *Genes Dev.* **12**, 2984–2991.
75. Vogel, K. S., Klesse, L. J., Velasco-Miguel, S., et al. (1999) Mouse tumor model for neurofibromatosis type 1. *Science* **286**, 2176–2179.
76. Luongo, C., Moser, A. R., Gledhill, S., and Dove, W. F. (1994) Loss of *Apc*+ in intestinal adenomas from *Min* mice. *Cancer Res.* **54**, 5947–5952.
77. Reilly, K. M., Loisel, D. A., Bronson, R. T., McLaughlin, M. E., and Jacks, T. (2000) *Nf1;Trp53* mutant mice develop glioblastoma with evidence of strain-specific effects. *Nat. Genet.* **26**, 109–113.
78. DeClue, J. E., Heffelfinger, S., Benvenuto, G., et al. (2000) Epidermal growth factor receptor expression in neurofibromatosis type 1-related tumors and NF1 animal models. *J. Clin. Invest.* **105**, 1233–1241.
79. Guo, H. F., The, I., Hannan, F., Bernards, A., and Zhong, Y. (1997) Requirement of *Drosophila* NF1 for activation of adenylyl cyclase by PACAP38-like neuropeptides. *Science* **276**, 795–798.
80. The, I., Hannigan, G. E., Cawley, G. S., et al. (1997) Rescue of a *Drosophila* NF1 mutant phenotype by protein kinase A. *Science* **276**, 791–794.
81. Guo, H. F., Tong, J., Hannan, F., Luo, L., and Zhong, Y. (2000) A neurofibromatosis-1-regulated pathway is required for learning in *Drosophila*. *Nature* **403**, 895–898.
82. Johnson, M. R., Look, A. T., DeClue, J. E., Valentine, M. B., and Lowy, D. R. (1993) Inactivation of the *NF1* gene in human melanoma and neuroblastoma cell lines without impaired regulation of GTP-Ras. *Proc. Natl. Acad. Sci. USA* **90**, 5539–5543.
83. The, I., Murthy, A., E., Hannigan, G. E., et al. (1993) Neurofibromatosis type 1 gene mutations in neuroblastoma. *Nat. Genet.* **3**, 62–66.
84. Andersen, L. B., Fountain. J. W., Gutmann, D. H., et al. (1993) Mutations in the neurofibromatosis 1 gene in sporadic malignant melanoma cell lines. *Nat. Genet.* **3**, 118–121.
85. Johnson, M. R., DeClue, J. E., Felzmann, S., et al. (1994) Neurofibromin can inhibit Ras-dependent growth by a mechanism independent of its GTPase-accelerating function. *Mol. Cell. Biol.* **14**, 641–645.
86. Zhu, Y., Ghosh, P., Chernay, P., Burns, D. K., and Parada, L. F. (2002) Neurofibromas in NF1: Schwann cell origin and role of tumor environment. *Science* **296**, 920–922.

15

Wilms' Tumor as a Model for Cancer Biology

Andrew P. Feinberg and Bryan R. G. Williams

1. Wilms' Tumor as a Model of Cancer Biology

Wilms' tumor of the kidney (WT) is the most common solid tumor of childhood, and it was first described in detail by Max Wilms' in 1899. WT is a paradigm of childhood cancer, because it has served as a model from four distinct perspectives. First, it was one of three tumors used by Knudson in the early 1970s as a model for understanding the epidemiology of childhood cancer. Second, WT has played a key role is the molecular biology of childhood cancer. WT was one of the first examples in which a tumor suppressor gene was mapped by somatic genetic alterations in the tumors, compared to normal tissues, i.e., somatic cell gene mapping. Third, WT was the tumor in which abnormal imprinting in cancer was discovered. As such, it has been a model for epigenetic alterations in cancer. Finally, WT is a model for understanding the pathology of childhood cancer. Pediatric solid tumors, unlike adult malignancies, typically reflect normal developmental stages of organogenesis. As such, Wilms' tumor has served as a developmental paradigm for nephrogenesis. The purpose of this chapter is to describe the evolution of WT as a model for cancer from each of these perspectives, and then to synthesize them in a unifying view that hopefully provides novel insights into the mechanism of cancer in general.

2. Wilms' Tumor as a Model of Cancer Epidemiology

Knudson and Strong proposed a hypothesis to explain the bimodal age distribution of WT *(1)*. Reviewing case reports of Wilms' tumor, they found that bilateral cancers arise in children at an earlier age than unilateral cancers. They proposed that WT arises by two sequential genetic events, or "hits" analogous to retinoblastoma. According to this model, patients with bilateral tumors are born with the first hit in the germline, leading to multiple and bilateral tumors occurring at a younger age. In contrast, patients with unilateral tumors are not born with a germline hit, and both mutations must arise in a single somatic cell lineage, accounting for the later age and unilateral distribution of tumors *(1)*. The distinction in location is between unilateral and bilateral, rather than between unifocal and multifocal, as tumors do not metastasize to the contralateral kid-

From: *Methods in Molecular Biology, Vol. 222: Tumor Suppressor Genes: Pathways and Isolation Strategies*
Edited by: Wafik S. El-Deiry © Humana Press Inc., Totowa, NJ

ney, but multifocal tumors might arise from a single clone. This two-hit model grew to accommodate an hypothesis by David Comings that hereditary cancers segregating as a dominantly transmitted trait could be caused by a functionally recessive tumor suppressor gene ***(2)***. Patients with cancer predisposition have transmitted one mutant allele vertically, requiring only a second hit to develop cancer. Comings's idea was incorporated into the two-hit model by the realization that the two hits were allelic. It is worth bearing in mind that Knudson's epidemiologic observations applied to three childhood cancers, the others being retinoblastoma ***(3)*** and neuroblastoma ***(4)***. The RB retinoblastoma gene fits the model exceptionally well, although there are complexities in the epidemiology of WT that are not fully explained by the model, as will be discussed later. A neuroblastoma gene fulfilling Knudson's model has eluded identification.

While Knudson's model explains the earlier onset of bilateral WT, it does not explain the age distribution of most WT. Breslow and colleagues have observed that bilateral tumors are relatively rare in WT, accounting for only approximately 3% of cases ***(5,6)***. In contrast, they have found that there are distinct pathologic differences between earlier and later-arising tumors, but a rationale for their distinct age of appearance has not been available until recently, as will be discussed under **Subheading 5.**

3. Wilms' Tumor as a Model of Cancer Genetics and Its Relationship to WT Epidemiology

A Wilms' tumor gene was the second tumor suppressor gene to be identified by positional cloning, i.e., using mapping information, rather than functional information, to identify candidate genes, and then identifying mutations within these genes in tumors. Interestingly, the first such gene was retinoblastoma, also studied by Knudson. A third gene implicated in neuroblastoma has not yet been identified. In particular, the WT1 gene on 11p13 was first mapped by identification of chromosomal microdeletions in patients with Wilms' tumor and mental retardation ***(7)***. Such patients, many of whom also show genitourinary malformations and aniridia, are an example of a contiguous gene syndrome in which multiple genes of unrelated function are deleted together because of their contiguity. In this case, the syndrome is termed WAGR, for Wilms'/aniridia/genitourinary_malformations/mental_retardation. In a convergence of epidemiology and molecular biology, WT was one of the first cancers that showed loss of allelic heterozygosity (LOH) of polymorphic markers, consistent with loss of a Knudsonian tumor suppressor gene ***(8)***. In one of the few triumphs of positional cloning to identify tumor suppressor genes, the late Bill Lewis identified a homozygous deletion in a WT ***(9)***, which was a guidepost for two laboratories to identify a zinc-finger transcription factor that showed mutations in other tumors ***(10,11)***.

This Wilms' tumor gene, WT1, is expressed at high levels during early renal development and is hypothesized to be involved in commitment of primitive nephroblasts to renal lineage-specific development, as well as in gonadal development ***(12)***. This hypothesis is based on the observation that mice lacking WT1 show gonadal and renal dysplasia ***(13,14)***. Furthermore, patients with the genetic disorder Denys-Drash (DDS) syndrome, or pseudohermaphroditism and Wilms' tumor, have germline mutations in WT1. However, the molecular pathways by which WT1 functions have not been worked out definitively. WT1 was found to bind to EGR1 sites, and to repress expression of the insulinlike growth factor II gene (IGF2) and the IGF1 receptor through these sites in transfection experiments. As described under **Subheading 4.**, IGF2 is of great importance in the epigenetics of WT. However, WT1 may have an effect on IGF2 opposite to that first described, and WT1 expression may also be regulated by IGF2. The physiologic functions of WT1 are complicated by the existence of up to 16 protein isoforms resulting from alternative splicing. The WT1 protein (52–54 kDa) contains a carboxy terminal DNA and RNA-binding domain with four zinc fingers of the Cys2-His2 type, and an amino-terminal transactivation domain rich in proline, glutamine, serine, and lysine that can mediate self-dimerization ***(15,16)***. Wilms' tumor and DDS mutations within the zinc-finger domain abrogate WT1 DNA-binding activity ***(17)***. Two alternative splice sites give rise to four WT1 isoforms. Alternative splicing of exon 5 results in the presence or absence of 17 amino acids in the amino-terminal domain but this is conserved only in mammals. Alternative splicing at the end of exon 9 determines the presence of three amino acids, lysine-threonine-serine (KTS), between zinc fingers 3 and 4 ***(18)***. Mutations within the +/–KTS splice junction upset the balance of WT1 isoforms, and cause the severe genitourinary abnormalities seen in Frasier syndrome ***(19)***. RNA editing changing proline to leucine at position 280, or the addition of 68 amino acids through the use of an alternative translation start, increases the number of potential isoforms to 16 ***(20)***. Further complexity is introduced by the posttranscriptional regulation of WT1 activity by phosphorylation ***(21,22)***.

A better understanding of the role of WT1 in development and cancer may come from more adequate modeling in the mouse. An essential role for WT1 in urogenital development was illustrated by the failure of mice carry a homozygous deletion of the WT1 gene to develop kidneys and gonads ***(13)***. Unfortunately, the mice die of a massive hemodynamic problem due to the incomplete formation of the epicardium and consequently have had restricted utility as a model. However, the analyses of mice that can produce only the WT1 –KTS or WT1 +KTS isoforms have clearly illustrated that the isoforms have specific and distinct activities in urogenital development ***(23)***. One overriding issue that studying these mice may help address is the physiologic relevance of a large number of putative WT1 –KTS target genes identified by in-vitro cotransfection studies. The +KTS WT1 isoform has also been implicated in RNA processing (reviewed in **ref.** *24*), but specificity in this process has not been determined. However, the results thus far from the mouse models indicating that the +KTS isoform is essential for male sex determination may lead to identification of affected transcripts. Potential candidate WT1-regulated genes include *Sry*, *DAX-1,* and *amphiregulin*. However, none of these is implicated in tumorigenesis. WT1-regulated genes implicated early in the pathway include the antiapoptotic gene *Bcl2*, the cell cycle arrest gene *p21*, and the growth-promoting *IGF2* and its receptor, *IGF1R*. Upsetting the balance of expression of these genes either through a failure of appropriate WT1 regulation or loss of imprinting (see below) likely contribute both to tumor initiation and progression.

Despite the incomplete understanding of the role of WT1, it does fulfill Knudson's hypothesis, in that patients with bilateral WT often show WT1 mutations in the germline, and such patients also show LOH in their tumors that includes the 11p13 region harboring WT1. Nevertheless, a conundrum in WT ***(25)*** biology is the fact that the region of LOH in WT more commonly involves 11p15 than 11p13. In most WT, LOH in fact does not involve WT1 at all. Furthermore, mutations of WT1 in WT are relatively rare, occurring in no more than 5% of sporadically occurring tumors. Thus a second Wilms' tumor suppressor gene has been hypothesized to involve 11p15, and it is commonly referred to as WT2 ***(25)***. To date, no gene on 11p15 has shown frequent mutations in WT with LOH of this region.

4. Wilms' Tumor as a Model of Cancer Epigenetics

Wilms' tumor was the first cancer in which altered genomic imprinting was observed ***(26)***. Imprinting is a parent of origin-specific gene silencing that leads to monoallelic, or preferential allelic expression, of the allele of a gene from a specific parent. Imprinting was suspected to play a role in WT when it was observed that LOH of *11p* virtually always involves the maternal allele ***(27)***. Sapienza proposed a model to explain this observation, in which an imprinted tumor suppressor gene undergoes LOH of the normally expressed, in this case, maternal allele ***(28)***. The problem with this model is that all people would effectively have lost one copy of a tumor suppressor gene, by virtue of imprinting-related gene silencing. Thus, all WT would appear to follow an epidemiologic model in which the first hit had already taken place, i.e., occur early and arise bilaterally, even though, as noted earlier, bilateral tumors are rare in WT. Sapienza suggested as a solution to this problem that imprinting arises aberrantly in some individuals, and as a mosaic trait in the kidney, but always involving silencing of the paternal allele when it arises ***(29)***. Thus, sporadically occurring, unilateral tumors would arise by a first epigenetic hit (imprinting related silencing) and a second genetic hit (LOH). At present, no gene fulfilling this hypothesis has been found, but it has not been ruled out either.

The first evidence for imprinting in cancer was quite unexpected and quite different than Knudson's model of gene silencing. We and others found that most WT undergo aberrant activation of the normally silenced copy of the imprinted IGF2 gene, as well as disrupted imprinting of the nearby imprinted H19 gene as well ***(26,30)***. The term we coined for this alteration is loss of imprinting, or LOI ***(26)***, and it connotes either the activation the normally silent copy of an imprinted growth-promoting gene, or the silencing of the normally active copy of an imprinted tumor suppressor gene. LOI has now been observed in most cancers, including the major malignancies of adults as well as other embryonal cancers of childhood ***(31–40)***. Although LOI was first observed in WT several years ago, a specific effect on gene expression has only recently been observed. Tumors with LOI show a doubling in expression of IGF2, consistent with activation of the normally silent maternal allele ***(41)***. IGF2 is normally expressed in the proliferating kidney during development ***(42)***. Given that IGF2 is a mitogen, its overexpression could lead to hyperproliferation and an expanded pool of nephroblasts subject to further carcinogenic alterations. IGF2 is also an important autocrine growth factor in a wide variety of malignancies ***(31–40)***, and LOI is likely to be a mechanism of its overexpression generally. In a mouse model of SV40 antigen transgene-induced cancer, LOI of *IGF2* is an early event in tumor progression ***(43)***. Furthermore, abrogation of LOI by deletion mutagenesis of an *IGF2* allele limits tumor progression ***(44)***. Thus, LOI is an important mechanism of cancer development, and WT is a model for understanding its role.

LOI of *IGF2* is also observed in the normal tissues of patients with Beckwith-Wiedemann syndrome (BWS), a disorder of prenatal overgrowth involving macroglossia, ear pits or creases, neonatal hypoglycemia, and midline abdominal wall defects, including umbilical hernia, diastasis recti, or omphalocele ***(45,46)***. BWS shows genetic heterogeneity, involving several imprinted genes on 11p15. Most patients show epigenetic alterations involving altered methylation of H19, or altered methylation of an antisense transcript termed LIT1 and lying within the K_VLQT1 gene ***(47)***. We have proposed a two-domain hypothesis, in which altered methylation of H19 leads to LOI and activation of the normally silent maternal allele of *IGF2*; and altered methylation of LIT1 leads to LOI and abnormal silencing of the normally active maternal allele of the $p57^{KIP2}$ gene, the latter of which is rarely mutated in BWS ***(48)***. Recently, we have found that altered methylation of LIT1 is specifically associated with midline abdominal wall defects and macrosomia in BWS ***(49)***, and we and others have found that altered methylation of H19 is specifically associated with cancer in BWS ***(49)***.

One mechanism of LOI in WT is altered methylation of a differentially methylated region (DMR) upstream of the *H19* gene, which permits activation of the maternal allele of *IGF2* by an enhancer shared with the H19 promoter and on the other side of the DMR, and which are normally separated by binding of the insulator factor CTCF to the unmethylated allele ***(50)***. Altered DNA methylation was first described in 1983 by Feinberg and Vogelstein ***(51)***, and thus LOI and altered DNA methylation are related epigenetic alterations. An intriguing observation in the case of colorectal cancer is that tumors with LOI also show microsatellite instability (MSI) in their tumors ***(52)***. Furthermore, patients with LOI in their colon tumors also show LOI in the normal mucosa, and in about 10% of adults in the general population, even though MSI is a property of the tumors specifically ***(52)***. Thus, it appears that LOI can arise as a field effect in the colon, and perhaps predispose to the development of cancers, particularly those MSI-positive tumors that are found in younger patients and in the difficult-to-detect right-sided colon cancers typically found to be MSI-positive. LOI is thus the first genetic alteration in normal cells that occurs at high frequency in cancer patients, and it may prove to be an important marker of neoplastic risk.

5. Wilms' Tumor as a Model of Cancer Pathology

A series of observations by Beckwith and colleagues led to a fundamental model of Wilms' tumor pathology. Beckwith observed that many WT were associated with premalignant nephrogenic rests (NR) that resembled the tumors with which they were associated. These NR were of two types. Intralobar nephrogenic rests (ILNR) showed heterotypic elements often found within WT. These heterotypic elements are often found normally in other developing tissues, such as striated muscle, cartilage, and adipocytes. In contrast, perilobar nephrogenic rests (PLNR) involve blastemal elements that resemble renal precursor cells in the developing nephron ***(53)***.

Beckwith hypothesized that the ILNR arose within the interior of the developing renal lobe, and indeed they are found near elements derived from the primitive ureteric bud, i.e., the collecting system of the kidney tubules. In contrast, PLNR are thought to arise from the periphery of the renal lobe, they do not show heterotypic elements, and indeed they are located near the renal cortex or along the periphery of the developing lobes ***(53)***. The tumors themselves also show ILNR-like or PLNR-like pathology in about two-thirds of cases, consistent with the idea that they arise from the rests ***(5,53)***. This is a more provocative idea than it may seem on the surface, as PLNRs are found in no less than 1% of newborn, otherwise normal infants ***(53)***. Thus, one must consider the possibility that at least this type of WT arises from precursor lesions common in the general population at birth. Such an idea is opposite to that suggested for common adult malignancies such

Table 1
Model of the Molecular Biology of Wilms' Tumor

Developmental stage	Nephrogenic commitment	Nephrogenic expansion
Age	Early	Late
Hereditary syndrome	Denys-Drash	Beckwith-Wiedemann
Nephrogenic rests	Intralobar	Perilobar
Kidney	Normal	Enlarged
Pathology	ILNR-like	PLNR-like
Mutations	WT1, "WT2"	None characteristic
LOH	11p13, 11p15	None
LOI	None	IGF2
Gene expression	Reduced WT1	Increased IGF2
Knudsonian	Yes	No
Primary mechanism	Genetic	Epigenetic
Kidney size	Normal	Enlarged

as colorectal cancer, in which the precursor lesion (adenoma) arises uncommonly and only after years of life and/or exposure to precipitating environmental factors. Finally, in another example of biological convergence in studies of WT, ILNR-like tumors were found to occur commonly in WAGR, and tumors with ILNR-like features show reduced expression of WT1 *(5,53)*. Similarly, PLNR-like tumors are commonly associated with BWS *(5,53)*, suggesting that the dichotomy of pathology in WT pathology extends to the genetics and epidemiology as well. Finally, several hereditary disorders predispose to WT. In two of these, the pathology of the tumors is characteristic. In Denys-Drash syndrome, tumors are ILNR-like; and in BWS, tumors are PLNR-like. This distinction is consistent with a difference in mechanism for cancer in the two disorders.

6. A Comprehensive Model for Understanding Wilms' Tumor Biology (Table 1)

There are several confusing and apparently contradictory observations in WT biology, and it is not immediately apparent how to reconcile the epidemiology, genetics, epigenetics, and pathology of WT in a coherent model. For example, in the epidemiology of WT, bilateral tumors are rare, and thus Knudson's model does not explain the bimodal age distribution of most WT. In the genetics of WT, most LOH does not include the WT1 gene on 11p13 but involves 11p15, even though no mutated gene has been identified within this region. In the epigenetics of WT, altered imprinting and methylation is found not only in the tumors, but also in nephrogenic rests, suggesting that epigenetic alterations may *precede* genetic changes. Finally, in the pathology of WT, the existence of ILNR and PLNR argue against a simple model of Wilms' tumorigenesis. Furthermore, PLNR are found in an astonishingly high frequency of otherwise normal individuals, as much as 1% of newborns. Such an abundance of premalignant lesions is not consistent with a conventional Knudsonian tumor suppressor gene, and it is at marked variance with the idea advanced by Vogelstein and others for adult tumors, that cancer arises from a series of individually rare events in a given somatic cell lineage.

We suggest that WT is fundamentally a dichotomous disorder. This idea is not entirely new, as support comes from several different venues, but it has not been viewed from all of them comprehensively before, to our knowledge. By this view, WT is a disorder involving disruption of normal nephrogenic development. This can happen at either of two stages. The first is just after nephroblasts have been formed from condensing mesenchyme, under the inductive influence of the ureteric bud. The second is during the period of maximum nephroblast proliferation during nephrogenic expansion. The first class of tumors, involving disruption of nephrogenic commitment, is similar to that seen in retinoblastoma. Even though the *RB* gene is ubiquitously expressed, it is critical for retinoblast development. Loss of the gene leads to a population of precursor cells that does not properly mature, and is a target for further carcinogenic mutations. These tumors thus fit Knudson's model well, with two mutations in the *WT1* gene, or potentially in *WT2* as well. In contrast, tumors arising during nephrogenic expansion are quite different, and do not fit Knudson's model. Here the defect is overexpansion of committed nephroblasts, again leading to an increased population of cells at risk of subsequent carcinogenic events. Here the mechanism is epigenetic, with LOI leading to overexpression of IGF2, with its attendant mitogenic effects.

It is important to give proper credit to those who have appreciated the distinction between the two types of tumors. Beckwith was first to raise the possibility of two developmental stages that account for pathologic differences in the tumors ***(53)***. He pointed out that the location of the rests within the tumors is consistent with this idea, because the maturing kidney migrates out from the region of the ureteric bud to the periphery of the developing lobe during development. Adjacent to the tumors themselves, the ILNR tend to lie near the collecting system and the PLNR near the cortex or the medulla, sinus, and caliceal wall within the kidney. Huang et al. first showed that most ILNR-like tumors exhibit reduced WT1 expression, and tumors with WT1 mutations are also ILNR-like ***(54)***. These authors pointed out that this alteration would mimic germline mutations in *WT1*, which are associated with renal dysplasia, and it could account for the presence of heterotypic elements within the ILNR and the ILNR-like tumors, as the precursor cells are not fully committed to renal development From an epidemiologic perspective, Breslow and colleagues recognized that tumors with PLNR arise later than those with ILNR ***(5,6)***. Recently, Ravenel et al. have found that PLNR-like tumors show LOI, arise at a later age than tumors with normal imprinting, and show increased expression of IGF2 ***(41)***.

An important implication of this dichotomous view of WT is that the later-arising tumors, which are PLNR-like and show LOI, likely follow a model of cancer quite different than that proposed by Vogelstein and colleagues for colorectal cancer ***(55)***. The implication of these "nephrogenic expansion" tumors is that there is a generalized epigenetic disturbance. Rather than a series of mutations leading gradually to the emergence of a tumor, the precursor lesion is epigenetic, involving LOI. Indeed, the PLNRs themselves show both altered gene expression and altered DNA methylation indicative of LOI. These alterations are widespread in the affected patients, often associated with nephromegaly, usually syndromically (BWS or idiopathic hemihypertrophy), but in some patients with sporadically occurring tumors as well. Given that PLNR are the lesions Beckwith observed in about 1% of newborns, epigenetic disturbances in fetal life would appear to predispose to cancer in a large proportion of the population. Fortunately, most of these lesions disappear over time. The encouraging news is that if these patients can be identified, the cancer risk might be amenable to chemoprevention, since epigenetic alterations are by definition reversible.

Acknowledgments

This work was supported by National Institutes of Health Grant CA65145.

References

1. Knudson, A. G. and Strong, L. C. (1972) Mutation and cancer: a model for Wilms" tumor of the kidney. *J. Natl. Cancer Inst.* **48,** 313–324.
2. Comings, D. E. (1973) A general theory of carcinogenesis. *Proc. Natl. Acad. Sci. USA* **70,** 3324–3328.
3. Knudson, A. G. (1971) Mutation and cancer: statistical study of retinoblastoma. *Proc. Natl. Acad. Sci. USA* **68,** 820–823.
4. Knudson, A. G., Jr. and Strong, L. C. (1972) Mutation and cancer: neuroblastoma and pheochromocytoma. *Am. J. Hum. Genet.* **24,** 514–532.
5. Breslow, N. E. and Beckwith, J. B. (1982) Epidemiological features of Wilms" tumor: results of the National Wilms" Tumor Study. *J Natl. Cancer Inst.* **68,** 429–436.
6. Breslow, N., Beckwith, J. B., Ciol, M., and Sharples, K. (1988) Age distribution of Wilms" tumor: report from the National Wilms" Tumor Study. *Cancer Res.* **48,** 1653–1657.
7. Riccardi, V. M., Sujansky, E., Smith, A. C., and Francke, U. (1978) Chromosomal imbalance in the aniridia-Wilms" tumor association: 11p interstitial deletion. *Pediatrics* **61,** 604–610.
8. Fearon, E. R., Vogelstein, B., and Feinberg, A. P. (1984) Somatic deletion and duplication of genes on chromosome 11 in Wilms' tumors. *Nature* **309,** 176–178.
9. Michalopoulos, E. E., Bevilacqua, P. J., Stokoe, N., Powers, V. E., Willard, H. F., and Lewis, W. H. (1985) Molecular analysis of gene deletion in aniridia-Wilms' tumor association. *Hum. Genet.* **70,** 157–162.
10. Haber, D. A., Buckler, A. J., Glaser, T., et al. (1990) An internal deletion within an 11p13 zinc finger gene contributes to the development of Wilms" tumor. *Cell* **61,** 1257–1269.
11. Gessler, M., Poustka, A., Cavenee, W., Neve, R. L., Orkin, S. H., and Bruns, G. A. (1990) Homozygous deletion in Wilms' tumours of a zinc-finger gene identified by chromosome jumping. *Nature* **343,** 774–778.
12. DiLella, A. G., Kwok, S. C. M., Ledley, F. D., Marvit, J., and Woo, S. C. (1986) Molecular structure and polymorphic map of the human phenylalanine hydroxylase gene. *Biochemistry* **25,** 743–749.
13. Kreidberg, J. A., Sariola, H., Loring, J. M., et al. (1993) WT-1 is required for early kidney development. *Cell* **74,** 679–691.
14. Glaser, T., Lane, J., and Housman, D. (1990) A mouse model of the aniridia-Wilms' tumor deletion syndrome. *Science* **250,** 823–827.
15. Moffett, P., Bruening, W., Nakagama, H., et al. (1995) Antagonism of WT1 activity by protein self-association. *Proc. Natl. Acad. Sci. USA* **92,** 11105–11109.
16. Reddy, J. C., Hosono, S., and Licht, J. D. (1995) The transcriptional effect of WT1 is modulated by choice of expression vector. *J. Biol. Chem.* **270,** 29976–29982.
17. Pelletier, J., Bruening, W., Kashtan, C. E., et al. (1991) Germline mutations in the Wilms" tumor suppressor gene are associated with abnormal urogenital development in Denys-Drash syndrome. *Cell* **67,** 437–447.
18. Rauscher, F. J., III (1993) The WT1 Wilms' tumor gene product: a developmentally regulated transcription factor in the kidney that functions as a tumor suppressor. *FASEB J.* **7,** 896–903.

19. Barbaux, S., Niaudet, P., Gubler, M. C., et al. (1997) Donor splice-site mutations in WT1 are responsible for Frasier syndrome. *Nat. Genet.* **17,** 467–470.
20. Bruening, W. and Pelletier, J. (1996) A non-AUG translational initiation event generates novel WT1 isoforms. *J. Biol. Chem.* **271,** 8646–8654.
21. Sakamoto, Y., Yoshida, M., Semba, K., and Hunter, T. (1997) Inhibition of the DNA-binding and transcriptional repression activity of the Wilms'' tumor gene product, WT1, by cAMP-dependent protein kinase-mediated phosphorylation of Ser-365 and Ser-393 in the zinc finger domain. *Oncogene* **15,** 2001–2012.
22. Ye, Y., Raychaudhuri, B., Gurney, A., Campbell, C. E., and Williams, B. R. (1996) Regulation of WT1 by phosphorylation: inhibition of DNA binding, alteration of transcriptional activity and cellular translocation. *EMBO J.* **15**, 5606–5615.
23. Hammes, A., Guo, J. K., Lutsch, G., et al. (2001) Two splice variants of the Wilms'' tumor 1 gene have distinct functions during sex determination and nephron formation. *Cell* **106,** 319–329.
24. Hastie, N. D. (2001) Life, sex, and WT1 isoforms—three amino acids can make all the difference. *Cell* **106,** 391–394.
25. Reeve, A. E., Sih, S. A., Raizis, A. M., and Feinberg, A. P. (1989) Loss of allelic heterozygosity at a second locus on chromosome 11 in sporadic Wilms'' tumor cells. *Mol. Cell. Biol.* **9,** 1799–1803.
26. Rainier, S., Johnson, L. A., Dobry, C. J., Ping, A. J., Grundy, P. E., and Feinberg, A. P. (1993) Relaxation of imprinted genes in human cancer. *Nature* **362,** 747–749.
27. Schroeder, W. T., Chao, L. Y., Dao, D. D., et al. (1987) Nonrandom loss of maternal chromosome 11 alleles in Wilms' tumors. *Am. J. Hum. Genet.* **40,** 413–420.
28. Scrable, H., Cavenee, W., Ghavimi, F., Lovell, M., Morgan, K., and Sapienza, C. (1989) A model for embryonal rhabdomyosarcoma tumorigenesis that involves genome imprinting. *Proc. Natl. Acad. Sci. USA* **86,** 7480–7484.
29. Sapienza, C. (1990) Parental imprinting of genes. *Sci. Am.* **263,** 52–60.
30. Ogawa, O., Eccles, M. R., Szeto, J., et al. (1993) Relaxation of insulin-like growth factor II gene imprinting implicated in Wilms'' tumour. *Nature* **362,** 749–751.
31. Rainier, S., Dobry, C. J., and Feinberg, A. P. (1995) Loss of imprinting in hepatoblastoma. *Cancer Res.* **55,** 1836–1838.
32. Zhan, S., Shapiro, D. N., and Helman, L. J. (1994) Activation of an imprinted allele of the insulin-like growth factor II gene implicated in rhabdomyosarcoma. *J Clin. Invest.* **94,** 445–448.
33. Vu, T. H., Yballe, C., Boonyanit, S., and Hoffman, A. R. (1995) Insulin-like growth factor II in uterine smooth-muscle tumors: maintenance of genomic imprinting in leiomyomata and loss of imprinting in leiomyosarcomata. *J. Clin. Endocrinol. Metab.* **80,** 1670–1676.
34. Hibi, K., Nakamura, H., Hirai, A., et al. (1996) Loss of H19 imprinting in esophageal cancer. *Cancer Res.* **56,** 480–482.
35. Hashomoto, K., Azuma, C., Koyama, M., et al. (1995) Loss of imprinting in choriocarcinoma. *Nat. Genet.* **9,** 109–110.
36. Jarrard, D. F., Bussemakers, M. J., Bova, G. S., and Isaacs, W. B. (1995) Regional loss of imprinting of the insulin-like growth factor II gene occurs in human prostate tissues. *Clin. Cancer Res.* **1,** 1471–1478.
37. Yee, D., Cullen, K. J., Paik, S., et al. (1998) Insulin-like growth factor II mRNA expression in human breast cancer. *Cancer Res.* **48,** 6691–6696.
38. El-Badry, O. M., Helman, L. J., Chatten, J., Steinberg, S. M., Evans, A. E., and Israel, M. A. (1991) Insulin-like growth factor II-mediated proliferation of human neuroblastoma. *J. Clin. Invest.* **87,** 648–657.
39. Lahm, H., Amstad, P., Wyniger, J., et al. (1994) Blockade of the insulin-like growth-factor-I receptor inhibits growth of human colorectal cancer cells: evidence of a functional IGF-II-mediated autocrine loop. *Int. J. Cancer* **58,** 452–459.

40. Leventhal, P. S., Randolph, A. E., Vesbit, T. E., Schenone, A., Windebank, A., and Feldman, E. L. (1995) Insulin-like growth factor-II as a paracrine growth factor in human neuroblastoma cells. *Exp. Cell Res.* **221,** 179–186.
41. Ravenel, J. D., Broman, K. W., Perlman, E. J., et al. (2001) Loss of imprinting of insulin-like growth factor-II (IGF2) gene in distinguishing specific biologic subtypes of Wilms' tumor. *J. Natl. Cancer Inst.* **93,** 1698–1703.
42. Hedborg, F., Holmgren, L., Sandstedt, B., and Ohlsson, R. (1994) The cell type-specific IGF2 expression during early human development correlates to the pattern of overgrowth and neoplasia in the Beckwith-Wiedemann syndrome. *Am. J. Pathol.* **145,** 802–817.
43. Christofori, G., Naik, P., and Hanahan, D. (1994) A second signal supplied by insulin-like growth factor II in oncogene-induced tumorigenesis. *Nature* **369,** 414–418.
44. Christofori, G., Naik, P., and Hanahan, D. (1995) Deregulation of both imprinted and expressed alleles of the insulin-like growth factor 2 gene during β-cell tumorigenesis. *Nat. Genet.* **10,** 196–201.
45. Weksberg, R., Shen, D. R., Fei, Y. L., Song, Q. L., and Squire, J. (1993) Disruption of insulin-like growth factor 2 imprinting in Beckwith-Weidemann syndrome. *Nat. Genet.* **5,** 143–150.
46. Steenman, M. J. C., Rainier, S., Dobry, C. J., Grundy, P., Horon, I. L., and Feinberg, A. P. (1994) Loss of imprinting of IGF2 is linked to reduced expression and abnormal methylation of H19 in Wilms" tumor. *Nat. Genet.* **7,** 433–439.
47. Lee, M. P., DeBaun, M. R., Mitsuya, K., et al. (1999) Loss of imprinting of a paternally expressed transcript, with antisense orientation to KVLQT1, occurs frequently in Beckwith-Wiedemann syndrome and is independent of insulin-like growth factor II imprinting. *Proc. Natl. Acad. Sci. USA* **96,** 5203–5208.
48. Lee, M. P., DeBaun, M., Randhawa, G. S., Reichard, B. A., and Feinberg, A. P. (1997) Low frequency of $p57^{KIP2}$ mutation in Beckwith-Wiedemann syndrome. *Am. J. Hum. Genet.* **61,** 304–309.
49. DeBaun, MR, Niemitz, EL, McNeil, ED, Brandenburg, SA, Lee, MP, and Feinberg, AP. (2002) Epigenetic alterations of H19 and LIT1 distinguish Beckwith-Wiedemann syndrome patients with cancer and birth defects. *Am. J. Hum. Genet.* **70**, 604–611.
50. Cui, H., Niemitz, E. L., Ravenel, J. D., et al. (2001) Loss of imprinting of insulin-like growth factor-II in Wilms" tumor commonly involves altered methylation but not mutations of CTCF or its binding site. *Cancer Res.* **61,** 4947–4950.
51. Feinberg, A. P. and Vogelstein, B. (1983) Hypomethylation distinguishes genes of some human cancers from their normal counterparts. *Nature* **301**, 89–92.
52. Cui, H., Horon, I. L., Ohlsson, R., Hamilton, S. R., and Feinberg, A. P. (1998) Loss of Imprinting in normal tissue of colorectal cancer patients with microsatellite instability. *Nat. Med.* **4,** 1276–1280.
53. Beckwith, J. B., Kiviat, N. B., and Bonadio, J. F. (1990) Nephrogenic rests, nephroblastomatosis, and the pathogenesis of Wilms" tumor. *Pediatr. Pathol.* **10,** 1–36.
54. Huang, A., Campbell, C. E., Bonetta, L., et al. (1990) Tissue, developmental, and tumor-specific expression of divergent transcripts in Wilms' tumor. *Science* **250,** 991–994.
55. Fearon, E. R. and Vogelstein, B. (1990) A genetic model for colorectal turmorigenesis. *Cell* **61,** 759–767.

16

Inadequate "Caretaker" Gene Function and Human Cancer Development

Theodore L. DeWeese and William G. Nelson

1. Introduction

Both genetic and environmental factors have been implicated in the pathogenesis of human cancers. Yet for certain cancers, genetic factors act as critical contributors, while for other cancers, environmental factors seem to predominate in cancer causation. With the completion of the human genome project and accompanying advances in genomic technologies, genetic descriptions of inherited cancer risk and genetic signatures of cancer development are becoming increasingly refined. Already, it is clear that germline inheritance of certain mutant genes markedly increases cancer risks independent of environmental exposures. For example, inherited mutations in *Rb*, encoding a master cell-cycle regulatory gene, leads to the development of retinoblastoma, an otherwise rare cancer of the eye in children ***(1,2)***. In genomic DNA from retinoblastoma cells, the inherited mutant *Rb* alleles are typically accompanied by somatic alterations affecting the wild-type allele, resulting in a complete absence of normal Rb function in the cancer cells ***(3,4)***. Mice carrying targeted disruptions in *Rb* are also prone to develop cancers, supporting the notion that *Rb* functions as a tumor suppressor gene ***(5,6)***. Exposure to environmental carcinogens also markedly increases cancer risks. For example, cigarette smoke promotes the development of lung cancer, bladder cancer, and other cancers in a dose-dependent manner ***(7)***. Polycyclic aromatic hydrocarbons, as well as other compounds found in cigarette smoke, also promote cancer development in laboratory animal models, supporting the designation of such compounds as carcinogens ***(8,9)***.

For the majority of human cancers, are genetic factors causative or is the environment responsible? In a recent study of cancer risks among 44,788 pairs of twins in Sweden, Denmark, and Finland ***(10)***, statistically significant effects of heritable factors were observed only for prostate cancer (42% of cases with a 95% confidence interval of 29–50%), colorectal cancer (35% of cases with a 95% confidence interval of 10–48%), and breast cancer (27% of cases with a 95% confidence interval of 4–41%). With these data, the authors concluded that the environment plays the principal role in causing most human cancers. What may also be likely is that if the environment has such a strong influence on human cancer development, then genes that determine interactions with

From: *Methods in Molecular Biology, Vol. 222: Tumor Suppressor Genes: Pathways and Isolation Strategies*
Edited by: Wafik S. El-Deiry © Humana Press Inc., Totowa, NJ

environmental factors may be critical heritable factors predisposing to cancer or critical targets for somatic alteration in cancer cells.

2. "Caretaker" Genes and the Pathogenesis of Human Cancer

Cancer cell DNA typically contains a plethora of alterations, including mutations, deletions, amplifications, translocations, and hypermethylated CpG islands, that affect the function of critical genes and contribute to the malignant phenotype ***(11,12)***. Gene defects in cancer cells may be inherited in the germline or may be acquired during cancer development. Historically, abnormal cancer genes have been classified as oncogenes, which, when activated by alteration, promote neoplastic transformation, and tumor suppressor genes, which when inactivated by alteration fail to prevent neoplastic transformation. More recently, increased attention has been afforded a new collection of abnormal cancer genes, termed "caretaker" genes, which when inactivated increase the rates at which somatic genome alterations appear, and by doing so, commensurately increase risks for cancer development ***(13)***. "Caretaker" genes identified thus far encode carcinogen-detoxification enzymes, DNA mismatch repair enzymes, DNA damage recognition and repair enzymes, and a variety of polypeptides responsible for maintaining chromosome integrity and the fidelity of chromosome segregation throughout DNA replication and mitosis. Defects in such "caretaker" genes typically appear early during the pathogenesis of cancer, often inherited in the germline or acquired at the earliest stages of cancer development. More direcly than oncogenes and tumor suppressor genes, "caretaker" genes potentially provide a clear link between genetic and environmental factors in cancer causation. By protecting the genome against environmental threats to genome integrity, "caretaker" genes serve as critical barriers to cancer caused by carcinogens. Inadequate "caretaker" function, resulting from inherited or acquired defects in "caretaker" genes, renders certain cells exquisitely vulnerable to genome damage inflicted by carcinogens and prone to cancer development.

Because loss of "caretaker" activity results in genetic instability, either by failing to suppress spontaneous genome accidents or by failing to correct genome damage inflicted by carcinogens, cells with defective "caretaker" genes have been thought to enjoy a selective growth or survival advantage during cancer pathogenesis *only* when additional genes have become altered (**Fig. 1A**). Can loss of "caretaker" function directly provide a selective growth or survival advantage? Natural selection of organisms in a given environment tends to favor genetic stability over genetic instability as a trait. When organisms display genetic instability, many of their progeny are genetic variants that do not reproduce successfully, leading to inefficiencies compromising population fitness. In contrast, in a changing environment, genetic instability likely offers a selective advantage as an organism trait. Although cancer cells arising in a target organ may not constitute an independent species subject to natural selection, such cells may perhaps be considered quasi-species influenced by selective pressures. Is the milieu of a target organ site a stable environment or a changing one? Before cancers arise, specific organ sites typically feature a constant temperature, a reliable source of nutrients via blood supply, and a steady supply of signals that regulate growth and survival. When cancers appear and grow, vascular perfusion becomes intermittent and inconsistent, and growth-regulatory signaling networks become compromised ***(14–17)***. Perhaps, such a site may be a suitable habitat for selecting cancer cells with genetic instability.

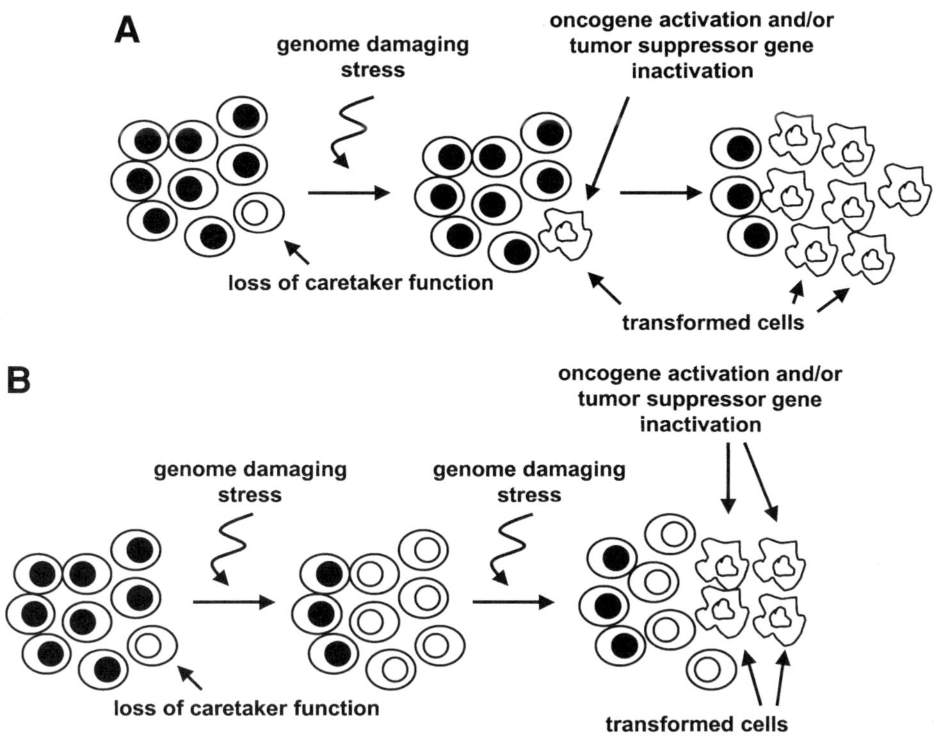

Fig. 1. (**A**) "Caretaker"-deficient cells are prone to genome damage that triggers oncogene activation or tumor suppressor gene inactivation, leading to growth of transformed cells. (**B**) "Caretaker"-deficient cells exhibiting damage "tolerance" enjoy a growth/survival advantage in response to genome damage, increasing the chance for oncogene activation or tumor suppressor gene inactivation, leading to growth of transformed cells.

One feature of inadequate "caretaker" function that threatens the growth and survival of cells with defective "caretaker" genes is that most genome-damaging stresses cause cell death as well as genome alterations. For this reason, cells that have lost "caretaker" activity might be expected to display a heightened sensitivity to both the cytotoxic effects and pro-mutagenic effects accompanying exposure to genome-damaging agents. Cancer pathogenesis triggered by genome damage in this setting must then be accompanied by some additional phenotypic change, whether the result of an adaptive response or of a new somatic genome alteration, that provides increased resistance to the cytoxicity of the damaging stress (**Fig. 1A**). For example, cancer cells arising in the face of such cytotoxic genome-damaging stresses might be predicted to contain alterations in genes that regulate cell survival. Apoptosis-regulating genes are known targets for somatic alteration during cancer pathogenesis ***(18,19)***. However, whether the frequency of alteration targeting such genes is different between cancers lacking "caretaker" genes vs cancers with apparently intact "caretaker" functions has not been established.

Recently, a curious phenotype has been identified among two families of "caretaker" genes that might contribute to cancer pathogenesis in the setting of genome damaging stresses. Cells with defects in *GSTP1*, encoding the π-class glutathione *S*-transferase (GST), or with defects in *MSH2*, encoding a DNA mismatch repair (MMR) enzyme, have been found to display oxidation damage "tolerance," manifest as an increased sensitivity

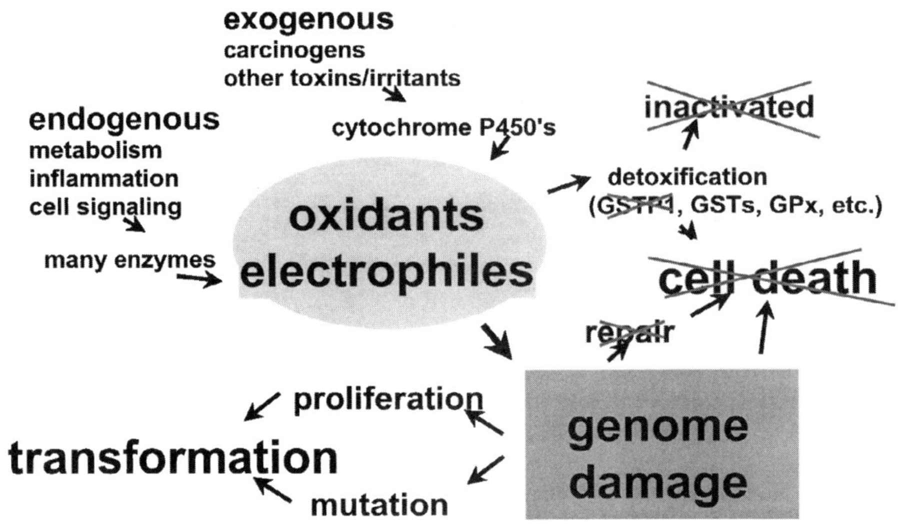

Fig. 2. Loss of *GSTP1* or *hMSH2* "caretaker" function leads to a genome damage "tolerance" phenotype: increased genome damage with decreased cell death upon exposure to oxidants or electrophiles.

to oxidative genome damage and mutations accompanied by an increased survival to oxidative genome damaging stress. This oxidation damage "tolerance" phenotype may be a mechanism by which genetic instability might be accompanied by a selective growth or survival advantage (**Fig. 1B**). In this chapter, we review the role of GSTs and MMR enzymes in cancer development and the evidence that defects in the genes encoding GSTs and MMR enzymes lead to an oxidation damage "tolerance" phenotype.

3. Glutathione S-Transferases as "Caretakers" and Human Cancer Development

Glutathione *S*-transferases (GSTs), enzymes capable of catalyzing conjugation reactions between glutathione, a peptide chemical-scavenging molecule, and a vast array of reactive chemical species, have long been thought to protect against cancer development by detoxifying carcinogens (**Fig. 2**) *(20)*. GSTs are dimeric enzymes with subunit polypeptides encoded by genes classified into several different gene families, including α, μ, π, and θ; the broad diversity in the composition of the GSTs likely provides great flexibility in substrate preferences. Different GST subunit polypeptides display different baseline expression patterns throughout various cell and tissue types, perhaps contributing intrinsic protection of these cells and tissues against chemical injury. In addition, many of the enzyme subunit genes stereotypically can be induced, via increased transcription, in response to chemical stresses *(21,22)*. Specific *cis* transcriptional regulatory sequences are required for stress induction *(23)*. New data have implicated the transcription factor Nrf2 in the induction response *(24)*. Whether present via intrinsic expression or via induced expression, *GST* genes comprise a critical barrier against reactive electrophiles and oxidants that threaten damage to the genome and other cellular macromolecules. This barrier may be essential for protection against chemical carcinogenesis: mice carrying disrupted *Gstp1/2* genes, encoding the murine π-class

GSTs, have been reported to manifest increased skin tumorigenesis upon topical exposure to 7,12-dimethylbenz anthracene (DMBA) ***(25)***, and mice carrying disrupted *Nrf2* genes have been reported to display increased gastric carcinogenesis upon exposure to benzo[a]pyrene ***(24)***. By catalyzing the detoxification of reactive species threatening genome damage, *GST* genes likely function as "caretaker" genes for cancers arising as a consequence of carcinogen exposure.

Inherited defects in human GST subunit genes confer a predisposition to cancer development upon carcinogen exposure. Some 40–50% of Caucasians are homozygous for null alleles for *GSTM1*, encoding a μ-class GST subunit polypeptide ***(26)***. Genetic epidemiology studies have associated nullizygosity for *GSTM1* with an increased risk for lung cancer and for bladder cancer, particularly among smokers ***(26–30)***. GSTM1, an enzyme capable of detoxifying cigarette smoke carcinogens by conjugation with glutathione, likely provides a measure of protection against cell and genome damage from cigarette smoke carcinogens ***(31,32)***. The combination of diminished GST "caretaker" activity and increased carcinogen exposure may be particularly dangerous. In one study, while nullizygosity for *GSTM1* conferred little increased risk for bladder cancer among nonsmokers, heavy smokers homozygous for null *GSTM1* alleles were at a six-fold increased risk for bladder cancer development ***(30)***. Approximately 15–20% of Caucasians homozygous for null alleles for *GSTT1*, encoding a θ-class GST subunit polypeptide ***(26)***. Recent genetic epidemiology studies have hinted at increased cancer risks associated with null-*GSTT1* alleles ***(26,33)***. No specific association with an identified carcinogen has yet been reported. Polymorphic variant alleles for *GSTP1*, encoding the π-class GST subunit polypeptide, have also been associated with increased cancer risks ***(34)***. As of yet, the mechanism by which the identified *GSTP1* allelic variants, found to encode polypeptides with altered enzyme activities ***(35)***, contribute to carcinogenesis has not been established. Nonetheless, genetic epidemiology studies of breast cancer development may provide a clue. In one study, homozygosity for the aa $105^{\text{ile to val}}$ *GSTP1* allele increased breast cancer risks 1.97-fold ***(36)***. However, in a follow-up study, women homozygous for low-activity *COMT* alleles that were also homozygous for aa $105^{\text{ile to val}}$ *GSTP1* alleles manifested an even more dramatically increased breast cancer risk ***(37)***. A model featuring redox cycling between estradiol and estrone, leading to the generation of oxidant species threatening genome damage, that otherwise might be controlled by adequate *COMT* and *GSTP1* function, has been proposed ***(37)***.

Somatic alterations in *GST* genes also appear to play a role in the pathogenesis of human cancers. Inactivation of *GSTP1*, by CpG island hypermethylation, has been reported in >90% of prostate cancers ***(38–43)***, in >85% of liver cancers ***(44)***, and in some 30% or more of breast cancers ***(45)***. In the normal prostate, GSTP1 is selectively expressed in basal epithelial cells, with little, if any enzyme detectable in nonstressed columnar epithelial cells. *GSTP1* expression appears strikingly induced in cells comprising proliferative inflammatory atrophy (PIA) lesions, the earliest identified precursor to prostate cancer ***(46)***. PIA lesions are characterized by inflammatory injury to the epithelium and exuberant epithelial cell proliferation/regeneration. PIA lesions appear to give rise directly to prostatic intraepithelial neoplasia (PIN) lesions ***(47)***. However, in PIN lesions, in contrast to PIA lesions, loss of GSTP1 expression, likely the result of *GSTP1* CpG island hypermethylation, is a characteristic finding ***(43)***. Prostatic carcinoma (PCA) cells are also characteristically devoid of GSTP1. Detailed molecular pathology analyses have discerned that either somatic *GSTP1* CpG island hypermethylation or other somatic genome alteration is found in all PCA cells in very nearly all PCA cases ***(48)***. Of note, despite the evidence that somatic *GSTP1* inactivation, detected in almost all PIN and PCA cells, appears selected during the pathogenesis of prostate cancer, *GSTP1* does not appear to act as a tumor suppressor gene. LNCaP PCA cells that express GSTP1, either as a result of treatment with 5-aza-C, an inhibitor of DNA methyltransferases that can reverses *GSTP1* CpG island hypermethylation and restores *GSTP1* function, or as a result of forced *GSTP1* cDNA expression, have been reported to grow at similar rates in vitro and in vivo when compared to unmodified LNCaP PCA cells that fail to express GSTP1 ***(48)***. Thus, rather than acting as a tumor suppressor gene, *GSTP1* appears more likely to function as a "caretaker" gene. For example, recent studies have suggested that *GSTP1* "caretaker" activity may provide a defense against ingested carcinogens present in "well-done" or "charred" meats ***(49)***. The heterocyclic aromatic amine carcinogen 2-amino-1-methyl-6-phenylimidazo[4,5-β]pyridine (PhIP) is known to cause mutations by adduction to DNA bases after metabolic activation by various cellular enzymes ***(50–54)***. Rats fed PhIP have been reported to display mutations in prostate DNA and develop PCA ***(55–57)***. When LNCaP PCA cells devoid of GSTP1 were exposed to metabolically activated PhIP, high levels of PhIP-DNA adducts were detected, while LNCaP PCA cells genetically modified to express GSTP1 exhibited substantial resistance to the formation of pro-mutagenic PhIP-DNA adducts ***(49)***. In this way, *GSTP1* "caretaker" activity may protect against cancer development driven by ingested "charred" meat carcinogens.

Model studies of the pathogenesis of hepatocellular carcinoma in a variety of different species confirm the "caretaker" role for GSTs. Normal rat hepatocytes, like normal prostatic columnar epithelial cells, do not express the rat π-class GST, GST-P. However, when rats are treated with various different hepatocarcinogens, hyperplastic liver nodules composed of cells containing very high GST-P levels typically appear ***(58–63)***. A few of these hyperplastic liver nodules progress to hepatocellular carcinoma, indicating that the high-GST-P-expressing hyperplastic liver nodules are cancer precursors. Nonetheless, the majority of the high-GST-P-expressing liver nodules regress over time, suggesting perhaps that GST-P "caretaker" activity might afford protection against further cell and genome damage. In support of this model, hepatocytes from rainbow trout (*Onchorhynchus mykiss*), fail to reliably increase GST expression upon exposure to hepatocarcinogens such as aflatoxin B_1 or 1,2-dimethylbenzanthracene ***(64)***. Rather, the carcinogen-treated fish develop both high-GST-expressing hyperplastic liver nodules and low-GST-expressing hyperplastic liver nodules. In support of a "caretaker" role for the fish GSTs during hepatocarcinogenesis, the low-GST-expressing hyperplastic nodules appear to be at higher risk for progression to hepatocellular carcinoma than high-GST-expressing liver nodules ***(64)***. The pathogenesis of human hepatocellular carcinoma may proceed via a similar mechanism. Hepatitis virus infection, a known risk for liver cancer development, has been reported to result in decreased GST levels in liver tissues ***(65)***. Hepatocellular carcinoma (HCC) cells arising in the setting of chronic hepatitis B or hepatitis C infection fail to express GSTP1, and like PCA cells, contain hypermethylated *GSTP1* CpG island alleles ***(44)***. *GSTP1* CpG island hypermethylation was also detected in tissues from human livers with cirrhosis and hepatitis ***(44)***. Augmentation of GST "caretaker" activity in liver tissues, using therapeutic inducers of GST activity, may prevent liver cancer development. Oltipraz, a GST inducer, has been shown to reduce aflatoxin B_1 damage when administered to a cohort at high risk for aflatoxin exposure and liver cancer development in China ***(66–68)***. All of these observations suggest that in the liver, loss of inducible GST "caretaker" activity tends to accelerate the pathogenesis of liver cancer upon carcinogen exposure, and activation of GST "caretaker" functions tends to attenuate carcinogen-associated hepatocarcinogenesis.

In human prostate cancers and liver cancers, all of the PCA cells and HCC cells display somatic *GSTP1* inactivation. How can loss of *GSTP1* "caretaker" functions be selected during the pathogenesis of these cancers? Early studies with human PCA cells and with murine hepatocytes have suggested that loss of π-class GST activity may provide a survival advantage in the face of certain cell- and genome-damaging stresses ***(69,70)***. When LNCaP PCA cells, devoid of GSTP1 activity, were challenged by prolonged exposures to oxidant stresses, the cells acquired high levels of oxidative DNA base damage and suffered some cell death ***(70)***. However, when GSTP1 expression was restored via *GSTP1* cDNA transfection, the resultant LNCaP PCA cells manifested markedly reduced levels of oxidized DNA bases and increased levels of cell death ***(70)***. Thus, loss of *GSTP1* "caretaker" function resulted in increased genome damage with decreased cell death. This oxidation "tolerance" phenotype may well be subject to selection during prostatic carcinogenesis, especially in a prostate milieu characterized by an abundance of inflammatory oxidants ***(46)***. Further evidence of damage "tolerance" associated with loss of *GST* genes has been recently reported in mice exposed to acetaminophen overdoses ***(69)***. Specifically, when compared to *Gstp1/2+/+* mice, *Gstp1/2−/−* mice displayed markedly less hepatocyte damage after high-dose acetaminophen administration ***(69)***. The mechanism underlying this acetaminophen "tolerance" phenotype was not established. However, after acetaminophen treatment, significant decrements in glutathione levels were seen in the livers of *Gstp1/2+/+* mice ***(69)***. A possible mechanism for diminished acetaminophen toxicity in *Gstp1/2−/−* mice may have been that less dramatic drops in gluthathione levels were evident in livers after acetaminophen treatment. Glutathione "bankruptcy" would thus render cells with intact GST "caretaker" activity prone to initiate cell death upon suffering extreme amounts of cell and genome damage. Other mechanisms for coupling damage to cell death via GST activity might include the toxification of reactive chemical species by π-class GSTs ***(71)*** or the possible effects of π-class GSTs on signal transduction pathways ***(72–74)***. Regardless of the mechanism, in both human PCA cells and murine hepatocytes, loss of π-class GST "caretaker" activity resulted a damage "tolerance" phenotype characterized by an increased survival of damaged cells.

4. DNA Mismatch Repair Enzymes as "Caretakers" in Human Cancer Development

DNA mismatch repair enzymes are polypeptides that have evolved to "proofread" DNA following replication in order to maintain DNA sequence fidelity ***(75,76)***. Along with DNA polymerase accuracy and exonuclease "proofreading," DNA mismatch repair helps control the rate at which mutations appear, particularly addition and deletion mutations, during cell proliferation ***(77)***. The mechanisms of DNA mismatch repair, as well as the specificities of DNA mismatch repair enzymes, are strikingly similar between species and appear highly conserved ***(78,79)***. Human homologs of the mismatch repair proteins found in bacteria, MutS and MutL, have been shown to exhibit activities characteristic of their bacterial counterparts ***(80–86)***. The human MutS homologs, hMSH, function as heterodimers, hMSH2-hMSH6 or hMSH2-hMSH3, acting primarily to bind and facilitate the repair of single-base mismatches and small insertions or deletions, and also assisting in the repair of larger insertions or deletions ***(77,87–89)***. The precise functions of the human homologs of MutL, hMLH1 and hPMS1, are not completely understood, but the enzymes can clearly participate with hMSH2 in binding DNA. The MutS homologs possess ATPase activity, similar to that of GTPase-activating proteins associated with G-protein signaling, which may act as a regulator of a critical molecular switch affecting the timing of events initiated by DNA mismatch recognition ***(90)***. Together, hMSH2, hMSH3, hMSH6, hMLH1, and hPMS1 are thought to function as a major cellular defense against replication errors. Defects in this DNA mismatch repair "caretaker" activity have been recognized to result in microsatellite repeat DNA instability and in the subsequent creation of insertion/deletion loop-type mismatched nucleotides ***(91)***. As such, inadequate DNA mismatch repair "caretaker" function leads to a predisposition to mutations and genomic instability accompanying DNA replication and cell division.

Inherited defects in human DNA mismatch repair genes can lead to the development of cancer ***(80,83,85)***. Early studies revealed that some of the somatic mutations present in genomic DNA from cancer tissues of patients with hereditary nonpolyposis colorectal cancer (HNPCC) resembled mutations typically present in nonhuman cells with defects in the MutHLS system ***(92)***. Fishel et al. demonstrated the presence of a T-to-C transition mutation, at a splice acceptor site, in the *hMSH2* germline gene present in affected members of two families with HNPCC ***(80)***, suggesting a role for *hMSH2* defects in the development of colon cancer. Using chromosome microdissection techniques, Leach et al. obtained and ordered a panel of polymorphic markers that permitted the definition of a 0.8-Mb portion of chromosome 2p16 as the likely locus of an HNPCC gene ***(83)***. An HNPCC gene candidate search in that genome region revealed a gene (*hMSH2*) homologous to MutS; germline *hMSH2* mutations were found in members of HNPCC kindreds, with germline and somatic *hMSH2* mutations present in cancer cells displaying replication error mutations, providing strong evidence that inherited *hMSH2* mutations might be responsible for HNPCC ***(83)***. Additional studies provided evidence that defective DNA mismatch repair was also involved in the development of other extracolonic tumors among members of HNPCC families ***(93)***. Germline defects in genes encoding DNA mismatch repair enzymes, accompanied presumably by somatic inactivation of the remaining wild-type allele, characteristically result in an HNPCC syndrome in which family members exhibit early development of proximal colon cancer only (Lynch syndrome I), or in an HNPCC syndrome in which family members develop endometrial cancer or stomach cancer in addition to colon cancer (Lynch syndrome II) ***(94,95)***. All cancers associated with HNPCC characteristically exhibit a "mutator" phenotype (associated with an accumulation of replication error mutations) and display DNA microsatellite repeat instability. Intriguingly, members of HNPCC families who have germline *MSH2* mutations appear prone to develop a different spectrum of extracolonic cancers than members of HNPCC families who have *hMLH1* mutations ***(96)***.

A number of animal studies have been performed in an attempt to define the role of mammalian MutS homologs as "caretakers" in the maintenance of genomic stability and as well as in tumor development. Using targeted gene disruption techniques, *Msh2−/−* mouse embryonic stem (ES) cells, and *Msh2−/−* mouse strains, were generated ***(97)***.

Msh2-deficient cells were found to lack DNA mismatch binding, to exhibit a "mutator" phenotype and microsatellite repeat instability, and to be tolerant to the toxicity associated with methylating agent exposure, all hallmarks of disrupted DNA mismatch repair ***(97)***. Mice carrying disrupted *Msh2* alleles developed lymphomas at an early age but did not display other significant functional or phenotypic abnormalities ***(98,99)***. The lymphomas developing in the *Msh2−/−* mice were of a specific type, similar to a human T-cell lymphoblastic lymphoma, with the expression of several genes, including *Tal-1*, *rhombotin-2*, and *Hox11*, in the mouse lymphoma cells reminiscent of genes stereotypically expressed in the human lymphoma cells. Of interest, *hMSH2* mutations have been identified in several of these human lymphomas, delineating a new pathway of human lymphomagenesis in which DNA mismatch repair plays a critical role ***(100)***.

Early reports suggested that inactivation of both *Msh2* alleles, with the subsequent loss of DNA mismatch repair, might be required for tumorigenesis in mice. However, de Wind et al. reported data indicating that *Msh2+/−* mice suffered increased numbers of spontaneous tumors when compared with *Msh2+/+* littermates ***(97)***. Furthermore, the spontaneous tumors in the *Msh2+/−* mice rarely displayed either loss of the second *Msh2* allele or microsatellite instability ***(97)***. In another study, Sansom et al. reported that the spontaneous mutation frequency (as determined by loss of function at the *Dlb-1* locus) in the intestine of *Msh2+/−* mice over 1 yr of age was significantly higher than that seen in the intestine of control *Msh2+/+* mice of a similar age ***(101)***. Together, these studies suggest the possibility of a gene dose effect for mutagenesis, and the possibility of an alternate mechanism for *Msh2*-related tumorigenesis not requiring loss of DNA mismatch repair. One possibility is that two *Msh2* alleles might be required to survey and correct genome damage, other than replication errors, that might arise as a consequence of as environmental exposures, or some other process, to inflict mutations leading to neoplastic transformation.

In support of a role for DNA mismatch repair enzymes in cellular responses to genome-damaging stresses, data have accumulated to indicate that cells with defects in genes encoding DNA mismatch repair proteins exhibit "tolerance" to injuries inflicted by a number of exogenous DNA-damaging agents when compared to cells with intact DNA mismatch repair. As an example, treatment of a variety of different DNA mismatch repair-deficient cells with alkylating agents such as N-methyl-N′-nitro-N-nitroguanidine has been found to result in substantially less cell killing than that observed for mismatch repair-competent cells ***(98,102–104)***. There is also evidence, although some of it conflicting, regarding the response of DNA mismatch repair-deficient cells to UV irradiation. Nonetheless, most of the analyses conducted suggest that intact mismatch repair is important in the transcription-coupled repair following UV exposures ***(105,106)***. Defective DNA mismatch repair, leading to inadequate or ineffective transcription-coupled repair, could conceivably lead to mutations upon exposure to carcinogens, promoting neoplastic transformation. By this mechanism, DNA mismatch repair may be critically important not only to replication error protection, but also to mutagenic carcinogen defenses. DNA mismatch repair proteins specifically recognize the 1,2-intrastrand d(GpG) crosslink produced by cisplatin; in addition, cells with defective DNA mismatch repair are relatively resistant to the cytotoxicity associated with cisplatin exposure ***(107)***. In all, these data strongly argue that both the cytotoxicity and mutagenicity of many DNA-damaging agents are, in part, modulated by DNA mismatch repair protein function. Furthermore, the "caretaker" functions of DNA mismatch repair enzymes may include the surveillance of genome damage as well as the "proofreading" of newly replicated DNA.

Many environmental factors have been implicated as contributors to human cancer development. During human colorectal carcinogenesis, the major promutagenic stresses faced by colonic epithelial cells may be chronic oxidative stress. Reactive oxygen species appear in many cells as a result of normal metabolic processes, with additional oxidants arising as a result of inflammation, of the actions of inducible enzymes, such as cyclooxygenase 2 (COX-2) and the microsomal oxidases, and of the actions of sex steroids and other hormones ***(108–110)***. Reactive oxygen species can inflict damage to many cellular components, including DNA ***(111)***; genome oxidation lesions include DNA strand breaks, oxidized DNA bases, and DNA–protein crosslinks, all of which can lead to mutations ***(111)***. The oxidation of DNA bases, such as the production of 8-hydroxyguanine (8-OH-G) from guanine, likely causes mutations via disruption of accurate base pairing during DNA replication ***(112–114)***. What is the impact of oxidative stress on tumorigenesis in the context of defective DNA mismatch repair? Does DNA mismatch repair "caretaker" activity extend to the preservation of genomic fidelity in the recognition and/or repair of oxidized DNA?

Clearly, DNA in living organisms is under constant attack by reactive oxygen species, leading to repeated oxidative injury, including the production of oxidized DNA bases ***(111)***. Guanine oxidation, for example, resulting in the formation of 8-8-OHG, can cause mutations during DNA replication if present in the template strand, as it tends to lead to the misincorporation of A opposite the oxidized guanine, resulting in G:C-to-T:A transversions ***(111)***. The involvement of DNA mismatch repair proteins in the modulation of such induced mutations is now being recognized. Several studies have documented the responses of DNA mismatch repair-deficient cells to exposure to ionizing radiation ***(115–117)***, a potent inducer of reactive oxygen species ***(112)***. Each study has reported that cells with defective DNA mismatch repair appear resistant to the toxicity of oxidative damage inflicted by ionizing radiation. As cell death following DNA damage by a mutagen constitutes a major barrier to neoplastic transformation, cells that fail to die as a consequence of significant mutagenic DNA injury would be prone to neoplastic transformation. Most interestingly, when either *Msh2+/−* or *Msh2−/−* ES cells were exposed to a prolonged (72 h) oxidant stress, the cells responded quite abnormally, exhibiting an oxidation "tolerance" phenotype ***(116)***. Both *Msh2+/−* and *Msh2−/−* ES cells, as compared to *Msh2+/+* ES cells, exhibited increased clonogenic survival, decreased apoptosis, increased oxidative genome damage (measured as the accumulation of 8-OH-G, 8-OH-A, and thymine glycol), and increased mutations following protracted, low-level ionizing radiation exposure ***(116)***. These data implicated DNA mismatch repair enzymes as critical participants in cellular responses to oxidative stress, a concept that has been solidified by a recent report which showed that in *S. cerevesiae*, MSH2–MSH6 complexes bind with high specificity to 8-OH-G:A mispairs ***(118)***. *S. cerevesiae* carrying MSH2 or MSH6, mutations in conjunction with mutations in *OGG1*, a gene encoding an 8-OH-G repair enzyme, exhibited synergistic increases in G:C-to-T:A transversion muations, suggesting that DNA mismatch repair enzymes perform critical roles in oxidative genome-damage repair ***(118)***. In further support of this concep, Wang et al. have shown that MSH2, MSH6, and MLH1 are present in large multisubunit protein complexes (termed BASCs) along with other important DNA repair proteins (BRCA-1, ATM, BLM, RAD50, RFC1, RFC2, and RFC4). These complexes have been proposed to play an important role as a sensor of DNA damage and DNA structural abnormalities and, perhaps, in coordinating and conducting repair ***(119)***.

All of the accumulated data suggest that defective "caretaker" activity resulting from DNA mismatch enzyme repair gene mutations can result not only in replication errors, but also in a reduced capacity to respond to carcinogen-induced genomic injury. Furthermore, defects in even one of two DNA mismatch repair enzyme gene alleles appear to be accompanied by a genome damage "tolerance" phenotype, evident upon prolonged exposure to reactive oxygen species. Thus, although inactivation of both mismatch enzyme repair alleles has been thought to be required at some stage for the development of colorectal cancers in HNPCC families, normal colon cells in HNPCC family members who are heterozygous for mutant DNA mismatch repair genes may also have defective damage recognition or repair capacity, or inadequate "caretaker" activity, specific to certain types of genomic injury that may contribute to the earliest stages of colonic carcinogenesis. As cells with at least one wild-type mismatch repair enzyme gene allele appear to contain sufficient mismatch repair capacity to permit repair of DNA mismatches in in vitro assays, and to prevent high mutation rates in nonstressed conditions ***(120)***, the oxidation "tolerance" phenotype exhibited by stressed cells with only one wild-type mismatch repair enzyme gene allele suggests that other "caretaker" functions performed by DNA mismatch repair are likely to require two intact and functioning mismatch repair enzyme gene alleles.

5. New Strategies for Cancer Prevention and Treatment Targeting "Caretaker" Genes

To interrupt cancer pathogenesis in the setting of defective "caretaker" gene function, several rational cancer prevention approaches might be considered, including: (a) *attenuation* or *abrogation* of genome damaging stresses via avoidance of exogenous carcinogens and/or reduction of endogenous carcinogenic (particularly oxidant) stresses, (b) *restoration* of defective "caretaker" function, and (c) *compensation* for inadequate "caretaker" activity via treatment with inducers of carcinogen detoxification activity. Several strategies for *attenuation* or *abrogation* of genome-damaging stresses are already under clinical evaluation for cancer prevention. For example, antioxidant micronutrients, including vitamin E, selenium, and carotenoids, such as lycopene, might be expected to reduce oxidant damage to DNA if used for cancer prevention ***(121–124)***. Each of these agents has reached human clinical studies. For prostate cancer, accompanied by somatic defects in *GSTP1* "caretaker" activity, a large (n = 32,400) randomized trial (The Selenium and Vitamin E Chemoprevention Trial; SELECT) is planned in the United States to ascertain whether selenium and vitamin E, alone or in combination, can reduce the incidence of prostate cancer in healthy men age 55 and older (age 50 and older for African-American men). *Restoration* of "caretaker" gene function may be feasible for some "caretaker" genes, such as *GSTP1* and *MLH1*, that can be somatically "silenced" by CpG island DNA hypermethylation ***(39,125)***. CpG island hypermethylation at the loci of "caretaker" genes in cancer precursor lesion cells and in cancer cells is likely maintained via the action of DNMT1, along with other DNA methyltransferases, that could be targeted for inhibition. For example, DNA methyltransferase inhibitors such as 5-aza-cytidine, 5-aza-deoxycytidine, and procainamide can restore GSTP1 expression in LNCaP PCA cells in vivo (X. Lin, K. Asgari, and W. G. Nelson,

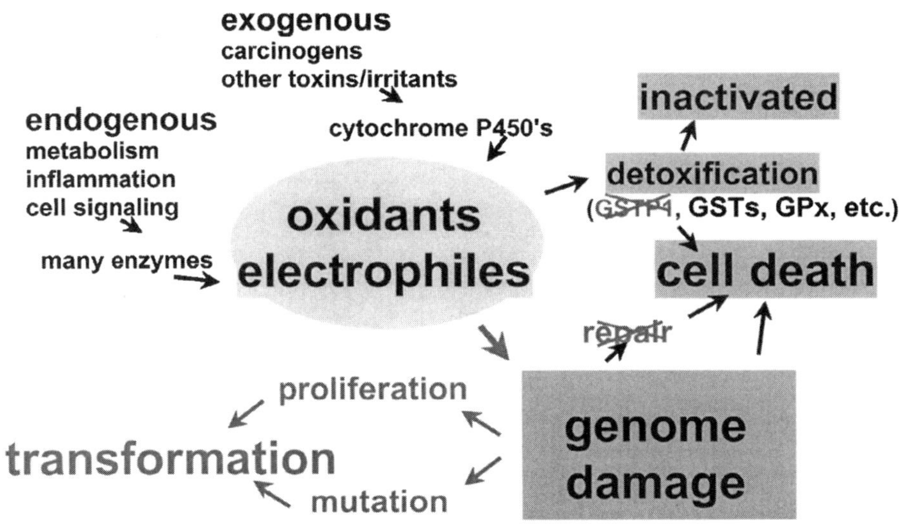

Fig. 3. Compensation for loss of *GSTP1* or *hMSH2* "caretaker" function via induction of carcinogen-detoxification enzyme gene expression.

unpublished data); these agents may be considered candidate cancer-preventation drugs. Chronic 5-aza-deoxycytidine treatment has been reported to attenuate intestinal carcinogenesis in mice with defective *Apc* alleles ***(126)***. In addition, data have been reported revealing interactions both between $^{5\text{-m}}$C-binding proteins and histone deacetylases (HDACs), and between DNA methyltransferases and HDACs, in effecting transcriptional repression ***(127–130)***. Perhaps, combinations of drugs active at inhibiting DNA methyltransferases and at inhibiting chromatin-remodeling enzymes might prove useful in restoring high-level "caretaker" gene expression to prevent life-threatening cancer development ***(130)***. The key issue for development of this approach, especially if it is to be used for cancer prevention, will be whether restoration of "caretaker" gene expression (as well as the expression of other "silenced" genes) can be accomplished with reasonable gene selectivity and with acceptable side effects.

Therapeutic *compensation* for inadequate "caretaker" gene function may hold the most promise for cancer prevention (**Fig. 3**). Glutathione *S*-transferases (GSTs), enzymes capable of catalyzing conjugation reactions between glutathione, a peptide chemical-scavenging molecule, and a vast array of reactive chemical species, have long been thought to protect against cancer development by detoxifying carcinogens ***(20)***. A key feature of the genes encoding GST-subunit polypeptides is that gene transcription can be induced, via an Nrf2-dependent mechanism, in response not only to injurious chemical stresses, but also to chemoprotective compounds ***(21,22)***. Augmentation of carcinogen-detoxification capacity, using a variety of such chemoprotective compounds, including isothiocyanates, 1,2-dithiole-3-thiones, terpenoids, etc., has been reported to prevent different cancers in different animal models by triggering the expression of carcinogen-detoxification enzymes ***(131)***. Most or all of these compounds likely act to prevent cancer by activating carcinogen-detoxification enzyme gene expression via the Nrf2-dependent transcription induction pathway, as oltipraz, an anti-schistosomal 1,2-dithiole-3-thione compound known to protect against benzo[a]pyrene gastric carcinogenesis in murine models, had no effect on gastric tumor formation in mice carrying disrupted *Nrf2* alleles ***(24)***. As mentioned previously, induction of GST "caretaker" activity in liver tissues, using oltipraz, a therapeutic inducer of GST activity, has been shown to reduce aflatoxin B_1 damage when administered to a human clinical study cohort at high risk for aflatoxin exposure and liver cancer development in China ***(66–68)***. Of perhaps even greater interest, many carcinogen-detoxification enzyme inducers have been detected in dietary components. For example, sulforaphane, an isothiocyanate that can trigger carcinogen-detoxification enzyme induction, is present in high amounts in cruciferous vegetables ***(132,133)***. Diets rich in carcinogen inducers such as sulforaphane have been associated with decreased cancer risks ***(134)***; clinical trials of carcinogen inducers are underway in efforts to prevent many different human cancers ***(135,136)***.

6. Conclusions

Both genetic and environmental factors contribute to the development of human cancers. "Caretaker" genes, encoding carcinogen-detoxification enzymes, DNA mismatch repair enzymes, DNA damage recognition and repair enzymes, and various polypeptides responsible for maintaining chromosome integrity and the fidelity of chromosome segregation throughout DNA replication and mitosis, may serve to protect against environmental carcinogenesis by preserving genome integrity in response to genome damaging stresses. Defects in two "caretaker" genes, *GSTP1* and *MSH2*, have been found to lead not only to increased genome damage, but also to decreased cellular toxicity (increased cell survival) in response to oxidative genome-damaging stresses. This genome-damage "tolerance" phenotype associated with defective "caretaker" gene function may lead to selection during cancer pathogenesis. Therapeutic strategies to *abrogate* genome damaging stresses, to *restore* "caretaker" gene function, or to *compensate* for inadequate "caretaker" gene function via induction of carcinogen-detoxification enzymes might be expected to attenuate tumorigenesis in the setting of defective "caretaker" genes.

References

1. Knudson, A. G., Jr. (1971) Mutation and cancer: statistical study of retinoblastoma. *Proc. Natl. Acad. Sci. USA* **68**, 820–823.
2. Knudson, A. G., Jr., Hethcote, H. W., and Brown, B. W. (1975) Mutation and childhood cancer: a probabilistic model for the incidence of retinoblastoma. *Proc. Natl. Acad. Sci. USA* **72**, 5116–5120.
3. Cavenee, W. K., Hansen, M. F., Nordenskjold, M., et al. (1985) Genetic origin of mutations predisposing to retinoblastoma. *Science* **228**, 501–503.
4. Friend, S. H., Bernards, R., Rogelj, S., et al. (1986) A human DNA segment with properties of the gene that predisposes to retinoblastoma and osteosarcoma. *Nature* **323**, 643–646.
5. Jacks, T., Fazeli, A., Schmitt, E. M., Bronson, R. T., Goodell, M. A., and Weinberg, R. A. (1992) Effects of an Rb mutation in the mouse. *Nature* **359**, 295–300.
6. Marino, S., Vooijs, M., van Der Gulden, H., Jonkers, J., and Berns, A. (2000) Induction of medulloblastomas in p53-null mutant mice by somatic inactivation of Rb in the external granular layer cells of the cerebellum. *Genes Dev.* **14**, 994–1004.
7. Doll, R. and Peto, R. (1981) The causes of cancer: quantitative estimates of avoidable risks of cancer in the United States today. *J. Natl. Cancer Inst.* **66**, 1191–1308.
8. Uematsu, K. and Huggins, C. (1968) Induction of leukemia and ovarian tumors in mice by pulse-doses of polycyclic aromatic hydrocarbons. *Mol. Pharmacol.* **4**, 427–434.

9. Deutsch-Wenzel, R. P., Brune, H., Grimmer, G., Dettbarn, G., and Misfeld, J. (1983) Experimental studies in rat lungs on the carcinogenicity and dose-response relationships of eight frequently occurring environmental polycyclic aromatic hydrocarbons. *J. Natl. Cancer Inst.* **71**, 539–544.
10. Lichtenstein, P., Holm, N. V., Verkasalo, P. K., et al. (2000) Environmental and heritable factors in the causation of cancer—analyses of cohorts of twins from Sweden, Denmark, and Finland. *N. Engl. J. Med.* **343**, 78–85.
11. Fearon, E. R. and Vogelstein, B. (1990) A genetic model for colorectal tumorigenesis. *Cell* **61**, 759–767.
12. Baylin, S. B. and Herman, J. G. (2000) DNA hypermethylation in tumorigenesis: epigenetics joins genetics. *Trends Genet.* **16**, 168–174.
13. Kinzler, K. W. and Vogelstein, B. (1997) Cancer-susceptibility genes. Gatekeepers and caretakers [news; comment] [see comments]. *Nature* **386**, 761, 763.
14. Sevick, E. M. and Jain, R. K. (1989) Geometric resistance to blood flow in solid tumors perfused ex vivo: effects of tumor size and perfusion pressure. *Cancer Res.* **49**, 3506–3512.
15. Jain, R. K. and Baxter, L. T. (1988) Mechanisms of heterogeneous distribution of monoclonal antibodies and other macromolecules in tumors: significance of elevated interstitial pressure. *Cancer Res.* **48**, 7022–7032.
16. Price, J. E., Naito, S., and Fidler, I. J. (1988) Growth in an organ microenvironment as a selective process in metastasis. *Clin. Exp. Metastasis* **6**, 91–102.
17. Radinsky, R. and Fidler, I. J. (1992) Regulation of tumor cell growth at organ-specific metastases. *In Vivo* **6**, 325–331.
18. Evan, G. and Littlewood, T. (1998) A matter of life and cell death. *Science* **281**, 1317–1322.
19. Adams, J. M. and Cory, S. (1998) The Bcl-2 protein family: arbiters of cell survival. *Science* **281**, 1322–1326.
20. Hayes, J. D. and Pulford, D. J. (1995) The glutathione S-transferase supergene family: regulation of GST and the contribution of the isoenzymes to cancer chemoprotection and drug resistance. *Crit. Rev. Biochem. Mol. Biol.* **30**, 445–600.
21. Rushmore, T. H., King, R. G., Paulson, K. E., and Pickett, C. B. (1990) Regulation of glutathione S-transferase Ya subunit gene expression: identification of a unique xenobiotic-responsive element controlling inducible expression by planar aromatic compounds. *Proc. Natl. Acad. Sci. USA* **87**, 3826–3830.
22. Rushmore, T. H. and Pickett, C. B. (1990) Transcriptional regulation of the rat glutathione S-transferase Ya subunit gene. Characterization of a xenobiotic-responsive element controlling inducible expression by phenolic antioxidants. *J. Biol. Chem.* **265**, 14648–14653.
23. Wasserman, W. W. and Fahl, W. E. (1997) Functional antioxidant responsive elements. *Proc. Natl. Acad. Sci. USA* **94**, 5361–5366.
24. Ramos-Gomez, M., Kwak, M. K., Dolan, P. M., et al. (2001) From the cover: sensitivity to carcinogenesis is increased and chemoprotective efficacy of enzyme inducers is lost in nrf2 transcription factor-deficient mice. *Proc. Natl. Acad. Sci. USA* **98**, 3410–3415.
25. Henderson, C. J., Smith, A. G., Ure, J., Brown, K., Bacon, E. J., and Wolf, C. R. (1998) Increased skin tumorigenesis in mice lacking pi class glutathione S-transferases. *Proc. Natl. Acad. Sci. USA* **95**, 5275–5280.
26. Rebbeck, T. R. (1997) Molecular epidemiology of the human glutathione S-transferase genotypes GSTM1 and GSTT1 in cancer susceptibility. *Cancer Epidemiol. Biomarkers Prev.* **6**, 733–743.
27. Seidegard, J., Pero, R. W., Markowitz, M. M., Roush, G., Miller, D. G., and Beattie, E. J. (1990) Isoenzyme(s) of glutathione transferase (class Mu) as a marker for the susceptibility to lung cancer: a follow up study. *Carcinogenesis* **11**, 33–36.
28. Kihara, M. and Noda, K. (1994) Lung cancer risk of GSTM1 null genotype is dependent on the extent of tobacco smoke exposure. *Carcinogenesis* **15**, 415–418.

29. McWilliams, J. E., Sanderson, B. J., Harris, E. L., Richert-Boe, K. E., and Henner, W. D. (1995) Glutathione S-transferase M1 (GSTM1) deficiency and lung cancer risk. *Cancer Epidemiol. Biomarkers Prev.* **4**, 589–594.
30. Bell, D. A., Taylor, J. A., Paulson, D. F., Robertson, C. N., Mohler, J. L., and Lucier, G. W. (1993) Genetic risk and carcinogen exposure: a common inherited defect of the carcinogen-metabolism gene glutathione S-transferase M1 (GSTM1) that increases susceptibility to bladder cancer. *J. Natl. Cancer Inst.* **85**, 1159–1164.
31. Kato, S., Bowman, E. D., Harrington, A. M., Blomeke, B., and Shields, P. G. (1995) Human lung carcinogen-DNA adduct levels mediated by genetic polymorphisms in vivo. *J. Natl. Cancer Inst.* **87**, 902–907.
32. Robertson, I. G., Jensson, H., Mannervik, B., and Jernstrom, B. (1986) Glutathione transferases in rat lung: the presence of transferase 7-7, highly efficient in the conjugation of glutathione with the carcinogenic (+)-7 beta, 8 alpha-dihydroxy-9 alpha, 10 alpha-oxy-7,8,9, 10-tetrahydrobenzo[a]pyrene. *Carcinogenesis* **7**, 295–299.
33. Landi, S. (2000) Mammalian class theta GST and differential susceptibility to carcinogens: a review. *Mutat. Res.* **463**, 247–283.
34. Strange, R. C. and Fryer, A. A. (1999) Chapter 19. The glutathione S-transferases: influence of polymorphism on cancer susceptibility. *IARC Sci. Publ.* **148**, 231–249.
35. Ali-Osman, F., Akande, O., Antoun, G., Mao, J. X., and Buolamwini, J. (1997) Molecular cloning, characterization, and expression in Escherichia coli of full-length cDNAs of three human glutathione S-transferase Pi gene variants. Evidence for differential catalytic activity of the encoded proteins. *J. Biol. Chem.* **272**, 10004–10012.
36. Helzlsouer, K. J., Selmin, O., Huang, H. Y., et al. (1998) Association between glutathione S-transferase M1, P1, and T1 genetic polymorphisms and development of breast cancer. *J. Natl. Cancer Inst.* **90**, 512–518.
37. Lavigne, J. A., Helzlsouer, K. J., Huang, H. Y., et al. (1997) An association between the allele coding for a low activity variant of catechol-O-methyltransferase and the risk for breast cancer. *Cancer Res.* **57**, 5493–5497.
38. Lee, W. H., Isaacs, W. B., Bova, G. S., and Nelson, W. G. (1997) CG island methylation changes near the GSTP1 gene in prostatic carcinoma cells detected using the polymerase chain reaction: a new prostate cancer biomarker. *Cancer Epidemiol. Biomarkers Prev.* **6**, 443–450.
39. Lee, W. H., Morton, R. A., Epstein, J. I., et al. (1994) Cytidine methylation of regulatory sequences near the pi-class glutathione S-transferase gene accompanies human prostatic carcinogenesis. *Proc. Natl. Acad. Sci. USA*, **91**, 1733–1737.
40. Millar, D. S., Ow, K. K., Paul, C. L., Russell, P. J., Molloy, P. L., and Clark, S. J. (1999) Detailed methylation analysis of the glutathione S-transferase pi (GSTP1) gene in prostate cancer. *Oncogene* **18**, 1313–1324.
41. Goessl, C., Krause, H., Muller, M., et al. (2000) Fluorescent methylation-specific polymerase chain reaction for DNA-based detection of prostate cancer in bodily fluids [In Process Citation]. *Cancer Res.* **60**, 5941–5945.
42. Santourlidis, S., Florl, A., Ackermann, R., Wirtz, H. C., and Schulz W. A. (1999) High frequency of alterations in DNA methylation in adenocarcinoma of the prostate. *Prostate* **39**, 166–174.
43. Brooks, J. D., Weinstein, M., Lin, X., et al. (1998) CG island methylation changes near the GSTP1 gene in prostatic intraepithelial neoplasia. *Cancer Epidemiol. Biomarkers Prev.* **7**, 531–536.
44. Tchou, J. C., Lin, X., Freije, D., et al. (2000) GSTP1 CpG island DNA hypermethylation in hepatocellular carcinomas. *Int. J. Oncol.* **16**, 663–676.
45. Esteller, M., Corn, P. G., Urena, J. M., Gabrielson, E., Baylin, S. B., and Herman, J. G. (1998) Inactivation of glutathione S-transferase P1 gene by promoter hypermethylation in human neoplasia. *Cancer Res.* **58**, 4515–4518.

46. De Marzo, A. M., Marchi, V. L., Epstein, J. I., and Nelson, W. G. (1999) Proliferative inflammatory atrophy of the prostate: implications for prostatic carcinogenesis. *Am. J. Pathol.* **155**, 1985–1992.
47. Putzi, M. J. and De Marzo, A. M. (2000) Morphologic transitions between proliferative inflammatory atrophy and high-grade prostatic intraepithelial neoplasia. *Urology* **56**, 828–832.
48. Lin, X., Tascilar M., Lee W.-H., et al. (in press) Evidence that GSTP1 CpG island hypermethylation may be selected during human prostatic carcinogenesis.
49. Nelson, C. P., Kidd, L. C., Sauvageot, J., et al. (2001) Protection against 2-hydroxyamino-1-methyl-6-phenylimidazo[4,5-b]pyridine cytotoxicity and DNA adduct formation in human prostate by glutathione S-transferase P1. *Cancer Res.* **61**, 103–109.
50. Nagaoka, H., Wakabayashi, K., Kim, S. B., et al. (1992) Adduct formation at C-8 of guanine on in vitro reaction of the ultimate form of 2-amino-1-methyl-6-phenylimidazo[4,5-b]pyridine with 2′-deoxyguanosine and its phosphate esters. *Jpn. J. Cancer Res.* **83**, 1025–1029.
51. Morgenthaler, P. M. and Holzhauser, D. (1995) Analysis of mutations induced by 2-amino-1-methyl-6-phenylimidazo[4,5-b]pyridine (PhIP) in human lymphoblastoid cells. *Carcinogenesis* **16**, 713–718.
52. Knize, M. G., Sinha, R., Rothman, N., et al. (1995) Heterocyclic amine content in fast-food meat products. *Food Chem. Toxicol.* **33**, 545–551.
53. Gross, G. A., Turesky, R. J., Fay, L. B., Stillwell, W. G., Skipper, P. L., and Tannenbaum, S. R. (1993) Heterocyclic aromatic amine formation in grilled bacon, beef and fish and in grill scrapings. *Carcinogenesis* **14**, 2313–2318.
54. Davis, C. D., Schut, H. A., and Snyderwine, E. G. (1994) Adduction of the heterocyclic amine food mutagens IQ and PhIP to mitochondrial and nuclear DNA in the liver of Fischer-344 rats. *Carcinogenesis* **15**, 641–645.
55. Shirai, T., Cui, L., Takahashi, S., et al. (1999) Carcinogenicity of 2-amino-1-methyl-6-phenylimidazo [4,5-b]pyridine (PhIP) in the rat prostate and induction of invasive carcinomas by subsequent treatment with testosterone propionate. *Cancer Lett.* **143**, 217–221.
56. Shirai, T., Sano, M., Tamano S., et al. (1997) The prostate: a target for carcinogenicity of 2-amino-1-methyl-6-phenylimidazo[4,5-b]pyridine (PhIP) derived from cooked foods. *Cancer Res.* **57**, 195–198.
57. Stuart, G. R., Holcroft, J., de Boer, J. G., and Glickman, B. W. (2000) Prostate mutations in rats induced by the suspected human carcinogen 2-amino-1-methyl-6-phenylimidazo[4,5-b]pyridine. *Cancer Res.* **60**, 266–268.
58. Farber, E. and Cameron, R. (1980) The sequential analysis of cancer development. *Adv. Cancer Res.* **31**, 125–226.
59. Farber, E. and Sarma, D. S. (1987) Hepatocarcinogenesis: a dynamic cellular perspective. *Lab. Invest.* **56**, 4–22.
60. Sato, K., Kitahara, A., Satoh, K., Ishikawa, T., Tatematsu, M., and Ito, N. (1984) The placental form of glutathione S-transferase as a new marker protein for preneoplasia in rat chemical hepatocarcinogenesis. *Gann* **75**, 199–202.
61. Satoh, K., Kitahara, A., Soma, Y., Inaba, Y., Hatayama, I., and Sato, K. (1985) Purification, induction, and distribution of placental glutathione transferase: a new marker enzyme for preneoplastic cells in the rat chemical hepatocarcinogenesis. *Proc. Natl. Acad. Sci. USA* **82**, 3964–3968.
62. Bannasch, P. (1986) Preneoplastic lesions as end points in carcinogenicity testing. I. Hepatic preneoplasia. *Carcinogenesis* **7**, 689–695.
63. Roomi, M. W., Ho, R. K., Sarma, D. S., and Farber, E. (1985) A common biochemical pattern in preneoplastic hepatocyte nodules generated in four different models in the rat. *Cancer Res.* **45**, 564–571.

64. Hayes, M. A., Smith, I. R., Rushmore, T. H., et al. (1990) Pathogenesis of skin and liver neoplasms in white suckers from industrially polluted areas in Lake Ontario. *Sci. Total Environ.* **94**, 105–123.
65. Zhou, T., Evans, A. A., London, W. T., et al. (1997) Glutathione S-transferase expression in hepatitis B virus-associated human hepatocellular carcinogenesis. *Cancer Res.* **57**, 2749–2753.
66. Wang, J. S., Shen, X., He, X., et al. (1999) Protective alterations in phase 1 and 2 metabolism of aflatoxin B1 by oltipraz in residents of Qidong, People's Republic of China. *J. Natl. Cancer Inst.* **91**, 347–354.
67. Jacobson, L. P., Zhang, B. C., Zhu, Y. R., et al. (1997) Oltipraz chemoprevention trial in Qidong, People's Republic of China: study design and clinical outcomes. *Cancer Epidemiol. Biomarkers Prev.* **6**, 257–265.
68. Kensler, T. W., He, X., Otieno, M., Egner, P. A., et al. (1998) Oltipraz chemoprevention trial in Qidong, People's Republic of China: modulation of serum aflatoxin albumin adduct biomarkers. *Cancer Epidemiol. Biomarkers Prev.* **7**, 127–134.
69. Henderson, C. J., Wolf, C. R., Kitteringham, N., Powell, H., Otto, D., and Park, B. K. (2000) Increased resistance to acetaminophen hepatotoxicity in mice lacking glutathione S-transferase Pi. *Proc. Natl. Acad. Sci. USA* **97**, 12741–12745.
70. DeWeese, T. L., Hanranhan, C., Veeraswamy, R., and Nelson, W. G. (in press) Oxidation tolerance associated with loss of GSTP1 function in human prostate cancer cells.
71. Diah, S. K., Smitherman, P. K., Townsend, A. J., and Morrow, C. S. (1999) Detoxification of 1-chloro-2,4-dinitrobenzene in MCF7 breast cancer cells expressing glutathione S-transferase P1-1 and/or multidrug resistance protein 1. *Toxicol. Appl. Pharmacol.* **157**, 85–93.
72. Wang, T., Arifoglu, P., Ronai, Z., and Tew, K. D. (2001) Glutathione S-transferase P1-1 (GSTP1-1) inhibits c-Jun N-terminal kinase (JNK1) signaling through interaction with the C terminus. *J. Biol. Chem.* **276**, 20999–21003.
73. Yin, Z., Ivanov, V. N., Habelhah, H., Tew, K., and Ronai, Z. (2000) Glutathione S-transferase p elicits protection against H2O2-induced cell death via coordinated regulation of stress kinases. *Cancer Res.* **60**, 4053–4057.
74. Adler, V., Yin, Z., Fuchs, S. Y., et al. (1999) Regulation of JNK signaling by GSTp. *EMBO J.* **18**, 1321–1334.
75. Wildenberg, J. and Meselson, M. (1975) Mismatch repair in heteroduplex DNA. *Proc. Natl. Acad. Sci. USA* **72**, 2202–2206.
76. Rydberg, B. (1978) Bromouracil mutagenesis and mismatch repair in mutator strains of *Escherichia coli*. *Mutat. Res.* **52**, 11–24.
77. Palombo, F., Iaccarino, I., Nakajima, E., Ikejima, M., Shimada, T., and Jiricny, J. (1996) hMutSbeta, a heterodimer of hMSH2 and hMSH3, binds to insertion/deletion loops in DNA. *Curr. Biol.* **6**, 1181–1184.
78. Kunkel, T.A. (1995) DNA-mismatch repair. The intricacies of eukaryotic spell-checking. *Curr. Biol.* **5**, 1091–1094.
79. Modrich, P. (1991) Mechanisms and biological effects of mismatch repair. *Annu. Rev. Genet.* **25**, 229–253.
80. Fishel, R., Lescoe, M. K., Rao, M. R., et al. (1993) The human mutator gene homolog MSH2 and its association with hereditary nonpolyposis colon cancer. *Cell* **75**, 1027–1038.
81. Horii, A., Han, H. J., Sasaki, S., Shimada, M., and Nakamura, Y. (1994) Cloning, characterization and chromosomal assignment of the human genes homologous to yeast PMS1, a member of mismatch repair genes. *Biochem. Biophys. Res. Commun.* **204**, 1257–1264.
82. Hughes, M. J. and Jiricny, J. (1992) The purification of a human mismatch-binding protein and identification of its associated ATPase and helicase activities. *J. Biol. Chem.* **267**, 23876–23882.

83. Leach, F. S., Nicolaides, N. C., Papadopoulos, N., et al. (1993) Mutations of a mutS homolog in hereditary nonpolyposis colorectal cancer. *Cell* **75**, 1215–1225.
84. Nicolaides, N. C., Papadopoulos, N., Liu, B., et al. (1994) Mutations of two PMS homologues in hereditary nonpolyposis colon cancer. *Nature* **371**, 75–80.
85. Papadopoulos, N., Nicolaides, N. C., Wei, Y. F., et al. (1994) Mutation of a mutL homolog in hereditary colon cancer. *Science* **263**, 1625–1629.
86. Fujii, H. and Shimada, T. (1989) Isolation and characterization of cDNA clones derived from the divergently transcribed gene in the region upstream from the human dihydrofolate reductase gene. *J. Biol. Chem.* **264**, 10057–10064.
87. Acharya, S., Wilson, T., Gradia, S., et al. (1996) hMSH2 forms specific mispair-binding complexes with hMSH3 and hMSH6. *Proc. Natl. Acad. Sci. USA* **93**, 13629–13634.
88. Drummond, J. T., Li, G. M., Longley, M. J., and Modrich, P. (1995) Isolation of an hMSH2-p160 heterodimer that restores DNA mismatch repair to tumor cells. *Science* **268**, 1909–1912.
89. Marsischky, G. T., Filosi, N., Kane, M. F., and Kolodner, R. (1996) Redundancy of Saccharomyces cerevisiae MSH3 and MSH6 in MSH2-dependent mismatch repair. *Genes Dev.* **10**, 407–420.
90. Gradia, S., Acharya, S., and Fishel, R. (1997) The human mismatch recognition complex hMSH2-hMSH6 functions as a novel molecular switch. *Cell* **91**, 995–1005.
91. Fishel, R., Ewel, A., Lee, S., Lescoe, M. K., and Griffith, J. (1994) Binding of mismatched microsatellite DNA sequences by the human MSH2 protein. *Science* **266**, 1403–1405.
92. Aaltonen, L. A., Peltomaki, P., Leach, F. S., et al. (1993) Clues to the pathogenesis of familial colorectal cancer. *Science* **260**, 812–816.
93. Green, R. C., Narod, S. A., Morasse, J., et al. (1994) Hereditary nonpolyposis colon cancer: analysis of linkage to 2p15-16 places the COCA1 locus telomeric to D2S123 and reveals genetic heterogeneity in seven Canadian families. *Am. J. Hum. Genet.* **54**, 1067–1077.
94. Lynch, H. T., Smyrk, T., and Lynch, J. F. (1996) Overview of natural history, pathology, molecular genetics and management of HNPCC (Lynch syndrome). *Int. J. Cancer* **69**, 38–43.
95. Kinzler, K. W. and Vogelstein, B. (1996) Lessons from hereditary colorectal cancer. *Cell* **87**, 159–70.
96. Scott, R. J., McPhillips, M., Meldrum, C. J., et al. (2001) Hereditary nonpolyposis colorectal cancer in 95 families: differences and similarities between mutation-positive and mutation-negative kindreds. *Am. J. Hum. Genet.* **68**, 118–127.
97. de Wind, N., Dekker, M., van Rossum, A., van der Valk, M., and te Riele, H. (1998) Mouse models for hereditary nonpolyposis colorectal cancer. *Cancer Res.* **58**, 248–255.
98. de Wind, N., Dekker, M., Berns, A., Radman, M., and te Riele, H. (1995) Inactivation of the mouse Msh2 gene results in mismatch repair deficiency, methylation tolerance, hyperrecombination, and predisposition to cancer. *Cell* **82**, 321–330.
99. Reitmair, A. H., Schmits, R., Ewel, A., et al. (1995) MSH2 deficient mice are viable and susceptible to lymphoid tumours. *Nat. Genet.* **11**, 64–70.
100. Lowsky, R., DeCoteau, J. F., Reitmair, A. H., et al. (1997) Defects of the mismatch repair gene MSH2 are implicated in the development of murine and human lymphoblastic lymphomas and are associated with the aberrant expression of rhombotin-2 (Lmo-2) and Tal-1 (SCL). *Blood* **89**, 2276–2282.
101. Sansom, O. J., Toft, N. J., Winton, D. J., and Clarke, A. R. (2001) Msh-2 suppresses in vivo mutation in a gene dose and lesion dependent manner. *Oncogene* **20**, 3580–3584.
102. Branch, P., Aquilina, G., Bignami, M., and Karran, P. (1993) Defective mismatch binding and a mutator phenotype in cells tolerant to DNA damage. *Nature* **362**, 652–654.
103. Kat, A., Thilly, W. G., Fang, W. H., Longley, M. J., Li, G. M., and Modrich, P. (1993) An alkylation-tolerant, mutator human cell line is deficient in strand-specific mismatch repair. *Proc. Natl. Acad. Sci. USA* **90**, 6424–6428.

104. Koi, M., Umar, A., Chauhan, D. P., et al. (1994) Human chromosome 3 corrects mismatch repair deficiency and microsatellite instability and reduces N-methyl-N′-nitro-N-nitrosoguanidine tolerance in colon tumor cells with homozygous hMLH1 mutation. *Cancer Res.* **54**, 4308–4312.
105. Reitmair, A. H., Risley, R., Bristow, R. G., et al. (1997) Mutator phenotype in Msh2-deficient murine embryonic fibroblasts. *Cancer Res.* **57**, 3765–3771.
106. Mellon, I., Rajpal, D. K., Koi, M., Boland, C. R., and Champe, G. N. (1996) Transcription-coupled repair deficiency and mutations in human mismatch repair genes. *Science* **272**, 557–560.
107. Duckett, D. R., Drummond, J. T., Murchie, A. I., et al. (1996) Human MutSalpha recognizes damaged DNA base pairs containing O6-methylguanine, O4-methylthymine, or the cisplatin-d(GpG) adduct. *Proc. Natl. Acad. Sci. USA* **93**, 6443–6447.
108. Ripple, M. O., Henry, W. F., Rago, R. P., and Wilding, G. (1997) Prooxidant-antioxidant shift induced by androgen treatment of human prostate carcinoma cells. *J. Natl. Cancer Inst.* **89**, 40–48.
109. Prescott, S. M. and White, R. L. (1996) Self-promotion? Intimate connections between APC and prostaglandin H synthase-2. *Cell* **87**, 783–786.
110. Gorsky, L. D., Koop, D. R., and Coon, M. J. (1984) On the stoichiometry of the oxidase and monooxygenase reactions catalyzed by liver microsomal cytochrome P-450. Products of oxygen reduction. *J. Biol. Chem.* **259**, 6812–6817.
111. Ames, B.N., Shigenaga, M. K., and Hagen, T. M. (1993) Oxidants, antioxidants, and the degenerative diseases of aging. *Proc. Natl. Acad. Sci. USA* **90**, 7915–7922.
112. Wood, M. L., Dizdaroglu, M., Gajewski, E., and Essigmann, J. M. (1990) Mechanistic studies of ionizing radiation and oxidative mutagenesis: genetic effects of a single 8-hydroxyguanine (7-hydro-8-oxoguanine) residue inserted at a unique site in a viral genome. *Biochemistry* **29**, 7024–7032.
113. Shibutani, S., Takeshita, M., and Grollman, A. P. (1991) Insertion of specific bases during DNA synthesis past the oxidation-damaged base 8-oxodG. *Nature* **349**, 431–434.
114. Breimer, L. H. (1990) Molecular mechanisms of oxygen radical carcinogenesis and mutagenesis: the role of DNA base damage. *Mol. Carcinog.* **3**, 188–197.
115. Fritzell, J. A., Narayanan, L., Baker, S. M., Bronner, C. E., et al. (1997) Role of DNA mismatch repair in the cytotoxicity of ionizing radiation. *Cancer Res.* **57**, 5143–5147.
116. DeWeese, T. L., Shipman, J. M., Larrier, N. A., et al. (1998) Mouse embryonic stem cells carrying one or two defective Msh2 alleles respond abnormally to oxidative stress inflicted by low-level radiation. *Proc. Natl. Acad. Sci. USA* **95**, 11915–11920.
117. Xu, X. S., Narayanan, L., Dunklee, B., Liskay, R. M., and Glazer, P. M. (2001) Hypermutability to ionizing radiation in mismatch repair-deficient, Pms2 knockout mice. *Cancer Res.* **61**, 3775–3780.
118. Ni, T. T., Marsischky, G. T., and Kolodner, R. D. (1999) MSH2 and MSH6 are required for removal of adenine misincorporated opposite 8-oxo-guanine in S. cerevisiae. *Mol. Cell* **4**, 439–444.
119. Wang, Y., Cortez, D., Yazdi, P., Neff, N., Elledge, S. J., and Qin, J. (2000) BASC, a super complex of BRCA1-associated proteins involved in the recognition and repair of aberrant DNA structures. *Genes Dev.* **14**, 927–939.
120. Parsons, R., Li, G. M., Longley, M. J., et al. (1993) Hypermutability and mismatch repair deficiency in RER+ tumor cells. *Cell* **75**, 1227–1236.
121. Gann, P. H., Ma, J., Giovannucci, E., et al. (1999) Lower prostate cancer risk in men with elevated plasma lycopene levels: results of a prospective analysis. *Cancer Res.* **59**, 1225–1230.
122. Giovannucci, E., Ascherio, A., Rimm, E. B., Stampfer, M. J., Colditz, G. A., and Willett, W. C. (1995) Intake of carotenoids and retinol in relation to risk of prostate cancer. *J. Natl. Cancer Inst.* **87**, 1767–1776.

123. Yoshizawa, K., Willett, W. C., Morris, S. J., et al. (1998) Study of prediagnostic selenium level in toenails and the risk of advanced prostate cancer [see comments]. *J. Natl. Cancer Inst.* **90**, 1219–1224.
124. Helzlsouer, K. J., Huang, H. Y., Alberg, A. J., et al. (2000) Association between alpha-tocopherol, gamma-tocopherol, selenium, and subsequent prostate cancer. *J. Natl. Cancer Inst.* **92**, 2018–2023.
125. Herman, J. G., Umar, A., Polyak, K., et al. (1998) Incidence and functional consequences of hMLH1 promoter hypermethylation in colorectal carcinoma. *Proc. Natl. Acad. Sci. USA* **95**, 6870–6875.
126. Laird, P. W., Jackson-Grusby, L., Fazeli, A., et al. (1995) Suppression of intestinal neoplasia by DNA hypomethylation. *Cell* **81**, 197–205.
127. Nan, X., Ng, H. H., Johnson, C. A., et al. (1998) Transcriptional repression by the methyl-CpG-binding protein MeCP2 involves a histone deacetylase complex. *Nature* **393**, 386–389.
128. Jones, P. L., Veenstra, G. J., Wade, P. A., et al. (1998) Methylated DNA and MeCP2 recruit histone deacetylase to repress transcription. *Nat. Genet.* **19**, 187–191.
129. Rountree, M. R., Bachman, K. E., and Baylin, S. B. (2000) DNMT1 binds HDAC2 and a new co-repressor, DMAP1, to form a complex at replication foci. *Nat. Genet.* **25**, 269–277.
130. Cameron, E. E., Bachman, K. E., Myohanen, S., Herman, J. G., and Baylin, S. B. (1999) Synergy of demethylation and histone deacetylase inhibition in the re-expression of genes silenced in cancer. *Nat. Genet.* **21**, 103–107.
131. Kensler, T. W. (1997) Chemoprevention by inducers of carcinogen detoxication enzymes. *Environ. Health Perspect.* **105 Suppl 4**, 965–970.
132. Zhang, Y., Talalay, P., Cho, C. G., and Posner, G. H. (1992) A major inducer of anticarcinogenic protective enzymes from broccoli: isolation and elucidation of structure. *Proc. Natl. Acad. Sci. USA* **89**, 2399–2403.
133. Zhang, Y., Kensler, T. W., Cho, C. G., Posner, G. H., and Talalay, P. (1994) Anticarcinogenic activities of sulforaphane and structurally related synthetic norbornyl isothiocyanates. *Proc. Natl. Acad. Sci. USA*, **91**, 3147–3150.
134. Spitz, M. R., Duphorne, C. M., Detry, M. A., et al. (2000) Dietary intake of isothiocyanates: evidence of a joint effect with glutathione S-transferase polymorphisms in lung cancer risk. *Cancer Epidemiol. Biomarkers Prev.* **9**, 1017–1020.
135. Shapiro, T. A., Fahey, J. W., Wade, K. L., Stephenson, K. K., and Talalay, P. (1998) Human metabolism and excretion of cancer chemoprotective glucosinolates and isothiocyanates of cruciferous vegetables. *Cancer Epidemiol. Biomarkers Prev.* **7**, 1091–1100.
136. Shapiro, T. A., Fahey, J. W., Wade, K. L., Stephenson, K. K., and Talalay, P. (2001) Chemoprotective glucosinolates and isothiocyanates of broccoli sprouts: metabolism and excretion in humans. *Cancer Epidemiol. Biomarkers Prev.* **10**, 501–508.

17

The Regulation of Tumor Suppressor Genes by Oncogenes

Paul Dent, Liang Qiao, Donna Gilfor, Michael Birrer, Steven Grant, and Paul B. Fisher

1. Introduction

Tumor suppressor genes (TSGs) and oncogenes represent the ying and the yang of cell growth, differentiation, and survival control. TSGs such as p53, the retinoblastoma (RB) gene product, and the cyclin kinase inhibitor (CKI) proteins p21 Cip-1/WAF1/mda6 (p21), p27 Kip-1 (p27), p16 INK4a (p16), and p19 ARF (p19), play the role of negative regulators of the cell cycle ***(1–5)***. In contrast, mutation of protooncogenes such as the epidermal growth factor receptor (EGFR) ***(6)***, ErbB2 (Neu) ***(7)***, Ras ***(8,9)***, Raf-1 ***(10)***, PTEN ***(11,12)***, MKP-1 ***(13)***, and c-Myc ***(14)*** can promote cell cycle progression, in part by abrogating the negative regulation of the cell cycle by tumor suppressor genes.

1.1. Signal Transduction Pathways Downstream of Protoncogenes

In a nontransformed cell, growth factors transmit signals to downstream pathways in a defined "on/off" manner via their receptors, leading to a tightly regulated proliferative response. Growth factor-induced activation of the "classical" MAPK (also called extracellular regulated kinase, ERK) pathway ***(15,16)***, the phosphatidyl inositol 3-kinase (PI3K) pathway ***(17,18)*** and the c-Jun NH_2-terminal kinase (JNK) pathway ***(19,20)*** have been proposed to play central roles in the proliferative responses of cells. In a transformed cell, however, signaling pathways are frequently found to be in a permanently active state. This is of note because, as mentioned above, EGFR, ErbB2, Ras, Raf-1, and PTEN all represent plasma membrane-associated or -proximal signaling molecules that are at the head of many signaling pathways, and whose functions can become modified during the process of transformation. Growth factor signaling from the EGFR, ErbB2, and Ras can promote the activation of multiple downstream signal transduction pathways, including the "classical" MAPK/ERK pathway, the PI3K pathway, and the JNK pathway. Truncation of the EGFR or ErbB2 results in enhanced basal activity of these receptors, leading to subsequent activation of the downstream pathways ***(21,22)***. Similarly, loss of GTPase function in Ha- or K-Ras proteins can promote activation of the MAPK/ERK, PI3K, and

From: *Methods in Molecular Biology, Vol. 222: Tumor Suppressor Genes: Pathways and Isolation Strategies*
Edited by: Wafik S. El-Deiry © Humana Press Inc., Totowa, NJ

JNK pathways ***(23,24)***. Downstream of the plasma membrane, truncation of Raf-1 can enhance MAPK/ERK pathway activity ***(25)***. Loss of PTEN expression and/or function will result in enhanced activity within the PI3K pathway, which may potentially also either enhance or suppress MAPK/ERK activity ***(26–29)***. More recently, loss of MKP-1 expression has been shown to correlate with advanced ovarian cancer; loss of MKP-1 expression will tend to enhance the activities of the MAPK/ERK and JNK pathways ***(30)***. Thus a variety of potential mechanisms exist within a transformed cell whereby mutant proteins can lead to a constitutive increase in the activity of multiple signaling pathways. These pathways act in a concerted fashion to manipulate cell growth and differentiation responses, including the control of TSG function.

1.2. Signal Transduction and CKI Expression in Nontransformed Cells

In a "normal" nontransformed cell, cell cycle progression through G1 phase requires growth factor-induced mitogenic signals transmitted down multiple signal transduction pathways, such as MAPK/ERK, PI3K, and JNK. These pathways control the expression of many proteins, including CKI and cyclin molecules, ultimately regulating the function of cyclin-dependent kinase (cdk) enzymes. The amount of CKI expression will ultimately determine whether a cell enters S phase or remains arrested in G1.

Two families of CKIs have been described in mammalian cells; the p21/27/57 family and the p15/16/18/19 INK4 family (e.g., **refs.** ***31,32***). P21 protein is a potent inhibitor of multiple cdks, which are required for both G1/S phase entry (e.g., cdk2 and cdk4) and also at the G2/M transition (e.g., cdk1, also called cdc2) ***(33)***. P21 binds directly to cdk catalytic subunits in complexes containing cyclins A, B, D, and E, and its overexpression has been shown to decrease cdk activity and to inhibit DNA synthesis ***(34–37)***. The INK4 family, in contrast to the p21 family, is more selective in the cdk enzymes to which it binds, binding only to cdk4 or cdk6, but not to cyclins or any of the other known cyclin-dependent kinases ***(38–41)***.

The regulation of CKI molecule expression is very complex; growth factor signaling from Ras through the MAPK/ERK pathway has been shown to increase expression of p16 and p21 (e.g., **ref.** ***42***), and via PI3K signaling to variably reduce expression of p27 (e.g., **refs.** ***43,44***). MAPK/ERK signaling can also increase expression of the alternate splice form of p16, termed p19 ARF ***(45)***. P19 can inhibit the negative regulator of p53, Mdm2, from binding to p53, thereby stabilizing p53 protein levels ***(46)***. Enhanced p53 protein expression, by this mechanism, will in turn have the potential to further enhance p21 levels. Both p19 and Mdm2 expression can be regulated by MAPK/ERK signaling ***(47)***. Hence, Mdm2 protein induced by MAPK/ERK signaling has the potential to reduce p53 levels in the absence of p19. However, in a variety of primary cells, MAPK/ERK signaling can increase protein levels of both Mdm2 and p19, which will thus be presumed to act in a dynamic balance to achieve a steady-state level of p53. Thus, of note a loss of p19 expression during transformation will tend to reduce p53, and by implication p21, protein expression.

In general, growth factor-induced signaling is believed to decrease expression of p27 and enhance expression of p21 and D- and E-type cyclins, leading to the activation of the G1 cdks, cyclin E:cdk2, and cyclin D:cdk4 ***(48,49)***. Several studies have shown that both MAPK/ERK and PI3K signaling can play roles in enhancing cyclin D1 and cyclin E expression, demonstrating that cdk activation is dependent on prior activations of signal transduction pathways ***(50,51)***. As noted previously, growth factor signaling via MAPK/ERK is proposed to enhance p21 levels and signaling via PI3K is proposed to reduce p27 expression ***(52)***. Thus, under normal conditions, in which proliferation occurs, p21 can often be found in complex with cyclin:cdk complexes, and only when more p21 (or p27) is added to these complexes does inhibition of cdk activity occur ***(53,54)***. These data suggest that p21 may exist bound to G1 cyclin:cdk complexes in monomeric and multimeric forms, and that only multimeric forms of p21 may inhibit cdk activity. The paradoxical effect of growth-promoting signals increasing CKI expression levels, such as p21, is due to the finding that low levels of CKI expression also appear to play an essential role in the formation and activation of the cyclin E:cdk2 and cyclin D:cdk4 complexes ***(55)***. Of note, it was recently demonstrated that in contrast to G1-phase cyclin:cdk complexes, a single molecule of p21 will inhibit the catalytic activity of a single molecule of the S-phase cyclin A:cdk2 complex, suggesting that it is the cyclin molecule that determines the stoichiometric impact of the p21 inhibitory effect ***(56)***.

Active cyclin D:cdk4, followed by active cyclin E:cdk2, phosphorylate and inactivate RB (and RB family members p107 and p130), leading to release of E2F transcription factors (e.g., **refs.** ***57,58***). Release of E2F factors permits cell cycle progression through the G1 restriction point with commitment to S phase. Thus it can be seen that the CKI molecules p16, p19, p21, and p27 act in concert to control cell cycle progression through G1 phase. Simplistically, this could be thought of as p21/p27 inhibiting cdk2 and cdk4 complexes, with p16 and p19 "topping up" the inhibitory effect to ensure appropriate growth control as required. Reality is more complex. For example, in addition to the impact of p19 on p53 and p21, overexpression of p16 can also blunt proliferative responses via two mechanisms. In the first, binding of p16 to cyclin D:cdk4: complexes blocks the ability of cdk4 to phosphorylate RB, thus inhibiting the release of E2F transcription factors essential for G1 progression ***(59,60)***. In cells that are RB null, enforced p16 expression cannot induce G1 arrest because E2F family factors are unbound and functionally active in the absence of RB expression ***(61)***. In cell systems in which basal levels of cyclin E are elevated, however, p16 overexpression has less ability to inhibit cell cycle progression ***(62)***. In the second, increased expression of p16 can cause a redistribution of p21 and p27 from cyclin D:cdk4 complexes to cyclin E:cdk2 complexes ***(63)***. Hence, in the absence of any alteration in p21 or p27 expression, increased p16 levels can increase association of p21 and p27 to cyclin E:cdk2 complexes to an extent that p21 and p27 inhibit cyclin E:cdk2 function. Thus enhanced basal expression of cyclin E has the potential to titrate the inhibitory effects of the redistributed p21 and p27.

Recent studies have also argued that in addition to cdk-mediated phosphorylation of RB in late G1 phase, an essential earlier growth factor-stimulated priming phosphorylation of RB is required for subsequent RB inactivation. This early RB phosphorylation is mediated via Raf-1, and is apparently independent of downstream MAPK/ERK signaling ***(64)***. Thus multiple sites of phosphorylation are required for complete inactivation of the TSG RB and E2F release. Of note, however, it can be envisaged that maintained p27 expression and/or too much p21 and p16 expression during this process will potentially lead to inhibition of cdk enzymes, reduced RB phosphorylation, and cause cell cycle arrest in G1 phase. Similarly, low expression of cyclin D or cyclin E can dampen cdk activity and enhance the inhibitory effects of CKI molecule expression, leading to RB dephosphorylation and a halt of cell cycle progression. Thus a normal cell has to carefully balance the relative expression levels of CKIs vs cyclin molecules during G1 progression to achieve cdk activation. It is probable that too much mitogenic stimulation in a nontransformed cell may lead to an overexpression of CKI molecules and promote growth arrest.

1.3. Signal Transduction and CKI Expression in Transformed Cells

In a transformed cell, multiple alterations may have occurred to cell signaling processes, and many of the "checks and balances" controlling proliferative responses are nonfunctional. Transformed cells frequently have acquired some form of autocrine loop, with increased expression of growth factors and the plasma membrane receptors to which they bind, e.g., transforming growth factor α (TGFα) and EGFR or ErbB2. Such autocrine loops will lead ultimately to enhanced downstream signaling pathway activity. As noted in sections under **Subheadings 1.1.** and **1.2.**, transformed cells also may have multiple mutational activation(s) of downstream signaling pathways via protooncogenes, e.g., Ras mutational activation, PTEN mutational inactivaμtion. Thus, as a generalization, transformed cells tend to have higher amounts of signal being propagated down their signaling pathways per se than nontransformed cells. At face value, this finding seems to suggest that a single mutation of—for example, Ras—would increase the activities of Raf-1 and MAPK/ERK, as well as PI3K and c-Akt, leading to increased cyclin levels, cdk activity, and RB phosphorylation. However, based on studies by several groups using inducible Raf constructs in nontransformed cells, prolonged intense activation of the MAPK/ERK pathway was found to cause growth arrest, whereas prolonged weak activation of the pathway promoted proliferation ***(65-67)***. This was correlated to superinduction of CKI molecules (*see* **Subheading 1.4.**). Thus an initial burst of intense MAPK/ERK activity, as may be induced following a mutation to Ras, would be expected to cause a rapid growth arrest, as was elegantly demonstrated in fibroblasts and epithelial cells ***(68)***. That other studies have also shown that Ras- and Raf-transformed cells have greater colony-forming abilities argues that a portion of these initially growth-arrested cells must have lost the function of the "arresting" mechanism ***(69)***. This argues strongly that for a cell to mount a proliferative response to prolonged intense MAPK/ERK signaling, a further "mutation" is required. For example, either a loss of CKI function/ inducibility, a loss of RB function, or potentially a gain of constitutive cyclin expression will be required. Collectively, it is these additional genetic events that are necessary for the initial mutation to have a plieotropic impact on cell function(s), leading to neoplastic transformation. Thus, loss of p53 and other transcription factors that modulate p21 expression levels can play a role in reducing the relative levels of p21 CKI molecules compared to cyclin D and cyclin E molecules, thus promoting cdk activation. In many cancers, loss of p16 and p27 expression can occur by deletion and/or methylation of transcriptional-promoting elements within these genes, also ultimately promoting cdk activation ***(70–72)***. Collectively, this leads to a relative diminished threshold that is required for proliferative signals, e.g., MAPK/ERK to stimulate growth, vs the possibility that too much signal will cause growth arrest. In addition and downstream of cdk enzymes, loss of RB function can largely remove the need for a profound cdk2/4 activation to achieve cell cycle progression from G1 into S phase, since loss of RB function allows permanent release of E2F factors and largely abrogates the G1 restriction point.

The release and activation of E2F transcription factors is a key step in the commitment of cells to S phase. E2F factors regulate the function of many genes, including that for the transcription factor c-Myc ***(73)***. The interactions between RB/E2F and c-Myc in the control of cell cycle progression are very complicated. c-Myc has been argued to be a direct downstream target for both MAPK/ERK and JNK signaling in several cells types, and furthermore, protein expression of c-Myc is often increased in neoplastic cells ***(74–76)***. The increased expression of c-Myc may be due to increased protein stabilization via MAPK/ERK signaling ***(74,75)***. Transcriptional activation of c-Myc can enhance proliferation by multiple mechanisms, including the recent findings of several groups showing that c-Myc can repress p21 promoter activity ***(34,77,78)***. Thus, in the presence of an activated MAPK/ERK pathway, an increase in basal c-Myc protein levels will profoundly reduce the ability of MAPK/ERK signaling to enhance p21 protein levels. Thus loss of p53 function together with a gain of MAPK/ERK and c-Myc functions will tend to reduce CKI expression and promote proliferation versus growth arrest. Clearly, it is evident that oncogenic transformation represents a complex alteration in the functions of multiple cell cycle control proteins and signaling pathways.

Recent work has revealed another connection between the MAPK signal pathways and the expression and activity of TSGs. The target of the MAPK cascades include the components of the AP-1 transcriptional complex, which in turn have been shown to modulate the expression of a number of TSGs. For instance, it has been well documented that increased c-Jun activity can transcriptionally repress the expression of p53 ***(79,80)***. In addition, there is evidence that there can be a direct transcriptional repressive effect of c-Jun on the p21 gene ***(81)***, although this may be cell type specific ***(78,82)***. Further, Jun B can transactivate the p16 gene ***(83)***. Since c-Jun and Jun B can form heterodimers and frequently antagonize one another actions, increases in the c-Jun/Jun B ratio (resulting from MAPK activity) can alter cell cycle regulation. Thus AP-1 can potentially modify cell cycle checkpoints mediated by p53, p21, or p16.

1.4. Regulation of CKI Expression by MAPK/ERK Signaling: a Temporal Relationship with Signaling Intensity Defines Proliferative vs Antiproliferative Effects

As noted above, a good example of how the function of a signaling pathway can be modified during the transformation process is the MAPK/ERK pathway. Studies performed shortly following the discovery and description of the MAPK/ERK pathway argued that prolonged elevation of pathway activity correlated with increased proliferation. For example, fibroblasts stably transfected with either oncogenic v-Ha-Ras or v-Raf were found to have elevated basal MAPK/ERK activity that correlated with increased proliferation and tumorigenic potential ***(68,69)***. Meanwhile, several investigators were also demonstrating that prolonged activation of the MAPK/ERK cascade can inhibit DNA synthesis, and these studies suggested that this was due to induction of CKI proteins ***(84,85)***. Treatment of PC12 cells with nerve growth factor (NGF) caused a sustained activation of B-Raf and the MAPK/ERK cascade with neuronal differentiation and cell cycle exit. This was blocked by the specific inhibitor of MEK1/2 activation, PD98059. NGF treatment of NIH 3T3 cells transfected with the high-affinity TrkA NGF receptor lead to the induction of p21, which was blocked by treatment with PD98059, and this was proposed to be the mechanism by which MAPK/ERK signaling was inhibiting proliferation. Treatment of several other established/transformed cell types with ionizing radiation, growth factors, or phorbol esters has been shown to induce prolonged activation of the MAPK/ERK cascade and cause induction of p21 and/or/both p27, leading to cell cycle arrest ***(86–88)***. The various cell types used in these studies were p53 wild type, p53 mutant, and p53 null. Thus induction of p21 protein levels by prolonged intense MAPK/ERK signaling is not always dependent on p53 function.

More recent studies, however, have suggested that the intensity and duration of MAPK/ERK signaling determines whether a cell proliferates, becomes senescent, or differentiates. In this regard, several groups have demonstrated that high levels of MAPK/ERK signaling for a short period of time, or low levels of sustained MAPK/ERK signaling, correlates with the induction of low amounts of CKI proteins and cyclins, e.g., p21 and cyclin D1, leading to increased cdk2/4 activity and DNA synthesis (e.g., **ref. *66***). In contrast, sustained high levels of MAPK/ERK signaling frequently induce high CKI levels, e.g., p16, p19, p21, p27 leading to growth arrest and potentially a senescence/differentiation response ***(66,67,89,90)***. As will be discussed under the following **Subheading 1.5.**, the relative ability of MAPK/ERK signaling to increase expression of the CKI p21 appears to be reduced in transformed hepatoma cells compared to primary hepatocytes. Thus the outcome of whether a given amount of MAPK/ERK signaling causes proliferation vs growth arrest changes during the process of transformation.

1.5. Regulation of CKI Expression in Primary Hepatocytes and Hepatoma Cells by MAPK/ERK Signaling; Proliferative vs Antiproliferative Effects

Partial hepatectomy or dissociation followed by primary culture triggers hepatocyte entry into the cell cycle ***(91,92)***. Maximal DNA synthesis in vivo occurs 12–36 h post partial hepatectomy (PHX) ***(93–95)***. Hepatocytes do not terminally differentiate and can enter and exit the cell cycle during cycles of liver regeneration. This is in contrast to other cell types, e.g., intestinal epithelial cells, chondrocytes, and keratinocytes, which undergo irreversible terminal differentiation and senescence processes.

Recently, several signaling pathways leading to increased DNA synthesis in primary hepatocytes have been shown to be the JNK, the p38, the PI3K, and the MAPK/ERK pathway ***(95–97)***. In addition, prolonged signaling by the MAPK/ERK pathway was shown to play a prominent role in causing cell cycle arrest in these cells ***(89,90)***. The ability of MAPK to cause cell cycle arrest in these studies was correlated with increased expression of the CKI proteins p21, and to a lesser extent p16. These findings will be presented in the following sections.

Hepatoma cells that are incapable of increasing p21 expression are known to be more tumorigenic in vivo than hepatoma cells that still retain the ability to express p21 ***(98)***. In part, this may be due to loss of p53 function, as noted previously, although several studies have reported that childhood hepatomas and early-stage adult liver cancers express functional p53 ***(99–101)***. A reduction in the ability of many cell types to increase p21 expression has also been suggested to be important in the process of transformation and differentiation. This may be due to a loss of transcription factor function(s), or a consequence of altered signaling via other pathways that regulate p21 expression ***(102–104)***. In further agreement with the importance of p21 expression in hepatocyte cell cycle control, it was recently shown that MAPK signaling had a reduced ability to increase p21 expression in hepatoma cells and that inducible overexpression of p21 could blunt liver regeneration ***(105–107)***. Together, these data suggest that regulation of p21 expression and function may play a pivotal role in both the regulation of liver regeneration and in hepatocellular transformation. Regulation of the p21 promoter appears to be complex, consisting of both potential positive and negative regulatory elements; multiple transcription factor-binding sites within the promoter have also been noted; potential regulatory transcription factors include p53; c-Jun, the glucocorticoid receptor family; Ets family; C/EBP family; Stat family; Sp family, and c-Myc (**refs. *88,89***). It is likely that multiple MAPK-dependent events may play a cell type- and growth status-specific role in modulating p21 expression, at the levels of transcription and posttranscriptional stabilization. Indeed, several studies have argued that the regulation of p21 protein levels by MAPK signaling is predominantly at the levels of increased mRNA stabilization and protein stabilization ***(78,88–90)***. The molecules/mechanisms that may control p21 mRNA stability, e.g., HuR and p21 protein stability, e.g., ubiquination, are less well described, and whether the functions of these mechanisms are reduced during the process of transformation remains to be determined.

2. Materials

2.1. Infection of Primary Mouse Hepatocytes by Poly-L-lysine Conjugated Adenovirus

1. HEPES-buffered saline (HBS): 20 m*M* HEPES; 150 m*M* NaCl; pH 7.3.
2. Filter sterilize stock poly-L-lysine (Sigma, poly-L-lysine hydrobromide, cat no. p7890): make into 0.1 mg/mL in HBS.
3. Filter-sterilize tissue culture medium (Williams E medium or DMEM).
4. Male C57BL/6J wild-type, p21-null, and p16-null mice (~4 mo old, ~30 g) that have had access to food and water ad libitum.
5. All inducible *Raf* constructs were kindly provided by Dr. M. McMahon (University of California, San Francisco, CA).
6. Recombinant p16 adenovirus was kindly provided by Dr. Prem Seth (National Institutes of Health, Bethesda, MD).
7. Anti-p42/44MAPK, also called ERK2/1 (sc-154AC), anti-β-actin (sc-1616), anti-cdk2 (sc-163AC), anti-cdk4 (sc-601AC), anti-p16 (sc-1207), anti-p27 (sc-528), and anti-p21 (sc-397-G and sc-817) were from Santa Cruz Biotechnology (Santa Cruz, CA).
8. Radiolabeled [γ-^{32}P]-ATP and ^{3}H-thymidine were from NEN.
9. Western immunoblotting was performed using the Enhanced Chemi-Luminescence (E.C.L.) system (Amersham).
10. Protein preparations and other reagents were as noted in **ref. *89***.
11. The specific MEK1/2 activation inhibitor was a gift from Parke Davis/Warner Lambert pharmaceuticals.

3. Methods

3.1. Poly-L-lysine-Coated Adenovirus: Generation

Replication-defective adenovirus was generated, and was conjugated to poly-L-lysine as described in **refs. *108*** and ***109***. Briefly, a nonreplicative adenovirus without insert was grown in 293 cells and purified on CsCl gradients and inactivated by exposure to UV irradiation. Purified virus was dialyzed against 2× filtered HBS (HBS = 150 m*M* NaCl/20 mM HEPES pH 7.3). The concentration of the virus was determined using UV spectrophotometric analysis. Freshly isolated adenovirus (1.4×10^{11} particles) was combined with poly-L-lysine (16 μ*M* final concentration) and 1-ethyl-3-(3-dimethylaminopropyl) carbodiimide at a final concentration of 130 μ*M*, in a final volume of 4 mL. The reaction mixture was incubated on ice for 4 h, followed by removal of unreacted components by ultracentrifugation (150 × 10^3 × *g*; 18 h on a CsCl gradient at a CsCl concentration of 1.35 g/mL). The conjugated adenovirus was dialysed vs HBS including 10% (v/v) glycerol, viral concentration determined using using UV spectrophotometric analysis, and stored at −80°C.

3.2. Infection of Primary Mouse Hepatocytes by Poly-L-lysine Conjugated Adenovirus (*see* Notes 1–5)

1. Freshly isolated mouse hepatocytes are plated in rattail collagen-coated 12-well plates at the density of 2 × 10^5 cells /well in medium. The cells are cultured for 3–4 h in a 37°C incubator before transfection. The plasmid and the virus need to be kept on ice once thawed.
2. In separate polystyrene tubes, mix 1 × HBS with poly-L-lysine, at the ratio of 13 μL of 1 × HBS + 2 μL of stock poly-L-lysine, total 15 μL per reaction; make appropriate number of tubes according to the experiment design and number of wells available; incubate at room temperature (RT) for 20–30 min.
3. In separate polystyrene tubes, mix 1 μL of plasmid DNA (0.5–1.5 μg), 3–5 μL of virus (300–500 MOI), and × μL of 1 × HBS to bring the total volume to 31 μL (*see* **Note 3**). Incubate in the dark at RT for 30 min (usually the reaction mixture should be covered with foil to prevent light exposure).
4. Mix the whole contents of HBS–poly-L-lysine into each DNA–virus mixture, and incubate at RT for 30 min.
5. Add the virus–DNA–poly-L-lysine mixture in **step 5** to each well (45 μl in each well of 12-well plate). Incubate for 4 h on a rocker in a CO_2 incubator.
6. Replace the medium with 3 mL of fresh culture medium. Allow 18–24 h after transfection/infection for the interested protein(s) to express (depending on the expression level of the target protein).
7. Proceed with any further treatment as desired.
8. Make the HBS and poly-L-lysine mixture *(mixture A):*

Component	Amount required in each well (μL)	Total volume to be made (μL)
1 × HBS	13	13 × *n*
Poly-L-lysine	2	2 × *n*
Total volume in each well of 12-well plate	13 + 2 = 15	15 × *n*

n = the number of wells to be transfected. For example, to transfect 7 of 12-well plates, *n* = 84. In total, you will need: 1 × HBS, 13 × 84 = 1092 (μL); poly-L-lysine: 2 × 84 = 168 (μL); total, 1260 (μl).

9. Make the plasmid and adenovirus mixture *(mixture B):*

Component	Amount required in each well	Total to be made
Plasmid	0.5 (μg)	0.5 × *n* (μg)
Adenovirus	2 (μL)	2 × *n* (μL)
1 × HBS	To bring the total volume to 31 μL	
Total volume in each well of 12-well plate	31 (μL)	31 × *n* (μL)

n = the number of wells to be transfected. For example, to transfect 7 of 12-well plates with pCMV (the concentration is 1 μg/μL), n = 84, in total, you will need: plasmid, 0.5 × 84 = 42 (μg), i.e., 42 μL of plasmid is to be added to the test tube; virus, 2 × 84 = 168 (μl); total volume, 31 × 84= 2604 (μL). Therefore, you will need HBS, 2604 − 168 − 42 = 2396 (μL).

10. Add mixtures A and B together, mix well to form the mixture C. Add 45 μL of mixture C to each well, leave on 37°C shaker for 4 h. Then put into regular 37°C incubator overnight, change the medium the following morning, and proceed with any other treatment.

3.3. Generation of a Recombinant Adenovirus: Recombination in Escherichia coli

We have generated recombinant adenoviruses using a novel methodology in bacteria. In this procedure, the full-length recombinant adenovirus genome is cloned in a plasmid, flanked by a rare cutter (*PacI*) restriction site and is generated using a recombination proficient *E. coli* strain (*BJ5183*) with the genotype *recBC sbcBC*. We developed a novel transfer plasmid using pZero2.1 (Invitrogen) and a plasmid containing the 35-kbp adenoviral genome pTG-CMV (kindly provided by Drs. Matthias Paul Wymann and Stefano Brenz Verca, University of Fribourg, Switzerland). Digestion of pZero2.1 with Aff III and Stu I was followed by insertion of a linker containing Pac I and Bgl II, forming the construct pZero-link. Digestion of pZero-link and pTG-CMV with Pac I, followed by fragment purification and annealing, produced the plasmid pZeroTG-CMV. The cDNA (gene) of interest was isolated and subcloned into pZeroTG-CMV (pZeroTG-CMV-cDNA). Colonies were selected using kanamycin. Recombination was achieved in the recombination proficient *E. coli* strain (*BJ5183*). The pZeroTG-CMV-cDNA plasmid (500 ng) was digested with Pac I and Bgl II; the pTG-CMV plasmid was cut with Cla I (1 μg), followed by cotransformation of *BJ5183* cells. Add each DNA to *BJ5183*-competent cells and sit on ice for at least 30 min. Heat-shock for 80 s, then put back on ice for 2 min. Add 250 mL SOC and incubate at 37°C for 1 h. Plate out 150 mL of the transformation media onto LB/glucose plates. Grow overnight at 37°C. Do not overgrow plate (important). Satellite colonies will form and must be avoided when picking colonies. After colonies grow, pick ~8 and grow all in 5 mL LB/ampicillin. Harvest in afternoon, allowing bacteria to grow at least 6 h. We use the Bio-Rad mini-prep kit to prepare plasmid DNA. *Note:* If you are not able to prepare DNA the same day as harvest, make sure to spin and pour off LB, then put pellet into −80°C freezer to store for later use. Prepare DNA from BJ5183 cells using Bio-Rad mini-prep kit (*see* **Note 6**). Bio-Rad protocol has been changed slightly: (1) Add 300 mL cell resuspension solution to bacterial pellet and resuspend pellet using a pipet or vortex. (2) Add 400 mL cell lysis solution with gentle rocking. Allow the solution to sit at room temperature until the lysate is clear. (3) Add 400 mL of the neutralization solution, again with gentle agitation, followed by chilling on ice for 5 min. (4) Clarify the solution by centrifugation at room temperature, discarding the pellet. (5) Add 200 mL of the binding matrix to the supernatant from the spin. (6) Follow the protocol as directed—wash steps. (7) Add 100 mL of TE to the binding matrix and filter. Spin and collect the DNA in a fresh tube. (8) Transform 5 mL into 50 mL XL I Blue cells. (9) Heat-shock for 42 s, followed by placement on ice for 2 min. (10) Add 250 mL SOC and incubate at 37°C for 1 h. (11) Plate out 150 mL onto LB/glucose plates. Grow overnight at 37°C. Pick five colonies the following morning and prepare each separately with Promega Wizard prep kit. Digest each DNA, along with control empty pTG-CMV with EcoR1, to confirm recombinant plasmid pTG-CMV-cDNA. The pTG-CMV-cDNA DNA was transfected into 80% confluent 293 cells using the $CaPO_4$-DNA coprecipitation technique. Cells were overlaid with agarose and plaques formed in 7–10 d. Plaques were isolated, the virus from each expanded, and protein expression from the relevant cDNA determined by Western blotting. The concentration of the recombinant adenovirus was assessed on the basis of the absorbency at 260 nm and on a limiting-dilution plaque assay. Recombinant adenoviruses are stored in small portions (~1 mL) at −80°C. A reduction in titer is observed after > 1 freeze–thaw cycle.

3.4. Infection of Cells with Recombinant Adenovirus

Prior to infection, cell number is determined. For infection, cells were incubated in a minimal volume of serum-free medium for the plate size, e.g., 0.5 mL per well of a 12-well dish. To this medium, the appropriate amount of recombinant adenovirus is added to give the required multiplicity of infection (m.o.i.). Cells are gently rocked for 4 h at 37°C in an incubator. At this time, medium can be ereplaced. Cells were infected with recombinant viruses at 30–100 m.o.i. and incubated at 37°C for an additional 24 h. To assess expression, we performed Western immunoblots 24 h after infection.

3.5. Preparation of Mouse Hepatocytes

Mice were anesthetized by intraperitoneal injection of sodium pentobarbital (50 mg/kg), and the lower thorax and abdomen was shaved to remove fur. A small (3-cm) vertical mid-line incision was made in the abdominal wall from just below the costal margin/xiphoid process. Hepatocytes were prepared by cannulation of the portal vein, collagenase perfusion of the liver, and washing in Dulbeco's Modified Eagle's Medium (DMEM) containing 5% (v/v) fetal calf serum as described.

3.6. Primary Culture, Hormonal Treatment, and Assay for DNA Synthesis in Cultures of Hepatocytes

Mouse hepatocytes were cultured on rat-tail collagen (Vitrogen)-coated plastic dishes (12 × 20 mm, 2 × 10^5 cells) in 1 mL phenol red-free DMEM in 5% (v/v) CO_2 supplemented with 50 n*M* insulin, 0.1 n*M* dexamethasone, 1 n*M* thyroxine. At this time, cells were infected with various adenoviruses according to the experimental protocol. For cells undergoing acute exposure, treatments occurred 90 min after plating. For adenoviral-infected cells, 4 h after infection, medium was replaced and hepatocytes were cultured in the same supplemented DMEM for 24 h. Hormonal treatment and/or protein kinase inhibitors (PD98059, 50 μ*M*) were added 24 h after the medium change (protein kinase inhibitors were added 30 min prior to further treatment). Twenty-four hours after infection, hepatocytes were treated, where indicted, for a further 0–36 h (60 h total) with 200 n*M* 4-hydroxytamoxifen. The activity of MAPK/ERK was determined prior to 4-hydroxytamoxifen addition, and 0–36 h after the start of treatment. Twenty seconds prior to terminating the experiment, medium was aspirated, followed by immediate homogenization. Cells were homogenized in 1 mL ice-cold homogenization buffer A (25 m*M* HEPES, pH 7.4, at 4°C, 5 m*M* EDTA, 5 m*M* EGTA, 5 m*M* benzamidine, 1 m*M* phenylmethyl sulphonylfluoride, 40-μg/mL pepstatin A, 1 μ*M* Microcystin-LR, 0.5 m*M* sodium vanadate, 0.5 m*M* sodium pyrophosphate, 0.05% (w/v) sodium deoxycholate, 1% [v/v] Triton X100, 0.1% (v/v) 2-mercaptoethanol), with trituration using a P1000 pipet to lyse the cells. Homogenates were stored on ice prior to clarification by centrifugation (4°C), and clarified aliquots were subjected to immunoprecipitation. For DNA synthesis assays, hepatocytes were isolated from wild-type, p21-, and p16-null mice and infected with ΔB-Raf:ER. Twenty-four hours after infection, hepatocytes were treated for 36 h with 200 n*M* 4-hydroxytamoxifen. Hepatocytes were cultured in the presence of 4 μCi ^{3}H-thymidine for a further 36 h, after which time cells were lysed with 0.5 *M* NaOH and DNA-precipitated with 12.5 % (w/v) TCA (final). Acid-precipitable material was transferred to glass fiber filters, washed with 5% (w/v) TCA, and ^{3}H-thymidine incorporation into DNA was quantified by liquid scintillation spectrometry.

3.7. Culture and Adenoviral Infection of HepG2 Cells

HepG2 cells were cultured in phenol red-free DMEM supplemented with 50 n*M* insulin, 0.1 n*M* dexamethasone, 1 n*M* thyroxine, in an identical manner to primary hepatocytes. In experiments assessing DNA synthesis, cells were cultured in phenol red-free DMEM supplemented with 5% (v/v) fetal calf serum (FCS). Cells were infected with adenoviruses as described for primary hepatocytes.

3.8. Immunoprecipitations from Homogenates

Fifty microliters of Protein A agarose (Ag) slurry (25 μL bead volume) was washed twice with 1 mL PBS containing 0.1% (v/v) Tween-20, and resuspended in 0.1 mL of the same buffer. Antibodies (2 μg, 20 μL) or serum (20 μL) were added to each tube and incubated (3 h, 4°C). Clarified hepatocyte homogenates (1.0 mL, 1 mg total protein) were mixed with Protein A–Ag-conjugated antibody in duplicate using gentle agitation (2.5 h, 4°C). Protein A–Ag was recovered by centrifugation, the supernatant discarded, and washed (10 min) sequentially with 0.5 ml buffer A (twice), PBS, and buffer B (25 m*M* HEPES, pH 7.4, 15 m*M* $MgCl_2$, 0.1 m*M* Na_3VO_4, 0.1% [v/v] 2-mercaptoethanol).

3.9. Assay of MAPK/ERK Activity

Immunoprecipitates were suspended in a final volume of 50 μL of buffer B containing 0.2 m*M* [γ-^{32}P]ATP (2000 cpm/pmol), 1 μ*M* Microcystin-LR, 0.5 mg/mL myelin basic protein (MBP), which initiated reactions and incubated at 37°C. After 20 min, 40 μL of the reaction mixtures were spotted onto 2-cm circles of P81 phosphocellulose paper (Whatman) and immediately placed into 180 m*M* phosphoric acid. Papers were washed four times (10 min each) with phosphoric acid, and once with acetone, and ^{32}P-incorporation into MBP was quantified by liquid scintillation spectroscopy. Preimmune controls were performed to ensure that MBP phosphorylation was dependent on specific immunoprecipitation of MAPK/ERK.

3.10. Western Blotting

Twenty-four hours after infection, in cells expressing ΔB-Raf:ER, hepatocytes were treated with 100 n*M* 4-hydroxytamoxifen. Hepatocytes were cultured for a further 6–36 h, after which the cells were lysed with either 10% (w/v) TCA or ice-cold homogenization buffer. TCA-precipitated protein was collected by centrifugation, washed once with

Table 1
MAPK Activity in Hepatocytes and Hepatoma Cells Expressing Inducible Raf Constructs After Treatment with Tamoxifen

Time (h) (+ TAM)	Primary hepatocyte ΔRaf-1(301):ER + TAM	Primary hepatocyte ΔA-Raf:ER + TAM	Primary hepatocyte ΔB-Raf:ER + TAM	Hep G2 ΔRaf-1(301):ER + TAM	Hep G2 ΔB-Raf:ER + TAM
0	1.0 ± 0.1	1.1 ± 0.1	1.1 ± 0.1	9.9 ± 0.8	10.1 ± 0.9
6	1.3 ± 0.2	1.4 ± 0.2	6.5 ± 0.5	10.3 ± 1.1	20.5 ± 1.8
12	1.2 ± 0.1	1.9 ± 0.2	9.4 ± 0.6	10.3 ± 1.4	28.6 ± 2.3
24	1.1 ± 0.1	2.1 ± 0.1	11.1 ± 0.6	10.0 ± 1.6	30.4 ± 3.5
36	1.2 ± 0.1	2.0 ± 0.2	10.7 ± 0.8	10.1 ± 1.0	30.2 ± 2.8

Hepatocytes and HepG2 cells were infected with inducible Raf constructs conjugated to poly-L-lysine adenovirus and/or with null recombinant adenovirus or a recombinant adenovirus expressing p21 antisense mRNA followed by culture as in **Subheading 3.** After 24 h, hepatocytes were treated with either vehicle control or with 4-hydroxytamoxifen for an additional 36 h. Cells were assayed for MAPK activity at the indicated times. Addition of 50 μ*M* PD98059 to the culture medium abolished the activation of MAPK (data not shown), in agreement with **ref. *89***. Data are expressed as -fold alterations in MAPK activity per mg cell protein in comparison to MAPK activity in ΔRaf-1(301):ER + TAM cells, prior to TAM addition at 0 time (defined as 1.00).

cold acetone, and resuspended in SDS-PAGE sample buffer prior to resolution on SDS-PAGE using 10–12% gels. Cells lysed in homogenization buffer were subjected to immunoprecipitation as described under the previous subheadings, prior to resuspension in SDS-PAGE sample buffer and resolution on SDS-PAGE using 10–12% gels. Gels were transferred to a 0.22-μm nitrocellulose filter and immunoblotting performed using the E.C.L. system (Amersham).

3.11. Data Analysis

Comparison of the effects of various treatments was performed using one way analysis of variance and a two-tailed *t*-test. Differences with a p value of <0.05 were considered statistically significant. Experiments shown are the means of individual points from multiple separate experiments (SEM).

3.12. Results

3.12.1. Prolonged Intense Activation of the MAPK Cascade, but not Prolonged Weak Activation, Increases $p21^{Cip\text{-}1/WAF1}$ and $p16^{INK4a}$ Expression in Primary Mouse Hepatocytes

Hepatocytes were isolated from wild-type, p16-null, and p21-null mice and infected with a construct to express an inducible estrogen receptor-B-Raf fusion protein (ΔB-Raf:ER), with a construct to express an inducible estrogen receptor-A–Raf fusion protein (ΔA-Raf:ER) or with an inducible estrogen receptor–Raf-1 kinase-inactive mutant of Raf-1 (ΔRaf-1:ER[301]). Cells were treated with 200 n*M* 4-hydroxytamoxifen and the activity of MAPK/ERK determined 0–36 h after treatment in immune complex kinase assays. Treatment of hepatocytes expressing ΔB-Raf:ER and ΔA-Raf:ER with 4-hydroxytamoxifen increased MAPK activity ~10-fold and ~2-fold, respectively (**Table

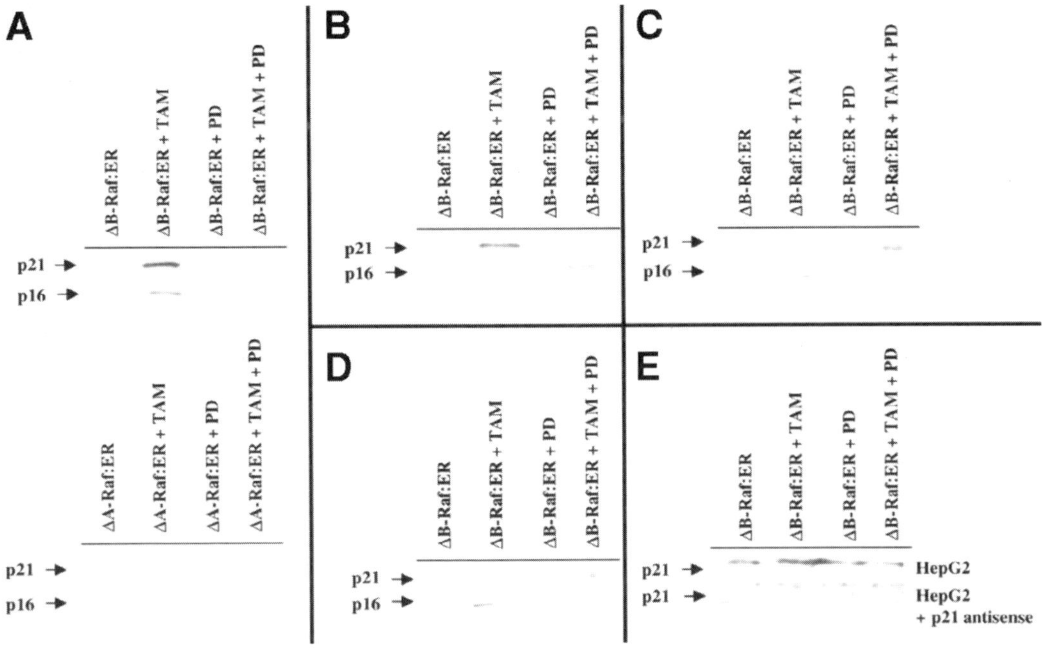

Fig. 1. Prolonged intense, but not prolonged weak, MAPK signaling increases p21 levels in primary hepatocytes and hepatoma cells: **(A)** wild-type hepatocytes; **(B)** p16 null hepatocytes; **(C)** p21-null hepatocytes; **(D)** wild-type hepatocytes with p21 antisense mRNA; **(E)** HepG2 cells and HepG2 cells with p21 antisense mRNA. Hepatocytes were infected with ΔB-Raf:ER or ΔA-Raf:ER, as indicated, poly-L-lysine adenovirus, and/or with either a null recombinant adenovirus or a p21 antisense mRNA recombinant adenovirus followed by culture as under **Subheading 3.** After 24 h, cells were treated with either vehicle control or with 4-hydroxytamoxifen for 36 h. Protein expression of p21 was determined by immunoblotting. Equal protein loading (200 μg) per lane. Lane 1, ΔB– or ΔA–Raf:ER + vehicle control; lane 2, ΔB– or ΔA–Raf:ER + 4-hydroxytamoxifen; lane 3, ΔB– or ΔA–Raf:ER + 4-hydroxytamoxifen + 50 μ*M* PD98059; lane 4, + 4-hydroxytamoxifen.

1). No increase in MAPK/ERK activity was observed in 4-hydroxytamoxifen-treated hepatocytes infected with kinase-inactive ΔRaf-1:ER(301) (**Table 1**). MAPK/ERK activation was blocked by incubation with the specific inhibitor of MEK1/2, PD98059 (50 μM). Prolonged MAPK/ERK activation induced by ΔB-Raf:ER caused an increase in p16 and p21 protein expression in wild-type hepatocytes, but did not alter p27 levels (**Fig. 1A**). No increase in p16 and p21 protein expression was observed in 4-hydroxytamoxifen-treated hepatocytes infected with either ΔA-Raf:ER or kinase-inactive ΔRaf-1:ER(301) (**Fig. 1A**) *(9)*.

Elevated MAPK/ERK activity induced by ΔB-Raf:ER increased p21 expression, but not p16 expression, in p16-null hepatocytes (**Fig. 1B**). Elevated MAPK/ERK activity induced by ΔB-Raf:ER increased p16 expression but not p21 expression, in p21-null hepatocytes (**Fig. 1C**). Expression of full-length human p21 antisense mRNA blunted the ability of mouse hepatocytes to increase p21 protein levels in response to elevated MAPK/ERK activity induced by ΔB-Raf:ER (**Fig. 1D**). The cross-species effectiveness of the antisense p21 likely relates to the extensive nucleic acid sequence conservation between mouse and human p21 genes. These data demonstrate that a prolonged 10-fold elevation of MAPK/ERK activity via ΔB-Raf:ER increases expression of p21 and p16 in primary hepatocytes. In contrast, a prolonged 2-fold activation of MAPK/ERK activity in primary hepatocytes by ΔA-Raf:ER does not induced either p21 or p16.

3.12.2. Prolonged Intense Activation of the MAPK Cascade Increases p21 Expression in HepG2 Hepatoma Cells

In a similar manner to primary hepatocytes, HepG2 hepatoma cells were infected with either a control recombinant adenovirus or a virus-expressing antisense p21 mRNA, together with a construct expressing ΔB-Raf:ER. Cells were treated with 4-hydroxytamoxifen and the activity of MAPK/ERK determined. Treatment of HepG2 cells expressing ΔB-Raf:ER with 4-hydroxytamoxifen increased MAPK/ERK activity ~4-fold after 6 h, which was also maintained for the following 36 h (**Table 1**). Of note, however, basal MAPK/ERK activity was ~10-fold higher in HepG2 cells than in primary hepatocytes. MAPK/ERK activation was blocked by incubation with the specific inhibitor of MEK1/2, PD98059 (50 μM). Prolonged MAPK/ERK activation increased p21 protein expression in control virus-infected cells, but not in cells expressing antisense p21 mRNA (**Fig. 1E**). Collectively, the data in **Table 1** and **Fig. 1** demonstrates that prolonged intense MAPK/ERK signaling can increase p21 protein levels in primary hepatocytes and HepG2 hepatoma cells.

3.12.3. Prolonged MAPK Activity Promotes DNA Synthesis in the Absence of p21 Expression

Hepatocytes isolated from wild-type, p16-null, and p21-null mice were infected to express either ΔB-Raf:ER, ΔA-Raf:ER, or ΔRaf:ER(301). In parallel, wild-type hepatocytes were also infected with a full-length human p21 antisense mRNA, shown to blunt the ability of mouse hepatocytes to increase mouse p21 protein levels in response to elevated MAPK/ERK activity. The ability of 4-hydroxytamoxifen treatment to alter ^{3}H-thymidine incorporation into DNA was determined over the following 36 h.

Prolonged intense MAPK/ERK activity induced by ΔB-Raf:ER reduced DNA synthesis in wild-type and p16-null hepatocytes, but increased DNA synthesis in p21-null hepatocytes and in hepatocytes infected with full-length human p21 antisense mRNA (**Table 2**). These data demonstrate that the ability of prolonged MAPK signaling to inhibit DNA synthesis in primary hepatocytes requires p21 expression. However, the relative ability of MAPK/ERK signaling to inhibit DNA synthesis was reduced in p16-null hepatocytes, arguing that enhanced p16 expression also plays an important role in MAPK/ERK-induced growth-arrest processes. In contrast to the intense MAPK/ERK activation induced by ΔB-Raf:ER, prolonged weak MAPK/ERK activity induced by ΔA-Raf:ER enhanced hepatocyte DNA synthesis ~2-fold. No effect on DNA synthesis was observed using the kinase-inactive mutant of Raf-1 construct with 4-hydroxytamoxifen treatment. Thus, prolonged weak MAPK/ERK activation enhances proliferation in primary hepatocytes.

Inhibition of MAPK/ERK activity caused a 37% reduction in basal DNA synthesis in HepG2 cells and prolonged intense MAPK/ERK activity reduced DNA synthesis in control virus-infected cells to 9% of control values. Expression of p21 antisense mRNA in HepG2 hepatoma cells blocked the ability of prolonged MAPK/ERK activation to increase p21 expression and permitted this stimulus to increase DNA synthesis by 31%

Table 2
DNA Synthesis in Hepatocytes Expressing Inducible Raf-Constructs After Treatment with Tamoxifen

	Wild-type + pTG-CMV	p16-null + pTG-CMV	p21-null + pTG-CMV	p16-null + pTG-p21 antisense
ΔA-Raf:ER + TAM	19,900 ± 800	nd	nd	nd
ΔB-Raf:ER + TAM	2,200 ± 200	4,500 ± 300	25,400 ± 1,900	26,900 ± 2,000
ΔA-Raf:ER + TAM + PD	10,100 ± 500	nd	nd	nd
ΔB-Raf:ER + TAM + PD	7,900 ± 500	8,100 ± 700	14,600 ± 2,500	13,200 ± 1,400
ΔA-Raf:ER + VEH	10,200 ± 600	nd	nd	nd
ΔB-Raf:ER + VEH	11,300 ± 700	10,900 ± 900	11,400 ± 1,200	11,300 ± 1,600
ΔA-Raf:ER + VEH + PD	8,700 ± 400	nd	nd	nd
ΔB-Raf:ER + VEH + PD	8,300 ± 500	8,100 ± 800	9,100 ± 750	8,900 ± 900

Hepatocytes were infected with inducible Raf constructs conjugated to poly-L-lysine adenovirus and/or with either null recombinant adenovirus or a p21 antisense recombinant adenovirus followed by culture as under **Subheading 3.** After 24 h, hepatocytes were continually incubated with ^{3}H-thymidine and treated with either vehicle control (VEH) or with 4-hydroxytamoxifen (TAM) for 36 h. In some experiments, cells were treated with 50 μ*M* PD98059 (PD). After 24 h, cells were lysed and ^{3}H-thymidine incorporation into DNA determined as under **Subheading 3.** Data are rounded to the nearest 100 cpm; nd, not determined.

(*see* **ref. *89***). These data support the view that the molecular mechanism by which prolonged MAPK/ERK signaling inhibits DNA synthesis in primary cultures of hepatocytes and in hepatoma cells is via increasing protein levels of p21.

In the Introduction, it was noted that transformed cells may be expected to have enhanced basal activities within multiple signaling pathways, such as the MAPK/ERK pathway, compared to nontransformed cells. In **Table 1,** HepG2 hepatoma cells had a ~10-fold higher basal MAPK/ERK activity than primary hepatocytes and were shown to have a high rate of proliferation, which was in part dependent on MAPK/ERK activity (**Table 2**). In itself, this is of little note. However, while ΔB-Raf:ER induced a prolonged, intense, ~10-fold increase in MAPK/ERK activity in primary hepatocytes, equivalent to the amount of basal MAPK/ERK activity in a hepatoma cell, this increase in MAPK/ERK activity *reduced* proliferation; i.e., the basal level of MAPK/ERK activity that promotes growth in a hepatoma cell is antiproliferative in a primary hepatocyte. Of note, a prolonged enhancement in MAPK/ERK activity within HepG2 cells could still enhance p21 levels to an extent whereby hepatoma cell proliferation was reduced ***(110)***. Thus the threshold at which MAPK/ERK signaling induces proliferation vs growth arrest changes during the process of transformation.

The hepatoma cell line chosen for this study, HepG2, expresses a wild-type p53 molecule, and thus the difference in the threshold for increased p21 expression is unlikely to be due, in this instance, to a loss of p53 function. Several other transcription factors that control p21 promoter function, such as C/EBPα and C/EBPβ, and whose activities are also controlled by MAPK/ERK signaling, often are found at reduced expression levels in transformed compared to nontransformed cells. Thus, a reduction in the expression levels of CKI regulatory transcription factors such as C/EBPα and C/EBPβ, prior to p53 mutation may also play an early role in the transformation process. Studies from both this group and those of Kunos and co-workers has argued that p21 pro-

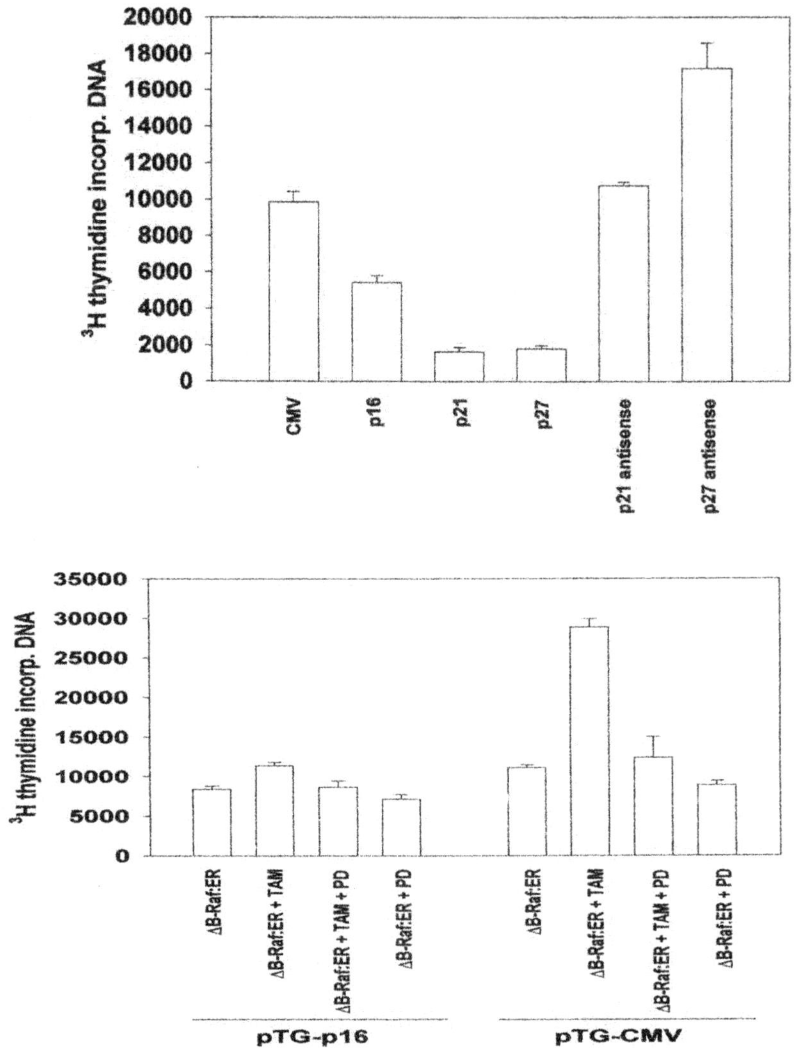

Fig. 2. Modulation of cell cycle progression in hepatocytes by overexpression of p16. **(A)** Wild-type hepatocytes were infected with recombinant adenoviruses to express p16, p21, p27, p21 antisense, or p27 antisense. Twenty-four hours after infection, hepatocytes were continually incubated with ^{3}H-thymidine for 24 h. After 24 h, cells were lysed and ^{3}H-thymidine incorporation into DNA determined as under **Subheading 3. (B)** p21-null hepatocytes were infected with inducible Raf constructs conjugated to poly-L-lysine adenovirus and/or with either null recombinant adenovirus or a p16 recombinant adenovirus followed by culture as under **Subheading 3.** After 24 h, hepatocytes were continually incubated with ^{3}H-thymidine and treated with either vehicle control or with 4-hydroxytamoxifen (TAM) for 36 h. After 24 h, cells were lysed and ^{3}H-thymidine incorporation into DNA determined as under **Subheading 3.**

tein expression in primary hepatocytes and HepG2 hepatoma cells is under control of MAPK/ERK signaling at multiple levels. MAPK/ERK signaling can increase p21 protein levels in these cells via increased transcription, increased mRNA stability, and increased protein stability. Thus it is also possible that the mechanisms by which MAPK/ERK signaling stabilizes p21 mRNA and p21 protein could be lost during the process of transformation.

3.12.4. Overexpression of p16 Blunts MAPK-Stimulated DNA Synthesis in Hepatocytes

Prolonged MAPK/ERK signaling increased expression of both p21 and p16, leading to growth arrest. The impact of p21 on inhibiting DNA synthesis was greater than that of p16. It is also known that enhanced expression of p16 can promote redistribution of p21 and p27 from Cdk4:cyclin D complexes to Cdk2:cyclin E complexes, leading to growth arrest ***(111)***. Thus the impact of enforced p16 overexpression on MAPK/ERK-stimulated DNA synthesis in primary hepatocytes was determined.

Overexpression of p16 in wild-type hepatocytes reduced basal DNA synthesis by ~50% and potentiated the ability of prolonged MAPK/ERK signaling to reduce growth (**Fig. 2A**). In contrast, when either p21 or p27 was overexpressed in these cells an ~90% inhibition in basal DNA synthesis was observed (**Fig. 2A**). When wild-type hepatocytes were infected with a recombinant adenovirus to express an antisense p27 mRNA, basal DNA synthesis was increased ~2-fold (**Fig. 2A**). The finding with loss of p27 function is in contrast to either p16-null, p21-null, or wild-type with antisense p21 mRNA hepatocytes, which did not show differences in basal proliferative potential (**Fig. 2A**). These data further argue that p21, and p27, are more potent inhibitors of hepatocyte cell cycle progression than p16. Overexpression of p16 in p21-null cells reduced basal DNA synthesis and was capable of abolishing the ability of prolonged MAPK/ERK signaling to enhance DNA synthesis in these cells (**Fig. 2B**). Collectively, these findings suggest that the relatively weak ability of MAPK/ERK signaling to increase p16 levels in hepatocytes may be linked to the dominant effect that p21 can have on MAPK/ERK-controlled cell cycle progression in these cells ***(112)***.

3.13. Conclusions on MAPK/ERK-Dependent Regulation of CKI Expression and the Impact of Their Expression on Cell Cycle Progression

Tumor suppressor genes play a vital role in the negative control of cell cycle progression. As noted in the introduction, when growth factors act on a cell to produce a mitogenic response, transient as well as low sustained activation of downstream signal transduction pathways are initiated, leading to the coordinated expression of cyclins and CKI molecules, permitting a cell to progress through G1 phase. Ultimately, activation of cyclin:cdk complexes permits the phosphorylation and inactivation of RB, allowing commitment for entry into S phase. In a variety of cell types, prolonged intense MAPK/ERK signaling has been proposed to inhibit cell cycle progression by inhibiting the functions of cyclin:cdk complexes. In some studies, the TSG p16 has been proposed to play the key role in growth arrest. In other studies, the TSG p21 has been proposed to be the essential cdk inhibitor. In other studies of p16, a G1 CKI has been proposed to play a role in senescence responses of cells in G2 phase. The most likely explanations for these similar yet different findings are cell type specific, in particular, differences in primary vs established cells, culture conditions, differences in the relative basal expression of cyclin D and cyclin E, differences in the relative basal expression of other CKI molecules, e.g., p27 and p57, and differences in the expression of regulatory transcription factors. For example, based on the data presented in this chapter, p21 appeared to be a more important growth inhibitory CKI than p16 in hepatocytes. However, in our hands primary hepatocytes in culture have high basal expression levels of cyclin E, and overexpression of cyclin E can potentially abrogate the growth-inhibitory effects of p16 expression. Furthermore, while prolonged MAPK/ERK signaling inhibited growth via p21, loss of p21 did not alter basal levels of DNA synthesis. In contrast, loss of p27 expression enhanced basal DNA synthesis and blunted the inhibitory effect of prolonged intense MAPK/ERK signaling. Hence p16, p21, and p27 act in concert with cyclin molecules to tightly control cyclin:cdk activities in G1 phase ***(112)***.

Thus multiple scenarios can be envisaged following even a single mutation, e.g., Ras, by which transformation takes place. Oncogenic mutation of Ras can activate the PI3K pathway, leading to lower p27 levels; mutation of Ras can also activate the MAPK/ERK pathway, leading to increased levels of p21 and p16/p19. In some systems, however, oncogenic Ras may increase expression levels of all of these CKI molecules. Depending on the relative expression of cyclins D and E, the modulation of CKI molecule expression by oncogenic Ras may potentiate growth if cyclin levels are constitutively high, or more probably inhibit proliferation if cyclin levels are lower. Thus, for oncogenic Ras to drive growth, a loss of CKI molecule function(s) must also occur. Multiple scenarios exist by which this could happen. For example, cells may decrease, in relative terms, their ability to induce p21, which may occur via a reduction or loss of function in multiple transcription factors, including p53 and members of the C/EBP family. Loss of the p16 locus is common in several tumor types and together with enhanced Mdm2 expression can result in constitutive downregulation of p53 (and presumably p21). Loss of RB function is also common in pancreatic cancer, which tends to abrogate the abilities of p16, p21, and p27 to halt G1 progression. Loss of p27 function is found in later-stage hepatomas. Mutational loss of PTEN function, combined with activation of Ras, will further lower p27 levels. Activation of Ras and mutation of PTEN, in a cell type-specific manner, may increase MAPK/ERK activity, thereby increasing p16 and p21 levels, or alternatively, PI3K signaling may suppress MAPK/ERK signaling, thereby suppressing p16 and p21 expression. Thus the impact of a single Ras mutation on cell cycle progression is dependent on the gain or loss of multiple additional factors.

In conclusion, the criteria that argue for certain signaling pathways and CKIs in the control of proliferation in one cell type may not be those reported by others in cells of a different lineage. This finding is most relevant when comparing primary cells vs established cell lines, including embryonic fibroblasts. Thus, in any assessment of oncogene vs CKI function, care is required in the conclusions drawn as to the relative importance of each CKI in the control of cell cycle progression in the face of oncogenic transformation.

4. Notes

1. The titer of virus is 4.5×10^{10} virus pfu/mL.
2. Plasmid DNA concentration should vary according to the culture area: in 12-well plates, 0.5 μg/well; in 24-well plates, 0.25 μg/well.
3. The amount of virus used depends on the virus concentration, and therefore is variable.
4. Conjugated virus should be titrated to maximize the efficiency.
5. All the virus-contaminated wastes (medium, tips, etc.) should be discarded into separate containers or bags, properly secured and autoclaved before disposal.
6. The Promega Wizard kit does not produce clean enough DNA for future transformation into XL1 Blue cells.

References

1. Malumbres, M., Ortega, S., and Barbacid, M. (2000) Genetic analysis of mammalian cyclin-dependent kinases and their inhibitors. *Biol. Chem.* **381**, 827–838.
2. Masciullo, V., Khalili, K., and Giordano, A. (2000) The Rb family of cell cycle regulatory factors: clinical implications. *Int. J. Oncol.* **17**, 897–902.
3. Bringold, F. and Serrano, M. (2000) Tumor suppressors and oncogenes in cellular senescence. *Exp. Gerontol.* **35**, 317–329.
4. Lowe, S. W. (1999) Activation of p53 by oncogenes. *Endocr. Relat. Cancer* **6**, 45–48.
5. Sherr, C. J. (2000) The Pezcoller lecture: cancer cell cycles revisited. *Cancer Res.* **60**, 3689–3695.
6. Antonyak, M. A., Moscatello, D. K., and Wong, A. J. (1998) Constitutive activation of c-Jun N-terminal kinase by a mutant epidermal growth factor receptor. *J. Biol. Chem.* **273**, 2817–2822.
7. Hynes, N. E. (1996) ErbB2 activation and signal transduction in normal and malignant mammary cells. *J. Mammary Gland Biol. Neoplasia* **1**, 199–206.
8. Shields, J. M., Pruitt, K., McFall, A., Shaub, A., and Der, C. J. (2000) Understanding Ras: "it ain't over 'til it's over." *Trends Cell Biol.* **10**, 147–154.
9. Sebti, S. M. and Hamilton, A. D. (2000) Farnesyltransferase and geranylgeranyltransferase I inhibitors in cancer therapy: important mechanistic and bench to bedside issues. *Expert Opin. Investig. Drugs* **9**, 2767–2782.
10. Kolch, W. (2000) Meaningful relationships: the regulation of the Ras/Raf/MEK/ERK pathway by protein interactions. *Biochem. J.* **351**, 289–305.
11. Maehama, T. and Dixon, J. E. (1999) PTEN: a tumour suppressor that functions as a phospholipid phosphatase. *Trends Cell Biol.* **9**, 125–128.
12. Vazquez, F. and Sellers, W. R. (2000) The PTEN tumor suppressor protein: an antagonist of phosphoinositide 3-kinase signaling *Biochim. Biophys. Acta* **1470**, M21–M35.
13. Yokoyama, A., Karasaki, H., Urushibara, N., et al. (1997) The characteristic gene expressions of MAPK phosphatases 1 and 2 in hepatocarcinogenesis, rat ascites hepatoma cells, and regenerating rat liver. *Biochem. Biophys. Res. Commun.* **239**, 746–751.
14. Lee, C. M. and Reddy, E. P. (1999) The v-myc oncogene. *Oncogene* **18**, 2997–3003.
15. Roovers K. (2000) Assoian RK integrating the MAP kinase signal into the G1 phase cell cycle machinery. *Bioessays* **22**(9), 818–826.
16. Kyriakis, J. M. (1999) Making the connection: coupling of stress-activated ERK/MAPK (extracellular-signal-regulated kinase/mitogen-activated protein kinase) core signalling modules to extracellular stimuli and biological responses. *Biochem. Soc. Symp.* **64**, 29– 48.
17. Vanhaesebroeck, B. and Alessi, D. R. (2000) The PI3K-PDK1 connection: more than just a road to PKB. *Biochem. J.* **346**, 561–576.
18. Dufner, A. and Thomas, G. (1999) Ribosomal S6 kinase signaling and the control of translation. *Exp. Cell Res.* **253**, 100–109.
19. Tibbles, L. A. and Woodgett, J. R. (1999) The stress-activated protein kinase pathways. *Cell. Mol. Life Sci.* **55**, 1230–1254.
20. Davis, R. J. (1999) Signal transduction by the c-Jun N-terminal kinase. *Biochem. Soc. Symp.* **64**, 1–12.
21. Moscatello, D. K., Holgado-Madruga, M., Emlet, D. R., Montgomery, R. B., and Wong, A. J. (1998) Constitutive activation of phosphatidylinositol 3-kinase by a naturally occurring mutant epidermal growth factor receptor. *J. Biol. Chem.* **273**, 200–206.
22. Norgaard, P., Law, B. K., Plovisson, H. S., and Moses, H. L. (1999) Farnesyltransferase inhibitor-induced regression of mammary tumors in TGF alpha and TGF alpha/neu transgenic mice correlates with inhibition of map kinase and p70s6 kinase phosphorylation. *Ann. N.Y. Acad. Sci.* **886**, 265–268.

23. Gire, V., Marshall, C., and Wynford-Thomas, D. (2000) PI-3-kinase is an essential anti-apoptotic effector in the proliferative response of primary human epithelial cells to mutant RAS. *Oncogene* **19**, 2269–2276.
24. De Ruiter, N. D., Wolthuis, R. M., van Dam, H., Burgering, B. M., and Bos, J. L. (2000) Ras-dependent regulation of c-Jun phosphorylation is mediated by the ral guanine nucleotide exchange factor-Ral pathway. *Mol. Cell. Biol.* **20**, 8480–8488.
25. Stofega, M. R., Yu, C. L., Wu, J., and Jove, R. (1997) Activation of extracellular signal-regulated kinase (ERK) by mitogenic stimuli is repressed in v-Src-transformed cells. *Cell Growth Differ.* **8**, 113–119.
26. Rommel, C., Clarke, B. A., Zimmermann, S., et al. (1999) Differentiation stage-specific inhibition of the Raf-MEK-ERK pathway by Akt. *Science* **286**, 1738–1741.
27. Liu, H., Kublaoui, B., Pilch, P. F., and Lee, J. (2000) Insulin activation of mitogen-activated protein (MAP) kinase and Akt is phosphatidylinositol 3-kinase-dependent in rat adipocytes. *Biochem. Biophys. Res. Commun.* **274**, 845–851.
28. Fukazawa, H. and Uehara, Y. (2000) U0126 reverses Ki-ras-mediated transformation by blocking both mitogen-activated protein kinase and p70 S6 kinase pathways. *Cancer Res.* **60**, 2104–2107.
29. Chaudhary, A., King, W. G., Mattaliano, M. D., et al. (2000) Phosphatidylinositol 3-kinase regulates Raf1 through Pak phosphorylation of serine 338. *Curr. Biol.* **10**, 551–554.
30. Manzano, R. G., Montuenga, L., Dayton, M., et al. (2001) CL100 has tumor suppressor properties in vivo and its downregulation is important in the progression of epithelial ovarian cancer. *Cancer Res.*, in press.
31. Harper, J. W., Elledge, S. J., Keyomarsi, K., et al. Inhibition of cyclin dependent kinases by p21. *Mol. Biol. Cell.* **6**, 387–400.
32. Serrano, M., Lin, A. W., McCurrah, M. E., Beach, D., and Lowe, S. W. (1997) Oncogenic Ras provokes premature cell senescence associated with accumulation of p53 and p16INK4a. *Cell* **88**, 593–602.
33. El-Deiry, W. S. (1998) p21/p53, cellular growth control and genomic integrity. *Curr. Top. Microbiol. Immunol.* **227**, 121–137.
34. Claassen, G. F. and Hann, S. R. (2000) A role for transcriptional repression of p21CIP1 by c-Myc in overcoming transforming growth factor beta -induced cell-cycle arrest. *Proc. Natl. Acad. Sci. USA* **97**, 9498–9503.
35. Kibbe, M. R., Li, J., Nie, S., et al. (2000) Inducible nitric oxide synthase (iNOS) expression upregulates p21 and inhibits vascular smooth muscle cell proliferation through p42/44 mitogen-activated protein kinase activation and independent of p53 and cyclic guanosine monophosphate. *J. Vasc. Surg.* **31**, 1214–1228.
36. Delgado, M. D., Vaque, J. P., Arozarena, I., et al. (2000) H-, K- and N-Ras inhibit myeloid leukemia cell proliferation by a p21WAF1-dependent mechanism. *Oncogene* **19**, 783–790.
37. Beier, F., Taylor, A. C., and LuValle, P. (1999) The Raf-1/MEK/ERK pathway regulates the expression of the p21(Cip1/Waf1) gene in chondrocytes. *J. Biol. Chem.* **274**, 30273–30279.
38. Serrano, M. (2000) The INK4a/ARF locus in murine tumorigenesis. *Carcinogenesis* **21**, 865–869.
39. Roussel, M. F. (1999) The INK4 family of cell cycle inhibitors in cancer. *Oncogene* **18**, 5311–5317.
40. Sharpless, N. E. and DePinho, R. A. (1999) The INK4A/ARF locus and its two gene products. *Curr. Opin. Genet. Dev.* **9**, 22–30.
41. Sherr, C. J. and Weber, J. D. (2000) The ARF/p53 pathway. *Curr. Opin. Genet. Dev.* **10**, 94–99.
42. Tombes, R., Auer, K. L., Brenz-Verca, S., et al. (1998) The mitogen-activated protein (MAP) kinase cascade can either stimulate or inhibit DNA synthesis in primary cultures of rat hepa-

tocytes depending upon whether its activation is acute/phasic or chronic. *Biochem. J.* **330**, 1451–1460.

43. Dijkers, P. F., Medema, R. H., Pals, C., et al. (2000) Forkhead transcription factor FKHR-L1 modulates cytokine-dependent transcriptional regulation of p27. *Mol. Cell. Biol.* **20**, 9138–9148.
44. Busse, D., Doughty, R. S., Ramsey, T. T., et al. (2000) Reversible G(1) arrest induced by inhibition of the epidermal growth factor receptor tyrosine kinase requires up-regulation of p27(KIP1) independent of MAPK activity. *J. Biol. Chem.* **275**, 6987–6995.
45. Inoue, K., Wen, R., Rehg, J. E., et al. (2000) Disruption of the ARF transcriptional activator DMP1 facilitates cell immortalization, Ras transformation, and tumorigenesis. *Genes Dev.* **14**, 1797–1809.
46. Weber, J. D., Kuo, M. L., Bothner, B., et al. (2000) Cooperative signals governing ARF-mdm2 interaction and nucleolar localization of the complex. *Mol. Cell. Biol.* **20**, 2517–2528.
47. Ries, S., Biederer, C., Woods, D., et al. (2000) Opposing effects of Ras on p53: transcriptional activation of mdm2 and induction of p19ARF. *Cell* **103**, 321–330.
48. Sherr, C. J. and Roberts, J. M. (1999) CDK inhibitors: positive and negative regulators of G1-phase progression. *Genes Dev.* **13**, 1501–1512.
49. Keyomarsi, K. and Herliczek, T. W. (1997) The role of cyclin E in cell proliferation, development and cancer. *Prog. Cell Cycle Res.* **3**, 171–191.
50. Gille, H. and Downward, J. (1999) Multiple ras effector pathways contribute to G(1) cell cycle progression. *J. Biol. Chem.* **274**, 22033–22040.
51. Sandhu, C., Donovan, J., Bhattacharya, N., Stampfer, M., Worland, P., and Slingerland, J. (2000) Reduction of Cdc25A contributes to cyclin E1-Cdk2 inhibition at senescence in human mammary epithelial cells. *Oncogene* **19**, 5314–5323.
52. Nakamura, N., Ramaswamy, S., Vazquez, F., Signoretti, S., Loda, M., and Sellers, W. R. (2000) Forkhead transcription factors are critical effectors of cell death and cell cycle arrest downstream of PTEN. *Mol. Cell. Biol.* **20**, 8969–8982.
53. Waga, S., Hannon, G. J., Beach, D., and Stillman, B. (1994) The p21 inhibitor of cyclin-dependent kinases controls DNA replication by interaction with PCNA. *Nature* **369**, 574–578.
54. Zhang, H., Hannon, G. J., and Beach, D. (1994) p21-containing cyclin kinases exist in both active and inactive states *Genes Dev.* **8**, 1750–1758.
55. Weiss, R. H., Joo, A., and Randour, C. (2000) p21(Waf1/Cip1) is an assembly factor required for platelet-derived growth factor-induced vascular smooth muscle cell proliferation. *J. Biol. Chem.* **275**, 10285–10290.
56. Adkins, J. N. and Lumb, K. J. (2000) Stoichiometry of cyclin A-cyclin-dependent kinase 2 inhibition by p21. *Biochemistry* **39**, 13925–13930.
57. Lipinski, M. M. and Jacks, T. (1999) The retinoblastoma gene family in differentiation and development. *Oncogene* **18**, 7873–7882.
58. Wazer, D. E. and Band, V. (1999) Molecular and anatomic considerations in the pathogenesis of breast cancer. *Radiat. Oncol. Investig.* **7**, 1–12.
59. Serrano, M., Gómez-Lahoz, E., DePinho, R. A., Beach, D., and Bar-Sagi, D. (1995) Inhibition of Ras-induced proliferation and cellular transformation by p16 INK4a. *Science* **267**, 249–252.
60. Serrano, M., Lee, H. -W., Chin, L., Cordon-Cardo, C., Beach, D., and DePinho, R. A. (1996) Role of the *INK4a* locus in tumor suppression and cell mortality *Cell* **85**, 27–37.
61. Trouche, D., Le Chalony, C., Muchardt, C., Yaniv, M., and Kouzarides, T. (1997) RB and hbrm cooperate to repress the activation functions of E2F1 *Proc. Natl. Acad. Sci. USA* **94**, 11268–11273.

62. Lukas, J., Herzinger, T., Hansen, K., et al. (1997) Cyclin E induced S phase without activation of the pRb/E2F pathway. *Genes Dev.* **11**, 1479–1492.
63. Grimison, B., Langan, A. T., and Sclafani, R. A. (2000) p16 INK4a tumor suppressor function in lung cancer cells involves cyclin dependent kinase 2 inhibition by Cip/Kip protein redistribution. *Cell Growth Different.* **11**, 507–515.
64. Wang, S., Ghosh, R. N., and Chellappan, S. P. (1998) Raf-1 physically interacts with Rb and regulates its function: a link between mitogenic signaling and cell cycle regulation. *Mol. Cell. Biol.* **18**, 7487–7498.
65. Woods, D., Parry, D., Cherwinski, H., Bosch, E., Lees, E., and McMahon, M. (1997) Raf-induced proliferation or cell cycle arrest is determined by the level of Raf activity with arrest mediated by p21Cip1 *Mol. Cell. Biol.* **17**, 5598–5611.
66. Lloyd, A. C., Obermuller, F., Staddon, S., Barth, C. F., McMahon, M., and Land, H. (1997) Cooperating oncogenes converge to regulate cyclin/cdk complexes. *Genes Dev.* **11**, 663–677.
67. Auer, K. L., Seth, P., Darlington, G., et al. (1998) Prolonged activation of the mitogen activated protein (MAP) kinase pathway promotes DNA synthesis in primary hepatocytes from $p21^{Cip-1/WAF1}$ knock out mice, but not in hepatocytes from $p16^{INK4a}$ knock out mice. *Biochem. J.* **336**, 551–560.
68. Pruitt, K., Pestell, R. G., and Der, C. J. Ras inactivation of the retinoblastoma pathway by distinct mechanisms in NIH 3T3 fibroblast and RIE-1 epithelial cells. *J. Biol. Chem.* **275**, 40916–40924.
69. Fukasawa, K. and Vande Woude, G. (1997) Synergy between the Mos/mitogen activated protein kinase pathway and loss of p53 function in transformation and chromosome instability *Mol. Cell. Biol.* **17**, 506–518.
70. Gonzalez, M., Mateos, M. V., Garcia-Sanz, R., et al. (2000) De novo methylation of tumor suppressor gene p16/INK4a is a frequent finding in multiple myeloma patients at diagnosis. *Leukemia* **14**, 183–187.
71. Miracca, E. C., Kowalski, L. P., and Nagai, M. A. (1999) High prevalence of p16 genetic alterations in head and neck tumours. *Br. J. Cancer* **81**, 677–683.
72. Tannapfel, A., Grund, D., Katalinic, A., et al. (2000) Decreased expression of p27 protein is associated with advanced tumor stage in hepatocellular carcinoma. *Int. J. Cancer* **89**, 350–355.
73. Logan, T. J., Moberg, K. H., and Hall, D. J. (1997) Multiprotein complex formation on the c-myc promoter. *Biochem. Mol. Biol. Int.* **43**, 945–953.
74. Sears, R., Leone, G., DeGregori, J., and Nevins, J. R. (1999) Ras enhances Myc protein stability. *Mol. Cell* **3**, 169–179.
75. Sears, R., Nuckolls, F., Haura, E., Taya, Y., Tamai, K., and Nevins, J. R. (2000) Multiple ras-dependent phosphorylation pathways regulate myc protein stability. *Genes Dev.* **14**, 2501–2514.
76. Noguchi, K., Kitanaka, C., Yamana, H., Kokubu, A., Mochizuki, T., and Kuchino, Y. (1999) Regulation of c-Myc through phosphorylation at Ser-62 and Ser-71 by c-Jun N-terminal kinase. *J. Biol. Chem.* **274**, 32580–32587.
77. Mitchell, K. O. and El-Deiry, W. S. (1999) Overexpression of c-Myc inhibits p21WAF1/CIP1 expression and induces S-phase entry in 12-O-tetradecanoylphorbol-13-acetate (TPA)-sensitive human cancer cells. *Cell Growth Differ.* **10**, 223–230.
78. Park, J. S., Boyer, S., Mitchell, K., et al. (2000) Expression of human papilloma virus E7 protein causes apoptosis and inhibits DNA synthesis in primary hepatocytes *via* increased expression of $p21^{Cip-1/WAF1}$. *J. Biol. Chem.* **275**, 18–28.
79. Shaulian, E., Schreiber, M., Piu, F., Beeche, M., Wagner, E. F., and Karin, M. (2000) The mammalian UV response. c-Jun induction is required for exit from p53-imposed growth arrest. *Cell* **103**, 897–907.

80. Schreiber, M., Kolbus, A., Piu, F., et al. (1999) Control of cell cycle progression by c-Jun is p53 dependent. *Genes Dev.* **13**, 607–619.
81. Wang, C. H., Tsao, Y. P., Chen, H. J., Chen, H. L., Wang, H. W., and Chen, S. L. (2000) Transcriptional repression of p21((Waf1/Cip1/Sdi1)) gene by c-jun through Sp1 site. *Biochem. Biophys. Res. Commun.* **270**, 303–310.
82. Kardassis, D., Papakosta, P., Pardali, K., and Moustakas, A. (1999) c-Jun transactivates the promoter of the human p21(WAF1/Cip1) gene by acting as a superactivator of the ubiquitous transcription factor Sp1. *J. Biol. Chem.* **274**, 29572–29581.
83. Passegue, E. and Wagner, E. F. (2000) JunB suppresses cell proliferation by transcriptional activation of p16(INK4a) expression. *EMBO J.* **19**, 2969–2979.
84. Jaiswal, R. K., Weissinger, E., Kolch, W., and Landreth, G. E. (1996) Nerve growth factor-mediated activation of the mitogen-activated protein (MAP) kinase cascade involves a signaling complex containing B-Raf and HSP90. *J. Biol. Chem.* **271**, 23626–23629.
85. Decker, S. J. (1995) Nerve growth factor-induced growth arrest and induction of p21Cip1/WAF1 in NIH-3T3 cells expressing TrkA. *J. Biol. Chem.* **270**, 30841–30844.
86. Park, J. S., Reardon, D. B., Carter, S., Fisher, P. B., Schmidt-Ullrich, R. K., and Dent, P. (1999) Mitogen activated protein (MAP) kinase pathway signaling is required for release/progression of cells through G_2/M after exposure to ionizing radiation. *Mol. Biol. Cell* **10**, 4231–4236.
87. Vrana, J. A., Kramer, L. B., Saunders, A. M., et al. (1999) Inhibition of PKC activator-mediated induction of p21CIP1 and p27KIP1 by deoxycytidine analogs in human leukemia cells:relationship to apoptosis and differentiation. *Biochem. Pharmacol.* **58**, 121–131.
88. Macloed, K. F., Sherry, N., Hannon, G., et al. (1995) p53-dependent and independent expression of p21 during cell growth, differentiation, and DNA damage. *Genes Dev.* **9**, 935–944.
89. Park, J. S., Qiao, L., Yang, M. Y., et al. (2000) A role for Ets and CCAATT (C/EBP) transcription factors in the MAPK-dependent increase in $p21^{Cip-1/WAF1}$ protein levels in primary hepatocytes. *Mol. Biol. Cell* **11**, 2915–2932.
90. Hu, P. P., Shen, X., Huang, D., Liu, Y., Counter, C., and Wang, X. F. (1999) The MEK pathway is required for stimulation of p21(WAF1/CIP1) by transforming growth factor-beta. *J. Biol. Chem.* **274**, 35381–35387.
91. Michalopoulos, G. K. and DeFrances, M. C. (1997) Liver regeneration. *Science* **276**, 60–66.
92. Diehl, A. M. and Rai, R. M. (1996) Liver regeneration 3: regulation of signal transduction during liver regeneration. *FASEB J.* **10,** 215–227.
93. Loyer, P., Cariou, S., Glaise, D., Bilodeau, M., Baffet, G., and Gugen-Guillouzo, C. (1996) Growth factor dependence of progression through G1 and S phases of adult rat hepatocytes in vitro. Evidence of a mitogen restriction point in mid-late G1. *J. Biol. Chem.* **271**, 11484–11492.
94. Westwick, J. K., Fleckenstein, J., Yin, M., et al. (1996) Differential regulation of hepatocyte DNA synthesis by cAMP in vitro in vivo. *Am. J. Physiol.* **271**, 780–790.
95. Talarmin, H., Rescan, C., Cariou, S., et al. (1999) The mitogen-activated protein kinase kinase/extracellular signal-regulated kinase cascade activation is a key signalling pathway involved in the regulation of G(1) phase progression in proliferating hepatocytes. *Mol. Cell. Biol.* **19**, 6003–6011.
96. Spector, M., Auer, K. L., Jarvis, D., et al. (1997) Differential regulation of the mitogen-activated protein and stress-activated protein kinase cascades by adrenergic agonists in quiescent and regenerating adult rat hepatocytes. *Mol. Cell. Biol.* **17**, 3556–3565.
97. Band, C. J., Mounier, C., and Posner, B. I. (1999) Epidermal growth factor and insulin-induced deoxyribonucleic acid synthesis in primary rat hepatocytes is phosphatidylinositol 3-kinase dependent and dissociated from protooncogene induction. *Endocrinology* **140**, 5626–5634.
98. Pu, H., Tsuji, T., Kondo, A., et al. (1997) Comparison of cellular characteristics between human hepatoma cell lines with wild-type p53 and those with mutant-type p53 gene. *Acta. Med. Okayama* **51**, 313–319.

99. Imai, Y., Oda, H., Arai, M., et al. (1996) Mutational analysis of the p53 and K-ras genes and allelotype study of the Rb-1 gene for investigating the pathogenesis of combined hepatocellular-cholangiocellular carcinomas. *Jpn. J. Cancer Res.* **87**, 1056–1062.
100. Chen, T. C., Hsieh, L. L., and Kuo, T. T. (1995) Absence of p53 gene mutation and infrequent overexpression of p53 protein in hepatoblastoma. *J. Pathol.* **176**, 243–247.
101. Teramoto, T., Satonaka, K., Kitazawa, S., Fujimori, T., Hayashi, K., and Maeda, S. (1994) p53 gene abnormalities are closely related to hepatoviral infections and occur at a late stage of hepatocarcinogenesis. *Cancer Res.* **54**, 231–235.
102. Olson, M. F., Paterson, H. F., and Marshall, C. J. (1998) Signals from Ras and Rho GTPases interact to regulate expression of p21Wafl/Cip1. *Nature* **394**, 295–299.
103. Hirai, A., Nakamura, S., Noguchi, Y., et al. (1998) Newly synthesized Rho A, not Ras, is isoprenylated and translocated to membranes coincident with progression of the G1 to S phase of growth-stimulated rat FRTL-5 cells. *J. Biol. Chem.* **272**, 13–16.
104. Serfas, M. S., Goufman, E., Feuerman, M., Gartel, A. L., and Tyner, A. L. (1997) p53-independent induction of p21WAF1/CIP1 expression in pericentral hepatocytes following carbon tetrachloride intoxication. *Cell Growth Different.* **8**, 951–961.
105. Albrecht, J. H., Meyer, A. H., and Hu, M. Y. (1997) Regulation of cyclin-dependent kinase inhibitor p21(WAF1/Cip1/Sdi1) gene expression in hepatic regeneration. *Hepatology* **25**, 557–563.
106. Wu, H., Wade, M., Krall, L., Grisham, J., Xiong, Y., and Van Dyke, T. (1996) Targeted in vivo expression of the cyclin-dependent kinase inhibitor p21 halts hepatocyte cell-cycle progression, postnatal liver development and regeneration. *Genes Dev.* **10**, 245–260.
107. Timchenko, N. A., Harris, T. E., Wilde, M., et al. (1997) CCAAT/enhancer binding protein alpha regulates p21 protein and hepatocyte proliferation in newborn mice. *Mol. Biol. Cell* **17**, 7353–7361.
108. Cristiano, R. J., Smith, L. C., Kay, M. A., Brinkley, B. R., and Woo, S. L. (1993) Hepatic gene therapy: efficient gene delivery and expression in primary hepatocytes utilizing a conjugated adenovirus-DNA complex. *Proc. Natl. Acad. Sci. USA* **90**, 11548–11552.
109. Cristiano, R. J., Smith, L. C., and Woo, S. L. (1993) Hepatic gene therapy: adenovirus enhancement of receptor-mediated gene delivery and expression in primary hepatocytes. *Proc. Natl. Acad. Sci. USA* **90**, 2122–2126.
110. Auer, K. L., Spector, M., Tombes, R. M., et al. (1998) Alpha-adrenergic inhibition of proliferation in HepG2 cells stably transfected with the alpha1B-adrenergic receptor through a p42MAPkinase/p21Cip1/WAF1-dependent pathway. *FEBS Lett.* **436**, 131–138.
111. Groth, A., Weber, J. D., Willumsen, B. M., Sherr, C. J., and Roussel, M. F. (2000) Oncogenic Ras induces p19ARF and growth arrest in mouse embryo fibroblasts lacking p21Cip1 and p27Kip1 without activating cyclin D-dependent kinases. *J. Biol. Chem.* **275**, 27473–27480.
112. Qiao, L., Leach, K., McKinstry, R., et al. (2001) Hepatitis B virus X protein increases expression of p21(Cip-1/WAF1/MDAG) and p27(Kip-1) in primary mouse hepatocytes, leading to reduced cell progression. *Hepatology* **34**, 906–917.

II

Identification of Tumor Suppressor Genes

18

Determination of Cancer Allelotype

Jennifer J. Ascaño and Steven M. Powell

1. Introduction

1.1. Evolution of Allelotyping Methodology

The term "allelotype" was first used by Vogelstein and colleagues ***(1)***, in analogy to karyotype, to describe a newly developed molecular analysis that surveyed chromosomal loss and/or aberration in a panel of human colorectal tumors. This study was the first to comprehensively screen 39 nonacrocentric chromosomal arms in surveying restriction fragment polymorphisms (RFLPs) using variable number tandem repeats (VNTRs) as probes. Simple tandem repeats, including VNTRs and microsatellite loci consisting of 1- to 5-bp repeats, have been found to be highly polymorphic and relatively stable among the human population, which facilitates the analysis of both maternal and paternal alleles. These markers have also been useful in linkage analysis when studying hereditary human disease ***(2–5)***.

Allelotyping was first made possible by the ability to enrich for neoplastic cells using cryostat sectioning method ***(6,7)*** and the development of these VNTR probes, which could be readily radiolabeled and used to identify restriction fragment length polymorphisms (RFLPs). These probes were used to hybridize to genomic DNA that had been cut with a specific restriction enzyme and transferred onto a nitrocellulose membrane (i.e., a Southern blot).

Polymerase chain reaction (PCR) amplification methods to detect polymorphisms have been widely used since they were first described by Weber et al. ***(8)***. Primers can be designed for microsatellite sequences on each arm of every chromosome (examples of representative markers are listed in **Table 1**). Since these discoveries, a variety of methods to visualize PCR products, ranging from radioactive to fluorescent labels, have arisen. This chapter describes three methods to label and PCR-amplify a microsatellite sequence. The first two methods utilize ^{32}P to radiolabel the PCR products, which are then electrophoresed on a standard 7% acrylamide vertical gel and subsequently visualized by autoradiography. The third method described uses a phosphoramidite-fluorescent label oligomer as one of the primers for PCR amplification. The resulting products can be run on a 373 DNA sequencer and computer-analyzed more accurately compared

From: *Methods in Molecular Biology, Vol. 222: Tumor Suppressor Genes: Pathways and Isolation Strategies*
Edited by: Wafik S. El-Deiry © Humana Press Inc., Totowa, NJ

Table 1
Microsatellite Markers for Comprehensive Allelotyping *(9)*

D1S1665 (1p21)	D8S261 (8p22)	D16S753 (16p11)
D1S552 (1p32)	D8S1106 (8p22)	D16S764 (16p12)
D1S1589 (1q24)	D8S1119 (8q21.3)	D16S3095 (16q21)
D1S549 (1q32)	D8S1179 (8q23)	D16S402 (16q24.2)
D2S2739 (2p13)	D9S1118 (9p12)	D17S947 (17p11)
D2S1400 (2p23)	D9S171 (9p21)	D17S974 (17p12)
D2S1384 (2q22)	D9S938 (9q21.3)	D17S1290 (17q22)
D2S1399 (2q32.1)	D9S934 (9q22.3)	D17S784 (17q25)
D3S1234 (3p14.2)	D10S191 (10p12.1)	D18S542 (18p11.1-3)
D3S2387 (3pter-24)	D10S2325 (10p12.3)	D18S976 (18p11.1-3)
D3S1764 (3q13.3)	D10S541 (10q22)	D18S46 (18q12.1-21.1)
D3S1262 (3q27)	D10S1237 (10q24.1)	D18S474 (18q12.3)
D4S2639 (4p15.2)	D11S1981 (11p14)	D19S586 (19p13.1-13.3)
D4S1601 (4p15.3)	D11S1999 (11p15.1)	D19S714 (19p13.1-13.3)
D4S1644 (4q26)	D11S860 (8q13.5)	D19S246 (19q13.2)
D4S2431 (4q31.3)	D11S4464 (8q23.1)	D19S589 (19q13.3)
D5S1470 (5p12)	D12S1042 (12p11.2)	D20S851 (20p12)
D5S807 (5p14)	D12S391 (12p12.1)	D20S482 (20p13)
D5S816 (5q21)	D12S78 (12q22)	D20S478 (20q11.2)
D5S393 (5q23)	D12S395 (12q24.1)	D20S480 (20q12)
D6S1017 (6p12)	D13S162 (13q21-31)	D21S1440 (21q21)
D6S1959 (6p21.3)	D13S170 (13q31)	D21S413 (21q22.1-22.3)
D6S1040 (6q21)	ERCC-5 (13q32-33)	D22S689 (22q11)
D6S305 (6q25-27)	D14S48 (14q24.3-32)	D22S683 (22q11.2)
D7S817 (7p14)	D14S306 (14q12-13)	
D7S1802 (7p15.1)	D15S643 (15q15)	
D7S820 (7q21.1)	D15S657 (15q25)	
D7S1824 (7q31.3)		

to the methods that use radiolabeling. Some recipes for stock supplies are different than the standard PCR protocols.

Regardless of the method, the interpretation of the data remains the same. The presence of specific microsatellite sequences is characterized and compared between DNA extracted from the tumor vs DNA from noncancerous tissue of the same patient. When DNA from noncancerous tissue is amplified and presents two distinctly sized alleles (reflecting differences from paternally vs maternally inherited alleles), that sample is labeled as "informative" (*see* **Fig. 1**). DNA extracted from the tumor of the same patient can then be amplified in the same manner and compared to analyze for the loss of either allele (e.g., loss of heterozygosity, LOH.) The respective material and paternal alleles can be analyzed with respect to allelic band intensity by a variety of methods including phosphoimaging or laser quantitation. The tumor samples may also exhibit abnormally sized allelic bands, which can indicate the presence of microsatellite instability (*see* **Subheading 1.4.**).

1.2. Loss of Genetic Material in Cancers Harboring Tumor Suppressor Genes

Chromosomal loss, detected by frequent allelic loss in simple tandem repeats, is thought to indicate the aberration and sequential inactivation of adjacent tumor sup-

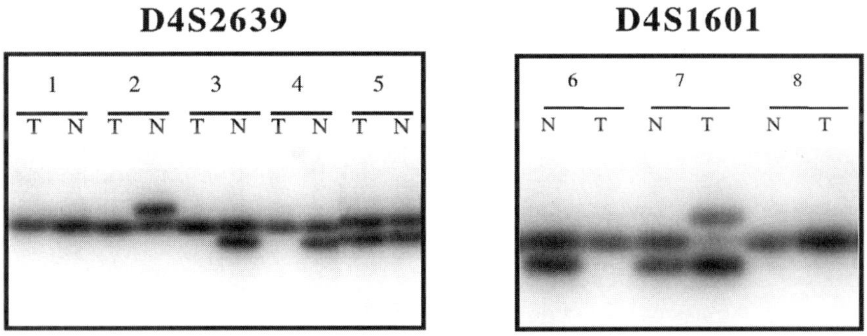

Fig. 1. Interpretation of representative data of PCR amplification of microsatellite loci. All lanes show the amplification products of DNA from tumor ("T") tissue and normal ("N") tissue from the eight different patients *(9)*. Lanes 1 and 8 are "noninformative," as the paternal allele is indiscernible from the maternal allele. Lanes 2–7 are "informative," as two distinctly sized alleles are present in normal tissue. Lane 5 is an example of retention of heterozygosity, as both alleles present in the normal tissue are also present in the tumor tissue of the sample patient. In contrast, lanes 2, 3, 4, and 6 are examples of loss of heterozygosity (LOH), as one allele that is present in normal tissue is no longer present in the tumor tissue. Lane 7 is an example of microsatellite instability, as one allele present in the normal tissue has shifted in molecular weight in the tumor tissue, indicating a defect in mismatch repair mechanisms.

pressor genes. These markers have since been used to explore the correlation between chromosomal loss and the development of several human malignancies. In earlier years, partial allelotyping analyses using markers representing only some of the chromosomal arms were performed on several human carcinomas, including carcinoma *in situ* in the bladder ***(10)*** and esophageal squamous cell carcinoma ***(11)***. As more polymorphic markers have been characterized, comprehensive allelotyping has become possible for a variety of different cancers. Among those studied are papillary thyroid cancer ***(12)***, salivary gland tumors ***(13)***, gastric cancer ***(9,14,15)***, pancreatic adenocarcinoma ***(16)***, nonfunctional pancreatic endocrine tumors ***(17)***, pancreatic acinar cell carcinoma ***(18)***, primary nasopharyngeal carcinoma ***(19,20)***, hepatocellular carcinoma ***(21)***, and intrahepatic cholangiocarcinoma ***(22)***, to name a few. These comprehensive studies have served to identify chromosomal regions that repeatedly exhibit aberrations in their respective tumors. Once identified, further extended studies can explore that chromosomal region in more detail and potentially ultimately identify the target of allelic loss harboring essential genetic information that has been inactivated.

1.3. Allelic Loss vs Allelic Imbalance in Cancer

True allelic loss of genetic material is detected by some molecular analyses such as Southern blotting of RFLP markers. However, other analyses such as PCR amplification of microsatellite markers generally can only suggest allelic imbalance without additional DNA analyses demonstrating either loss or amplification of one allele relative to the other. PCR amplification is not quantitative enough by itself to determine the mechanism of altered allelic ratio in tumors vs paired normal samples. Thus, unless the tumor sample is of the highest homogeneous quality, without any contaminating normal tissue present to amplify a signal, additional studies such as Southern blot or FISH analyses are required to define loss of amplification as the source of altered allelic ratios in tumor samples.

1.4. Microsatellite Instability

Instability of the repeated sequences within microsatellite loci, aptly named "microsatellite instability," has also been observed in human cancers, including colon cancer ***(23–26)***, gastric cancer ***(27–41)***, and uterine cancer ***(42)***. Length variation of the repeated sequences at these microsatellite loci, both contraction and expansion, are observed as abnormal-sized allelic bands (*see* **Fig. 1**). As microsatellite sequences are known to be relatively stable, a high rate of microsatellite instability is now recognized to be indicative of a deficiency of genetic products responsible for DNA mismatch repair. Mutations that inactivate these MMR genes have been found to be inherited among pedigrees of HNPCC ***(43,44)***. Microsatellite instability has since been used as a marker for this specific hereditary colon cancer predisposition trait and high risk of cancer (reviewed in **ref.** ***45***).

2. Materials

2.1. End Labeling of the PCR Primers with ^{32}P

1. Microsatellite primers (20 μ*M*) (for representative examples, *see* **Table 1**).
2. T4 kinase (10 U/μL) (NEB cat. no. 201S).
3. 10× Kinase buffer (NEB cat. no. 201S).
4. [γ-^{32}P]ATP (150 μCi/μL).
5. dH_2O.
6. Pharmacia Microspin G-25 (cat. no. 27-5325).
7. Boehringer Master Mix (cat. no. 1636 103).
8. DNA template (25 μg/μL)

2.2. Internal Labeling of the PCR Primers with ^{32}P

1. dNTP set (100 m*M* solutions) (Amersham cat. no. 27-2035-01).
2. Tris-HCl (1 *M*, pH 8.8, and another solution at pH 7.5).
3. $(NH_4)_2SO_4$ (2 *M*).
4. β-Mercaptoethanol.
5. Purified bovine serum albumin (BSA) (10 mg/mL) (NEB cat. no. 007-BSA).
6. Nuclease-free dH_2O.
7. $MgCl_2$.
8. EDTA (0.2 *M*, pH 7.5).
9. Microsatellite primers (*see* **Table 1**).
10. Amplitaq (Perkin Elmer cat no. N808-0153).
11. $^{32}\alpha$-P-dCTP (ICN cat no. 39011X.2).

2.3. Gel Electrophoreses and Autoradiography

1. Acrylamide.
2. bis-Acrylamide.
3. Formamide (Sigma cat. no. F7503).
4. Urea.
5. Tris-HCl (1 *M*, pH 7.5).
6. Trizma base.
7. Boric acid.
8. EDTA (Whitaker cat. no. 16-004Y).
9. Bromophenol blue.
10. Xylene cyanol.

11. Methanol.
12. Acetic acid.
13. Owl rig vertical electrophoreses apparatus, including plates.
14. Glass shield: nontoxic glass coating (Genomyx cat. no. GLSO4B901).
15. Filter paper (Bio-Rad cat. no. 1650962).
16. Gel dryer (Bio-Rad model 583).
17. Flat gel loading tips.
18. TEMED.
19. Ammonium persulfate.
20. Film (Kodak X-OMAT AR).

2.4. Flourescent Labeling of Primers

1. 96-Well PCR plates (Robbins, cat. no. 1055-00-0).
2. GENUNC tape (Nunc cat. no. 2-32689).
3. Flat gel loading tip.
4. Glass plates:
   a. Notched, optically pure glass plate 16.75 in (PE cat. no. 401069).
   b. Plain, optically pure glass plate 17.75 in (PE cat. no. 401069).
5. Pair of spacers 16.75 in (PE cat. no. 400571).
6. 24-Well square-tooth comb (PE cat. no. 401444) (24-cm well-to-read, nonstretch config.).
7. 36-Well square-tooth comb (PE cat. no. 401497).
8. Gel tape from CBS scientific, GT-72-20 Permacel stretch wide tape.
9. Magnetooptical disk (Online Distribution, cat. no. DGR256MB).
10. 373 DNA sequencer and accompanying ABI Prism GeneScan Analysis Software (Perkin Elmer).
11. Microsatellite primers (*see* **Table 1**).
12. 2 m*M* dNTP mix.
13. Formamide.
14. EDTA.
15. 10× PCR buffer.
16. DMSO.
17. Amplitaq.
18. Tris-HCl.
19. Boric acid.
20. EDTA.
21. Acrylamide.
22. Bis-Acrylamide.
23. Urea.
24. 0.22-µm filter.
25. Ethanol.
26. TEMED.
27. Ammonium persulfate.

3. Methods

3.1. PCR Amplification of Microsatellite Loci

3.1.1. Kinase End Label of the PCR Primer with ^{32}P

1. Prepare the following primer mix for each sample:

Forward primer (20 µM)	10 µL
10× Kinase buffer	4.0 µL

dH_2O	4.5 µL
$^{32}\gamma$P-ATP (150 µC/µL)	1.0 µL*
T4 kinase (10 U/µL)	0.5 µL
	20 µL total

*Adjust to obtain optimal activity.

2. Incubate at 37°C for 18 min.
3. Incubate at 68°C for 5 min.
4. Clean up column using Pharmacia MicroSpin G-25:
   a. Resuspend buffer by vortexing, loosen the cap, and snap off bottom closure.
   b. Spin off packing buffer for 2 min at 2500 xg.
   c. Place column into a clean 1.7-mL screw-cap tube.
   d. Load product on to center of column.
   e. Spin for 2 min at 2500 xg.
5. Add 9 µL of the reverse primer to the mix.
6. Measure activity of 0.5 µL in a scintillation counter.
7. Prepare the following PCR reaction for each sample (*see* **Notes 1** and **2**):

Boehringer master mix	5 µL
DH_2O	3 µL
γ-^{32}P map pair mix (from above)	1 µL
DNA template (25 µg/µL)	1 µL
	10 µL total

3.1.2. Internally Label PCR Product via Radiolabeled "C" Integration (see **Note 3***)*

1. Prepare LoC mix: In a 1.7 mL tube, add 100 µL of 10 m*M* solutions of dATP, dGTP, and dTTP. Add 2.5 µL of dCTP for a total volume of 302.5 µL of solution. Always keep solution on ice and store at –20°C.
2. Prepare 10× RDA buffer: In a sterile 100-mL bottle, add 33.5 mL of Tris-HCl (1 *M*, pH 8.8, 25°C), 4.0 mL of $(NH_4)_2SO_4$ (2 *M*), 352 µL of β-mercaptoethanol, and 5.0 mL of purified BSA (10 mg/mL). Bring the solution up to a final volume of 50 mL by adding 7.15 mL of nuclease-free dH_2O. Make 1-mL aliquots and store at –20°C.
3. Prepare the LoTE buffer (3 m*M* Tris-HCl/0.2 m*M* EDTA): In a sterile 1-L bottle, add 3 mL of Tris-HCl (1 *M*, pH 7.5), 1 mL of EDTA (0.2 *M*, pH 7.5), and 900 mL of dH_2O. Adjust the pH to 7.5 and bring to volume up to 1 L with dH_2O. Store solution at 4°C.
4. Prepare the following PCR reaction for each template (*see* **Notes 1** and **2**):

LoC	0.2 µL
RDA buffer	0.8 µL
$MgCl_2$ (50 m*M*)	0.4 µL
LoTE	5.6 µL
Forward primer	0.15 µL
Reverse primer	0.15 µL
Taq	0.5 µL
[α-^{32}P]dCTP (ICN cat. no. 39011X.2)	0.2 µL
DNA template (25 µg/µL)	2.0 µL
	10 µL

3.1.3. Amplification of the PCR Product

1. Place reaction mixture in a thermocycler and incubate at the following temperatures: an initial denaturation step at 94°C for 2 min, followed by 35 cycles of 94°C for 30 s, then the

appropriate annealing temperature for 30 s (*see* **Note 4**), and 72°C for 60 s. Incubate at a final temperature of 72°C for 5 min.

3.1.4. Running a 7% Analytical Acrylamide Gel

1. Prepare 10× TBE (Tris-borate electrophoreses) buffer: In a sterile 2-L bottle, add 216 g of Trizma base, 110 g of boric acid, and 11.7 g of EDTA to approximately 1800 mL of dH_2O. Adjust pH as needed (it should be between 8.1 and 8.3). Bring volume up to 2 L with dH_2O.
2. Prepare a 40% acrylamide stock solution: In a dark bottle, add 95 g of acrylamide and 5 g of bis-acrylamide. Bring volume up to 250 mL with dH_2O. Place solution over gentle heat and stir until completely dissolved. Store at 4°C.
3. Prepare a 7% acrylamide working solution: In a dark bottle, add 175 mL of the 40% acrylamide stock solution from **step 2.**, 320 mL of formamide, 336 g of urea, and 100 mL of 10× TBE.
4. Prepare the gel-fixing solution: In a sterile 2-L bottle, add 100 mL of methanol and 100 mL of acetic acid to 1800 mL of dH_2O (*see* **Note 5**).
5. Prepare the formamide loading buffer: In a sterile 100-mL bottle, add 90 mL of formamide, 10 mL of 10× TBE, 20 mg of bromophenol blue, and 20 mg of xylene cyanol. Mix the solution until thoroughly dissolved and filter solution through into a clean 100-mL bottle. Store at 4°C for up to 1 yr or at –20°C for long-term storage.

3.1.5. Pouring the Gel

1. Prepare the glass plates to cast a 7% acrylamide gel. When washing your plates, designate one side of each plate as "rough" by placing a small piece of tape. This side should always face outside, whereas the "smooth" side faces the gel. Also, designate the notched plate as "top" and the plain gel as "bottom." The "smooth" side of the "top" plate should be treated with a siliconizing agent (glass shield, see materials) after it has been washed and dried. This will facilitate the separation of the plates after the samples have been run.
2. Place the bottom plate, rough side down. Moisten the spacers, and align them on both sides. Be sure that the smooth side is completely dry by wiping it down with a Kimwipe. Place the top plate on the bottom plate; the two smooth sides should face one another. Place the top plate so that the notched half overhangs the bottom plane, leaving the bottom plate's smooth side exposed. This exposed area can be used as a "pool" in which you can pour the gel mix.
3. Place 80 mL of the 7% acrylamide gel solution in a clean squirt bottle. Add 36 µL of TEMED and 475 µL of fresh 10%APS into 80 ml of 7% acrylamide gel solution. Pour the gel immediately (*see* **Note 6**). The gel is cast by the capillary action of the space between the plates, pulling the liquid toward the notched top. If bubbles form as the gel mix is traveling, pull gently on the notched half of the top plate (up toward you) and gently knock the glass in the area. This should be done carefully so as not to disturb the rest of the gel. When bubbles form early in the run (when the liquid has traveled less than halfway), gentle knocking can usually get rid of the bubbles.
4. Once the gel mix has traveled along the length of the bottom plate, begin to slide the top plate back so that it is aligned with the bottom. Do this slowly, making sure there is enough gel mix left in the "pool" to cast this bottom third.
5. Place three pairs of clamps, tightening the bottom, middle, and top of the plates. *Do not* place the clamp on the notched area.
6. Insert the comb, straight edge in (the teeth should face out) and clamp the plates down around the comb. Allow the gel to polymerize for at least 1 h.
7. When the gel is ready, take off the clamps and clean off the plates using dH_2O and Kimwipes.

8. Wet the comb with 1× TBE and pull it gently out. Clean out the area, clearing out any extra acrylamide. Slowly insert the comb, teeth side in, until the teeth first reach the acrylamide front. This will maximize the volume you can load.
9. Fill the lower buffer chamber of the Owl rig with 1× TBE. Set the gel in place, adjust the knobs tightly, and add 1× TBE to the upper chamber.

3.1.6. Electrophorese the PCR Products

1. Prepare the samples by adding 2 µL of formamide loading buffer to the PCR products.
2. Heat the samples to 80°C for 3 min and then immediately chill on ice.
3. Load approximately 3.5–5.0 µL of the sample onto your gel using a flat gel tip (*see* **Note 7**).
4. Electrophorese samples at 2200 V and ~80 W for approximately 1.5 h. The gel should be heated evenly at a constant temperature, being careful not to exceed 50°C (see **Note 8**).
5. Separate the glass plates using a spatula; the gel should easily be separated from the top plate and remain on the bottom plate. Place the whole bottom plate, with the gel, into a tub of the gel-fixing solution for 15–20 min (*see* **Note 9**).
6. Transfer gel by gently placing a sheet of filter paper (Bio-Rad) directly onto the gel, being careful to avoid air bubbles (*see* **Note 10**). Cover the gel with plastic wrap, applying gentle pressure again to get rid of any air bubbles. Do not fold over the edges.
7. Place the wrapped gel into the gel dryer (Bio-Rad model 583.) Allow to dry for 2 h at 50°C.
8. Expose the dried gel to film (Kodak X-OMAT AR) overnight (*see* **Note 11**).

3.2. ABI Instrument Detection

3.2.1. Fluorescent Labeling of Microsatellite Markers by PCR

1. Prepare the primers to be used: For example, MP primers (Research Genetics) come as 8-µm stocks, 200 µL each (one F—forward, one R—reverse); one is labeled, one is not. To use 0.5 µL of each one per reaction, make a primer mix of equal amounts of F + R, then use 1 µL of mix per reaction, i.e., 50 µL of each stock.
2. Prepare the dNTP solutions. This can done in two ways: dilute 25 or 2.5 m*M* stocks already made for regular PCR to 2m*M* (10×); or make from original Pharmacia by using known (ones for which the OD has been measured) 100 m*M* stocks of separate dNTPs: mix 60 µL of each stock together with 2760 µL dH_2O to make 3 mL of a 2m*M* stock dNTP mix.
3. Prepare stop buffer: To a sterile 100-mL bottle, add 38 mL Formamide and 1.6 mL of EDTA (0.5 M). Store at –20°C.
4. Prepare the following PCR reaction for each template (*see* **Note 12**):
   a. Label and/or cut to size the 96-well plate.
   b. Pipette 50 ng (1µL) of the DNA template to appropriately labeled wells. Add 1 µL of H_2O to the "negative control" well. Set the plate aside.
   c. Prepare the following reaction for each template (*see* **Notes 1** and **2**):

DI H_2O	4.94 µL
10× PCR buffer	1.0 µL
2 m*M* dNTP mix	1.0 µL
DMSO (*see* **Note 13**)	1.0 µL
Primer mix	1.0 µL
Amplitaq	0.06 µL
	9.0 µL

   d. Mix reaction well and spin solution down quickly, keeping as dark as feasible. Aliquot 9 µL of the reaction to each well.
5. Pipet 20 µL of PCR oil over the reaction mix and cover rows with GENUNC tape.

6. Remove the white grid and thermistor from the PCR thermocycler block and set the plate in wells. Use the plastic lid/foam lid cover and shut lid (will be tight).
7. Run program: You will use a simulated tube control, which requires you to enter a calibration factor (CAL=). For this recipe, CAL=300 (10× the total volume in the well, including oil, *see* **Note 14**). A typical program would be to incubate the samples for 35 cycles of denaturing temperature of 93°C for 20 s, then to the appropriate annealing temperature for 30 s, and a final extension temperature of 70°C for 30 s; followed by an extension temperature of 70°C for 5 min.
8. Add 40 μL of stop buffer to samples (*see* **Note 15**), under oil, and freeze at –20°C (*see* **Note 16**).
9. Pooling reaction to run on gel: It is possible to run reactions together as long as the PCR products you want to analyze are either very different size ranges or are labeled with different colors (HEX, FAM, TET). Also added is an internal size standard labeled with ROX standards, plus more stop buffer if needed for volume. Per lane, add 0.5 μL ROX std., 1 μL each reaction, plus enough stop buffer to make total volume = 4 μL if it is less. Only 3 μL can be loaded (i.e., for one sample with two products labeled; pool 0.5 μL ROX350, 1 μL PCR product 1, 1 μL PCR product 2, and 1.5 μL stop).
10. To run samples on gel, heat to 80°C for 3 min, then immediately chill on ice. Use flat gel-loading tip.

3.2.2. Pouring and Running a Gel for the ABI Instrument

1. Prepare a 6% acrylamide, 8 *M* urea gel: To a sterile dark bottle, add 960 g of urea, 200 mL of 10× TBE (*see* **Note 17**), and approximately 600 mL of dH_2O. Heat gently until dissolved into solution. Add 300 mL of 40% acrylamide. Bring volume up to 2 L and filter solution (using 0.22 μm filter) into a clean dark bottle.
2. Prepare glass plate to cast the 6% acrylamide gel (*see* **Note 18**): Place spacers on outside edges of notched plate and lay plain plate on top (freshly cleaned insides together), aligning edges. Using no clamps, place one continuous piece of permacel tape around sides and bottom of plates. Be sure there are no wrinkles. Pull tape tight (tape stretches) and seal to glass sides, being careful to seal beveled edges well. Put an extra layer of tape around the bottom, and then reinforce the bottom corners with one horizontal layer at each. Gel is now ready to pour.
3. For one gel, pour 80 mL of the filtered mix into a squirt bottle and add 400 μL of fresh 10%APS and 40 μL TEMED.
4. Pour gel (taking same precautions as under **Subheading 3.1.5.**), get rid of bubbles, and insert comb by pushing it all the way to the glass and then running a weighing spatula between the top of the comb and the glass to measure them away from each other. Allow to polymerize for a minimum of 1 h and a maximum of 18 h (*see* **Notes 6** and **19**).
5. Before use, clean all excess gel material from the glass plates on both sides with DI H_2O. Then clean the bottom 2/3 area (where the laser will read). The plates should be clean, clear, and streak-free when washed with dH_2O and Kimwipes.
6. When placing the gel in the 373 DNA sequencer, set plates as far right as possible (*see* **Note 20**). Close the laser stop bar (use a lint magnet on inside facing glass and inside the stop bar, the outside facing glass) and lock down.
7. Follow the instructions on the Genescan setup card from here. Remember to change the running time on the 373, to use a sample sheet, and to give the collection file a different name than the sample sheet.
8. Plate-check before adding 1× TBE to upper chamber, check for leaks and plate-check again, then add buffer to lower chamber. Clean wells with syringe and load the products using the flat gel-loading tip.

9. (Follow card) Press Genescan run on 373, wait for voltage to come on, and PRESS COLLECT BUTTON on GS collection screen. Can leave over night or even over the weekend.
10. To disassemble, unplug electric leads, release stop bar, and lift out gel and upper chamber attached to carry to a sink to dump. Clean plates and retape to store.
11. Magnetooptical drive must be on before the computer for it to recognize it.

4. Notes

1. Make 1–2 extra reactions worth of mix to allow for pipet error.
2. Keep DNA polymerase at –20°C until needed. Keep on ice while in use, and return to freezer immediately.
3. This is an alternative method for radiolabeling the PCR product.
4. Most primers will anneal at 50°C, although the optimal annealing temperature for each individual primer may need to be tested.
5. Always pour acid into water. The gel-fixing solution can used for 3–4 gels before being discarded.
6. Pour some of the solution (~1 mL) into a scintillation vial to ensure the gel has polymerized correctly before proceeding. After allowing the gel to polymerize for approximately 1 h, it should be consistent and adhere strongly when probed with a spatula. If the gel does not stick and instead feels too liquidy, do not proceed and pour a fresh gel using a fresh acrylamide solution and APS.
7. Volume will vary according to the size of the comb used, but as little as 3.5 μL is usually sufficient for a signal.
8. Running time may again vary, although allowing the leading dye to travel ~20 cm from the top of the well is sufficient for proper separation of informative alleles.
9. After the gel has been soaked appropriately, it can easily slip off the bottom plate. Therefore care should be taken when taking it out of the solution.
10. Allow the paper to absorb for a few moments, becoming completely saturated with liquid, and then carefully pull the paper off. It is helpful to check a corner first before proceeding, ensuring that the gel has indeed adhered to the paper. However, if the gel has been properly treated in the gel fixing solution, it should transfer easily.
11. Exposure time will vary with the current activity of ^{32}P used.
12. The most important thing to remember is to keep stock primers at as low a light level as possible while working with them and in the dark when not in use.
13. Substituting DMSO with dH_2O can yield more product.
14. The program will not start running until you enter this (unlike regular PCR), so do not forget. Also, use a few degrees lower a melting temperature than you would have planned, i.e., 93°C if you plan 95°C.
15. Amount of buffer to add is 40 μL, but (per ABI) this can differ according to the strength of the fluorescent label, e.g., HEX is twice as strong as FAM, so add 40 μL to a HEX reaction, 20 μL to a FAM reaction.
16. This seems to be important for equal mixing, as opposed to adding stop and pooling immediately.
17. 10× TBE must be prefiltered with a 0.22-μm filter.
18. Plates must be clean and free from possible fluorescing material, i.e., dust, lint, soap film, EtOH film, etc. Use Alconox or Micro soap to wash and DI H_2O to rinse. Use Kimwipes to dry "insides" of plates clean and streak-free. Use "magnetic" lint attractor brush to clear off any lint, etc.
19. Wells are better if gel is allowed to polymerize overnight.
20. By doing this, you will keep all lanes in the read area. You may occasionally lose lane 1 if this is not done.

Acknowledgment

This work was supported by National Cancer Institute grant CA67900-06.

References

1. Vogelstein, B., Fearon, E. R., Kern, S. E., et al. (1989) Allelotype of colorectal carcinomas. *Science* **244**, 207–211.
2. Jeffreys, A. J., Wilson, V., and Thein, S. L. (1985) Hypervariable "minisatellite" regions in human DNA. *Nature* **314**, 67–73.
3. Nakamura, Y., Leppert, M., O'Connell, P., et al. (1987) Variable number of tandem repeat (VNTR) markers for human gene mapping. *Science* **235**, 1616–1622.
4. Dib, C., Faure, S., Fizames, C., et al. (1996) A comprehensive genetic map of the human genome based on 5,264 microsatellites. *Nature* **380**, 152–154.
5. Hearne, C. M., Ghosh, S., and Todd, J. A. (1992) Microsatellites for linkage analysis of genetic traits. *Trends Genet.* **8**, 288–294.
6. Goelz, S. E., Hamilton, S. R., and Vogelstein, B. (1985) Purification of DNA from formaldehyde fixed and paraffin embedded human tissue. *Biochem. Biophys. Res. Commun.* **130**, 118–126.
7. Fearon, E. R., Feinberg, A. P., Hamilton, S. H., and Vogelstein, B. (1985) Loss of genes on the short arm of chromosome 11 in bladder cancer. *Nature* **318**, 377–380.
8. Weber, J. L. and May, P. E. (1989) Abundant class of human DNA polymorphisms which can be typed using the polymerase chain reaction. *Am. J. Hum. Genet.* **44**, 388–396.
9. Yustein, A. S., Harper, J. C., Petroni, G. R., Cummings, O. W., Moskaluk, C. A., and Powell, S. M. (1999) Allelotype of gastric adenocarcinoma. *Cancer Res.* **59**, 1437–1441.
10. Rosin, M. P., Cairns, P., Epstein, J. I., Schoenberg, M. P., and Sidransky, D. (1995) Partial allelotype of carcinoma in situ of the human bladder. *Cancer Res.* **55**, 5213–5216.
11. Shibagaki, I., Shimada, Y., Wagata, T., Ikenaga, M., Imamura, M., and Ishizaki, K. (1994) Allelotype analysis of esophageal squamous cell carcinoma. *Cancer Res.* **54**, 2996–3000.
12. Califano, J. A., Johns, M. M., 3rd, Westra, W. H., et al. (1996) An allelotype of papillary thyroid cancer. *Int. J. Cancer* **69**, 442–444.
13. Johns, M. M., 3rd, Westra, W. H., Califano, J. A., Eisele, D., Koch, W. M., and Sidransky, D. (1996) Allelotype of salivary gland tumors. *Cancer Res.* **56**, 1151–1154.
14. Tamura, G., Sakata, K., Nishizuka, S., et al. (1996) Allelotype of adenoma and differentiated adenocarcinoma of the stomach. *J. Pathol.* **180**, 371–377.
15. Sano, T., Tsujino, T., Yoshida, K., et al. (1991) Frequent loss of heterozygosity on chromosomes 1q, 5q, and 17p in human gastric carcinomas. *Cancer Res.* **51**, 2926–2931.
16. Seymour, A. B., Hruban, R. H., Redston, M., et al. (1994) Allelotype of pancreatic adenocarcinoma. *Cancer Res.* **54**, 2761–2764.
17. Rigaud, G., Missiaglia, E., Moore, P. S., et al. (2001) High resolution allelotype of nonfunctional pancreatic endocrine tumors: identification of two molecular subgroups with clinical implications. *Cancer Res.* **61**, 285–292.
18. Rigaud, G., Moore, P. S., Zamboni, G., et al. (2000) Allelotype of pancreatic acinar cell carcinoma. *Int. J. Cancer* **88**, 772–777.
19. Shao, J. Y., Wang, H. Y., Huang, X. M., et al. (2000) Genome-wide allelotype analysis of sporadic primary nasopharyngeal carcinoma from southern China. *Int. J. Oncol.* **17**, 1267–1275.
20. Lo, K. W., Teo, P. M., Hui, A. B., et al. (2000) High resolution allelotype of microdissected primary nasopharyngeal carcinoma. *Cancer Res.* **60**, 3348–3353.
21. Roncalli, M., Borzio, M., Bianchi, P., and Laghi, L. (2000) Comprehensive allelotype study of hepatocellular carcinoma. *Hepatology* **32**, 876.

22. Kang, Y. K., Kim, Y. I., and Kim, W. H. (2000) Allelotype analysis of intrahepatic cholangiocarcinoma. *Mod. Pathol.* **13**, 627–631.
23. Thibodeau, S. N., Bren, G., and Schaid, D. (1993) Microsatellite instability in cancer of the proximal colon. *Science* **260**, 816–819.
24. Aaltonen, L. A., Peltomaki, P., Leach, F. S., et al. (1993) Clues to the pathogenesis of familial colorectal cancer. *Science* **260**, 812–816.
25. Ionov, Y., Peinado, M. A., Malkhosyan, S., Shibata, D., and Perucho, M. (1993) Ubiquitous somatic mutations in simple repeated sequences reveal a new mechanism for colonic carcinogenesis. *Nature* **363**, 558–561.
26. Peltomaki, P., Lothe, R. A., Aaltonen, L. A., et al. (1993) Microsatellite instability is associated with tumors that characterize the hereditary non-polyposis colorectal carcinoma syndrome. *Cancer Res.* **53**, 5853–5855.
27. Lin, J. T., Wu, M. S., Shun, C. T., et al. (1995) Microsatellite instability in gastric carcinoma with special references to histopathology and cancer stages. *Eur. J. Cancer* **31A**, 1879–1882.
28. Leung, W. K., Kim, J. J., Kim, J. G., Graham, D. Y., and Sepulveda, A. R. (2000) Microsatellite instability in gastric intestinal metaplasia in patients with and without gastric cancer. *Am. J. Pathol.* **156**, 537–543.
29. Nakashima, H., Inoue, H., Mori, M., Ueo, H., Ikeda, M., and Akiyoshi, T. (1995) Microsatellite instability in Japanese gastric cancer. *Cancer* **75**, 1503–1507.
30. Renault, B., Calistri, D., Buonsanti, G., Nanni, O., Amadori, D., and Ranzani, G. N. (1996) Microsatellite instability and mutations of p53 and TGF-beta RII genes in gastric cancer. *Hum. Genet.* **98**, 601–607.
31. Rhyu, M. G., Park, W. S., and Meltzer, S. J. (1994) Microsatellite instability occurs frequently in human gastric carcinoma. *Oncogene* **9**, 29–32.
32. Rugge, M., Shiao, Y. H., Guido, M., and Bovo, D. (1999) Microsatellite instability and gastric cancer subtypes. *Hum. Pathol.* **30**, 108–109.
33. Sepulveda, A. R., Santos, A. C., Yamaoka, Y., et al. (1999) Marked differences in the frequency of microsatellite instability in gastric cancer from different countries. *Am. J. Gastroenterol.* **94**, 3034–3038.
34. Schneider, B. G., Pulitzer, D. R., Brown, R. D., et al. (1995) Allelic imbalance in gastric cancer: an affected site on chromosome arm 3p. *Genes Chromosomes Cancer* **13**, 263–271.
35. Semba, S., Yokozaki, H., Yasui, W., and Tahara, E. (1998) Frequent microsatellite instability and loss of heterozygosity in the region including BRCA1 (17q21) in young patients with gastric cancer. *Int. J. Oncol.* **12**, 1245–1251.
36. Shiao, Y. H., Bovo, D., Guido, M., et al. (1999) Microsatellite instability and/or loss of heterozygosity in young gastric cancer patients in Italy. *Int. J. Cancer* **82**, 59–62.
37. Strickler, J. G., Zheng, J., Shu, Q., Burgart, L. J., Alberts, S. R., and Shibata, D. (1994) p53 mutations and microsatellite instability in sporadic gastric cancer: when guardians fail. *Cancer Res.* **54**, 4750–4755.
38. Suzuki, H., Itoh, F., Toyota, M., et al. (1999) Distinct methylation pattern and microsatellite instability in sporadic gastric cancer. *Int. J. Cancer* **83**, 309–313.
39. Wang, Y., Shinmura, K., and Sugimura, H. (1997) [Microsatellite instability in gastric cancer with varied structure]. *Rinsho Byori* **45**, 651–655.
40. Wang, Y., Ke, Y., Ning, T., et al. (1998) [Studies of microsatellite instability in Chinese gastric cancer tissues]. *Zhonghua Yi Xue Yi Chuan Xue Za Zhi* **15**, 155–157.
41. Wu, M. S., Lee, C. W., Shun, C. T., et al. (1998) Clinicopathological significance of altered loci of replication error and microsatellite instability-associated mutations in gastric cancer. *Cancer Res.* **58**, 1494–1497.
42. Nishimura, M., Furumoto, H., Kato, T., Kamada, M., and Aono, T. (2000) Microsatellite instability is a late event in the carcinogenesis of uterine cervical cancer. *Gynecol. Oncol.* **79**, 201–206.

43. Leach, F. S., Nicolaides, N. C., Papadopoulos, N., et al. (1993) Mutations of a mutS homolog in hereditary nonpolyposis colorectal cancer. *Cell* **75**, 1215–1225.
44. Fishel, R., Lescoe, M. K., Rao, M. R., et al. (1993) The human mutator gene homolog MSH2 and its association with hereditary nonpolyposis colon cancer. *Cell* **75**, 1027–1038.
45. Maehara, Y., Oda, S., and Sugimachi, K. (2001) The instability within: problems in current analyses of microsatellite instability. *Mutat. Res.* **461**, 249–263.

19

Epidemiological Approaches to the Identification of Cancer Predisposition Genes

Timothy R. Rebbeck

1. High- and Low-Penetrance Genes

Commonly occurring forms of human cancer have a complex multifactorial etiology that involves the interaction of inherited genotypes and endogenous or exogenous environmental exposures. The ability to understand the inherited genetic causes of cancer in human populations therefore requires appropriate study designs and analytical methods.

In order to understand inherited genetic susceptibility to develop cancer in the general population, it is useful to think about more than one class of susceptibility genes. For simplicity, two classes of cancer susceptibility genes may be considered (**Table 1**). First, there may be a few genes with allelic variants that confer a high degree of risk to the individual. These genes will be referred to here as high-penetrance genes. Relatively few individuals in the population carry risk-increasing genotypes at these loci. Therefore, the population attributable risk (i.e., the proportion of cancer in the population that may be explained by these genotypes) is low. Because of the large magnitude of effects these genotypes have on cancer risk, one hallmark of high-penetrance genes is the creation of a Mendelian (usually autosomal dominant) pattern of cancer, sometimes involving multiple cancer sites that form a cancer syndrome. Germline mutations in high-penetrance tumor suppressor genes include *CDKN2a* in hereditary melanoma, *BRCA1* or *BRCA2* in hereditary breast and ovarian cancer, and *APC* in familial adenomatous polyposis of the colon.

Second, there are genes with relatively common disease-associated allele frequencies that confer a small to moderate cancer risk. However, since these variants are carried by a large number of individuals, the population attributable risk of cancer explained by these genes may be high. The majority of commonly occurring cancer in the general population occurs in the absence of a strong cancer family history. These tumors are likely to be caused by the interaction of many genetic and environmental factors. These genes will be denoted here as low-penetrance (**Table 1**), based on the relatively smaller size of their effects on an individual's cancer risk. Classes of genes usually considered to be low-penetrance include those involved in steroid hormone metabolism, DNA repair, immune surveillance, and carcinogen metabolism.

From: *Methods in Molecular Biology, Vol. 222: Tumor Suppressor Genes: Pathways and Isolation Strategies*
Edited by: Wafik S. El-Deiry © Humana Press Inc., Totowa, NJ

Table 1
High- and Low-Penetrance Genes

		Class of gene	
		High penetrance	Low penetrance
Characteristics	Variant allele frequency	Rare	Common
	Size of effects (individual)	Large	Small
	Number of genes	Few	Many
	Attributable risk (population)	Small	Large
Usual analytical methods		Segregation analysis Linkage analysis	Association studies
Typical sampling design		Family-based	Case–control, cohort

The different patterns of risk conferred by high- and low-penetrance genes also dictate that different analytical approaches may be used to identify them. Because high-penetrance genes often confer hereditary patterns of cancer risk, family-based study designs that make use of Mendelian principles may be an efficient and worthwhile approach to gene identification. However, low-penetrance genes typically do not confer patterns of hereditary cancer risk. As a result, methods that rely on the observation of Mendelian patterns of disease in families are unlikely to be successful in identifying low-penetrance genes responsible for the disease.

As suggested in **Table 1**, low-penetrance genes may explain a larger proportion of cancer in the general population than major genes. An estimate of the proportion of cancer attributable to a major gene and a polygene can be undertaken using the formula $100\% \times p \times (R - 1)/[p \times (R - 1) + 1]$ (*1*), where R is the risk conferred by the variant, and p is the variant allele frequency in the general population. Assuming a typical major gene variant confers a large relative risk of $R = 8$, and has a rare frequency of $p = 0.005$, the resulting attributable risk is approximately 3%. In contrast, an estimate of 20% attributable risk results if we assume a low risk ($R = 1.5$) but a common variant genotype frequency ($p = 0.5$) for a typical polygene variant. While these estimates are at best hypothetical approximations, they suggest that a sizable proportion of cancer may be explained by the effect of low-penetrance genotypes. These estimates further support the idea that studies of commonly occurring cancer etiology may need to consider both high and low penetrance genes to explain genetic susceptibility to develop cancer in the general population.

The distinction between high and low penetrance genes as shown **Table 1** is an oversimplification of the relationship between allele frequency and size of effects. Based on empirical observations, genetic variants with large, deleterious effects on cancer risk tend to be rare, while variants having smaller effects on risk tend to be more common. Furthermore, the conceptual model presented in **Table 1** does not preclude the possibility that a high penetrance may also behave as a low penetrance gene. For example, variants in the *APC* gene are associated with familial adenomatous polyposis of the colon. Germline *APC* variant carriers have an extremely high lifetime risk of developing colon cancer (reviewed by Foulkes, **ref.** *2*). In addition to these high penetrance manifestations, *APC* variants may also confer cancer risks consistent with the behavior of a low-penetrance genes. For example, the I1307K variant in the *APC* gene, found in approximately 7% of

U.S. Ashkenazi Jews, has recently been reported to confer a slightly increased risk of colon cancer ***(3)*** with an odds ratio effect of 1.5 (95% CI: 1.12–1.97). Thus, while some variants in a gene have characteristics of major genes because they confer extremely high cancer risk, other variants in the same gene may have effects that confer small increases in cancer risk.

2. Possible Explanations for Genotype–Cancer Relationships

The statistical inference of a genotype–cancer relationship can be explained by at least three phenomena. First, linkage disequilibrium (defined here as the nonrandom association of alleles in a population, e.g., Pr[AB]/Pr[A] Pr[B], where A and B are alleles at a locus) may explain an observed statistical association between genotype and disease. For example, the observed association may simply represent the fact that the genotype being studied is in linkage disequilibrium with the true causative allele or genotype at another locus. Second, spurious associations can be induced by improper study design or analysis methods. The most widely discussed of these potential errors is that of population stratification, described below. Third, there may be a true functional association between a genotype and disease, whereby these genotypes are causally related to the disease etiology. This is the type of relationship that is sought in genetic and molecular epidemiologic studies. A central goal of the methodologies described here is to maximize the chance that a causal relationship between genotype and disease will be found.

3. Genetic Epidemiologic Strategies for Cancer Gene Identification

Genetic epidemiologic approaches use the principles of Mendelian inheritance to evaluate the co-inheritance of a disease with a disease-causing gene. The gene being modeled is an unmeasured entity that is specified quantitatively assuming Mendelian models. The main goals of these analyses are (a) to determine the mode of inheritance of the putative disease gene (e.g., in complex segregation analysis) and (b) to identify the genomic location of that gene (e.g., genetic linkage analysis). The data required for these methods involve related individuals including pairs of relatives, nuclear families, or extended pedigrees. The statistical methods of linkage analysis are therefore aimed at determining the relationship of the putative (unmeasured) disease gene with disease, and the physical relationship of the unmeasured disease gene with a "marker" gene with a known genomic location.

Figure 1 presents an overview of the hierarchical strategies that may be applied in the identification of high-penetrance cancer susceptibility genes (for a more detailed description, *see* **ref. *4***). Generally, the steps that are undertaken begin with the evaluation of aggregation of disease in families (e.g., do some families have more disease than expected?) followed by an evaluation of patterns of inheritance by using methods of segregation analysis (i.e., does the disease appear to follow a Mendelian pattern of inheritance?), and finally linkage analytical methods that can be used to identify the genomic location of specific genes responsible for the observed pattern of disease in families. Subsequent to the localization of a relevant genomic region, molecular cloning techniques are used to isolate the specific DNA sequence associated with the cause of the disease in question.

It is notable that this hierarchical approach relies primarily on patterns of disease in families to infer the existence of a causative gene, and that little or no information about the structure or function of these genes is available until the gene is cloned and studied

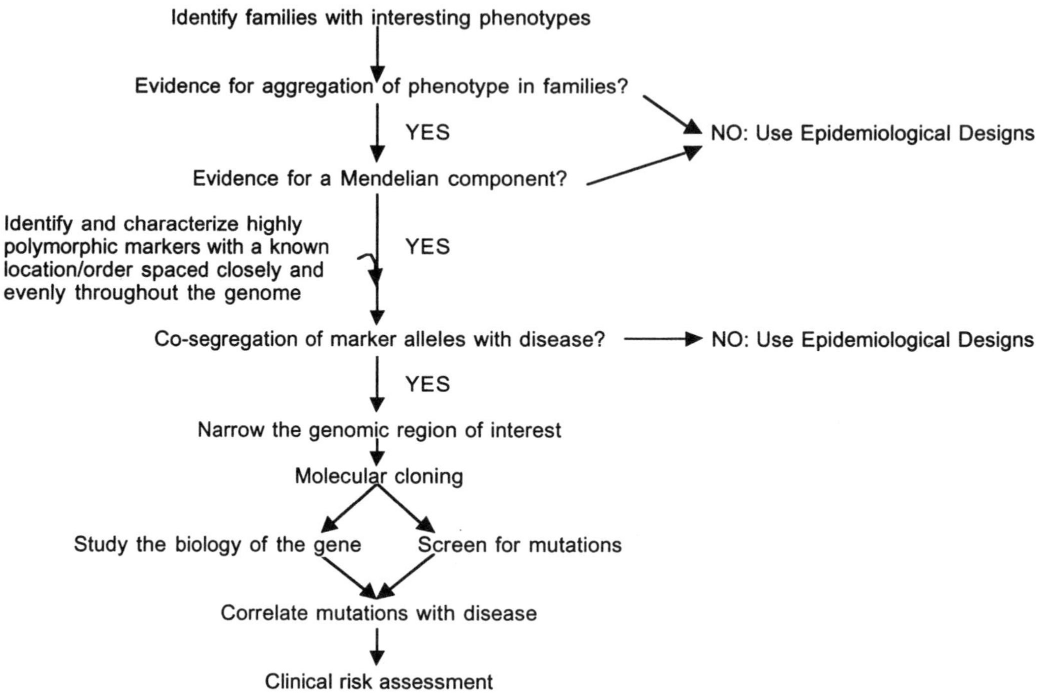

Fig. 1. Hierarchy of approaches used to identify high-penetrance genes.

using methods of molecular biology. With the completion of the human genome sequence and the characterization of genes and germline genetic variants, these modeling approaches take on a different meaning: less reliance on identifying genes, and more value in modeling the relationship of these genes with disease itself.

3.1. Studies of Mendelian Segregation

Analyses that determine a disease aggregates in families provide limited information about the existence of a single gene as an explanation for this aggregation. Familial aggregation can arise from the sharing of either genes or environments, or both, among relatives. Therefore, additional steps are required before it can be concluded that a single gene is responsible for the pattern of cancer in families.

An important step in this process is to evaluate whether the pattern of disease in a family is consistent with a Mendelian pattern of inheritance. For some cancer syndromes, this requires little in the way of statistical methods because the pattern is clearly consistent with the inheritance of a single high-penetrance gene. Examples include hereditary retinoblastoma, Li-Fraumeni syndrome, and others. However, the majority of cancer does not follow a simple pattern of Mendelian inheritance, and sophisticated statistical models have been developed that can be used to evaluate objectively the evidence that a single gene may be acting to cause disease.

3.1.1. Classical Segregation Analysis Approaches

Classical methods have been developed that model whether the probability of an offspring of a particular mating type will be affected with the disease (**ref. 5**, pp. 236–242).

These methods are based on the expected proportion (i.e., based on Mendelian principles) of affected children of a given mating type ***(6)***. Although they are useful in some circumstances, these methods are limited in their ability to account for the potentially complex patterns of disease and its etiology. In particular, these models become complicated when the parental mating type is not known (as is often the case). In addition, correction for ascertainment of study subjects may be required to obtain unbiased estimates of the proportions of interest. Determination and correction for ascertainment in these studies can be difficult. Because more flexible models of segregation anlaysis are available, these models will not be described in detail here.

3.1.2. Likelihood-Based Segregation Analysis Models

Complex segregation analysis has been developed to account for the observed patterns of disease in family data. The principle of these likelihood-based methods is to specify a statistical model for which the observed data are most likely. Thus, likelihood-based approaches model past events with known outcomes. The first step in specifying these models is to specify a likelihood function ***(7)***. For known, fixed data points x, and a parameter or set of parameters ρ that you want to estimate, it is possible to write:

$$L_n(\theta) = f(x_1, x_2, \ldots, x_n | \theta) = f(x_1 | \theta), f(x_2 | \theta), \ldots, f(x_n | \theta) = \Pi_i f(x_1 | \theta)$$

where $x_1, x_2, \ldots, x_n$ are usually assumed to represent a random sample of data, and θ represents parameters of interest, as described below. It is common to work in log likelihood terms so that computations can be made in terms of sums of log likelihoods, i.e., $\log L_n(\theta) = \log[\Pi_i f(x_i; \theta)] = T_i \log f(x_i; \theta)$. The maximum likelihood estimator (MLE) of θ is the value that maximizes $L_n(\theta)$ or $\log L_n(\theta)$. To find the MLE, take first derivative of $\log L_n(\theta)$, set this derivative equal to zero, and solve the equation for θ.

This principle can be applied to pedigree data to write the likelihood of the observed pedigree data given some parameters that describe the family. Three components can be used to describe a pedigree completely, as depicted in **Fig. 2** ***(8)***. First, it is assumed that a single factor (possibly a gene) exists that is causative of disease. A prior probability (PRIOR) Pr(g) of this factor is the probability that a randomly chosen individual will have factor g. These models generally assume that random mating has occurred with respect to the disease gene of interest. Second, the models may specify the transmission (TRANS) of this factor across generations ***(9)***. The model thus considers a mode of inheritance that may or may not be consistent with a single gene effect. For example, if we set TRANS = Pr (offspring genotype|father genotype, mother genotype) = τ, then the following would model a factor that is consistent with a major gene such that τ_1 = Pr (AA|AA, AA) = 1.0, τ_2 = Pr (AA|AA, Aa) = 0.5, and τ_3 = Pr (AA|aa, aa) = 0.0. Models that specify the values of the τ's can be used to determine if the pattern of transmission is consistent with a single gene.

Finally, it is important to model the relationship of genotype to phenotype using a penetrance function (PEN), which is defined as Pr (phenotype|genotype). For example, for an autosomal dominant disorder with complete penetrance, we may specify Pr (Affected|g) = 1 if g = AA, Aa, and 0 if g = aa, assuming that A is the allele associated with disease. However, more complex models can be specified that allow the penetrance function to consider age-specific penetrance, incomplete penetrance, etc. Similarly, the penetrance function can be used to evaluate whether the disease is consistent with dominant, recessive, or other modes of genetic inheritance.

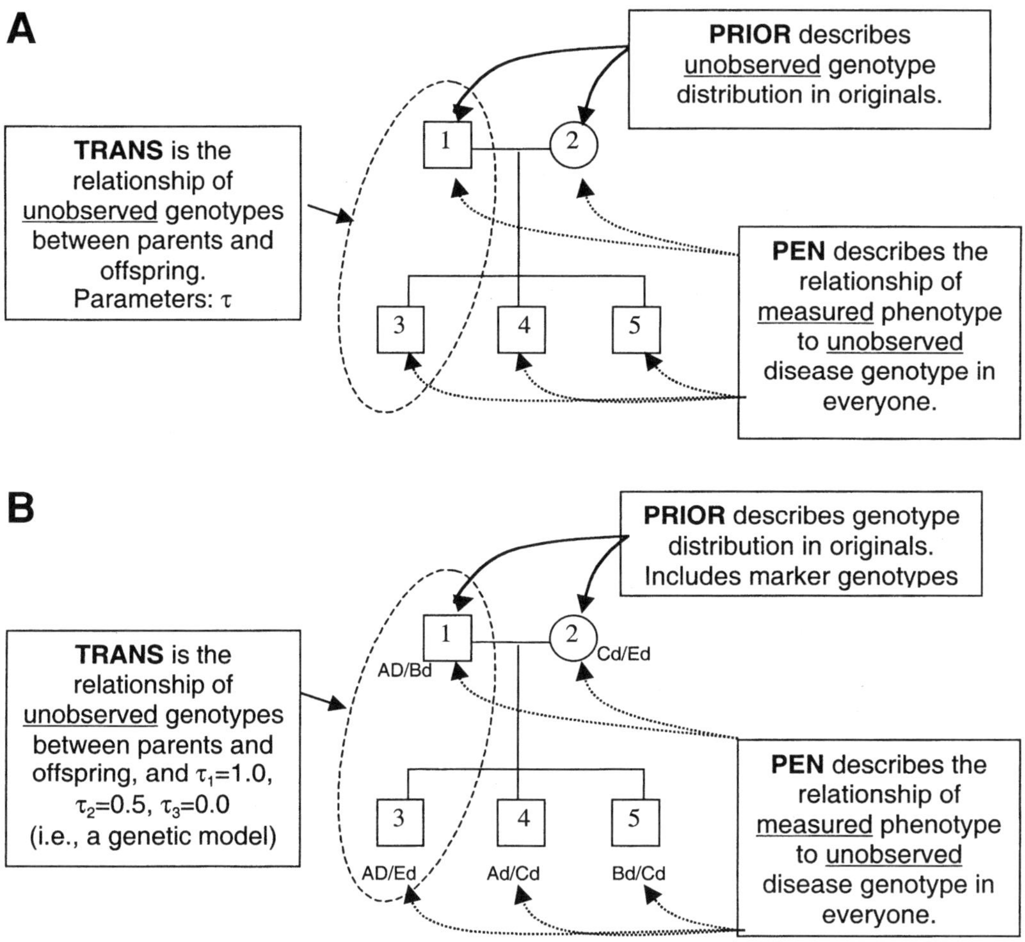

Fig. 2. Pedigree likelihoods: **(A)** Components of the pedigree likelihood used in segregation analysis; **(B)** Components of the pedigree likelihood used in linkage analysis.

Inferences from these models are made by comparing the likelihoods resulting from two or models using a variety of statistics that assess which model best fits the data. This best fitting model can then be used to determine whether a single gene, environmental exposure, or some other combination of factors explains the pattern of disease in families.

There are many advantages to undertaking a segregation analysis. First, the models provide information about whether a single gene can be detected that explains the pattern of disease in families, and thus may provide impetus for future genetic studies of the disease, or redirection of efforts in other directions. For example, the segregation analysis can be used to identify specific types of families that may be more likely to have a single gene etiology, and would thus be more amenable to genetic linkage analysis (*see* **Subheading 3.2.**). Segregation analysis models can also estimate the frequency of disease-causing mutations in the population, and possibly determine the mode of inheritance of the genetic disorder. In both cases, this information may be valuable in the genetic characterization of disease, and may be required for subsequent modeling studies using genetic linkage analysis.

3.2. Studies of Genetic Linkage

Once there is evidence for a genetic etiology for a disease (possibly coming from formal segregation analyses), families are identified that are likely candidates for having such a gene. These families tend to have more clearly Mendelian patterns of disease, and may have patterns that include an earlier age of diagnosis or other clinical features that hallmark apparently hereditary forms of the disease. Rarely, patterns of multiple cancers can be observed that are consistent with hereditary syndromes.

The goal of genetic linkage analysis is to track the co-segregation of disease with genomic DNA biomarkers across relatives. Two general approaches to linkage analysis exist using Identity by State (IBS) and Identity by Descent (IBD) methods, as described below.

3.2.1. IBS Approaches

IBS occurs when two alleles are identical on inspection, regardless of their ancestral source. Alleles that are IBS are not always IBD. Methods that use IBS to infer genetic linkage rely on the principle that relatives with the disease may share genes IBS more frequently than expected for family members with a particular degree of relationship ***(10)***. This approach is valuable because it does not require that a penetrance model or mode of inheritance be specified, and these models can be efficient in that they may require only affected individuals.

Using Mendel's laws it is possible to predict the probability that two relatives will share an allele IBS. IBS linkage analysis approaches simply compare the observed distribution of alleles among relatives with specific degrees of relationship to the expected distribution. If the observed and expected distributions are not different from one another, then the null hypothesis (no linkage) is true.

In order to determine this objectively, it is possible to create a test statistic that evaluates whether the observed sharing of alleles is greater than that expected under the null hypothesis. For example, compute Z_{ij} over all pairs of typed relatives i and j such that $Z_{ij} = \frac{1}{4}\{[\delta_{11}; f(p_1)] + [\delta_{12}; f(p_1)] + [\delta_{21}; f(p_2)] + [\delta_{22}; f(p_2)]\}$, where δ is 1 if an allele is shared and 0 if not, and $f(p)$ is some weighting function of the allele frequency ***(11)***.

3.2.2. IBD Approaches

IBD models require that identical alleles have been inherited from a common ancestor. Alleles that are IBD are always IBS. The principle used in these approaches is to track the co-segregation of alleles from ancestors to descendants assuming specific models of Mendelian inheritance. These models may required data on both affected and unaffected relatives, and usually require pedigree data with information about parental genotypes for the individuals being studied. However, analysis can be done using affecteds only, pairs of relatives, or whole pedigrees.

The principle of this approach is to measure the likelihood that a genetic marker with a known genomic location co-segregates with an inferred (i.e., unknown, unmeasured) disease gene. If evidence for linkage is found with a particular genetic marker, it is assumed that the marker lies nearby the disease gene of interest.

The most common IBD approach is that of lod score linkage analysis using likelihood-based models similar to those described above ***(12,13)***. The lod score is defined as

$$Z(\theta) = \log_{10}\frac{L(\text{family data}|\theta)}{L(\text{family data}|\theta = ½)}$$

where θ is a recombination fraction defined as the probability that a gamete produced by a parent is a recombinant, L is the likelihood specified in a manner similar to that described above for segregation analysis. Inferences about the presence or absence of genetic linkage can be made as follows. A value of $Z > 3$ provides evidence for linkage, while a value of $Z < -2$ provides evidence against linkage. Analyses fall into two general categories: two-point and multipoint analyses. Two-point analysis estimates a single recombination fraction θ between two loci (two markers, or a marker and a disease gene). Multipoint analysis is used to determine the relative locations of multiple loci.

Although this type of linkage analysis can provide the strongest evidence for a causal relationship of a single gene with disease, and can be extremely flexible in terms of modeling the relationship of the gene to its associated disease, the lod score method is also prone to violations of underlying assumptions and may be limited in its ability to account completely for the complexity of the disease process itself. These models are sensitive to model misspecification, errors in determining marker genotypes, and missing pedigree data. These approaches also need to (and in many cases are able to) account for the fact that a disease may be etiologically heterogeneous, when the disease is explained by different genes in different families.

Lod score-based linkage analysis generally proceeds in two phases: a hypothesis generating phase (in which initial indications for linkage are explored) and a hypothesis testing phase (in which initial linkages are confirmed and the specific location of a disease gene is narrowed down to a small region of a particular chromosome relative to known markers). The initial hypothesis generating phase of linkage analysis usually involves two-point analyses, in which a single recombination fraction θ between two loci (two markers, or a marker and a disease gene) is estimated. This often involves computing a lod scores for a range of values of θ (e.g., 0, 0.01, 0.05, 0.1, 0.2, 0.3, 0.4), and may involve using iterative methods to estimate the maximum lod score and the associated value of θ that maximizes the likelihood $L(\theta)$.

Once a linkage has been identified by this two-point analysis, refinement of the location of the putative gene of interest is accomplished by the use of multipoint linkage analysis to determine the relative locations of multiple marker loci and the disease gene. Although many algorithms exist for this type of analysis, multipoint lod scores can be obtained by simultaneously estimating the relative location of three or more loci, one of which may be a disease locus. A general problem with this approach is that for n loci, there are ½ $n!$ orders that may need to be considered. Thus, location scores can be computed that determine the location of a putative disease locus relative to a fixed map of marker loci. This approach requires that only n orders be considered. For example, for a fixed map of three markers (M_1, M_2, M_3) and a disease locus (D), determine first $\log_{10} L$ for each order (i.e., $L_1 = D\text{-}M_1\text{-}M_2\text{-}M_3$; $L_2 = M_1\text{-}D\text{-}M_2\text{-}M_3$; $L_3 = M_1\text{-}M_2\text{-}D\text{-}M_3$; $L_4 = M_1\text{-}M_2\text{-}M_3\text{-}D$). Then, use $\log_{10}(L_i) - \log_{10}(L_j) = t > 0$ to determine whether the odds that i is a better order than j are $10^t : 1$, and conclude which order is best if $t \geq 3$. Using this information, a multipoint map can be produced with location scores to identify the most likely location of the disease gene.

The previous description implies that a targeted approach to linkage analysis can be undertaken that involves specific genomic regions. However, there is often no *a priori* information about the putative location of a disease gene, and a genome-wide scan may be required. While either IBS or IBD approaches may be used, genome-wide scans generally involve a two-step approach that uses many (often 300 or more) highly polymorphic markers. First, two-point analyses are undertaken using a set of "anchor" markers that are spaced evenly throughout the human genome. Potential linkages are evaluated, often with potential linkages being considered even though lod < 3.0. Second, areas containing possible linkages are reanalyzed using more closely spaced markers in the region(s) of interest to confirm the presence of linkage, and possibly to undertake multipoint analysis to refine the location of the putative disease gene. Positive evidence for linkage, including the evaluation of obligate recombinant events, may narrow the region of interest to a level at which molecular cloning techniques can be applied to isolate the DNA sequence responsible for the observed linkage.

3.2.3. Transmission Disequilibrium Test Methods

The transmission disequilibrium (TDT) approach is a means of evaluating linkage (i.e., $\theta < \frac{1}{2}$) between alleles and disease using a family-based approach ***(14)***. Briefly, the TDT approach compares chromosomes (i.e., alleles) rather than individuals in a matched design in which the 'cases' consist of those chromosomes that have been transmitted from parent to affected offspring, while the 'controls' consist of those chromosomes that have not been transmitted from parent to affected offspring. The data used for this approach are represented in the following table:

		Nontransmitted allele		
		M	m	Total
Transmitted allele	M	w	x	$w + x$
	m	y	z	$y + z$
Total		$w + y$	$x + z$	$2n$

A standard McNemar's test can be used to evaluate, over a series of such parent–offspring transmissions, whether there is evidence for deviation from Mendelian expectations.

The TDT approach is a valid test for genetic linkage in the presence of association. That is, linkage will only be detected if association is also present. This approach has also been extended in a variety of ways to allow for flexibility in modeling and in terms of which siblings can be included in the analysis to reconstruct missing parental genotypes ***(15)***. The model can also be extended to incorporate analysis of genotype–environment interactions ***(16–18)***. Similarly, extensions of the original TDT method have been developed that incorporate information about siblings ***(15)***.

3.2.4. Linkage Methods Specific to the Identification of Tumor Suppressor Genes

The power to detect genetic linkage for tumor suppressor genes can be improved if the use of loss of constitutional heterozygosity (LOH) in tumor tissue is considered. A central goal of tracking marker segregation across multiple generations in a family is to establish the gametic phase of alleles. However, knowledge of gametic phase can also be gained by knowing which allele(s) have been lost in a tumor. Information about LOH can therefore be used to augment information about gametic phase gained from segregation of alleles. In particular, for cancers with a late age of onset, for which information about older generations may not be available, LOH data can substantially improve the power to detect genetic linkage.

Approaches for the incorporation of LOH data in both lod score linkage analysis and affecteds-only approaches ***(18–23)*** have been proposed. The value of LOH data in linkage analyses depends with what certainty gametic phase can be inferred ***(18)***, as well as how often LOH is observed in tumors ***(23)***. Therefore, use of LOH data in genetic linkage analysis can be of value in identifying tumor suppressor genes.

4. Association Studies

Genetic association studies tend to involve measured genes (usually candidate genes with a known or hypothesized function) that are involved in more complex disorders that do not necessarily follow a Mendelian pattern of inheritance. These approaches generally employ unrelated individuals in case–control or cohort study designs, although family members are used in some association study designs. The statistical methods of association studies are aimed at determining the relationship of a measured genotype with the occurrence of disease.

Unlike high-penetrance genes, which may confer Mendelian dominant patterns of disease inheritance consistent with tumor suppressor genes, most low-penetrance genes act by mechanisms other than tumor suppression. For example, most low-penetrance genes are involved in the metabolism of environmental carcinogens (e.g., members of the cytochrome P450 family) or steroid hormones, are involved in DNA damage and repair pathways, or function in other metabolic pathways relevant to the generation of somatic mutations, maintenance of DNA integrity, or control of cell growth. However, polymorphic variation in tumor suppressor genes (e.g., missense variants rather than protein truncating mutations) may confer a slight increase in cancer risk. In contrast to the methods described above, which would be used to identify high-risk effects of mutations in a tumor suppressor gene, a very different set of analytical tools is required to identify the relatively weaker associations of low-penetrance variants.

4.1. Haplotype Relative Risk or Affected Family-Based Controls Methods

In order to minimize the potential for population structures leading to a false positive association, a number of approaches have been used that employ data on relatives of cases. One of these approaches is the haplotype relative risk (HRR) or affected family-based controls (AFBAC) method. Falk and Rubinstein ***(24)*** originally proposed the HRR method comparing the $2n$ parental alleles that are transmitted to affected offspring against the $2n$ parental alleles not transmitted to affected offspring as a means of detecting allelic associations with disease. This method was similarly presented by Thomson ***(25)***, and is calculated by using the data shown in the following table:

	M allele	*m* allele	Total
Transmitted allele	a	$2n - a$	$2n$
Nontransmitted allele	b	$2n - b$	$2n$
Total	$a + b$	$4n - a - b$	$4n$

This approach controls for ethnicity, and thus should alleviate the potential for population stratification as an explanation for an association between a gene and disease. Ewens and Spielman ***(26)*** have shown that this approach tests association in the presence of linkage, and is not a valid test of genetic linkage. In addition, Ewens and Spielman ***(26)*** have shown that this test may not be a valid test of association because it requires random mating and the absence of population admixture for at least two generations prior to the sampling of affected offspring. Because of these limitations, the HRR/AFBAC methods have not been as widely applied as the TDT or case–control methodologies described elsewhere in this chapter.

4.2. Other Approaches Using Relatives as Controls

A number of study designs have been proposed that use related controls to compare with cases. Penrose ***(27)*** and others have proposed the use of unaffected siblings as controls ***(28–29)***. The use of relatives as controls limits the potential for population stratification (see above). Gaudermann et al. ***(30)*** have shown that sibling or cousin controls are less efficient than unrelated controls for detecting the main effect of genotype on disease, but are more efficient for the detection of interactions involving genotypes. However, this relative efficiency decreases as the genotype frequency increases, so that the use of relative controls may be advantageous to only some situations in which the genotypes of interest are very rare. Related designs have been proposed using siblings ***(31–34)*** or parents (the haplotype relative risk or AFBAC method ***[24–25,35]***) as controls. The TDT method, described above, is an additional variation on the use of relative controls that tests for linkage in the presence of association.

Use of relatives is limited because an association may be explained by linkage disequilibrium and not a true disease-causing effect. Another disadvantage is that using unaffected siblings or parents as controls may lead to an increased sample size requirement, possibly because of overmatching on exposures ***(34)***. Relative controls are also more costly to identify and collect that unrelated controls. The primary advantage of these methods is that the alleles studied are 'matched' on ethnicity. Therefore, this approach has the advantage of eliminating the potential for spurious associations due to the effects of population stratification (i.e., confounding due to ethnicity). However, for late-onset diseases such as cancer, this may be an impractical approach, and it has been shown that population stratification among mixed Caucasian populations may not be a significant issue in most study settings ***(36)***. Nonetheless, it may not be possible to characterize ethnicity adequately in many settings. If this is the case, these relative-based methods are appropriate analytical choices.

4.3. Studies of Association Using Unrelated Individuals

Studies using candidate genes are amenable to traditional epidemiological study designs including case–control and cohort studies. In general, the same principles that are required for a well-designed and well-conducted epidemiologic study are also relevant for a molecular epidemiologic study relating candidate genes to disease. Concerns about study biases from traditional epidemiology are similarly relevant to candidate gene association studies. As these designs are discussed in detail elsewhere ***(37)***, they will not be discussed further here. Although the case–control and cohort designs are effective and appropriate for the study of genotype-disease associations, other nontraditional study designs have emerged to accomplish genotype–disease associations. Some of these are described in the subsequent sections.

4.3.1. Case–Control and Cohort Designs

4.3.1.1. Studies of Interaction

An important aspect of association studies is the ability to detect interactions among genotypes and between genotypes and environments. In its simplest sense, interaction can be defined as deviations from independence between two or more factors. Interaction effects can be measured on additive or multiplicative scales. For risks R (measured, for example, by risk or odds ratios), additive interactions can be defined as $R_{ge} = R_g + R_e - 1$, and multiplicative interactions can be defined as $R_{ge} = R_g \times R_e$.

Generally speaking, interactions can be generated using either joint or stratified models. For the simplest case of an exposure with two levels (exposed and unexposed) and a genotype with two levels (a low-risk genotype and a high-risk genotype), a joint model would consider a single reference group defined by the genotype and exposure (usually those who are unexposed and have the low-risk genotype), which is compared with the remaining genotype–exposure combinations (e.g., those who are unexposed and have the high-risk genotype, those who are exposed and have the low-risk genotype, and those who are exposed and have the high-risk genotype). Three odds or risk ratios would therefore be estimated. A stratified model would consider the effect of exposure in the low-risk genotype and in the high-risk genotype separately, giving two odds or risk ratios. These models are readily extended beyond two levels of exposure or genotype. In each case, formal tests of interaction should be generated.

4.3.1.2. Special Considerations When Making Inferences from Association Studies

Although the general statistical considerations for genotype–disease association studies are not different than those of 'traditional' association studies, there are some considerations that are specific to associations involving genotype data. A commonly discussed concern with case–control or cohort-based association studies is that of population stratification. Population stratification is a form of confounding by ethnicity, in which populations with different genotype frequencies and different disease risks are combined analytically to produce a spurious positive association. There are four major determinants of population stratification. First, allele frequencies must vary by ethnicity. Second, cancer rates must vary by ethnicity. Third, variation in allele frequencies and cancer rates must track together across populations (e.g., in those populations in which cancer rates are higher, susceptibility allele frequencies must also be higher). Finally, population stratification is more likely to be of concern when ethnicity cannot be easily determined, and thus cannot be dealt with analytically (e.g., by matching or statistical adjustment).

A hypothetical example of population stratification is presented in **Fig. 3**. In this example, the frequency of disease and the frequency of the genetic variant of interest differ between populations, but there is no association of genotype and disease within either population. However, if the two populations are mixed, and analysis is undertaken ignoring the fact that these are distinct populations, a spurious association is induced. Analyzing populations that are ethnicity or racially heterogeneous may result

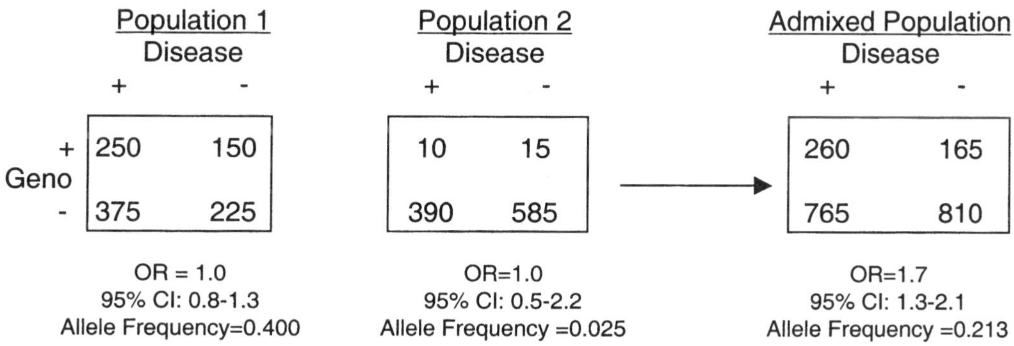

Fig. 3. Hypothetical example of population stratification inducing a spurious genotype–disease association.

in associations that suffer from population stratification. Two general approaches have been proposed to deal with this potential problem. The first is to use family-based studies in which all individuals being analyzed are related, and thus "matched" for ethnicity. The second is a recognition that population stratification is a manifestation of confounding by ethnicity, and can be corrected by using standard tools of epidemiology that deal with confounding in general, namely, matching or statistical adjustment. Wacholder et al. ***(36)*** and Millikan ***(38)*** considered typical values for parameters used in association studies (background genotype frequency rates, etc.), and reported that population stratification is not likely to represent a major problem in typical association studies involving Caucasian populations, or populations of African Americans and Caucasians in which these ethnicities can be known and accounted for analytically. In particular, Wacholder et al. ***(36)*** reported that the larger the number of ethnicities that are considered, the less likely it is that population stratification is likely to affect study results.

As with other measures used in epidemiologic association studies, misclassification in exposure assessment or genotype determination can affect the statistical power (and therefore the sample size required to detect an association). For example, Garcia-Closas et al. ***(39)*** estimated that a 5% misclassification in genotype would inflate the sample size required to detect an OR of 2 from 720 to 900 cases, and could result in a bias in the odds ratio estimate toward the null hypothesis. Although this problem is well known in epidemiology, Increased misclassification leading to decreased power, increased study cost, and biased point estimates should be considered in assessing genotype–disease associations.

A final consideration is whether it is appropriate to use alleles or genotypes in association with disease. Clearly, it is attractive to consider analyses with a sample size that is doubled because alleles are counted rather than genotypes (i.e., individuals). Under Hardy-Weinberg equilibrium (HWE) conditions, analysis of alleles or genotypes is equivalent. However, when HWE conditions are not met, χ^2 tests and OR estimates of association may be biased ***(40)***. Therefore, tests for HWE may be valuable prior to undertaking associations. In addition, it may be valuable to consider that the concept of "risk" is for individuals (i.e., genotypes), not alleles in isolation. Therefore, depending on the research question of interest, some inferences may be better made using genotype information.

4.3.2. Case–Case Designs

Case–case designs are a potentially efficient approach to evaluate potential genotype–genotype or genotype–environment ***(41–43)*** or genotype–genotype ***(5)*** interactions. For the simple situation in which there is a binary genotype and a binary exposure (or two binary genotypes), the following contingency table can be drawn:

		Genotype	
		−	+
Exposure	−	*w*	*x*
	+	*y*	*z*

A measure of the relationship between these two factors is given by the case-only odds ratio $\psi = wz/xy$, which measures the departure from a multiplicative joint effect of the genotype and the exposure. This model assumes that the genotype and exposures are independent in the source population from which cases arose (or the two genotypes are not in linkage disequilibrium). That is, there is independence of the factors in controls, had they been measured. This approach also assumes that there is no differential selection of cases based on factors of interest. It has not been formally evaluated whether this assumption will usually be met. Using this measure, we can infer a multiplicative interaction if $\psi = 1$, a more than multiplicative interaction if $\psi > 1$, and a less than multiplicative (e.g., additive) interaction if $\psi < 1$.

The advantages of this approach are the reliance on cases with disease only, which are often readily available and thus an efficient study sample. Because there are no controls, issues of appropriate control selection, matching, etc., are not a consideration. However, careful definition of cases, including the use of proper inclusion/exclusion criteria, is important. The primary disadvantage of this approach is that it does not provide a direct measure of disease risk, since all study subjects have the disease. Begg and Zhang ***(41)*** have specified the relationship between ψ and the true relative risk of disease exists. Schmidt and Schaid ***(44)*** have noted that ψ is not identical to the odds ratio obtained from a case–control study, since the case–control design compares cases with unaffected individuals (i.e., controls), while the case-only study compares affecteds against a general reference population. Other disadvantages of this approach include the fact that this method provides only a measure of departures from multiplicativity of two risk ratios, not departures from additivity, and it does not evaluate the marginal effect of individual factors (genes or environments separately) on disease risk.

4.3.3. Psuedo-Data Approaches

A number of sampling designs has been proposed the use various combinations of genotype and/or exposure data from a study in which cases and controls are available, but there are limited resources, logistical concerns, or ethical limitations in generating genotype data on all study subjects. For example, Umbach and Weinberg ***(17)*** have proposed a series of log-linear models that can be used to efficiently evaluate genotype–environment interactions, as described below.

4.4.3.1. De-linked Controls

Using a log-linear model, estimates of genotype–environment interaction (but not the main effects of genotype or environment separately) can be obtained from data in which both cases and controls are genotyped, but the genotype information is 'delinked' from controls. This provides only marginal genotype frequencies (i.e., no link between an individual control and their genotype), and the genotype itself is not linked to their exposure. This model can be defined as:

$$\log u_{dge} = \mu_0 + \alpha_0 E + \beta_0 G + \gamma_0 GE + \mu_1 D + \alpha_1 DE + \beta_1 DG + \gamma_1 DGE$$

where D, G, and E denote the disease, genotype, and environment, respectively; μ_0, μ_1, α_0, β_0, γ_0 = joint genotype–exposure distribution in controls, μ_1 = fitted marginal totals for cases and controls compared to observed data, α_1 = disease–exposure relationship, β_1 = disease–genotype relationship, and γ_1 = genotype–environment interaction relationship with disease. A central assumption of this design is that genotype and exposure are independent in controls. When this is the case, $\gamma_0 = 0$ and

$$\log u_{dge} = \mu_0 + \alpha_0 E + \beta_0 G + \mu_1 D + \alpha_1 DE + \beta_1 DG + \gamma_1 DGE$$

Therefore, valid estimates of the genotype–environment interaction are possible without using individual information about genotypes of controls. This approach may be valuable when there are concerns about linking genetic information to an individual.

4.3.3.2. Exposure Data Only in Controls

Using the same model structure as described above, we can define

$$\log u_{dge} = \mu_0 + \alpha_0 E + \mu_1 D + \alpha_1 DE + \beta_1 DG + \gamma_1 DGE$$

where genotype can be coded as a (0, 1) variable in cases and missing in controls. As above, the interaction effects involving genotype can be estimated, but genotype main effects are not estimable. This design may be valuable when there are limitations in resources available to genotype all study participants or when there are ethical concerns about genotyping controls.

4.3.3.3. Independent Control Groups

If genotype data are available from one control group, and exposure data from another, it is possible to estimate genotype–environment interaction effects using two separate control groups using the models described in the previous sections. The major concern with this approach is that a poor choice of genotyped 'controls' can induce bias, particularly if the genotyped control group and the exposure-assessed control group are not comparable, and even more so if one or more of the control groups has not been drawn from the population out of which the cases arose.

5. Summary

The goal of appropriate epidemiologic approaches to identify cancer susceptibility genes is to maximize the potential of identifying causal relationships and minimize the potential that spurious relationships will be found. It is similarly important to target the appropriate approach for the identification of specific types of genes. Germline mutations in high-penetrance genes often confer Mendelian patterns of inheritance, and can be optimally identified by segregation and genetic linkage analysis methods. In contrast, low-penetrance genes may not confer Mendelian patterns of inheritance, and may be involved in cancer etiology only through complex interactions with other genes and exposures. Association studies may be more appropriate means of determining a role for these genes in cancer etiology.

References

1. Lilienfeld, A. M. and Lilienfeld, D. E. (1980) *Foundations of Epidemiology*. Oxford University Press, New York, pp. 346–347.
2. Foulkes, W. D. (1995) A tale of four syndromes: familial adenomatous polyposis, Gardner syndrome, attenuated APC and Turcot syndrome. *Q. J. Med.* **88** (12), 853–863.
3. Redston, M., Nathanson, K. C., Yuan, Z. Q., et al. (1998) The APC I1307K polymorphism and breast cancer risk. *Nat. Genet.,* **20**, 13–14.
4. Khoury, M. J., Beaty, T. H., and Cohen, B. H. (1993*) Fundamentals of Genetic Epidemiology.* Oxford University Press, New York.
5. Khoury, M. J. and Flanders, W. D. (1996) Nontraditional epidemiologic approaches in the analysis of gene-environment interaction: case-control studies with no controls. *Am. J. Epidemiol.* **144**, 207–213.
6. Elandt-Johnson, R. (1971) *Probability Models and Statistical Methods in Genetics.* Wiley, New York.
7. Edwards, A. W. F. (1978) *Likelihood: An Account of the Statistical Concept of Likelihood and Its Applications to Scientific Inference.* Cambridge University Press, London, UK.
8. Elston, R. C. and Stewart, J. (1971) A general model for the genetic analysis of pedigrees. *Hum. Hered.* **21**, 523–542.
9. Cannings, C., Thompson, E. A., and Skolnick, M. H. (1978) Probability functions on complex pedigrees. *Adv. Appl. Prob.* **10**, 26–91.
10. Penrose, L. S. (1938) Genetic linkage in graded human characters. *Ann. Eugen.* **8**, 233–237.
11. Weeks, D. E. and Lange, K. (1988) The affected pedigree member method of linkage analysis. *Am. J. Hum. Genet.* **42**, 315–322.
12. Haldane, J. B. S. and Smith, C. A. B. (1947) A new estimate of the linkage between genes for colour-blindness and haemophilia in man. *Ann. Eugen.* **14**, 10–31.
13. Morton, N. E. (1955) Sequential tests for the detection of linkage. *Am. J. Hum. Genet.* **7**, 277–318.
14. Spielman, R. S., McGinnis, R. E., and Ewens, W. J. (1993) Transmission test for linkage disequilibrium. The insulin gene region and insulin-dependent diabetes mellitus (DDM). *Am. J. Hum. Genet.* **52**, 506–516.
15. Spielman, R. S. and Ewens, W. J. (1996) The TDT and other family-based tests for linkage disequilibrium and association. *Am. J. Hum. Genet.* **59**, 983–989.
16. Maestri, N. E., Beaty, T. H., Hetmanski, J., et al. (1997) Application of transmission disequilibrium tests to nonsyndromic oral clefts: including candidate genes and environmental exposures in the models. *Am. J. Med. Genet.* **73**(3), 337–344.
17. Umbach, D. M. and Weinberg, C. R. (2000) The use of case-parent triads to study joint effects of genotype and exposure. *Am. J. Hum. Genet.* **66**(1), 251–261.
18. Rebbeck, T. R., Lustbader, E. D., and Buetow, K. H. (1994) Somatic allele loss in genetic linkage analysis of cancer. *Genet. Epidemiol.* **11**, 419–429.
19. Steichen-Gersdorft, E., Gallion, H. H., Ford, D., et al. (1994) Familial site-specific ovarian cancer is linked to BRCA1 on 17q12-21 *Am. J. Hum. Genet.* **55**, 870–875
20. Teare, M. D., Rohde, K., Santibanez Koref, M. (1994) The use of loss of constitutional heterozygosity data to ascertain the location of predisposing genes in cancer families. *J. Med. Genet.* **31**, 449–452.
21. Lustbader, E. D., Rebbeck, T. R., and Buetow, K. H. (1995) Using loss of heterozygosity data in affected pedigree member linkage tests. *Genet. Epidemiol.* **12**, 339–350.
22. Rhode, K., Teare, M. D., Scherneck, S., and Santibanez Koref, M. (1995) A program using loss of constitutional heterozygosity data to ascertain the location of predisposing genes in cancer families. *Hum. Hered.* **45**, 337–345.

23. Rohde, K, Teare, M. D., and Santibanez Koref, M. (1997) Analysis of genetic linkage and somatic loss of heterozygosity in affected pairs of first-degree relatives. *Am. J. Hum. Genet.* **61**(2), 418–422.
24. Falk, C. T. and Rubinstein, P. (1987) Haplotype relative risks: an easy reliable way to construct a proper control sample for risk calculations. *Am. J. Hum. Genet.* **51**(Pt 3), 227–233.
25. Thomson, G. (1988) HLA disease associations: models for insulin dependent diabetes mellitus and the study of complex human genetic disorders. *Annu. Rev. Genet.* **22**, 31–50.
26. Ewens, W. J. and Spielman, R. J. (1995) The transmission/disequilibrium test, history, subdivision and admixture. *Am. J. Hum. Genet.* **57**, 455–464.
27. Penrose, L. S. (1938) Genetic linkage in graded human characters. *Ann. Eugen.* **8**, 233–237.
28. Goldstein, A. M., Hodge, S. E., and Haile, R. W. (1989) Selection bias in case-control studies using relatives as the controls. *Int. J. Epidemiol.* **18**, 985–989.
29. Andrieu, N. and Goldstein, A. M. (1996) Use of relatives of cases as controls to identify risk factors when an interaction between environmental and genetic factors exists. *Int. J. Epidemiol.* **25**, 649–657.
30. Gauderman, W. J., Witte, J. S., and Thomas, D. C. (1999) Family-based association studies. *J. Natl. Cancer Inst. Monogr.* **26**, 31–37.
31. Curtis, D. (1997) Efficient strategies for genome scanning with affected sib pairs. *Am. J. Hum. Genet.* **62**(1), 204–7.32.
32. Boehnke, M. and Langefeld, C. D. (1998) Genetic association mapping based on discordant sib pairs: the discordant-alleles test *Am. J. Hum. Genet.* **62**(4), 950–961.
33. Risch, N. and Teng, J. (1998) The relative power of family-based and case-control designs for linkage disequilibrium studies of complex human diseases I. DNA pooling. *Genome Res.* **8**(12), 1273–1288.
34. Schaid, D. J. and Rowland, C. (1998) Use of parents, sibs, and unrelated controls for detection of associations between genetic markers and disease. *Am. J. Hum. Genet.* **63**(5), 1492–1506.
35. Terwilliger, J. D. and Ott, J. (1992) A haplotype-based "haplotype relative risk" approach to detecting allelic associations. *Hum. Hered.* **42**(6), 337–346.
36. Wacholder, S., Rothman, N., and Caporaso, N. (2000) Population stratification in epidemiologic studies of common genetic variants and cancer: quantification of bias. *J. Natl. Cancer Inst.* **92**(14), 1151–1158.
37. Breslow, N. E. and Day, N. E. (1980). *Statistical Methods in Cancer Research: Vol. 1—The Analysis of Case-Control Studies.* IARC, Lyon, France.
38. Millikan, R. C. (2001) Population stratification in epidemiologic studies of common genetic variants and cancer: quantification of bias. *J. Natl. Cancer Inst.* **93**(2), 156–158.
39. Garcia-Closas, M., Rothman, N., and Lubin, J. (1999) Misclassification in case-control studies of gene-environment interactions: assessment of bias and sample size. *Cancer Epidemiol. Biomarkers Prev.* **8**(12), 1043–1050.
40. Sasieni, P. D. (1997) From genotype to genes: doubling the sample size. *Biometrics* **53**, 1253–1261.
41. Begg, C. B. and Zhang, Z. F. (1994) Statistical analysis of molecular epidemiology studies employing case-series. *Cancer Epidemiol. Biomarkers and Prev.* **3**, 173–175.
42. Piegorsch, W. W., Weinberg, C. R., and Taylor, J. A. (1994) Non-hierarchical logistic models and case-only designs for assessing susceptibility in population-based case-control studies. *Stat. Med.* **13**, 153–162.
43. Yang, Q., Khoury, M. J., Sun, F., and Flanders, W. D. (1998) Case-only design to measure gene–gene interaction. *Epidemiology* **10**, 167–170.
44. Schmidt, S. and Schaid, D. J. (1999) Potential misinterpretation of the case-only study to assess gene-environment interaction. *Am. J. Epidemiol.* **150**, 878–885.

20

Characterization of Translocations in Human Cancer

Keri Fair and Michelle M. Le Beau

1. Introduction

Recurring chromosomal abnormalities are associated with distinct subtypes of leukemia or lymphoma that have unique morphologic, immunophenotypic, and clinical features, such as response to therapy ***(1–4)***. Thus, cytogenetic analysis of an individual's malignant cells plays a major role in the diagnosis and subclassification of a hematologic neoplasm. Examples of the clinical-cytogenetic subtypes of acute myeloid leukemia (AML) include the t(15;17) in acute promyelocytic leukemia (AML-M3), and the t(8;21) in acute myeloblastic leukemia with maturation (AML-M2) ***(1–3)***. A number of recurring abnormalities have also been recognized in solid tumors. At present, there are more than 600 recurring translocations associated with human cancers ***(1–3,5)***. The vast majority of these have been identified in the hematologic malignant diseases; however, some are associated with mesenchymal tumors, particularly the pediatric small round blue cell tumors, e.g., Ewing sarcoma and rhabdomyosarcoma. Although epithelial tumors represent the most common cancers (lung, breast, and colon carcinomas), only a few recurring abnormalities have been identified in these diseases.The Mitelman Database of Chromosome Aberrations in Cancer provides a registry of the abnormal karyotypes identified in greater than 37,000 neoplasms (>100,000 aberrations) ***(5)***. Another relevant database has been developed by the Cancer Chromosome Aberration Project, which integrates cytogenetic and genomic databases relevant to cancer (http://www.ncbi.nlm.nih.gov/CCAP).

Molecular analysis has revealed that the recurring chromosomal abnormalities result in the altered function of oncogenes or tumor suppressor genes; thus, chromosomal aberrations represent genetic mutations that are involved in the pathogenesis and progression of human tumors ***(1,2,6,7)***. Translocations typically act as dominant, or dominant negative, genetic traits, in that altering a single allele is sufficient to alter cell fate. The genes that are located at translocation breakpoints encode proteins that participate in a variety of normal cellular functions, including growth-factor receptor activation and signal transduction; however, the largest class of genes encodes transcription factors.

There are several mechanisms by which chromosomal translocations result in altered gene function. The first is deregulated gene expression. This mechanism is characteristic of the translocations in the lymphoid neoplasms that involve the immunoglobulin

From: *Methods in Molecular Biology, Vol. 222: Tumor Suppressor Genes: Pathways and Isolation Strategies*
Edited by: Wafik S. El-Deiry © Humana Press Inc., Totowa, NJ

genes in B-lineage tumors, or the T-cell receptor genes in T-lineage tumors ***(1,2,7)***. These rearrangements result in inappropriate expression (either overexpression or aberrant expression in a tissue that does not normally express the gene) of the partner gene involved in the translocation, typically with no alteration in its protein structure. The second, and most common mechanism, is the expression of a novel fusion protein, resulting from the juxtaposition of coding sequences that are normally located on different chromosomes ***(1,2,7)***. Such fusion proteins are tumor-specific in that the fusion gene does not exist in nonmalignant cells; thus, the detection of such a fusion gene, fusion transcript, or fusion protein can be important in diagnosis, in identifying residual disease or relapse, and in the development of tumor-specific therapy. An example is the BCR/ABL fusion protein, resulting from the t(9;22) in chronic myelogenous leukemia; Gleevec, an ABL tyrosine kinase inhibitor, and has led to impressive responses in patients with this disease ***(8)***.

Several experimental techniques can be used for the characterization of translocations in human tumors, including (a) conventional cytogenetic analysis, (b) fluorescence *in situ* hybridization (FISH), (c) polymerase chain reaction (PCR) and reverse transcription PCR (RT-PCR), and (d) real-time PCR. This chapter provides protocols for conventional cytogenetic analysis and FISH detection of chromosomal abnormalities.

1.1. Characterization of Translocations Using Cytogenetic Analysis

The use of cytogenetic analysis to detect recurring translocations in malignant diseases requires the analysis of metaphase chromosomes obtained from dividing cells. Mitogens that induce cells to divide in culture are rarely used in cancer cytogenetics, because stimulation of the division of normal cells may interfere with the analysis of spontaneously dividing malignant cells. In the leukemias and lymphomas, cytogenetic analysis is typically performed on bone marrow aspirates or biopsies, peripheral blood, pleural or ascites fluid, or lymph node or tumor mass biopsies. For solid tumors, the analysis is performed on tumor biopsies following short-term culture. Detailed protocols for processing specimens from patients with leukemia, lymphoma, or a solid tumor have been described previously ***(9,10)***. To increase the numbers of cells in metaphase, cells are accumulated in mitosis using Colcemid, followed by sequential exposure to a hypotonic solution to increase cell volume, and fixation. Chromosome preparations are generated by dropping the fixed cell suspension onto microscope slides, and are stained using histologic stains. Cytogenetic analysis is performed by microscopy, and images are recorded by photomicroscopy, or digitally using an automated karyotyping system.

1.2. Characterization of Translocations Using FISH

FISH techniques have been reviewed elsewhere, and are summarized here ***(11,12)***. FISH is based on the same principle as Southern blot analysis, namely, the ability of single-stranded DNA to anneal to complementary DNA. In the case of FISH, the target DNA is the nuclear DNA of interphase cells, or the DNA of metaphase chromosomes affixed to a microscope slide. FISH can also be performed using bone marrow or peripheral blood smears, or fixed and sectioned tissue. The test probe is labeled through incorporation of a chemically modified reporter nucleotide covalently bound to biotin or digoxigenin, or directly conjugated to fluorochromes. After the test probe anneals to its complementary sequences, the probe is detected via the attachment of a fluorochrome to molecules, such as streptavidin or anti-digoxigenin antibodies, which have a high affinity for the reporter nucleotide. The end result is visualized by fluorescence microscopy, and appears as a brilliantly colored signal at the hybridization site. The high sensitivity and specificity of FISH, combined with the speed with which the assays can be performed, has made FISH a powerful adjunct to cytogenetic analysis. In addition, FISH can be performed with interphase cells, thus eliminating the need for dividing cells.

A variety of probes are used to detect chromosomal abnormalities by FISH; a partial list of probes commercially available for cancer diagnostics is given in **Table 1**. A number of these probes are directly conjugated to fluorochromes, which has the advantage of eliminating the probe detection step. Directly labeled probes also provide strong signal with low background. FISH probes can be divided into three groups: (a) probes that identify a specific chromosome structure, such as the centromere; (b) probes that hybridize to multiple chromosomal sequences, such as whole-chromosome painting probes; and (c) probes that hybridize to unique DNA sequences, e.g., translocation-specific probes.

FISH strategies for the detection of recurring translocations take advantage of the ability to detect hybridized probes derived from the breakpoints of specific translocations using multiple fluorochromes ***(13)***. In the conventional method, a *single fusion signal* is detected via the hybridization of dual-color locus-specific probes to their target sequences at the junction of the translocation breakpoints on one of the derivative chromosomes. Probes of this type have a relatively high number of false-positive fusion signals (1.5–5.6%), as a result of the close proximity and random juxtaposition of target chromosomes in interphase nuclei. The sensitivity of single-fusion FISH probes is 5–10% (mean number of fusion signals in normal samples plus two standard deviations), thus limiting their use in the detection of minimal residual disease ***(14)***.

The next generation of translocation-specific FISH probes, *extra-signal (ES) probes*, was introduced recently (Vysis, Inc, http://www.vysis.com). With the two-color ES probes, DNA sequences flanking the breakpoints on the involved chromosomes are brought together to form a yellow fusion signal; however, part of the DNA sequences recognized by one of the probes remains at the original site, giving rise to an extra signal in a single color. False-positive signals can be distinguished from genuine fusion signals resulting from the translocation by virtue of the absence of the extra signal. The application of these probes results in an improvement of sensitivity, and can detect translocation-positive cells in <1% of nuclei in a sample ***(15)***.

A third type of probe is a *dual-fusion signal* probe, e.g., the *BCR/ABL* probe marketed by Ventana Medical Systems, http://www.ventanamed.com. In this case, the probes span the involved genes; cells with the translocation have one red and one green signal observed on the normal homologs, and two yellow fusion signals observed on both derivative homologs. This type of probe has a substantially lower rate of false positive cells, and the cutoff value for a positive result is ~1% ***(16)***.

A fourth type of probe is referred to as a *two-color break-apart probe*. An example is the *MLL* probe marketed by Vysis, Inc., to detect the recurring translocations of 11q23. In two-color break-apart probes, DNA sequences from the 5′ and 3′ regions of a single gene are differentially labeled and detected with red and green fluorochromes. In the germline configuration, a yellow fusion signal is observed, whereas separate red and green signals are observed when the sequences are separated as a result of a translocation. The sensitivity of this type of probe is exceedingly high (cutoff value of ~1.7%), with very high specificity.

Table 1
FISH Probes to Detect Chromosomal Abnormalities[a]

Disease[b]	Abnormality	Probe	Format[c]	Vendor[d]
AML-M2	t(8;21)	RUNX1/ETO	Two-color dual fusion	Vysis
AML-M4Eo	inv(16)/t(16;16)	CBFB	Two-color break-apart	Vysis, Ventana
AML-M3	t(15;17)	PML/RARA	Two-color single fusion	Vysis, Ventana
AML	t(11q23)	MLL	Two-color break-apart Single color	Vysis, Ventana
AML/MDS	−5/del(5q)	EGR1/D5S721	Two-color deletion	Vysis
AML/MDS	−7/del(7q)	D7S522/CEP7	Two-color deletion	Vysis
AML/MDS	del(20q)	D20S108	Single color	Vysis
AML/MDS	+8	CEP8	Single color	Vysis, Cytocell, Ltd
CML	t(9;22)	BCR/ABL	Two-color single fusion	Vysis
		BCR/ABL ES	Two-color extra signal	Vysis
		MBCR/ABL	Two-color single fusion	Ventana
		BCR/ABL	Two-color dual fusion	Ventana
	+8	CEP8	Single color	Vysis, Cytocell, Ltd
	i(17q)	HER2/CEP17	Two color (17q/centromere)	Vysis, Ventana
ALL	t(12;21)	TEL/AML1	Two-color extra signal	Vysis
ALL	t(11q23)	MLL	Two-color break-apart	Vysis
ALL	t(9;22)	BCR/ABL ES	Two-color extra signal	Vysis
		MBCR/ABL	Two-color single fusion	Ventana
		mBCR/ABL	Two-color single fusion	Ventana
	t(8;14)	IGH/MYC	Two-color dual fusion	Vysis
	−9/del(9p)	p16/CEP9	Two-color (9p/centromere)	Vysis
CLL	+12	CEP12	Single color	Vysis
	del(13q)	D13S25/ D13S319	Single color	Vysis
NHL	t(14q32)	IGH	Two-color break-apart	Vysis

continued

2. Materials

2.1. Culture of Hematopoietic Cells for Cytogenetic or FISH Analysis

1. Standard culture medium: 90% RPMI 1640, 10% heat-inactivated fetal bovine serum, 10 m*M* HEPES (1 mL per 100 mL), 100 U/mL penicillin/100 μg/mL streptomycin (1 mL per 100 mL), 37°C, pH 7.2–7.3. All reagents are obtained from Invitrogen Life Technologies. Prepared media may be stored for up to 7 d at 4°C, and used within this time period.
2. Unopette test (Becton Dickinson).

2.2. Preparation of Metaphase Cells for Cytogenetic or FISH Analysis

1. KaryoMAX Colcemid (10 μg/mL) in PBS (Invitrogen Life Technologies, 10 mL): Dilute with 10 mL phosphate-buffered saline (PBS). The resulting concentration of the Colcemid stock solution is 5 μg/mL.

Table 1 *(continued)*
FISH Probes to Detect Chromosomal Abnormalities[a]

Disease[b]	Abnormality	Probe	Format[c]	Vendor[d]
	t(8;14)	IGH/MYC	Two-color dual fusion	Vysis
	t(14;18)	IGH/BCL2	Two-color dual fusion	Vysis
	t(3q27)	BCL6	Two-color break-apart	Vysis
NHL-ALCL	t(2;5)	ALK	Two-color break-apart	Vysis
MCL	t(11;14)	CCND1/IGH	Two-color dual fusion	Vysis
Myeloma	−13/del(13q)	LSI RB1	Single color	Vysis
		LSI D13S19	Single color	Vysis
	−17/del(17p)	LSI p53	Single color	Vysis
		LSI D13S19	Single color	Vysis
	t(14q32)	IGH	Two-color break-apart	Vysis
Miscellaneous				
Bone marrow transplants		CEPX/CEPY	Single color or two color	Vysis
Subtelomeric-each chromosomal arm			Single color Two-color	Vysis, Cytocell, Ltd

[a]Modified from **ref.** ***12***.

[b]ALCL, anaplastic large cell lymphoma; ALL, acute lymphoblastic leukemia; AML, acute myeloid leukemia; CLL, chronic lymphocytic leukemia; CML, chronic myelogenous leukemia; MCL, Mantle cell lymphoma; MDS, myelodysplastic syndrome; NHL, non-Hodgkin's lymphoma.

[c]In two-color break-apart probes, DNA sequences from the 5′ and 3′ regions of a single gene are labeled and detected with red and green fluorochromes. In the germline configuration, a yellow fusion signal is observed, whereas separate red and green signals are observed when the sequences are separated as a result of a translocation.

With two-color fusion probes, DNA sequences flanking the breakpoints of the involved genes are brought together to form either one (single-fusion probes) or two (dual-fusion probes) yellow fusion signal(s).

With the two-color extra-signal probes, DNA sequences flanking the breakpoint on the partner chromosomes are brought together to form a yellow fusion signal; however, part of the DNA sequences recognized by one of the probes may remain at the original site, giving rise to an extra signal in a single color.

[d]Vendors of probes for detecting translocations and other abnormalities in human tumors include Vysis, Inc. (http://www.vysis.com), Cytocell (http://www.cytocell.co.uk), and Ventana Medical Systems (http://www.ventantmed.com).

2. Dulbecco's phosphate-buffered saline at 37°C (Invitrogen Life Technologies cat. no. 14040).
3. Hypotonic KCl (0.075 *M*, 0.56%) at 37°C: Potassium chloride, certified ACS grade, Fisher, 5.6 g, deionized distilled water, 1.0 L. Store the hypotonic solution at 37°C, or at 4°C.
4. Fixative: 3:1, absolute methanol:glacial acetic acid, 3 parts Methyl alcohol anhydrous AR, low acetone, ACS grade, Mallinckrodt (VWR). 1 part acetic acid, glacial, Baker analyzed reagent (VWR), Mix fixative in a glass-stoppered flask or bottle immediately before initial use. Store at room temperature. Mix by swirling the flask immediately before each use. Fixative should be prepared fresh for each harvest, and stored for a maximum of 2 h.

2.3. Preparation of Slides for Cytogenetic or FISH Analysis

1. Fixative (*see* **Subheading 2.2.**).
2. 95% ethyl alcohol.
3. Glass microscope slides.

2.4. Trypsin-Giemsa Banding of Metaphase Cells

1. 10% hydrogen peroxide: 17 mL of 30% hydrogen peroxide, Baker analyzed (VWR); 33 mL of double distilled H_2O. Just prior to banding, mix the two reagents together well in a Coplin jar. The jar of 10% H_2O_2 should be covered when not in use.
2. Isotonic saline diluent (VWR). Adjust pH to 6.8.
3. Trypsin solution: 100 mg trypsin powder, Difco 1:250 (VWR), 100 mL Isoton (pH adjusted to 6.8). Stir on a magnetic stirrer for 1–2 h. Allow the solution to cool to room temperature (RT) before banding. Unused solution may be stored at 4°C for 2–3 d, and warm to RT prior to use.
4. Gurr buffer solution: 1 Gurr buffer tablet, pH 6.8 (Bio-Medical Specialties), 1 L double-distilled H_2O. Stir on a magnetic stirrer until tablet has dissolved. Adjust pH to 6.8. Solution may be stored at room temperature for up to 6 mo.
5. Giemsa stain solution: 1.5 mL Gurr's Improved R66 Giemsa Stain (Bio-Medical Specialties), 50 mL Gurr Buffer Solution (pH adjusted to 6.8). Mix both reagents well in a Coplin jar. The thin layer of film/undissolved material that rises to the surface may be skimmed off with a Kimwipe, or a clean microscope slide. Working solutions of stain should be made daily.
6. Seven Coplin jars.

2.5. Fluorescence In Situ Hybridization

1. 20× SSC: 353 g sodium citrate, 701.3 g NaCl. Bring volume to 4 L with double-distilled H_2O, and adjust pH to 7.0.
2. 2× SSC: 100 mL 20× SSC:900 mL double-distilled H_2O, adjust pH to 7.0.
3. 4× SSC: 200 mL 20× SSC:800 mL double-distilled H_2O, adjust pH to 7.0.
4. RNAase: 20× stock solution: 2 mg/mL in 2× SSC. Dissolve 400 mg in 200 mL 2× SSC. Boil at 100°C for 10 min, cool, and aliquot. Store at –20°C. Working solution: Prepare 200 mL RNAse by diluting 10 mL 20× RNase with 190 mL 2× SSC, pH 7.0 (final concentration 100 μg/mL).
5. 70% Formamide/30% 4× SSC: 35 mL formamide (deionized):10 mL 20× SSC:5 mL double-distilled H_2O, pH 7.0. Use deionized nucleic acid-hybridization grade formamide for the denaturation step; molecular biology-grade formamide may be used for posthybridization washes.
6. Triton-X 100: 0.1% in 4× SSC.
7. 4,6-Diamidino-2-phenylindole (DAPI) counterstain: Dissolve 10 μg DAPI in 50 mL 2× SSC (final concentration 200 ng/mL); filter solution into a Coplin jar wrapped in foil or black tape. Store at 4°C.
8. PDD antifade: Dissolve 300 mg *p*-phenylenediamine dihydrochloride in 30 mL pH 7.0 Dulbecco's PBS. Adjust pH to 8.0. Combine 20 mL PDD stock solution with 80 mL glycerol, and store this working solution at –20°C in the dark.
9. 70%, 80%, and 95% ethanol.
10. Coplin jars, glass staining boxes, and inserts.

3. Methods

3.1. Culture of Hematopoietic Cells for Cytogenetic or FISH Analysis (see Note 1)

1. Centrifuge the bone marrow or blood sample at 800 × *g* for 8–10 min. Under the laminar-flow hood, draw off the buffy coat layer in a volume of ~1 mL, and place in a sterile 15-mL centrifuge tube. Bring the volume to 5–10 mL using medium.

2. Enumerate the cells using the Unopette test (lyses red cells to facilitate cell counting) and a hemacytometer; it may be necessary to prepare a dilution (usually 1:5 or 1:10).
3. Using the cover on the capillary pipet, puncture the diaphragm of the Unopette reservoir. Fill capillary tube by touching the tip to the sample. Squeeze reservoir slightly, cover opening of the capillary pipet, and insert pipet securely into reservoir neck. Release pressure on reservoir and squeeze reservoir gently 2 or 3 times to rinse capillary bore.
4. Mix diluted sample thoroughly by inverting assembly. Convert the assembly to a dropper by withdrawing pipet from reservoir and reseal securely in a reversed position. Gently squeeze sides of reservoir and discard first 3–4 drops. Fill both hemacytomer chambers.
5. Allow the cells to settle for a few minutes, and enumerate the mononuclear cells in ten 1-mm squares by scoring cells in the 4 corners and middle squares in each of the top and bottom chambers (100× magnification). Score cells lying on the borderlines of each square by selecting 2 of the 4 borderlines and uniformly scoring cells lying on these sides only, e.g., score cells on top line and left side only.
7. Calculate cell number using the following formula: No. of cells in 10 squares $\times 10^3$ (hemacytometer factor or number of cells/mm^3) $\times$ 20 (Unopette dilution factor) or No. cells $\times$ 20,000 = No. cells/mL of sample.
8. Short-term cultures are initiated by incubating 10×10^6 cells in 10 mL of culture medium (1×10^6 cells/mL) at 37°C (5% CO_2/95% air, humidified atmosphere) for 24–72 h. Incubate culture flasks upright with caps slightly loose. If there are sufficient cell numbers, initiate 4 cultures (two flasks can be harvested after 24 h, and the remaining two flasks after 48 or 72 h of incubation).

3.2 Preparation of Metaphase Cells for Cytogenetic or FISH Analysis (see Note 2*)*

1. To accumulate cells in mitosis, add Colcemid (0.1 mL per 10 mL culture, final conc. 0.05 µg/mL) to each flask, and mix cell suspension by gently swirling the flask. Incubate for 10–25 min. If there are two cultures, use 10 min for one flask, 25 min for the second flask; use 25 min if there is only a single flask.
2. Transfer each cell suspension to a 15-mL tube, rinse each flask with 3 mL PBS (37°C) and transfer PBS to the corresponding tube.
3. Centrifuge the tubes at 800–900 rpm for 8 min.
4. Decant the supernatant. Gently resuspend each pellet by tapping the bottom of the tube, then add 10 mL of prewarmed hypotonic KCl (0.075 *M*). Resuspend by inserting a Pasteur pipet to the bottom of the tube and bubbling air gently through the solution. Incubate the tubes at 37°C for 8 min.
5. Centrifuge at 800–900 rpm for 8 min, and decant the supernatant. Gently resuspend each pellet by tapping the bottom of the tube. While agitating the tube, add fixative. This is accomplished by adding 5 drops while rotating the rim of the tube, and allowing the fixative to run down the tube. Tap the bottom of the tube gently as the fixative mixes with the cell suspension. Repeat twice until tube contains ~1 mL. Process each tube in this fashion, then bring volume of fixative in each tube to 10 mL. Mix the cell suspension by bubbling air through a Pasteur pipet. Allow tube to sit at ambient temperature for 10 min.
6. Centrifuge for 8 min at 800–900 rpm. After centrifugation, decant the supernatant, and gently resuspend each pellet by tapping the bottom of the tube. Add 10 mL of fixative, and mix each cell suspension by bubbling air through a Pasteur pipet.
7. Repeat **step 6.** until the supernatant is clear (2–3 times).
8. The cells may be used to prepare slides immediately, or the tubes may be stored at –20°C for up to 10 yr.

3.3. Preparation of Slides for Cytogenetic or FISH Analysis (see Note 3)

1. After harvesting, cells can be stored at –20°C indefinitely, or slides can be prepared immediately. In either case, the cells must be washed in fixative 3–6 times. The number of washes depends on how clean the pellet is, and how clear the fixative is after harvesting.
2. Centrifuge tubes at approximately 1000 rpm for 8 min to pellet the cells. The supernatant is decanted, and the cell pellet is gently resuspended by tapping the tube. Add 10 mL of fixative to the tube, and mix the cell suspension by bubbling air through a Pasteur pipet. This washing process promotes improved chromosome spreading and morphology, and removes cellular debris.
3. After the last wash, centrifuge sample and discard the supernatant. Resuspend cell pellet in enough fixative to produce a slightly milky cell suspension (the volume of fresh fixative required is approximately equal to 10 times the volume of the cell pellet). Care should be taken to dilute the sample correctly, since underdilution can result in an overly dense slide and poor spreading of metaphase cells.
4. Clean glass microscope slides by immersing in 95% ethanol followed by wiping with a Kimwipe or other lint-free cloth.
5. Place one drop of fixative on the top of the slide (below frosted edge); the fixative will spread and cover the surface of the slide. Care should be taken not to place more than *one* drop of fixative onto the dry slide, as too much fixative may cause cell nuclei and metaphase cells to accumulate along the edges of the slide.
6. Hold the slide at a 60° angle over a steam bath, and drop 4–6 drops of the cell suspension from a Pasteur pipet held 18–24 in above the slide, beginning at the top of the slide and moving downward. Immediately place the slide over a steam bath created by running hot tap water (45–50°C) into a pan. Two glass rods, placed 1.5 in apart across the rim of the pan, can be used to hold the slides above the water level in the pan.
7. The slides are removed from the pan when the fixative has dried from the upper surface (2–3 min). Examine 1–2 slides/culture under phase contrast microscopy to evaluate the quality of the metaphase cells. Adjust slide dropping conditions as needed (*see* **Notes**).
8. Slides should be aged at room temperature or in a drying oven (55–60°C) for 1–3 days prior to staining. Slides to be used for FISH or spectral karyotyping analysis should be stored at 4°C in a slide box with a desiccant (Drierite) and sealed with tape.

3.4. Trypsin-Giemsa Staining (see Notes 4 and 5)

1. Age air-dried slides for 1–3 d at room temperature or in a drying oven (50–60°C), then place on a slide warmer (55–60°C) for 2–4 h. Age fresh slides by placing on a 55–60°C slide warmer for 4 h prior to staining. Allow slide to cool to room temperature or until slightly warm and start banding procedure immediately. Do not allow slides to sit at room temperature after removal from the slide warmer.
2. Just prior to banding, prepare a series of Coplin jars containing the following solutions:
   Jar 1: 10% hydrogen peroxide (17 mL 30% H_2O_2:33 mL ddH_2O).
   Jar 2: Double-distilled water.
   Jar 3: Trypsin solution (50 mL).
   Jar 4: 45 mL Isoton (pH 6.8) with 5 mL fetal bovine serum (serum will inactivate trypsin).
   Jar 5: 50 mL Isoton.
   Jar 6: 1.5 mL Giemsa stain in 48.5 mL Gurr buffer (pH 6.8).
   Jar 7: 50 mL Gurr buffer (pH 6.8).
3. Immerse slide in 10% hydrogen peroxide for 15 s, agitating continuously, then rinse in the double-distilled water jar (jar 2), and drain. The hydrogen peroxide removes cytoplasm that may be covering metaphase chromosomes, thereby permitting uniform exposure of the chro-

matin to the trypsin treatment. This will result in more consistent staining of slides prepared from different samples.

4. Immerse slide in the trypsin solution for 10–25 s. The slide should be agitated continuously by lifting the slide out of the solution and replacing it carefully. *The trypsin time may vary from sample to sample; therefore, one slide from each culture should be tested to determine the optimal exposure time at the start of staining.* A single slide may be used to test two trypsin times by immersing the bottom half 2–10 s longer than the top portion of the slide.
5. Quickly dip the slide 5–7 times in jar 4, containing Isoton and FBS.
6. Dip slide 5–7 times in Isoton (jar 5).
7. Stain in buffered Giemsa solution for 8 min.
8. Drain slide and rinse by dipping in Gurr buffer (jar 7). Blot back of slide dry, invert slide by placing the frosted end down, and allow slide to dry on slide rack in a vertical position. *Do not* blot front of the slide.
9. The test slides are examined under the microscope after staining to select the optimal length of trypsin treatment and staining period. Increase or decrease time as needed.
10. Change solutions after 20 slides are processed.
11. Slides can be coverslipped for permanent storage using Pro-Texx mounting medium. However, we do not routinely mount slides with permanent coverslips because such slides do not give good results consistently when additional procedures are applied, e.g., FISH of previously banded slides.

3.5. Fluorescence In Situ Hybridization (see Note 6*)*

The following protocol is designed for directly labeled, commercially available probes, e.g., centromere-specific (CEP), locus-specific (LSI), or whole-chromosome painting (WCP) probes labeled with SpectrumGreen or SpectrumOrange (Vysis, Inc., Downers Grove, IL). Detailed protocols for labeling and FISH of genomic/cDNA probes have been described elsewhere ***(11)***.

1. Slides are prepared for FISH as described under **Subheadings 2.3.** and **3.3.**, and are stored at 4°C in a sealed box with a desiccant (Drierite) for up to 5 yr. Allow slides to warm to room temperature. Examine the slides under a microscope using phase contrast, and locate the area to be hybridized (select an area of 22 × 22 mm that contains a sufficient density of interphase nuclei or metaphase cells). Optional: Mark the hybridization area by scratching the slide with a glass etcher. It is possible to hybridize more than one area per slide.
2. Prepare RNAse solution (100 µg/mL in 2× SSC) in a glass staining box, and warm to 40°C in a water bath. Place slides in the RNAse solution for 30–60 min. Note: This step is omitted for G-banded slides, or slides being hybridized for the second time.
3. Remove slides and wash in 2× SSC for 2 min (room temperature, pH 7.0). Repeat three times, followed by dehydration in a graded ethanol series for 2 min each in 70%, 80%, and 95% ethanol. Air-dry slides.
4. Denature DNA by placing slides in a solution of 70% formamide/30% 4× SSC, pH 7.0, at 72–75°C for 2 min (3 min for G-banded slides). Dehydrate the slides in a graded ethanol series for 2 min each in 70%, 80%, and 95% ethanol. Air-dry slides and place on a slide warmer at 40°C.
5. Warm the commercially available probe and hybridization buffer (provided by the manufacturer) to room temperature. For 22 × 22 mm hybridization area, mix 1 µL of probe, 2 µL of sterile tissue-culture-grade H_2O, and 7 µL of the appropriate hybridization buffer (10 µL total). Use 0.3–0.5 µL of probe, 2.5 µL of hybridization buffer, and 0.5 µL of H_2O (3–3.5 µL total) for 15-mm circular hybridization areas.
6. Vortex probe mixture for 15 min.

7. Denature probe mixture at ~75°C for 5 min. Note: Some probes, e.g., CEPX/CEPY dual-color probe from Vysis, Inc., are provided as a premixed, single-stranded probe. In this case, eliminate the denaturation step, and use 10 µL for a 22 × 22-mm hybridization area, or 3 µL for a 15-mm circular hybridization area.
8. Preanneal probe mixture for 15–20 min at 37°C.
9. Apply probe mixture to hybridization area on slide. Place a 22 × 22-mm square or 15-mm round cover slip over the solution. Avoid air bubbles, which interfere with hybridization.
10. Seal the edges with rubber cement applied using a 5- or 10-mL syringe, and place into a sealed humid chamber, such as a sealed plastic container, in a 40°C water bath overnight.
11. Remove the rubber cement from the slides (no more than 5 slides at a time) and place the slides in a Coplin jar containing 0.4 × SSC at 37°C for 2 min, after which the cover slips should slide off.
12. Transfer the slides to a Coplin jar containing 0.4 × SSC/0.3%Triton-X at 73°C, cover and agitate for 10 s. Soak slides for 3 min.
13. Transfer the slides to a Coplin jar containing 2 × SSC/0.1%Triton-X at room temperature for 1 min. Rinse under running distilled water, and counterstain in DAPI for 30–60 s. Apply cover slip with PDD mounting medium.

4. Notes

1. Errors in hemacytometer counting can occur as a result of inaccuracies in preparing dilutions, incomplete mixing of cells prior to filling the hemacytometer, and overfilling or underfilling of the chamber. A major source of error is overdilution or an overly concentrated sample. Accuracy is greatest when the cell number/square is between 30 and 45 (300–450/10 squares). Therefore, samples should be diluted or concentrated to achieve this range.
2. Processing of anchorage-dependent cells. Tumor cell lines, or primary cultures of tumors may be processed using the procedures outlined under **Subheading 3.2.** with the following modifications. After exposure to Colcemid (**step 1.**), the cells are removed from the surface of the flask using trypsin-EDTA, following standard procedures for passaging anchorage-dependent cells. Continue with **steps 3–8**.
3. If the metaphase cells are *underspread*, one or more of the following steps should be applied:
   a. Perform 2–5 additional washes in fixative,
   b. Increase the angle at which the slide is held while dropping the suspension,
   c. Increase the distance between the slide and the pipet,
   d. Dilute the cell suspension by adding fixative.

   If the metaphase cells are *overspread*, one or more of the following steps should be applied:
   a. Decrease the angle at which the slide is held while dropping the suspension,
   b. Decrease the distance between the slide and the pipette.
4. If the cells are *undertrypsinized*, the chromosomes will have poor differentiation of bands with uniform, darkly stained chromatids. If the cells are *overtrypsinized*, the chromosomes will appear fuzzy, chewed or frayed, and bloated.
5. Variability in metaphase spreading and length of chromosomes will influence the optimal trypsin exposure period. The trypsin time may vary for metaphase cells from different tissues. Also, temperature and pH (the enzyme is most active at a slightly basic pH, pH 8.0) of the solution can affect trypsin time. If the room temperature is warmer than usual (25–27°C), a shorter trypsin time may be necessary (the activity of the enzyme increases with higher temperatures). These and other variables can result in over- or undertrypsinization. Therefore, at least one slide/culture should be assessed by examining 3–10 cells with different

spreading and chromosome length. It is also important to examine cells from different parts of the slide, as chromosome spreading and banding may differ along the length of the slide.

6. Although hybridization of unstained slides is optimal, FISH may be performed on slides that have been stained previously using trypsin-Giemsa banding, quinacrine banding, and some R-banding methods. Stained slides ≤3 mo old yield better results than older slides. Slides that have already been processed for FISH may be rehybridized using a second probe, following the modifications described under **Subheading 3.5.** It is preferable to use a different fluorochrome to detect the second probe, particularly when the initial probe was a centromere-specific probe. In most instances, the initial probe is removed during the denaturation step; however, the presence of residual probe may result in weak background signal.

References

1. LeBeau, M. M. and Larson, R. A. (1999) Cytogenetics and neoplasia, in *Hematology: Basic Principles and Practice*, 3rd ed. (Hoffman, R., Benz, E. J., Shattil, S. J., et al., eds.). Livingstone, New York, pp. 848–870.
2. Heim, S. and Mitelman, F. (eds.) (1995) *Cancer Cytogenetics*, 2nd ed. Wiley-Liss, New York.
3. Mitelman, F. (2000). Recurrent chromosome aberrations in cancer. *Mut. Res.* **462,** 247–253.
4. Grimwade, D., Walker, H., Oliver, F., et al. (1998) The importance of diagnostic cytogenetics on outcome in AML: analysis of 1,612 patients entered into the MRC AML 10 trial. *Blood* **92,** 2322–2333.
5. Mitelman, F., Johansson, B., and Mertens, F. (eds.), Mitelman Database of Chromosome Aberrations in Cancer (2001) http://cgap.nci.nih.gov/Chromosomes/Mitelman.
6. Look, A. T. (1997) Oncogenic transcription factors in the human acute leukemias. *Science* **278,** 1059–1063.
7. Rabbits, T. (1994) Chromosomal translocations in human cancer. *Nature* **372,** 143–149.
8. Druker, B. J., Talpaz, M., Resta, D. J., et al. (2001) Efficacy and safety of a specific inhibitor of the BCR-ABL tyrosine kinase in chronic myeloid leukemia. *N. Engl. J. Med.* **344,** 1031–1037.
9. Roulston, D. and Le Beau, M. M. (1997) Cytogenetic analysis of hematologic malignant diseases, in *The AGT Cytogenetics Laboratory Manual*, 3rd ed. (Barch, M. J., Knutsen, T., and Spurbeck, J., eds.). Lippincott-Raven, Philadelphia, pp. 325–374.
10. Thompson, F. H. (1997) Cytogenetic methods and findings in human solid tumors, in *The AGT Cytogenetics Laboratory Manual*, 3rd ed. (Barch, M. J., Knutsen, T., and Spurbeck, J., eds.). Lippincott-Raven, Philadelphia, pp. 375–415.
11. Espinosa R., III, and Le Beau, M. M. (1996) Gene Mapping by FISH, in *Methods in Molecular Biology Vol 68: Gene Isolation and Mapping Protocols* (Boultwood, J., ed.). Humana, Totowa pp. 53–76.
12. Gozzetti, A. and Le Beau, M. M. (2000) Fluorescence in situ hybridization: uses and limitations. *Sem. Hematol.* **37,** 320–333.
13. Nederlof, P. M., van der Flier, S., Wiegant, J., et al. (1990) Multiple fluorescence in situ hybridization. *Cytometry* **11,** 126–131.
14. Dewald, G. W., Schad, C. R., Christenson, E. R., et al. (1993) The application of fluorescent *in situ* hybridization to detect M*bcr/abl* fusion in variant Ph chromosomes in CML and ALL. *Cancer Genet. Cytogenet.* **71,** 7–14.
15. Dewald, G. W., Wyatt, W. A., Juneau, A. L., et al. (1998) Highly sensitive fluorescence in situ hybridization method to detect double BCR/ABL fusion and monitor response to therapy in chronic myeloid leukemia. *Blood* **91,** 3357–3365.
16. Sinclair, P. B., Green, A. R., Grace, C., and Nacheva, E. P. (1997) Improved sensitivity of BCR-ABL detection: A triple probe three-color fluorescence in situ hybridization system. *Blood* **90,** 1395–1402.

21

Positional Approaches to Cancer Genetics

Gavin P. Robertson, Linda Sargent, and Mark A. Nelson

1. Introduction

Loss of genetic material accumulates during tumor development as cancer cells select for the physical removal or functional inactivation of genes whose encoded proteins regulate normal cellular behavior. The hallmark indication for this type of gene, now termed a tumor suppressor gene, is a genetic event resulting in loss or deletion of chromosomal material. Position-oriented approaches have taken advantage of the intimate involvement of tumor suppressor gene inactivation through deletion to localize, identify, and demonstrate the involvement of these genes in carcinogenesis. Since deletion of genetic material exceeds genomic amplifications in most cancer types, loss of gene function appears to play a prominent role in tumor formation. However, despite the intimate involvement of tumor suppressor genes in the neoplastic process, relatively few of these genes have been identified, leaving the identity of the majority unknown. Thus, the use of position-based strategies for mapping and isolating tumor suppressor genes remains a prime tool in their identification.

In this chapter we discuss the positional approaches used in the localization and identification of tumor suppressor genes. This process can be broadly divided into the following 4 steps leading to the identification and validation of these genes: (a) genome-wide surveys to identify the chromosomal location of genes potentially involved in tumor development; (b) refinement and/or confirmation of the chromosomal region of loss or deletion; (c) cloning genes from consistent regions of loss; and (d) validating the candidate gene as a tumor suppressor through the identification of truncating or missense mutations as well as by growth and tumor suppressing assays.

Any strategy that detects genome-wide loss of genetic material could conceivably be used as a position-oriented approach to map the location of potential tumor suppressor genes. At the chromosomal level of resolution, traditional cytogenetic analysis (*see* **Note 1**), spectral karyotyping (SKY), and comparative genomic hybridization (CGH) have been used extensively for these purposes. All three of these chromo-

From: *Methods in Molecular Biology, Vol. 222: Tumor Suppressor Genes: Pathways and Isolation Strategies*
Edited by: Wafik S. El-Deiry © Humana Press Inc., Totowa, NJ

some-based approaches have a unique set of strengths and weaknesses and are therefore quite complementary in defining regions of loss in the genome. Comprehensive analyses based on these strategies and performed on collections of tumors have identified nonrandom areas of loss within particular cancers, and areas of loss common to multiple tumor types. In certain instances, the identification of these areas as "hot spots" for loss of genomic material has facilitated the identification of cancer suppressor genes at these sites.

Once a chromosomal region has been identified as a target of deletion, the region is narrowed by mapping using fluorescence *in situ* hybridization (FISH) or by searching for smaller regions of loss in tumors in comparison to normal tissue from the same patient. The latter process is termed loss of heterozygosity (LOH) and employs a variety of molecular strategies to detect regional loss in tumors by comparison to normal tissue from the same patient. These techniques initially used restriction fragment length polymorphism (RFLP), but now commonly employ polymerase chain reaction (PCR)-based strategies to identify each chromosomal allele present in the heterozygous condition from normal tissue and loss of one allele in tumor material, indicating deletion. The PCR-based protocol is discussed below. Like the cytogenetic approaches described above, nonrandom loss of a region or allele can infer disruption of a tumor suppressor gene and thereby indicate its chromosomal location. The identification of "hot-spot" areas of LOH has also facilitated the identification of several cancer suppressor genes.

Ideally, a small region of homozygous loss can be identified in the tumor material using LOH, and a physical map spanning the region can be constructed from which candidate genes can be cloned. Strategies such as exon trapping or the use of microarray analysis can aid in the identification of genes mapping to a particular regions lost from the tumor genome. The uses of arrayed genes or expressed sequence tags (ESTs) has the further advantage of identifying genes regulated by the tumor suppressor, thereby unraveling the biology of the gene. Another extension of this approach examines directly genes or ESTs that occur in regions of homozygous or hemizygous deletion for the presence of mutations. As more genes and expressed sequences are identified and mapped to subchromosomal regions that are frequently deleted in cancers, determining their involvement in tumor progression will become an increasingly important strategy in the identification of this class of cancer-causing genes. In many instances only a small portion of sequence need be obtained, and then the intact gene can be pieced together and then screened for the presence of mutations. On-line databases such as those of The National Center for Biotechnology Information (http://www.ncbi.nlm.nih.gov/) or The Institute for Genome Research (http://www.tigr.org/) are useful when using this approach.

Candidate cancer suppressor genes identified by either approach must be shown to be biologically relevant to the pathogenesis of the cancer being studied. Therefore, in addition to the identification of large-scale deletions in tumors, further confirmation is provided by the discovery of nonsense and missense mutations that either truncate the protein or affect its normal cellular function in primary tumors and cell lines. Approaches such as direct sequencing of single-stranded conformational polymorphism or coupled in-vitro transcription and translation (also called the protein truncation test) can aid in the identification of mutations.

2. Materials

2.1. Giemsa Trypsin Banding

1. Gurr buffer tablets (pH 6.8): BDH Chemicals, 50-tablet lots.
2. Gurr buffer stock solution: 1 Gurr tablet and 1000 mL of sterile distilled water.
3. Hanks' balanced salt solution, with phenol red (1×) (Ca/Mg-free): Gibco, 500-mL lots.
4. Leischman's stain (Gurr): BDH Chemicals, 25-g lots.
5. Leischman's stain stock solution: 0.8 g Leischman's stain and 500 mL methyl alcohol. Incubate at 37°C overnight and filter.
6. Leischman's stain working solution: 10 mL Leischman's stain stock solution and 30 mL Gurr buffer stock solution. Good for only 2 h.
7. Methyl alcohol anhydrous: Mallinckrodt AR, 4-L lots.
8. Sodium phosphate dibasic: Columbus Chemical Industries, ACS powder, 500-g lots.
9. Sodium phosphate dibasic (0.4 *N*): 5–6 g sodium phosphate dibasic and 100 mL sterile distilled water.
10. Trypsin-EDTA (10×): Sigma Chemical Co (cat. no. T-9395), 100-mL lots. Aliquoted into 5 mL batches and frozen at −20°C.
11. Trypsin EDTA working solution: 5 mL trypsin EDTA and 50 mL Hanks' with 2 drops of 0.4 *N* Na_2HPO_4. Good for 2 working days.

2.2. SKY Analysis

1. Applied Spectral Imaging Probe mix of fluorescently labeled spectral probes: SKY/M10 for mouse and SKY/H10 for human.
2. MilliQ Deionized water for all solutions.
3. 0.01 *M* HCL: Add 0.5 mL of 1 *M* HCL to 49.5 mL deionized (DI) H_2O in a Coplin jar and heat to 37°C.
4. Pepsin stock: Make 100 mg/mL in sterile H_2O. Store at −20°C. Add 6–15 µL to 0.01 *M* HCL solution.
5. Phosphate-buffered saline (PBS) 1× (Gibco/BRL): Stored at room temperature (RT). Make up 3 Coplin jars at RT.
6. 1× PBS/MgCl:. Add 50 mL of 1.0 *M* $MgCl_2$ to 950 mL of 1× PBS. Make up 1 Coplin jar at RT.
7. 1% Formaldehyde: Add 2.7 mL of 37% formaldehyde to 100 mL of 1× PBS/$MgCl_2$. Make up 1 Coplin jar at RT
8. Ethanol series: Make up 2 Coplin jars each of 70%, 80%, and 100% ethanol, one at RT and one at 4°C.
9. Denaturation solution: Add 35 mL formamide, Fluka deionized cat. no. 47671, Sigma; 10 mL deionized H_2O and 5 mL 20× SSC. pH to 7.0. Put aliquot of 100 µL/slide in a tube and warm to 72°C.
10. Water baths: One bath at 72°C and one at 37°C.
11. 50% Formamide/2× SSC: Add 15 mL 20× SSC, 60 mL DI H_2O, and 75 mL formamide, pH to 7.0. Make up 3 Coplin jars and warm to 45°C.
12. 1× SSC: Add 12.5 mL of 20× SSC to 237.5 mL of DI H_2O. Make up 2 Coplin jars warmed to 45°C.
13. 4× sodium chloride sodium citrate (SSC)/0.1% NP-40: Add 100 mL of 20× SSC, 400 mL DI H_2O, and 0.5 mL NP-40 and mix well. Make up 6 Coplin jars and warm to 45°C. Denature probe (vial #1) 10 µL/slide in a microfuge tube in a floating rack at 72°C for 7 min. Then transfer to 37°C for at least 1 h. Vectashield mounting medium with DAPI and antifade, Vector cat. no. H-1200.

2.3. Pretreatment of Slides for CGH

Use MilliQ DI water for all solutions.

1. RNase A: To remove RNA from the target slides. Boehringer cat. no. 109169, 10-mg stock solution: 20 mg/mL sterile water, boil for 15 min, cool to RT, make aliquots, store at −20°C.
2. 5 g Pepsin (Sigma): Stock solution: 10% = 100 mg/mL, dissolve in sterile water, keep on ice, make 50-L aliquots, store at −20°C.
3. PBS/$MgCl_2$: 50 mL of 1*M* $MgCl_2$ + 950 mL of 1× PBS.
4. 1% Formaldehyde: Formamide from Fluka, cat no. 47671. Prepare in 1× PBS/$MgCl_2$. Add 2.7 mL of 37% formaldehyde to 97.3 mL of 1× PBS/$MgCl_2$.
5. 2× SSC: 50 mL 20× SSC, 450 mL H_2O, room temperature.

2.4. Postfixation for CGH

1. PBS at room temperature.
2. 1% Formaldehyde: Add 2.7 mL of 37% formaldehyde to 97.3 mL of 1× PBS/$MgCl_2$.

2.5. CGH

Use MilliQ DI water for all solutions.

1. CGH nick translation kit (nick translation enzyme, 10× nick translation buffer, dTTP, dCTP, dATP dGTP, nuclease-free water, unlabeled and Spectrum Green-labeled control DNA, Vysis cat. no. 32-801-3000).
2. Spectrum Red dUTP, Vysis cat. no. 30-803400.
3. Spectrum Green dUTP, Vysis cat. no. 30-803200.
4. Human COT-1 DNA 32-800028.
5. Salmon sperm DNA, 1 mg/mL.
6. 3 *M* sodium acetate.
7. 100% ethanol.
8. Nick translation kit with direct-labeled nucleotides, Vysis cat. no. 32-801-3000.
9. LSI/WCP hybridization buffer, Vysis cat. no. 30-804826 (dextran sulfate, 2× SSC, and formamide, pH 7.0).

2.6. Hybridization

2.6.1. Solutions for CGH

1. 70% Fluka formamide: 35 mL formamide, 10 mL of 20× SSC, 5 mL distilled H_2O, pH to 7.0, heat for 30 min to 74°C.
2. Ethanol, 70%, 80%, 100%.

2.7. CGH Detection

1. 0.4× SSC 0.3% NP 40: To 10 mL 20× SSC, add H_2O so that final volume is 500 mL, heat for 30 min at 74°C.
2. 2× SSC 0.1% NP-40: 50 mL 20× SSC: 450 mL H_2O, 0.5 mL NP-40, room temperature.

Washing of the slide should be done with extensive agitation of the slide by hand, because of the large amounts of DNA.

2.8. Less Stringent Wash Solutions for CGH

1. 50% Formamide, Fluka brand from Fisher Scientific, cat. no. BP228-100.

2. Formamide/SSC: 30 mL 20× SSC, 120 mL sterile water, 150 mL formamide. Adjust pH to 7.4 (200 mL FA/SSC requires 500 µL of 1*M* HCl). Heat for 30 min at 43°C.
3. 2X SSC 0.1% NP40: 50 mL 20X SSC, 450 mL distilled water, 0.5 mL NP 40. Heat to 37°C for 30 min.
4. Vectashield mounting medium with DAPI and antifade (Vector).

2.9. Labeling of BAC, P1, or Yac Clones

1. Nick translation kit (nick translation enzyme, 10× nick translation buffer, dTTP, dCTP, dATP dGTP, nuclease-free water, unlabeled and Spectrum Green-labeled control DNA, Vysis cat. no. 32-801-3000.
2. Texas Red dUTP, Molecular Probes cat. no. C-7608.
3. Spectrum Green dUTP, Vysis cat. no. 30-803200.
4. CGH hybridization reagents (CGH hybridization buffer, 20× SSC, NP40, and DAPI II counterstain Vysis cat. no. 32-801023.
5. Human COT-1 DNA, Vysis cat. no. 32-800028.
6. Nick translation kit with direct-labeled nucleotides, Vysis cat. no. 32-801-3000.
7. Vectashield mounting medium with DAPI and antifade, Vector cat. no.
8. LSI/WCP hybridization buffer, Vysis cat. no. H-1000 (dextran sulfate, SSC, and formamide pH 7.0).

2.10. Bac Hybidization

1. Vysis Nick Translation kit.
2. 70% Formamide/2× SSC.
3. 70%, 80%, 100% ethanol.
4. Hybridization buffer.

2.11. PCR-Based Copy

1. PCR buffer.
2. dNTP mix.
3. Flourescently labeled STS.
4. Automated DNA sequence analyzer.

2.12. Exon Trapping

1. Genomic DNA for this protocol can be obtained from cosmids, plasmids, phage, yacs, Bacs, or Pacs.
2. Restriction enzymes: BstXI and one or more of the following for cloning in vector pSPL3: BamHI, SstI, NotI, XhoI, XmaIII, PstI.
3. PSPL3 (Gibco)
4. 7.5 *M* ammonium acetate.
5. 100% ethanol.
6. TE buffer, pH 8.
7. Calf intestinal alkaline phosphatase (CIP).
8. T4 DNA ligase.
9. Competent *Escherichia coli.*
10. LB medium and plates with 100 µg/mL ampicillin.
11. COS-& cells.
12. DMEM supplemented with L-glutamine, pen/strep, and 10%10% FBS.
13. Cationic lipid (your choice).
14. Trizol (Gibco).

15. PBS lacking calcium and magnesium.
16. DEPC-treated water.
17. 20 μ*M* primer SA2:5′-ATCTCAGTGGTATTTGTGAGC-3′.
18. 5× first-strand buffer.
19. 0.1 M DTT.
20. 10 m*M* dNTP mix in DEPC water.
21. 200 U/μ*M* MLVH reverse transcriptase.
22. 2 U/μL Rnase H.
23. Taq DNA polymerase and 10× buffer.
24. 50 m*M* $MgCl_2$.
25. 20 μ*M* primer dUSA4:5′-CUACUACUACUACACCTGAGGAGTGAATTGGTCG-3′.
26. 20 μM primer SUSD2:
27. 5′-CUACUACUACUAGTGAACTGCACTGCACTGTGAAAGCTGC-3′.
28. 1.5% Agarose gel.
29. 6-Well tissue culture plates.
30. Polystyrene microfuge tubes.
31. TA cloning kit.

2.13. Microarray

1. dNTP mix: 25 m*M* dA/G/TTP, 10 m*M* dCTP. For 100-μL volume, use 25 μL 100 m*M* dATP, 25 μL 100 m*M* dGTP, 25 μL 100 m*M* dTTP, 10 μL 100 m*M* dCTP, 15 μL sterile dH_2O.

2.14. Slide Wash Solutions

1. Wash 1: .5× SSC, 0.01% SDS.
2. Add 12.5 mL 20× SSC and 0.5 mL 10% SDS into a total vol. of 500 mL dH_2O (487 mL dH_2O). Measure volumes in grad. cylinder, *not* wash bottles.
3. Wash 2: 0.06× SSC, 0.01% SDS.
4. Add 1.5 mL 20× SSC and 0.5 mL 10% SDS into a total vol. of 500 mL dH_2O (498 mL dH_2O). Measure volumes in grad. cylinder, *not* wash bottles.
5. Wash 3: 0.06× SSC.
6. Add 1.5 mL 20× SSC into a total vol. of 500 mL dH_2O (498.5 mL dH_2O), and filter with a 0.22-μm filter. Measure volumes in grad. cylinder, not wash bottles.

2.15. Preparation of 6% Polyacrlyamide Gels (100-mL volume) for SSCP (see Notes 3 and 4)

1. 20 mL Acrylamide (49:1).
2. 6 mL 10× TBE.
3. 200 μL Ammonium persulfate (10%).
4. 20 μL TEMED.
5. qs to 100 mL ddH_2O.

2.16. Preparation of Stop Solution (50-mL volume) for SSCP

1. 10 mL 95% Formamide.
2. 10 mg Xylene cyanol.
3. 10 mg Bromophenol blue.
4. 200 μL 0.5 *M* EDTA.
5. 10 μL 10 *M* NaOH.
6. qs to 50 mL ddH_2O.
7. Mix completely and aliquot into 1.5-mL Eppendorf tubes. Store at −20°C.

2.17. PTT Analysis

1. QIA quick PCR columns (Qiagen, Santa Clarita, CA).
2. TNT T7 Quick Coupled Transcription/Translation reaction (Promega, Madison, WI).
3. 0.3 m*M* magnesium acetate.
4. 10–15% discontinuous SDS-PAGE gel.
5. Amplify (Amersham Corp. Arlington, IL).
6. Phosphoimager and screens.

3. Methods

The analysis of chromosomes by the individual banding pattern allows a genome-wide scan of the karyotype. Staining methods allow the unambiguous identification of each individual chromosome. The study of cytogenetic changes in malignancies has been especially useful in the understanding of the pathogenesis of the disease. Genetic alterations in cancer commonly occur in chromosomal regions that regulate growth and development. Identifying these chromosomal regions is the first step in determining the critical genes for the initiation and progression of cancer. The analysis of these changes is the first step of isolating the specific genes that are involved in the progression of cancer. The identification of cytogenetic changes in tumors is limited by quality of the chromosome morphology. This technique is most powerful when it is done in combination with molecular cytogenetic methods.

3.1. Giemsa Trypsin Banding (see Note 1)

This procedure involves the proteolytic treatment of the chromosomes to produce a differential staining in the chromosomes arms. This treatment allows pairing of homologs and the detection of chromosomal rearrangements. The biochemical nature of the reactions involved has not yet been completely defined.

1. Immerse aged slide in working trypsin solution for 20–90 s at 20–25°C.
2. Rinse in tap water.
3. Immerse in working Leischman's stain for 1.5–2.5 min.
4. Rinse in tap water.
5. Dry the stained slide by blowing the compressed air at moderate force over the surface of the slide.
6. Examine the slide for quality of bands and staining. If slide is adequate for analysis and photography, place on slide warmer.
7. Adjust trypsin treatment times and/or stain times to find the best banding for the meatphase preparation.
8. Banded slides should be left on slide warmer for 5–10 min, then cover-slipped and allowed to dry before screening and analysis.

3.2. Spectral Karyotyping (SKY)

Nonrandom chromosome abnormalities associated with the development of tumors can target genes that are altered during carcinogenesis. Specific chromosome break points have also been associated with genes that confer susceptibility as well as resistance to cancer. Karyotypic analysis of tumor metaphase preparations is hindered by low mitotic index and complicated translocations as well as small rearrangements that are difficult to identify by traditional chromosome banding. Although comparative genomic hybridization has made it possible to characterize the loss and gain of chromosomal material, this method cannot identify translocations. The accuracy of the identification of karyotypic alterations has been greatly increased by the development of spectral karyotyping. Spectral karyotyping is a novel imaging method that combines spectroscopy and imaging. This type of approach allows analysis of the full spectrum of light at all pixels of the image. The spectral-based approach can simultaneously identify 100% of the chromosomes in a metaphase spread. The loss, gain, or translocation of any chromosome in a metaphase spread can be characterized in one experiment ***(2)***.

The spectral paints used to detect the chromosomes are prepared using flow-sorted chromosomes labeled with a combination of 5 fluorochromes by degenerate oligonucleotide-primed PCR ***(2,3)***. All 46 chromosomes are labeled a unique color, which is analyzed and karyotyped using computer imaging and Applied Spectral Imaging software (ASI, Carlsbad, CA). The analysis of the slides is done using a microscope attached to a spectral cube (Applied Spectral Imaging) and a filter cube (SKY1, Chroma Technology, Brattleboro, VT). The attached spectral cube and filter allow for the simultaneous excitation of all dyes and the measurement of their emission spectra. The spectral measurements are analyzed using ASI software. A classification color is assigned based on the best match for each chromosome. A color image is then created in which every pixel is displayed in the color that corresponds to the chromosome-specific emission spectrum ***(4)***. The spectral classification is the basis for chromosome identification and SKY.

1. Warm 50 mL of 0.01 *M* HCL with 6 μL pepsin stock (may need up to 15 μL depending on cytoplasm on slide) to 37°C in a Coplin jar.
2. Incubate slides in pepsin solution for 1–3 min for human and up to 1 min for mouse.
3. Wash slides in 1× PBS at room temperature for 5 min. Do a second wash in PBS for 5 min at room temperature.
4. Wash slides in 1× PBS/$MgCl_2$ at room temperature for 5 min.
5. Place slides in a Coplin jar with 1% formaldehyde for 10 min at room temperature.
6. Wash slides in 1× PBS for 5 min, agitating the slide during the wash.
7. Dehydrate slides in 70%, 80%, and 100% ethanol for 2 min each. Air-dry slide completely.

3.2.1. Denaturation

1. Warm denaturation solution (70% formamide/2× SSC, pH 7) to 72°C in a tube in a water bath. Warm the slide in a hybridization oven to 72°C.
2. Add 100 μL denaturation solution warmed to 72°C directly onto slide, add coverslip, and place in a hybridization oven at 72°C for 30–45 s. Shake off the cover slip and immediately put slide in 70% ethanol. It is important to do only one slide at a time.
3. Place slides in ice-cold 70%, 80%, and 100% ethanol for 2 min each, then air-dry completely.
4. Add 10 μL of denatured probe (vial #1) to the hybridization area and cover with a 22 × 22-cm glass cover slip, being sure the probe spreads evenly. Immediately seal edges with rubber cement. Let slide on slide warmer for 4 h to overnight for rubber cement to cure.
5. Dampen paper towels with water and place the folded towels at the bottom of a Petri plate to assemble the humidified chamber. Place Q-tips over the paper towels, then lay the slides directly over the Q-tips. Seal the Petri plate with aluminum foil. Transfer slides to a humidified chamber at 37°C for 72 h (from time the probe is added).

3.2.2. Wash/Detection After Hybridization

1. Remove slides from humidified chamber and carefully remove cover slip.
2. Transfer slides to a Coplin jar with 50% formamide/2× SSC, prewarmed to 45°C for 5 min. Repeat this twice for 3 full washes, 5 min each.
3. Wash slides in a Coplin jar of 1× SSC prewarmed to 45°C for 5 min. Agitate the slides by hand for each wash. Repeat this once for 2 full washes.
4. Wash slides in a Coplin jar with 4× SSC/0.1% NP-40 prewarmed to 45°C for 2 min.
5. Pipette 60–80 μL of blocking reagent (vial #2) to hybridization area and cover with a 24 × 30-mm cover slip, being sure it spreads evenly. Lay flat and incubate at 37°C for 30 min in a humidified chamber.
6. Carefully remove cover slip and add 60–80 μL of buffer 1 (vial #3) and cover slip. Incubate for 45 min in a humidified chamber at 37°C.
7. Carefully remove cover slip and wash slides in 4× SSC/0.1% NP-40 prewarmed to 45°C for 5 minutes. Repeat this with fresh wash for 3 full washes with agitation.
8. Add 60–80 μL of buffer 2 (vial #4) to hybridization area and cover slip. Incubate at 37°C in a humidified chamber for 45 min. *Repeat step 8 with new wash.*
9. Pipet about 20 μL of DAPI/antifade solution (vial #5) over the hybridization area, being sure to completely cover area evenly with no bubbles. Apply 24 × 60-mm cover slip to cover whole hybridization area. Store at −20°C until ready to view.

3.3. Comparative Genomic Hybridization Introduction (see Note 2*)*

Neoplastic transformation is a complex process involving both positive and negative regulatory elements. Determination of loss of heterozygosity by allelotype studies alone underestimates the loss of genetic material. Using computer imaging systems, the degree of chromosomal loss and gain in frozen and archival tumor samples can be measured by comparative genomic hybridization (CGH). The degree of loss and gain of chromosomal material of archival samples can be estimated within 10 megabases using CGH software *(6)*. The CGH analysis can be followed by further FISH analysis of the common break-points with specific BAC, cosmid, and YAC probes *(6)*. The precise mapping of the alteration of a single gene can be accomplished using small probes, possibly using YAC, BAC, and cosmid clones *(7)*. The defined break-points identified in tumor cells can then be followed back to earlier changes both in dividing populations and in interphase cells using the specific genetic region probes within those regions.

3.3.1. Pretreatment of Slides

1. Equilibrate slides in 2× SSC at RT.
2. RNase A treatment:
   a. Dilute the RNase A stock 1:200 (in 2× SSC).
   b. Apply 100 μL to 24 × 60-mm^2 cover slip, touch slide to cover slip, incubate at 37°C for 45 min.
   c. Remove cover slips and wash 3 times for 5 min in 2× SSC, RT, with shaking (Coplin jar).
3. Pepsin treatment:
   a. Prepare solution: Make 0.01 *M* HCl by adding about 1 mL of 1 *M* HCl to 99 mL of dH_2O, prewarmed at 37°C. Add 10–50 (μ) 1 pepsin first, then add 100 mL prewarmed 0.01 *M* HCl. Mix well and adjust pH to 2.0 with sodium hydroxide.
   b. Incubate slides at 37°C in Coplin jar for 4–10 min.
   c. Wash 2 times for 5 min each in 1× PBS, at room temperature, with shaking.
   d. Wash 1 time for 5 min at room temperature with 1× PBS/$MgCl_2$.

3.3.2. Pretreatment Procedure

1. Make solution of 1% formaldehyde in 1× PBS/$MgCl_2$.
2. Incubate for 10 min at room temperature.
3. Wash 1 time for 5 min in 1× PBS, at room temperature, with shaking.
4. Dehydrate slides in 70%, 90%, and 100% ethanol for 3 min each.
5. Air-dry slides.

3.3.3. Probe Preparation

1. In a microfuge tube cooled on ice, combine the following:
   a. 1–3 µg DNA.
   b. 2.5 µL of 0.2 m*M* Spectrum Green dUTP or Spectrum Orange/Red dUTP.
   c. 5 µL of 0.1 m*M* dTTP.
   d. 10 µL of dNTP mix (0.3 m*M* dGTP, 0.3 m*M* dATP, 0.3 m*M* dCTP).
   e. 5 µL of 10× nick translation buffer
   f. 10 µL of nick translation enzyme cocktail
2. Vortex briefly.
3. Incubate 6–12 h (usually 8 h for YACs and BACs) at 15°C.
4. Stop reaction by heating to 70°C for 10 min—incubation and stop can be done in a thermal cycler.
5. Chill on ice or store at 4°C until ready to use probe.
6. Determine probe size by running 8–10 µL of the sample on a 2% agarose gel with ethidium bromide. The majority of the smear should be between 500 and 3000 bp for best hybridization; if the probe is larger, you can add more enzyme and incubate at 15°C for longer and stop reaction by 70°C for 10 min and rerun sample on a gel.
   It is important that the length of the test and the control DNA are the same.
7. From the smear, determine the amount of product that is required for a probe (usually 10–30 µL of product).
8. To a small centrifuge tube, add: 100 ng/labeled DNA product (usually 10–30 µL), 3 µL COT-1 DNA, for blocking, 0.5 µL salmon sperm, for blocking, 0.1 vol of 3 *M* sodium acetate, 2.5 vol of 100% ethanol.
9. Vortex briefly and incubate on Dry Ice for 15–30 min. Spin at >12,000 rpm for 30 min at 4°C. Carefully pour off supernatent. Dry pellet completely in a speed vacuum centrifuge for 5–8 min.
10. To the pellet add 3 µL nuclease-free water and 7 µL Vysis hybridization buffer (dextran sulfate, SSC, and formamide, pH 7.0).
11. Mix well and allow resuspending at least overnight at −20°C before hybridizing.

3.3.4. Assembly of Probe Mixture

1. Combine in a 1.5-mL tube:
   a. 10 µL (200 ng) Spectrum Green test DNA (nick translated).
   b. 1 µL (100 ng) Spectrum Red total genomic reference DNA.
   c. 10 µL (10 µg) Cot-1 DNA.
   d. 0.5 µL (10 mg/mL) Salmon sperm.
2. Add 2.1 µL (0.1 vol) 3 *M* sodium acetate. Then add 52.5 µL (2.5 vol) of 100 % ethanol and vortex briefly. Let stand on Dry Ice for 15–30 min.
3. Centrifuge at 12,000 rpm at 4°C for 30 min.
4. Pour off supernatent and dry pellet for 10–15 min under vacuum at ambient temperature.
5. Resuspend pellet in 3 µL nuclease-free H_2O and 7 µL hybridization buffer.
   Store at −20°C at least overnight.

3.3.5. Procedure

1. Select slides with well-spread, long chromsomes with very little cytoplasm. Choose slide and mark hybridization area with diamond pen. The area is 22 × 22 cm.
2. Warm in a H_2O bath set to 74 ± 1°C with a Coplin jar of denaturation solution (70% formamide).
3. Immerse slides in denaturation solution for 3–5 min.
4. Dehydrate slides in ethanol for 2 min each, 70%, 80%, 100% ethanol.
5. Denature probe by placing it in a float in the 74°C water bath for at least 5 min.
6. Completely dry slides by wiping bottom and then put on slide warmer.
7. Apply 10 µL of probe to hybridization area while on slide warmer.
8. Apply cover slip and be sure probe spreads evenly.
9. Immediately seal with rubber cement, make sure it is completely sealed, and place immediately in the incubator. Place slides in a humidified box at 37°C for 2–4 d.

3.3.6. Washing Slide, Stringent Wash

1. Warm the water bath and a Coplin jar with a solution of 0.4× SSC/0.3% NP to 74 ± 1°C for at least 30 min.
2. Prepare a Coplin jar of room-temperature 2× SSC/0.1% NP-40.
3. Remove rubber cement and cover slip from slides carefully and immediately place in Coplin jar at 74°C of 0.4× SSC/0.3% NP-40 for 1–3 s and agitate slide.
4. Put slide in the room-temperature 2× SSC/0.1% NP-40 for 5–60 s with agitation to remove the probe. Metaphase preparations with a lot of cytoplasm will require the longer washing time.
5. Air-dry slide in darkness.
6. Apply 10 µL of DAPI II counterstain (Vectashield mounting medium with DAPI and antifade, Vector) to hybridization area and cover slip.
7. Detection of the losses and gains requires a fluorescent microscope, a sensitive camera, and computer software to quantitate the changes as detailed by Kallioniemi et al., 1992 ***(8)***.

3.3.7. Procedure for Less Stringent Wash

1. Make 3 Coplin jars of 50% formamide/2× SSC pH 7.0 and warm to 43°C in a H_2O bath. Make 3 Coplin jars of 2× SSC and warm to 37°C in a H_2O bath. Make 1 Coplin jar of distilled H_2O at room temperature.
2. Place slides in first 50% formamide jar at 43°C for 5 min. Repeat for 3 full washes.
3. Place slides in a 2× SSC at 37°C for 5 min. Repeat for 3 full washes in fresh 2× SSC.
4. Rinse slides in distilled deionized room temperature H_2O three times.
5. Apply 19 µL DAPI II counterstain (Vectashield).

3.4. Refinement of the Region of Consistent Abnormality. Mapping by Fluorescent In Situ Hybridization (FISH)

BACs or PACs containing an average of 100 kbp can be useful in refining regions lost during the tumorigenic process. They may be particularly useful in mapping regions of deletion or chromosome alteration by labeling DNA from these reagents with fluorescent tags and hybridizing to metaphase chromosomes obtained from cell lines or tumor material to map consistently altered regions of the genome. Narrowing the region of consistent abnormality can aid in the identification of candidate cancer suppressor genes by refining the search to a single BAC or PAC.

3.4.1. Procedure for Nick Translation of BAC, YAC, P1, or Cosmid Clones

1. In a microfuge tube cooled on ice, combine the following:
   a. 1–3 µg Extracted P1, BAC, cosmid, or YAC DNA. Prepare 0.2–1-µg/µL solution of extracted DNA in Tris-EDTA (10 m*M* Tris base, 1 m*M* EDTA, pH 8.5) buffer.
   b. 2.5 µL of 0.2 m*M* Spectrum Green dUTP or Spectrum Orange/Red dUTP.
   c. 5 µL of 0.1 m*M* dTTP.
   d. 10 µL each of dNTP mix, 0.3 m*M* dGTP, 0.3 m*M* dATP, 0.3 m*M* dCTP.
   e. 5 µL of 10× nick translation buffer.
   f. 10 µL of nick translation enzyme cocktail.
2. Vortex briefly.
3. Incubate 6–12 h (usually 8 h for YACs and BACs) at 15°C.
4. Stop reaction by heating to 70°C for 10 min.
5. Remove unincorporated nucleotides by gel filtration using a Sephadex G-50 spin column equilibrated in 10 m*M* Tris base, 1 m*M* EDTA, and 0.1% SDS (the SDS prevents the probe from sticking in the column).
6. Chill on ice or store at 4°C until ready to make probe. Incubations and stop reactions can be done in a thermal cycler.
7. Determine probe size by running 8–10 µL of the sample on a 2% agarose gel with ethidium bromide. The majority of the smear should be around 300 and 3000 bp for best hybridization. If the probe is larger than 300 bp, you can add enzyme and incubate at 15°C for longer and stop reaction by 70°C for 10 min and rerun sample on a gel. Probe fragments that are larger may produce bright fluorescent speckles across the hybridization area.

3.4.2. FISH Hybridization with BAC, P1, YAC, and Cosmid Region-Specific Probes

1. Mark slide hybridization area.
2. Denature slide in 70% formamide/2× SSC at 78°C for 7 min.
3. Make probe hybridization mixture from the nick translation reaction of about 10–20 µL nick translation product per slide.
4. Dehydrate slides in ethanol series (70%, 80%, and 100%) for 2 min each.
5. During dehydration denature probe mix at 78°C for 7 min (optional: pre-anneal the probe for 1–2 hours at 37°C prior to hybridization).
6. Immediately apply probe (10–20 µL) to hybridization area, apply a 22 × 22-cm glass cover slip, and seal with rubber cement.
7. Incubate at 37°C for 36–72 h.

3.4.2.1. Wash

1. Wash slides in 50% formamide/2× SSC at 45°C for 5 min each. Do 3 full washes in fresh solution in clean jars.
2. Wash slides in 0.1× SSC at 60°C for 3 full washes at 5 min each.
3. Wash slides in 4× SSC/0.1% NP-40 for a dip (1–3 s) at 45°C. Dip in RT tap water. Then air-dry.
4. Apply DAPI (Vectashield mounting medium with DAPI and antifade, Vector) to cover the hybridization area and cover slip.

3.5. Loss of Heterozygosity (LOH) to Map Homozygous Deletion

LOH is a molecular approach used to identify regions of consistent loss or to narrow a region of deletion in tumor material in comparison to normal tissue from the same individual. Mapping regions of loss using this approach has successfully localized nonfamilial cancer suppressor genes to particular chromosomes and further refined the region of deletion to chromosomal bands or even smaller genetic regions. Under ideal circumstances, a small region of homozygous loss can be identified in the tumor genome and a physical map spanning the region can be constructed from which candidate genes can be cloned. Illustrative examples demonstrating the success of this approach are the cloning of INK4A/p16 in 9p21, RB1 in 13q14, and PTEN in 10q23. These successes demonstrate the usefulness of LOH in cancer suppressor gene identification for the future.

The techniques used to detect LOH have traditionally been molecular-based approaches such as restriction fragment length polymorphism (RFLP) or PCR-based amplification of polymorphic tandem repeats (STS, sequence-tagged-sites) to identify the alleles present in each respective chromosome. RFLP utilizes the normal variation in sequence that exists between chromosomal alleles to detect changes in the ability of restriction enzymes to cut DNA. Usually the variation is observed as fragments of different size produced after digesting DNA with a restriction enzyme. The PCR-based approach identifies polymorphic variation that occurs in the number of tandem repeats present in the maternally and paternally derived chromosomes. Isotopic labels have been widely used for both strategies; however, the latter technique now commonly employ fluorescently labeled PCR primers, with the product being visualized using fluorescence detectors. Fluorescently labeled STS primer sets or MapPairs are commercially available from Research Genetics, Inc. (http://www.resgen.com/).

STSs or MapPairs generally amplify under uniform PCR conditions; however, specific reaction conditions for each marker are described in GeneMap 99 (http://www.ncbi.nlm.nih.gov/genemap99/), from the Genome Data Base (http://www.gdb.org/) or can be obtained from a reference cited therein. Markers spaced at desired intervals along the length of the chromosome can be obtained through the GDB or GeneMap 99. Each marker should be accessed twice and examined by two independent evaluators prior to determining LOH.

Any standard procedure for DNA extraction from tumor samples and matched normal lymphocytes can be used to isolate DNA of sufficient quality for use in this procedure.

1. Prepare the following buffer and set up the PCR reaction as follows:
   a. Buffer conditions: 1.5 m*M* $MgCl_2$, 50 m*M* KCl, 10 m*M* Tris-HCL, pH 9.3.
   b. Protocol: 10 ng template: 5 p*M* of each primer, 4 n*M* of each dNTP, 0.025 U/µL of Taq. Total volume 20 µL (*see* **Note 9**).
2. Heat reaction to 95°C for 5 min.
3. PCR thermalcycler program: 94°C for 40 s; 55/56°C for 30 s; 72°C for 40 s. Do 35 cycles followed by 2 min at 72°C.
4. From 1 to 5 µL of PCR reaction can be run on an ABI sequence analyzer. Fluorescent size markers should be added to the samples and used as an internal control to determine the approximate location and size of the PCR products.

3.6. Cloning Candidate TSG Genes from the Region of Consistent Abnormality

3.6.1. Exon Trapping

Exon trapping or exon amplification identifies expressed DNA sequences present in a segment of genomic DNA by selecting for functional splice sites found in that genomic DNA ***(10)***. No prior knowledge is required regarding tissue-specific gene expression, and the protocol is amenable to complex genomes. An added advantage of this approach is that it can identify constitutive exons and alternative exons but cannot be used to identify genes that lack introns.

The procedure requires the use of an "exon trap" vector called pSPL3 (Gibco), which contains an artificial mini gene consisting of three parts. The first is a segment of the simian virus 40 (SV40) containing an origin of replication and promoter. The second elements are two splicing-competent exons flanking an intron sequence that contains a multiple cloning site. The third part is an SV40 polyadenylation site.

The portion of genomic DNA to be screened for the presence of exons is inserted into a restriction site in the multiple cloning site followed by transfection into a mammalian cell line such as monkey COS-7 cells. Transcription of RNA occurs from the SV40 promoter and the transcript undergoes splicing driven by the machinery of the host cell. Any exon contained in the genomic fragment becomes attached between the upstream and downstream minigene exons. RT-PCR with primers that are specific for the minigene exons are used to confirm successful exon trapping. Trapped exons produce a PCR product that is larger than that of vector alone. The identity of the inserted product can be determined through sequencing using primers specific to the minigene sequence. Sequencing, followed by a database search, is the most direct way to obtain information from the trapped sequences. Since trapped exons are also putative cDNAs, they can be used as probes on cDNA libraries, Southern blots, Northern blots, or for FISH.

Exon trapping has proven to be an important part of positional cloning. Demonstrations of the successful use of exon trapping include the identification of at least 30 genes, including the Menkes disease gene, the NF-2 gene, and the HD gene ***(5,10–15,18)***.

Day 1–3: DNA Preparation and Subcloning

1. Digest approximately 1 µg of both genomic and pSPL3 DNA in separate microfuge tubes with the same restriciton enzyme or enzyme combinations. The pSPL3 multiple cloning site contains unique restriction sites for EcoRI, SstI, XhoI, NotI, ZmaIII, PstI, bamHI, and EcoRV. Single or double digests can be used (your choice).
2. Phenol extract DNA followed by ethanol precipitation (1/2 vol of 7.5 *M* ammonium acetate and 2 vol of 100% ethanol). Wash twice using 70% ethanol.
3. Dissolve the pellet in TE buffer, pH 8.0. The final concentration should be ~250 ng/mL.
4. If only a single enzyme is used, pSPL3 should be dephosphorylated with CIP (according to supplier's protocols). This step is not necessary if two different restriction enzymes are used.
5. Subclone genomic fragments into pSPL3 using T4 DNA ligase according to supplier's protocol.
6. Transform *E. coli* with ligated DNA.
7. Incubate 500 µL of cells with 5 mL of LB containing ampicillin overnight at 37°C.
8. Plate 10 and 100 µL of ligation reaction onto LB/ampihicillin plates to determine the degree of ligation. Place at 37°C overnight.
9. Isolate plasmid containing insert from the 5-mL culture by any standard miniprep procedure.

Day 4–5: Lipid-Mediated Transfection

10. On the afternoon before transfection (~15 h prior), plate COS-7 cells in DMEM supplemented with 10% FBS into 6-well plates to achieve 70–80% confluency (~4–6 × 10^5 cells).
11. Add an optimum amount of lipid (determined from supplier's protocol) reagent to 100 µL of serum-free medium in a sterile polystyrene tube, mix gently, and incubate at room temperature for 5 min.

12. For each transfection, add 1 to 100 μL of serum-free medium in separate polystyrene tubes. Combine lipid and DNA solutions with gentle mixing and incubate at room temperature for 15 min.
13. Aspirate medium from COS-7 cells and add 2 mL of 37°C serum-free DMEM.
14. Add an additional 1 mL of serum-free DMEM to the lipid/DNA mix.
15. Remove medium from the plates and add the entire lipid/DNA mix of one tube to the plate; incubate for 6 h at 37°C.
16. Following the 6-h incubation, add an additional 1 mL of DMEM medium supplemented with 10–20% FBS and incubate an additional 24 h at 37°C.
17. Rinse cells with PBS lacking calcium and magnesium then add 1 mL of Trizol (Gibco). Prepare RNA according to manufacture's protocol. Resuspend in 50 μL DEPC-treated water.

Day 6: RT-PCR

18. Add the following components in a polypropylene microfuge tube: 2 μg RNA, 1 μL 20 m*M* oligonucleotide SA2, and DEPC water to 8 μL.
19. Heat to 70°C for 10 min, ice for 10 min, and centrifuge briefly.
20. Add 3 μL 5× reaction buffer: 2 μL 0.1 *M* DTT, 1 μL mixed dNTP stock (10 m*M* of each dNTP).
21. Mix gently, centrifuge briefly, and equilibrate for 2 min to 37°C.
22. Add 1 μL MMLV H-RT and mix gently.
23. Incubate at 37°C for 1 h, then incubate at 55°C for 5 min.
24. Add 1 μL of RNase H, mix gently, and incubate at 55°C for 10 min.
25. Add 15 μL of DEPC water (final volume 30 μL).

Primary PCR

26. Combine the following reagents in a polypropylene tube:
    a. 5 μL of RT reaction mix.
    b. 5 μL of 10× Taq buffer.
    c. 1.5 μL 50 m*M* $MgCI_2$.
    d. 1 μL dNTP mix (10 m*M* of each in sterile water).
    e. 2.5 μL 20 m*M* primer SA2.
    f. 2.5 μL 20 m*M* primer SD6.
    g. Sterile water to 47.5 μL final volume.
27. Heat reaction to 95°C for 5 min, then add 2.5 μL of Taq (1 U/mL).
28. Run the following amplification cycles: 6 cycles of 1 min at 95°C, 1 min at 60°C, 5 min at 72°C; 1 cycle of 10 min at 72°C, followed by a soak at 55°C.
29. Add 25 U BstXI and incubate overnight at 55°C. This step eliminates vector-only sequence and sequences using the cryptic splice-donor site.
30. Add an additional 5 U of BstXI, incubate 2 h more at 55°C.

Secondary PCR

31. Combine the following reagents in a polypropylene PCR tube:
    a. 1 μL of a 1:10 dilution of the primary PCR product.
    b. 78 μL of sterile water.
    c. 10 μL of 10× Taq DNA polymerase.
32. Heat to 958C for 5 min, then add 3 μL of Taq DNA polymerase.
33. Perform the following amplification cycle: 30 cycles of 30 s at 558C, 2 min at 728C; 1 cycle of 10 min at 728C, followed by a soak at 48C.
34. Electrophorese 10 μL of PCR product with dye in a 1.5% agarose gel and identify reactions that contain PCR products.

35. PCR products containing exon can be sequenced using the AD2 and UAP primers and the PCR product can be cloned using the TA cloning kit (according to supplier's protocol, Promega).

3.7 Microarray Analysis

A major obstacle in postional cloning of TSG is identifying the specific mutated gene from within a large physical contig. The completion of the human genome has incited technologic advances that are geared toward efficiently exploiting these sequences and the associated physical resoureces to identify human disease genes. Here we describe the application of DNA microarray technology to a defined genomic region (physical map) to identify: (a) exons without *a priori* sequence data and (b) the TSG based on differential gene expression in a tumor. Rather than arrayed cDNA elements, this approach utilizes an arrayed genomic library within the genetic interval of interest. The feasbility of this approach has been demonstrated by Stephen et al. *(7)* who used such a strategy to identify the Niemann-Pick type C gene.

3.7.1. Total RNA Isolation with Trizol (see Note 10)

Prep. notes: Turn centrifuge on to 4°C.

1. For a confluent monolayer of cells, use 5 mL of Trizol for each T75 flask or 10 mL of Trizol for each T175 flask (generally yields 1 mg total RNA).
2. Lyse cells by repeated pipetting, making sure that the mixture has been sufficiently homogenized. *This is an important step.*
3. Incubate at room temperature for 5 min.
4. At this point the homogenized sample can be stored at −80°C (for up to 1 mo) for later use.
5. Add 0.2 mL of chloroform per 1 mL of Trizol (ex.: 5 mL of Trizol:1 mL of chloroform).
6. Shake tubes vigorously by hand for at least 15 s.
7. Incubate at room temperature for 3 min.
8. Spin samples at no more than 12,000*g* for 15 min at 4°C.
9. Transfer the clear aqueous phase to a fresh tube.
10. Add 0.5 mL of isopropanol per 1 mL of Trizol (ex.: 5 mL of Trizol:2.5 mL of isopropanol).
11. Invert tubes several times.
12. Incubate at room temperature for 10 min.
13. Spin at no more than 12,000 × g for 10 min at 4°C.
14. A gel pellet should now be visible on the bottom of the tube.
15. Discard the supernatant and wash the pellet with 1 mL of 75% ethanol in DEPC H_2O per 1 mL of Trizol (ex.: 5 mL of Trizol:5 mL of 75% EtOH).
16. Mix the sample by vortexing and centrifuge at no more than 7500 × *g* for 5 min at 4°C.
17. Remove the supernatant and briefly air-dry the pellet.
18. Resuspend the pellet in DEPC H_2O and keep on ice (store long-term at −80°C). Try to resuspend so that final conc. is ≥ 2 μg/μL (in general use 100–300 μL).
19. Read samples at 260 nm to determine approximate RNA concentration and run 1 μg on a 1% agarose gel with ethidium bromide (look for the presence of 28s and 18s bands).

3.7.2. Poly A+ Isolation with Oligotex (Oligotex mRNA batch Protocol)

Prep. notes: Set centrifuge to 4°C, put Oligotex suspension in 37°C block, heat another block to 70°C (heat OEB or DEPC H_2O here), and place buffer OBB in 37°C incubator (to dissolve precipitant.)

Ex.: Preparation of 1 mg total RNA, starting material at 10 μg/μL.

20. Thaw total RNA on ice.
21. 100 μL of total RNA (for 1 mg total) + 400 μL of RNase-free water for a total volume of 500 μL.
22. Add 500 μL buffer OBB (make sure this solution has not precipitated upon storage; if it has, heat at 37°C to dissolve).
23. Add 55–85 μL of Oligotex suspension (for 1 mg of total RNA, 70 μL works well)—make sure to vortex suspension just before adding.
24. Mix the contents by inverting the tube.
25. Incubate the sample for 5 min at 70°C (linearizes mRNA).
26. Make sure the cap is still tight, and invert the tube to mix. Incubate at room temperature for 10 min (mRNA attaches to beads).
27. Spin for 2 min at 15,000 *g* at 4°C.
28. Remove the supernatant by pipetting (at this point it is okay to leave 50 μL of supernatant in the tube).
29. Resuspend the pellet in 1 mL of buffer OW2 by pipetting and vortexing.
30. Spin for 2 min at 15,000 *g* at 4°C.
31. Remove the supernatant by pipetting (once again, it is okay to leave 50 μL of supernatant in the tube).
32. Again resuspend the pellet in 1 mL of buffer OW2 by pipetting and vortexing.
33. Spin for 2 min at 15,000 *g* at 48°C.
34. This time, remove all supernatant (be careful not to disturb the pellet; use a p10 pipet to remove the residual supernatant).
35. Resuspend the pellet in 30 lL DEPC H20 (at 708C) by repeated pipetting.
36. Spin for 2 minutes at 15,000 *g* at 4°C.
37. Transfer the supernatant containing the eluted poly A+ RNA to a clean RNase-free tube and place on ice (we usually remove only 25 μL at this point, so as not to disturb the pellet).
38. Again wash the pellet in 70°C DEPC H_20 to elute RNA, but this time use 25 μL.
39. Spin for 2 min at 15,000 *g* at 4°C.
40. Transfer the remaining supernatant to the tube on ice containing the eluted poly A+ RNA.
41. Finally, spin this tube for 2 min at 15,000 *g* at 4°C to pellet any residual Oligotex particles and transfer this solution to another clean RNase-free tube and place on ice (for long-term storage place at −80°C).
42. Read the samples at 260 and 280 nm to determine concentration and purity. The 260 nm/280 nm ratio should between 1.7 and 2.0 (a ratio of 2.0 represents a pure preparation). If not, the procedure can be repeated on the sample by following the miniprep protocol (using 15 μL of Oligotex).

3.7.3. Probe Preparation

This procedure can be performed in an Eppendorf thermocycler, heat blocks, hybridization oven.

Prep. notes: mRNA conc. must be at least 0.35 μg/μL; if it is not, lyophilize in a speed vacuum briefly to volume below 11.4 μL. Heat block to 70°C, turn on 42°C instrument.

43. Use 4 μg RNA.
44. Add DEPC H_2O to a volume of 11.4 μL.
45. Add 1 lg of oligo dT (2 lL stock at 0.5 mg/mL).
46. Heat 5 min at 70°C (turn block to 65°C), place on ice for 30 s, flash-spin.
47. Incubate 10 min at 25°C.
48. Add 14.6 μL of Cy3 or Cy5 master mix (per reaction tube).
49. Add 6 μL of 5× 1st strand buffer: μL 0.1 *M* DTT, 3 μL Cy3 or Cy5 dCTP, 0.6 μL dNTPs (25 m*M* dA/G/TTP and 10 m*M* dCTP), 2 μL RNasin (Gibco), 14.6 μL.

50. Once master mix has been added to mRNA, keep covered from light.
51. Add 2 μL of Superscript RT II.
52. Incubate for 2 h at 42°C.
53. Spin down tubes after incubation.
54. Pool the two tubes (Cy3 and Cy5 labeled).
55. Add 2.65 μL of 25 m*M* EDTA (to stop reaction).
56. Add 3.3 μL of 1 *M* NaOH (to degrade RNA strands).
57. Incubate for 10 min at 65°C.
58. Add 3.3 μL 1 *M* HCl (to neutralize reaction).
59. Add 5 μL 1 *M* Tris, pH 6.8.
60. Can store temporarily at −80°C at this point, if necessary.

3.7.4. Probe Purification

Perform all steps in minimal light.

3.7.4.1. Microcon 30

Label 2 collection tubes per probe.

61. Insert column into supplied tube.
62. Add 400 μL of dH_2O.
63. Apply the probe mixture (about 80 μL) onto the column and mix well by pipetting.
64. Spin at room temperature (25°C) for 8 min at 12,000 *g* (*see* **Note 11**).
65. Remove column from assembly, add an additional 20 μL of dH_2O, and place it upside down in a new tube (cut cap off of this tube).
66. Spin for 1 min at 15,000 rpm to collect probe (cap strap should now face outward).
67. Remove column and store tube on ice.

3.7.4.2. QIAquick Nucleotide Removal Kit

Label column, elution tube, 0.5-mL final tube.

68. Add 5 vol of buffer PN to 1 vol of the reaction sample (ex.: If 50 μL of sample is recovered from the microcon, use 250 μL of buffer PN).
69. Transfer this solution to a QIAquick spin column in a 2-mL collection tube.
70. Spin at room temperature for 1 min at 6000 rpm.
71. Discard flow-through and place column back into the same tube.
72. Add 750 μL of buffer PE (be sure to add ethanol to buffer PE before use).
73. Spin at room temperature for 1 min at 6000 rpm.
74. Discard flow-through and place column back into the same tube.
75. Spin at room temperature for 1 min at 13,000 rpm to remove residual ethanol.
76. Place column in a clean 1.5-mL tube.
77. Add 30 μL of dH_2O to the center of the column.
78. Let stand for 2 min at room temperature.
79. Spin at room temperature for 1 min at 13,000 rpm.
80. Remove column and transfer probe to 0.5-mL tube for lyophilization.
81. Lyophilize sample in a speed vacuum until dry (about 1 min per microliter of sample); do not overdry the pellet.
82. Wrap tube containing the sample in foil and store in the dark at −20°C for temporary storage (<24 h), −80°C for long-term storage.

3.7.5. Slide Fixation

For non-amino-linked DNAs printed onto silane-coated slides.

83. Once dry, UV crosslink (0.3 J or 3000 μJ × 100) using Stratalinker.
84. Rinse in 0.1 % SDS for 1 min.
85. Rinse 2× in dH_2O for 1 min.
86. Air-dry.

3.7.6. Hybridization

Perform all steps in minimal light.

87. Denature slide (DNA side facing up) in a beaker of 95°C water for 3 min.
88. Place in a 50-mL tube containing ice cold ethanol for 15 seconds.
89. Spin-dry at 500 × *g* for 1 min (place slide in a 50-mL tube with a Kimwipe in the bottom).
90. Add 30 μL of 2× SSC to each of the hybridization chamber's lower grooves.
91. Set denatured slide into chamber DNA side up and seal for the time being.
92. Retrieve probe from −20°C and resuspend pellet in (for slides printed 10/00):
    a. 9.0 μL of dH_2O.
    b. 5.0 μL of 20 × SSC.
    c. 4.0 μL of 1% SDS.
    d. 2.0 μL of Cot-1 DNA (10 mg/mL).
93. Transfer mixture to a screw-cap tube (do not place probe mixture on ice once resuspended, as the SDS will precipitate).
94. Boil 3 min.
95. Immediately flash-spin sample and apply solution to the slide resting in the hybridization chamber, using the reference slide as a guide.
96. Gently place a cover slip over the array, try to avoid large air bubbles (small bubbles are not a concern because they will disappear).
97. Seal the hybridization chamber and carefully place it in a 64°C water bath overnight (14–18 h).

3.7.7. Washing

Perform all steps in minimal light.

98. Remove hybridization cassette containing slide from water bath, and place slide in the first wash. Carefully remove the cover slip from the slide by continuously dipping into the wash tube (or remove the cover slip from the tube if it has already come off).
99. All wash steps are performed at room temperature on a rocker in 50-mL conical tubes:
    First wash: 5 min, 0.5× SSC, 0.01% SDS.
    Second wash: 5 min, 0.06× SSC, 0.01% SDS.
    Third wash: 2× 2 min, 0.06× SSC filtered.
100. Spin-dry at 500 × *g* for 1 min (place slide in a 50-mL tube with a Kimwipe in the bottom).
101. Store the slide in a slide box until scanned (ScanArray/GenePix).

3.8. Database Searches for Homology to Human Genes or ESTs

When the region of loss or consistent abnormality has been narrowed to approximately 1 cM, genes or EST mapping can be identified by sequencing databases according to the following outline.

1. Either Blast sequence the region or search databases to identify gene ESTs from these regions.

2. Retrieve sequences of known STS from Genbank and Blast them against genomic sequences. Various Blast databases can be found at http:www.ncbi.nlm.nih/BLAST/blast_databases.html. For additional information see NCBI Blast tutorial and Blast course.
3. When genomic sequence is available, the following type of database search can be conducted:
   a. Mask the retrieved genomic sequences. The presences of particular types of repetitive elements and vector sequences can distort the Blast results and gene prediction. In particular, L1 elements are often predicted as genes.
   b. To avoid these problems, prescreen for repetitive elements within the sequences using a program such as RepeatMasker or Censor. These programs are designed to replace sequence segments that match any of the elements common to your organism (i.e., Alu) with the same number of asterisks or N's.
      RepeatMasker1: http://aurora.bwh.harvard.edu/cg9-bin/RepeatMasker.cgi
      RepeatMasker2: http://ftp.genome.washington.edu/cgi-bin/RepeatMasker
      VecScreen: http://www.ncbi.nlm.nih.gov/VecScreen/VecScreen.html
4. 5× First-strand buffer (250 m*M* Tris-CI, pH 8.3, 375 m*M* KCI, 15 m*M* $MgCl_2$). This buffer is supplied with the SuperScript Reverse Transcriptase: 0.1 *M* DTT, 1 μL 10 m*M* dNTP mix in DEPC water, 1 μL Rnasin (Rnase inhibitor from Promega), DEPC water to 19 μL.
5. For prediction of potential genes or ORFs in the retrieved genomic sequences, use a gene prediction program such as FGENESH, which can be found in the Sanger center. However, proceed with caution when using these programs, since gene prediction programs are designed to predict and can give false predictions. Therefore, be sure to compare several predictions using programs based on different prediction algorithms.
6. Genes and ESTs can be screened for genomic sequences, by Blasting masked sequences against the dEST database. When genomic sequences are not available or the region of consistent alteration is too large, the following steps can be helpful in identifying candidate genes.
   a. Check candidates in GenMap 99 and other relevant mapping databases.
   b. Assembled sequences of candidates, based on mapped EST clusters, can be retrieved from other assembly databases such as:
      DOTS (http://.cbil.upenn.edu/DOTS*/dotsweb?page5blast)
      MIPS (http://www.mips.biochem.mpg.de/proj/human/human_blast.html)
      TBI (http://www.dfkx-heidelberg.de/tbi/Welcome.html)
      TIGR (http://www.ncbi.nlm.nih.gov/BLAST.theblast.html)
      STACK (http://ziggy.sanbi.ac.za/stack/stacksearch.html)
   c. The analysis and evaluation of candidiate genes and EST clusters can be based on the following criteria:
      The map position of the clone (GenMap 99)
      Expression in relevant tissues (UniGene)
      Protein similarities (ExPASy Proteomics tools)
   d. In order to find available genomic sequence and homologies of specific candidates to other genes, Blast sequence of chosen candidates against relevant databases.

There are several advantages to using available genomic sequence. First, when genomic sequence is available, new polymorphic STSs can be made to further refine the region of consistent alteration. Second, when genomic sequence is available, it is easier to determine the candidate genomic structure and organization (i.e., intron/exon borders), which can be helpful in designing primers for screening genomic DNA. Once a candidate gene is identified, the next key step is to identify disease-causing alterations associated with the tumor suppressor gene.

3.9. Testing Candidate Genes: Mutational Analysis of the Putative TSG

Direct detection of mutations is not simple, for many reasons that are related both to genes themselves and to current technologies. It is the scanning of large genes for base changes that is the rate-limiting step. Complete sequencing of large genes can also be time-consuming, costly, and tedious, and mutations can be missed. The alternative approach is to use scanning methods that allow rapid analysis of exon and intron boundaries, then to use limited sequencing to confirm and identify the mutation in a fragment. This approach avoids the labor and cost of sequencing large numbers of negative samples. A variety of methods for detecting unknown mutations currently exist. This section covers single-strand conformation polymorphism (SSCP) analysis, the protein truncation tests, and direct DNA sequencing.

3.9.1. Search for Mutations by PCR-SSCP

3.9.1.1. Preparation of Samples for SSCP analysis

It is critical to use optimized PCR conditions that minimize unwanted extra products because these can result in artifact bands that interfere with interpretation of the SSCP results. DNA samples and PCR should be done according to standard methods; however, the following pointers could help in generating the best SSCP results.

1. Use only highly purified, salt-free template DNA in the PCR reaction.
2. Use primers that contain no partial mismatches in the target sequence.
3. Optimize reagent and primer concentrations for each amplification reaction.
4. Determine thermal cycler settings that eliminate nonspecfic priming.
5. Use the minimum number of PCR cycles to obtain a sufficient quantity of DNA, usually 30 cycles (or fewer) on 100 ng of genomic DNA.

The primers used for amplification can be radiolabled with ^{32}P prior to PCR, or ^{32}P-dCTP can be added to the PCR cocktail (5 μCi) before aliquotting to individual reaction tubes. Following PCR amplification, it is a good idea to check the success of the PCR products on a 3% agarose gel to ensure amplification (*see* **Note 3**).

3.9.1.2. Gel Preparation and Pouring

1. Glass plates must be clean and free of soap residues or dried gel. To remove residues, apply ethanol to both plates and wipe dry.
2. To ensure that the gel will not stick to the glass plates, treat one of the plates with Gel Slick solution or a similar antistick product. (If the plates were previously silanized, the coating must be removed completely prior to applying a fresh coat of antistick solution).
3. Assemble the glass plates according to the manufacturer's instructions (use 0.4-mm spacers).
4. Either polyacrylamide or MDE matrix can be used to cast gels.
5. Pour into sequencing gel apparatus and allow to solidify. For 10% glycerol gels, add 5.0 mL of pure glycerol to the mixture and q.s with ddH_2O to 100 mL.
6. Place all samples on ice, remove 5.0 μL of hot sample, and mix with 45 μL of stop solution.
7. Heat-denature at 94°C for 2 min.
8. Transfer samples directly to ice. (Quick-spin samples to ensure that all the product is at the bottom of the tube).

3.9.1.3. Electrophoresis

1. Rinse the top of the gel thoroughly with running buffer.
2. Prerun gel for at least 20 min, no longer than an hour. Gels to be run at 4°C should be placed at 4°C for 1 h prior to rerunning and rerun at 4°C.
3. Load 10 μL of sample on to the gel and electrophorese at 20 W for 5 h at room temperature.
4. The gels should be removed promptly and placed on transfer paper (Whatman 3MM filter paper), covered with saran wrap, and dried at 80°C for 30 min.
5. Perform autoradiography or phosphor imaging analysis using standard techniques (*see* **Note 8**).

3.9.2. Protein Truncation Test (Coupled In-Vitro Transcription and Translation) (see **Notes 5** *and* **6***)*

Coupled in-vitro transcription and translation (also known as the protein truncation test; PTT) is a convenient single-tube reaction to determine if mutations or deletions produce proteins that differ in size when compared to the normal product. This procedure is highly effective for detecting mutations that lead to the termination of mRNA translation and subsequently protein truncation. The types of mutations that can be detected are: nonsense mutations in which there is a single nucleotide substitution that produces a stop codon (TGA, TAA, or TAG); as frameshift mutations in which one or more nucleotides are either inserted or deleted if the number of bases is not divisible by 3, the altered reading fame frequently results in a stop codon or mutations at a splice site. PTT has been especially successful for detecting mutations in genes in which the frequency of missense mutations is low.

3.9.2.1. PTT or Coupled In-Vitro Transcription and Translation (*see* **Notes 12** *and* **13**)

1. First-strand cDNA synthesis (start with 1–5 μg of total RNA or 50–500 ng of mRNA).
2. Combine the following components into a nuclease-free microcentrifuge tube: 1 μL or 50–250 ng of random primers, 1 μL of oligo dT primers (stock solution 20 umo), 1–5 μg total RNA, sterile distilled water to 12 μL.
3. Heat mixture to 70°C for 10 min and chill on ice.
4. Collect the contents of the tube by brief centrifugation and add:
   a. 5× first-strand buffer (250 m*M* Tris-Cl, pH 8.3, 375 m*M* KCl, 15 m*M* $MgCl_2$). This buffer is supplied with the SuperScript Reverse Transcriptase.
   b. 2 μL 0.1 *M* DTT.
   c. 1 μL 10 m*M* dNTP mix in DEPC water.
   d. 1 μL Rnasin (RNase inhibitor from Promega).
   e. DEPC water to 19 μL.
5. Mix contents of the tube gently and incubate at 42°C for 2 min.
6. Add 1 μL (200 U) of SUPERSCRIPT II and mix by gently swirling the pipet tip in the tube.
7. Incubate 50 min at 42°C. Inactivate the reaction by boiling for 10 min. Spin down contents.
8. Add the following to a PCR reaction tube for a final reaction volume of 100 μL: 10 μL 10× PCR buffer (200 m*M* Tris-HCl [pH 8.4], 500 m*M* KCl), 3 μL 50 m*M* $MgCl_2$, 2 μL 10 m*M* dNTP mix, 2 μL amplification primer 1 (10 μ*M*), 2 μL amplification primer 2 (10 μ*M*), 1 μL Taq DNA polymerase (2–5 U/μL), 2 μL cDNA (from first-strand reaction), 80 μL autoclaved, distilled water. Mix gently. Heat reaction to 94°C, 45 s at 55–65°C (depending on the primer sequence in your gene), and 1 min at 72°C. A final 10-min cycle at 72°C followed by a soak at 4°C is also common.

9. Purify the PCR product using the QIAquick columns (Qiagen, Santa Clarita, CA) according to the supplier's recommendations.
10. After amplification and purification, 100 ng of product is used in a TNT T7 Quick Coupled Transcription/Translation reaction (Promega, Madison, WI) with the addition of 0.3 m*M* magnesium acetate. For detection, a labeled amino acid is included. The label can be either a radionucleotide such as ^{35}S, which is visualized by autoradiography, or biotin for detection by chemiluminescence (see Promega TNT Quick Coupled Transcription/Translation protocol for details.
11. The resultant proteins are run out on a SDS-PAGE gel for sizing against normal control products and protein markers.

3.9.3 Search for Mutations by Direct DNA Sequencing

Direct sequencing of PCR products based on the Sanger dideoxy chain-termination method is a final step of any mutation scanning procedure, as all the methods described above are capable of detecting mutations with varying efficiencies, but none defines precisely the nature of the change. Major improvements have been achieved with recent automated capillary electrophoresis instruments and multicolor fluorescent detection.

1. Optimize PCR conditions for the region of interest in the putative TSG sequence. Amplify PCR products from samples of interest (i.e., cell lines, archival tumor material, blood samples, etc.) under the optimal conditions (total volume is 50 μL).
2. Verify PCR products by 2% agarose-EtBr gel electrophoresis.
3. Purify PCR products by using GFX PCR, DNA, and Gel Band Purification Kit (Amersham Pharmacia). Or, extract PCR product bands from low-melting-point agarose by using the same kit. Elute DNA in an appropriate volume of sterile nuclease-free water.
4. Check the concentration of the purified PCR products on 2% agarose-EtBr gel using Precision Molecular Mass Standard (Bio-Rad).
5. Adjust PCR products to an appropriate concentration (10 ng/100 bases) so that 5 μL can be used per reaction.
6. Sequence samples with the dideoxy termination method on an ABI 373A DNA sequencer.
7. Align sequences using CLUSTAL W (EMBL, http://www.ebi.ac.uk).
8. 1 min in (the radioactive room), inject the solution of probe to the bag that has the prehybridization solution and membrane, hybridize overnight.

4. Notes

1. For GTL banding the following may be helpful:
   a. Times of trypsin treatment vary with each case, the age of the slides to be banded, and the technique used to make the slides.
   b. Chromosomes are undertreated when they stain homogeneously; when banding is present, but bands appear to blur into each other; when banding is present, but a darkly stained bar is noticeable between the chromatids.
   c. Chromosomes are overtreated when they appear swollen, with a "crust" around the chromatids; when they appear fuzzy or have a cobwebbed appearance.
   d. Due to variation in the stages of condensation of chromosomes within a specimen, some spreads may appear over- or undertreated, while others appear perfect for analysis. It is the technologist's judgment as to whether treatment or staining time is right for the particular length and morphology of chromosome he or she are working with.
2. If the amount of tumor DNA is not sufficient for CGH, the DNA can be amplified by degenerated oligonucleotide-primer (DOP)-PCR ***(3)***.
3. The primers used for amplification can be radiolabled with ^{32}P prior to PCR, or [^{32}P]dCTP (5 μCi) can be added to the PCR cocktail before aliqouting to individual reaction tubes. Fol-

lowing PCR amplification, it is a good idea to check the success of the PCR products on a 3% agarose gel to ensure amplification.

4. We have found that a 30% stock solution of acrylamide with a 49:1 (acrylamide:bis) ratio works extremely well for SSCP analysis. Store the solution in a foil-covered bottle. Prior to preparing the stock solution, treating the acrlyamide with amberlite solution can improve the quality of the acryl amide. Add dissolved acryl amide to a beaker containing 100 g of dry Amberlite MB-1 resin and stir overnight in the cold room. Filter through Whatmen #1 filter paper.
5. The PTT protocol is useful for detecting truncating mutations, i.e., disease-causing and not missense mutations, which often represent non-disease-related sequence variation. Large stretches of coding sequence (up to 5 kb) can be screened, but a 1.3–1.6-kb cDNA yields the best results. This protocol works equally well with large single exons derived from a DNA template or from multiple exons using an RNA template. The latter can significantly reduce the workload. The length of the truncated protein pinpoints the position of the mutation, thereby facilitating its confirmation by sequencing analysis. If RNA is used as a template, any abnormalities of message splicing can potentially be detected. Compared to other mutation detection techniques, PTT can detect mutated alleles present at 5–10% in a sample.

   This technique is not applicable to genes in which there are low levels of truncating mutations. For example, *APC, BRCA1, BRCA2,* and *Dystrophin* all have approximately 90–95% truncating mutations, but *NFI* and *RTS* have only 50% and 10% truncating mutations, respectively. Unfortunately, archival material is usually DNA and not RNA, which further limits the usefulness of this technique. Transcripts carrying truncating mutations can also be highly labile, but this can be overcome by using DNA-based PTT. A final limitation of this technique is that it cannot detect mutations occurring outside the coding region that regulate expression and RNA stability.
6. Troubleshooting PTT:
   a. False negatives due to a failure to amplify the mutated allele or only detecting very small deletion/insertions or missense mutations might be caused by (1) mutations in the primer binding site; (2) very small in-frame deletions/insertions undetectable by the mobility shifts; (3) mutation may only occur germline or be present as somatic mosaicism; (4) large insertions, translocations, and inversions that enlarge the region under analysis beyond amplifiable length. The solution is to use different percentage gels and overlapping primer sets.
   b. False positives or alternate splicing give rise naturally to different size transcripts and can cause artifacts during RT or PCR procedures. These usually disappear in the huge amount of correctly amplified fragments, but problems can occur if the errors happen in the first 1–3 rounds of amplification, producing substantial amounts of artifactual products. A solution is to perform two independent RT and PCR reactions and check at each stage of the process to allow for identification of this potential problem.
7. In some tumors, a situation occurs in which there is physical loss of one allele of the gene with the second allele remaining wild type. This condition, which is termed haploinsufficiency, results in the suppressor protein being expressed at reduced levels that are insufficient to block tumor progression. Under these circumstances, mRNA expression analysis becomes important and protocols such as semiquantitative RT-PCR and Northern blot analysis are useful for quantifying gene expression levels. The ultimate validation of tumor suppressor gene function is reintroduction of the putative cancer suppressor gene into cancer-derived cell lines lacking endogenous expression and observing a reversion to a less tumorigenic, more "normal" phenotype. Protocols describing the functional testing of tumor suppressor genes, including transfection studies, are discussed in detail in other chapters of this volume.

8. Troubleshooting for SSCP analysis can be found at www.bioproducts.com/technical/sscp-analysiswithmdegelsolution.shtmL.
9. For each PCR reaction, to determine LOH occurring in tumor DNA, a control should be run using normal lymphocyte DNA from the same patient.
10. If using 13-mL round-bottom tubes, spin at 9000 *g*.
11. Note that the cap strap should be aligned toward the center of the rotor.
12. The cDNA can now be used as a template for PCR amplification to produce a product for use in the PTT assay. A specifically designed tailed sense primer should contain the following regions: a 5′-end containing a T7 RNA-polymerase promoter sequence that is needed to facilitate the in-vitro production of RNA, followed by a 5–7-bp spacer, and finally a eukaryotic translation initiation sequence (Kozak sequence), which includes an ATG start codon that will facilitate the initiation of protein synthesis. The 3′ portion of the primer will contain gene-specific sequence, allowing amplification reads inframe from the ATG. The gene specific 3′ primer should have a similar annealing temperature to that of the 5′ primer.
13. Large deletions, duplications, and splicing mutations may be detected by agarose gel electrophoresis at this stage.

References

1. Hack, M. A. and Lance, H. J. (eds.) (1980). *ACT Laboratory Manual,* pp. 38–62.
2. Schröck E., du Manior, S., Veldman, T., et al. (1996). Multicolor spectral karyotyping of human chormosomes. *Science* **273**, 494–497.
3. Telenius, H., Pelmear, A. H., Tunnacliffe, A., et al. (1992). Cytogenetic analysis by chromosome painting using DOP-PCR amplified flow-sorted chromosomes. *Genes Chromosomes Cancer* **4**, 257–263.
4. Veldman, T., Vignon, C., Schröck, E., Rowley, J. D., and Ried, T. (1997). Hidden chromosome abnormalities in haematological malignancies detected by multicolor spectral karotyping. *Nat. Genet.* **4**, 406–410.
5. Vulpe, C., Levinson, B., Whiteny, S., Packman, S., and Gitschier, J. (1993) Isolation of a candidate gene for Menkes disease and evidence that it encodes a copper-transporting ATPase. *Nat. Genet.* **3**, 7–13.
6. Thompson, C. T. and Gray, J. W. (1993). Cytogenetic profiling using fluorescence in situ hybridization (FISH) and comparative genomic hybridization (CGH). *J. Cell. Biochem.* **17**, 139–143.
7. Joos, S., Fink, T. M., Rätsch, A., and Lichter, P. (1994). Genomic mapping and chromosome analysis: the potential of fluorescence *in situ* hybridization. *J. Biotechnol.* **35**, 135–153.
8. Ried, T., Schröck, E., Ning, Y., and Wienberg, J. (1998) Chromosome painting: a useful art. *Hum. Mol. Genet.* **7**, 1619–1626.
9. Kallioniemi, A., Kallioniemi, O. P., Sudar, D., et al. (1992). Comparative genomic hybridization for molecular cytogenetic analysis of solid tumors. *Science* **258**, 818–821.
10. Buckler, A. J., Chang, D. D., Graw, S. L., et al. (1991) Exon amplification: a strategy to isolate mammalian genes based on RNA splicing. *Proc. Natl. Acad. Sci. USA* **88**, 4005–4009.
11. Church and Buckler (1999) Gene amplification by exon amplification. *Meth. Enzymol.* **303**, 83–99.
12. Church, D. M., Banks, L. T., Roger, A. C., et al. (1993) Identification of human chromosome 9 specific genes using exon amplification. *Hum. Mol. Genet.* **2**, 1915–1920.
13. Church, D. M., Stotler, C. J., Rutter, J. L., Murrell, J. R., Trofatter, J., A., and Buckler, A. J. (1994) Isolation of genes from complex sources of mammalian genomic DNA using exon amplification. *Nat. Genet.* **6**, 98–105.

14. Datson, N. A., vand de Vosse, E., Dauwerse, H. G., Bout, M., van Ommen, G. J., and den Dunnen, J. T. (1994) Specific isolation of 3′-terminal exons of human genes by exon trapping. *Nucleic Acids Res.* **22**, 4148–4153.
15. Datson, N. A., Duyk, G. M., Van Ommen, J. B., and Den Dunnen, J. T. (1996) Scanning for genes in large genomic regions: cosmid-based exon trapping of multiple exons in a single product. *Nucleic Acids Res.* **24**, 1105–1111.
16. Den Dunnen et al. (1999) Cosmid-based exon trapping. *Meth. Enzymol.* **303**, 100–110.
17. Stephen, D. A., Chen, Y., Jiang, Y., et al. (2000). Positional cloning utilizing genomic DNA microarrays: the Nieman-Pick Type C gene as a model system. *Mol. Genet. Metab.* **70**, 10–18.
18. Den Dunnen, J. T. and Can Ommen, G.-J. B. (1999) The protein truncation test: a review. *Hum. Mutat.* **14**, 95–102.

22

Hybrid Capture of Putative Tumor Suppressor Genes

Bryan L. Betz and Bernard E. Weissman

1. Introduction

During the last part of the twentieth century, research on human cancer increasingly focused on the molecular basis of this disease. These studies have identified many facets of cellular transformation, including aberrant cell cycle regulation, inhibition of programmed cell death or apoptosis, impaired DNA damage repair systems, genomic instability, and altered signal transduction in an increasing number of pathways. However, we have only begun the formidable task of identifying all the genes responsible for these cellular and genetic alterations. Over the past four decades, investigators have developed multiple models to investigate the processes of malignant transformation, including transformation of cells both in vitro and in the animal by chemical carcinogens or oncogenic viruses. Immortal human and rodent tumor cell lines have proved especially amenable for identification of oncogenes and tumor suppressor genes. In order to map and identify functional tumor suppressor genes directly, we and others have used the technique of microcell hybridization to transfer chromosomes from normal human cells into human tumor cell lines *(1,2)*. By this approach, we first showed that introduction of a normal human chromosome 11 into a Wilms' tumor cell line causes a complete suppression of tumorigenicity *(3)*. After our initial study, multiple laboratories have established the validity of this experimental system by the demonstration that transfer of specific human chromosomes into a variety of different human tumor cell lines results in the suppression of tumorigenic potential *(4–9)*. In this chapter, we outline a general strategy for mapping the location of tumor suppressor genes by microcell-mediated chromosome transfer (MMCT). Once the chromosomal site of a functional tumor suppressor gene becomes apparent, investigators can use other techniques to lead to the isolation of the operative gene, including positional cloning (**Fig. 1**). Thus, one can search for tumor suppressor genes armed with an established functional assay for verification of candidate genes.

1.1. General Strategy

The MMCT technique can provide a powerful tool for the identification of tumor suppressor genes. However, we have found several pitfalls and caveats in our own studies

From: *Methods in Molecular Biology, Vol. 222: Tumor Suppressor Genes: Pathways and Isolation Strategies*
Edited by: Wafik S. El-Deiry © Humana Press Inc., Totowa, NJ

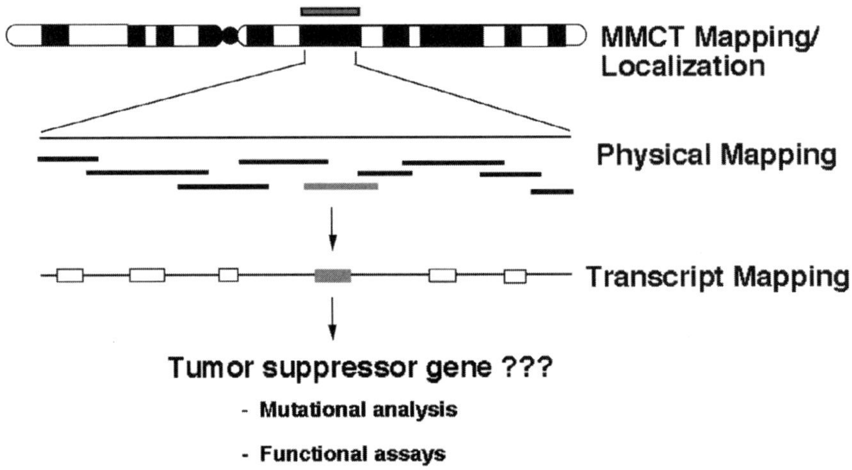

Fig. 1. Overall strategy for identification of tumor suppressor genes by MMCT. Suppression of transformed phenotypes such as anchorage-independent growth, immortality, and tumorigenicity after chromosome transfer of a normal chromosome into a tumor cell indicates the presence of a functional tumor suppressor gene. Transfer of smaller chromosome fragments narrows the tumor suppressor region to <2 Mb. At that point, traditional positional cloning approaches would lead to the identification of candidate tumor suppressor genes. Identification of the operative tumor suppressor gene would rely on its ability to suppress the transformed phenotype in a similar fashion to the whole chromosome.

that can lead to potential problems in data interpretation. Therefore, we suggest the use of several guiding principles for these types of experiments.

1. *The presence of introduced chromosomes must be documented by both cytogenetic and molecular means.* Clonal variation in tumor cell lines can often lead to the presence of an extra copy of a chromosome. Finding the expected number of chromosome copies in a putative microcell hybrid by karyotypic analysis would not control for this caveat. In contrast, the presence of a number of polymorphic markers from the transferred chromosome by molecular techniques does not assure the introduction of an intact chromosome. Breakage of the transferred chromosome occurs frequently during MMCT, resulting in transfer of small fragments. If this occurs, only the genetic information close to the site of the selection marker may be retained after growth in the selection medium. Finally, if the transferred chromosome is maintained in a cell line of a different species than the recipient, the investigator must show that the microcell hybrid did not receive genetic information from the donor cell line. If one finds rodent material in a human cell line after MMCT, one must control for this additional and substantial genetic information. We also suggest that investigators carry out these characterizations at the time of transformation and tumorigenicity assays. Continued growth of microcell hybrids in tissue culture may rapidly select the presence of tumor suppressor genes. If one waits five passages after molecular and cytogenetic analysis before inoculation into nude mice or soft agar, a significant population of segregants may appear in the population.
2. *Transfer of only one chromosome does not control adequately for nonspecific effects of this technique.* In other words, the isolation of nontumorigenic microcell hybrids containing the same chromosome does not always indicate the presence of a tumor suppressor gene on that chromosome. We have observed cases where the introduction of a number of different chromosomes leads to tumor suppression. In other instances, we have found that single-cell clones of a tumor cell line display a broad range of tumorigenic potentials, including nontumori-

genic clones. Thus, one should always produce microcell hybrids with at least one other normal chromosome that does not affect the transformed properties of the cancer cell line.

3. *If possible, investigators should use a selection system that allows for the addition and subtraction of the chromosome of interest.* The use of such reagents provides an obvious benefit. One can validate the presence of tumor suppressor activity on a chromosome transfer by demonstrating a reexpression of the transformed phenotype after loss of that chromosome. For some of our studies, we have used chromosomes from normal cells that contain a translocation between an autosome and the long arm of the X chromosome, site of the *HPRT* gene. Thus, we can select for the presence of the chromosome using growth medium containing HAT reagents and its loss by growth in medium containing 6-thioguanine (6TG). However, these X; autosome translocations are limited. Therefore, one group has developed libraries of human donor chromosomes that possess selectable markers such as the hygromycin-HSV/TK fusion gene ***(10,11)***. This allows for positive selection in hygromycin and negative selection in gancyclovir.
4. *One must always characterize the cells that actually formed tumors or soft agar colonies for chromosome content.* Reconstitution of tumors or soft agar colonies into tissue culture allows the investigator to confirm the presence of the transferred chromosome. Again, propagation of the cells before inoculation can allow loss of material from the transferred chromosome. If only 10% of the cells have lost the tumor suppressor gene, they may form tumors as rapidly as the parent may.
5. *We have come to rely mainly on positive results when mapping tumor suppressor genes.* Several studies have provided evidence for alterations in expression of genes on donor chromosomes, especially in cases of human chromosomes propagated in a rodent cell line ***(12)***. Thus, a chromosome may contain a complete copy of a tumor suppressor gene yet fail to express it upon chromosome transfer. Investigators should use caution in interpreting a lack of tumor suppressor activity after transfer as an indication of the absence of a tumor suppressor activity.

2. Materials

The general microcell hybridization method is shown in **Fig. 2**. Donor cell lines are treated with colcemid for 48 h to induce micronuclei formation. At that time, microcells are isolated by either centrifugation of flasks or by Percoll gradients. The microcells are then fused with the recipient cell lines using polyethylene glycol (PEG), divided into tissue culture dishes and allowed to incubate overnight. Selection media can be added the next day.

2.1. Microcell Preparation

Two general methods exist for the preparation of microcells. One approach uses cells attached to plastic or glass dishes or flasks ***(1,13,14)***. The second protocol uses Percoll gradients to isolate microcells ***(15)***. We have used the former technique almost exclusively for our tumor suppressor mapping studies. Virtually all donor cell lines containing human chromosome for MMCT are derived from mouse L-cells. The following cell technique works well for these cell lines.

1. T-25 flasks: Nunc cat. no. 1 52094 (*see* **Note 1**).
2. Filter units, gaskets, and nucleopore filters: 25-mm Swinnex filter units, cat. no. SX0002500.
3. Silicone gaskets: 25-mm Swinnex gaskets, cat. no. SX0002501.
4. Polycarbonate filters:VWR Scientific, 8 µm, cat. no. 28158-806; 5-µm, cat. no. 28158-668; 3-µm, cat. no. 28158-602. Assemble a filtration unit comprised of three individual Swinnex

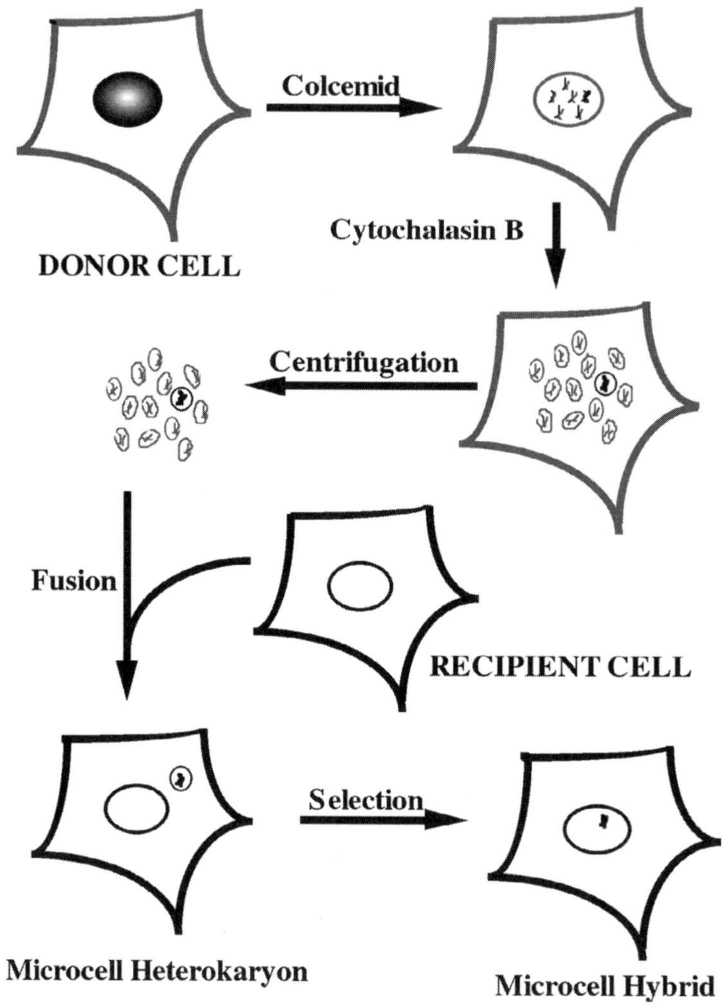

Fig. 2. Schematic of microcell-mediated chromosome transfer protocol. Microcell fusion consists of five essential steps: (1) micronucleation of the donor cells; (2) enucleation of the donor cell population; (3) isolation of the microcells; (4) fusion of the microcells to the recipient cells; and (5) isolation of microcell hybrids by growth in selection medium.

filters units stacked in series, containing an 8-µm filter (top), a 5-µm filter (middle) and a 3-µm filter (bottom). The filter should be inserted into the filter units so that the shiny side makes contact with the cells first, then sterilize by autoclaving for 30 min with slow exhaust (*see* **Note 2**).

5. Colcemid: 10-µg/ml stock, GibcoBRL cat. no. 15210-040.
6. Cytochalasin B: Sigma cat. no. C6762. Prepare a stock solution of 2 mg/mL in dimethylsulfoxide (DMSO) and store at −20°C. (*see* **Note 3**).

2.2. Microcell Fusion

1. Polyethylene glycol (PEG): Boehringer Mannheim cat. no. 783641 (*see* **Note 4**).
2. Phytohemmaglutin P(PHA-P): Difco cat. no. 3110-56-4 or Sigma cat. no. L9132. Prepare a 100-µg/mL stock in unsupplemented tissue culture medium. Sterilize by filtration through a 0.22-µm filter and aliquot 5-10 mL in 15-mL plastic tubes. Store frozen at −20°C.

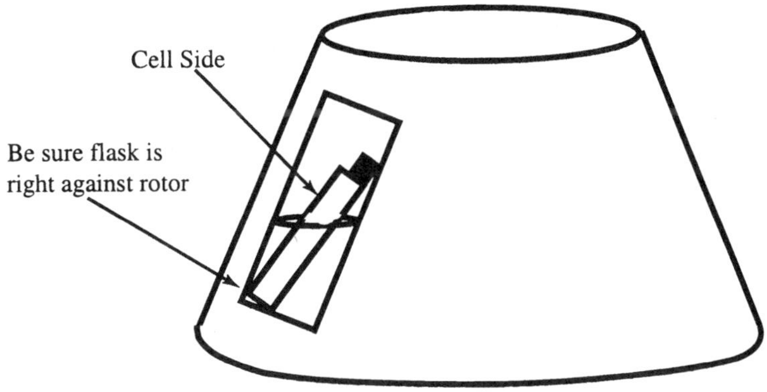

Fig. 3. Diagram of proper setup for Nunc T-25 flasks in GSA rotor.

3. Methods

We have generally used cell lines that attach to plastic as recipient cell lines for MMCT. Therefore, we have presented our standard fusion protocol in this chapter. However, one can use a suspension cell line as a recipient for chromosome transfer. In this instance, one follows a standard hybridoma fusion protocol with the additional step of using the PHA-P treatment to attach the microcells to the recipient suspension cells *(17)*.

3.1. Preparation of Microcells

1. Grow donor cells in 6 Nunc T-25 flasks until they reach approximately 90% confluence. Remove medium from cells and add medium containing 0.2-µg/mL colcemid (100 µL of 10-µg/mL stock in 5 mL) to each of the 6 Nunc T-25 flasks. Incubate cells at 37°C for 48 h (*see* **Note 5**).
2. After 48 h, aspirate medium from Nunc flasks of donor cells. Fill flasks completely to the top (including neck) with prewarmed cytochalasin B media (10 µg/mL). Incubate flasks upright in a 37°C incubator for 20 min. Wrap tops of tightly capped flasks with parafilm to prevent leakage during centrifugation.
3. Place the 6 Nunc flasks into wells of a Sorvall GSA rotor containing 100 mL water per well. Flasks should be oriented in wells such that the cell side faces up and the cap rests against the inside side of the well (**Fig. 3**). It is critical that the bottom of flask is tight against the bottom outer corner of the well, otherwise the flasks will crack. In this orientation the flask is situated at an angle in the well, with only the cap and two end edges touching the rotor. The rotor should be prewarmed to room temperature and its wells wiped with 70% ethanol prior to use. The flasks should be centrifuged at 24,000 *g* for 60 min.
4. Following centrifugation, gently decant cytochalasin B medium from Nunc flasks (this medium may be sterile filtered and reused 4–5 times). At this point you should see a pellet in each corner at the bottom of the flask (they may be different sizes).
5. Collect microcell pellets by adding 1.5 mL unsupplemented medium to each Nunc flask. One should try to resuspend the pellets only to avoid contamination of cellular debris. Resuspend each pellet with a 2-mL pipet and combine them in a plastic 15-mL conical centrifuge tube. Spin 10 min at approximately 1500 *g*.
6. Aspirate medium from pellet, gently and thoroughly break apart the pellet by tapping the tube, and resuspend microcells in 10 mL unsupplemented medium. Filter suspension through the Swinnex 8–5–3 µm filter stack using a syringe. Use a low amount of pressure to avoid

breaking the filters (*see* **Note 6**). Centrifuge filtered microcells for 15 min at approximately 1500 *g*. A small pellet should be visible at this point.

3.2. Fusion

1. Rinse recipient and control flasks with unsupplemented medium and remove medium by aspiration. Add 2 mL of PHA medium (100 μg/mL) to control flask. Resuspend microcell pellet in 2 mL PHA and add this to recipient flask. Lay flasks flat for 2 min, then transfer to a 37°C incubator for 15 min. At this step you should be able to see microcells attached to the recipient cells under the microscope.
2. Aspirate off PHA medium and add 2 mL of 50% PEG to each T-25 flask for exactly 1 min. Aspirate off PEG and wash cells quickly and sequentially 5 times with 5 mL of unsupplemented medium. Add complete medium containing 10% serum and incubate flasks overnight.

3.3. Selection

1. After overnight incubation, the cells in each T-25 flask are removed by trypsin-EDTA treatment and divided equally into eight 100-mm tissue culture dishes. The dishes are allowed to incubate overnight.
2. The next day, remove medium and add fresh medium with appropriate concentration of selective agent (*see* **Note 7**). Feed plates twice per week with fresh selection medium.
3. When colonies reached a size of 500–1000 cells (1–3 wk), they can be isolated and expanded for characterization (*see* **Note 8**).

3.4. Results of MMCT Studies

The use of MMCT has led to the mapping of functional tumor suppressor genes on the majority of human chromosomes. Functional activities include suppression of tumor growth in animals ***(18)***, suppression of immortality ***(19)*** and suppression of metastatic potential ***(20,21)***. In many cases, functional studies have implicated the presence of multiple tumor suppressor genes on the same chromosome ***(5,22–24)***. However, in many cases, loss of heterozygosity (LOH) studies had already implicated these regions in human cancer development of progression. The LOH studies followed by laborious positional cloning led eventually to the identification of the operative tumor suppressor gene. In some cases, MMCT studies led directly to the identity of a tumor suppressor gene, especially in the cases of suppressors of metastasis ***(25–28)***. Thus, the use of chromosome transfer has given important functional evidence to support the location of tumor suppressor genes identified by cytogenetic, LOH, and other molecular methods. However, they have not fulfilled the promise of rapid isolation of novel tumor suppressor genes.

3.5. Future Directions

Will MMCT prove useful for future investigations into tumor suppression? Although this technique has not led to the isolation of a large number of tumor suppressor genes, it still retains utility for other purposes. Transfer of a tumor suppressor gene in its native genomic structure still gives the most accurate accounting of its normal regulation. This fact may become increasingly important as chromatin remodeling, histone modification, and epigenetic regulation emerge as neoplastic processes ***(29–35)***. In addition, methods for modification of the donor chromosome in a targeted fashion have recently become available. The finding that the DT40 chicken cell line retains a high level of homologous recombination has opened new avenues of investigation ***(36)***. That report demonstrated that a human chromosome retained in the DT40 cell line became amenable to gene targeting ***(36)***. Thus, one can delete a gene on a human chromosome using an approach similar to that used in the production of knockout mice ***(37)***. Alternatively, one can insert a new telomere at a specific site in the chromosome by homologous recombination leading to a loss of genetic information distal to that site ***(38)***. Oshimura and his colleagues have developed a library of human chromosomes for MMCT with either the maternally derived or paternally derived chromosome ***(39)***. They have also shown that certain human chromosomes may be successfully propagated in transgenic mice with expression of human genes ***(40–42)***. Therefore, manipulation of single chromosomes has evolved from simple mapping experiments to more exquisite applications in solving the riddle of human cancer development.

4. Notes

1. These are the only flasks that will withstand the centrifugation step. Other brands will crack under the pressure. Alternative approaches include the use of a plastic retainer to enforce other brands of plastic flasks ***(13)*** or plastic "bullets" in centrifuge tubes ***(16)***.
2. It is critical that you sterilize by autoclaving for 30 min with slow exhaust so that the filters do not crack.
3. Cytochalasin B is considered a mutagen. Follow proper safety protocols.
4. We prefer to buy the PEG as a ready-to-use 50% solution. Although it is more expensive, it avoids potential problems with toxicity associated with laboratory preparations of PEG solutions.
5. We use this concentration for mouse L-cell donor lines, the most commonly used cells for routine MMCT. However, colcemid concentrations may vary for other cell types, due to differences in toxicity. One can optimize colcemid concentrations for other donor cell lines using the protocol of Killary and Fournier ***(16)***.
6. Filter breakage will allow any intact donor cells to pass through, leading to contaminating colonies after fusion and selection.
7. Cell lines vary widely in their sensitivity to different selection reagents. We highly recommend a cytotoxicity assay to determine the optimal concentration for your cell line. We prefer the lowest dose that kills the cells. Remember, you are introducing only one copy of the gene coding for the selection agent. Furthermore, the transferred chromosome must be integrated into the recipient cell nucleus and start producing enough enzyme to neutralize the selection agent before microcell hybrids become resistant. Therefore, unlike gene transfections, we do not recommend placing the recipient cells into high doses of the selection agent right after fusion.
8. Any method can be used. We prefer to use glass-cloning rings. We also recommend that colonies be expanded in a gradual manner by first transferring to a 24-well plate followed by a T-25 flask. We always grow our cells under selective pressure to maintain the transferred chromosome.

References

1. Fournier, R. E. K. and Ruddle, F. H. (1977) Microcell-mediated transfer of murine chromosomes into mouse, Chinese hamster, and human somatic cells. *Proc. Natl. Acad. Sci. USA* **74,** 319–323.
2. Ege, T. and Ringertz, N. R. (1974) Preparation of microcells by enucleation of micronucleated cells. *Exp. Cell. Res.* **87,** 378–382.

3. Weissman, B. E., Saxon, P. J., Pasquale, S. R., Jones, G. R., Geiser, A. G., and Stanbridge, E. J. (1987) Introduction of a normal human chromosome 11 into a Wilms' tumor cell line controls its tumorigenic expression. *Science* **236,** 175–180.
4. Koi, M., Morita, H., Yamada, H., Satoh, H., Barrett, J. C., and Oshimura, M. (1989) Normal human chromosome 11 suppresses tumorigenicity of human cervical tumor cell line SiHa. *Mol. Carcinogen.* **2**, 12–21.
5. Nihei, N., Ichikawa, T., Kawana, Y., et al. (1996) Mapping of metastasis suppressor gene(s) for rat prostate cancer on the short arm of human chromosome 8 by irradiated microcell-mediated chromosome transfer. *Genes Chromosomes Cancer* **17**, 260–268.
6. Oshimura, M., Kugoh, H. M., Shimizu, M., et al. (1989) Multiple chromosomes carrying tumor suppressor activity, via microcell-mediated chromosome transfer, for various tumor cell lines, in *Genetic Basis for Carcinogenesis: Tumor Suppressor Genes and Oncogenes* (Knudson, A. G., Stanbridge, E. J., Sugimura, T., Terada, M., and Watanabe, S., eds.), Scientific Societies Press, Tokyo, pp. 279–257.
7. Satoh, H., Lamb, P. W., Dong, et al. (1993) Suppression of tumorigenicity of A549 lung adenoacarcinoma cells by human chromosomes 3 and 11 introduced via microcell-mediated chromosome transfer. *Mol. Carcinogen.* **7,** 157–164.
8. Tanaka, K., Kikuchi-Yanoshita, R., Muraoka, M., Konishi, M., Oshimura, M., and Miyaki, M. (1996) Suppression of tumorigenicity and invasiveness of colon carcinoma cells by introduction of normal chromosome 8p12-pter. *Oncogene* **12,** 405–410.
9. Yamada, H., Wake, N., Fujimoto, S., Barrett, J. C., and Oshimura, M. (1990) Multiple chromosomes carrying tumor suppressor activity for a uterine endometrial carcinoma cell line identified by microcell-mediated chromosome transfer. *Oncogene* **5,** 1141–1147.
10. Speevak, M. D., Berube, N. G., McGowan-Jordan, I. J., Bisson, C., Lupton, S. D., and Chevrette, M. (1995) Construction and analysis of microcell hybrids containing dual selectable tagged human chromosomes. *Cytogenet. Cell Genet.* **69,** 63–65.
11. Trott, D. A., Cuthbert, A. P., Todd, C. M., Themis, M., and Newbold, R. F. (1995) Novel use of a selectable fusion gene as an "in-out" marker for studying genetic loss in mammalian cells. *Mol. Carcinogen.* **12,** 213–224.
12. Kucherlapati, R. and Shin, S. I. (1979) Genetic control of tumorigenicity in interspecific mammalian cell hybrids. *Cell* **16,** 639–648.
13. Koi, M., Shimizu, M., Morita, H., Yamada, H., and Oshimura, M. (1989) Construction of mouse A9 clones containing a single human chromosome tagged with neomycin-resistance gene via microcell fusion. *Jpn. J. Cancer Res.* **80,** 413–418.
14. Saxon, P. J., Srivatsan, E. S., Leipzig, G. V., Sameshima, J. H., and Stanbridge, E. J. (1985) Selective transfer of individual human chromosomes to recipient cells. *Mol. Cell. Biol.* **5,** 140–146.
15. Stubblefield, E. and Pershouse, M. (1992) Direct formation of microcells from mitotic cells for use in chromosome transfer. *Somat. Cell. Mol. Genet.* **18,** 485–491.
16. Killary, A. M. and Fournier, R. E. (1995) Microcell fusion. *Meth. Enzymol.* **254,** 133–152.
17. Nguyen, T. M. and Morris, G. E. (1996) Production of panels of monoclonal antibodies by the hybridoma method. *Meth. Mol. Biol.* **66,** 377–389.
18. Anderson, M. J. and Stanbridge, E. J. (1993) Tumor suppressor genes studied by cell hybridization and chromosome transfer. *FASEB J.* **7,** 826–833.
19. Leung, J. K. and Pereira-Smith, O. M. (2001) Identification of genes involved in cell senescence and immortalization: potential implications for tissue ageing. *Novartis Found. Symp.* **235,** 105–110.
20. Yoshida, B. A., Sokoloff, M. M., Welch, D. R., and Rinker-Schaeffer, C. W. (2000) Metastasis-suppressor genes: a review and perspective on an emerging field. *J. Natl. Cancer Inst.* **92,** 1717–1730.
21. Ichikawa, T., Hosoki, S., Suzuki, H., et al. (2000) Mapping of metastasis suppressor genes for prostate cancer by microcell-mediated chromosome transfer. *Asian J. Androl.* **2,** 167–171.

22. Nihei, N., Ohta, S., Kuramochi, H., et al. (1999) Metastasis suppressor gene(s) for rat prostate cancer on the long arm of human chromosome 7. *Genes Chromosomes Cancer* **24,** 1–8.
23. O'Briant, K., Jolicoeur, E., Garst, J., Campa, M., Schreiber, G., and Bepler, G. (1997) Growth inhibition of a human lung adenocarcinoma cell line by genetic complementation with chromosome 11. *Anticancer Res.* **17,** 3243–3251.
24. Sabbioni, S., Negrini, M., Possati, L., et al. (1994) Multiple loci on human chromosome 11 control tumorigenicity of BK virus transformed cells. *Int. J. Cancer* **57,** 185–191.
25. Lee, J. H., Miele, M. E., Hicks, D. J., et al. (1996) KiSS-1, a novel human malignant melanoma metastasis-suppressor gene. *J. Natl. Cancer Inst.* **88,** 1731–1737.
26. Yang, X., Welch, D. R., Phillips, K. K., Weissman, B. E., and Wei, L. L. (1997) KAI1, a putative marker for metastatic potential in human breast cancer. *Cancer Lett.* **119,** 149–155.
27. Dong, J.-T., Lamb, P. W., Rinker-Schaeffer, C. W., Isaacs, J. T., and Barrett, J. C. (1994) Cloning and characterization of a putative metastasis suppressor gene on human chromosome 11p11.2-13 for prostatic cancer. *Proc. Am. Assoc. Cancer Res.* **35,** 186.
28. Seraj, M. J., Samant, R. S., Verderame, M. F., and Welch, D. R. (2000) Functional evidence for a novel human breast carcinoma metastasis suppressor, BRMS1, encoded at chromosome 11q13. *Cancer Res.* **60,** 2764–2769.
29. Versteege, I., Sevenet, N., Lange, J., et al. (1998) Truncating mutations of hSNF5/INI1 in aggressive paediatric cancer. *Nature* **394,** 203–206.
30. Jones, P. A. and Laird, P. W. (1999) Cancer epigenetics comes of age. *Nat. Genet.* **21,** 163–167.
31. Wong, A. K., Shanahan, F., Chen, Y., et al. (2000) BRG1, a component of the SWI-SNF complex, is mutated in multiple human tumor cell lines. *Cancer Res.* **60,** 6171–6177.
32. Zhang, H. S., Gavin, M., Dahiya, A., et al. (2000) Exit from G1 and S phase of the cell cycle is regulated by repressor complexes containing HDAC-Rb-hSWI/SNF and Rb-hSWI/SNF. *Cell* **101,** 79–89.
33. Strobeck, M. W., Knudsen, K. E., Fribourg, A. F., et al. (2000) BRG-1 is required for RB mediated cell cycle arrest. *Proc. Natl. Acad. Sci. USA* **97,** 7748–7753.
34. Decristofaro, M. F., Betz, B. L., Rorie, C. J., Reisman, D. N., Wang, W., and Weissman, B. E. (2001) Characterization of SWI/SNF protein expression in human breast cancer cell lines and other malignancies. *J. Cell. Physiol.* **186,** 136–145.
35. Muraoka, M., Konishi, M., Kikuchi-Yanoshita, R., et al. (1996) p300 gene alterations in colorectal and gastric carcinomas. *Oncogene* **12,** 1565–1569.
36. Dieken, E. S., Epner, E. M., Fiering, S., Fournier, R. E. K., and Groudine, M. (1996) Efficient modification of human chromosomal alleles using recombination-proficient chicken/human microcell hybrids. *Nat. Genet.* **12,** 174–182.
37. Horike, S., Mitsuya, K., Meguro, M., et al. (200) Targeted disruption of the human LIT1 locus defines a putative imprinting control element playing an essential role in Beckwith-Wiedemann syndrome. *Hum. Mol. Genet.* **9,** 2075–2083.
38. Kuroiwa, Y., Shinohara, T., Notsu, T., et al. (1998) Efficient modification of a human chromosome by telomere-directed truncation in high homologous recombination-proficient chicken DT40 cells. *Nucleic Acids Res.* **26,** 3447–3448.
39. Kugoh, H., Mitsuya, K., Meguro, M., Shigenami, K., Schulz, T. C., and Oshimura, M. (1999) Mouse A9 cells containing single human chromosomes for analysis of genomic imprinting. *DNA Res.* **6,** 165–172.
40. Shinohara, T., Tomizuka, K., Takehara, S., et al. (2000) Stability of transferred human chromosome fragments in cultured cells and in mice. *Chromosome Res.* **8,** 713–725.
41. Kuroiwa, Y., Tomizuka, K., Shinohara, T., et al. (2000) Manipulation of human minichromosomes to carry greater than megabase-sized chromosome inserts. *Nat. Biotechnol.* **18,** 1086–1090.
42. Tomizuka, K., Shinohara, T., Yoshida, H., et al. (2000) Double trans-chromosomic mice: maintenance of two individual human chromosome fragments containing Ig heavy and kappa loci and expression of fully human antibodies. *Proc. Natl. Acad. Sci. USA* **97,** 722–727.

23

Approaches to Proteomic Analysis of Human Tumors

Mamoun Ahram and Michael R. Emmert-Buck

1. Introduction

The involvement of tumor suppressor genes (TSG) in cancer initiation and progression is well documented in several tumor types such as colon (APC, p53) and breast (BrcA1, BrcA2) cancers. Loss of heterozygosity of distinct chromosomal regions, which are thought to harbor as yet unidentified TSGs, have also been linked to many cancer types. Traditional research approaches such as positional cloning have greatly assisted in elucidation of these genetic factors, and continue to be critical in efforts to understand the nature and role of TSGs in human cancers. In parallel, molecular profiling is a new strategy for analysis of tumors that has emerged based on the information provided by the Human Genome Project and the development of several high-throughput technologies. This new research concept utilizes global measurements of mRNA and protein expression patterns in tumor cells and their normal counterparts in search of the genetic culprits responsible for tumorigenesis ***(1)***.

Based on current knowledge, the biologic functions of tumor suppressor genes are thought to range widely, from protecting genome stability, to modulating proliferation, differentiation, survival, and finally to mediating invasion and metastasis. However, much remains to be learned regarding the roles of TSGs in normal cellular physiology and tumor development. The field of "proteomics" is an important component of molecular profiling and likely will be extremely valuable in efforts to increase our knowledge of tumor suppressor gene function. Proteomics may be defined as the global identification, measurement, and analysis of the complete set of proteins of a biological system, referred to as the proteome. Although proteomic efforts began in the 1970s with the establishment of two-dimensional gel technology, the field has not kept pace with the rapid developments in genomic research. This is owing in large measure to the technical difficulties associated with protein purification and analysis. These include the lack of a well-defined set of "hybridization reagents" (i.e., similar to expressed sequence tagged clones for mRNA microarray-based measurements) for each protein, and the absence of a protein amplification technology such as polymerase chain reaction (PCR). Nonetheless, there has been a recent revival of interest in proteomics, fueled by both the completion of the Human Genome Project as well as the development of promising new

From: *Methods in Molecular Biology, Vol. 222: Tumor Suppressor Genes: Pathways and Isolation Strategies*
Edited by: Wafik S. El-Deiry © Humana Press Inc., Totowa, NJ

technologies that may permit global protein measurements to become a reality. The purpose of the present chapter is to provide readers with background information on proteomic tools and their general applications in studying cancer, with an emphasis on their utility for identifying tumor suppressor genes and investigating biological functions. Detailed methodologies of the approaches discussed can be found referenced between the articles. The chapter begins with a short section describing tissue microdissection, a developing technology that has become increasingly valuable in both genomic and proteomic studies of human tumor specimens.

2. Tissue Microdissection

Clinical tissue specimens represent a critical source of information on the DNA, mRNA, and protein profiles associated with cancer. Moreover, these samples permit investigators to examine features of tumorigenesis (premalignant lesions, tumor endothelium, host response) that are not easily studied in model systems such as cultured cell lines. However, there are many challenges associated with studying clinical samples, based primarily on the fact that the majority of the pathological features are microscopic in nature ***(1)***. Thus, methods that permit investigators to specifically procure cell types of interest from a tissue sample have become increasingly important. Our group at the National Cancer Institute utilizes tissue microdissection to recover cell types of interest from tissue samples. This approach permits investigators to examine a histologic section of specimens microscopically, assess the pathology that is present, and specifically recover phenotypically defined cell populations.

Methodologies to perform tissue microdissection have evolved significantly over the past decade, and include the recent advent of laser-based technologies, such as laser capture microdissection (LCM) ***(2,3)***. LCM was developed at the National Institutes of Health through a cooperative research and development agreement (CRADA) with Arcturus Engineering (Mountain View, CA) ***(2)***. LCM works by covering a histologic section with a thermoplastic film and procuring the cells of interest with a near-infrared laser. The thermal film is located on the underside of a test tube cap that is placed onto 0.5-mL microcentrifuge tube after dissection. Damage to the molecular contents of the tissue during dissection is minimal due to the brief laser pulse (5 ms) and limited amount of energy that traverses the film and reaches the tissue. Commercial LCM instruments permit imaging of the tissue before and after cell procurement, allowing for complete documentation of the dissection process. LCM has proven to be a simple, accurate, and rapid method for tissue microdissection, and has been widely utilized in several studies of tumor suppressor genes ***(4–12)***. As will be discussed, LCM can be coupled with proteomic tools to examine the proteome of normal, premalignant and tumor cells.

3. Proteomic Tools

3.1. Two-Dimensional Gel Electrophoresis

At present, the most widely used technique for global protein analysis is two-dimensional polyacrylamide gel electrophoresis (2D-PAGE). 2D-PAGE separates proteins on the basis of isoelectric charge (first dimension) and molecular weight (second dimension) ***(13,14)***. Following protein extraction in a buffer composed of nonionic detergents (e.g., urea, thiourea) and reducing agents (e.g., DDT, β-mercaptoethanol), proteins are first separated according to charge. Carrier ampholytes were initially used for the first dimension; however, this method often resulted in inconsistent protein separation. The recent utilization of immobilized pH gradient (IPG) gels for isoelectric separation of proteins has significantly improved the reproducibility of protein focusing ***(15–17)***. Moreover, the addition of narrow range IPG strips (e.g., pH 4–7, 5–8, 6–9, 6–11) has allowed for increased resolution and, thus, identification of a larger number of proteins. For the second dimension separation, proteins are separated by size on a polyacrylamide gel.

The major strength of 2D-PAGE is the ability to generate a two-dimensional "protein fingerprint" of a biologic sample. Many hundreds to several thousands of proteins can be analyzed simultaneously, depending on the system employed. Although it has been an extremely useful method, 2D-PAGE does have significant shortcomings. For example, extraction and separation of both hydrophobic membrane proteins and highly basic proteins is challenging, although some success has been made recently by modifying extraction buffers ***(16–20)***. Another limitation is the difficulty identifying low- and moderate-abundant proteins. Considering that a cellular proteome consists of approximately 6000–10,000 proteins, 2D-PAGE reveals only a subset of the total protein content, primarily those that are high and moderate-high in abundance levels. Silver nitrate has been the primary staining method for 2D-PAGE due to its relatively good sensitivity for detection (0.1–1 ng). However, the need to fix proteins in the gel prior to staining limits downstream applications such as mass spectrometry-based sequencing. Alternative staining protocols that do not require in-gel fixation include fluorescent labeling with compounds such as Sypro Ruby ***(21)***. The latter has a detection sensitivity of 1–10 ng. Radioactive protein labeling improves sensitivity, but is expensive and requires special handling ***(22)***. For more information on the protocols that our group uses for 2D-PAGE, please refer to: http://cgap-mf.nih.gov/Protocols/DNARNAProteomicAnalysis/Proteomics/2Dpage.html.

3.2. Mass Spectrometry

Mass spectrometry (MS) is the primary tool for identifying proteins after 2D-PAGE separation. Proteins are digested by a protease such as trypsin and then identified by MS. There are two main approaches to MS-based protein analysis (for review, *see* **refs.** ***23*** and ***24***). In the first, termed matrix-assisted laser desorption/ionization (MALDI), the masses of peptide fragments are calculated by measuring their acceleration or time-of-flight (TOF) distribution following laser-induced ionization. Since the generated peptides are unique for each protein, identity can be predicted by computational searches of existing public databases.

The second approach utilizes MS as a protein sequencing method. An initial peptide fragment (parent ion) is isolated from a protein and further fragmented (daughter ions), followed by mass determination of the resulting peptides. This method is referred to as collision-induced fragmentation (CID) or tandem MS (MS/MS) ***(25–27)***. The molecular masses of peptides from a digestion of a protein and those of the fragments represent a unique fingerprint. With an appropriate protein sequence database and search program (e.g., PepFrag or ProFound, at http://prowl.rockefeller.edu), protein identity and amino acid sequence can be determined. Although this technique is more complex than MALDI-TOF, it is particularly useful in studies of novel proteins with an unknown sequence.

Two-dimensional-PAGE and MS have been applied to discover new markers of colorectal cancer and hepatocellular carcinoma ***(28)***, bladder carcinoma ***(29,30)*** as well as other tumor types (for review, *see* **ref. *31***). LCM has been used in combination with 2D-PAGE to define the protein profiles of epithelial cells, both normal and cancerous, of the prostate ***(32)*** and esophagus ***(33)***. These efforts have led to the discovery of several differentially expressed proteins, including annexin I ***(34)***. Downregulation of 14-3-3 protein in breast carcinoma cell lines was discovered using a similar approach ***(35)***. With regard to TSGs, proteomic efforts have examined the effect of specific genes such as p53 and APC on global patterns of protein expression ***(36–38)***. Future studies of this sort have great potential to provide insight into the network of protein–gene and protein–protein interactions that involve TSGs.

3.3. Surface-Enhanced Laser Desorption Ionization

Surface-enhanced laser desorption ionization (SELDI) is a new technology for protein analysis that is based on TOF of proteins following analysis on affinity-based protein chips (for review, *see* **ref. *39***) This method is based on binding total cellular protein content to a small area (1–2 mm^2) of a chip with a specified surface affinity (e.g., hydrophobic, cationic, anionic, aliphatic). Laser activation results in ionization and release of the captured proteins that travel with an inverse correlation to their molecular weight. Measurement of the TOF of the released proteins or peptides is illustrated either graphically as peaks, with every peak representing a protein molecule, or as a gel-like image similar to a first-dimension PAGE gel.

Two major advantages of SELDI are the small sample size required for analysis (microliters) and the speed with which multiple samples can be studied. Several studies have illustrated the potential of this technology for analysis of tumor samples. In our group, combination of LCM and SELDI revealed differences in protein fingerprints between microdissected normal, premalignant, and malignant prostate cells ***(40)***. A follow-up SELDI-based study of multiple cancer types (breast, colon, and ovary) identified protein markers that were specific for each cancer. This study also illustrated the high sensitivity of SELDI to detect protein profiles from small numbers of cells ***(41)***. In addition to cell samples, body fluids may also be examined by SELDI for the presence of cancer-specific markers. For example, reproducible protein changes were identified in the urine of patients with bladder cancer ***(42)***.

There are several potential uses of SELDI technology for tumor suppressor gene research. The affinity of a tumor suppressor protein for binding specific DNA sequences may be determined using SELDI chips. Similarly, protein–protein interactions can be studied by immobilizing a protein(s) of interest on a chip and analyzing the binding partners that are present in a cellular lysate. Additionally, antibody-based affinity chips may be used for investigating specific proteins. This was demonstrated by a study that verified the presence of defensin, a protein marker specific for bladder cancer, using an antibody-bound chip ***(42)***.

4. Tissue Reagents

4.1. Tissue Array

Proteomic studies are capable of identifying large numbers of disease-associated protein alterations, including candidate tumor suppressor proteins that show decreased expression levels in cancer. To validate data sets derived from these global, discovery-based efforts, technologies need to be developed that permit protein alterations to be validated in large numbers of clinical tissue specimens. Recently, researchers have developed "tissue arrays," a high-throughput methodology that permits investigation of chromosomal alterations (fluorescent *in situ* hybridization), RNA expression (*in situ* hybridization), and protein levels (immunohistochemistry) ***(43,44)***. Small tissue samples (0.6 mm) are removed from specific regions of a paraffin-embedded block and arrayed into a "recipient" paraffin block. The new block is subsequently processed into histology slides, each containing up to several hundred tissue samples. Serial sections are generated and then utilized for DNA, mRNA, or protein analyses, depending on the needs of the investigator.

Tissue microarrays have been applied successfully to measure amplification of genes in breast cancer ***(45–47)***. Alternatively, Bubendorf et al. Used tissue array technology to measure protein expression of several genes that initially showed mRNA deregulation in prostate cancer xenografts by cDNA microarray. In a separate study, expression levels of the tumor suppressor gene candidate, Nkx3.1, was evaluated in prostate tissue in a large sample set using tissue arrays ***(48)***. This study showed a correlation between significantly decreased Nkx3.1 expression and progression of prostate cancer, in particular hormone-refractory tumors and metastatic lesions. The large number of samples that were studied and showed decreased levels of Nkx3.1 provided strong support for the hypothesis that Nkx3.1 functions as a tumor suppressor gene in prostate cancer.

4.2. Protein Arrays Under Development

The success of nucleic acid-based array systems for global measurements of transcript levels has spurred significant interest in applying similar strategies to protein-based studies. Several attempts have been made to establish protein arrays, each with a variable amount of success (for review, *see* **refs.** ***49–51***). Several different methodologies have been employed. For example, purified proteins, antibodies, or small ligands can be spotted on a chip and used as "bait" to analyze cellular lysates. These arrays have utility for multiple applications such as determination of protein–protein or protein–ligand interaction, as well as for analyzing differential expression of proteins in multiple cell samples. MacBeath and Schreiber developed a multipurpose array system consisting of immobilized proteins spotted on glass microscope slides ***(52)***. These arrays were used for identification of protein–protein and protein–small molecule interactions. In a separate approach, bacterial lysates were spotted at high density on microarray filters to facilitate large-scale discovery of bacterial clones expressing a protein of interest ***(53)***. Specific protein arrays such as "mitogenesis pathway chips" or "prosurvival/apotosis chips" are being developed for use both in the laboratory and in clinical settings ***(54)***. These arrays are composed of various surface ligands (antibodies, phages, proteins/peptide, nucleic acids) in order to determine the signaling pathway associated with tumorigenesis. A new technology, termed "reverse-phase protein arrays" (RPA), spots LCM-derived cell lysates on nitrocellulose membranes ***(55)***. Each array contains cell populations that were procured directly from tissue specimens. Currently, these arrays are being used for high-throughput analysis of several different cancer types, including measuring expression levels of several tumor suppressor proteins.

5. Integration of Clinical Studies and Laboratory Methods

5.1. Example: Yeast Two-Hybrid System

To understand fully the role that TSGs play in the development and progression of human cancer, it is important to integrate information from both clinical and laboratory research efforts. This permits discoveries made in the laboratory to be evaluated in patient specimens, and observations made in the clinic to be evaluated mechanistically. As an example, protein–protein interactions represent a critical component of cellular regulation. A powerful technique that can be used to analyze these interactions is the yeast two-hybrid (Y2H) system ***(56)***. This method was used by several groups to discover potentially important binding partners of tumor suppressor genes, including classical tumor suppressor proteins such as p53 ***(57,58)***, Bcl-2 ***(59)***, NF 2 ***(60)***, BrcA1 and 2 ***(61,62)***, WT-1 ***(63)***, PTEN ***(64)***, Rb ***(65)***, p16INK4a ***(66)***, and APC ***(67)***. However, the fact that two proteins can interact in an in-vitro setting is not definitive evidence that they interact in vivo. Therefore, the current and developing proteomic analysis tools described above have a critically important role in (in)validating observations made in the laboratory. Moreover, the nature of tumor suppressor protein interactions can be analyzed more thoroughly. For example, are normal protein–protein interactions disrupted during tumorigenesis, and if so, at what pathologic stage does this occur?

6. Bioinformatics

The development of high-throughput nucleic acid and protein analysis technologies is resulting in a dramatic growth of information. Thus, there is a pressing need for computational-based methods that can analyze large biologic data sets. For proteomic studies, several databases have been established (for review, *see* **ref. *68***). For example, the Expert Protein Analysis Program (ExPASy) was developed by the Swiss Institute of Bioinformatics and is publicly available at http://www.expasy.ch. At this site, images of 2D gels from various biologic systems can be viewed, and one can also obtain information on protein properties such as proposed function, amino acid sequence, and protein structure. Celis and colleagues have developed an excellent proteomic database that can be viewed at http://biobase.dk/cgi-bin/celis; ***(69)***. The SWISS-PROT group has developed an informational database that contains annotation and description of all known proteins as part of the Human Proteome Initiative (HPI). The site is available at http://www.ebi.ac.uk/swiss prot/hpi/hpi.html ***(70)***.

7. Summary

The completion of the Human Genome Project and the successful use of high-throughput array formats for mRNA measurements has spurred a revival of interest in the field of proteomics. Clinical cancer specimens will be a critical component of proteomic studies of cancer, both for discovering new insights into TSG function, and for validation of discoveries that are made in the laboratory. Global protein analysis methods hold great potential for facilitating the discovery of novel tumor suppressor genes, as well as providing a better understanding of the biochemical role of these proteins.

References

1. Emmert-Buck, M., Strausberg, R., Krizman, D., et al. (2000) Molecular profiling of clinical tissue specimens: feasibility and applications. *Am. J. Pathol.* **156**, 1109–1115.
2. Emmert-Buck, M. R., Bonner, R. F., Smith, P. D., et al. (1996) Laser capture microdissection. *Science* **274**, 998–1001.

3. Gillespie, J. W., Ahram, M., Best, C. J., et al. (2001) The role of tissue microdissection in cancer research. *Cancer J.* **7**, 32–39.
4. Brown, M. R., Chuaqui, R., Vocke, C. D., et al. (1999) Allelic loss on chromosome arm 8p: analysis of sporadic epithelial ovarian tumors. *Gynecol. Oncol.* **74**, 98–102.
5. Chung, T. K., Cheung, T. H., Lo, W. K., et al. (2000) Loss of heterozygosity at the short arm of chromosome 3 in microdissected cervical intraepithelial neoplasia. *Cancer Lett.* **154**, 189–194.
6. Boni, R., Matt, D., Voetmeyer, A., Burg, G., and Zhuang, Z. (1998) Chromosomal allele loss in primary cutaneous melanoma is heterogeneous and correlates with proliferation. *J. Invest. Dermatol.* **110**, 215–217.
7. Emmert-Buck, M. R., Lubensky, I. A., Dong, Q., et al. (1997) Localization of the multiple endocrine neoplasia type I (MEN1) gene based on tumor loss of heterozygosity analysis. *Cancer Res.* **57**, 1855–1858.
8. Romagnoli, S., Roncalli, M., Graziani, D., et al. (2001) Molecular alterations of Barrett's esophagus on microdissected endoscopic biopsies. *Lab. Invest.* **81**, 241–247.
9. Shivapurkar, N., Sood, S., Wistuba, II, et al. (1999) Multiple regions of chromosome 4 demonstrating allelic losses in breast carcinomas. *Cancer Res.* **59**, 3576–3580.
10. Shivapurkar, N., Virmani, A. K., Wistuba, II, et al. (1999) Deletions of chromosome 4 at multiple sites are frequent in malignant mesothelioma and small cell lung carcinoma. *Clin. Cancer Res.* **5**, 17–23.
11. Tannapfel, A., Benicke, M., Katalinic, A., et al. (2000) Frequency of p16(INK4A) alterations and K-ras mutations in intrahepatic cholangiocarcinoma of the liver. *Gut* **47**, 721–727.
12. Werness, B. A., Parvatiyar, P., Ramus, S. J., et al. (2000) Ovarian carcinoma in situ with germline BRCA1 mutation and loss of heterozygosity at BRCA1 and TP53. *J. Natl. Cancer Inst.* **92**, 1088–1091.
13. O'Farrell, P. Z. and Goodman, H. M. (1976) Resolution of simian virus 40 proteins in whole cell extracts by two-dimensional electrophoresis: heterogeneity of the major capsid protein. *Cell* **9**, 289–298.
14. Klose, J. (1975) Protein mapping by combined isoelectric focusing and electrophoresis of mouse tissues. A novel approach to testing for induced point mutations in mammals. *Humangenetik* **26**, 231–243.
15. Sanchez, J. C., Rouge, V., Pisteur, M., et al. (1997) Improved and simplified in-gel sample application using reswelling of dry immobilized pH gradients. *Electrophoresis* **18**, 324–327.
16. Gorg, A., Obermaier, C., Boguth, G., et al. (2000) The current state of two-dimensional electrophoresis with immobilized pH gradients. *Electrophoresis* **21**, 1037–1053.
17. Wildgruber, R., Harder, A., Obermaier, C., et al. (2000) Towards higher resolution: two-dimensional electrophoresis of *Saccharomyces cerevisiae* proteins using overlapping narrow immobilized pH gradients. *Electrophoresis* **21**, 2610–2616.
18. Rabilloud, T., Adessi, C., Giraudel, A., and Lunardi, J. (1997) Improvement of the solubilization of proteins in two-dimensional electrophoresis with immobilized pH gradients. *Electrophoresis* **18**, 307–316.
19. Santoni, V., Rabilloud, T., Doumas, P., et al. (1999) Towards the recovery of hydrophobic proteins on two-dimensional electrophoresis gels. *Electrophoresis* **20**, 705–711.
20. Chevallet, M., Santoni, V., Poinas, A., et al. (1998) New zwitterionic detergents improve the analysis of membrane proteins by two-dimensional electrophoresis. *Electrophoresis* **19**, 1901–1909.
21. Berggren, K., Chernokalskaya, E., Steinberg, T. H., et al. (2000) Background-free, high sensitivity staining of proteins in one- and two-dimensional sodium dodecyl sulfate-polyacrylamide gels using a luminescent ruthenium complex. *Electrophoresis* **21**, 2509–2521.
22. Vuong, G. L., Weiss, S. M., Kammer, W., et al. (2000) Improved sensitivity proteomics by postharvest alkylation and radioactive labelling of proteins. *Electrophoresis* **21**, 2594–2605.

23. Blackstock, W. P. and Weir, M. P. (1999) Proteomics: quantitative and physical mapping of cellular proteins. *Trends Biotechnol.* **17**, 121–127.
24. Humphery-Smith, I. (1998) Proteomics: from small genes to high-throughput robotics. *J. Protein Chem.* **17**, 524–525.
25. Biemann, K. (1992) Mass spectrometry of peptides and proteins. *Annu. Rev. Biochem.* **61**, 977–1010.
26. Hunt, D. F., Yates, J. R., 3rd, Shabanowitz, J., Winston, S., and Hauer, C. R. (1986) Protein sequencing by tandem mass spectrometry. *Proc. Natl. Acad. Sci. USA* **83**, 6233–6237.
27. Qin, J., Herring, C. J., and Zhang, X. (1998) De novo peptide sequencing in an ion trap mass spectrometer with 18O labeling. *Rapid Commun. Mass Spectrom.* **12**, 209–216.
28. Jungblut, P. R., Zimny-Arndt, U., Zeindl-Eberhart, E., et al. (1999) Proteomics in human disease: cancer, heart and infectious diseases. *Electrophoresis* **20**, 2100–2110.
29. Celis, J. E., Celis, P., Ostergaard, M., et al. (1999) Proteomics and immunohistochemistry define some of the steps involved in the squamous differentiation of the bladder transitional epithelium: a novel strategy for identifying metaplastic lesions. *Cancer Res.* **59**, 3003–3009.
30. Celis, J. E., Rasmussen, H. H., Vorum, H., et al. (1996) Bladder squamous cell carcinomas express psoriasin and externalize it to the urine. *J. Urol.* **155**, 2105–2112.
31. Alaiya, A. A., Franzen, B., Auer, G., and Linder, S. (2000) Cancer proteomics: from identification of novel markers to creation of artifical learning models for tumor classification. *Electrophoresis* **21**, 1210–1217.
32. Ornstein, D. K., Gillespie, J. W., Paweletz, C. P., et al. (2000) Proteomic analysis of laser capture microdissected human prostate cancer and in vitro prostate cell lines. *Electrophoresis* **21**, 2235–2242.
33. Emmert-Buck, M. R., Gillespie, J. W., Paweletz, C. P., et al. (2000) An approach to proteomic analysis of human tumors. *Mol. Carcinog.* **27**, 158–165.
34. Paweletz, C. P., Ornstein, D. K., Roth, M. J., et al. (2000) Loss of annexin 1 correlates with early onset of tumorigenesis in esophageal and prostate carcinoma. *Cancer Res.* **60**, 6293–6297.
35. Vercoutter-Edouart, A. S., Lemoine, J., Le Bourhis, X., et al. (2001) Proteomic analysis reveals that 14-3-3sigma is down-regulated in human breast cancer cells. *Cancer Res.* **61**, 76–80.
36. Araki, N., Morimasa, T., Sakai, T., et al. (2000) Comparative analysis of brain proteins from p53-deficient mice by two- dimensional electrophoresis. *Electrophoresis* **21**, 1880–1889.
37. Minowa, T., Ohtsuka, S., Sasai, H., and Kamada, M. (2000) Proteomic analysis of the small intestine and colon epithelia of adenomatous polyposis coli gene-mutant mice by two-dimensional gel electrophoresis. *Electrophoresis* **21**, 1782–1786.
38. Cole, A. R., Ji, H., and Simpson, R. J. (2000) Proteomic analysis of colonic crypts from normal, multiple intestinal neoplasia and p53-null mice: a comparison with colonic polyps. *Electrophoresis* **21**, 1772–1781.
39. Merchant, M. and Weinberger, S. R. (2000) Recent advancements in surface-enhanced laser desorption/ionization–time of flight-mass spectrometry. *Electrophoresis* **21**, 1164–1177.
40. Paweletz, C. P., Liotta, L. A., and Petricoin, E. F., 3rd. (2001) New technologies for biomarker analysis of prostate cancer progression: Laser capture microdissection and tissue proteomics. *Urology* **57**, 160–163.
41. Paweletz, C. P., Gillespie, J. W., Ornstein, D. K., et al. (2000) Rapid protein display profiling of cancer progression directly from human tissue using protein biochip. *Drug Develop. Res.* **49**, 34–42.
42. Vlahou, A., Schellhammer, P. F., Mendrinos, S., et al. (2001) Development of a novel proteomic approach for the detection of transitional cell carcinoma of the bladder in urine. *Am. J. Pathol.* **158**, 1491–1502.
43. Kononen, J., Bubendorf, L., Kallioniemi, A., et al. (1998) Tissue microarrays for high-throughput molecular profiling of tumor specimens. *Nat. Med.* **4**, 844–847.

44. Schraml, P., Kononen, J., Bubendorf, L., et al. (1999) Tissue microarrays for gene amplification surveys in many different tumor types. *Clin. Cancer Res.* **5**, 1966–1975.
45. Barlund, M., Forozan, F., Kononen, J., et al. (2000) Detecting activation of ribosomal protein S6 kinase by complementary DNA and tissue microarray analysis. *J. Natl. Cancer Inst.* **92**, 1252–1259.
46. Bubendorf, L., Kononen, J., Koivisto, P., et al. (1999) Survey of gene amplifications during prostate cancer progression by high-throughout fluorescence in situ hybridization on tissue microarrays *Cancer Res.* **59**, 803–806.
47. Richter, J., Wagner, U., Kononen, J., et al. (2000) High-throughput tissue microarray analysis of cyclin E gene amplification and overexpression in urinary bladder cancer. *Am. J. Pathol.* **157**, 787–794.
48. Bowen, C., Bubendorf, L., Voeller, H. J., et al. (2000) Loss of NKX3.1 expression in human prostate cancers correlates with tumor progression. *Cancer Res.* **60**, 6111–6115.
49. Cahill, D. J. (2001) Protein and antibody arrays and their medical applications. *J. Immunol. Meth.* **250**, 81–91.
50. Kodadek, T. (2001) Protein microarrays: prospects and problems. *Chem. Biol.* **8**, 105–115.
51. Emili, A. Q. and Cagney, G. (2000) Large-scale functional analysis using peptide or protein arrays. *Nat. Biotechnol.* **18**, 393–397.
52. MacBeath, G. and Schreiber, S. L. (2000) Printing proteins as microarrays for high-throughput function determination. *Science* **289**, 1760–1763.
53. Lueking, A., Horn, M., Eickhoff, H., Bussow, K., Lehrach, H., and Walter, G. (1999) Protein microarrays for gene expression and antibody screening. *Anal. Biochem.* **270**, 103–111.
54. Bichsel, V. E., Liotta, L. A., and Petricoin, E. F., 3rd. (2001) Cancer proteomics: from biomarker discovery to signal pathway profiling. *Cancer J.* **7**, 69–78.
55. Paweletz, C. P., Charboneau, L., Bichsel, V. E., et al. (2001) Reverse phase protein microarrays which capture disease progression show activation of pro-survival pathways at the cancer invasion front. *Oncogene* **20**, 1981–1989.
56. Fields, S. and Song, O. (1989) A novel genetic system to detect protein-protein interactions. *Nature* **340**, 245–246.
57. Li, B. and Fields, S. (1993) Identification of mutations in p53 that affect its binding to SV40 large T antigen by using the yeast two-hybrid system. *FASEB J.* **7**, 957–963.
58. Iwabuchi, K., Li, B., Bartel, P., and Fields, S. (1993) Use of the two-hybrid system to identify the domain of p53 involved in oligomerization. *Oncogene* **8**, 1693–1696.
59. Kasof, G. M., Goyal, L., and White, E. (1999) Btf, a novel death-promoting transcriptional repressor that interacts with Bcl-2-related proteins. *Mol. Cell. Biol.* **19**, 4390–4404.
60. Gronholm, M., Sainio, M., Zhao, F., Heiska, L., Vaheri, A., and Carpen, O. (1999) Homotypic and heterotypic interaction of the neurofibromatosis 2 tumor suppressor protein merlin and the ERM protein ezrin. *J. Cell Sci.* **112**, 895–904.
61. Sharan, S. K. and Bradley, A. (1998) Functional characterization of BRCA1 and BRCA2: clues from their interacting proteins. *J. Mammary Gland Biol. Neoplasia* **3**, 413–421.
62. Marston, N. J., Richards, W. J., Hughes, D., Bertwistle, D., Marshall, C. J., and Ashworth, A. (1999) Interaction between the product of the breast cancer susceptibility gene BRCA2 and DSS1, a protein functionally conserved from yeast to mammals. *Mol. Cell. Biol.* **19**, 4633–4642.
63. Little, N. A., Hastie, N. D., and Davies, R. C. (2000) Identification of WTAP, a novel Wilms' tumour 1-associating protein. *Hum. Mol. Genet.* **9**, 2231–2239.
64. Wu, X., Hepner, K., Castelino-Prabhu, S., et al. (2000) Evidence for regulation of the PTEN tumor suppressor by a membrane- localized multi-PDZ domain containing scaffold protein MAGI-2. *Proc. Natl. Acad. Sci. USA* **97**, 4233–4238.
65. Durfee, T., Mancini, M. A., Jones, D., Elledge, S. J., and Lee, W. H. (1994) The amino-terminal region of the retinoblastoma gene product binds a novel nuclear matrix protein that co-localizes to centers for RNA processing. *J. Cell Biol.* **127**, 609–622.

66. Yang, R., Gombart, A. F., Serrano, M., and Koeffler, H. P. (1995) Mutational effects on the p16INK4a tumor suppressor protein. *Cancer Res.* **55**, 2503–2506.
67. Su, L. K., Burrell, M., Hill, D. E., et al. (1995) APC binds to the novel protein EB1. *Cancer Res.* **55**, 2972–2977.
68. Persson, B. (2000) Bioinformatics in protein analysis. *EXS* **88**, 215–231.
69. Celis, J. E., Gromov, P., Ostergaard, M., et al. (1996) Human 2-D PAGE databases for proteome analysis in health and disease: http://biobase.dk/cgi-bin/celis. *FEBS Lett.* **398**, 129–134.
70. O'Donovan, C., Apweiler, R., and Bairoch, A. (2001) The human proteomics initiative (HPI). *Trends Biotechnol.* **19**, 178–181.

24

Representational Difference Analysis of Gene Expression

James M. Bugni and Norman R. Drinkwater

1. Introduction

The technique of representational difference analysis (RDA) allows the selective amplification of DNA fragments that differ greatly in abundance between two samples. The method was originally developed by Lisitsyn and co-workers for detecting differences between complex genomes ***(1)***, and has been used to identify genomic deletions (including putative tumor suppressor genes), genomic amplifications, genetic polymorphisms, and viral insertions ***(2)***. RDA was later adapted for application to cDNA by Hubank and Schatz ***(3)***, and like other techniques for the comparative analysis of gene expression, including microarray hybridization ***(4)***, differential display ***(5)***, and serial analysis of gene expression ***(6)***, RDA has been used to identify genes deregulated in cancers and cancer cell lines ***(7–9)***. Some major advantages of RDA relative to other approaches include the potential for identifying novel or rare transcripts that differ greatly in expression between two samples.

The principle underlying RDA is depicted in **Fig. 1.** In order to identify transcripts that are significantly more abundant in a test sample relative to a control sample, cDNA is prepared from each, digested with restriction endonuclease to generate small fragments, and the fragments are ligated to adapter oligonucleotides. Primers complementary to the adapters are used to generate representations of the cDNA populations by polymerase chain reaction (PCR). The adapters are removed from both samples by restriction enzyme digestion and a new set of oligonucleotide adapters is ligated to the product from the test sample (designated Tester). An aliquot of the Tester is mixed with a large excess of the control sample (designated Driver), and the duplexes are melted and allowed to reanneal. These samples are subjected to PCR using primers complementary to the Tester adapters allowing the exponential amplification of only those cDNA fragments that are unique to (or highly abundant in) the Tester relative to the Driver. After removing the adapters, this product may be subjected to further rounds of RDA until a discrete set of fragments derived from cDNAs differentially expressed in the test sample are obtained. The roles of the cDNA representations from the test and control samples may be reversed in order to isolate cDNA fragments that are much less abundant in the test sample than the control.

From: *Methods in Molecular Biology, Vol. 222: Tumor Suppressor Genes: Pathways and Isolation Strategies*
Edited by: Wafik S. El-Deiry © Humana Press Inc., Totowa, NJ

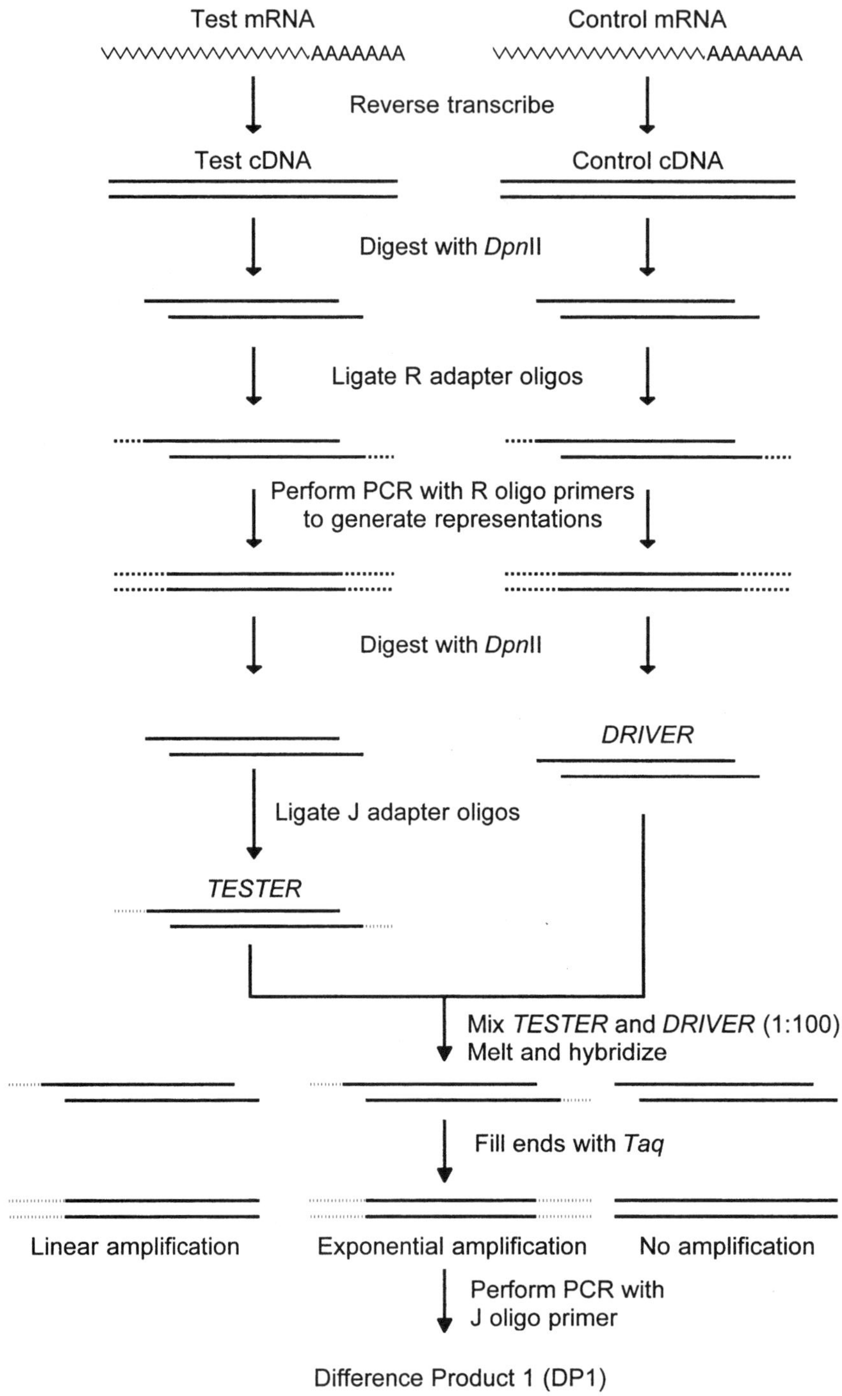

Fig 1. Schematic diagram of RDA procedure. The protocol for RDA of cDNA described by Hubank and Schatz *(3)* is diagrammed. See text for further details of the method.

This chapter describes the cDNA RDA protocol developed by Hubank and Schatz ***(3)*** with minor modifications as noted below. Furthermore, as differences in the quality of starting populations of cDNA can increase the rate of false positives, a protocol for the preparation and monitoring of cDNA synthesis is given, in addition to a rapid, comprehensive method for the analysis of difference products.

2. Materials

1. SuperscriptII reverse transcriptase (Gibco BRL, Bethesda, MD).
2. RNasin (Promega, Madison, WI).
3. Restriction enzymes (*Dpn*II, *Hae*III, *Hinf*I, *Mbo*I), mung bean nuclease (MBN), Taq polymerase, *Escherichia coli* DNA polymerase I and its Klenow fragment, T4 DNA ligase, *E. coli* DNA ligase.
4. Oligo dT15.
5. RDA adapters and PCR oligos:
   a. R-Bgl-12 5′-GATCTGCGGTGA-3′
   b. R-Bgl-24 5′-AGCACTCTCCAGCCTCTCACCGCA-3′
   c. J-Bgl-12 5′-GATCTGTTCATG-3′
   d. J-Bgl-24 5′-ACCGACGTCGACTATCCATGAACA-3′
   e. N-Bgl-12 5′-GATCTTCCCTCG-3′
   f. N-Bgl-24 5′-AGGCAACTGTGCTATCCGAGGGAA-3′

   Dissolve 24-mers to 2-µg/µL stocks and 12-mers to 1-µg/µL stocks in water and store at –20°C (*see* **Note 1**).
6. Universal primers: T7 promoter primer, T3 promoter primer, KS primer (Stratagene, La Jolla, CA).
7. pBluescript (Stratagene; La Jolla, CA).
8. RNeasy, QiaexII gel purification kit, and Oligotex beads (Qiagen, Valencia, CA).
9. Dithiothreitol (DTT).
10. Tris-buffered phenol (pH 7.5), and chloroform. Also prepare stocks of phenol:chloroform (1:1) (P:C), and phenol:chloroform:isoamyl alcohol (25:24:1) (P:C:IAA).
11. Trichloroacetic acid (TCA).
12. Buffers for enzymatic reactions:
    a. Second-strand synthesis buffer (10×): 250 m*M* Tris, pH 7.5, 250 m*M* KCl, 200 m*M* $(NH_4)_2SO_4$, 35 m*M* $MgCl_2$.
    b. T4 DNA ligase buffer (10×): 500 m*M* Tris (pH 7.8), 100 m*M* $MgCl_2$, 100 m*M* DTT, 10 m*M* ATP. Store frozen in 10-µL aliquots to avoid repeated freezing and thawing and loss of ATP.
    c. PCR reaction buffer (5×): 335 m*M* Tris-HCl, pH 8.8, 20 m*M* $MgCl_2$, 80 m*M* $(NH_4)_2SO_4$, and 166-µg/mL bovine serum albumin (BSA) ***(1)***.
    d. EE×3 buffer: EPPS 30 m*M* EPPS (N-[2-hydroxyethyl]piperazine-N′-3-propanesulfonic acid), pH 8.0; 3 m*M* EDTA.
    e. *Dpn*II buffer (10×): 1 *M* NaCl, 0.5 *M* Bis Tris-HCl (pH 6.0), 100 m*M* $MgCl_2$, 10 m*M* DTT.
    f. Klenow buffer (10×): 100 m*M* Tris-HCl (pH 7.5), 50 m*M* $MgCl_2$, 75 m*M* DTT.
    g. MBN buffer (10×): 500 m*M* sodium acetate (pH 5.0), 300 m*M* NaCl, 10 m*M* $ZnSO_4$.
    h. Restriction buffer (10×): 100 m*M* Tris-HCl (pH 7.9), 100 m*M* $MgCl_2$, 500 m*M* NaCl, 10 m*M* DTT.
    i. Colony PCR buffer (10×): 100 m*M* Tris-HCl (pH 8.3), 15 m*M* $MgCl_2$, 500 m*M* KCl.
13. dNTPs (4 m*M*).
14. Yeast tRNA (dissolve to 5 mg/mL in TE and store at –20°C).

15. Glycogen. Dissolve to 1 mg/mL in water and refrigerate.
16. [α-^{32}P]dCTP (10 mCi/mL).
17. Sephadex G50 microspin columns (AP Biotech, Piscataway, NJ).

3. Methods

Subheading 3.1. describes an effective method for cDNA synthesis and analysis. **Subheading 3.2.** describes the protocol for RDA, in which iterative cycles of hybridization and PCR are performed to amplify cDNA fragments representing differentially expressed genes. Finally, **Subheading 3.3.** describes a rapid method for screening difference products.

3.1. Synthesis and Analysis of cDNA

Prepare total RNA using RNeasy, then purify mRNA by poly A+ selection using oligotex beads, both by the manufacturer's (Qiagen) instructions (*see* **Note 2**).

3.1.1. cDNA Synthesis

1. Add the following reagents to a 0-5 mL tube on ice:
   a. 5 µg mRNA in 15 µL water.
   b. 5 µL 10 mM DTT.
   c. 3 µL RNasin (3 units; dilute 1 µL of the 10-units/µL stock 10-fold in water on ice).
   d. 10 µL 5× SuperscriptII buffer (provided by the manufacturer).
   e. 2.5 µL oligo dT15 (1.25 µg).
2. Bring the reaction up to 36 µL with water. In a thermal cycler, heat to 70°C for 5 min, and cool to 25°C with a 20 min ramp time. Then add:
   a. 3 µL RNasin (3 U).
   b. 5 µL BSA (0.5 µg).
   c. 5 µL 10 m*M* dNTPs.
   d. 2 µL (200 U) SuperscriptII reverse transcriptase.
   Incubate at 42°C for 1 h.
3. Add the following reagents to the first-strand synthesis reaction:
   a. 26 µL water.
   b. 10 µL 10× second-strand synthesis buffer.
   c. 2.5 µL *E. coli* DNA pol I (25 U).
   d. 0.5 µL *E. coli* DNA ligase (5 U).
   e. 1 µL RNase H (3 U).
   f. 10 µL 0.1 *M* DTT.
   Incubate at 16°C for 3 h.

3.1.2. Radioactive Tracer Reactions to Monitor cDNA Synthesis

Because the technique of RDA is sensitive to differences between samples in the size distribution of cDNA, prepare radioactive tracer reactions to monitor the quality of the cDNA.

1. Remove 10% of the cDNA synthesis reactions (may be done for both the first- and second-strand reactions, but need only be performed for second-strand synthesis) to a new tube and add 0.5 µL [α-^{32}P]dCTP (1 mCi/mL, diluted 10-fold from the original stock). Because evaporation is a concern with small volumes, add mineral oil to the tubes, and incubate in parallel with first- and second-strand syntheses.

2. Dilute tracer reactions to 50 µL with TE and purify on Sephadex G50 spin columns. Use 10 µL for subsequent electrophoresis.
3. Label λ*Hind* III markers by adding to a 0.5-mL tube:
   a. 18 µL water.
   b. 1 µL λ *Hind* III (1 µg/ µL).
   c. 1 µL [α-^{32}P]dCTP (1 mCi/mL).
   d. 2.5 µL 10× Klenow buffer.
   e. 1 µL each of 1 m*M* dATP and 1 m*M* dGTP.
   f. 1 µL Klenow fragment (5 U).

   Incubate for 1 h at room temperature, then purify on a Sephadex G50 spin column. Dilute 100-fold and use 10 µL for subsequent electrophoresis.
4. Prepare a 1.4% agarose gel in 50 m*M* NaCl, 1 m*M* EDTA, and equilibrate in an alkaline electrophoresis buffer (30 m*M* NaOH, 1 m*M* EDTA). Add 2× alkaline loading dye (20 m*M* NaOH, 20% glycerol, 0.025% bromphenol blue) to samples and electrophorese at 5 V/cm until the dye has migrated about 7 cm. Neutralize the gel by soaking in at least 5 gel volumes 5% TCA for 30 min. Place the entire gel in a tray on plastic wrap and cover with a wetted piece of nylon membrane and two wetted pieces of filter paper (Whatmann 3MM). Also cover with a stack of paper towels and a weight, then allow the gel to dry overnight. Remove the paper towels, wrap the dried gel and membrane in plastic wrap, and expose to a phosphorimager screen for approximately 1 h. Use ImageQuant (Molecular Dynamics, Sunnyvale, CA) or equivalent software to compare the relative sizes and amounts of cDNA synthesized in the two samples to be subtracted (**Fig. 2**) (*see* **Note 3**).

3.2. Representational Difference Analysis

This section describes the selective amplification of fragments present in greater abundance in the Tester sample than in the Driver (**Fig. 1**).

3.2.1. Digest cDNA and Add R-Adapters

1. Digest approximately 2 µg cDNA with 3 µL *Dpn*II (50,000 U/mL) and 10 µL 10× *Dpn*II buffer in 100 µL at 37°C for 2 h. Dilute to 200 µL with TE.
2. Extract twice with equal volumes phenol:chloroform:isoamyl alcohol (P:C:IAA) and once with chloroform:isoamyl alcohol (C:IAA). Add 2 µg glycogen carrier, and precipitate with 50 µL 10 *M* NH_4OAc and 650 µL ethanol on ice for 20 min. Centrifuge at full speed for 15 min. (Perform all centrifugations in a refrigerated centrifuge.) Wash the pellet with 70% ethanol, air-dry, and resuspend in 20 µL TE (*see* **Note 4**).
3. Ligate R-Bgl adapters to the cut cDNA by combining:
   a. 12 µL digested cDNA (1.2 µg).
   b. 4 µL R-Bgl-24 (2 µg/µL).
   c. 4 µL R-Bgl-12 (1 µg/µL).
   d. 6 µL 10× T4 DNA ligase buffer.
   e. 31 µL water.

   Heat to 50°C and cool to 10°C with a 40-min ramp time. Then add 3 µL T4 DNA ligase (1200 U) and incubate overnight at 15°C. Dilute the ligations to 200 µL with TE.

   In this and all subsequent ligation reactions, the 12-mer has 8 bases of complementarity to the 24-mer, producing a 4-base, 5′ overhang that serves as a "sticky end" for the ligation. Because the oligos are not phosphorylated, only the 24-mer is ligated to the end of the cDNA fragments, and the 12-mer is melted away in the subsequent reaction.

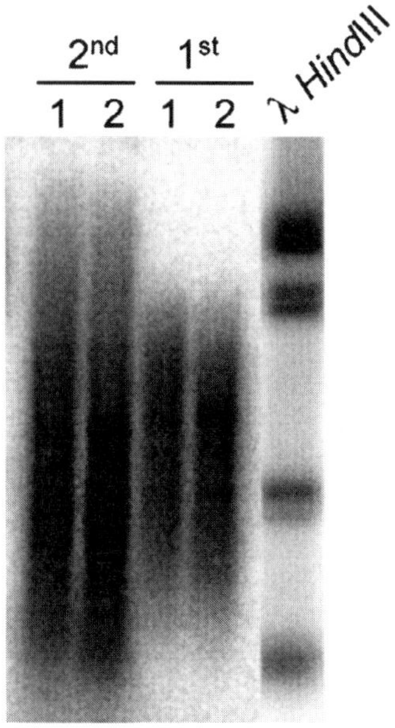

Fig. 2. Gel analysis of tracer reactions to monitor cDNA synthesis. First- and second-strand synthesis reactions were monitored as described (**Subheading 3.1.2.**) for samples to be subtracted against one another. Note that the cDNA size range and overall intensity of the products is similar for both samples.

3.2.2. Generating Representations

1. Prepare trial amplifications by using various amounts of the diluted ligation as template. To a series of 0.5-mL thin-walled PCR tubes add the following:
   a. 0.2–5 µL R-ligated cDNA (For template volumes less than 2 µL it is convenient to perform dilutions, then add 2–5 µL).
   b. 40 µL 5× PCR reaction buffer.
   c. 17 µL dNTPs (4 m*M* each).
   d. 2 µL R-Bgl-24 primer.
   e. 70 µL water.

   Combine all reagents (minus the template) in a "master mix" for the number of reactions plus one (*see* **Note 5**). Add the appropriate amount of cocktail to each tube and then add the diluted R-ligated cDNA. Overlay each tube with mineral oil (not necessary when using a thermal cycler equipped with a heated lid).
2. Heat the samples to 72°C for 3 min to melt away the 12-mer. Then add (also in master mix format):
   a. 1 µL Taq polymerase.
   b. 68 µL water.
3. Incubate for 20 cycles of 5 min at 72°C, 1 min at 95°C, and 3 min at 72°C, then heat to 72°C for 10 min.
4. Cool to 4°C, and electrophorese 5 µL of each reaction on a 1.8% agarose gel. The representation should contain a smear of amplicons ranging in size from 150 bp to 1 kb.
5. Scale up the synthesis of representations using the amount of template that provides the highest product-to-template ratio (*see* **Note 6**). Combine the products from 4 tubes into a 1.5-mL microcentrifuge tube. It does not matter if a small amount of mineral oil is carried over, as it will be extracted in the next step. Extract each tube twice with 700 µL P:C:IAA and once with C:IAA. Precipitate on ice with 75 µL 3*M* NaOAc (pH 5.3) and 800 µL iso-

propanol for 20 min. Centrifuge at full speed for 15 min, wash the pellet with 70% ethanol, and resuspend the representation in 150 μL (to a final concentration of 0.5 μg/μL). Combine representations from the same cDNA into a single tube (*see* **Note 7**).

3.2.3. Preparation of Driver

Digest approximately 300 μg of each representation by combining 600 μL of the representation with 70 μL 10× *Dpn*II buffer and 15 μL *Dpn*II (50,000 U/mL). Incubate at 37°C for at least 4 h. Extract twice with P:C:IAA and once with C:IAA. Precipitate with 70 μL 3 *M* NaOAc (pH 5.3) and 700 μL isopropanol on ice for 20 min. Centrifuge at full speed 15 min, wash with 70% ethanol, and resuspend in 500 μL water. Quantify by spectrophotometry and adjust the concentration to 0.5 μg/μL. This digested product is the Driver.

3.2.4. Preparation of Tester

Tester is prepared by gel purifying the digested representation away from the adapters and adding new (J-oligo) adapters.

1. Prepare a 1.2% agarose gel and load 20 μg of the digested representation in a total of 100 μL (diluted with TE and containing 1× loading dye). Run the gel until the bromphenol blue has migrated approximately 2 cm. Excise the portion of the gel containing the amplicons and purify using Qiaex beads (Qiagen, Valencia, CA) (*see* **Note 8**), and quantify by spectrophotometry.
2. Ligate the J-oligos to the gel-purified representation by performing **step 3** of **Subheading 3.2.1.** using 4 μL gel-purified representation (2 μg) and 39 μL water. Perform the ligation at 14°C. Following the ligation, dilute to 10 ng/μL by adding 120 μL TE. The J-ligated representation is the Tester.

3.2.5. Prepare Tester and Driver Hybridization

1. To a 0.5-mL tube add:
   a. 80 μL Driver (40 μg).
   b. 40 μL of the J-ligated Tester (0.4 μg).
2. Extract once with 120 μL phenol and once with chloroform. Precipitate by adding 30 μL 10 *M* NH_4OAc and 380 μL ethanol, and cool to –70°C for 10 min. Follow this precipitation by warming to 37°C for 1 min to prevent salt precipitation. Centrifuge at full speed for 15 min and wash the pellet twice with 70% ethanol. Dry the pellet and resuspend in 4 μL EE×3 buffer by pipetting up and down at least 20 times, warming to 37°C for 5 min, and further pipetting (at least 10 times).
3. Overlay the Tester:Driver mixture with 35 μL mineral oil and denature at 98°C for 5 min. Cool to 67°C and immediately add 1 μL 5 *M* NaCl directly into the aqueous phase. Incubate at 67°C for 20 h.
4. Remove most of the mineral oil. Dilute the Tester:Driver hybrids stepwise by adding 8 μL TE containing 5 μg/μL yeast tRNA (pipet at least 10 times), 25 μL TE (pipet at least 10 times), and 362 μL TE (mix by vortexing).

3.2.6. Generating the First Difference Product (DP1)

1. Prepare 4 independent PCR reactions for each subtraction by adding to each tube:
   a. 60 μL water.
   b. 40 μL 5× PCR buffer.
   c. 17 μL 4 m*M* dNTPs.
   d. 20 μL diluted hybridization reaction.

2. Overlay with mineral oil and incubate at 72°C for 3 min then add:
   a. 30 μL water.
   b. 1 μL Taq polymerase.
3. Incubate another 5 min at 72°C and add:
   a. 30 μL water.
   b. 2 μL J-Bgl-24.
   Taq polymerase will fill in the ends on hybrids containing Tester molecules.
4. Incubate in a thermal cycler for 10 cycles at 95°C for 1 min and 70°C for 3 min, then heat to 72°C for 10 min.
5. Combine the products of the four reactions in a 1.5-mL tube. Extract twice with P:C:IAA and once with C:IAA. Precipitate by adding 2 μg glycogen, 75 μL 3 *M* NaOAc, and 800 μL isopropanol on ice for 20 min. Centrifuge at full speed for 15 min and wash the pellet with 70% ethanol. Resuspend the pellet in 40 μL 0.2× TE.
6. Digest the PCR product with mung bean nuclease to remove linearly amplified products by adding to a 0.5-mL tube:
   a. 20 μL amplified DNA.
   b. 4 μL 10× MBN buffer.
   c. 14 μL water.
   d. 2 μL MBN (10 U/μL).
   Incubate at 30°C for 35 min. Stop the reaction by adding 160 μL 50 m*M* Tris HCl, pH 8.9, and heat inactivate the nuclease at 98°C for 5 min (*see* **Note 9**).
7. Perform the second round of amplification, again with 4 tubes per subtraction by adding to new tubes:
   a. 20 μL MBN-treated DNA.
   b. 80 μL water.
   c. 40 μL 5× PCR buffer.
   d. 17 μL 4 m*M* dNTPs.
   e. 1 μL J-Bgl-24 (1 μg/μL).
8. Incubate at 95°C for 1 min. Cool to 80°C and add:
   a. 40 μL water.
   b. 1 μL Taq polymerase.
9. Incubate in thermal cycler for 18 cycles at 95°C for 1 min and 70°C for 3 min, then heat to 72°C for 10 min.
10. Combine the products from the 4 tubes and electrophorese 5 μL of each subtraction on a 1.8% agarose gel to visualize the bands. The first difference product will consist of a diffuse mixture of amplicons in the 150 bp-to-1 kb range and does not appear appreciably different from the starting representation (**Fig. 3**).
11. Extract twice with P:C:IAA, and once with C:IAA. Precipitate with 75 μL 3 *M* NaOAc and 800 μL isopropanol on ice for 20 min. Centrifuge at full speed 15 min and wash with 70% ethanol. Resuspend the pellet in 100 μL TE (approximately 0.5 μg/μL). This product is the first difference product (DP1).

3.2.7. Generating DP2

1. To change the oligo adapters from the J-oligos to the N-oligos, add to a 0.5-mL tube:
   a. 40 μL (DP1), approximately 20 μg.
   b. 15 μL 10× *Dpn*II buffer.
   c. 3 μL *Dpn*II (150 U).
   d. 92 μL water.

Incubate at 37°C for 4 h. Extract twice with P:C:IAA and once with C:IAA. Precipitate with 33 μL 3 *M* NaOAC, and 800 μL ethanol at –20°C for 20 min. Centrifuge at full speed for 15 min, wash with 70% ethanol, dry the pellet, and resuspend it in 40 μL TE. Quantify the product by spectrophotometry and adjust the concentration to 0.5 μg/μL with TE.

2. Dilute 5 μL of the digested DP1 10-fold with TE (to a final concentration of 50 ng/μL). Ligate N-oligos by performing **step 2** of **Subheading 3.2.4.** using 4 μL digested DP1. Dilute the ligation product to a concentration of 1.25 ng/μL with TE.
3. To generate the second difference product, mix 40 μL (50 ng) of the N-ligated DP1 with 80 μL Driver (40 μg). Perform the same reactions that were used to generate DP1 (i.e., all steps in **Subheading 3.2.6.** with the only difference being the use of 72°C rather than 70°C for the extension reactions in **steps 1** and **3**). Electrophorese 5 μL of each subtraction. In the second difference product, one should expect distinct bands with a less evident background smear (**Fig. 3**).

3.2.8. Generating DP3

To generate the third difference product, digest the N-ligated products of DP2 and ligate the J-oligos. Dilute the J-ligated DP2 to 1 ng/μL and dilute 10 μL of this solution 100-fold in TE containing 30-ng/μL yeast tRNA (final DP2 concentration = 10 pg/μL). Perform the hybridization with 100 pg (10 μL) of the J-ligated DP2 and 80 μL (40 μg) of Driver. Generate DP3 by performing all steps of **Subheading 3.2.6.**, using 22 cycles of amplification for **step 3** rather than 18. Electrophorese an aliquot of the products on an agarose gel. At this stage, distinct bands should be visible with little or no background smear (**Fig. 3**). These bands can be gel purified and cloned (*see* **Note 10**).

3.3. Cloning and Characterizing the Difference Products

3.3.1. Gel Purification and Cloning Fragments (see **Note 11***)*

It is convenient to clone the *Dpn*II-digested difference products into the *Bam*HI site of pBluescript.

1. Digest DP3 by adding to a 0.5-mL tube:
   a. 30 μL DP3 (15 μg).
   b. 57 μL water.
   c. 10 μL 10× *Dpn*II buffer.
   d. 3 μL *Dpn*II (150 U).

   Electrophorese 5 μg of the product on a 1.3% gel. Because most regions of each lane contain some product, we have found it most convenient to gel purify the products in approximately 3 or 4 size ranges from 200 up to 750 bp. Perform gel purification using QiaexII beads and elute twice in 30 μL 0.2× elution buffer.
2. Shotgun clone products purified from different size ranges (*see* **Note 12**) by combining:
   a. 3 μL (15 ng) vector (pBluescript, digested with *Bam*HI and treated with phosphatase).
   b. 5 μL gel-purified DP3.
   c. 1.5 μL 10× T4 DNA ligase buffer.
   d. 0.5 μL T4 DNA ligase.
   e. 5 μL water.

   Incubate at 16°C overnight.
3. Transform high-efficiency, chemically competent *E. coli* DH5α cells by standard methods (5 μL ligation with 100 μL cells). Grow cells on Xgal/IPTG/ampicillin plates.

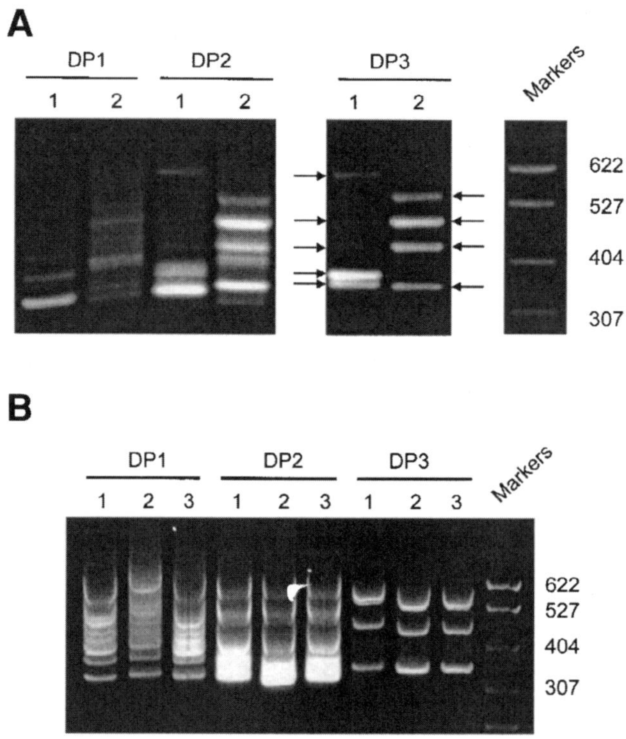

Fig. 3. Gel analysis of RDA difference products. **(A)** Difference products from a two-way subtraction to find genes upregulated (subtraction 1) and downregulated (subtraction 2) by growth hormone in the mouse liver. Note the emergence of distinct bands with successive rounds of selective PCR. Fragments indicated by the arrows in DP3 represent differentially expressed genes. **(B)** Three independent Tester samples were prepared from a single cDNA sample and subtracted against the same Driver. Note that the products in the three independent subtractions are nearly identical, demonstrating the reproducibility of the technique. Although three major bands are seen in DP3, using the method described in under **Subheading 3.3.**, six amplified fragments were identified in each subtraction.

3.3.2. Amplify the Inserts Directly from Colonies

1. Pick white colonies from the plates with a sterile toothpick and swirl each colony in 10 µL water in a PCR tube (use the same toothpick to innoculate a culture for glycerol storage).
2. Add the following reagents (in "master mix" format):
   a. 2.5 µL 10× colony PCR buffer.
   b. 1 µL T7 promoter primer (10 µ*M*).
   c. 1 µL T3 promoter primer (10 µ*M*).
   d. 2.5 µL dNTPs (1 m*M* each).
   e. 10 µL water.

   Heat the mixture to 95°C for 6 min, incubate in a thermal cycler for 35 cycles at 95°C for 35 s, 58°C 35 s, and 72°C 50 s, and heat to 72°C for 7 min.

3.3.3. Colony PCR Fingerprinting

The amplification products are digested with *Dpn*II alone to determine (a) the size of the insert and (b) whether the insert is chimeric. Products are also digested with a cocktail of restriction enzymes to "fingerprint" the clones. This approach provides a convenient method for exhaustively analyzing RDA difference products, while reducing the number of clones requiring DNA sequence analysis.

1. Digest products with *Dpn*II by adding to a fresh tube:
   a. 5 µL amplification reaction.
   b. 1.5 µL 10× *Dpn*II buffer.
   c. 0.2 µL *Dpn*II.
   d. 8.5 µL water.

   Incubate at 37°C for 1 h, and electrophorese 10 µL of the reaction on a 1.8% agarose gel. Analyze the products visually, and record insert sizes. Exclude clones that generated more than one insert.
2. Also remove 5 µL of the PCR reaction to a new tube and add:
   a. 1.5 µL 10× restriction buffer.
   b. 0.2 µL of the following enzymes: *Hin*fI, *Hae*III, and *Mbo*I (*Mbo*I is an isoschizomer of *Dpn*II that has higher activity in this restriction buffer).
   c. 8 µL water.

   Digest at 37°C for 1 h. Electrophorese the products on a 7% polyacrylamide gel (5 V/cm for 4 h). Multiple, differentially expressed genes can produce cDNA fragements of the same size. However, they will not yield the same pattern of fragments following digestion with multiple restriction enzymes. These patterns are easily scored visually to determine which clones are from the same cDNA fragment.
3. Choose two amplified fragments from each fingerprint that was represented more than once among the clones. Prepare the samples using a PCR purification kit (Qiagen) (elute with 50 mL 0.2× elution buffer) and determine the DNA sequence by standard methods using the nested KS promoter primer. Northern analysis, differential hybridization, and quantitative PCR are among the techniques that can be used to verify differential expression of the genes identified. Furthermore, the purified fragments amplified directly from colonies can be used to supplement the Driver sample in subsequent RDA experiments to prevent their amplification and allow the detection of additional, differentially expressed genes ***(3,10)***.

3.4. Perspectives

We have performed RDA, EST sequencing, and hybridization to microarrays to analyze differential gene expression between the livers of wild-type and growth hormone-deficient mice (Bugni et al., in preparation). Using RDA, we identified 6 genes (including 3 novel genes) that were not identified by the other methods and that were previously not known to be growth hormone-regulated. The technique of RDA is biased toward identifying genes with large differences in expression, and we typically measured greater that 10-fold differences in expression for the genes we found by this method. In summary, RDA is a relatively rapid technique that can be used to identify rare or novel transcripts that differ greatly in abundance and can complement other techniques for assessing differential gene expression.

4. Notes

1. Although these oligonucleotide sets yield good results, other oligos that have less self-complementarity have been used ***(11,12)***. Furthermore, HPLC purification of primers may reduce the incidence of false positives when starting with small amounts of RNA ***(11)***.
2. We have used both ultracentrifugation methods and Qiagen RNeasy to purify total RNA. Two-step purification of mRNA (total, then poly A+) is preferable to one-step purification

from cell lysates, as small amounts of DNA contamination can greatly increase the number of false positives. If one-step purification is used, DNase treatment is recommended prior to generation of the representations (**Subheading 3.2.**). We have found that poly A+ purification using Qiagen oligotex beads is useful because no ribosomal RNA bands are evident by ethidium bromide staining after a single round of selection. Ribosomal RNA is removed much less efficiently with oligo-dT cellulose. Starting with 5 μg of poly A+ RNA yields good results, but the technique of RDA has been performed with much less ***(9,11,12)***.

3. Different amounts of cDNA template are tested prior to the generation of representations (**Subheading 3.2.**). For this reason, the absolute amount of cDNA used for RDA is less important than the fact that the cDNAs from the samples to be subtracted are very similar in quality.
4. Several organic extractions and subsequent DNA precipitations are performed in the course of this protocol. Sound technique is essential. When performing phenol extractions, vortex vigorously for at least 30 s, and centrifuge at full speed for 1 min before removing the aqueous phase. After pelleting DNA precipitates, much of the DNA is not visible on the side of the tube well above the pellet. To maximize yields, resuspend the DNA by repeated pipetting down the side of the tube. This procedure is especially important in the steps of **Subheading 3.2.5.** Some investigators have replaced these steps with the use of gel-filtration spin columns ***(12)***.
5. We have found it convenient to add reagents, including Taq polymerase, in master mix format. Take caution to avoid cross-contamination of samples.
6. We have consistently used 1- or 2-μl templates for generating representations, and we have found that 24 independent reactions per sample is convenient and sufficient to produce enough representation for the entire subtraction.
7. At this point it is not necessary to quantify precisely the representation prior to digestion. However, it is imperative to quantify by spectrophotometry all nucleic acid samples that are used for subtraction. This requirement applies to the digested representations, the Tester, and each difference product (after digestion) that is mixed with the Driver.
8. When preparing the preparative gel, the wells in which the representations are loaded should be at least 1 cm long, to avoid overloading and smearing of the bands. If a minigel comb of this type is not available, one can tape together teeth from a standard comb. This latter technique also allows markers to be loaded in standard-sized wells. Also, ensure that there is sufficient space (~1 cm) between lanes from different representations to avoid cross-contamination. When excising gel fragments, use a low-intensity, high-wavelength UV source to minimize damage to the representations. In fact, at this step the only concern is to gel-purify the amplicons away from the digested linkers. For this reason, one can quickly make a slice between the linkers and the amplicons. Then place a piece of aluminum foil under the lanes containing representations to prevent their continued UV exposure, and use a ruler and the markers as a guide for making additional slices in the gel. Also, by using a low-energy light source, some products in the representation may not be seen. However, it is not problematic to gel-purify the region of the representation up to ~1.5 kb. Although the Qiagen spin columns for gel purification are more convenient than the QiaexII beads, we recommend the beads because they provide a higher yield of product.
9. Modifications of the procedure have eliminated MBN digestion ***(9,12)***.
10. Ratios of Tester to Driver or difference products to Driver can be varied empirically to optimize RDA for different applications ***(7,10,12)***.
11. Differentially expressed genes have been identified by screening DP2 products ***(7,8,13–15)***, but the incidence of false positives is higher. In one experiment, we found that only 3 of 7 DP2 clones that appeared more than once (using the method described under **Subheading 3.3.**) were differentially expressed, as opposed to 6 of 6 clones screened from DP3. The lack of specificity of DP2 products representing differentially expressed genes is not problematic when the products are screened by hybridization to arrays. Early-round difference products have been used as probes ***(13,14)*** or as targets ***(7,15)*** for array analysis. With the latter method it is still possible to clone novel genes.

12. Do not shotgun clone the entire mixture of products from DP3. A shotgun method will favor the cloning of very short products. For example, in one experiment we shotgun cloned *Dpn*II-digested difference products without gel purification. By gel analysis, nearly all of the visible products were greater than 200 bp in length, but after shotgun cloning, nearly all of the clones contained inserts under 150 bp in length.

Acknowledgments

We thank Andrea Bilger and Jennifer Drew for their critical comments on this manuscript. The authors' work was supported by grants CA07175, CA09135, and CA22484 from the National Cancer Institute; J.M.B. was also supported by funds from the Cremer Foundation.

References

1. Lisitsyn, N., Lisitsyn, N., and Wigler, M. (1993) Cloning the differences between two complex genomes. *Science* **259**, 946–951.
2. Lisitsyn, N. A. (1995) Representational difference analysis: finding the differences between genomes. *Trends Genet.* **11**, 303–307.
3. Hubank, M. and Schatz, D. G. (1994) Identifying differences in mRNA expression by representational difference analysis of cDNA. *Nucleic Acids Res.* **22**, 5640–5648.
4. Schena, M., Shalon, D., Davis, R. W., and Brown, P. O. (1995) Quantitative monitoring of gene expression patterns with a complementary DNA microarray. *Science* **270**, 467–470.
5. Liang, P. and Pardee, A. B. (1995) Recent advances in differential display. *Curr. Opin. Immunol.* **7**, 274–280.
6. Velculescu, V. E., Zhang, L., Vogelstein, B., and Kinzler, K. W. (1995) Serial analysis of gene expression. *Science* **270**, 484–487.
7. Ismail, R. S., Baldwin, R. L., Fang, J., et al. (2000) Differential gene expression between normal and tumor-derived ovarian epithelial cells. *Cancer Res.* **60**, 6744–6749.
8. Graveel, C. R., Jatkoe, T., Madore, S. J., Holt, A. L., and Farnham, P. J. (2001) Expression profiling and identification of novel genes in hepatocellular carcinomas. *Oncogene* **20**, 2704–2712.
9. Welford, S. M., Gregg, J., Chen, E., et al. (1998) Detection of differentially expressed genes in primary tumor tissues using representational differences analysis coupled to microarray hybridization. *Nucleic Acids Res.* **26**, 3059–3065.
10. Hubank, M. and Schatz, D. G. (1999) cDNA representational difference analysis: a sensitive and flexible method for identification of differentially expressed genes. *Meth. Enzymol.* **303**, 325–349.
11. O'Neill, M. J. and Sinclair, A. H. (1997) Isolation of rare transcripts by representational difference analysis. *Nucleic Acids Res.* **25**, 2681–2682.
12. Pastorian, K., Hawel, L., and Byus, C. V. (2000) Optimization of cDNA representational difference analysis for the identification of differentially expressed mRNAs. *Anal. Biochem.* **283**, 89–98.
13. Geng, M., Wallrapp, C., Muller-Pillasch, F, Frohme, M., Hoheisel, J. D., and Gress, T. M. (1998) Isolation of differentially expressed genes by combining representational difference analysis (RDA) and cDNA library arrays. *Biotechniques* **25**, 434–438.
14. Kim, S., Zeller, K., Dang, C. V., Sandgren, E. P., and Lee, L. A. (2001) A strategy to identify differentially expressed genes using representational difference analysis and cDNA arrays. *Anal. Biochem.* **288**, 141–148.
15. Wallrapp, C., Muller-Pillasch, F., Micha, A., et al. (1999) Strategies for the detection of disease genes in pancreatic cancer. *Ann. N.Y. Acad. Sci.* **880**, 122–146.

25

Antisense Libraries to Isolate Tumor Suppressor Genes

Adi Kimchi

1. Introduction

1.1. Apoptosis and Tumor Suppressor Genes

Apoptosis has become a subject that draws tremendous attention and research efforts in the cancer field, since it has a major impact on tumor initiation, progression, and metastasis. At various stages during the course of tumor development, cells are subjected to stressful conditions that trigger different types of programmed cell death, including the classic type of apoptosis. As a consequence, mutations leading to inhibition of apoptosis confer a selective advantage to cells. In premalignant cells, activation of oncogenes and the consequent hyperproliferation provoke a cellular response that leads to elimination of those cells by apoptosis. Subsequently, transformed cells in the tumor microenvironment are under constant selective death pressure, due to lack of oxygen (hypoxia), depletion of growth/survival factors, attacks by the immune system and often death by anoikis that results from loss of cell–matrix interactions. At later stages, when metastasizing tumor cells enter into circulation, they encounter many additional death-inducing signals, such as superoxides, nitric oxide (NO), killing cytokines, and mechanical shearing forces. Thus, all along the multistage process of tumorigenesis, induction of apoptosis functions as a tumor suppressor mechanism from which cells have to escape in order to survive (reviewed in **refs.** ***1–3***). This means that tumor cells should benefit from mutations that either inactivate proteins that positively mediate programmed cell death (PCD) or alternatively abnormally activate antiapoptotic genes.

The first classical example that established the concept that genes in the apoptotic machinery are mutated in cancer was documented with the cloning of *Bcl-2*. The gene resides at the site of the (8;14) chromosomal translocation characteristic of follicular B-cell lymphoma, a genomic rearrangement that activates its function. The Bcl-2 protein protects from cell death in cell culture systems, and promotes lymphomagenesis in transgenic mice models ***(4,5)***. The second well established example is the *p53* gene, whose pro-apoptotic functions have been thoroughly studied ever since they were first documented ***(6)***. Inactivating mutations of *p53* are frequently found in a wide range of human tumors. The inactivation of *p53*, by deletions or mutations, reduces the sensitivity of cells to apoptosis triggered by oncogene activation, hypoxia, telomere erosion, changes in cell

From: *Methods in Molecular Biology, Vol. 222: Tumor Suppressor Genes: Pathways and Isolation Strategies*
Edited by: Wafik S. El-Deiry © Humana Press Inc., Totowa, NJ

adhesion, and DNA-damaging agents, thus providing a powerful positive selection at the different stages of tumor development (reviewed in **ref.** *7*). These two well-studied examples provided the milestones for establishing the link between apoptosis and cancer. Yet, in light of the complexity of the molecular networks of apoptosis and the diversity of stress signals operating in the multistep process of tumorigenicity, it was obvious that the two first examples reflect the edge of an iceberg and that inactivation or loss of many other pro-apoptotic genes may be directly implicated in cancer development. Here we describe the rescue of a novel pro-apoptotic gene that displays tumor suppressor activity, and present the approach that was developed in order to clone and identify such genes.

1.2. Developing Function-Based Gene Cloning Selections in Cell Cultures for Trapping Novel Death-Promoting Genes

As in every field in modern molecular biology, including apoptotsis, the powerful genetic tools available in *Caenorhabditis elegans* and *Drosophila* resulted in the discovery of several critical genes ***(8,9)***. Yet, in light of the higher complexity of molecular networks in mammalian systems, it was obvious that systematic screens for mammalian orthologues of nematode and *Drosophila* genes may not be sufficient. It was therefore necessary to develop in parallel unbiased function-based genetic selections that could be performed in mammalian systems. The necessity for such screens was further sharpened in light of the second role that some of the death-promoting genes exert in mammals, i.e., their potential tumor suppressive activity discussed above. Different types of programmed cell death can be initiated and brought to completion in cells that grow in culture, suggesting that the molecular network(s) could be studied under these relatively simple in-vitro conditions. The main obstacle was the lack of available genetic tools for hitting and rescuing the relevant genes in these mammalian cell cultures.

One of the predictions made in the dissection of complex molecular networks is that the specific inactivation of a gene which controls an important "junction" within the network should weaken the biologic outcome. In the case of PCD, the inactivation should slow down the death of cells. The strategy that was developed in our laboratory, to confront the challenge of function-based gene cloning in cell cultures, was based on the three classical principles of genetic screens: (a) random inactivation of gene expression; (b) selection of the phenotype of interest; and (c) rescue of gene(s) whose inactivation was responsible for the alteration in phenotype (**Fig. 1**). The phenotypic selection was relatively simple. It consisted of selection of cells that display prolonged survival, as compared to their neighbors, due to a decreased sensitivity to the apoptotic stimulus that is continuously present in the cell culture. In other words, we used the powerful positive selection that stems from the growth advantage conferred on cells that carry the specific gene inactivation. The random gene inactivation was achieved by transfection with antisense cDNA libraries, to block gene expression instead of mutating the DNA itself. We named our strategy TKO, for Technical Knock Out, and defined it as a general methodology for function-based gene isolation in mammalian cells ***(10)***.

The first instance of system validation (i.e., the ability to isolate from a complex antisense cDNA library a rate-limiting death-protective cDNA fragment) was documented in our laboratory in 1991 ***(10)***. The subsequent comprehensive application of the TKO selection over the following years has resulted in the isolation and identification of half a dozen novel genes that control different steps in PCD (reviewed in **refs.** *11* and *12*). These novel

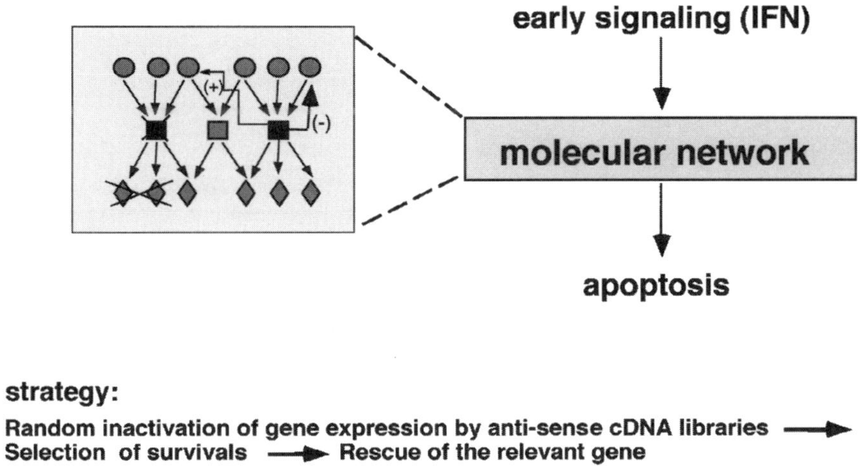

Fig. 1. Identifying the central junctions (■) within the apoptotic network by a functional approach to gene cloning.

positive mediators of PCD were named DAP genes (for Death Associated Proteins). Three of these genes—DAP-kinase, DAP-5, and DAP-3—were extensively studied over the past few years at various levels, including detailed structure/function analysis of the proteins, their intracellular localizations, and modes of action ***(13–19)***. Additionally, DAP-kinase exhibited many characteristics of a tumor suppressor gene, as will be detailed later on.

This chapter will refer to two main topics: (a) description of the technology behind the TKO selection and of the specific precautions and critical considerations that made the strategy feasible and successful; and (b) description of the experimental approaches that determined whether an isolated death-promoting gene has tumor suppressive activity. DAP-kinase will provide the proof of principle for the hypothesis, which argues that application of the TKO selection in various types of programmed cell death can be used as a functional genetic approach toward isolating novel tumor suppressor genes.

2. Technical Details on the TKO Strategy

The TKO strategy was designed with the intention of establishing a technology that will directly target genes whose function is rate limiting within a given molecular network. The type of selection pressure applied in order to yield the phenotype of interest is tailored according to the class of genes one wishes to select. To rescue genes that participate in PCD, the selection consisted of pooling survivals that resisted (at least partially) the apoptotic pressure of an external stimulus. The selection should be powerful enough to detect an event whose frequency should be around 10^{-5} to 10^{-4} (assuming that the numbers of actively expressed genes is approximately 30,000). The initial goal was to be able to perform these selections in vitro in mammalian cell cultures. In principle, the approach had to rely on two main technologies, both directed at overcoming impediments to the use of genetic methods in cultured cells. One is the use of antisense RNAs expressed from cDNA libraries as "virtual mutagens" for trapping genes based on their function. This would allow identification of genes that control central "junctions" in the molecular network underlying a certain biologic process, since a reduction in their

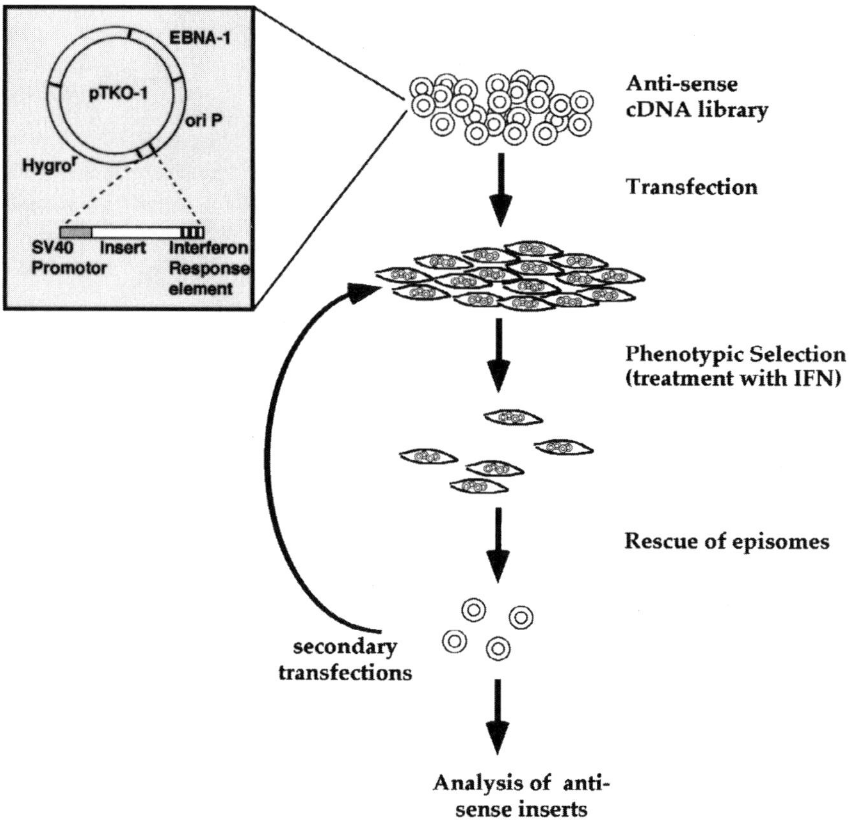

Fig. 2. Description of the technical knock-out (TKO) strategy.

expression is rate limiting and weakens significantly the biologic response. The second technology was aimed at developing efficient strategies for introduction and recovery of complex cDNA libraries, to allow multiple rounds of phenotypic selection and rapid screens for identification of the functionally relevant genes. These two goals were achieved by:

1. Preparing specific antisense cDNA libraries that should hit preferentially the relevant genes.
2. Developing an expression vector that provides an efficient gene transfer system and thus the capability to represent a complex library in a single transfection event.
3. Upgrading its efficiency by working out the best transcription cassette to drive maximal anti-sense RNA expression during the selection procedure.
4. Calibrating the selection to give minimal background of "spontaneous" resistant cells so that the ratio between functional cDNA fragments and false positive cDNAs is the highest.
5. Easy and rapid recovery approaches for the rescue of the transfected cDNAs, which also enable multiple rounds of phenotypic selection to enrich the functionally relevant elements.

The details on the different steps and of the logic behind them are discussed below under **Subheadings 2.1.–2.4.** The general outline of the TKO selection that was eventually used successfully is illustrated in the scheme in **Fig. 2**. It was based on HeLa cells that were transfected with an antisense cDNA expression library cloned into an Epstein-Barr virus (EBV)-based episomal vector (named pTKO-1). Cultures were then subjected to selection by applying interferon-γ, an efficient killing cytokine in the HeLa cell line. Surviving cells were pooled and the episomes were extracted. The rescued episomes were then subjected to second rounds of transfections by testing each individual episome for its capability to protect cells to some extent from interferon-γ killing. The cDNA inserts carried by the episomes that were positively scored, based on these criteria, were further sequenced and analyzed.

2.1. Choice of the Cell Culture System, of the Killing Cytokine, and Preparation of the Antisense cDNA Library

2.1.1. The Killing Cytokine

Interferon-γ was chosen as the preferred external trigger mainly because in some cell systems it induces a biphasic pattern of responses, i.e., proliferation arrest followed by PCD ***(13)***. Our initial hypothesis was that such interesting scenarios may provide a unique system to study within a single genetic constellation the points where the cell cycle inhibitory and apoptotic mechanisms diverge. In retrospect, all the rescued genes were part of the cell death machinery, yet the fact that death by interferon-γ displayed many characteristics of type II PCD offered an opportunity to rescue genes that underlie these less studied subcellular features of PCD. Another reason to choose this killing cytokine stemmed from the fact that much information was available at the time we started the approach on the early signaling of interferons. This information was used to efficiently increase the antisense RNA expression levels from the transfected cDNA library during the selection, as detailed below.

2.1.2. The Target Cells

Many cell lines were systematically screened. The choice was based on two main parameters: (a) cells that display high efficiency of transfection, to allow maximal representation of the library within a population whose size does not exceed 5×10^7 cells (a feasible cell number to work with); and (b) the cells must display a maximal sensitivity to interferon-γ to yield a minimal background of "spontaneous" resistant cells (between 10^{-5} and 10^{-4} survival after long-term treatment with interferon-γ). This was an important prerequisite for reducing the fraction of false positive cDNA fragments later on during the selection procedure. Eventually, the HeLa cells that fulfilled these requirements were chosen as the preferred target cells.

2.1.3. Construction of the Antisense cDNA Library

The library was prepared from RNA derived from a mixture of nontreated and interferon-γ-treated HeLa cells (at 1, 2, 4, 12, 24, and 48 h after treatment) to include both constitutive and inducible mRNAs. To this end, oligo (dt)-selected mRNA was reverse transcribed; double-stranded cDNA was generated and further ligated to specific linkers for the unidirectional cloning. The ligation to the polyadenylated end of the cDNA generated a *Hind*III site, whereas at the 5′ end of the cDNA a *Bam*HI site was generated. The presence of the *Hind*III site in the proximity of the promoter in the vector enabled cloning of cDNA fragments into the expression vector in antisense orientation. The library consisted of 10^5 independent clones.

2.2. Choice of Expression Vector to Drive the Antisense RNA Expression

We have chosen an EBV-based, self-replicating, episomal shuttle vector to express a directional antisense cDNA library (**Fig. 2**). The episomal vector had several advantages over vectors that integrate into the genome:

1. It reduced the background of nonrelevant phenotypic alterations occurring as a result of random integrations into DNA.
2. The episomes were easily rescued from the transfected cells by a simple DNA extraction procedure (the Hirt procedure), with no need for other genetic manipulations. This yielded a rapid and convenient way to perform multiple rounds of phenotypic selection and also solved the issue of plasticity of tissue culture cells, since only the individual cDNA fragments that transduced the phenotypic change in subsequent transfections were scored as real positives.
3. The episomal vector accumulated at multiple copies per cell in the stable transfectants, resulting in high expression levels, which, as detailed below, was critical for efficient reduction in protein levels.

2.3. Screening for the Best Transcription Cassette to Yield Maximal Antisense RNA Expression During the Selection Procedure

The unidirectional strategy in the antisense orientation increased the probability of acquiring "recessive mutations" due to loss of expression. At the initial planning stage, several promoters driving reporter gene expression were screened by using transient assays in Hela cells treated with interferon-γ. Many of the tested promoters (e.g., LTR) were strongly attenuated by the continuous exposure to interferon-γ and therefore were not suitable for the project. Eventually the SV40 promoter was chosen, and the promoter cassette of the vector was manipulated to confer high inducible expression levels of antisense RNAs during the phenotypic selection. This was achieved by introducing an interferon-responsive enhancer element into the vector to increase expression during the selection by the killing cytokine (**Fig. 2**; *see also* **ref. *10***). In retrospect, the latter manipulation was found to be critical for the success of the functional cloning, since expression of high antisense RNA levels during the phenotypic selection was essential to efficiently reduce the protein levels ***(13,16)***. Moreover, the chances of cloning genes belonging to the basic cell death machinery, as opposed to those that function proximal to the interferon-γ receptors, were increased by making the selection dependent on this specific enhancer element.

2.4. First Cycle of Plasmid Rescue from Survival Cells and Subsequent Rounds of Transfections to Select the Ultimate Functional Fragments

The antisense cDNA library was introduced into 5×10^7 Hela cells, followed by the double selection with interferon-γ and the hygromycin B drug carried by the pTKO-1 vector. The initial cell density and the concentration of the drugs were calibrated in pilot transfections with a control vector and were found to strongly influence the appearance of false positives. The chosen protocol started with a cell density of 1500 cells/cm^2; selection with 200–1000 U/mL interferon-γ plus 200 μg/mL hygromycin B for 28 days. During the selection period, fresh medium and drugs were supplemented twice a week and interferon was removed from the culture medium a few days before the final harvest of cells. Stable transfected colonies that survived after long-term selection with the cytokine were pooled and the episomal DNA was extracted from them by the method of Hirt ***(10)***. The episomal fraction was subjected to the restriction enzyme DpnI, which cleaves only input nonreplicated DNA. The latter step further enriched the fraction of episomal vectors that have undergone replication cycles in the HeLa cells. These were then shuttled into bacteria and further analyzed individually by retransfections into HeLa cells and examining whether as single episomes they could protect the HeLa cells from death by interferon-γ. The EBV plasmids which in several repetitive transfections were positively scored, using the above-mentioned long-term clonability assays, were then individually tested in other death assays to confirm and further characterize the nature of the death protection that they each confer to cells.

2.5. Notes Concerning the TKO Selection

The TKO strategy may be tailored to other types of phenotype selections in cell cultures for the rescue of genes that control replicative senescence, cell cycle arrest, or suppressors of tumorigenicity. However, this should be carefully designed taking into account all the above-mentioned precautions and advises elaborated for the intereron selections. Also there exists a possibility to work out these selections in vivo in mice. Obviously, other transcription enhancer elements to drive the antisense RNA expression should be used, depending on the nature of the selection procedure. Other vectors that fulfill the necessary prerequisites could be developed as well, especially if in-vivo selections will be developed. Finally, it should be mentioned that from the analysis of 8 functional fragments that we have already rescued so far, 6 turned out to be short antisense fragments that correspond to internal regions within the cDNA (250–400 bp in length). They corresponded to internal *Hind*III–*Bam*HI regions in the genes rather than extending from the poly-A tail as originally planned. The enrichment of such fragments in the selection means that these are the most efficient elements. Therefore it is recommended that in the future the design of novel antisense cDNA libraries should consist of short internal fragments .

2.6. The Long Journey from the Isolation of a Functional Antisense cDNA Fragment to the Notion that a Critical Death-Promoting Gene Has Been Rescued—Focusing on DAP-Kinase as an Example

The rescue of an antisense cDNA fragment that to some extent protects cells from death is the beginning of long-term research aimed at establishing the exact role of that particular gene within the apoptotic network. We will briefly summarize this multistep procedure, and illustrate its outcome in one well-studied example, DAP-kinase. A general outline illustrating how one proceeds from the starting point toward the notion that a critical death-promoting gene with a defined function has been isolated is shown in **Fig. 3**.

The first step is obviously the sequencing of a functional antisense cDNA fragment and isolation of the corresponding full-length cDNA clone. One of the short antisense mRNAs that protected HeLa cells to some extent from PCD induced by interferon-γ corresponded to a novel gene that we named DAP-kinase ***(13)***. Sequence analysis of the full-length cDNA was very informative even before the recombinant protein was expressed and studied in detail. It predicted that DAP-kinase is a potential Ca^{2+}/calmodulin (CaM)-regulated serine/threonine kinase with an interesting multidomain structure that is unique in the field of protein kinases. The sequence also predicted that the protein carries 8 ankyrin repeats, and a conserved death domain ***(20)***. It has been speculated that the 8 ankyrin repeats, as well as the death domain, may mediate interactions with putative effector proteins, or influence the specificity and/or stability of kinase–substrate interactions.

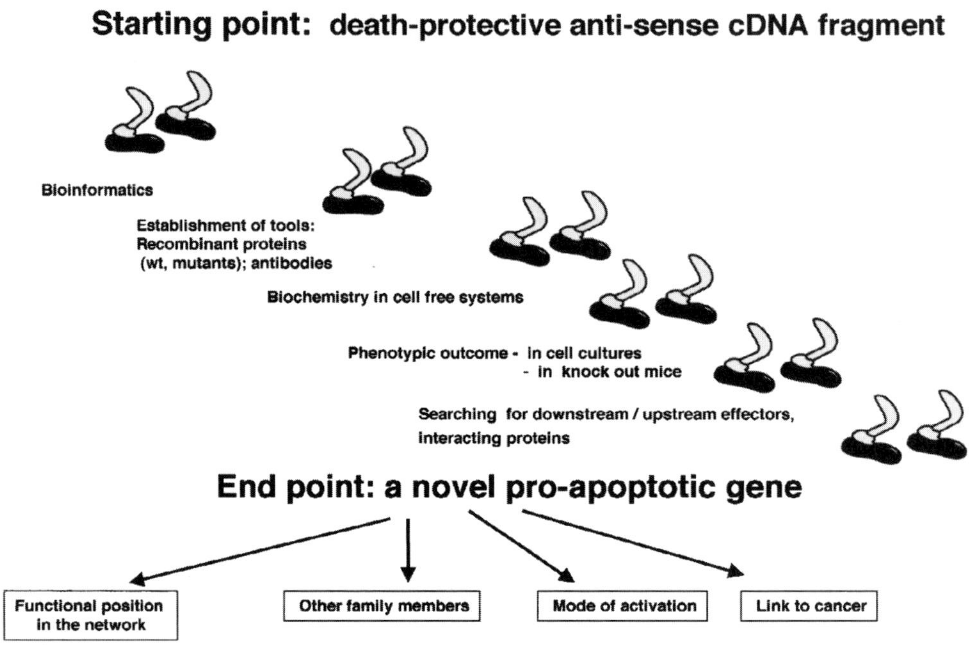

Fig. 3. Outline of the different steps in the characterization of a gene, rescued by a functional approach to gene cloning.

The second step is to construct the molecular tools to study the protein experimentally. This includes the production of the recombinant wild-type protein, design of loss-of-function and gain-of-function mutants, and production of specific antibodies. In the case of DAP-kinase, the recombinant protein, 160 kDa in size, was produced in baculovirus and mammalian cell systems for the structure/function analysis. Specific monoclonal antibodies were generated and used to follow the expression and intracellular localization of the protein by immunostaining and biochemical fractionation of cells. These experiments established that DAP-kinase is localized to the cytoskeleton, specifically in association with the actin microfilaments *(14)*. The cytoskeleton-binding region was then grossly mapped by testing experimentally a series of deletion mutants *(14)*.

The third step is to develop cell-free systems, if possible, to study the biochemical function of the protein. In the case of DAP-kinase we developed the conditions for assaying the kinase activity, using myosin light-chain (MLC) as a substrate. These kinase assays helped us to reveal interesting features concerning the activation of DAP-kinase. They indicated that the enzyme's active site is inhibited by the CaM-regulatory segment and that this inhibition is relieved upon binding of Ca^{2+}/CaM. They also revealed a unique mechanism of autoinhibition of kinase activity, which is controlled by a specific autophosphorylation.

In the next step, the function of the protein in cell cultures and in the context of the entire organism has to be studied. In the case of DAP-kinase, the cellular effects were extensively analyzed by several independent approaches. Activated forms of the kinase were ectopically expressed in cells, and conversely, dominant-negative mutants of the kinase were tested for possible protection against different apoptotic stimuli *(15)*. Based on these studies it was concluded that DAP-kinase is involved in various physiologic scenarios in which elimination of cells is carried out. These include death induced by interferon-γ, Fas, TNF-α, detachment from extracellular matrix, and oncogene activation *(13,15,21,22)*. Current efforts to dissect the subcellular events that are being wired to DAP-kinase, using electron microscopy, indicate that the kinase mediates membrane blebbing and autophagocytosis—two phenotypic outcomes on which very little is known from the molecular point of view. The generation of DAP-kinase-deficient mice is the classical strategy to study the function of a gene in vivo. These mice were recently established in our laboratory; they are viable and fertile, and their phenotype is currently investigated in detail.

Finally, one has to identify upstream and downstream proteins that regulate/mediate the function of the studied gene in order to start mapping it functionally within the complex network. In the case of DAP-kinase, one of the main challenges is to identify the entire range of DAP-kinase's physiologic substrates and to understand how their phosphorylation impinges upon the biochemical pathways through which DAP-kinase executes its pro-apoptotic actions. Another challenge is to identify the proteins that interact directly with DAP-kinase, either via the ankyrin repeats or the death domain—two classical modules that mediate protein–protein interactions.

All the aforementioned steps build the profile of a novel gene that was originally selected based on its rate-limiting function in the network. Studies of its mode of activation and mapping its functional position within the network with respect to other known genes are critical steps for the precise evaluation of its importance (**Fig. 3**). Another interesting issue is the search for close members, which in the case of DAP-kinase led to the discovery of a novel family *(23,24)*. Then the possible link to cancer should be studied as a different aspect of its function. The latter parameter was extensively studied in the case of DAP-kinase, as detailed below.

3. The Link between DAP-Kinase and Cancer—Harnessing the Power of the TKO Selection in the Search for Tumor Suppressor Genes

Obviously, one of the challenges was to find out whether the DAP-kinase gene is a potential tumor suppressor subjected to loss-of-function mutations in cancer. To this end, a few independent directions were undertaken in the past few years, based on functional assays and tumor screens. The functional assays were performed both in cell cultures and in mouse model systems, which together assessed the potential activity of this protein in suppressing different stages of tumorigenicity. In parallel, the status of the gene in human tumor specimens was analyzed and correlations were made with respect to the aggressiveness of the disease in several cases. These three independent lines of research are summarized below.

3.1. DAP-Kinase in Oncogene-Induced P53 Activation—Channeling P53 Toward Apoptosis

One of the first apoptotic checkpoints in the multistep process of tumorigenicity is turned on early during transformation of primary cells in response to unbalanced hyperproliferative signals. Induction of apoptosis at this stage is critical for the elimination of oncogene-bearing cells, and in many systems, it is carried out in a p53-dependent manner *(25)*. Oncogene-induced activation of p53 is mediated by p19ARF—a regulator of p53 stability that is necessary for responses to mitogenic signals but not for responses to DNA damage *(26)*.

Once the various aspects of the antitumorigenic functions of DAP-kinase started to emerge in our laboratory, it became of interest to investigate whether DAP-kinase specifically operates in this early apoptotic checkpoint. A basic finding that initiated the interest in this direction was the strong suppression that activated DAP-kinase exerted on transformation of primary fibroblasts by Myc and Ras or E1A and Ras *(22)*. Most important, this inhibition of foci formation by the introduction of activated DAP-kinase was completely abrogated in p53−/− fibroblasts, indicating that functional p53 was absolutely required. A detailed molecular dissection of the process revealed that DAP-kinase increased p53 protein levels, as well as several p53-responsive genes, in a p19ARF-dependent manner. Moreover, the p53-dependent reduction in foci number resulted from DAP-kinase-induced apoptosis, which involved caspase activation and DNA degradation *(22)*.

The finding that ectopically expressed DAP-kinase killed primary mouse embryonic fibroblasts in a p19ARF/p53-dependent manner immediately raised the following intriguing questions: Does it reflect the involvement of endogenous DAP-kinase in the well-studied safeguard mechanism that is turned on in response to oncogenes? Is endogenous DAP-kinase one of the missing components that couples oncogenes to p19ARF/p53 activation and is responsible for channeling these fibroblasts towards apoptosis? Subsequent experiments indeed confirmed that endogenous DAP-kinase functions in this specific apoptotic checkpoint to modulate p53 responses. First, it was found that endogenous DAP-kinase protein is upregulated in response to c-Myc or E2F-1 activation. Second, and most important, the inactivation of DAP-kinase either functionally by a dominant negative mutant, or genetically by targeted mutation, attenuated the apoptotic responses to c-Myc or E2F-1. These results imply that DAP kinase is an intrinsic component along the p19ARF/p53 pathway that is activated by oncogenes to induce apoptosis.

3.2. DAP Kinase and Metastasis—Affecting Late Steps in Tumorigenesis

A survey of cancer cell lines derived from various human tumors has shown that mRNA and protein expression of DAP-kinase was frequently lost in these cells *(27)*. This initial screen provided the first hint that DAP-kinase expression may be lost in the course of malignant transformation.

An interesting paradigm emerged from the study of mouse lung carcinoma cell lines in experimental metastasis assays. It was found that whereas all the nonmetastatic clones deriving from Lewis lung carcinoma still expressed normal levels of DAP-kinase, the highly metastatic clones were all DAP-kinase negative. DAP-kinase expression was then restored into the highly metastatic cells, yielding clones expressing the transgene at physiologic levels, with no changes in the overall pattern of their growth in vitro. The metastatic activity of these genetically manipulated cells was examined by intravenous injection of the cells into syngeneic mice. Strikingly, restoration of physiologic levels of DAP-kinase into the highly metastatic Lewis carcinoma cells suppressed their ability to form lung metastases in mice *(21)*. Both the number and the size of lung lesions were strongly reduced. Conversely, rare lung lesions, which were spontaneously selected in mice following injection of the original poorly metastatic cells, had lost endogenous DAP kinase expression at high frequency. Together, these experiments suggested that loss of DAP-kinase expression provides a positive selective advantage during the formation of lung metastases. The transgene also delayed local tumor growth, yet with lower efficiency than the suppression of the metastatic activity ***(21)***. A detailed dissection of the system revealed that DAP-kinase-mediated suppression of metastasis results, at least in part, from the increased sensitivity that it conveys to the various death-inducing stimuli that metastasizing cells encounter in their long way from the primary tumor toward the secondary lesion. It was therefore proposed that loss of DAP-kinase expression confers a selective advantage to cancer cells and may play a causative role in tumor progression toward the more aggressive metastasizing cells ***(21)***.

In conclusion, the metastatic assays and the in-vitro function-based assays in primary fibroblasts attributed tumor suppressive properties to DAP-kinase in at least two different apoptotic checkpoints that operate in the course of tumor development. These assays propose that loss of DAP-kinase may confer a selective advantage in the early oncogene-activated apoptotic checkpoint, by attenuating the p53 responses, and during the late stages of metastasis by reducing the sensitivity of the metastasizing cells to different stresses including detachment from extracellular matrix. The obvious question that emerged from these experiments was whether DAP-kinase is serving as a bona-fide tumor suppressor gene in humans. To this aim, *in situ* screens for testing the status of DAP-kinase gene in fresh human tumor specimens were undertaken simultaneously by a number of groups in the past few years, demonstrating loss of DAP-kinase expression in many types of cancers, as summarized below.

3.3. Loss of DAP-Kinase Expression in Human Tumor Specimens

An observation that we made a few years ago in the cancer cell lines that we screened facilitated some of the subsequent *in situ* measurements of DAP-kinase in cancer patients. We found that in some (but not all) of the DAP-kinase-minus cancer cell lines, DAP-kinase expression could be restored by the DNA-demethylating drug 5-aza-2′-deoxycytidine ***(27)***. This indicated that aberrant DNA methylation was one of the mechanisms responsible for turning off this gene. Later, a CpG island at the 5′UTR of DAP-kinase was found to be a target for hypermethylation ***(28)***. The biologic relevance for this methylation was demonstrated in the Burkitt's lymphoma cell line Raji, in which DAP-kinase is fully methylated and not expressed. In these cells, demethylation by treatment with 5-aza-2′-deoxycytidine restored their apoptotic sensitivity to IFN-γ ***(28)***.

Promoter methylation is considered the main epigenetic modification in mammals, and abnormal methylation of CpG island in the promoter regions of genes leads to transcriptional silencing. The methylation status of DAP-kinase's 5′UTR was analyzed in DNA extracted from primary normal and tumor samples, using methylation-specific polymerase chain reaction (MSP) developed by Herman and his co-workers. This analysis revealed high incidence of methylations in B-cell malignancies ***(28)***, thyroid lymphoma ***(29)***, non-small-cell lung carcinoma ***(30–32)***, head and neck cancer ***(33,34)***, multiple myeloma ***(35)***, and breast and colon carcinoma cancers (J. Kissil and A. Kimchi, unpublished data). All the normal tissues had unmethylated copies exclusively. The highest frequency of DAP-kinase methylation was detected in B-cell lymphoma. All Burkitt's lymphoma tested and 84% of B-cell non-Hodgkin's lymphomas were hypermethylated in the DAP-kinase CpG island ***(28)***. Two of these studies also linked DAP-kinase methylation status with disease prognosis. In the head and neck cancer study,

18% of the tested patients displayed promoter methylation of DAP-kinase. Within this group, hypermethylation of DAP-kinase correlated with the presence of lymph node metastases and advanced disease stage. In non-small-cell lung cancer (NSCLC), a retrospective study found hypermethylation of the DAP-kinase promoter in 44% of the tumors, and a strong association between hypermethylation of the DAP-kinase promoter and poor disease-specific survival prospects ***(31)***.

These multiple studies collectively indicate that the DAP-kinase gene is abnormally methylated in a significant portion of human tumors, providing the first proof that inactivation of this gene is critical for the development of these common malignancies. However, as is the case for other tumor suppressor genes, and as predicted from the 5′-aza-2′-deoxycytidine experiments discussed above, DNA methylation is not an exclusive way to inactivate DAP-kinase. Loss or inactivation may also result from mutations, deletions, or other DNA rearrangements. A few LOH cases were already identified in our initial screen of breast and colon carcinoma cancers once DAP-kinase gene was mapped in our laboratory to chromosome 9, region q21.3-q22.3 (J. Kissil and A. Kimchi, unpublished data). Obviously, within the next few years an extensive search for intragenic mutations/deletions will follow these studies.

Altogether, the functional approaches discussed under **Subheadings 3.1.** and **3.2.** and the tumor screens performed so far provide solid support for the notion that the TKO selection has yielded a fascinating new tumor suppressor gene that will be in the center of many cancer studies in the future.

References

1. Kaufmann, S. H. and Gores, G. J. (2000) Apoptosis in cancer: cause and cure. *Bioessays* **22**, 1007–1017.
2. Lowe, S. W. and Lin, A. W. (2000) Apoptosis in cancer. *Carcinogenesis* **21**, 485–495.
3. Wyllie, A. H., Bellamy, C. O., Bubb, V. J., et al. (1999) Apoptosis and carcinogenesis. *Br. J. Cancer*, **80 Suppl 1**, 34–37.
4. McDonnell, T. J., Deane, N., Platt, et al. (1989) bcl-2-Immunoglobulin transgenic mice demonstrate extended B cell survival and follicular lymphoproliferation. *Cell* **57**, 79–88.
5. Strasser, A., Harris, A. W., Vaux, D. L., et al. (1990) Abnormalities of the immune system induced by dysregulated bcl-2 expression in transgenic mice. *Curr. Top. Microbiol. Immunol.* **166**, 175–181.
6. Yonish-Rouach, E., Resnitzky, D., Lotem, J., Sachs, L., Kimchi, A. and Oren, M. (1991) Wild-type p53 induces apoptosis of myeloid leukaemic cells that is inhibited by interleukin-6. *Nature* **352**, 345–347
7. Gottlieb, T. M. and Oren, M. (1998) p53 and apoptosis. *Semin. Cancer Biol.* **8**, 359–368.
8. Horvitz, H. R. (1999) Genetic control of programmed cell death in the nematode *Caenorhabditis elegans*. *Cancer Res.* **59**, 1701s–1706s.
9. Steller, H. (1995) Mechanisms and genes of cellular suicide. *Science* **267**, 1445–1449.
10. Deiss, L. P. and Kimchi, A. (1991) A genetic tool used to identify thioredoxin as a mediator of a growth inhibitory signal. *Science* **252**, 117–120.
11. Kimchi, A. (1998) DAP genes: novel apoptotic genes isolated by a functional approach to gene cloning. *Biochim. Biophys. Acta* **1377**, F13–F33.
12. Kissil, J. L. and Kimchi, A. (1998) Death-associated proteins: from gene identification to the analysis of their apoptotic and tumour suppressive functions. *Mol. Med. Today* **4**, 268–274.

13. Deiss, L. P., Feinstein, E., Berissi, H., Cohen, O., and Kimchi, A. (1995) Identification of a novel serine/threonine kinase and a novel 15-kD protein as potential mediators of the gamma interferon-induced cell death. *Genes Dev.* **9**, 15–30.
14. Cohen, O., Feinstein, E., and Kimchi, A. (1997) DAP-kinase is a Ca^{2+}/calmodulin-dependent, cytoskeletal-associated protein kinase, with cell death-inducing functions that depend on its catalytic activity. *EMBO J.* **16**, 998–1008.
15. Cohen, O., Inbal, B., Kissil, J. L., et al. (1999) DAP-kinase participates in TNF-alpha- and Fas-induced apoptosis and its function requires the death domain. *J. Cell Biol.* **146**, 141–148.
16. Kissil, J. L., Deiss, L., Bayewitch, M., Raveh, T., Khaspekov, G., and Kimchi, A. (1995) Isolation of DAP3, a novel mediator of interferon-gamma-induced cell death. *J. Biol. Chem.* **270**, 27932–27936.
17. Kissil, J. L., Cohen, O., Raveh, T. and Kimchi, A. (1999) Structure–function analysis of an evolutionary conserved protein, DAP3, which mediates TNF-alpha- and Fas-induced cell death. *EMBO J.* **18**, 353–362.
18. Levy Strumpf, N., Deiss, L. P., Berissi, H., and Kimchi, A. (1997) DAP-5, a novel homolog of eukaryotic translation initiation factor 4G isolated as a putative modulator of gamma interferon-induced programmed cell death. *Mol. Cell. Biol.* **17**, 1615–1625 .
19. Henis-Korenblit, S., Strumpf, N. L., Goldstaub, D., and Kimchi, A. (2000) A novel form of DAP5 protein accumulates in apoptotic cells as a result of caspase cleavage and internal ribosome entry site-mediated translation. *Mol. Cell Biol.* **20**, 496–506.
20. Feinstein, E., Wallach, D., Boldin, M., Varfolomeev, E., and Kimchi, A. (1995) The death domain: a module shared by proteins with diverse cellular functions. *Trends. Biochem. Sci.* **20**, 342–344.
21. Inbal, B., Cohen, O., Polak-Charcon, S., et al. (1997) DAP kinase links the control of apoptosis to metastasis. *Nature* **390**, 180–184.
22. Raveh, T., Droguett, G., Horwitz, M. S., DePinho, R. A., and Kimchi, A. (2001) DAP-kinase activates a p19ARF/p53-mediated apoptotic checkpoint to suppress oncogenic transformation. *Nat. Cell Biol.* **3**, 1–7.
23. Inbal, B., Shani, G., Cohen, O., Kissil, J. L., and Kimchi, A. (2000) Death-associated protein kinase-related protein 1, a novel serine/threonine kinase involved in apoptosis. *Mol. Cell Biol.* **20**, 1044–1054.
24. Kogel, D., Prehn, J. H. and Scheidtmann, K. H. (2001) The DAP kinase family of pro-apoptotic proteins: novel players in the apoptotic game. *Bioessays* **23**, 352–358.
25. Lowe, S. W. (1999) Activation of p53 by oncogenes [In Process Citation]. *Endocr. Relat. Cancer* **6**, 45–48.
26. Sherr, C. J. (1998) Tumor surveillance via the ARF-p53 pathway. *Genes Dev.* **12**, 2984–2991.
27. Kissil, J. L., Feinstein, E., Cohen, O., et al. (1997) DAP-kinase loss of expression in various carcinoma and B-cell lymphoma cell lines: possible implications for role as tumor suppressor gene. *Oncogene* **15**, 403–407.
28. Katzenellenbogen, R. A., Baylin, S. B., and Herman, J. G. (1999) Hypermethylation of the DAP-kinase CpG island is a common alteration in B-cell malignancies. *Blood* **93**, 4347–4353.
29. Nakatsuka, S., Takakuwa, T., Tomita, Y., Miwa, H., Matsuzuka, F., and Aozasa, K. (2000) Role of hypermethylation of DAP-kinase CpG island in the development of thyroid lymphoma. *Lab. Invest.* **80**, 1651–1655.
30. Esteller, M., Sanchez-Cespedes, M., Rosell, R., Sidransky, D., Baylin, S. B., and Herman, J. G. (1999) Detection of aberrant promoter hypermethylation of tumor suppressor genes in serum DNA from non-small cell lung cancer patients. *Cancer Res.* **59**, 67–70.
31. Tang, X., Khuri, F. R., Lee, J. J., et al. (2000) Hypermethylation of the death-associated protein (DAP) kinase promoter and aggressiveness in stage I non-small-cell lung cancer. *J. Natl. Cancer Inst.* **92**, 1511–1516.

32. Kim, D. H., Nelson, H. H., Wiencke, J. K., et al. (2001) Promoter methylation of DAP-kinase: association with advanced stage in non-small cell lung cancer. *Oncogene* **20**, 1765–1770.
33. Sanchez-Cespedes, M., Esteller, M., Wu, L., et al. (2000) Gene promoter hypermethylation in tumors and serum of head and neck cancer patients. *Cancer Res.* **60**, 892–895.
34. Rosas, S. L., Koch, W., da Costa Carvalho, M. G., et al. (2001) Promoter hypermethylation patterns of p16, O6-methylguanine-DNA-methyltransferase, and death-associated protein kinase in tumors and saliva of head and neck cancer patients. *Cancer Res.* **61**, 939–942.
35. Ng, M. H., To, K. W., Lo, K. W., et al. (2001) Frequent death-associated protein kinase promoter hypermethylation in multiple myeloma. *Clin. Cancer Res.* **7**, 1724–1729.

26

Genetic Suppressor Elements in the Characterization and Identification of Tumor Suppressor Genes

Igor B. Roninson and Andrei V. Gudkov

The subject of the present chapter is the genetic suppressor element (GSE) methodology, a functional genomics platform for identifying and characterizing genes involved in different cellular phenotypes, and the applications of this methodology to the study of tumor suppressor genes (TSG). In this overview, we will first describe the principles and basic elements of GSE selection and then concentrate on the methods and examples for using this methodology to identify and analyze TSGs. For detailed "cookbook" protocols of methods used in GSE selection, we refer the reader to previously published methodologic works *(1,2)*.

1. Methods of GSE Selection

1.1. What Is a GSE?

The GSE concept *(3,4)* arose from two general methodologies that had been widely used for targeted inhibition of gene function: antisense nucleic acids and dominant negative mutant proteins. Antisense RNA or oligonucleotides targeting a specific mRNA vary widely in their efficacy, and even today there are no reliable algorithms to predict which nucleic acid sequences will be efficient as antisense inhibitors. The general principle of designing protein variants that would act as dominant negative mutants *(5)* is based on the inactivation of some essential domains within a protein, leaving other domain(s) intact. The intact domains of a mutant protein remain capable of binding to their corresponding ligands, but inactivation of the other domains renders the protein nonfunctional. When expressed in a cell, such nonfunctional proteins will compete with their intact progenitor for binding to the molecules that are recognized by the intact domains. Domain inactivation may even render the binding of the mutant protein nearly irreversible, since such a protein can no longer complete its normal cycle of interactions. Such irreversible binding would result in a strong dominant negative effect. Selective inactivation of protein domains can be achieved either by point mutations or by truncation. In an extreme scenario, the bulk of the protein is removed, leaving only a single domain that retains its capacity for binding the appropriate counterpart. The application

From: *Methods in Molecular Biology, Vol. 222: Tumor Suppressor Genes: Pathways and Isolation Strategies*
Edited by: Wafik S. El-Deiry © Humana Press Inc., Totowa, NJ

of this principle to a specific target protein, however, runs into the same problem as the design of efficient antisense inhibitors: with very few exceptions, it is impossible to predict which, if any, segment of the protein would act as a functional binding domain and therefore a dominant inhibitor.

Both of the above-described types of genetic inhibitors can be generated from the mRNA sequence of the target gene and expressed in the cell using a suitable expression vector, either in antisense orientation (for antisense RNA inhibition) or in sense orientation (for dominant negative protein fragments). We have used the term GSE to designate both sense- and antisense-oriented gene fragments that can inhibit or modify the function of the target gene when expressed in a cell. We then hypothesized that both types of functional GSEs can be generated by random fragmentation of the DNA of the target gene and identified by function-based selection of fragments that confer the desired cellular phenotype. Such function-based GSE selection should make it possible to develop genetic inhibitors for the selected targets, delineate protein functional domains, and identify genes involved in various complex phenotypes.

Over the years, this hypothesis proved to be correct and applicable to a wide variety of systems. **Table 1** lists selected examples of the application of GSE technology to different biologic problems (other than TSG). These examples were chosen to illustrate different types of targets, vectors, and GSE selection procedures, which are discussed in the next section.

1.2. Basic Elements of GSE Selection

A generalized scheme of GSE selection is depicted in **Fig. 1.** This process starts with the selection of target DNA and its random fragmentation (**A**). The resulting fragments are then supplied with adaptors suitable for the expression of protein fragments (**B**) and ligated into an expression vector, to generate a random-fragment library (**C**). This library is then introduced into the recipient cell type (**D**), which is then subjected to selection for the desired phenotype (**E**). Library-derived transgenic DNA fragments are then recovered from the selected cells (**F**). The recovered fragments may be used (in the form of an enriched library) for another round of selection, or subjected directly to sequencing (**G**) and functional testing (**H**). Functional testing allows one to identify biologically active fragments that can then be properly designated as GSEs. Sequence analysis shows whether the selected GSEs are sense- or antisense-oriented, and from which part of the mRNA or protein they are derived. If the target DNA (**A**) comprised more than one gene, the GSE sequence will pinpoint the genes that are likely to be involved in the biologic process under study.

The following is a general overview of the basic elements of GSE selection.

1.2.1. Target DNA and the Fragmentation Procedure

If the goal of GSE selection is to develop specific inhibitors for the gene under study or to identify the functional domains of the corresponding protein, a single cloned gene or cDNA is used as the starting material for GSE selection. The next level of target complexity is provided by viral genomes, which are used to develop GSEs that block viral infection or specific viral functions. When the goal of GSE selection is to identify genes involved in a specific cellular phenotype, the starting material represents a mixture of multiple genes. In the simplest case, a mixture of cloned cDNAs of many candidate genes

Table 1
Examples of GSE Selection (Other than TSGs)

Target DNA and vector type	Selection strategies	GSEs isolated	GSE phenotypes	References
From single genes:				
Human topoisomerase IIα cDNA; retroviral vector	Etoposide resistance in HeLa cells	Sense and antisense fragments from different regions of cDNA	Resistance to different topoisomerase II-interactive drugs	***30***
Human DAP-kinase cDNA; episomal plasmid vector	Protection of human HeLa cells from interferon-induced apoptosis	Sense fragments from different regions of cDNA	Inhibition of interferon-induced and TNF-induced apoptosis and of DAP-kinase activity	***11***
Human BCL2 cDNA; retroviral vector	FACS selection for apoptosis-specific changes in cellular DNA content in human AA2 leukemia cells treated with dexamethasone or vincristine	A single sense-oriented GSE	Decreased apoptotic cell death in several types of tumor cells	***28***
Human MDR1 cDNA; retroviral vector	FACS selection for increased accumulation of Rhodamine 123; selection of floating dead cells in the presence of low-dose colchicine	Sense and antisense fragments from different regions of cDNA	Inhibition of MDR1-mediated efflux; sensitization to MDR1-transported drugs	***26***
From mixtures of a limited number of genes:				
E. coli ribosomal operon (rrnB); bacterial plasmid vector	Erythromycin resistance in *E. coli*	Sense-oriented fragments from the same region of 23S rRNA, encoding a novel peptide	Resistance to erythromycin	***61***
cDNA mixture of 19 immediate-early mouse genes; IPTG-inducible plasmid vector	BrdU suicide selection for IPTG-dependent growth arrest in mouse NIH 3T3 cells	Sense-oriented GSEs from three genes	Growth inhibition	***23***

(continued)

Table 1 *(continued)*
Examples of GSE Selection (Other than TSGs)

Target DNA and vector type	Selection strategies	GSEs isolated	GSE phenotypes	References
From viral genomes:				
Bacteriophage lambda DNA; bacterial plasmid vector	Resistance to lysis by phage lambda in *E. coli*	Multiple sense and antisense fragments from different lambda genes	Inhibition of lytic infection suppression of lysogen rescue (some GSEs); transcriptional inhibition of LamB receptor (full-length and truncated Ea8.5 gene)	***3***
Full-length HIV-1 cDNA; retroviral vector	Selection against HIV induction in a latently infected OM10.1 cell line (FACS-based selection for CD4 retention); selection against HIV infection in CEM-ss cells (based on CD4 positivity and negativity for viral p24 antigen)	Multiple sense and antisense fragments from different HIV genes	Resistance to provirus induction in OM10.1 and HIV infection in normal CEM-ss cells	***25***
From yeast genome:				
S. cerevisiae genome; yeast GFP-fusion plasmid vector	Resistance to α-factor in *S. cerevisiae*	Sense fragments from multiple genes	Bypass of α-factor-induced growth arrest	***12***
From total mammalian cDNA:				
Normalized cDNA of human HeLa cells; retroviral vector	Resistance to aphidicolin in human HT1080 cells	Sense and antisense fragments from four different genes	Combination of four GSEs produces resistance to multiple DNA-interactive drugs and increases expression of genes involved in DNA repair and stress response	***16,62***

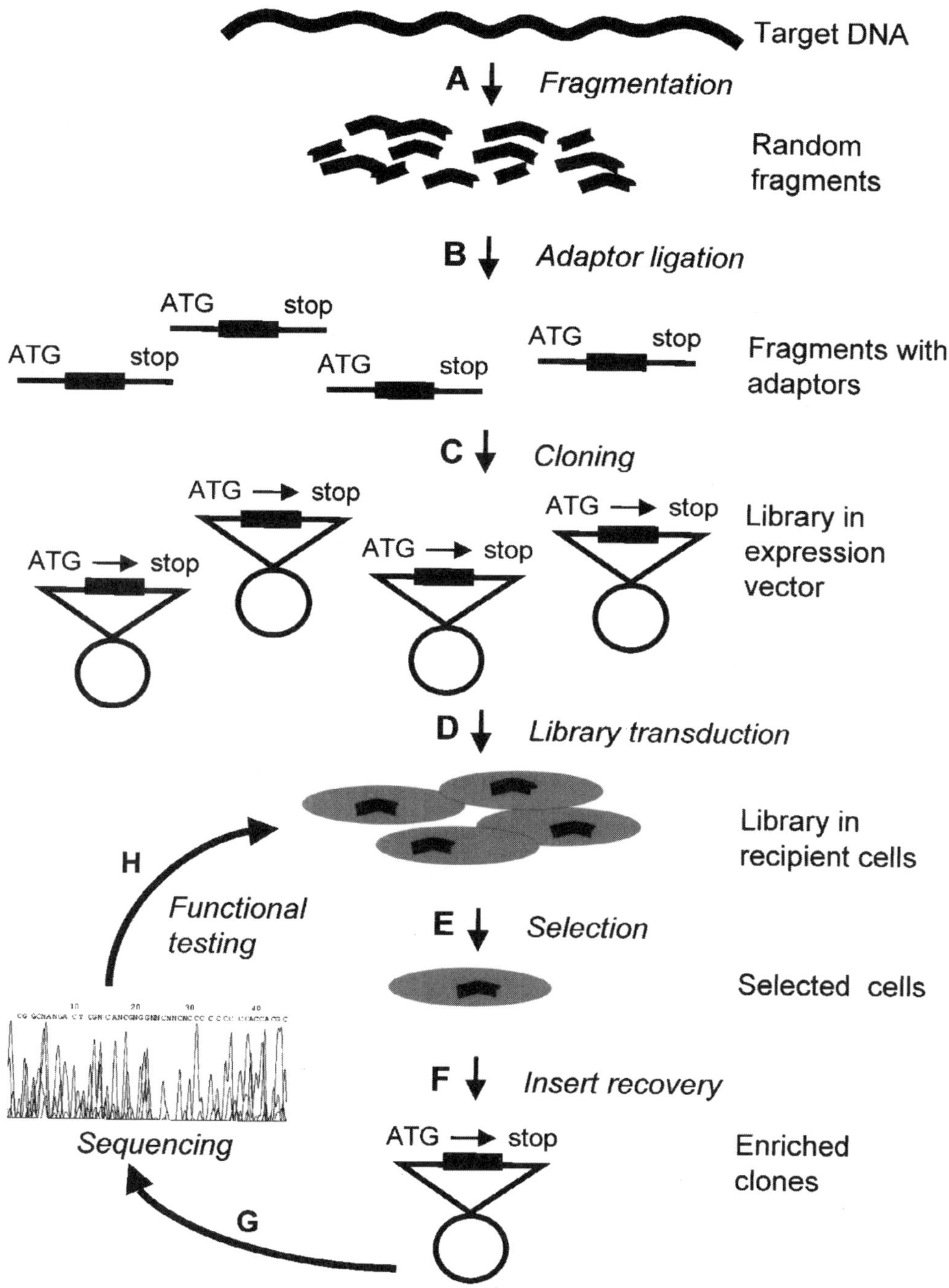

Fig. 1. General scheme of GSE selection. See text for details.

is manually put together prior to fragmentation. In the case of prokaryotes and lower eukaryotes, the whole genome can be used as the starting material (technical limitations of library construction and delivery have so far prevented the use of total mammalian DNA as the target for GSE selection). The most complex libraries have been derived from fragmented cDNA populations of mammalian cells. Such libraries were prepared after the steps of subtraction (to enrich for cDNA sequences overrepresented in a specific cell type) or normalization (to reduce the redundancy of cDNA sequences), which allows one to decrease the size of a library that contains all the mRNA sequences expressed in a cell.

The target DNA is subjected to random fragmentation (as opposed to restriction enzyme digestion), to obtain the widest possible permutation of the endpoints of the encoded peptides and RNA molecules. The size of DNA fragments used for GSE selection typically ranges from 50 to 500 bp. This range was selected from rough estimates for the size of a typical protein domain and inhibitory antisense RNA molecules. The procedure that has been used in most cases for random fragmentation of single cDNA clones, viral genomes, or mixtures of cDNA clones involves digestion with DNase I in the presence of Mn^{2+}. T4 DNA polymerase and Klenow fragment are then used to fill in the ends prior to adaptor ligation.

When cellular cDNA is used as the starting material for GSE selection, random fragmentation is incorporated into the cDNA synthesis procedure. The generation of random fragments is assured by (a) fragmenting the poly (A)+ RNA template (by boiling) prior to cDNA synthesis, leading to random termination of cDNA fragments, and (b) priming cDNA synthesis with oligonucleotides that carry a stretch of random sequence at their 3′ ends, providing for random initiation. Following the ligation of adaptors (*see* **Subheading *1.2.2.***), the cDNA fragment population is amplified by PCR, using adaptor-derived primers. Once a population of adaptor-tagged cDNA fragments has been generated, this population can be used for normalization or subtraction. A cDNA subtraction protocol used for GSE selection from TSG associated with breast cancer ***(6)*** was based on a modification of the procedure known as representational difference analysis (RDA) ***(7)***, and is described under **Subheading 2.3.2**. The cDNA normalization procedure that we have used to generate libraries from different mammalian cell lines ***(1,8)*** is based on the relatively laborious C_0t fractionation protocol of Patanjali et al. ***(9)***. Simpler procedures have since been developed for cDNA subtraction and normalization. In particular, the suppression subtractive hybridization (SSH) technique of Diatchenko et al. ***(10)*** can be used for both subtraction and normalization. This procedure can in principle be modified to switch from restriction fragments to random cDNA fragments for subtraction and normalization.

1.2.2. Adaptor Design

The principal aim of adding adaptors to random fragments prior to cloning is to provide translation initiation and termination codons that would allow for the expression of sense-oriented cDNA fragments as peptides. Different types of adaptors used for this purpose are described in **ref. *1***. Our initial protocols used a single adaptor that contained a single ATG start codon in the sense orientation, three stop codons in all reading frames in antisense orientation, and a single restriction site for nondirectional cloning. More recently, we have used different adaptors for directional cloning into different restriction sites, with one adaptor containing three start codons, and the other adaptor other three stop codons (in all three reading frames), to enable directional cloning in the appropriate orientation. The adaptor design, however, lends itself to additional purposes. For example, studies that were specifically oriented toward peptide-encoding GSEs have used adaptors encoding the FLAG epitope ***(11)***, the green fluorescent protein ***(12)***, or penetratin peptide for transmembrane delivery ***(13)***.

Another modification of the adaptor design that we have used more recently for normalized cDNA libraries has been developed to ease the determination of the orientation of GSEs that are derived from unknown genes. In this protocol, reverse transcription is initiated not with a random primer (as in the earlier methods) but with a primer that carries eight random nucleotides at its 3′ end, coupled to a defined sequence tag. This tag comprises translation termination codons in all three reading frames and a restriction site. Double-stranded cDNA fragments carrying this adaptor at their 3′ ends are then ligated to another adaptor that contains translation initiation codons in all three reading frames. This ligation is followed by PCR with the primers that correspond respectively to the second adaptor and to the tag sequence in the reverse transcription primer. As a result, the cDNA fragments acquire the start and the stop codons in the orientation that corresponds to the original mRNA sequence, making it possible to deduce the original orientation of the fragment from the position of these adaptors.

1.2.3. Expression Vectors and Library Size

The choice of expression vector is determined by the nature of recipient cells, the size of the library that should be delivered into the cells (see below), and other requirements (e.g., the need for an inducible promoter). GSE selections in mammalian cells have been carried out using conventional and episomal plasmid vectors, as well as retroviral vectors. All the GSE selections carried out in our laboratories in mammalian cells have used exclusively retroviral vectors. This choice has been dictated by several considerations. (a) Retroviral vectors are still the only delivery system that can be used for stable transduction into a very high fraction (up to 100%) of recipient cells. (b) High-complexity libraries can be generated by standard cloning procedures in retroviral plasmid vectors, and these libraries can be easily converted into a representative mixture of recombinant retroviruses by transient transfection of retrovirus-packaging derivatives of 293 cells, such as BOSC 23 ***(14)***. (c) Retroviral infection, unlike most DNA transfection procedures or adenoviral infection, is essentially nonstressful for the recipient cells, avoiding stress-related artifacts in subsequent selection. (d) Integrated proviruses lend themselves to several convenient rescue procedures (*see* **Subheading 1.2.6.**). (e) Retroviral vectors frequently integrate at more than one copy per infected cell ***(15)***. This makes it possible to select combinations of GSEs that have a synergistic effect on the target phenotype ***(16)***.

Retroviral vectors, however, also have certain disadvantages. (a) Some cell types are not very efficiently infected by retroviral vectors (*see* **Subheading 1.2.4.**). (b) A more serious problem concerns the general design of retroviral vectors. Such vectors (e.g., LNCX ***[17]*** used in many GSE selections) typically contain two promoters (in the same orientation), including the retroviral long terminal repeat (LTR). One of the promoters drives the expression of the cloned gene, and the other a selectable marker (e.g., neo). We have found that selection for the marker gene is associated with frequent rearrangement or silencing of the promoter driving the gene of interest ***(15)***. This problem is likely to be even more pronounced when the promoter driving GSE expression is positioned in antisense orientation to the LTR, as is the case for inducible retroviral vectors, such as LNXCO3 ***(18)***. In the latter vectors, expression of one promoter was shown to inhibit gene expression from the other promoter, presumably through antisense-mediated inhibition ***(19)***. We therefore recommend not selecting cells for the marker gene prior to GSE selection, unless required by low infection efficiency.

The size of the random fragment library is determined by the nature and the size of the target DNA. The usual goal of GSE selection from a single gene or a small number of genes is to obtain a near-comprehensive set of functional cDNA domains. In our experience, an extensive set of GSEs can be derived from a single cDNA clone if the random fragment library contains 5000–10,000 clones/kb of cDNA. When large-scale cDNA libraries are used, the goals of selection are limited to obtaining at least one GSE for all or almost all the genes associated with the selected phenotype, without a comprehensive coverage of each individual gene. The frequency of selectable GSEs for individual genes, in our experience, varies from 1/100 to 1/500, depending on the gene and the selection protocol. Assuming that a normalized cDNA library from a typical cell line contains about 10,000 genes at approximately equal representation, a library of $1–5 \times 10^7$ clones should contain selectable GSEs for essentially all the genes expressed in the recipient cells.

1.2.4. Recipient Cells

The choice of recipient cells is dictated by specific needs of each selection procedure and the source of the library. Most of GSE selections in mammalian cells have been conducted in immortal cell lines, but some TSG-targeting selections were also carried out in primary rodent embryo fibroblasts ***(20)*** (*see* **Subheading 2.3.1.**). A promising novel class of recipient cells that may be particularly useful for human TSG-based GSE selection strategies is the telomerase (hTERT)-immortalized primary cells. Aside from their immortality, hTERT-immortalized cell lines maintain all the properties of untransformed cells, including normal karyotype, contact inhibition, inability to grow in soft agar or form tumors in animals, and the ability to undergo accelerated senescence ***(21,22)***. Recipient cell lines can be subjected to additional modifications prior to library transduction, such as transfection with repressor or transactivator proteins that are required for regulated promoters ***(23)***.

An essential practical issue for GSE selection from retroviral libraries is the susceptibility of recipient cells to retroviral infection, since some cell lines show relatively poor infection efficiency. In some cases, this is due to low levels of retroviral receptor on the cell surface. This problem usually can be corrected by transfecting recipient cells with the gene for the murine ecotropic receptor ***(24)***. Such transfection also makes it possible to infect nonmurine cells with ecotropic retroviruses (which can be used under BL1 biosafety conditions) instead of amphotropic or pantropic viruses that require BL2 containment. Some cell lines (especially those derived from hematopoietic cells) have low efficiency of proviral integration, making it impossible to achieve very high transduction rates. Nevertheless, infection rates that are achievable with leukemic cells are still sufficient for complex GSE selections ***(25)***.

1.2.5. Selection Principles

The design of a selection strategy is the most critical part of a GSE project, requiring great care and precision. Although some of the selected GSEs proved to be extremely powerful (e.g., p53-derived GSE56), the inhibitory effect of a GSE in most cases is much weaker than that of gene knockout or of disabling mutations in both alleles of the target gene. The selection conditions therefore should be designed to the least possible stringency, allowing for inevitable background. Furthermore, the expression levels of the same GSE will vary in different infected cells, depending on the provirus integration site and on the effects of other viruses that may be co-transduced in the same cell. This means that a GSE is unlikely to be selected unless it is present in at least 20–50 cells prior to the onset of selection. This is best achieved by maximizing the efficacy of retroviral transduction and using the largest possible number of infected cells for the selection. To reconcile both requirements (relatively mild selection conditions and a large number of cells infected with each viral clone), most of our past selections were carried out over two or more rounds of infection, selection, and recovery. As described under **Subheading 1.2.6.**, the use of high-throughput sequencing may allow one to identify candidate GSEs without additional rounds of selection.

The most frequently used GSE selection strategy in mammalian cells has been based on improved cell survival after exposure to cytotoxic drugs or apoptogenic agents (*see* **Table 1**). It is also possible, however, to select "conditionally lethal" GSEs that are detrimental to cell growth under specific conditions. For example, isolation of floating dead cells that die in the presence of normally noncytotoxic doses of colchicine has been used to select GSEs from the cDNA of the multidrug transporter MDR1 (P-glycoprotein) ***(26)***. A selection strategy with great potential for the study of oncogenes has been used by Pestov and Lau ***(23)*** to isolate growth-inhibitory GSEs from a mixture of 17 proliferation-associated genes. This selection was based on the use of an isopropyl-β-thio-galactoside (IPTG)-inducible plasmid vector for GSE expression and bromodeoxyuridine suicide selection technique ***(27)*** for cells that are growth-arrested in the presence of IPTG. The same selection strategy has been scaled up through the use of IPTG-inducible retroviral vectors ***(18)*** and used to isolate growth-inhibitory GSEs from a normalized cDNA fragment library, as an approach to identifying the complex of genes that are essential for cell growth (T. Primiano, B. D. Chang, M. Baig, S. Fellars, S. Axenovich, T. A. Holzmayer, and I. B. Roninson, unpublished).

Some of the most flexible methods of selection, which are particularly suitable to the isolation of GSEs with modest phenotypic effects, are based on the use of fluorescence-activated cell sorter (FACS) to separate cell populations with quantitatively different levels of specific proteins (detectable by immunofluorescence staining) or other fluorescent tags. FACS selection has been used to isolate GSEs from HIV-1 cDNA, on the basis of their ability to interfere with HIV-mediated inhibition of CD4 receptor or the expression of a viral antigen in infected cells ***(25)***. FACS selection was also used as an independent approach to isolate GSEs from MDR1 cDNA, based on the ability of such GSEs to inhibit the efflux of P-glycoprotein-transported fluorescent vital dye Rhodamine-123 ***(2,26)***. FACS selection of cells with altered DNA binding of acridine orange or with increased membrane permeability to propidium iodide has also been used for GSE selection from the apoptosis inhibitor gene *BCL-2* ***(28)***. Many other FACS-based selection strategies can be readily designed for specific purposes, such as the isolation of GSEs that modify the activity of promoters driving the expression of GFP or some other fluorescent marker.

Specific selection strategies for the study of TSG are described under **Subheading *2.2.***

1.2.6. Recovery of Integrated Fragments

Whereas recovery of plasmid vectors is a straightforward procedure in prokaryotic and yeast cells, the rescue of integrated retroviral inserts is one of the more artifact-prone steps in GSE selection in mammalian cells. The most commonly used and generally applicable recovery method is polymerase chain reaction (PCR) amplification of cDNA inserts from genomic DNA, using primers that correspond to vector sequences flanking the inserts ***(30)***. The PCR products are then recloned into the expression vector to generate a GSE-enriched library. Given the small size of GSE fragments, their PCR amplification is an efficient procedure, which can even be used with partially fragmented DNA of cells undergoing apoptosis ***(28)***. The PCR conditions, however, should be carefully monitored ***(1)***, to avoid the loss of representativity and artifactual fusion of PCR products. The same type of PCR amplification and recloning can be carried out using cDNA rather than genomic DNA from the selected cells as a PCR template ***(29)***. The latter modification has an apparent advantage that it may enrich for functional GSEs by favoring actively transcribed proviruses.

Several alternative procedures avoid the relatively laborious step of recloning the mixture of recovered PCR products. One of these methods is based on the use of a retroviral vector that carries *Escherichia coli* plasmid replication origin sequence and contains *lox* sites in its LTRs. Integrated proviral DNA is excised from genomic DNA of the selected cells by *Cre* recombinase and used to transform competent bacterial cells ***(31)***. Several other protocols recover the integrated proviruses in the form of infectious virus rather than plasmids. (a) Selections can be carried out directly on retrovirus-packaging cells, which secrete the infectious virus into culture media ***(8)***. (b) The selected cells can be subjected to polyethylene glycol-mediated fusion with a retrovirus-packaging cell line, leading to a release of infectious virus ***(6)***. (c) Infectious virus can also be rescued by transfection with expression plasmids carrying retroviral genes *gag-pol* and *env* (our unpublished data). (d) Full-length proviral sequences can be amplified from genomic DNA by long-range PCR, followed by direct transfection of retrovirus-packaging cells with the PCR product. This procedure is quick and maintains very well the representativity of the rescued proviral population, but it is more sensitive to DNA degradation and impurities than conventional short-range PCR ***(33)***.

Although all of the above methods speed up consequent rounds of selection relative to the common protocol (PCR amplification and recloning of inserts), it now appears that the latter procedure has a previously unappreciated advantage. This method provides an extensive library of inserts from the selected cells in a form that is ready for high-throughput sequencing. As described in the next section, such sequencing may facilitate the process of identifying functional GSEs.

1.2.7. Sequencing of Recovered Fragments

Technological advances in automated DNA sequencing, which have made it possible to obtain the human genome sequence long before the original projections, are changing the experimental philosophy in many other molecular biologic applications. The decreased cost and labor involved in the sequencing of multiple clones allow us to improve the quality control and greatly speed up the process of GSE selection. High-throughput sequencing of several hundred clones of the original random fragment library can now be used to verify sequence representation and to confirm the presence of the appropriate adaptors. Most important, however, it is now possible to sequence a large number of clones from a GSE-enriched library generated by PCR amplification and recloning from the selected cells. Analysis of the sequence distribution allows one to identify the clones that have been enriched by selection, even if the enriched clones constitute a small part of the library. Sequence-verified clones that represent different genes or domains can then be used directly for functional testing. Through this approach, we have been able to isolate a set of growth-inhibitory GSEs from a normalized cDNA library after only one round of selection (T. Primiano, B. D. Chang, M. Baig, S. Fellars, S. Axenovich, T. A. Holzmayer, and I. B. Roninson, unpublished).

1.2.8. Functional Testing of Candidate GSEs

Verifying individual GSEs by their functional activity is the final and most difficult step of GSE selection. Such testing is usually conducted by transducing the tested GSEs and the control vector into the recipient cells and testing the transduced cell populations by the same procedure that has been used to select the GSEs. In a previous overview ***(1)***, we have made general recommendations for GSE testing, including careful matching of the GSE-transduced and control populations, maintenance of the same conditions for cell culture throughout the process of selection and testing, and analyzing the GSE-transduced population as early as possible after transduction. Paradoxically, we have observed in several projects that functional activity of confirmed GSEs is no stronger and sometimes weaker than the activity of a mixed GSE-enriched library. The solution to this paradox came from the demonstration that multiple proviruses integrate in the same cell under the common infection conditions ***(15)***, and that co-infection of cells with several weak GSEs produces a synergistic increase in their functional activity ***(16)***. We suggest therefore that functional testing of clones enriched by GSE selection be carried out not only individually, but also with a mixture of all the enriched clones. If individual clones show little or no functional activity but the mixture of clones is active, active clone combinations can then be identified through combinatorial approaches. The identification of cooperating GSEs not only proves their functionality but also suggests possible functional interactions between the genes that give rise to the cooperating GSEs.

1.3. Validation and Analysis of the GSE Targets

With a proper design of the selection system and screening protocol, it is relatively easy to isolate a set of GSEs with the desired biologic activity. Successful GSE selection, however, is only the first step in a long process of identifying the molecular mechanisms of the studied phenotype. So far there is no high-throughput procedure that would allow one to effectively follow numerous GSEs all the way toward the mechanisms of their activity.

Although a GSE is most likely to cause its biologic effect by counteracting the gene from which it is derived, this may not be necessarily true for all cases. It is therefore essential to verify the GSE target, especially if GSEs are used for cloning new genes. One of the most general and powerful approaches to prove that the gene from which a GSE is derived is indeed the target of the GSE is to isolate other GSEs with similar properties representing different portions of the target. This approach was used successfully to prove that a GSE inducing etoposide resistance and representing an antisense fragment of ubiquitous kinesin heavy chain (uKHC) mRNA in fact targets uKHC. A new

GSE library was constructed from randomly fragmented uKHC cDNA, and multiple GSEs conferring etoposide resistance and encoding both antisense RNA and truncated uKHC-derived proteins were isolated ***(29)***. Another approach involves generation and testing of rationally designed dominant negative mutants side by side with the GSE. A similarity between the activities of such mutants and a GSE argues for the parental gene being the target of the GSE. This approach was also used in the studies of uKHC-mediated drug response ***(29)***. Another important confirmation of the specificity of the GSE effect may be provided by finding cases of deregulation of the GSE-corresponding gene in naturally arising cellular models with a GSE-related phenotype. Such evidence has been found in the case of TSG-derived GSEs, in which the gene corresponding to the GSE was lost or downregulated in human tumor cells ***(6)***, and in the case of a GSE inducing drug resistance, in which the corresponding gene was found to be downregulated in independently isolated drug-resistant cell lines ***(8)***.

2. Applications of GSE Methodology to the Study of TSGs

2.1. Variety of TSG Functions

A TSG has been defined as "a genetic element whose loss or inactivation allows a cell to display one or another phenotype of neoplastic growth deregulation" ***(34)***. In accordance with this definition, if the inhibition of a certain gene promotes any step in the process of neoplastic transformation, such a gene should be regarded as a candidate tumor suppressor. A TSG can therefore be identified by its ability to give rise to GSEs that promote various aspects of the neoplastic phenotype. TSGs control cell growth in different ways. In particular, the multifunctional regulator p53 acts as a positive regulator of transient growth arrest (quiescence), permanent growth arrest (senescence), and apoptosis ***(35,36)***. Through its latter capacity, p53 and its partners in the pathway (for example, ARF) mediate cellular sensitivity to apoptosis-inducing agents and resistance to transformation by dominant oncogenes ***(37)***. On the other hand, the breast tumor suppressors BRCA1 and BRCA2 act to maintain genome stability by controlling DNA repair and centrosome duplication ***(38)***. Similarly, the tumor suppressor ATM is involved in control of DNA repair and the G2/M checkpoint ***(39)***. The variety of biologic functions that are associated with TSG makes it possible to design various strategies for GSE selection. In the following sections, we discuss several established strategies and provide examples of GSE selection from TSG. These examples are summarized in **Table 2**.

2.2. Strategies for GSE Selection from TSGs

Functional identification of TSGs is not an easy task. Tumor phenotypes associated with the loss or inactivation of TSGs are recessive, since they cannot be detected in the presence of the wild-type gene. This makes it impossible to apply a powerful combination of positive selection and gene transfer techniques directly to the identification of tumor suppressors. This obstacle can be overcome by the use of the GSE approach. Since the loss of TSG activity induces transformed phenotype, TSG-derived GSEs are expected to act as dominant oncogenes in gene transfer protocols, providing a strategy for effective isolation of inhibitory elements for known TSGs and for the identification of new TSGs. Isolation of transforming GSEs is therefore a general approach toward cloning and functional analysis of TSGs.

Table 2
Examples of GSE Selection from TSGs

Target DNA	Selection strategies	GSEs isolated	GSE phenotypes	References
Rat p53 cDNA	Focus formation in MEF; etoposide resistance in MEF	Sense fragments from four different regions of p53 cDNA, with different biologic activities	Lifespan extension of primary MEF; cooperation with activated ras in transformation; resistance to etoposide; inhibition of transcriptional transactivation by p53	***20***
Human and mouse p53 cDNAs	Cisplatin resistance in A2780 human ovarian carcinoma	Sense and antisense fragments from different regions of p53 cDNA	Cisplatin resistance; decreased damage-induced growth arrest and apoptosis; GSE-based synthetic peptide inhibitor of p53 developed	***48***
Mouse p53 cDNA (antisense fragments)	Resistance to p53-mediated growth inhibition in an immortalized MEF cell line	Antisense fragments from two regions of p53	Bypass of p53-induced growth arrest; lifespan extension of primary MEF	***31***
Mouse NIH 3T3 cDNA, normalized	Focus formation in MEF	GSE TR6 from unknown gene	Increased cell density	***53***
		GSE TR7 from phosphotidyl inositol receptor	Increased cell density	***53***
	Resistance to p53-mediated apoptosis in C8 cells	GSE4 from unknown gene	Resistance to etoposide and γ-irradiation; Anchorage independence	*
	Resistance to etoposide in NIH 3T3-derived packaging cells	GSE VPC from ubiquitous kinesin heavy chain (*uKHC*) (two other GSEs from unknown genes isolated)	uKHC: resistance to DNA-damaging drugs; sensitization to microtubule-depolymerizing agents; lifespan extension of primary MEF	***8,29***
Subtracted cDNA library of normal versus malignant human mammary cells	Tumorigenicity of mouse NMuMG cells	GSE M3 from ING1, novel candidate tumor suppressor	Anchorage independence; lifespan extension of primary MEF; focus formation	***6***

* V. S. Ossovskaya, A. Budanov, A. V. G., and P. M. Chumakov, unpublished data.

Isolation of transforming GSEs is based on the principles of genetic screening for transforming genetic elements established in the field of dominant oncogenes, as well as the general rules of GSE selection:

1. Selectable phenotypes should be chosen by the likelihood that they would be controlled to a large extent by a single gene.
2. The selection protocol should be stringent enough to minimize the background but relaxed enough to allow the selection of relatively weak GSEs.
3. Every clone in the random fragment library should be delivered to numerous cells, since only rare GSEs are strong enough to induce selectable phenotype in each transduced cell.

Neoplastic transformation requires a combination of several genetic alterations in one cell, and even such a potent oncogene as activated *ras* cannot transform a normal cell alone ***(40)***. It is therefore impossible to expect a single GSE to be sufficient to induce complete transformed phenotype on its own. In fact, all function-based selections of transforming genetic elements have been based so far on the use of a selection system already predisposed to transformation, i.e., a system in which cooperating genetic events have already occurred. For example, NIH 3T3 cells, a classical cellular system for oncogene selection, are susceptible to transformation with *ras*, probably because of the inactivation of the *p53* pathway ***(41)***. In another example, transgenic mice expressing *c-myc* under the control of an immunoglobulin promoter have enabled the identification of oncogenes *pim-1* and *bmi-1*, which are activated by murine leukemia virus in lymphomas that appeared after infection ***(42)***. Hence, isolation of GSEs inducing neoplastic transformation requires the use of cellular systems that are already predisposed for transformation. GSE-mediated suppression of TSGs in such a system could then cooperate with the preexisting genetic alterations in establishing the transformed phenotype.

Methods for GSE selection from TSG include classical in-vitro transformation assays such as focus formation, anchorage independence, or tumorigenicity. An alternative approach is to isolate GSEs that induce less complicated cellular phenotypes that are also associated with carcinogenesis but, unlike full transformation, do not require cooperation of several genetic events. Immortalization of primary cell cultures and suppression of apoptosis are examples of such selections.

2.2.1. Immortalization of Senescent Cell Cultures

Replicative senescence is a potent anticancer mechanism limiting proliferation of normal cells. The following general principles are especially important for the selection of immortalizing GSEs.

1. The random fragment library should be delivered into actively proliferating cell cultures, since retroviral integration requires DNA replication.
2. The transduced culture should not be too far from reaching senescence. This allows one to avoid random fluctuations in clone representation due to the drift caused by natural variability in the rates of proliferation of different cells within the population.
3. Overgrowing cell clones should be picked as early as possible after the majority of cells in the culture reach senescence, to avoid losses of GSEs in less rapidly growing clones.
4. The rate of spontaneous immortalization should be carefully controlled in parallel cell cultures transduced with the insert-free vector (uninfected cultures should never be used as a control, since exposure of target cells to virus-containing media from packaging cells may by itself alter cell growth).

5. It is extremely useful to establish conditions of selection in pilot experiments. Such experiments will use artificial mixtures of insert-free virus (in excess) and small amounts of a virus expressing known immortalizing genes (*c-myc, E6, SV40 T-antigen*, etc.).
6. If the spontaneous immortalization rate is high, additional rounds of selection should be applied until the selected virus will reliably increase the frequency of cells with prolonged lifespan relative to the control population.
7. On the other hand, the number of rounds of selection should be as small as possible to avoid loss of weaker GSEs.

The strictness of the control of replicative senescence varies among different species: appearance of immortal clones is an extremely rare event in human and in chicken cell cultures, while rodent cells have a much higher incidence of spontaneous immortalization. Hence, viral oncogenes that immortalize rodent fibroblasts can only prolong the lifespan of human and chicken cell cultures ***(43,44)***. Therefore, specific protocols of selection of immortalizing GSEs should be adjusted according to the properties of different cell systems. High and unpredictable background of spontaneous immortalization in rodent cells always requires the use of several rounds of GSE selection. This can be avoided if screening is done in human or chicken cells.

2.2.2. In-Vitro Transformation

The same general principles are applicable to the selections aimed at the isolation of GSEs that induce focus formation in monolayers of quasi-normal cell cultures or anchorage-independent growth in semisolid media. Immortal cell lines retaining contact inhibition of growth and suitable for transformation with dominant oncogenes can be used for this purpose. Mouse NIH 3T3 and NMuMG cells of fibroblast and breast epithelial origin, respectively, provide excellent examples of such cell systems that have been used successfully for the isolation of transforming GSEs ***(6)***.

2.2.3. Inhibition of Apoptosis

Apoptosis is a general mechanism of negative growth regulation that provides an arena for the activity of different TSGs. Inhibition of apoptosis is an excellent selectable phenotype for isolating GSEs, since in many cases it can be achieved by suppressing a single component of the apoptotic pathway. Although the above-listed general principles of GSE selection are also applicable to this selection, it has some specific features. Thus, the majority of established cellular systems for the induction of apoptosis leave relatively high backgrounds of surviving cells. This results in low enrichment for the desired antiapoptotic GSEs after a single round of selection. This problem can be resolved by optimizing the selection system for low background of survivors (usually less than 1%, although the exact parameters depend on the frequency of antiapoptotic elements in the library). Another solution is to apply several subsequent treatments to the same library-transduced population. This approach, however, should be used with caution, since it can easily lead to the isolation of naturally occurring apoptosis-resistant cell variants. Monitoring the frequency of survivors in a control cell population transduced with the insert-free vector is essential to detect a raise in the spontaneous background of apoptosis-resistant cells. Such an increase should indicate that it is time to stop apoptosis-inducing treatments.

2.2.4. Selection for In-Vivo Tumorigenicity

Although tumor formation by originally nontumorigenic cells is in theory the ideal selection that reproduces the natural course of transformation, its value for GSE selection is limited to libraries with relatively low complexity and therefore high frequency of the desired elements. The reason for this limitation is the unavoidable massive death of cells injected in vivo (s.c., i.p., or even more dramatically, i.v.), leading to a great decrease in the size of library-transduced population that is subject to selection. Regardless of this problem, this approach has been used successfully for the isolation of transforming GSEs and subsequent identification of a candidate TSG *ING1* ***(6)*** (see below).

2.2.5. Selection of Transforming GSEs in a Whole Mouse?

Comprehensive screening of a GSE library requires large numbers of target cells and a very high efficiency of viral transduction. It is often difficult to fit the above requirements with the GSE selections done in vitro. Another serious limitation of cell cultures is that the success of selection depends on specific properties of the recipient cells, which can reveal biologic activity of only a minor fraction of GSEs. Both of these problems could be overcome if selection were done in the animal rather than in vitro. In this case, a random-fragment library could theoretically be exposed to an almost unlimited number of cells of all possible differentiation lineages. It is noteworthy that the majority of dominant oncogenes were first identified exactly in this way: by natural in-vivo selection of spontaneous tumorigenic variants of chicken and mouse retroviruses ***(45)***. However, it would be unrealistic to expect that a single GSE will be sufficient to induce tumor formation in a normal animal. Functional selection of transforming GSEs in vitro requires the use of cells that already carry cooperating genetic alterations. In principle, a similar strategy could be applied for the tumorigenic GSE selection in vivo, if one could use transgenic mice with high predisposition to cancer (such as *p53*-knockout mice or transgenic mice expressing dominant oncogenes). This approach remains a theoretical possibility awaiting experimental testing. Its success depends strongly on the effective delivery of GSE libraries in vivo, which might require the use of replication-competent virus vectors or helper viruses.

2.3. Examples of GSE Selection from TSGs

2.3.1. Isolation and Use of p53-Derived GSEs

Three groups have carried out GSE selection using p53 TSG as the target and fragmented p53 cDNA as a source of GSEs. In the first of these studies ***(20)***, a retroviral random fragment library of rat p53 cDNA was delivered to mouse embryo fibroblasts (MEFs). Two different strategies were applied to isolate GSEs (*see* **Fig. 2A**). One set of GSEs was isolated by selecting early-passage MEFs for focus formation in culture. The second strategy involved isolation of GSEs that induce etoposide resistance in MEFs, thus interfering with p53-mediated apoptosis. Unexpectedly, these two selections yielded nonoverlapping sets of sense-oriented GSEs, corresponding to different domains of p53. GSEs from different regions of p53 were tested for functional activity in different assays, which included the ability to cooperate with the activated *Ha-ras* oncogene in transformation of rat embryo fibroblasts (REFs), to suppress p53-mediated transcriptional activation, and to inhibit apoptosis. This analysis is illustrated in **Fig. 2B**.

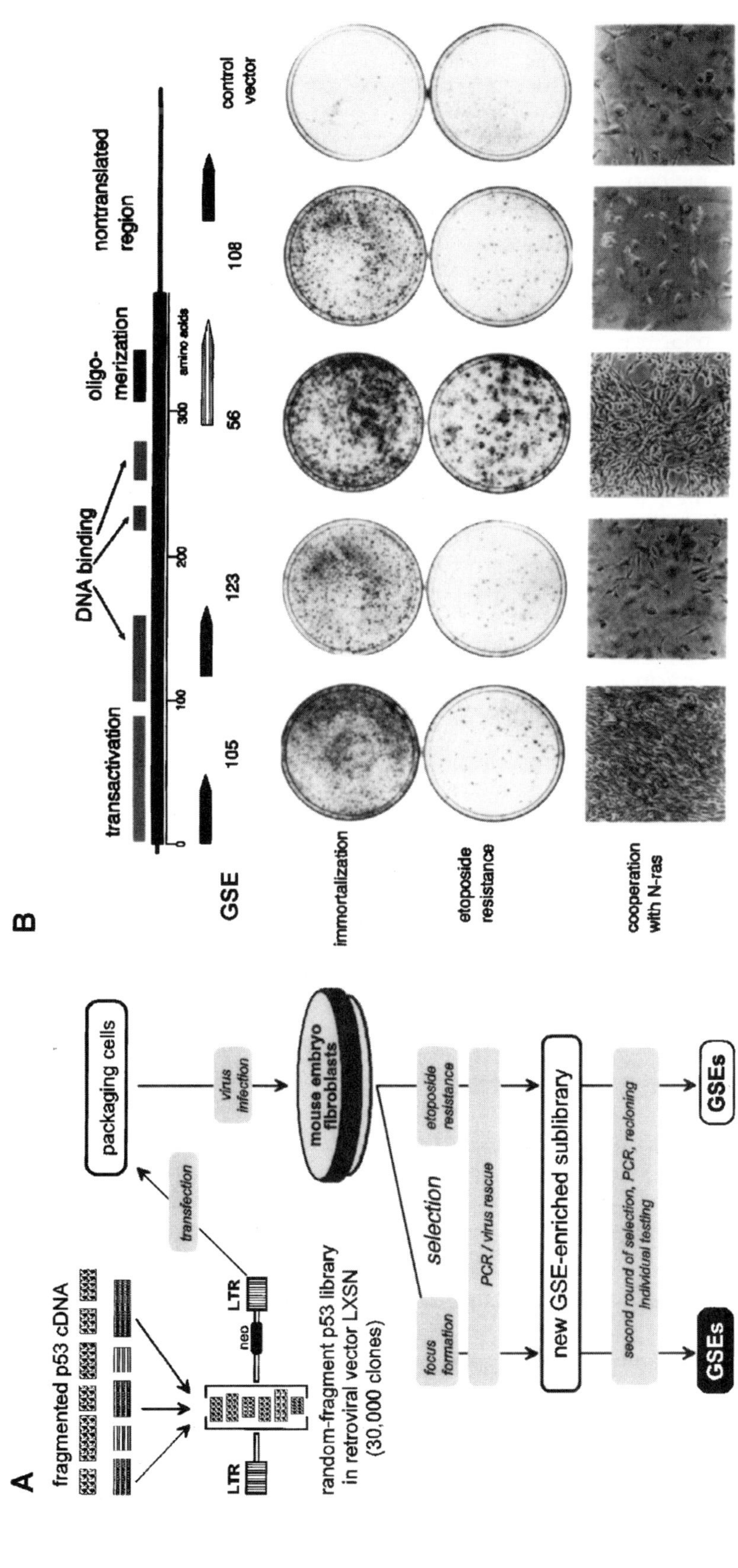

Fig. 2. Scheme of GSE selection from p53 **(A)** and comparison of biologic activity of GSEs derived from different portions of p53 cDNA **(B)** (reproduced from **ref. 20**). See text for details.

Only one cluster of the tested GSEs, typified by GSE56 (**Fig. 2B**), inhibited all the tested functions of p53. GSE56 has now become a widely used genetic inhibitor of p53 ***(46,47)***. GSEs derived from the other domains of p53 showed selective inhibition of some but not all p53 functions, with different GSEs showing different spectra of activity (**Fig. 2**). The distinct functions of the isolated GSEs indicate differences in the biologic activity of the corresponding domains of p53.

In the second study ***(48)***, GSEs from human and mouse p53 cDNAs were selected for the ability to confer cisplatin resistance in human A2780 ovarian adenocarcinoma cells, which are highly prone to apoptosis. This selection yielded both sense- and antisense-oriented GSEs. The sequence of a short sense-oriented GSE, containing only 11 amino acids of human p53, was used to generate a synthetic peptide. This peptide was found to confer resistance to cisplatin and to inhibit sequence-specific DNA binding of the p53 protein. The ability to design biologically active peptides and oligonucleotides based on the information derived from GSE selection provides another general application of the GSE methodology. In the most recent study ***(31)***, random antisense-oriented fragments of mouse p53 cDNA were introduced into an immortalized MEF cell line, carrying a thermolabile p53 mutant. A set of antisense-oriented GSEs was selected for the ability to overcome p53-induced growth arrest in this cellular system. The selected fragments were also shown to promote MEF immortalization.

2.3.2. GSE Selection from a Subtracted Library of cDNA Sequences Downregulated in Breast Cancer: Isolation and Properties of ING1

A more complex selection ***(6)*** was aimed at identifying TSGs that are expressed in normal mammary epithelium but downregulated in human breast cancers. A cDNA population specific for normal breast cells was prepared using a modified DNA subtraction procedure known as representational difference analysis (RDA) ***(7)***. cDNA was generated from poly(A)$^+$ RNA of quasi-normal mammary epithelial cells (nontumorigenic human mammary cell line 184A1) and from eight human breast carcinoma cell lines. cDNA preparations from the carcinoma cell lines were digested with Sau3A and ligated with different synthetic adaptors for PCR amplification. cDNA from untransformed cells was ligated to a different adaptor and used as a driver for several rounds of DNA subtraction. Subtracted fractions from each carcinoma cell line were combined and used as a probe for screening of a conventional cDNA library. About 300 phage clones corresponding to differentially expressed genes were isolated. Their cDNA inserts were combined, fragmented by DNase I, and cloned into a retroviral vector, producing a library of about 10^6 clones. The above library was transduced into mouse breast epithelial cell line NMuMG and used to isolate GSEs that endow these cells with in-vivo tumorigenicity. Two retroviral clones were isolated from tumors formed in nude mice by the library-transduced cells. Testing in vitro showed that both viruses induced anchorage-independent growth of NMuMG cells, as well as focus formation in NIH 3T3 cells.

The stronger of the two GSEs was found to be an antisense oriented fragment of the 3′-untranslated region of a mRNA encoding a polypeptide designated p33, a novel nuclear protein, expression of which was inhibited by the GSE. The p33-encoding gene was named ING1 (ING1 stands for "INhibitor of Growth"). Overexpression of ING1 caused growth arrest in different cell lines. ING1 was rearranged in one of four tested neuroblastoma cell lines and its expression was reduced in several breast carcinomas ***(6)***.

ING1 expression was found to be increased in senescent human diploid fibroblasts, and anti-ING1 GSE prolonged the lifespan of these cells in culture ***(32)***. In addition, ING1 GSE reduced the apoptosis response in some cellular systems ***(49)***. Association of ING1 with growth suppression, replicative senescence, anchorage dependence, and apoptosis allowed us to define ING1 as a candidate TSG. Further characterization of biologic properties of ING1 and ING1-derived GSE demonstrated that this gene is involved in the p53 pathway ***(50)***. ING1 is expressed as three alternative transcripts that encode two proteins, $p37^{ing1}$ and $p31^{ing1}$. The second protein is a truncated version of the first protein, lacking 104 N-terminal amino acids. Strikingly, we have found that the shorter variant of ING1 acts as a p53 activator, whereas the longer variant acts as a repressor of p53 ***(51)***. ING1 interacts physically with the p53–MDM2–ARF complex (S. S. Kwek and A. V. G., in preparation), and its longer form participates in HDAC1 transcriptional repressor complex ***(52)***.

2.3.3. Selection of Transforming GSEs from a Library of Normalized Murine cDNA Fragments

We have used a large normalized cDNA library from murine NIH 3T3 cells to select GSEs that induced a transformed phenotype in mouse cells in vitro ***(53)***. This library was transduced into early-passage MEFs or into spontaneously immortalized Swiss 3T3 cells, which were then selected for the formation of transformed foci. Three GSEs were isolated by this selection (**Table 2**). Two of the GSEs represent unknown genes, and the third corresponds to a fragment of the *phosphotidyl inositol class 3 receptor*, a gene with previously determined tumor suppressor activity ***(54)***.

A GSE that suppresses p53-mediated apoptosis was isolated from the same normalized NIH 3T3 library, by selection in E1A and RAS-transformed MEF cell line C8. This cell line is extremely sensitive to apoptosis caused by DNA damage or detachment from substrate ***(55)***, and the apoptosis program in these cells is strongly dependent on p53 ***(56)***. Two rounds of selection for the survival of etoposide yielded several clones. The GSE that displayed the strongest activity was characterized in more detail. This GSE contained a 551-bp insert that does not correspond to any known mouse gene, nor does it contain motifs related to any known sequences. This GSE blocks etoposide-induced apoptosis and provides anchorage-independent growth in methylcellulose almost as efficiently as *bcl-2* and dominant-negative *p53* (V. S. Ossovskaya, A. Budanov, A. V. G., and P. M. Chumakov, unpublished data).

2.3.4. A Drug Resistance-Inducing GSE Indicates a TSG Function of a Motor Protein

A GSE selection that was not originally aimed at TSG has indicated that a known motor protein has properties of a potential tumor suppressor. In that study ***(8)***, we used the normalized NIH 3T3 cDNA library to select GSEs that induce resistance to etoposide in NIH 3T3 cells and in a NIH 3T3-derived packaging cell line. After two rounds of selection, we isolated three GSEs that were able to increase the resistance of NIH 3T3 cells to etoposide. Two GSEs carried cDNA fragments of unknown genes, and one was derived from a gene that belongs to the superfamily of kinesin motor proteins ***(57)***. The latter GSE is an antisense-oriented fragment of the ubiquitously expressed kinesin heavy-chain (uKHC) cDNA. Decreased expression of the uKHC gene was found in four

HeLa-derived cell lines that had been selected independently for etoposide resistance, suggesting a role for KHC suppression in the resistant phenotype. GSE selection from full-length uKHC cDNA, together with the use of rationally designed mutants *(29)*, indicated that the C-terminus corresponding to the tail domain of uKHC was a determinant of the effect of kinesin on drug resistance. The C-terminal domain is known to interact with kinesin light chains, which are believed to mediate the link between kinesin and its ligands. Drug resistance associated with uKHC deregulation was not limited to etoposide but involved other DNA-damaging drugs with different mechanisms of toxicity, including doxorubicin, cisplatin, and camptothecin *(8)*. On the other hand, uKHC deregulation was also associated with increased sensitivity to microtubule-depolymerizing agents *(29)*. This association of resistance to some drugs with collateral sensitivity to others may potentially be exploited in cancer chemotherapy.

Unexpectedly, the uKHC GSE not only affected drug resistance and sensitivity, but also promoted the immortalization of primary MEF and prolonged the lifespan of human diploid fibroblasts in culture *(8)*. This combination of biologic effects of kinesin is indicative of a TSG function and closely resembles the functions of p53. We therefore investigated whether kinesin could play a role in the p53 pathways. Specifically, kinesin is known to be involved in cellular secretion, presumably by transporting secretory vesicles between ER, Golgi and plasma membrane *(58)*. On the other hand, p53 is known to control stress-dependent cellular export of growth-inhibitory factors *(59)*. This function possibly determines the "bystander effect" associated with p53 gene therapy *(60)*. We therefore hypothesized that deregulation of kinesin could affect p53-dependent secretion of growth inhibitors, thus increasing the overall survival of the treated cell population. This hypothesis was tested using cells transduced with dominant negative mutants of kinesin and poliovirus protein 3A that acts as an inhibitor of secretion (K. Gurova, A. Komarov, and A. V. Gudkov, in preparation). This analysis showed that DNA damage response involves p53-dependent kinesin-mediated secretion of growth-inhibitory factors, which contribute to the overall sensitivity of a cell population to the treatment. Thus, the initial discovery of kinesin-derived drug resistance-inducing GSE has led to an important conclusion that the sensitivity of a cell population to genotoxic stress is determined not only by the response of the affected cells, but also by a newly described "natural bystander effect," which is mediated through the kinesin function.

In conclusion, applications of the GSE approach to the field of TSGs have provided abundant new information on the functional organization of p53 and on the nature of proteins involved in the p53-regulated pathways. GSE selection has also yielded several new candidate TSGs. This methodology can be readily applied to the study of other TSGs and to the selection of GSEs with transforming activity in different types of normal tissues. The GSE-based research strategy is likely to increase our knowledge of TSGs and TSG-mediated pathways in different types of human cancer.

Acknowledgments

We wish to thank all our collaborators in GSE studies for their valuable contributions to the development and applications of the GSE methodology. Our work on GSEs was supported by National Institutes of Health Grants CA62099, CA56736 (I.B.R.), CA60730, CA75179 (A.V.G.), and a contract from PPD Discovery, Inc. (I.B.R.).

References

1. Gudkov, A. V. and Roninson, I. B. (1997). Isolation of genetic suppressor elements (GSEs) from random fragment cDNA libraries in retroviral vectors, in *Methods in Molecular Biology, Vol. 69: cDNA Library Protocols* (Cowell, I. G. and Austin, C. A., eds.) Humana Press, Totowa, NJ, pp. 221–240.
2. Roninson, I. B., Zuhn, D., Ruth, A., and de Graaf, D. (1998) Isolation of altered-function mutants and genetic suppressor elements of multidrug transporter P-glycoprotein by expression selection from retroviral libraries. *Meth. Enzymol.* **292**, 225–248.
3. Holzmayer, T. A., Pestov, D. G., and Roninson, I. B. (1992) Isolation of dominant negative mutants and inhibitory antisense RNA sequences by expression selection of random DNA fragments. *Nucleic Acids Res.* **20**, 711–717.
4. Roninson, I. B., Gudkov, A. V., Holzmayer, T. A., et al. (1995) Genetic suppressor elements: new tools for molecular oncology. *Cancer Res.* **55**, 4023–4028.
5. Herskowitz, I. (1987) Functional inactivation of genes by dominant negative mutations. *Nature* **329**, 219–222.
6. Garkavtsev, I., Kazarov, A., Gudkov, A. V., and Riabowol, K. (1996) Suppression of the novel growth inhibitor $p33^{ING1}$ promotes neoplastic transformation. *Nat. Genet.* **14**, 415–420.
7. Lisitsyn, N., Lisitsyn, N., and Wigler, M. (1993) Cloning the differences between two complex genomes. *Science* **259,** 946–951.
8. Gudkov, A. V., Kazarov, A. R., Thimmapaya, R., Axenovich, S., Mazo, I., and Roninson, I. B. (1994) Cloning mammalian genes by expression selection of genetic suppressor elements: association of kinesin with drug resistance and cell immortalization. *Proc. Natl. Acad. Sci. USA* **91**, 3744–3748.
9. Patanjali, S. R., Parimoo, S., and Weissman, S. M. (1991) Construction of a uniform abundance (normalized) cDNA library. *Proc. Natl. Acad. Sci. USA* **88**, 1943–1947.
10. Diatchenko, L., Lau, Y-F. C., Campbell, A. P.,et al. (1996) Suppression subtractive hybridization: A method for generating differentially regulated or tissue-specific cDNA probes and libraries. *Proc. Natl. Acad. Sci. USA* **93**, 6025–6030.
11. Raveh, T., Berissi, H., Eisenstein, M., Spivak, T., and Kimchi, A. (2000) A functional genetic screen identifies regions at the C-terminal tail and death-domain of death-associated protein kinase that are critical for its proapoptotic activity. *Proc. Natl. Acad. Sci. USA* **97**, 1572–1577.
12. Caponigro, G., Abedi, M. R., Hurlburt, A. P., Maxfield, A., Judd, W., and Kamb, A. (1998) Transdominant genetic analysis of a growth control pathway. *Proc. Natl. Acad. Sci. USA* **95**, 7508–7513.
13. Mittelman, J. M. (2001) Generation of gene-supressing peptides from genetic suppressor elements. Ph. D. Thesis, University of Illinois at Chicago.
14. Pear, W. S., Nolan, G. P., Scott, M. L., and Baltimore, D. (1993) Production of high-titer helper-free retroviruses by transient transfection. *Proc. Natl. Acad. Sci. USA* **90**, 8392–8396.
15. Schott, B., Iraj, E. S., and Roninson, I. B. (1996) Effects of infection rate and selection pressure on gene expression from an internal promoter of a double-gene retroviral vector. *Somat. Cell Mol. Genet.* **22**, 291–309.
16. Levenson, V. V., Lausch, E., Kirschling, D. J., et al. (1999) A combination of genetic suppressor elements produces resistance to drugs inhibiting DNA replication. *Somat. Cell Mol. Genet.* **25**, 9–26.
17. Miller, A. D. and Rosman, G. J. (1989) Improved retroviral vectors for gene transfer and expression. *Biotechniques* **7**, 980–986.
18. Chang, B. D. and Roninson, I. B. (1996) Inducible retroviral vectors regulated by lac repressor in mammalian cells. *Gene* **183**, 137–142.

19. Kandel, E. S., Chang, B. D., Schott, B., Shtil, A. A., Gudkov, A. V., and Roninson, I. B. (1997) Applications of green fluorescent protein as a marker of retroviral vectors. *Somat. Cell Mol. Genet.* **23**, 325–340.
20. Ossovskaya, V. S., Mazo, I. A., Chernov. M. V., et al. (1996) Use of genetic suppressor elements to dissect distinct biological effects of separate p53 domains. *Proc. Natl. Acad. Sci. USA* **93**, 10309–10314.
21. Jiang, X. R., Jimenez, G., Chang, E., et al. (1999) Telomerase expression in human somatic cells does not induce changes associated with a transformed phenotype. *Nat. Genet.* **21**, 111–114.
22. Morales, C. P., Holt, S. E., Ouellette, M., et al. (1999) Absence of cancer-associated changes in human fibroblasts immortalized with telomerase. *Nat. Genet.* **21**, 115–118.
23. Pestov, D. G. and Lau, L. F. (1994) Genetic selection of growth-inhibitory sequences in mammalian cells. *Proc. Natl. Acad. Sci. USA* **91**, 12549–12553.
24. Albritton, L. M., Tseng, L., Scadden, D., and Cunningham, J. M. (1989) A putative murine ecotropic retrovirus receptor gene encodes a multiple membrane-spanning protein and confers susceptibility to virus infection. *Cell* **57**, 659–666.
25. Dunn, S. J., Park, S. W., Sharma, V., et al. (1999) Isolation of efficient antivirals: genetic suppressor elements against HIV-1. *Gene Ther.* **6**, 130–137.
26. Zuhn, D. L. (1996) Genetic suppressor elements inhibiting P-glycoprotein function. Ph. D. Thesis, University of Illinois at Chicago.
27. Stetten, G., Davidson, R. L., and Latt, S. A. (1977) 33258 Hoechst enhances the sensitivity of the bromodeoxyuridine-light method of isolating conditional lethal mutants. *Exp. Cell Res.* **108**, 447–452.
28. Tarasewicz, D. (1995) Inhibition of BCL-2 with genetic suppressor elements. Ph. D. Thesis, University of Illinois at Chicago.
29. Axenovich, S. A., Kazarov, A. R., Boiko, A. D., Armin, G., Roninson, I. B., and Gudkov, A. V. (1998) Altered expression of ubiquitous kinesin heavy chain results in resistance to etoposide and hypersensitivity to colchicine: mapping of the domain associated with drug response. *Cancer Res.* **58**, 3423–3428.
30. Gudkov, A. V., Zelnick, C., Kazarov, A. R., et al. (1993) Isolation of genetic suppressor elements, inducing resistance to topoisomerase II-interactive cytotoxic drugs, from human topoisomerase II cDNA. *Proc. Natl. Acad. Sci. USA* **90**, 3231–3235.
31. Carnero, A., Hudson, J. D., Hannon, G. J., and Beach, D. H. (2000) Transdominant genetic analysis of a growth control pathway. *Nucleic Acids Res.* **28**, 2234–2241.
32. Garkavtsev, I. and Riabowol, K. (1997) Extension of the replicative life span of human diploid fibroblasts by inhibition of the p33ING1 candidate tumor suppressor. *Mol. Cell. Biol.* **17**, 2014–2019.
33. Schott, B., Kandel, E. S., and Roninson, I. B. (1997) Efficient recovery and regeneration of integrated retroviruses. *Nucleic Acids Res.* **25**, 2940–2942.
34. Weinberg, R. A. (1991) Tumor suppressor genes. *Science* **254**, 1138–1146.
35. Gottlieb, T. M. and Oren, M. (1996) p53 in growth control and neoplasia. *Biochim. Biophys. Acta* **1287**, 77–102.
36. Gottlieb, T. M. and Oren, M. (1998) p53 and apoptosis. *Semin. Cancer Biol.* **8**, 359–368.
37. Lowe, S. W. (1999) Activation of p53 by oncogenes. *Endocr. Relat. Cancer* **6**, 45–48.
38. Zheng, L., Li, S., Boyer, T. G., and Lee, W. H. (2000). Lessons learned from BRCA1 and BRCA2. *Oncogene* **19**, 6159–6175.
39. Kastan, M. B. and Lim, D. S. (2000) The many substrates and functions of ATM. *Nat. Rev. Mol. Cell Biol.* **1,** 179–186.
40. Land, H., Parada, L. F., and Weinberg, R. A. (1983) Tumorigenic conversion of primary embryo fibroblasts requires at least two cooperating oncogenes. *Nature* **304,** 596–602.

41. Serrano, M., Lin, A. W., McCurrach, M. E., Beach, D., and Lowe, S. W. (1997) Oncogenic *ras* provokes premature cell senescence associated with accumulation of p53 and p16INK4a. *Cell* **88**, 593–602.
42. van Lohuizen, M., Verbeek, S., Scheijen, B., Wientjens, E., van der Gulden, H., and Berns, A. (1991) Identification of cooperating oncogenes in E mu-myc transgenic mice by provirus tagging. *Cell* **65**, 737–752.
43. Lima, L. and Macieira-Coelho, A. (1972) Parameters of aging in chicken embryo fibroblasts cultivated in vitro. *Exp. Cell Res.* **70**, 279–284.
44. White, A. E., Livanos, E. M., and Tlsty, T. D. (1994) Differential disruption of genomic integrity and cell cycle regulation in normal human fibroblasts by the HPV oncoproteins. *Genes Dev.* **8,** 666–677.
45. Weinberg, R. A. (1995) The molecular basis of oncogenes and tumor suppressor genes. *Ann. N.Y. Acad. Sci.* **758**, 331–338.
46. Chang, B. D., Xuan, Y., Broude, E. V., et al. (1999) Role of p53 and p21waf1/cip1 in senescence-like terminal proliferation arrest induced in human tumor cells by chemotherapeutic drugs. *Oncogene* **18**, 4808–4818.
47. Rokhlin, O. W., Gudkov, A. V., Kwek, S. S., Glover, R. A., Gewies, A. S., and Cohen, M. B. (2000) p53 is involved in tumor necrosis factor-alpha-induced apoptosis in the human prostatic carcinoma cell line LNCaP. *Oncogene* **19**, 1959–1968.
48. Gallagher, W. M., Cairney, M., Schott, B., Roninson, I. B., and Brown, R. (1997) Identification of p53 genetic suppressor elements which confer resistance to cisplatin. *Oncogene* **14**, 185–193.
49. Helbing, C. C., Veillette, C., Riabowol, K., Johnston, R. N., and Garkavtsev, I. (1997) A novel candidate tumor suppressor, *ING1*, is involved in the regulation of apoptosis. *Cancer Res.* **57**, 1255–1258.
50. Garkavtsev, I. V., Grigorian, I. A., Ossovskaya, V. S., Chumakov, P. M., and Gudkov, A. V. (1998) A candidate tumor suppressor $p33^{ING1}$ cooperates with p53 in cell growth control. *Nature* **391**, 295–298.
51. Zeremski, M., Hill, J. E., Grigorian, I. A., et al. (1999) Structure and regulation of the mouse *ing1* gene: three alternative transcripts encode two PHD finger proteins that have opposite effects on p53 function. *J. Biol. Chem.* **274**, 32171–32181.
52. Skowyra, D., Zeremski, M., Neznanov, N., et al. (2001) Differential association of products of alternative transcripts of the candidate tumor suppressor *ING1* with the mSin3/HDAC1 transcriptional corepressor complex. *J. Biol. Chem.* **276**, 8734–8739.
53. Mazo, I. A. (1997) Isolation of genetic suppressor elements promoting neoplastic transformation. Ph. D. Thesis, University of Illinois at Chicago.
54. Fischer, G. A., Clementi, E. Raichman, M., Sudhof, T., Ullrich, A., and Meldolesi, J. (1994) Stable expression of truncated inositol 1,4,5-trisphosphate receptor subunits in 3T3 fibroblasts. Coordinate signaling changes and differential suppression of cell growth and transformation. *J. Biol. Chem.* **269**, 19216–19224.
55. Nikiforov, M. N., Ossovskaya, V. S., Hagen, K., Lowe, S., Deichman, G. I., and Gudkov, A. V. (1996) p53 modulation of anchorage independent growth and experimental metastasis. *Oncogene* **13**, 1709–1719.
56. Lowe, S. W., Ruley, H. E., Jacks, T., and Housman, D. E. (1993) p53-dependent apoptosis modulates the cytotoxicity of anticancer agents. *Cell* **74**, 957–967.
57. Vale, R. D. and Fletterick, R. J. (1997) The design plan of kinesin motors. *Annu. Rev. Cell Dev. Biol.* **13**, 745–777.
58. Reilein, A. R., Rogers, S. L., Tuma, M. C., and Gelfand, V. I. (2001) Regulation of molecular motor proteins. *Int. Rev. Cytol.* **204,**179–238.
59. Komarova, E. A., Diatchenko, L., Rokhlin, O. W., et al. (1998) Stress-induced secretion of growth inhibitors: a novel growth regulatory function of p53. *Oncogene* **17**, 1089–1096.

60. Qazilbash, M. H., Xiao, X., Seth, P., Cowan, K. H., and Walsh, C. E. (1997) Cancer gene therapy using a novel adeno-associated virus vector expressing human wild-type p53. *Gene Ther.* **4**, 675–682.
61. Tenson, T., DeBlasio, A., and Mankin, A. (1996). A functional peptide encoded in the Escherichia coli 23S rRNA. *Proc. Natl. Acad. Sci. USA* **93**, 5641–5646.
62. Levenson, V. V., Davidovich, I. A., and Roninson, I. B. (2000). Pleiotropic resistance to DNA-interactive drugs is associated with increased expression of genes involved in DNA replication, repair and stress response. *Cancer Res.* **60**, 5027–5030.

27

Cross-linking Subtractive Hybridization to Identify Tumor Suppressor Genes

Gen Sheng Wu

1. Introduction

Cell division is tightly controlled by both positive and negative growth factors. Disruption of this balance leads to the onset of neoplasm *(1)*. For example, a group of proteins, mostly encoded by oncogenes, act as positive factors to promote cell growth and proliferation. Tumor suppressor gene-encoding proteins, on the other hand, act as brakes to slow or stop cell growth and proliferation. Cancer is associated with activation of oncogenes and inactivation of tumor suppressor genes, which results in uncontrolled cellular proliferation and luck of adequate cell death.

A number of methods have been developed to identify the differential gene expression, including tumor suppressor gene expression at various conditions. Examples of such methods include subtractive hybridization screening *(2,3)*, differential display-polymerase chain reaction (DD-PCR) *(4,5)*, and microarray screening *(6,7)*. Each method has its advantages and limitations. In this chapter, I will detail the subtractive hybridization methodology.

1.1. Principle of Cross-linking Subtractive Hybridization

Subtractive hybridization technology has been widely used in biomedical research and several important genes, such as p21 and KILLER/DR5, were isolated by this method *(8,9)*. Essentially, this method is used to compare the transcripts between two experimental settings. It includes several steps (Fig. 1). The first step is to prepare high-quality mRNA. Two RNA resources are used in subtractive hybridization, tester mRNA and driver mRNA. Tester mRNA is isolated from the cells or tissues in which you will expect to have increased transcripts. The driver (control) mRNA usually serves as a control. The second step is to construct a cDNA library in which you wish to identify increased RNA transcripts in your experimental setting. This step involves standard molecular technology. For example, tester mRNA is reverse-transcribed into a single cDNA, followed by second-strand synthesis. The resulting double-strand DNA is ligated with EcoRI adapters and cloned into a λ vector. The vectors are then packaged and transformed into the host

From: *Methods in Molecular Biology, Vol. 222: Tumor Suppressor Genes: Pathways and Isolation Strategies*
Edited by: Wafik S. El-Deiry © Humana Press Inc., Totowa, NJ

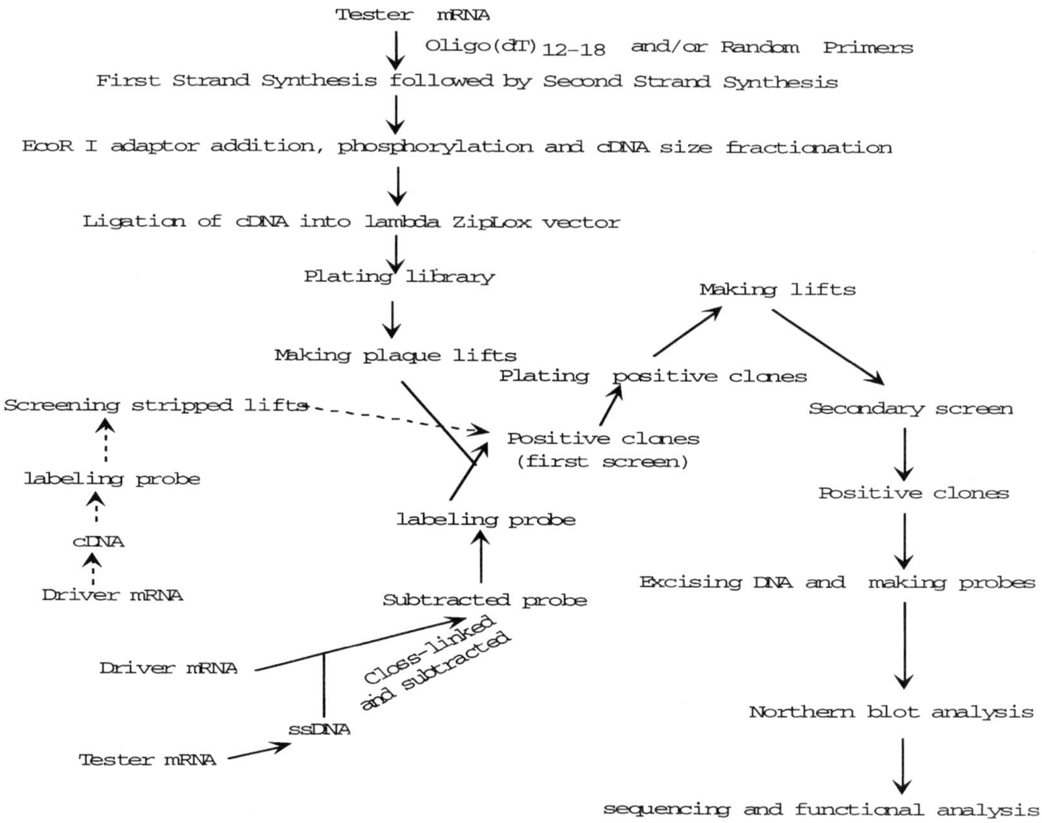

Fig. 1. Summary of cDNA library construction, probe preparation, and subtractive hybridization screen. A cDNA library is made with tester mRNA by first- and second-strand synthesis, followed by ligation of cDNA into a λ vector and library plating. To make subtracted probes, tester mRNA is reverse-transcribed into ssDNA, and then hybridized with driver mRNA, followed by cross-linking, and subtraction. Control probe is made directly from driver mRNA by labeling cDNA from tester mRNA, as shown by the dashed arrows. It should be pointed out that positive clones are those plaques that have stronger signal when hybridized with subtracted probes than when hybridized with control probes.

cells. The library is titrated and plated. The phage plaques containing representations of cDNA population are blotted onto the nitrocellular member lifts. The third step is to make subtracted probe, as well as control probe. Here a chemical cross-linking method is introduced to make subtracted probe ***(10)***. cDNA reverse transcripted from the tester mRNA is hybridized with the driver mRNA and the resulting cDNA–mRNA complexes are chemically cross-linked. The unhybridized cDNA population is isolated, which serves as subtracted probes. Two methods are described in the context of labeling cDNA probes; the first is to label cDNA starting with mRNA by a reverse transcription reaction; the other is to label cDNA by the random primer method. The fourth step is to screen a cDNA library. Usually, two rounds of screening are required. Lifts containing cDNA are hybridized with subtracted probe, washed, and exposed to X-ray film. After completion of screening with subtracted probe, the lifts are stripped and reprobed with control probe. The signal is observed after exposure of the lifts on the film. Once you have finished library screening, the following step (step 5) is to identify positive clones. Because subtracted probe enriches the representation of the transcripts of interest. The last step is to verify the identified putative positive clones. For this step, the most commonly used method is Northern blot analysis. Positive plaques are those that are associated with a stronger signal when hybridized with a subtracted probe as compared with the control probe. Once you have confirmed differential expression between driver and tester mRNAs, the positive clones can be characterized by sequencing and functional analysis. A number of methods can be used to characterize genes functionally. Readers are encouraged to refer to these methods in other chapters of this book.

2. Materials

2.1. cDNA Library Construction

1. DEPC-H_2O.
2. Oligo$(dT)_{12\text{-}16}$ primer (0.5 μg/μL).
3. 5× First-strand buffer: 250 m*M* Tris-HCL, pH 8.3, 375 m*M* KCl, 15 m*M* $MgCl_2$.
4. 0.1 *M* dithiothreitol (DTT).
5. RNasin (40 U/μL).
6. 10 m*M* dNTP mix: 10 m*M* each dATP, dCTP, dGTP, dTTP.
7. α^{32}P-dCTP (10 μCi/μL).
8. SuperScript II RT (200 U/μL).
9. 20 m*M* EDTA.
10. 10% (w/v) TCA in 1% sodium pyrophosphate.
11. 7.5 *M* NH_4OAc.
12. Absolute ethanol.
13. 75% ethanol.
14. 5× Second-strand buffer: 100 m*M* Tris-HCl, pH 6.9, 450 m*M* KCl, 23 m*M* $MgCl_2$, 0.75 m*M* β-NAD^+, 50 m*M* $(NH4)_2SO_4$.
15. *Escherichia coli* DNA ligase (10 U/μL).
16. *E. coli* DNA polymerase I (10 U/μL).
17. *E. coli* RNase H (2 U/μL).
18. T4 DNA ligase (5 U/μL).
19. Phenol:chloroform:isoamyl alcohol (20:24:1).
20. 5× Adapter buffer: 330 m*M* Tris-HCl, pH 7.6, 50 m*M* $MgCl_2$, 5 m*M* ATP.
21. EcoRI (Not I) adapters (1 μg/μL): the sequence is as follows:

5′-AATTCGCGGCCGCGTCGAG-3′

3′-____GCGCCGGCGCAGCTGp-5′

EcoRI Not I EcoRI

22. T4 polynucleotide kinase (10 U/μL).
23. 5× T4 DNA ligase buffer: 250 m*M* Tris-HCl, pH 7.6, 50 m*M* $MgCl_2$, 5 m*M* ATP, 5 m*M* DTT, 25% (w/v) PEG 8000.
24. cDNA size fraction columns (Sephacry S-500) (Gibco/BRL).
25. TEN buffer: 10 m*M* Tris-HCl, pH 7.5, 0.1 m*M* EDTA, 25 m*M* NaCl.
26. TE: 10 m*M* Tris–HCl, pH 7.6, 1 m*M* EDTA.

2.2. Ligation of cDNA into a λZipLox Vector

1. 10× DNA ligase buffer.
2. λZipLox, EcoR I Arms (250 μg/mL) (Gibco/BRL).

3. T4 DNA ligase (1 U/μL).
4. 0.1 *M* DTT.

2.3. In-Vitro Packaging and Transformation of E. coli

1. LB-agar plate.
2. LB-agar plate with ampicillin (50 μg/mL).
3. LB-agar plate with kanamycin (50 μg/mL).
4. LB top agar: 6 g agarose, 16 g LB broth, in 800 mL of water, sterile.
5. BACTi Y1090 (ZL) strain.
6. BACTi DH10B (ZIP) strain.
7. GigoPack kit (Stratagene).
8. Phase dilution buffer (PDB): 50 m*M* Tris-HCl, pH 7.9, 100 m*M* NaCl, 10 m*M* $MgCl_2$, 0.01% (w/v) gelatin.
9. Chloroform.
10. 20% maltose (sterile).
11. 10 m*M* $MgSO_4$.

2.4. Plating cDNA Library and Preparation of Lifts

1. 150-nm agar plate.
2. Nitrocellular filters (NEN).

2.5. Preparation of Probes and Hybridization

1. Olig $dT_{12\text{-}16}$.
2. 5× First-strand buffer.
3. 0.1 *M* DTT.
4. RNasin (40 U/μL) (Gibco/BRL).
5. 10 m*M* dNTP (dATP, dCTP, dGTP, and dTTP).
6. MMLV Superscript II (200 U/μL) (Gibco/BRL).
7. 0.5 *N* HCl.
8. Glycogen (20 mg/mL) (Roche Diagnostic).
9. 2 *M* NaClO4.
10. CCSB (chemical cross-linking subtraction buffer) or 1× subtractive hybridization buffer: 0.5 *M* NaCl, 25 m*M* HEPES (pH 7.5), 5 m*M* EDTA, 1% sodium dodecyl sulfate (SDS).
11. 10 m*M* 2,5-diaziridinyl-1,4-benzoquinone (DZQ).
12. Bovine serum albumin (BSA) (10 mg/mL).
13. Oligo-labeling buffer with dNTP (OLB): 125 m*M* Tris-HCl, 12.5 m*M* MgCl2, 25 m*M* 2-mercaptoethanol, 150 μg/mL oligodeoxyribonucleotide primers (random octamers), 10 m*M* dATP, 10 m*M* dCTP, 10 m*M* DGTP, 10 m*M* dTTP.
14. α-^{32}P-dCTP (3000 Ci/mmol, 10 mCi/mL).
15. Sequenase (3 U/μL) (USB Corporation).
16. Placental DNA (2 mg/mL).
17. 20× sodium chloride sodium citrate (SSC).
18. Blotto 10: 20 g of SDS, 10 g powdered milk, 120 g PEG 8000, 1000 mL 10× SSPE, 200 mL formamide, 20 mL salmon sperm DNA (24 mg/mL), and 680 mL H_2O.
19. Wash solution 1: 0.1× SSC, 0.1% SDS.
20. Wash solution 2: 0.3× SSC, 0.1% SDS.

3. Methods

3.1. cDNA Library Construction

We have used the SuperScript™ Choice System Kit (GIBCO/BRL) to construct a cDNA library and give us a satisfactory result. The following is the method, essentially based on the manufacturer's instruction with some modifications.

3.1.1. First-Strand cDNA Synthesis (see **Notes 1** and **2**)

The mRNA that is used to construct a cDNA library that contains transcripts you expect to increase in your experimental system.

1. Combine the following components in a sterile 1.5-mL microcentrifuge tube:
   a. 2–4 µL Poly $(A)^+$ RNA (2 µg).
   b. 2 µL Oligo $dT_{12\text{-}18}$ (1 µg of 0.5 µg/µL).
   c. DEPC H_2O to 8 µL.
2. Heat the mixture to 70°C for 10 min and then quickly cool the tube on ice for 2 min, followed by a brief spin.
3. Add the following to the reaction tube.
   a. 4 µL 5× first-strand buffer.
   b. 2 µL of 0.1 *M* DTT.
   c. 0.5 µL Rnasin (40 U/µL).
   d. 1 µL of 10 m*M* dNTP mix.
   e. 0.5 µL of α-^{32}P-dCTP.
   f. H_2O to 18 µL.

   Mix gently and spin briefly, and incubate at 37°C for 2 min, then add 2 µL SuperScript II RT (200 U/µL). Incubate at 37°C for 1 h. Collect the reaction by a brief spin and place the tube on ice.
4. Remove 2 µL from the reaction to a new 1.5-µL Eppendorf tube containing 43 µL of 20 m*M* EDTA (pH 7.5) and 5 µL of yeast tRNA. Spot duplicate10 µL of 50-µL diluted sample onto 2 glass fiber filters. Dry one of the filters at room temperature or under a heat lamp; this filter is used to determine the yield of the first-strand cDNA.
5. Wash the second filter with 10% (w/v) TCA containing 1% (w/v) sodium pyrophosphate 3 times (5 min for each). After washing with 95% ethanol for 2 min, dry the filter at room temperature or under a heat lamp.
6. Count the radioactivity of both filters in standard scintillant with ^{32}P channel to determine the amount of ^{32}P in the reaction and incorporated ^{32}P in the first strand.
7. Add 15 µL of 7.5 *M* NH_4OAc and 90 µL of absolute ethanol to the remaining 30 µL of diluted reaction. Mix thoroughly and centrifuge at room temperature for 20 min at 14,000*g*.
8. Remove the supernatant carefully, and wash the pellet with 0.5 mL of 70% ethanol (–20°C). Centrifuge at 14,000*g* for 5 min and remove the supernatant.
9. Leave the cDNA pellet at room temperature for 20 min to dry, and proceed to the next section.

3.1.2. Second-Strand Synthesis

1. Add the following reagents into the remaining18 µl of the first strand reaction in order. All the steps must be carried out on ice, otherwise indicated.
   a. 18 µL First-strand reaction.
   b. 93 µL DEPC-H_2O.
   c. 30 µL 5× second-strand buffer.
   d. 3 µL of 10 m*M* dNTP.

   e. 1 µL *E. coli* DNA ligase (10 U/µL).
   f. 4 µL *E. coli* DNA polymerase I (10 U/µL).
   g. 1 µL *E. coli* RNase H (2 U/µL).
2. Mix the tube gently and spin briefly, and then incubate the tube at 16°C for 2 h.
3. Add 2 µL of T4 DNA polymerase (5 U/µL), mix well, and continue to incubate at 16°C for another 5 min.
4. Place the tube on ice and add 10 µL of 0.2 m*M* EDTA to stop the reaction.
5. Add 250 µL of phenol:chloroform:isoamyl alcohol (25:24:1), vortex thoroughly, and centrifuge at 14,000× *g* for 4 min at room temperature. Carefully remove the upper aqueous layer, and transfer to a new 1.5-mL Eppendorf tube.
6. Add 70 µL of 7.5 *M* NH_4OAc and 0.5 mL of absolute ethanol (–20°C). Vortex thoroughly and precipitate cDNA by centrifuging at room temperature at 14,000× *g* for 20 min.
7. Carefully remove the supernatant and add 0.5 mL of 70% ethanol (–20°C). Centrifuge at 14,000× *g* for 5 min and carefully remove the supernatant.
8. Dry the cDNA at room temperature for 20 min and resuspend the cDNA in 18 µL of DEPC-H_2O.

3.1.3. Addition of EcoRI (NotI) Adapter

1. Add the following reagents into above **step 8:**
   a. 10 µL 5× Adapter.
   b. 10 µL EcoRI (Not I) adapters.
   c. 7 µL of 0.1 *M* DTT.
   d. 5 µL T4 DNA ligase.
   e. 50 µL Total.
2. Mix gently and spin briefly, and then incubate the tube at 16°C overnight (more than 16 h).
3. Inactivate the reaction by heating the tube at 70°C for 10 min.
4. Spin briefly, leave the reaction on ice, and proceed to the next step of phosphorylation.

3.1.4. Phosphorylation of EcoRI-Adapted cDNA

1. Add 3 µL of T4 polynucleotide kinase to the above **step 4.**
2. Mix gently and incubate the tube at 37°C for 30 min.
3. Heat the reaction at 70°C for 10 min.
4. Spin briefly and leave the reaction on ice.

3.1.5. Column Chromatography (see **Note 3**)

This procedure is to separate larger cDNA fragments from smaller fragments and adapters and ensure that larger cDNA fragments represent in the cDNA library.

1. Place one cDNA size fraction column in a support. Remove the top cap and the bottom cap, respectively, and allow the liquid to drain.
2. Apply 0.8 mL of TEN buffer (10 m*M* Tris-HCI, pH 7.5, 0.1 m*M* EDTA, 25 m*M* NaCl) onto the column. Let it drain completely. Repeat this step 3 more times for a total of 3.2 mL.
3. Label 20 sterile 1.5-mL microcentrifuge tubes from 1 to 20, place them in a rack, and place the washed column in these labeled tubes starting at number 1.
4. Add 97 µL of TEN buffer to the adapter-ligated reaction in the above **step 4** and mix well.
5. Add the all sample to the center of the top frit and leave it to drain into the bed and collect all the effluent into tube 1.
6. Add 100 µL of TEN buffer to the column and collect the effluent into tube 2.

Table 1
Example of Analysis of cDNA from the cDNA Size Fractionation*

	A	B	C	D	E
No.	Fraction volume (μL)	Total volume (μL)	Cerenkov counts (cpm)	Amount of cDNA (ng)	Concentration of cDNA (ng/μL)
1	129	129	54		
2	101	230	44		
3	17.5	247.5	31		
4	30	277.5	40		
5	13	290.5	43		
6	33	323.5	180		
7	34	357.5	5,539		
8	36	393.5	26,955		
9	27	420.5	43,316	918	34
10	15	435.5	25,742.5	546	36.4
11	40	475.5	794,866	1,685	42.1
12	28	503.5	61,505	1,304	46.6
13	18	521.5	30,868	654	36.3
14	26	547.5	45,336	961	37
15	17	564.5	54,557		
16	33	597.5	36,926		
17	8	605.5	8,730		
18	10	615.5	7,712		
19	19	634.5	12,635		
20	19.5	654	15,524		

*Left column is tube number. Column **A** is the actual volume for each fraction. Column **B** is the cumulative elution volume by adding one fraction after another. Column **C** is the radioactive counts corresponding to each fraction. Column **D** is the amount of cDNA (ng) obtained with the equation as shown in the Notes. Column **E** is the concentration of cDNA (9 ng/μL) based on the numbers in columns **A** and **D**.

7. Add 100 μL of TEN buffer, and collect single-drop (about 35-μL) fractions into individual tubes, one drop per tube. Continue to add another 100 μL of TEN buffer until the totals of 18 drops from tube 3 to tube 20 are collected.
8. Measure the volume using an automatic pipet, one tip for each tube. Record each volume in A column of **Table 1**. Add each elution volume together and place them in column **B**.
9. Measure radioactivity for each tube in a scintillation counter and record Cerenkov counts in column C of **Table 1**.
10. For each fraction in which the Cerenkov counts exceed background, calculate the amount of cDNA using the equation in **Note 4**.
11. Divide each cDNA amount in column **D** by each fraction volume in column **A** and record number in column **E**.
12. Because the plasmid vector or λ vector ligation reactions require 10 and 50 ng of cDNA, respectively, it usually takes equal volume of each fraction from fractions 9 to 14, totaling 50 ng of cDNA if λ vector is used in the ligation, to a new 1.5-mL tube.
13. Add 0.5 and 2 vols of NH_4OAc and absolute ethanol (–20°C) and vortex thoroughly, followed by centrifuging at 14,000× *g* at room temperature for 20 min.
14. Remove the supernatant carefully and overlay the pellet with 0.5 mL of 70% ethanol (–20°C). Spin at 14,000× *g* at room temperature for 5 min and discard the supernatant carefully.

15. Leave the cDNA at room temperature to evaporate trace ethanol. It usually takes about 30–45 min.
16. Dissolve the pellet in 10 µL of TEN buffer if you ligate it into a plasmid vector. For λ vector, the dried pellet is ready to go to the next step. The following procedure is to ligate cDNA into λ vector.

3.1.6. Ligation of cDNA into a λZipLox Vector (see **Note 5**)

1. Prepare 5× ligase buffer. Do not use 5× regular ligation buffer. Dilute 10 µL of 10× ligase buffer supplied with the λZipLox kit, with 10 µL of 100 m*M* DTT.
2. Add reaction components as follows:
   a. 50 ng above dried cDNA.
   b. 1 µL distilled water.
   c. 1 µL of 5× DNA ligase buffer.
   d. 2 µL λZipLox Arm, Not I-Sal1 (250 ng/µL).
   Mix well until the pellet is completely dissolved.
3. Add 1 µl (1 U) of T4 ligase and mix gently by pipetting up and down.
4. Leave the ligation reaction at room temperature overnight.
5. It is ready for in-vitro packaging.

3.1.7. In-Vitro Packaging and Transformation

3.1.7.1. Preparation of Bacterial Strain

Streak 50 µL of Y1090 (ZL) on L-Agar plate, and 50 µL of either DH10B or DH12S on kanamycin plates. Grow them at 37°C overnight.

3.1.7.2. Packaging of λZipLox in Vitro Using GigoPack from Stratagene

1. Remove the appropriate number of packing extracts from a –80°C freezer and place the extracts on Dry Ice.
2. Quickly thaw the packaging extract between fingers until just beginning to thaw.
3. Add the ligated DNA immediately to the packaging extract and place on ice.
4. Stir to mix well. You may pipet, but do not introduce air bubbles.
5. Spin the tube quickly for a few seconds to ensure that all contents are at the bottom of the tube.
6. Incubate the tube at room temperature (about 20°C) for 2 h.
7. Add 0.5 mL of phase dilution buffer.
8. Add 20 µL of chloroform and mix gently.
9. Spin briefly to sediment the debris.
10. Supernatant is now ready to be titered.

3.1.8. Phage Titration

1. Prepare L-agar plates.
2. Prepare bacterial cultures: Take a single clone of Y1090ZL strain and place in 10 mL of LB broth with 0.2 mL of 20% maltose in the absence of antibiotics.
3. Grow at 37°C with OD_{600} of 1.0 for 4–6 h, or you may grow them overnight at 30°C with shaking at 200 rpm.
4. Spin the bacteria at 500× *g* for 10 min.
5. Resuspend the pellet in 5 mL of sterile 10 m*M* $MgSO_4$.
6. Dilute the bacteria to an OD_{600} with sterile 10 m*M* $MgSO_4$.
7. Prepare LB-top agar (sterile) and leave it in water bath (48°C).

8. Place L-agar plate at 37°C to prewarm.
9. Prepare serial dilutions of the packaged phages in PDB. Dilute the packaged phage at 0, 1/5, 1/10, 1/50, 1/100, 1/500, $1/10^3$, $1/10^4$, $1/10^5$, and a tube without phage as a negative control. Each tube contains 20 μL.
10. Add 100 μL of host cells in tubes containing diluted phages.
11. Incubate tubes at 37°C for 10 min.
12. Add 4 mL of LB top agar (48°C) into tubes and mix gently.
13. Pour it on prewarmed LB plates immediately and make sure that top agar covers all over the plate.
14. Incubate the plates at 37°C overnight.
15. Count the plaques and determine the titer in plaque-forming units per milliliter (pfu/mL).

3.1.9. Examining the Inserts in the Packaging Phages

1. Pick up 20 individual plaques with a pasteur pipet with s rubber bulb and place in 1.5-mL tubes.
2. Add 250 μL of phage dilution buffer.
3. Vortex 10 s and let sit at room temperature for 5 min.
4. Add 25 μL of DH10B ZIP bacteria.
5. Let sit at room temperature for 5 min.
6. Plate on L-agar plates with ampicillin.
7. Grow them at 37°C overnight.
8. Pick up one colony from each plate and grow them up in 10 mL LB broth.
9. Miniprep plasmid.
10. Digest with EcoRI and run on 1% agarose gel to check the size of the inserts.

3.2. Plating the cDNA Library (see Note 6)

1. You may amplify the constructed cDNA library, or plate the cDNA library directly, based on titration of the library.
2. Plate cDNA library in prewarmed 150-nm agar plates, which must be made and sit at a cold room (4°C) at least 2 d in advance. Each plate contains 50,000 or less phage particles.
3. Incubate plates at 37°C until plaques reach a diameter of about 1.5 mm.
4. Place plates at 4°C for at least 1 d to allow the top agar to solidify.

3.3. Preparation of Plaque Lifts

1. Number nitrocellular filters (NEF978 from NEN) with a soft-lead pencil.
2. Take the plates from 4°C to room temperature and place the filters onto the surface of top agar and leave it on 2 min.
3. Peel the filter off with a blunt-ended forceps and proceed the filters as described elsewhere *(11)*.

3.4. Preparation of Subtracted Probes (see Note 7)

This procedure involves three different steps. The first step is to synthesize single-strand cDNA and then construct a cDNA library. The second step is to hybridize cDNA with driver mRNA, and the last step is to label the cDNA with α-^{32}P-dCTP.

3.4.1. Preparation of Probes Using the Random Primer Labeling System

1. Tester cDNA synthesis: Use mRNA, identical to that used to construct the cDNA library, and add the following components:

   a. 2 μg Tester mRNA.
   b. 2 μL of 1 μg oligo $dT_{12\text{-}16}$ (0.5 μg/μl).
   c. dH_2O to 11.5 μL.
2. Mix well and heat 70°C for 10 min, then cool immediately on ice for 2 min and spin quickly.
3. Add the following components:
   a. 4 μL of 5× First-strand buffer.
   b. 2 μL of 0.1 *M* DTT.
   c. 0.5 μL RNasin (40 U/μL).
   d. 1 μL of 10 m*M* dNTP mix.
4. Mix gently and leave at 37°C for 2 min.
5. Add 2 μL of MMLV Superscript II (200 U/μL) and incubate at 37°C for 1 h.
6. Inactivate the reaction by adding 21 μL of 1 *M* NaOH at 55°C for 15 min.
7. Add the following components to precipitate cDNA:
   a. 42 μL of 0.5 *N* HCl.
   b. 516 μL TE.
   c. 3 μL Glycogen.
   d. 200 μL of 2 *M* $NaClO_4$.
   e. 400 μL of 2-Propanol.
8. Spin 15 min at 14,000*g*.
9. Wash the pellet with 70% ethanol.
10. Resuspend the pellet in 10 μL of sterile water.
11. Assume that the concentration of this synthesis cDNA is 100 ng/μL and use 5 μL of this to do subtractive hybridization.

3.4.2. Labeling Probes Using the Random Labeling System

3.4.2.1. Preparation of Subtractive Probe

1. 10 μg of driver (control) mRNA and 5 μL of above cDNA.
2. Add DEPC-H_2O to 200 μL.
3. Add 100 μL of 10 *M* NH_4AC and 0.7 mL of 100% of ethanol.
4. Centrifuge at 14,000*g* 10 min at 4°C.
5. Wash the pellet with 70% ethanol and dry it by Speedvac.
6. Resuspend the pellet in 10 μL of 1× subtractive hybridization buffer
7. Spin quickly.
8. Add 1 drop of mineral oil.
9. Heat to 85°C for 5 min and then place at 68°C for overnight (22–24 h).
10. Add 40 μL of RNase-free dH_2O.
11. Remove 50 μL of the above to a new Eppendorf tube.
12. Add 150 μL of 100% absolute ethanol.
13. Spin for 5 min at room temperature.
14. Wash the pellet with 0.9 mL of 70% ethanol.
15. Remove the supernatant carefully.
16. Dry by Speedvac.
17. Resuspend the pellet in 50 μL of CCSB.
18. Incubate 3 min at 68°C and move it to 45°C.
19. Add 1 μL of 10 m*M* DZQ.
20. Incubate at 45°C for 20 min.
21. Add 150 μL of TE, 100 μL of 10 *M* NH_4Ac, 3 μL of glycogen, and 0.7 mL of 100% ethanol, respectively.
22. Spin at room temperature for 5 min.

23. Wash the pellet with 0.9 mL of 70% ethanol.
24. Dry by Speedvac.
25. Resuspend the dried pellet in 20 µL of dH_2O and store at –70°C or use immediatedly for the next step.

3.4.2.2. Labeling Subtractive cDNA

1. You may label substractive cDNA in duplicate. At the end, combine and purify them:
   a. 3 µL Subtracted cDNA.
   b. 3 µL TE.
   c. 1 µL BSA.
   d. 4 µL OLB
   e. 10 µL of α-^{32}P-dCTP.
   f. 2 µL Sequenase (3 U/µL in sequenase enzyme dilution buffer).
   Mix thoroughly and incubate 30 min at room temperature.
2. Add 170 µL of TE9, 100 µL of 10 *M* NH_4OA_C, 3 µL of glycogen, and 0.7 mL of 100% ethanol, respectively.
3. Centrifuge at 14,000× *g* for 20 min at room temperature.
4. Discard the supernatant.
5. Wash the pellet with 1 mL of 70% ethanol.
6. Speedvac to dry the pellet.
7. Resuspend in 50 µL TE.
8. Add 100 µL of placental DNA (2 mg/mL) and then add 50 µL of 20× SSC and mix well.
9. Boil the labeling cDNA for 2 min and place it at 68°C.
10. Meanwhile, take 2 µL out of 200 µL and measure CPM using standard liquid scintillation counting.
11. Calculate the CPM.

3.4.3. Labeling Subtractive cDNA Using the Reverse Transcription Reaction

3.4.3.1. Incorporation of ^{32}P into the Single-Strand Tester cDNA

Usually, start with three individual tubes in triplicate and at the end pool them and measure their radiolabeled cDNA activity.

1. For 2 µg tester mRNA, add 2 µL of 1 µg oligo $dT_{12\text{-}16}$ (0.5 µg/µL), and then add dH_2O to 11.5 µL.
2. Mix well and heat at 70°C for 10 min, then cool immediately on ice for 2 min and spin quickly.
3. Add the following components:
   a. 4 µL of 5× First-strand buffer.
   b. 2 µL of 0.1 *M* DTT.
   c. 0.5 µL of RNasin (40 U/µL).
   d. 1 µL of 10 m*M* dNTP (dATP, dGTP, dTTP) mix.
   e. 8 µL of α-^{32}P-dCTP.
4. Mix gently and place at 37°C for 2 min.
5. Add 2 µL of MMLV Superscript II (200 U/µL) and incubate at 37°C for 1 h.
6. Inactivate the reaction by adding 21 µL of 1 *M* NaOH at 55°C for 15 min.
7. Add the following components to precipitate cDNA:
   a. 42 µL of 0.5 *N* HCl.
   b. 516 µL of TE.
   c. 3 µL of Glycogen.

   d. 200 μL of 2 *M* $NaClO_4$.
   e. 400 μL of 2-propanol.
8. Spin 15 min at 14,000 *g*.
9. Wash the pellet with 70% ethanol.
10. Resuspend the pellet in 10 μL of sterile water.
11. Pool three tubes, with totaling 30 μL.

3.4.3.2. Subtracted Probe Preparation

The whole procedure for making the subtractive probe is the same as described above, except starting with ^{32}P-labeled first-strand cDNA from reverse transcription reaction.

1. 10 μg of driver (control) mRNA and 30 μL of cDNA.
2. Add DEPC-H_2O to 200 μL.
3. Add 100 μL of 10 *M* NH_4OAc and 0.7 mL of 100% of ethanol.
4. Centrifuge at 14,000 *g* for 10 min at 4°C.
5. Wash the pellet with 70% ethanol and dry by Speedvac.
6. Resuspend the pellet in 10 μL of 1× subtractive hybridization buffer.
7. Spin quickly.
8. Add 1 drop of mineral oil.
9. Heat to 85°C for 5 min and then place at 68°C for overnight (22–24 h).
10. Add 40 μL of RNase-free dH_2O.
11. Remove 50 μL to a new Eppendorf tube.
12. Add 150 μL of 100% absolute ethanol.
13. Spin for 5 min at room temperature.
14. Wash the pellet with 0.9 mL of 70% ethanol.
15. Remove the supernatant carefully.
16. Dry the pellet by Speedvac.
17. Resuspend the pellet in 50 μL of CCSB.
18. Incubate for 3 min at 68°C and move it to 45°C.
19. Add 1 μL of 10 m*M* DZQ.
20. Incubate it at 45°C for 20 min.
21. Add 150 μL of TE, 100 μL of 10 *M* NH_4OAc, 3 μL of glycogen, and 0.7 mL of 100% ethanol, respectively.
22. Spin at room temperature for 5 min.
23. Wash the pellet with 0.9 mL of 70% ethanol.
24. Dry the pellet by Speedvac.
25. Resuspend the dried pellet in 20 μL of dH_2O and store at –70°C or use immediately for hybridization.

3.5. Subtractive Hybridization with Subtracted Probe

1. Prewarm hybridization buffer Blotto 10 at 60°C.
2. Place all plaque lifts in a large hybridization bag.
3. Add 100 mL of Blotto 10 buffer in the bag and prehybridize it at 60°C for 2–4 h.
4. Prepare hybridization buffer: Add labeled probe, as described above, and 2.5 mL of salmon sperm DNA (2 mg/mL) in 50 mL of prewarmed Blotto 10 buffer, mix well. Note that the radioactivity should be at least 0.7×10^6 cpm/mL or more.
5. Pour prehybridization buffer out and replace with probe-containing buffer.
6. Hybridize at 60°C for overnight (about 18–24 h) with gentle shaking.
7. Take out filters and place in 1 L of wash solution 1 (0.1× SSC, 0.1% SDS). Make sure that filters are kept separate from each other.

8. Wash twice for 5 min at room temperature with gentle shaking.
9. Wash once for 45–60 min at 65°C with 0.3× SSC and 0.1% SDS wash solution with gentle shaking.
10. Wash twice for 5 min at room temperature with 0.1× SSC and 0.1% SDS wash solution with gentle shaking.
11. Take plaque filters out and blot them on 3-mm filters briefly.
12. Wrap them in plastic and place them on Kodak X-OMAT AR film with an intensifying screen and expose the film overnight (*see* **Note 8**).
13. Hybridize plaque lifts with labeled driver probe.

3.6. Stripping Hybridized Lifts

1. Heat 0.1× SSC buffer and 0.1% SDS wash solution to boiling.
2. Place the lifts in a tray and pour boiled 0.1× SSC and 0.1% SDS buffer over the lifts and wash 10 min at room temperature with gentle shaking.
3. Repeat above step, until the filters are not hot any more.
4. Proceed with the prehybridization step for another hybridization or store the lifts in plastic wrap at –4°C.

3.7. Label Control (Driver) Probe

Driver cDNA probe can be labeled using the random labeling system or the reverse transcription reaction as described above.

3.8. Prehybridization and Hybridization of Lifts with Labeled Driver Probe

Follow the procedure as described above. Expose to the X-ray film as described above.

3.9. Picking Positive Plaques (see Notes 9 and 10)

1. Use the fluorescent or radioactive markers to align the autorads with the probed filters and mark the needle holes in the filters, which enables the filters to be aligned with the film.
2. Identify positive spots that appear in duplicate filters with plaque shape and size; those positive spots appear as stronger signals on the filters screening, with subtracted probe over control (driver) probe.
3. Use a Pasteur pipet equipped with a rubber bulb to pick up individual potential positive plaques in phage dilution buffer and proceed to secondary screen.

3.10. Secondary Screen

1. Prepare the positive plaques identified in the first round screen and plate them as described above.
2. Make the plaque lifts as described above.
3. Make the subtractive probe and control (driver) probe individually as described above.
4. Screen the plaque lifts with labeled subtracted and driver probes separately.
5. Identify the positive spots that have stronger signals on the filters hybridized with subtracted probe over the control probe. At this point, some of the filters have stronger and uniform signals in subtracted probe over control probe, because the plaques are enriched from the first round screen.

3.11. Characterization of the Identified Positive Plaques (see Note 11)

1. Excise the λ phage DNA into plasmid as described previously.
2. Grow up bacteria and prepare plasmid DNA as described elsewhere.

3. Digest the plasmid with EcoRI enzyme because the vector used in constructing the library contains this unique site, and then isolate the insert.
4. Label the inserts with ^{32}P and perform Northern blot analysis to confirm the differential expression between the driver mRNA and tester mRNA.
5. Sequence the inserts to determine the identity of the inserts.
6. Functionally characterize them by determining their effects on the cell cycle and apoptosis using FACS analysis, colony-formation assays, and apoptosis assays, which are described in other chapters of this book.

4. Notes

1. Starting mRNA materials: The quality of starting RNAs is very important. If starting mRNA contaminates genomic DNA, it causes a lot of problems including high background with a number of false positive clones. Therefore, we recommend treating mRNA with Rnase-free Dnase and then analyzing by formaldehyde agarose gel electrophoresis and ethidium bromide staining. 18S and 28S rRNA should be visible. In addition, it is a precaution to change gloves frequently to avoid Rnase contamination during the course of library construction and probe preparation.
2. Either oligio $dT_{12\text{-}18}$ or random primers can be used for the first-strand synthesis. Since some of the oligio $dT_{12\text{-}18}$ synthesized cDNAs may not have 5' coding sequence and some of the random primer synthesized cDNA may not contain the full-length cDNA, the mixture with random primer and oligio $dT_{12\text{-}18}$ is recommended in the first-strand synthesis.
3. To use column chromatography to collect fragments, it is important to let the column dry between 100-µL aliquots. It is also recommended not to take fractions beyond 600 µL, which contains smaller cDNA fragments.
4. The following equation is used to calculates the amount of in each fraction cDNA:

$$\text{Amount of dscDNA (ng)} = \frac{(\text{Cerenkov cpm}) \times 2 \times (4 \text{ pmol dNTP/pmol dCTP}) \times (1000 \text{ ng/}\mu\text{L dscDNA})}{\text{SA (cpm/pmol dCTP)} \times (1515 \text{ pmol dNTP/}\mu\text{g ds cDNA})}$$

5. If you intend to ligate the cDNA into a plasmid vector, there are several vectors as options, such as pSV-SPORT vector (Gibco/BRL).
6. For the plating library, L-agar plates are made and solidify at 4°C at least 2 d earlier; only solidified plates can be used for plating the cDNA library.
7. There are two approaches for preparing the subtracted probes as described in this chapter. The first one is to subtract cDNAs synthesized directly from the tester mRNAs with the driver mRNAs. The resultant tester cDNAs will be labeled with α-^{32}P-dCTP by random primer labeling reaction. Although this approach is commonly used, in some cases, it may not give high labeling efficiency to provide enough CPM count for the cDNA library screening. To avoid this, we recommend the second approach, in which α-^{32}P-dCTP is incorporated directly into the first strand of the tester cDNA by reverse transcription reaction. The resulting cDNAs will be hybridized with driver mRNA to remove cDNAs that express at the same levels between the tester and driver mRNAs.
8. Expose several films with different exposure times and use the one with best exposure time. Also, use the fluorescent or radioactive markers to help align the developed autorad with the filters.
9. Although positive plaques can be easily identified by hybridizing with control probe and subtracted probe, it may be difficult to pick up positive plaques if the plaques merge with each other. To avoid this, do not plate too many plaques in a plate, especially for the second round of screening. We recommend plating less than 50,000 plaques in a 150-mm dish.

10. It helps to identify positive plaques if duplicated plaque lifts are made and they are hybridized at same time. Only both lifts hybridized with probes are considered for positive clones if the differential expression is observed between control probe and subtracted probes.
11. It is very important to verify the differential expression of identified positive clones by Northern blot analysis before any further characterizations are made, since some of the identified clones may be false positive, as observed by using all screening methods such as differential display. Once verified, truly positive clones will be further characterized by sequencing or by functional analysis.

References

1. Vogelstein, B. and Kinzler, K. W. (1993) The multistep nature of cancer. *Trends Genet.* **9**, 138–141.
2. Hedrick, S. M., Cohen, D. I., Nielsen, E. A., and Davis, M. M. (1984) Isolation of cDNA clones encoding T cell-specific membrane-associated proteins. *Nature* **308**, 149–153.
3. Rebagliati, M. R., Weeks, D. L., Harvey, R. P., and Melton, D. A. (1985) Identification and cloning of localized maternal RNAs from Xenopus eggs. *Cell* **42**, 769–777.
4. Liang, P. and Pardee, A. B. (1992) Differential display of eukaryotic messenger RNA by means of the polymerase chain reaction. *Science* **257**, 967–971.
5. Welsh, J., Chada, K., Dalal, S. S., Cheng, R., Ralph, D., and McClelland, M. (1992) Arbitrarily primed PCR fingerprinting of RNA. *Nucleic Acids Res.* **20**, 4965–4970.
6. Lennon, G. G. and Lehrach, H. (1991) Hybridization analyses of arrayed cDNA libraries. *Trends Genet.* **7**, 314–317.
7. Kafatos, F. C., Jones, C. W., and Efstratiadis, A. (1979) Determination of nucleic acid sequence homologies and relative concentrations by a dot hybridization procedure. *Nucleic Acids Res.* **7**, 1541–1552.
8. El-Deiry, W. S., Tokino, T., Velculesue, V. E., et al. (1993) WAF1, a potential mediator of p53 tumor suppression. *Cell* **75**, 817–825.
9. Wu, G. S., Burns, T. F., McDonald, E. R., et al. (1997) KILLER/DR5 is a DNA damage-inducible p53-regulated death receptor gene. *Nat. Genet.* **17**, 141–143.
10. Hampson, I. N., Pope, L., Cowling, G. J., and Dexter, T. M. (1992) Chemical cross linking subtraction (CCLS): a new method for the generation of subtractive hybridisation probes. *Nucleic Acids Res.* **20**, 2899.
11. Sambrook, J., Fritsch, E. F., and Maniatis, T. (1989) *Molecular Cloning: A Laboratory manual*, 3rd ed. Cold Spring Harbor Laboratory, Cold Spring Harbor, NY.

28

Suppression Subtractive Hybridization for Identification and Functional Analysis of Tumor Suppressor Genes

Iuri D. Louro, Evans C. Bailey, and J. Michael Ruppert

1. Introduction

Alterations leading to inactivation of tumor suppressor genes are often associated with reduced levels of tumor suppressor gene expression. Specific tumors or cell lines may exhibit deletion of one or both gene copies, promoter methylation, splice-site mutations, nonsense mutations that induce premature translational termination and destabilize mRNA transcripts, or a combination of these. Such mutations result in complete absence or partial reduction in levels of tumor suppressor mRNA. Therefore, cDNA subtraction has been frequently used for identification of candidate tumor suppressor genes ***(1–7)***. As multiple tumor suppressor genes function in signaling pathways that regulate expression of oncogene transcription factors ***(8)***, subtraction methods are also useful for identification of downstream targets of these pathways. For example, embryo cells can be derived from mice engineered to be deficient in both alleles of a tumor suppressor. Then primary cell cultures can be analyzed by comparison with littermate control cells for identification of candidate target genes.

The technique of suppression subtractive hybridization (SSH) identifies transcripts that are differentially expressed between two cell populations ***(9,10)*** (Clontech Laboratories, Inc., Palo Alto, CA). Briefly, the method involves generation of "tester" cDNA fragments by ligation of adaptors, sequential subtractive hybridization steps, and then a 3′ end-fill and polymerase chain reaction (PCR) step to identify hybrids that contain specific adaptor sequences on each end.

Following reverse transcription and second-strand synthesis, cDNA from the two populations is digested using a restriction enzyme that cleaves at a 4-base recognition sequence to leave blunt ends. Portions of this cDNA are set aside as driver. To prepare tester, aliquots of cDNA are ligated to one of two nonphosphorylated adapters, such that the longer strand of an adapter is ligated to the 5′ ends of the cDNA, while the shorter strand remains only noncovalently associated. These two adapted tester fractions are individually mixed with driver, heat denatured, and allowed to anneal. Only molecules that remain single-stranded (ss) in this first hybridization step will be identified by PCR at the end of the entire procedure. The technique therefore simultaneously normalizes

From: *Methods in Molecular Biology, Vol. 222: Tumor Suppressor Genes: Pathways and Isolation Strategies*
Edited by: Wafik S. El-Deiry © Humana Press Inc., Totowa, NJ

(concentrations of abundant and rare transcripts become more equal) and enriches for differentially expressed transcripts. The two (subtracted and normalized) tester fractions are then mixed and incubated (without denaturation), allowing ss cDNA molecules modified with distinct adaptor-derived squences to anneal. Typically, the technique results in a "library" of several thousand molecules that can be rescued by PCR. This is accomplished by filling in the 3′ ends prior to the first denaturation step in the PCR reaction, followed by nested PCR. Hybrids containing distinct adapters at each end (i.e., derived from different tester fractions) amplify exponentially. Hybrids containing the same adaptor sequence at each end (largely representing abundant tester transcripts that annealed in the first hybridization step) fail to amplify, due to formation of long inverted repeats that form panhandle-like structures and prevent binding of primers during PCR (hence the term suppression). Driver-driver hybrids do not contain binding sites for PCR primers, and residual ss cDNA molecules as well as tester-driver hybrids contain a primer at only one end and therefore amplify linearly. Typically, 10–40% of amplified molecules derive from differentially expressed transcripts.

Our laboratory has used the technique in parallel with microarray analysis to identify targets of transforming oncogenes, and found considerable overlap among transcripts identified by the two methods ***(11)***. In addition, SSH successfully identified multiple transcripts that were not detected by microarray, and serves well as a rapid, reliable, and inexpensive adjunct to supplement the results obtained by array-based approaches.

2. Materials

1. Bio-Spin chromatography columns and end caps (Bio-Rad, Hercules, CA, cat. no. 732-6008, caps cat. no. 731-1660).
2. 15-mL Conical-bottom tube (Corning cat. no. 25319-15, dimensions 17 mm × 118 mm, polypropylene).
3. Microcentrifuge tubes (USA Scientific, Ocala, FL, cat. no. 1615-5500, 1415-8799, 1.5-mL, snap-cap or screw-cap polypropylene).
4. Parafilm (American National Can, Greenwich, CT, cat. no. PM992).
5. PCR tubes (PE Applied Biosystems, Foster City, CA , cat. no. N801-0540, thin-walled).
6. Spin-X tubes (Costar, Cambridge, MA, cat. no. 8161, with 0.2-µm cellulose acetate filter).
7. Clean bench (Laminar Flow Work Station, Marietta, OH, Thermo Forma).
8. Pipetman (Rainin, Woburn, MA, models P2, P10, P20, P200, P1000, dedicated for RNA work and PCR setup).
9. Thermal cycler (PE Applied Biosystems, GeneAmp PCR System 9700 or equivalent).
10. Water purification system (Millipore, Bedford, MA, Synthesis A10 or equivalent).
11. cDNA Synthesis Kit (Stratagene, La Jolla, CA, cat. no. 200401).
12. PCR Select cDNA subtraction kit (Clontech, Palo Alto, CA) (*see* **Note 1**).
13. Zero Background/Kan cloning kit (pZero-2) (Invitrogen, Carlsbad, CA, cat. no. K2600-01).
14. Deoxynucleoside triphosphates (dNTPs) (Promega, Madison, WI, cat. no. U120A, U121A, U122A, U123A).
15. Chloroform (Mallinckrodt, Paris, KY, cat. no. 4440).
16. Diethylpyrocarbonate (DEPC) (Sigma, St. Louis, MO, D5758).
17. EDTA, disodium salt (JT Baker, Phillipsburg, NJ, cat. no. 4040).
18. Glycogen, 20 mg/mL (Roche, Indianapolis, IN, cat. no. 901393).
19. Mineral oil, high-quality (Sigma, cat. no. M3516).
20. Oligo-dT cellulose (Stratagene, cat. no. 300138).
21. Tris base (JT Baker, cat. no. 4109).

22. Sodium acetate (Mallinckrodt, cat. no. 7364).
23. Sodium chloride (Mallinckrodt, cat. no. 7581).
24. Sodium hydroxide (JT Baker, cat. no. 3722).
25. NEB1 10× restriction enzyme reaction buffer (New England Biolabs, Beverly, MA).
26. Phenol (United States Biochemical, Cleveland, OH, cat. no. US75830).
27. Elution buffer: 10 m*M* Tris-HCl, 1 m*M* EDTA, pH 7.4. Store at room temperature (RT).
28. High-salt buffer: 10 m*M* Tris-HCl, 1 m*M* EDTA, 0.5 *M* NaCl, pH 7.4. Store at RT.
29. Low-salt buffer: 10 m*M* Tris-HCl, 1 m*M* EDTA, 0.1 *M* NaCl, pH 7.4. Store at RT.
30. Sample buffer: 10 m*M* Tris-HCl, 1 m*M* EDTA, 3 *M* NaCl, pH 7.4. Store at RT.
31. 0.2 *M* EDTA, pH 8.0 (DEPC-treated). Store at RT.
32. 5 *M* sodium chloride (DEPC-treated, autoclaved). Store at RT.
33. 5 *M* sodium hydroxide. Store at RT.
34. LoTE: 3 m*M* Tris-HCl, pH 7.5, 0.2 m*M* EDTA (autoclaved). Store at RT.
35. PC9: phenol/chloroform/TRIS-EDTA, pH 8.9. Store at −20°C.
36. 2 *M* potassium acetate, pH 5.0 (DEPC-treated, autoclaved). Store at RT.
37. TAE electrophoresis buffer (1×): 40 m*M* Tris-acetate, pH 7.9, 5 m*M* sodium acetate, 1 m*M* EDTA. Store a 25× solution at RT.
38. TE9: 20 m*M* EDTA, 10 m*M* NaCl, 0.5 *M* Tris-HCl, pH 8.9 (autoclaved). Store at RT.
39. Water, DEPC-treated: DEPC H_2O, incubate overnight at 37°C in a 0.1% (0.1 gr DEPC per 100 mL) solution, autoclave, and store at RT.

3. Methods

3.1. Preparation of Total RNA and Selection of Poly A$^+$ mRNA (see Note 2*)*

1. Total RNA is prepared using the acid guanidinium thiocyanate/phenol/chloroform extraction technique as described ***(12)***. Dissolve in DEPC H_2O, quantitate by absorption at 260 nm (1 O.D. = 40 µg/mL), and store at –70°C.
2. Poly A$^+$ mRNA is selected twice using the same oligo dT-cellulose column. Weigh out 33 mg oligo-dT cellulose (~0.2–0.25 mL packed volume when hydrated) into a sterile, RNAse-free 15-mL Corning tube containing 5 mL of 0.1 *N* NaOH. Rock end over end at 25°C for 15 min.
3. Remove the tab from the bottom of a Bio-Spin minicolumn, put an end cap in place over the outlet, and add 0.3 mL of 0.1 *N* NaOH to the column.
4. Pour cellulose slurry into the column. In this and subsequent steps, always allow fluid to drain until the buffer level is just above the cellulose bed, avoiding introduction of air bubbles. Remove column cap and let drain. Fill and drain once using DEPC H_2O. Fill and drain twice using high-salt buffer. Cap the tip and check that the pH of the last few drops is below 8.0 using pH paper (*see* **Note 3**).
5. Dilute 1.0–4.0 mg total RNA to 1.0 mL total volume by adding H_2O, LoTE, or elution buffer.
6. Heat to 65°C for 5 min. Cool on ice for 5 min, then add 0.2 mL sample buffer (i.e., final [NaCl] = 0.5 *M*). Preheat elution buffer to 65°C for subsequent use.
7. Apply RNA to column in 0.2-mL aliquots, collecting flow-through (FT) in a 1.5-mL tube on ice. Wash with ~1 col vol (0.25 mL) high-salt buffer, and pool the FT.
8. Heat the FT at 65°C for 5 min, cool on ice for 5 min, reapply to column as above, and then discard.
9. Wash column twice with 0.6 mL high-salt buffer. Wash 3 times with 0.6 mL low-salt buffer.
10. Apply 0.25 mL hot elution buffer, collect the eluate in a clean tube, and store on ice. Repeat 3 times and pool eluates for a total volume of 1 mL.
11. Fill and drain twice with high-salt buffer and perform the second selection by starting at **step 6** above. After second elution, decant eluted poly-A$^+$ mRNA to fresh tube if contaminating

oligo-dT cellulose is visible. Check the O.D. 260 of a small aliquot diluted 1:10. Yield is typically 1–4%.

12. Add 1/10 vol of 2 *M* KOAc, pH 5.0. Aliquot to three 1.5-mL tubes, add 1.1 mL of 100% ethanol (EtOH), mix, and incubate overnight at −80°C. Centrifuge 30 min at 15,000 × g at 40°C to precipitate. Wash with 70% EtOH, air dry 5 min, and resuspend in DEPC H_2O.

3.2. Preparation of Tester and Driver cDNAs

poly A^+ mRNA derived from the cells or tissues to be compared (referred to as Samples 1 and 2) is converted into double-stranded cDNA and then subsequently digested with RsaI (*see* **Note 4**). "Tester" fractions from each cDNA are prepared by adapter ligation. Subtractions are performed both in the forward (Sample 2 tester subtracted with Sample 1 driver) and reverse (Sample 1 tester subtracted with Sample 2 driver) directions. This allows one to detect both induced and repressed transcripts, and also provides a control for differential screening of clones. Clontech recommends parallel preparation of a control driver skeletal muscle cDNA using mRNA provided in the PCR Select Kit, and a control tester prepared by admixing a small quantity of ϕX174 DNA cut with HaeIII. These controls are not discussed further, but may be helpful to include.

1. Label four 0.2–0.5-mL polypropylene microcentrifuge tubes: Tester 1, Driver 1, Tester 2, Driver 2. Combine the following in each:
   a. 4 μL Poly A^+ RNA (2 μg) from Sample 1 or 2.
   b. 1 μL cDNA synthesis primer (10 μ*M*, from Clontech PCR-Select Kit, sequence as described *(9)*).
2. Incubate at 70°C for 2 min in a thermal cycler. Cool on ice for 2 min. Centrifuge briefly.
3. Prepare the following mixture on ice (total volume 32 μL):
   a. 5.80 μL 10× first-strand buffer (Stratagene cDNA Synthesis Kit).
   b. 3.48 μL 10 m*M* dNTP's (unmodified nucleotides from Promega).
   c. 21.6 μL DEPC-treated water (Stratagene cDNA Synthesis Kit).
   d. 1.16 μL RNAse block inhibitor (Stratagene cDNA Synthesis Kit).
4. Mix by vortexing briefly, then spin briefly in a microcentrifuge.
5. Add 6.90 μL of the above mixture to each reaction tube.
6. Mix gently, incubate at RT for 10 min.
7. Add 0.6 μL MMLV-RT (50 U/μL). The total reaction volume is 12.5 μL.
8. Incubate at 37°C for 1 h. Wrap the lid using Parafilm and submerge tube completely to prevent condensation on upper surfaces. Alternatively, incubate in a thermal cycler with a heated lid or using one drop of mineral oil to prevent condensation.
9. To synthesize the second strand while avoiding formation of hairpins and undigestable, unclonable cDNAs, keep all samples and reagents ice-cold. Place the first-strand reaction on ice and allow to cool prior to adding the second-strand reagents. All second-strand components are from the Stratagene cDNA Synthesis Kit. Add all of these, except the RNAse H and DNA polymerase I, as a master mix (through the oil interface, if oil is present) and pipet up and down gently. Lastly, add the Rnase H and DNA polymerase through the oil interface, and gently pipette up and down again.
   a. 8 μL 10× second-strand buffer.
   b. 2.4 μL second-strand dNTP mixture.
   c. 51.9 μL sterile distilled water (DEPC water not required).
   d. 0.8 μL of RNAse H (1.5 U/μL).
   e. 4.4 μL of DNA polymerase I (9.0 U/μL).

10. Quickly vortex, spin briefly, and incubate 2.5 h at 15.5°C. Do not allow the temperature to go above 16°C at any time. Following the 2.5-h incubation, place the tube on ice. Blunt the cDNA termini by adding (from the Stratagene cDNA Synthesis Kit):
    a. 9.2 µL of blunting dNTP mix.
    b. 0.8 µL of cloned Pfu DNA Polymerase (2.5 U/µL).
    Quickly vortex and spin. Incubate at 72°C for 30 min exactly.
11. To stop the reaction, add 4 µL 20× EDTA/glycogen mix (from Clontech PCR Select Kit). Add 1 vol (~0.1 mL) of PC9 (*see* **Note 5**). Vortex 2 and centrifuge at 14,000 *g* for 2 min at RT. Decant the top layer and place in a clean 0.5–1.5-mL microcentrifuge tube. Add 1 vol (~0.1 mL) chloroform/isoamyl alcohol (CIA, 24/1), vortex 2 min, and centrifuge 2 min. Decant to a clean tube. Add 0.1 vol (~10 µL) of 3 *M* sodium acetate (NaOAC, pH 5.2); vortex well and add 2.5 vol (250 µL) 100% ethanol. Mark one side of the tube as an indicator of where the pellet will form. Vortex well and incubate 10 min on ice. Spin in a microcentrifuge at 14,000 *g* for 15 min at 4°C, orienting the tube with the lid hinge in the same orientation for all spins.
12. Pipet off the supernatant and wash with 500 µL 70% ethanol, being careful not to disturb the pellet. Centrifuge at 14,000 *g* for 2 min at RT with the pellet oriented precisely as in the previous centrifugation. Aspirate the ethanol wash and air dry for 10 min to evaporate residual alcohol. Dissolve in 50 µL water. Transfer 6 µL to a clean tube and store at –20°C for use as a control for analysis of restriction enzyme digestion (below).

3.3. RsaI Digestion

1. Mix the following:
   a. 43.5 µL ds cDNA.
   b. 5.0 µL (or use NEB1 reaction buffer) 10× RsaI reaction buffer.
   c. 1.5 µL RsaI (10 U/µL).
2. Mix well and centrifuge briefly, then incubate at 37°C for 1.5 h. Analyze the efficiency of digestion by electrophoresis of uncut (2.5-µL) and RsaI-digested (5-µL) cDNAs in 1% agarose (detailed methods are described in the Clontech PCR Select manual). The size range should shift from 0.5–10.0 kb to 0.1–2 kb. If the size distribution is not reduced after RsaI digestion, repeat the PC9 and CIA extractions, ethanol precipitation, and digestion.
3. Add 2.5 µL of 20× EDTA/glycogen mixture to terminate the reaction. Extract sequentially in PC9 and CIA and then ethanol precipitate and wash as described under **Subheading 3.2.11**. Resuspend Tester tubes in 5.5 µL H_2O and store at –20°C. Resuspend Driver tubes in 3.0-µL H_2O.

3.4. Preparation of Adapters and Adapter-Ligated Tester cDNA

We obtained good results using a version of adapters as described ***(9)***. Newer versions of the SSH kit contain modified adaptors that may work equally well. The following primers were synthesized and cartridge-purified (Life Technologies, Rockville, MD), although gel purification may be helpful as recommended ***(9)***:

L1: 5′ gtaatacgactcactatagggctcgagcggccgcccgggcaggt 3′

S1: 5′ acctgccc 3′

L2: 5′ tgtagcgtgaagacgacagaaagggcgtggtgcggagggcggt 3′

S2: 5′ accgccctccg 3′

Prepare Adapter 1 by mixing equal moles of primers L1 and S1 and add 5 *M* NaCl, 0.2 *M* EDTA, and 1 *M* Tris-HCl, pH 7.5, to obtain each primer in the range of 50–100 µ*M* in a solution of 50 m*M* NaCl, 1 m*M* EDTA, 10 m*M* Tris-HCl, pH 7.5. Similarly prepare

Adapter 2 by mixing primers L2 and S2. Heat primer mixtures to 65°C for 10 min in a water bath, then allow to cool to room temperature over a period of at least 2 h (use a Styrofoam container and lid filled with water at an initial temperature of 65°C, or place tubes in a prewarmed heating block and allow to cool slowly on the lab bench). Dilute a working stock to 10 μ*M* (i.e., 5–10-fold) using ice-cold LoTE, and store at –20°C. Thaw when needed by incubation in a 16°C water bath, then store on ice. Avoid warming to RT.

3.4.1. Adapter Ligation (cDNA to be used for driver does not get ligated to adapters)

1. Dilute 1 μL of each RsaI-digested Tester sample from **Subheading 3.3.3.** with 5 μL of H_2O. Prepare a ligation master mix on ice using components from the PCR Select Kit per reaction:
   a. 3.6 μL Sterile water.
   b. 2.4 μL 5× ligation buffer.
   c. 1.2 μL T4 DNA ligase (400 U/μL).
2. For each of the two Samples (i.e., N = 1 or 2), prepare Tester N1 by ligating sample Tester N1 cDNA to adapter 1, and Tester N2 by ligating to adapter 2 (*see* **Table 1**). Assemble ligation reactions on ice.
3. In a fresh microcentrifuge tube, mix 2 μL of Tester N1 and 2 μl of Tester N2 ligation reactions. This is your Unsubtracted Tester Control. Repeat for all samples. Centrifuge briefly and incubate all reactions at 16°C overnight. Stop the reactions by adding 1 μL of EDTA/glycogen mix. Heat tubes at 72°C for 5 min to inactivate ligase. Briefly centrifuge. Label tubes Adaptor-Ligated Tester cDNAs. Remove 1 μL of each Unsubtracted Tester Control and dilute into 1 mL of H_2O (these diluted samples will be used later for PCR). Store all samples frozen. Perform the ligation efficiency analysis as described in the PCR-Select Manual (Clontech).

3.5. Subtractive Hybridization and Normalization

1. For analysis of Samples 1 and 2 in both forward and reverse directions, mix Driver 1 with Tester 2-1 (ligated to Adaptor 1) and Tester 2-2 (ligated to Adaptor 2). Mix Driver 2 with Testers 1-1 and 1-2. This corresponds to four hybridization reactions. Prewarm the 4× hybridization buffer at RT for 15 min. Check that no visible precipitates are present. Vortex gently. Combine the following in 0.2–0.5-mL tubes (compatible with your thermal cycler):

	Tester Hybridization N1	Tester Hybridization N2
RsaI-digested Driver cDNA	0.75 μL	0.75 μL
Adaptor 1-Ligated Tester N	1.5 μL	—
Adaptor 2-Ligated Tester N	—	1.5 μL
4× Hybridization buffer	0.75 μL	0.75 μL

2. Overlay samples with one drop of mineral oil and centrifuge briefly.
3. Incubate at 98°C for 1.5 min in a thermal cycler. Incubate at 68°C for 24 h.
4. Add fresh, denatured driver to each tube by mixing the following for each pair of Tester Hybridizations:
   a. Driver cDNA: 1.5 μL or all remaining in tube.
   b. 4× Hybridization buffer: 0.52 μL.
5. Check volume with P10 pipet (don't discard tip), add water to make the volume 2.1 μL. Mix and centrifuge briefly. Using the same tip, aliquot 1.0 μL each to two clean screw-cap tubes and overlay each with a drop of mineral oil.
6. Incubate at 100°C for 1.5 min to denature. Microfuge, vortex briefly, and incubate the tube on a heat block at 75°C for 2 min.

Table 1
Generation of Tester N1 and Tester N2 from Sample N (repeat for each sample).

Component	Tester N1 (µL)	Tester N2 (µL)
Diluted cDNA from Tester N	2	2
Adaptor 1 (10 µ*M*)	2	—
Adaptor 2 (10 µ*M*)	—	2
Master mix	6	6
Final volume	10	10

7. Add whole volume (including oil) of each tube to one of the two Tester Hybridizations. Spin briefly, mix by flicking the tube several times, then spin briefly again. Incubate for another 24 h at 68°C in the thermal cycler.

3.6. Hybridization of Subtracted Tester Fractions

Mix the two tester populations in one tube, spin briefly, then mix by flicking the tube and spin briefly again. Incubate at 68°C for 48 h. Add 0.2 mL of dilution buffer to each hybridization reaction and mix by pipetting. Incubate at 68°C for 7 min in the thermal cycler and store at –20°C. This constitutes a library of molecules enriched for differentially expressed sequences that can be rescued by PCR (*see* **Note 6**).

3.7. Fill-in and PCR

1. PCR primers (5 µM) and reaction conditions are as described ***(9)***:

   P1, 5′ gtaat acgac tcact atagg gc 3′

   P2, 5′ tgtag cgtga agacg acaga a 3′

   PN1, 5′ tcgag cggcc gcccg ggcag gt 3′

   PN2, 5′ agggc gtggt gcgga gggcg gt 3′
2. The template for PCR is a 1-µL aliquot taken from one of the following: the libraries resulting from the forward and reverse subtractions (total volume ~0.2 ml), or an aliquot of the diluted Unsubtracted Tester Controls for each of the two Samples (*see* **Subheading 3.4.1.3.**). Additional controls recommended by Clontech may be added as necessary for troubleshooting.
3. Aliquot 1 µL of each diluted hybridization product (**step 3.6**) and each Unsubtracted Tester Control (**step 3.4.1.3.**) into a PCR tube.
4. Prepare a Master Mix as follows (volumes shown are per reaction; multiply volumes by the number of PCR reactions and then by 1.2 to allow for lost volume during pipetting):
   a. 18.5 µL Sterile H_2O.
   b. 2.5 µL of 10× PCR reaction buffer.
   c. 0.5 µL dNTP mix (10 m*M*).
   d. 1.0 µL PCR primer 1 (5 µ*M*).
   e. 1.0 µL PCR primer 2 (5 µ*M*).
   f. 0.5 µL of 50× Advantage cDNA polymerase mix.
   g. Total volume 24.0 µL.

   Mix well by vortexing and briefly centrifuge. Aliquot 24 µL of Master Mix into each of the reaction tubes prepared in **step 3.7.3**. Overlay with mineral oil if recommended for your thermal cycler. The optimal parameters may vary for different thermal cyclers. We use a Perkin

Elmer model 9700 with a heated lid (no mineral oil), and using the following profile: 75°C × 5 min × 1 cycle (extends 3′ ends and creates primer binding sites), then 94°C × 25 min × 1 cycle (denatures DNA), then [94°C × 10 min, 66°C × 30 min, 72°C × 90 min] × 27 cycles, then 72°C × 5 min.

5. Following completion of this primary PCR profile, dilute 3 µL of each reaction in 27 µL H_2O. Aliquot 1 µL of this dilution (corresponding to 0.1 µL of the primary PCR yield) into a PCR tube.
6. Prepare a Master Mix for the secondary PCR as follows (multiply volumes by the number of PCR reactions and then by 1.2):
   a. 18.5 µL Sterile H_2O.
   b. 2.5 µL of 10× PCR reaction buffer.
   c. 1.0 µL Nested PCR primer PN1 (5 µ*M*).
   d. 1.0 µL Nested PCR primer PN2 (5 µ*M*).
   e. 0.5 µL dNTP Mix (10 µ*M*).
   f. 0.5 µL of 50× Advantage cDNA Polymerase Mix.
   g. Total volume 24.0 µL.
7. Mix well by vortexing and briefly centrifuge. Aliquot 24 µL of Master Mix into each of the reaction tubes prepared in **step 3.7.5**. The following profile was used in a Perkin Elmer 9700 thermal cycler: [94°C × 10 s, 68°C × 30 s, 72°C × 90 s] × 25 cycles; 72°C × 5 min.
8. Analyze 5 µL of secondary PCR reaction on a 1.5% agarose gel. Standard techniques for agarose gel electrophoresis and Southern blot preparation have been described ***(13)***. A smear composed of numerous bands should be evident in the products of subtracted samples. Southern blots prepared from these gels allow comparison of housekeeping gene expression in subtracted tester with Unsubtracted Tester Control. The secondary PCR product of forward and reverse subtractions should exhibit greatly reduced expression of these genes. If any known transcripts exhibit differential expression in the two starting samples, confirm that this positive control is enriched in the appropriate sample. Use approximately equal masses of secondary PCR product (e.g., 1 µg of DNA) to prepare Southern blots.

3.8. Cloning of PCR Products

1. To clone SSH products, purify the secondary PCR reaction by sequential extractions in PC9, ethanol precipitate, and resuspend in LoTE. Digest for 2 h with RsaI enzyme (10 µ/µg) in 1× restriction enzyme buffer (NEB1, New England Biolabs).
2. Purify products > 50 bp from a 1.5% TAE-agarose gel as follows: Freeze the gel slice 20 min) at –70°C, then thaw at 37°C for 5 min. Centrifuge up to 0.4 gr of agarose in a Spin-X tube for 15 min at Rtemp. Rotate tubes 180° in microcentrifuge rotor, then centrifuge for 10 min. The second spin increases the yield slightly. Dilute the filtrate to 0.4 mL using LoTE, and ethanol precipitate using 2.0 µL of glycogen (20 mg/mL) as carrier. Resuspend the washed pellet in 10 µL LoTE, and quantitate 0.5 µL on an agarose gel.
3. Ligate 50 ng insert to 50 ng of pZero-2 at RT in 10 µL of 1× ligase buffer containing 0.5 µL T4 ligase (Invitrogen/Life Technologies). Extract with PC9, ethanol precipitate, and electoporate into competent TOP-10 *E. coli* (provided in the Zero Background kit). Cells are rendered competent for electroporation as described ***(14)***.

3.9. Identification of Differentially Expressed Transcripts

Typically, 10–40% of SSH products detect differentially expressed transcripts by Northern blot. Several investigators have identified high-throughput methods for identification of such fragments, generally using complex hybridization probes prepared from gel-purified product of secondary PCR reactions to identify differentially expressed clones bound to filters (e.g., dot blots, Southern blots) ***(10)***. However, we experienced high background using such probes. We therefore prepared Northern blots by loading one entire gel with RNA from a starting sample, and cutting the filter into many strips. Methods for Northern blot analysis are as described ***(15)***. Pairs of these strips (corresponding to Sample 1 and Sample 2 RNAs) were analyzed using radiolabeled SSH fragments purified from pZero-2 plasmids by agarose gel electrophoresis. Three rounds of hybridization and autoradiography identified 30 induced transcripts out of a total of 120 probes analyzed, with fold induction (as determined by analysis of timed exposures) estimated as 3.0 to >20.

4. Notes

1. Modifications to the current version of the Clontech PCR Select Kit include the use of an older version of adapters and primers, altered hybridization times, volumes, and DNA concentrations, and others. We obtained better results using these modifications, although this conclusion is based on a small number of experiments.
2. Use standard RNAse precautions. Use clean, dedicated pipets with autoclaved tips and tubes. Treat Bio-Spin columns and caps, stir bars, and bottle caps with DEPC (0.1% solution in water, incubate overnight at 37°C, then autoclave and dry). Bake all glassware for 8 h at 338°F. To prepare 0.1 L RNAse-free 1 *M* Tris-HCl, pH 7.4 (density 1.04 gr/mL), first tare a baked 500-mL bottle. Add 83 gr of DEPC H2O, 24.2 g of Tris-HCl base (unopened bottle preferred), and 8.00 gr concentrated hydrochloric acid. Stir to dissolve, then add DEPC H_2O to 104 gr. Pour 5 mL into a scintillation vial, check the pH, and then discard this aliquot. Autoclave 30 min at 250°F.
3. Some batches of cellulose drain very slowly. If draining by gravity is time consuming, load RNA by gravity and then drain wash solutions and elution buffer by centrifugation for 2 min at 270 × *g*. Use clean forceps to place the entire column inside a 15-mL tube that contains a 2-mL microfuge tube as a spacer at the bottom.
4. The choice of starting material is critical. Best results are obtained when Samples 1 and 2 are as closely matched as possible. For example, it would be reasonable to compare morphologically normal tissue from wild-type mice vs the same strain carrying a genetic lesion (e.g., gene knockout or transgene). Comparison of morphologically normal tissue with morphologically abnormal tissue may identify many differences that are unrelated to the pathway of interest. In our studies we used retroviral transduction to derive mass populations of RK3E cells ***(16)*** using vector-only or oncogene expression vectors, cultured these in parallel for just 10 d in selective media, then prepared RNA. Analysis of late-passage transformed cell lines identified many nonspecific transcriptional alterations that confounded identification of real target genes (unpublished data).
5. Prepare PC9 by mixing 160 mL phenol, 213 mL chloroform, and 107 mL TE9; shake and let phases separate at 4°C for 2 h; repeat this extraction 1 time; discard aqueous layer (top layer) and store at −20°C. To extract samples: thaw PC9 at RT. Dilute sample to 400 µL with TE9 and add 1 vol (400 µL) of PC9. Vortex for 1 min, microcentrifuge for 2 min at maximum speed, and transfer the upper aqueous layer to a new microcentrifuge tube. Repeat extraction with PC9. Extract one final time with chloroform/isomyl alcohol, and transfer the upper aqueous layer to a new microcentrifuge tube.
6. We find it useful to assess the number of clones obtained in the final library. Such information helps in determining the appropriate volume of library to analyze for differentially expressed clones. For example, if you were planning to array 500 clones, you might start with at least 5000 clones so as to limit redundant analysis of a single clone that was copied multiple times during PCR. Take 1 µL and do serial 10-fold dilutions, analyzing each diluted sample by primary and secondary PCR reactions and agarose gel electrophoresis as described above. Count the number of clones in an appropriate lane and back-calculate to determine the number of

clones in the entire library. When comparing oncogene-transduced RK3E cells with vector control cells we obtained between 1000 and 12,000 clones in several subtractions. *(11)*

Acknowledgments

This work was supported by National Institutes of Health Grant CA65686 (to J.M.R.) and by a CAPES grant from the Brazilian government (to I.D.L.).

References

1. Zou, Z., Anisowicz, A., Hendrix, M. J., et al. (1994) Maspin, a serpin with tumor-suppressing activity in human mammary epithelial cells. *Science* **263**, 526–529.
2. Rutberg, S. E., Lee, E. J., Hansen, L. H., Glick, A. B., and Yuspa, S. H. (1997) Identification of differentially expressed genes in chemically induced skin tumors. *Mol. Carcinogen.* **20**, 88–98.
3. Stubbs, A. P., Abel, P. D., Golding, M., et al. (1999) Differentially expressed genes in hormone refractory prostate cancer: association with chromosomal regions involved with genetic aberrations. *Am. J. Pathol.* **154**, 1335–1343.
4. Yu, Y., Xu, F., Peng, H., et al. (1999) NOEY2 (ARHI), an imprinted putative tumor suppressor gene in ovarian and breast carcinomas. *Proc. Nat. Acad. Sci. USA* **96**, 214–219.
5. Chen, H. M., Schmeichel, K. L., Mian, I. S., Lelievre, S., Petersen, O. W., and Bissell, M. J. (2000) AZU-1: a candidate breast tumor suppressor and biomarker for tumor progression. *Mol. Biol. Cell* **11**, 1357–1367.
6. Hildebrandt, T., Preiherr, J., Tarbe, N., Klostermann, S., Van Muijen, G. N., and Weidle, U. H. (2000) Identification of THW, a putative new tumor suppressor gene. *Anticancer Res.* **20**, 2801–2809.
7. Russo, J. and Russo, I. H. (2001) The pathway of neoplastic transformation of human breast epithelial cells. *Radiat. Res.* **155**, 151–154
8. Taipale, J. and Beachy, P. A. (2001) The Hedgehog and Wnt signaling pathways in cancer. *Nature* **411**, 349–354.
9. Diatchenko, L., Lau, Y. F., Campbell, A. P., et al. (1996) Suppression subtractive hybridization: a method for generating differentially regulated or tissue-specific cDNA probes and libraries. *Proc. Natl. Acad. Sci. USA* **93**, 6025–6030.
10. Jin, H., Cheng, X., Diatchenko, L., Siebert, P. D., and Huang, C. C. (1997) Differential screening of a subtracted cDNA library—a method to search for genes preferentially expressed in multiple tissues. *Biotechniques* **23**, 1084–1086.
11. Louro, I.D., Bailey, E.C., Li, X., et al. (2002) Comparative gene expression profile analysis of GLI and c-MYC in an epithelial model of malignant transformation. *Cancer Res.* **62**, 5867–5873.
12. Chomczynski, P. and Sacchi, N. (1987) Single-step method of RNA isolation by acid guanidinium thiocyanate-phenol-chloroform extraction. *Anal. Biochem.* **162**, 156–159.
13. Brown, T. (2001) Analysis of DNA sequences by blotting and hybridization, in *Current Protocols in Molecular Biology*. (Ausubel, F., Brent, R., Kingston, R. E., et al. eds.), Wiley, New York.
14. Zabarovsky, E. R. and Winberg, G. (1990) High efficiency electroporation of ligated DNA into bacteria. *Nucleic Acids Res.* **18**, 5912.
15. Brown, T. and Mackey, K. (2001) Analysis of RNA by Nothern and slot blot hybridization, in *Current Protocols in Molecular Biology*. (Ausubel, F., Brent, R., Kingston, R. E., et al. eds.), Wiley, New York.
16. Ruppert, J. M., Vogelstein, B., and Kinzler, K. W. (1991) The zinc finger protein GLI transforms rodent cells in cooperation with adenovirus E1A. *Mol. Cell. Biol.* **11**, 1724–1728.

29

SAGE as a Strategy to Isolate Cancer-Related Genes

Kathy Boon and Gregory J. Riggins

1. Introduction

Serial Analysis of Gene Expression (SAGE) is a sequence-based approach that enables the comprehensive and quantitative analysis of gene expression within any cell type or tissue ***(1)***. The method is based on three main principles. First is the generation of a single 10-bp 'tag' sequence that contains sufficient information to uniquely identify a transcript, provided that the tag is obtained from a defined position within each transcript. Second is the concatenation of the tags, allowing the serial analysis of multiple transcripts in one single sequence run. Third is the quantification of the number of times a particular tag is observed in a certain SAGE tag population (SAGE library), which provides direct information on the expression level of the corresponding transcript.

Numerous reviews have been published describing the technology ***(2–5)***. The basic procedure has remained almost the same since its development and many studies have been performed, generating a comprehensive analysis from a diversity of tissues. A detailed protocol can be obtained through the SAGE Home Page from the Johns Hopkins Oncology Center (http://www.sagenet.org). The technology is patented by Johns Hopkins University and licensed to Genzyme Molecular Oncology (Framingham, MA) but freely available to academia and nonprofit organizations for research purposes. Further information on the license agreements for commercial applications can be obtained directly from Genzyme (www.genzyme.com/sage/welcome.htm).

Various modifications to the original technique have been reported to reduce the amount of polyadenylated RNA required for library construction. The use of small amounts of sample to generate a SAGE library is commonly referred to as micro-SAGE. One major improvement in most micro-SAGE protocols has been the use of streptavidine-coated polymerase chain reaction (PCR) tubes or magnetic beads to efficiently capture polyadenylated mRNA from as few as 5000 cells or microdissected material and allowing subsequent steps to occur in the same tube ***(6–9)***. The benefits of this approach are the increased efficiency of certain enzymatic steps and the alternative to generate a gene expression profile of limiting amounts of cells or tissue. The recent development of the i-SAGE kit from Invitrogen (http://www.invitrogen.com/SAGE) has made the

From: *Methods in Molecular Biology, Vol. 222: Tumor Suppressor Genes: Pathways and Isolation Strategies*
Edited by: Wafik S. El-Deiry © Humana Press Inc., Totowa, NJ

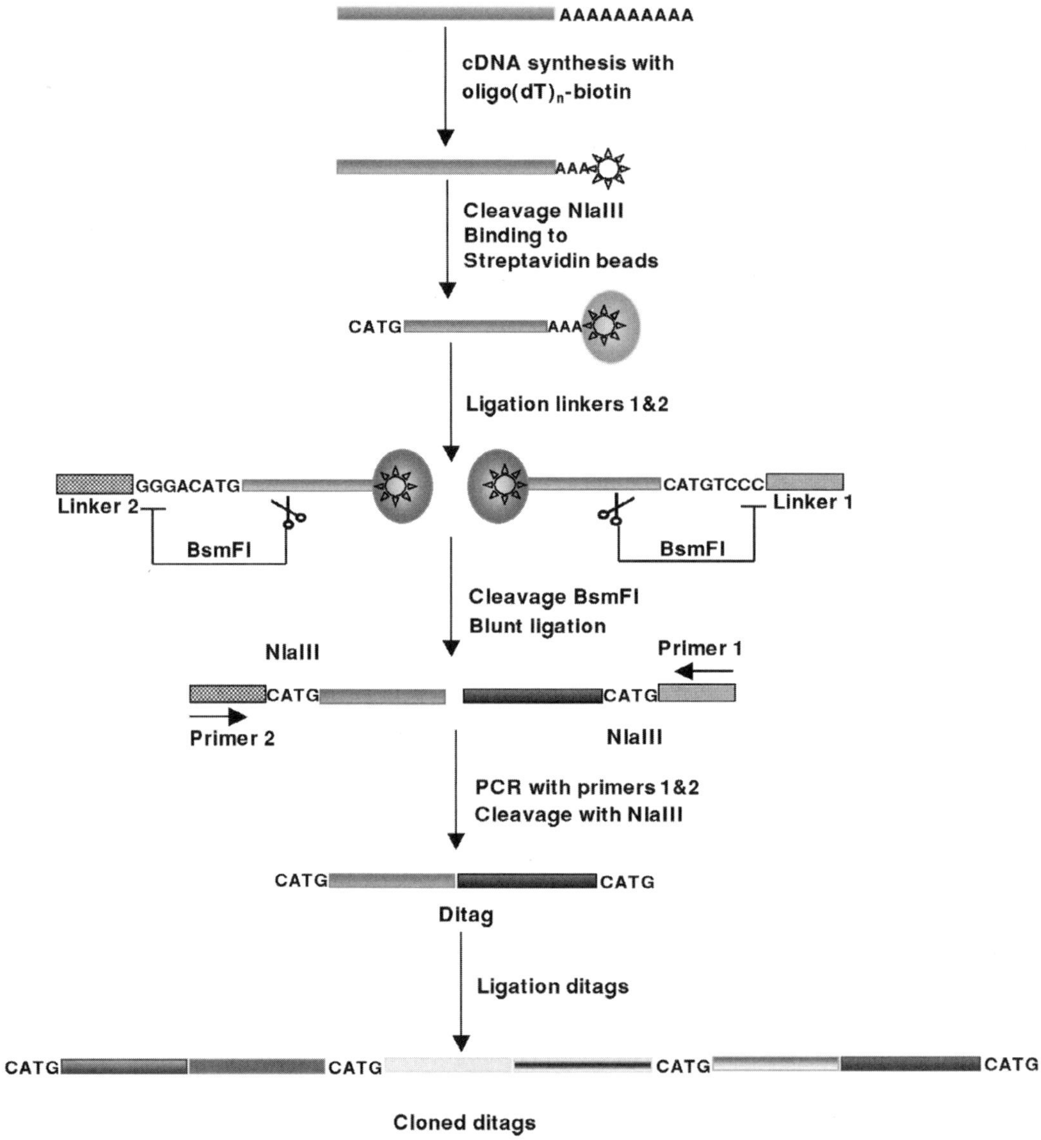

Fig. 1. Schematic representation of the standard SAGE protocol. Gene expression is quantified in a population of cells by isolating a transcript from the expressed genes. These tags are paired into ditags, ligated to form concatemeres, and cloned into a sequencing vector for efficient counting on an automated sequencer. Tag counts from each tissue type are stored for comparison to other cell population.

construction of SAGE libraries less demanding, in particular for those unfamiliar with SAGE library construction.

A schematic representation of the standard SAGE method is outlined in **Fig. 1**. Briefly, the construction of a SAGE library starts with the purification of polyadenylated mRNA, which subsequently is converted into double-stranded cDNA using a biotin-labeled oligo-dT primer during the first-strand synthesis for the recovery of the cDNA. A frequent-cutting anchoring enzyme, usually *Nla* III, defines the position in the transcripts from which the sequence tags are derived. After digestion of the generated cDNA with the anchoring enzyme, the 3′-terminal cDNA fragments are bound to streptavidine-coated beads. Thereafter an oligonucleotide linker containing recognition sites for a type II restriction enzyme (tagging enzyme) is ligated to the bound cDNA fragments. Type II enzymes cut at a defined distance away from the recognition site. The SAGE tags are then released from the bound cDNA by cleavage with the tagging enzyme (usually *BsmF* I) and dimerized by a tail-to-tail blunt-end ligation. The linked ditags (102 bp) are then amplified and digested with the same anchoring enzyme used for the initial digestion of the double-stranded cDNA. The resulting ditag fragments of 26–28 bp all have cohesive ends and can therefore be linked together to form concatemers. The size of these long serial molecules is strictly dependent on the quality and the ligation conditions applied. Cloning of the concatemers in a suitable vector yields a library of clones for sequencing. Each clone contains multiple ditags separated by the recognition sites for the anchoring enzyme. The efficiency of this serial process depends not only on the length of the cloned inserts, but also on the quality and read length of the subsequent sequencing. The combination of high-quality sequencing from a well-constructed library can generate >2800 tags from a single 96-capillary sequencing run.

2. Materials

1. FastTrack 2.0 (Invitrogen, Carlsbad, CA, cat. no. K1593-02) is needed only for the standard SAGE protocol. For micro-SAGE use the mRNA Direct Kit (Dynal, cat. no. 610.11).
2. cDNA Synthesis System from Life Technologies (cat. no. 18267-013).
3. SuperScript II RT is used in the micro-SAGE protocol (Life Technologies, cat. no. 18064-014).
4. Gel-purified 5-biotinylated oligo-dT (Standard SAGE protocol only).
5. Glycogen (Boehringer Mannheim, Indianapolis, IN, cat. no. 901-393).
6. 100% and 75% Ethanol.
7. LoTE buffer: 3m*M* Tris-HCl, 0.2 m*M* EDTA, pH 7.5.
8. Restriction enzyme *Nla* III (New England BioLabs, Beverly, MA, cat. no. 125S).
9. DNAse and RNAse free BSA (New England BioLabs).
10. 7.5 *M* ammonium acetate.
11. Phenol chloroform (PC8): 480 mL phenol, 320 mL 500 m*M* Tris-HCl (pH 8.0) and 640 mL chloroform (*see* **Note 1**).
12. MPC S magnet (Dynal, cat. no. 120.20).
13. Streptavidine Dynabeads M-280 (standard SAGE protocol only, Dynal, Lake Success, NY, cat. no. 112.05).
14. 2× B+W buffer: 10 m*M* Tris-HCl, pH 7.5, 1 m*M* EDTA, 2 *M* NaCl.
15. Linkers 1A, 1B, 2A, and 2B (*see* **Table 1** for sequences).
16. Primers 1, 2, M13Forward, and M13Reverse (*see* **Table 1** for sequences).
17. 3 m*M* Tris-HCl, pH 7.5 (micro-SAGE protocol only).
18. High-concentration T4 DNA ligase (Life Technologies cat. no. 15224-041).
19. Restriction enzyme *BmsF* I (New England BioLabs cat. no. 572S).
20. Klenow (USB cat. no. 27-0929-01).
21. 100 m*M* dNTPs (Life Technologies cat. no. 18427-013).
22. Taq DNA polymerase recombinant (Life Technologies cat. no. 10342-020).
23. DMSO (Sigma, St. Louis, MO, cat. no. D-2650).
24. 10× PCR buffer: 166 m*M* $(NH_4)_2SO_4$, 670 m*M* Tris-HCl, pH 8.8, 67 m*M* $MgCl_2$, 100 m*M* β-mercaptoethanol. Aliquot and store at –20°C.
25. DNA ladders: 1 kbp, 100 bp, 20 bp, and 10 bp.

Table 1.
Primer and Linker Sequences

Linker 1A	5′-TTT GGA TTT GCT GGT GCA GTA CAA CTA GGC TTA ATA GGG ACA TG -3′
Linker 1B	5′-TCC CTA TTA AGC CTA GTT GTA CTG CAC CAG CAA ATC C[amino mod. C7] -3′
Linker 2A	5′-TTT CTG CTC GAA TTC AAG CTT CTA ACG ATG TAC GGG GAC ATG -3′
Linker 2B	5′-TCC CCG TAC ATC GTT AGA AGC TTG AAT TCG AGC AG[amino mod. C7]-3′
Primer 1	5′-GGA TTT GCT GGT GCA GTA CA -3′
Primer 2	5′-CTG CTC GAA TTC AAG CTT CT -3′
M13Forward	5′-GTA AAA CGA CGG CCA GT -3′
M13Reverse	5′-GGA AAC AGC TAT GAC CAT G -3′
Biotinylated oligo dT	5′ [biotin]T_{18}

Linkers 1A, 1B, 2A, 2B, Primers 1 and 2, and the biotinylated oligodT should be obtained gel-purified. High quality is crucial to several steps in the SAGE method.

26. X-cell Surelock Mini-Cell system for Novex minigels (Invitrogen cat. no. EI0001) or similar system.
27. 12% Acrylamide stock solution in 1× TAE buffer using 40% acrylamide (19:1 acrylamide:bis, Bio-Rad cat. no. 161-0144).
28. 8% Acrylamide stock solution in 1× TAE buffer using 40% acrylamide (37.5:1 acrylamide:bis, Bio-Rad cat. no. 161-0148).
29. 10% Ammonium persulfate (APS).
30. Tetramethylethylenediamine (TEMED).
31. T4 polynucleotide kinase (10 U/µL, NEB cat. no. 201S).
32. 50-mL Oak Ridge centrifuge tubes (cat. no. 3114-0050).
33. 50-mL conical tubes.
34. 10× loading buffer: 0.2% bromphenol blue, 0.2% xylene cyanol, 50% glycerol in 1× TAE buffer.
35. SYBR Green I (Sigma cat. no. S9430).
36. Standard 0.5-, 1.5-mL microcentrifuge tubes and 2.0-mL round-bottom tubes.
37. Spin X columns (Costar cat. no. 8160).
38. Restriction enzyme *Sph* I (NEB).
39. pZeroO-1 cloning kit (Invitrogen cat. no. K2500-01).
40. Electromax DH10Bs cells (Life Technologies cat. no. 15224-041).
41. Platinum Taq DNA polymerase (Life Technologies cat. no. 10966-034).
42. Agarose and DNA sequencing equipment.
43. Ready-made 20% Novex gel (Invitrogen cat. no. EC6315).
44. Streptavidin (Sigma cat. no. S-4762).

3. Methods

3.1. Standard SAGE Protocol

3.1.1. mRNA Isolation

Total RNA can be prepared from tissue or cell line of choice using standard methods such as Trizol reagent (Life Technologies, Gaithersburg, MD) or guanidium isothiocynate RNAgent (Promega, Madison, WI). Gel electrophoresis and/or Northern blotting should be applied to ensure that the quality of the total RNA is high. For this method 500 µg to 1 mg of total RNA is needed. A variety of methods are available for the isolation of polyA$^+$ RNA. Good results have been obtained using FastTrack 2.0 (Invitrogen cat. no. K1593-02). At least 5 μg of polyA$^+$ RNA is required. In cases that the amount of RNA is limiting, the micro-SAGE procedure should be considered.

3.1.2. cDNA Synthesis

1. Double-stranded cDNA synthesis is carried out according to the cDNA Synthesis System (Life Technologies). Replace the oligo-dT supplied in the kit by 2.5 μg of a gel-purified 5′-biotinylated oligo-dT, and follow the manufacturer's instructions. For testing biotinylation of biotin-oligo dT, *see* **Note 2**. For best recovery of the synthesized cDNA, add 60 μg glycogen to the final precipitation step.
2. Wash pellet twice with 75% EtOH, air-dry, and resuspend cDNA in 20 μL LoTE.

3.1.3. Cleavage of Biotinylated cDNA with the Anchoring Enzyme

1. The cDNA (4 μg) is digested with the anchoring enzyme *Nla* III for 1 h at 37°C in a 200-μL reaction volume, containing 50 U *Nla* III, 20 μg bovine serum albumin (BSA) and 1× restriction buffer 4 (NEB).
2. The enzyme reaction is inactivated by phenol extraction (PC8) followed by ethanol precipitation (200 μL sample, 5 μL glycogen (20 mg/mL), 133 μL 7.5 *M* ammonium acetate, and 777 μL 100% EtOH).
3. Spin in microcentrifuge at full speed for 15 min.
4. Wash pellet twice with 75% EtOH, air-dry, and resuspend cleaved cDNA in 20 μL LoTE buffer.

3.1.4. Binding Biotinylated cDNA to Magnetic Beads

1. Take 200 μL Streptavidine Dynabead M-280 slurry and divide over 2 microcentrifuge tubes.
2. Immobilize beads on a magnet and remove supernatant.
3. Wash beads once with 200 μL 1× B+W buffer, mix, magnet, and remove wash.
4. Add 100 μL 2× B+W buffer to each tube, 90 μL H_2O and 10 μL cleaved cDNA from **Subheading 3.1.3., step 4.** Incubate for 15 min at room temperature by gentle mixing.
5. Wash as described 3 times with 200 μL 1× B+W buffer, once with 200 μL LoTE, and proceed immediately to the following step.

3.1.5. Ligating Linkers to Bound cDNA

1. For the preparation of the linkers, *see* **Notes 3** and **4**.
2. Remove the LoTE from the beads and add 5 μL (1 μg) of the annealed linkers 1A,B to tube 1 and 5 μL (1 μg) of the annealed linkers 2A,B to tube 2.
3. Add 25 μL LoTE and 8 μL 5× ligase buffer (Gibco-BRL) to each tube. Heat for 2 min at 50°C and let sit at room temperature for 15 min.
4. To each tube add 10 U T4 DNA ligase and incubate mixtures for 2 h at 16°C.
5. After ligation, wash tubes 3 times with 200 μL 1× B+W buffer; during the last wash, transfer to a new tube.
6. Wash once more with 200 μL 1× B+W buffer and twice with 200 μL 1× Reaction Buffer 4 (NEB).

3.1.6. Release of cDNA Tags Using Tagging Enzyme

1. After removing the last wash from the beads (**Subheading 3.1.5., step 6**), the linked cDNA is released from the beads by digestion with the tagging enzyme *BmsF* I for 1 h at 65°C in a reaction volume of 100 μL, consisting of 2 U of *BmsF* I, 20 μg BSA, and 1× restriction buffer 4.

2. This time, collect the supernatant containing the released linked tags.
3. Perform a PC8 extraction and ethanol-precipitate the linked tags (200 µL sample, 5 µL glycogen [20 mg/mL], 133 µL 7.5 *M* ammonium acetate, and 1000 µL 100% EtOH).
4. Spin in microcentrifuge at full speed for 30 min at 4°C.
5. Wash pellets twice with 75% EtOH, air-dry, and resuspend in 10 µL LoTE.

3.1.7. Blunt Ending of the Released Linked Tags

1. The blunt ending reaction is performed with Klenow (USB) for 30 min at 37°C in a reaction volume of 50 µL, containing 3 U Klenow, 10 µg BSA, 1 m*M* dNTPs, and 1× second-strand buffer (Gibco-BRL, cDNA Synthesis Kit).
2. Increase the volume of both tubes with 150 µL LoTE.
3. Extract with 200 µL PC8 and ethanol-precipitate the blunt-ended linked tags (200 µL sample, 5 µL glycogen [20 mg/mL], 133 µL 7.5 *M* ammonium acetate, and 1000 µL 100% EtOH).
4. Spin in microcentrifuge at full speed for 30 min at 4°C.
5. Wash pellets twice with 75% EtOH, air-dry, and resuspend in 6 µL LoTE.

3.1.8. Ligation of Linked Tags to Form Ditags

1. The ditags are formed by ligating the two blunt-ended linked tags together in a total volume of 6 µL containing 2 µL of each of the blunt-ended samples 1 and 2, 4 U of T4 DNA ligase and 1× ligase buffer (Life Technologies). Incubate overnight at 16°C.
2. The negative control consists of 1 µL of each blunt-ended sample and no ligase in a reaction volume of 6 µL. Incubate overnight at 16°C.
3. After ligation, increase the volume by adding 14 µL of LoTE. Aliquot the mixture and keep at −20°C.

3.1.9. PCR Amplification of Linked Ditags (see **Note 5**)

1. The amplification of the ditags is optimized by testing different dilutions of the template for both the ligation and the negative control (0-, 25-, 50-, 100-, and 200-fold dilutions).
2. Of each dilution, 1 µL is used as a template in a 50 µL PCR reaction containing 5 U Taq DNA polymerase (Life Technologies), 6% dimethyl sulfoxide (DMSO), 1.5 m*M* dNTPs, 26.5 µ*M* of each Primer 1 and Primer 2 and 1× PCR buffer. The following PCR conditions are optimized for the Hybaid Multiblock System 0.2G: an initial denaturation step for 1 min at 94°C, amplification for 27 cycles of 30 s at 94°C, 1 min at 55°C, and 1 min at 70°C, and a final extension of 5 min at 70°C. The negative control should be amplified for 35 cycles.
3. Test 10 µL of each reaction on a 12% TAE-PAGE gel (*see* **Note 6**), using a 100-bp and a 20-bp ladder as a marker. The amplified ditags should be 102 bp in size (**Fig. 2**). A background band of equal or less intensity occurs at 80 bp. The negative control should not show any amplification at the ditag length of 102 bp even at 35 cycles.
4. After establishing the best conditions for each sample, usually 300 50-µL PCR reactions are performed for the isolation of the ditags.

3.1.10. Isolation of Linked Ditags

1. Next, pool your PCR reactions into a 50-mL conical tube.
2. Add an equal volume of PC8, mix well, and spin in a swing-out rotor at 4000 *g* for 15 min.
3. Transfer the aqueous phase to a new tube and ethanol-precipitate the DNA from the 11.5-mL solution by adding 5.1 mL of 7.5 *M* ammonium acetate, 191 µL glycogen, and 38.3 mL of 100% EtOH. Vortex vigorously and spin in fixed angle rotor at 10,000 *g* for 30 min. At this step the precipitation can be stored at −20°C. Best results are obtained if precipitation is performed in 50-mL Oak Ridge centrifuge tubes.

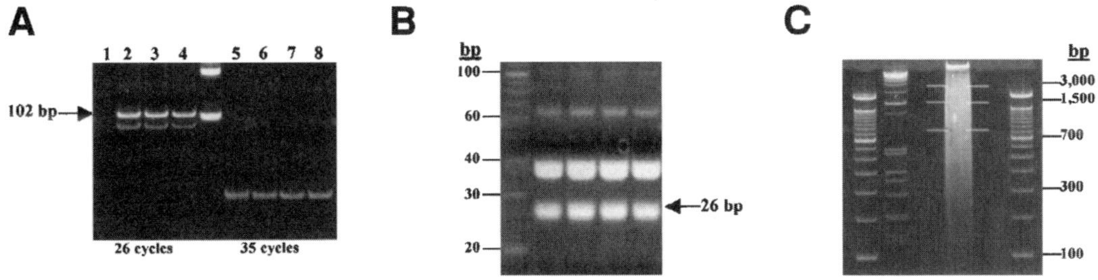

Fig. 2. SYBR Green I-stained polyacrylamide gels showing examples of: **(A)** PCR amplification of the ditags: *Lane 1*, PCR reaction mix; *lanes 2–4*, different dilutions of the ligated ditag (1/10, 1/100, 1/200); *lanes 5–8*, various dilutions of the corresponding negative control. **(B)** After digestion with *Nla* III, the 22–26 bp ditags are excised from gel and purified in **step 11**. **(C)** The isolated ditags are ligated to form the concatemeres. The region from 800 to 1500 bp is usually excised, purified, and cloned in pZero (**Subheading 3.1., step 12**).

4. Wash pellets with 5 mL 75% EtOH, air-dry, and resuspend in 400 µL LoTE buffer.
5. Add 40 µL of 10× loading buffer and load PCR product onto four 1.5-mm 12% TAE-PAGE gels using 2D-Well combs (*see* **Note 6**). Use a mixture of a 100-bp and a 20-bp ladder as a marker on each gel. Run gels at 160 V (xylene cyanol dye runs between 40 and 50 bp in this gel composition).
6. After electrophoresis, stain the gels with SYBR Green at a 1:10,000 dilution in 1× running buffer for a minimum of 15 min.
7. Excise the 102-bp band and load equally onto twelve 0.5-mL tubes, pierced twice with a 21-gage needle, and insert in a 2.0-mL round-bottom tube.
8. Spin at full speed for 5 min to fragment the gel pieces.
9. Discard the 0.5-mL tubes and add 250 µL LoTE buffer and 50 µL of 7.5 *M* ammonium acetate to each 2.0-mL tube. Vortex and place the tubes for 20 min at 65°C.
10. Remove the polyacrylamide gel by transferring the mixture into 12 Spin X columns (Costar) and spinning at full speed for 5 min.
11. Ethanol-precipitate eluates in twelve 1.5-mL microcentrifuge tubes (300 µL sample, 5 µL glycogen [20 mg/mL], 133 µL 7.5 *M* ammonium acetate, and 1000 µL 100% EtOH).
12. Spin in a microcentrifuge at full speed for 15 min.
13. Wash twice with 75% EtOH, air-dry pellets, and resuspend each pellet in 10 µL LoTE.
14. Pool the samples together and quantify the total amount of DNA. At this stage the yield should be between 10 and 20 µg.

3.1.11. Purification of 26-bp Ditags (see **Note 7***)*

1. Digest PCR product with *Nla* III for 1 h at 37°C in a reaction volume of 400 µL consisting of 400 U of *Nla* III, 40-µg BSA, and 1× reaction buffer 4 (NEB).
2. Phenol extract and ethanol precipitate (200 µL sample, 66 µL 7.5 *M* ammonium acetate, 5 µL glycogen, 825 µL 100% EtOH). Place in Dry Ice/methanol bath for 15 min or place at −80°C overnight before spinning.
3. Spin in a microcentrifuge at full speed for 30 min at 4°C.
4. Wash pellets once with cold 75% EtOH and air-dry shortly. Resuspend each pellet in 20 µL ice-cold TE (not LoTE) buffer.
5. Place on ice, add 4 µL of a 10× loading buffer, load sample onto four lanes of a 12% 1.5-mm TAE-PAGE gel, and run at 160 V. Use a 10-bp ladder as a marker.
6. Excise the ditag band that runs at 22–26 bp (**Fig. 2**) and place the two gel fragments containing the cut out bands in one 0.5-mL tube.

7. Elute ditags as described for the 102-bp band, except that the incubation should be performed at 37°C instead of 65°C (**Subheading 3.1.10, steps 6–12**).
8. Resuspend each pellet in 2.5 μL cold TE buffer (7.5 μL total), keeping the tube on ice.

3.1.12. Ligation of 26-bp Ditags to Form Concatamers

1. The purified ditags can be ligated to form large concatameres in a 10-μL reaction mix consisting of 7 μL ditags, 5 U T4 DNA ligase (Life Technologies), and 1× ligation buffer. The incubation time is crucial and can vary between different samples. Usually this reaction is carried out for 70 min at 16°C.
2. Heat the sample for 5 min at 65°C, place on ice, and add 2 μL of a 10× loading buffer.
3. The entire mixture is run in a single lane of an 8% 1.0-mm TAE-PAGE gel (*see* **Note 6**). Use a 100-bp and a 1-kb ladder as a marker and run gel at 100 V.
4. Stain the gel with SYBR Green I and cut out the region between 800 and 1500 bp.
5. Purified concatemers as described for the 102-bp band under **Subheading 3.1.10**. Resuspend the pellet in 6 μL LoTE.

3.1.13. Cloning of Concatamers (see **Note 8**)

1. The concatemers can be cloned into an *Sph* I-cleaved pZeroO-1 vector (Invitrogen). Digest 1 μg pZero with *Sph* I for 20 min at 37°C in a 10-μL reaction volume consisting of 10 U *Sph* I (NEB) and 1× reaction buffer 2 (NEB). Do *not* incubate longer than 20 min.
2. Phenol-extract and ethanol-precipitate (200 μL sample, 5 μL glycogen [20 mg/mL], 133 μL 7.5 *M* ammonium acetate, and 1000 μL 100% EtOH).
3. Spin in a microcentrifuge at full speed for 15 min.
4. Resuspend pellet in 30 μL LoTE (final concentration 33 ng/μL).
5. Ligate the concatemers in the *Sph* I-cleaved pZero in a 10-μL reaction mix containing 6 μL concatemers sample, 33 ng digested vector, 5 U T4 DNA ligase (Life Technologies), and 1× ligase buffer (Life Technologies). Incubate overnight at 16°C.
6. After the ligation is completed, increase the volume up to 200 μL. Phenol-extract ligation mix and ethanol-precipitate as described under **Subheading 3.1.13., step 2.**
7. Spin in a microcentrifuge at full speed for 15 min.
8. Wash pellets 4 times with 75% EtOH, air-dry pellet, and resuspend in 10 μL LoTE.
9. Transfect 1 μL of each ligation mix into Electromax DH10Bs cells (Life Technologies). Plate one-tenth of the transfected bacteria onto Zeocin-containing plates (100 μg/mL) and grow for 16 h at 37°C.
10. Analyze colonies by colony PCR. Set up a 25-μL PCR reaction containing 2.5 U platinum Taq DNA polymerase (Life Technologies), 6% DMSO (Sigma), 0.5 m*M* dNTPs (Life Technologies), 26.5 μ*M* each of M13-forward and M13-reverse primers, and 1× PCR buffer. The following PCR conditions are optimized for the Hybaid OmnE PCR machine. Initial heat step for 2 min at 95°C, amplification for 30 cycles of 30 s at 95°C, 1 min at 56°C, 1 min at 70°C, and a final extension of 5 min at 70°C.
11. Run 5 μL of the colony PCR reaction on a 1.5% agarose gel. For efficient sequencing of a SAGE library the average insert size of the clones should be >650 bp, and a cloning efficiency of >80% is recommended (**Fig. 3**).

3.1.14. Sequencing

Sequencing can be performed in several different ways and is dependent on the availability of automated sequencing. Both capillary and slab-gel automated sequencers have been successfully employed. For direct sequencing using four-dye-labeled fluorescence terminators, PCR products can either be purified (various methods can be used) or diluted

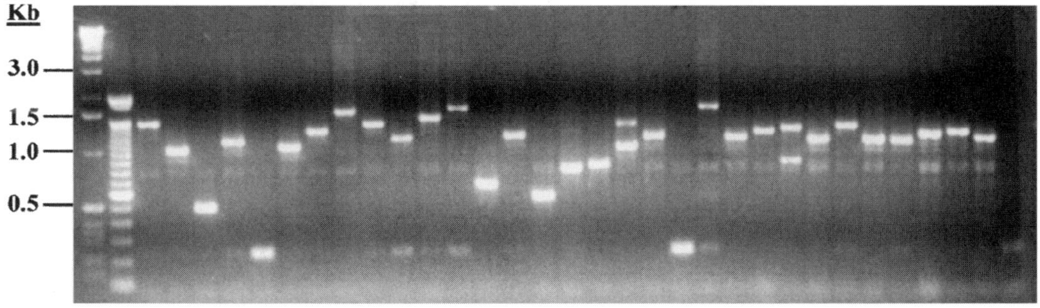

Fig. 3. Ethidium-stained 1.5% agarose gel showing an example of a colony PCR analysis of a good-quality SAGE library in which over 80% of the clones have an insert size of > 650 bp.

(correct dilution is dependent on the chemistry used and needs to be determined first). Good results have been obtained using the BigDye Primer Cycle Sequencing Ready Reaction kit (M13 or REV cat. no. 403050/1, Applied Biosystems) according to the manufacturer's instructions. Sequence text files in standard ASCII format (with a '.seq' filename extension) can be analyzed directly using the SAGE 2000 software version 4.12.

3.2. Micro-SAGE Protocol

The Micro SAGE method ***(9)*** is suitable for small samples (down to ~50 µg tissue or ~50,000 cells). Good results have been obtained starting with as little as 1–5 µg of total RNA (or 100 ng of polyA$^+$ RNA). Total RNA or polyA$^+$ RNA can be isolated using the protocols described for the standard SAGE method (**Subheading 3.1.**), although starting with total RNA has been the most efficient approach in our hands. Care has to be taken not to lose beads during the various steps. Furthermore, do not pipet the beads up and down during the washes, since the beads will stick to the tips. The beads can be mixed efficiently by 'flicking' the tube with your finger or a slow-speed vortex. Addition of BSA to most of the buffers is essential to reduce the sticking of the beads, which can lead to losses and inefficient washes.

3.2.1. Binding mRNA to Dynabeads oligo-dT

1. Thoroughly resuspend Dynabeads oligo-dT (Dynal mRNA Direct Kit) and transfer 100 µL to a 1.5–mL microcentrifuge tube.
2. After placing the tube next to the magnet in the magnet holder, remove supernatant and wash beads with 500 µL lysis/binding buffer (kit).
3. Adjust volume of the required amount of total RNA up to 1 mL with lysis/binding buffer, mix well, and add immediately to the beads after removing the last wash. Incubate for 5 min at room temperature by gentle mixing.
4. Using the magnet, remove the supernatant.
5. Wash the beads 2 times with 1 mL Washing buffer A LiDS (kit) supplemented with 20 µg/mL glycogen, once with 1 mL Washing buffer B (kit) also supplemented with 20 µg/mL glycogen, and 4 times with 100 µL 1× first-strand buffer (cDNA Synthesis System, Life Technologies).

3.2.2. cDNA Synthesis

1. Mix together 54 µL H_2O, 18 µL 5× First-strand buffer (cDNA synthesis kit), 9 µL 100 m*M* DTT (kit), and 4.5 µL 10 m*M* dNTPs (kit), and add this mix to the beads after removing the

last wash of **Subheading 3.2.1**. Incubate for 2 min at 42°C. Add 3 μL SuperScript II RT (Life Technologies) and incubate for 1 h at 42°C, mix intermittantly by gently flicking the tube.

2. Place on ice and add the components of the second-strand synthesis one by one as follows: 302 μL cold H_2O, 75 μL 10× second-strand buffer, 15 μL 10 m*M* dNTPs, 3 μL *Escherichia coli* DNA ligase, 12 μL *E. coli* DNA polymerase I, and 3 μL *E. coli* RNAseH. Mix and incubate for 2 h at 16°C. Mix beads every 15 min.
3. Stop the reaction by placing on ice and adding 100 μL of 200 m*M* EDTA, pH 8.0.
4. Wash the beads once with 500 μL 1× B+W buffer (with 1% SDS) and resuspend the beads in 200 μL of the same buffer. Heat sample for 20 min at 75°C.
5. At this point the beads get very sticky. Wash beads 4 times with 200 μL 1× B+W buffer with 200 μg/mL BSA (NEB) and once with 200 μL 1× reaction buffer 4 (NEB) with 200 μg/mL BSA.
6. Transfer the beads to a new microcentrifuge tube and repeat the last wash.

3.2.3. Cleavage of Bound cDNA with the Anchoring Enzyme

1. The bound cDNA is digested with the anchoring enzyme *Nla* III (NEB) for 1 h at 37°C by resuspending the beads in 170 μL LoTE buffer, 5 μL (50 U) *Nla* III, 4 μL (40 μg) BSA (NEB), and 20 μL 10× restriction buffer 4 (NEB).
2. Wash the beads twice with 200 μL 1× B+W buffer (supplemented with 0.1% SDS and 200-μg/mL BSA). If the BSA precipitates, heat the buffer for a few min at 65°C.
3. Then wash 4 times with 200 μL 1× B+W buffer (with 200-μg/mL BSA) and twice with 200 μL 1× ligase buffer (Life Technologies). During the last wash, transfer 100 μL of sample into two new microcentrifuge tubes.

3.2.4. Ligating Linkers to Bound cDNA

1. For preparation of the linkers, *see* **Notes 3** and **4**.
2. Remove the last wash from the beads and add 1 μL (20 ng) of the annealed linkers 1A,B to tube 1 and 1 μL (20 ng) of the annealed linkers 2A,B to tube 2.
3. Add 6 μL LoTE and 2 μL 5× ligase buffer (Life Technologies) to each tube. Heat for 2 min at 50°C and let sit at room temperature for 15 min.
4. To each tube add 1 μL (5 U) T4 DNA ligase (Life Technologies) and incubate mixtures for 2 h at 16°C. Mix beads intermittently.
5. After ligation, wash tubes once with 200 μL 1× B+W buffer (with 0.1% SDS and 200 μg/mL BSA) and transfer to new tubes.
6. Pool tubes 1 and 2 together in order to minimize loss in subsequent steps. Repeat the previous wash step.
7. Wash 4 times with 200 μL 1× B+W (with 200-μg/mL BSA) and twice with 200 μL 1× reaction buffer 4 (NEB).

3.2.5. Release of cDNA Tags Using Tagging Enzyme

1. After removing the last wash, the linked cDNA is released by resuspending the beads in 170 μL LoTE buffer, 20 μL 10× reaction buffer 4 (NEB), 4 μL (40 μg) BSA, and 2 μL (2 U) of *BmsF* I (NEB). Incubate mixture for 1 h at 65°C.
2. Collect the supernatant containing the released linked tags by centrifugation for 2 min at 14,000 *g* and transfer to a new tube.
3. Wash beads with 40 μL LoTE and pool the supernatants together (total volume 240 μL).
4. Extract with equal volume PC8 and ethanol-precipitate the linked tags (240 μL sample, 5 μL glycogen [20 mg/mL], 133 μL 7.5M ammonium acetate, and 1000 μL 100% EtOH).
5. Spin in a microcentrifuge at full speed for 30 min at 4°C.

6. Wash pellet twice with 75% EtOH, air-dry, and resuspend cleaved cDNA in 10 μL LoTE buffer.

3.2.6. Blunt Ending of the Released Linked Tags

1. The blunt ending reaction is performed with Klenow (USB) for 30 min at 37°C in a reaction volume of 50 μL, containing 3 U Klenow, 10 μg BSA, 0.5 m*M* dNTPs, and 1× second-strand buffer (Life Technologies, cDNA synthesis kit).
2. Increase the volume with 190 μL LoTE (total volume 240 μL), extract with equal volume of PC8.
3. Remove 200 μL of the aqueous phase and transfer to the ligase-plus tube.
4. The rest (~40 μL) is transferred to the ligase-minus tube.
5. Ethanol-precipitate both samples (200 μL sample, 5 μL glycogen [20 mg/mL], 133 μL 7.5 *M* ammonium acetate, and 1000 μL 100% EtOH).
6. Spin in a microcentrifuge at full speed for 30 min at 4°C.
7. Wash twice with 75% EtOH, air-dry and resuspend the ligase-plus reaction in 5 μL LoTE and the ligase-minus reaction in 3 μL LoTE buffer.

3.2.7. Ligation of Linked Tags to Form Ditags

1. The 102-bp ditags are formed by the ligation of the blunt-ended linked tags. Add 5 μL of a 2× ligation plus mix (2.5 μL 3 m*M* Tris-HCl, pH 7.5, 3 μL 5× ligase buffer [Life Technologies], and 2 μL (10 U) T4 DNA ligase [Gibco-BRL] to the ligase-plus tube.
2. Add 3 μL of a 2× ligation minus mix (2.25 μL 3 m*M* Tris-HCl, pH 7.5, 1.5 μL 5× ligase buffer [Life Technologies]) to the ligase-minus tube.
3. Incubate both tubes overnight at 16°C.
4. After ligation, increase the volume up to 20 μL by adding 10 or 14 μL of LoTE to the ligase-plus and ligase-minus tube, respectively.
5. Aliquot the ligase-plus mixture and keep at −20°C.

3.2.8. Return to the Standard SAGE Method

Proceed with Subheadings 3.1.9.–3.1.14. of the standard SAGE method.

3.3. SAGE Bioinformatics

The most challenging part of SAGE is the analysis of the generated digital data. The PC-based application SAGE 2000 version 4.12 is freely available for noncommercial use through the SAGE Home Page posted by the Kinzler-Vogelstein Laboratory at Johns Hopkins Oncology Center (http://www.sagenet.org). The SAGE software finds *Nla* III sites (or other anchoring enzymes) within the raw sequence files, extracts the ditags, removes any duplicate ditags, counts the occurrence of each tag, and provides a report containing a list of tags with their corresponding expression levels. The exclusion of duplicate ditags eliminates possible PCR bias during the amplification step. The program has many useful tools such as normalization, comparison of multiple SAGE libraries, and a statistical function based on the Monte Carlo approach, which allows the determination of *p*-values for the observed differences. In addition, tags derived from the linker sequences and 1-bp variations of these linker tags can be excluded from any type of analysis. The linker tags are an artifact of the SAGE library construction, but do not influence the relative abundance of the transcripts.

3.3.1. Tag to Gene Mapping

The matching of the tags to the appropriate reference sequence can be done in different ways. Most common is to download the latest update of the GenBank database, which can be obtained from the National Center for Biotechnology Information (NCBI) FTP site (ftp://ncbi.nlm.nih.gov/genbank/), and to subtract the tags from the sequence files using the SAGE 2000 software. In order to be able to do this, the files must be converted from UNIX format into DOS format beforehand. Linking of a list of tags to the generated GenBank database reveals the corresponding gene information. Tag-to-gene mappings from reference sequences and generated by the SAGE software are available at the SAGE home page (http://www.sagenet.org). These files can be exported as Microsoft Access files for use in a relational database, which greatly improves SAGE data management and analysis. SAGE tags can also be mapped to UniGene clusters through the SAGEmap web site ***(10)*** of the Cancer Genome Anatomy Project (CGAP), making the identification of a gene from a differentially expressed tag easier (http://www.ncbi.nlm.nih.gov/SAGE/). The public database, SAGEmap, is maintained by the NCBI and is a resource for SAGE analysis that contains, to date, over 3 million SAGE tags from 92 different malignant and normal cell types. On-line tools ***(11)*** allow statistical-based comparisons between libraries using the 'xProfiler', or data can be downloaded for local analysis from the NCBI SAGEmap FTP site. Full and reliable tag-to-gene mappings for human, rat, and mouse, based on *Nla* III or *Sau*3A, are available for download from the SAGEmap FTP site (ftp://ncbi.nlm.nih.gov/pub/sage/). These text files can be imported into Microsoft Access and used in a relational database.

Although the use of the GenBank database in combination with the tag-to-gene mappings from SAGEmap certainly improves the tag assignments, a significant number of these assignments are incorrect. Manual screening is still needed to verify the tag-to-gene assignments. A proper SAGE tag is derived from a sequence that contains a poly-adenylation signal and/or poly-adenylation tail and is located downstream from the *Nla* III site closest to the poly-adenylation tail. Well-characterized mRNA sequences, do not always contain a poly-adenylation signal and/or polyadenylation tail; therefore the extracted tag is not necessarily the one adjacent to the most 3′ *Nla* III site. Among EST sequences, those with a poly-adenylation signal and/or a poly-adenylation tail and annotated as 3′-sequences are the most reliable. Hence, when the extracted tags from cDNA equal the ones extracted from 3′-EST sequences, the tags are called most reliable. When the tags are different, it is most likely that the cDNA did not contain the 3′-untranslated region including the poly-adenylation signal. To determine the correct tag of that particular gene it is best to align the sequences present in the corresponding Unigene cluster. Another possible explanation for multiple tags for one gene is the occurrence of alternative splicing of a transcript. In that case, one has to consider the frequencies to establish the most likely tag for a gene.

Recently the CGAP SAGE project created SAGE Genie, a new web site for the analysis and presentation of SAGE data (http://CGAP.nci.nih.gov/SAGE). An enhanced link between SAGE tag and gene is based on an experimental sample of 6.8 million human SAGE tags, which was used to evaluate custom databases derived from GenBank, EST sequences, and other public transcript sequence databases. In this way SAGE Genie is able to automatically identify a tag from primary transcripts and/or alternatively poly-adenylated transcripts and at the same time screen for experimental artifacts. These informatics provide a rapid and intuitive view of expression levels in the human body or brain, displayed on the SAGE Anatomic Viewer ***(12)***.

3.4. SAGE Applications in Cancer

SAGE analysis generates large digital gene expression profiles, representing the entire set of genes expressed within the system analyzed. Such a profile is defined as a transcriptome. SAGE has been shown to be an excellent tool in defining various transcriptomes ***(13,14)***. Analysis of SAGE data among many human tissues has already allowed characterization of constitutively expressed genes that are common to all gene expression profiles, as well as those genes expressed specifically in certain tissues/cells ***(13)***. Global expression changes in cancer were first assayed by comparison of colon carcinoma to normal colon ***(15)***. Similar profiles have been generated for lung tumors, brain tumors, Hodgkin's lymphoma, breast, prostate, and ovarian cancer ***(16–20)***.

The complex expression changes that characterize the difference between normal and malignant changes are too numerous for the expression profiles to be immediately useful. However, this wealth of digital data has many interesting practical applications. The expression level of any specific gene can be visualized in many other normal or cancer tissues without the need to perform an actual Northern blot analysis. Most important, the results from any new experiment are comparable directly to existing gene expression profiles, which can lead to the immediate identification of genes that are expressed or absent in other tissues, representing potential prognostic or diagnostic markers ***(18,21)***. The tissue inhibitor of metalloproteinase type I (TIMP-1) was shown by SAGE analysis to be overexpressed in pancreatic cancer cells but not in normal cells or cancerous colon cells. The TIMP-1 gene by itself was not a good prognostic factor, but in combination with other pancreatic cancer markers the sensitivity of pancreatic cancer detection was significantly increased ***(22)***. The PGP9.5 ubiquitin hydralose was reported in another SAGE study to be specifically associated with lung cancer development, increasing its prognostic value ***(23)***.

SAGE is particularly well suited for pathway analysis. A well-controlled expression system, which provides an isogenic background, should reveal only those few transcriptional changes related to the experimental manipulation of the cells. An example of such an application was initially presented by the analysis of SAGE profiles of rat embryo fibroblast cells harboring a temperature-sensitive p53 mutation ***(24)***. Similarly, SAGE was applied to a human system in which cells expressing functional p53 resulted in apoptosis ***(25)***. A large fraction of the new-found downstream targets of p53 could be categorized into a functional gene family explaining the apoptotic phenotype. Another example of the utility of SAGE for pathway analysis is the study of pancreatic cells with an inducible expression system for the tumor suppressor APC (adenomatous polyposis coli). The APC gene is inactivated in most colorectal cancers and can partially regulate cell growth through the β-catenin/Tcf-4 transcription complex. The expression of the c-MYC oncogene was decreased by the activation of wild-type APC, implying that overexpression of c-MYC in colorectal tumors is mediated through an increased β-catenin/Tcf4 activity ***(26)***. Cyclin-dependent kinase 4 (CDK4) was identified as a transcriptional target of the c-MYC oncogene using SAGE, providing a direct link between the oncogenic effect of c-MYC and cell cycle regulation ***(27)***. A recent SAGE comparison of neuroblastoma cells expressing the N-myc oncogene with the corresponding parental line revealed an unexpected set of genes with a possible role in ribosome biogenesis and protein synthesis. The results indicate that N-myc can function as a regulator of cell growth and might in part explain the aggressiveness of N-myc-amplified tumors ***(28)***. These studies demonstrate how SAGE can further elucidate biologic pathways implicated in cancer.

3.5. Conclusions

One of the most interesting applications of SAGE has been the identification of cancer-related genes. Well-designed comparisons of the comprehensive gene expression profiles of tumors and normal tissues can yield a subset of genes that are potential diagnostic markers or therapeutic targets. Early detection is needed for many cancers, for which surgery prior to metastasis is the only effective treatment. Genes that are highly expressed in cancer cells, especially those with a secreted gene product, could be measured in blood and other body fluids. Analysis of gene expression profiles of responsive vs nonresponsive tumor samples could lead to a better understanding of the mechanism of action of certain chemotherapeutic agents and might lead eventually to a more specify and therefore more effective treatment.

The better-controlled experiments that characterize pathway analysis yield a more manageable set of differentially expressed genes and have the advantage of finding new genes, but they can also point to a previously unsuspected but well-known gene. SAGE experiments can be designed by disruption, overexpression or reintroduction of a known TSG in cultured cells. The application of SAGE to study tumor suppressor genes can help reveal the downstream genes controlled at a transcriptional level, and provide insight into their function.

4. Notes

1. Phenol chloroform (PC8): Add, in sequence, 480 mL phenol (warm to 65°C), 320 mL 500 mM Tris-HCl (pH 8.0), 640 mL chloroform, shake, and place at 4°C. After 2–3 h, shake again. After another 2–3 h, aspirate aqueous layer. Aliquot and store at −20°C.
2. Testing biotinylation of biotin-oligo dT: Obtain biotin-oligo dT gel purified from oligo-synthesis company. Test biotinylation by adding to several hundred nanograms of biotin-oligo dT to 1 μg streptavidin (Sigma cat. no. S-4762). Incubate several min at room temperature. Both the oligo alone and bound to streptavidin are run on a 20% Novex gel (Invitrogen). If the oligo is well biotinylated, the entire amount of oligo should be shifted to higher molecular weight in the lane containing the streptavidin. Alternatively, increasing amounts of oligo (from several hundred nanograms to several micrograms) can be incubated with and without separate aliquots of 100 μL of Dynabeads (Dynal). After 15 min, the beads are separated from the supernatant using a magnet, the supernatant is removed, and DNA quantitation is performed at OD 260. At low amounts of oligo, when bead-binding capacity is not saturated, the ratio of unbound oligo to the total oligo will indicate the percent of oligo that is not biotinylated.
3. Phosphorylation linkers: The linkers, described in Table 1, should be obtained gel purified. Dilute linkers 1B and 2B to 350 ng/μL. Add to a tube labeled 1B, 9 μL linker 1B (350 ng/μL), and to another tube labeled 2B, 9 μL linker 2B (350 ng/μL). In addition, add to each of the tubes 6 μL LoTE, 2 μL 10× kinase buffer (NEB), 2 μL 10 mM ATP, 1 μL T4 polynucleotide kinase (10 U/μL, NEB). Incubate the reaction for 30 min at 37°C. Heat-inactivate at 65°C for 10 min.

4. Annealing linkers: Dilute linkers 1A and 2A to 350 ng/μL. Add 9 μL linker 1A to the 20-μL phosphorylation mix of linker 1B (final concentration 200 ng/μL) and 9 μL linker 2A to the 20-μL phosphorylation mix linker 2B (final concentration 200 ng/μL). Heat samples for 2 min at to 95°C, then place for 10 min at 65°C, followed by 10 min at 37°C and finally 20 min at room temperature. Store at −20°C. The efficiency of the phosphorylation reaction should be tested by self-ligation of about 200 ng of each linker pair. Analyze the self-ligation on a 12% TAE-PAGE gel. The phosphorylated linkers should allow linker–linker dimers (80–100 bp) to form after ligation, while unphosphorylated linkers will prevent self-ligation. Only linker pairs that self-ligate >70% should be used in further steps.
5. The 102-ditag-containing PCR product can fail to form for a number of reasons. Even though there are normally other background PCR products formed and seen on the gel, the 102-bp band should be clearly visible and at least the same intensity as the other products. Starting materials such as the RNA or cDNA may be insufficient or degraded, so first run a gel to check the integrity of each of these steps, running a standard of known concentration for comparison. The linkers can often be a problem, so check the linkers as described above. Degradation of the Nla III enzyme has been a problem in the past, and we now store our Nla III at −80°C and ask that the enzyme be shipped on Dry Ice. Cutting a standard, such as a plasmid, known to contain one or more Nla III sites, can check the quality of the enzyme. Another potential problem is that if the dNTP concentration is too high in the PCR reaction, this can inhibit the formation of the 102-bp band. Therefore, check to make sure that dilutions have been made properly and see if a 102-bp band can be formed using serial dilutions of the dNTPs. Also, do not attempt too many PCR cycles; this will not only result in the potential loss of the 102-bp band, but in duplicate ditags. Run a range of cycle numbers but in no case over 30 cycles. Finally, if you have made any 'modifications' to the protocol, try it as written.
6. Gel electrophoresis: We are currently using the X-cell Surelock Mini-Cell system for Novex minigels. Prepare a 12% acrylamide stock solution in 1× TAE buffer using 40% acrylamide (19:1 acrylamide:bis), and store at 4°C. A different acrylamide:bis ratio is used for the 8% gels (40% polyacrylamide; 37.5:1). For one 1.0-mm gel use 7.5 mL stock solution, add 7.5 μL 10% ammonium persulfate (APS), and 5 μL TEMED. Let the gel polymerize for at least 30 min.
7. If the 102-bp product will not digest with Nla III, test to make sure your Nla III is active, or try increasing the amount of enzyme. If you can obtain some 102-bp product that you know is good, try cutting this as a control. In many cases we have observed, the Nla III site may not exist in the 102-bp bands that were purified. One potential reason for this is that exonuclease contamination may have digested the sensitive CATG overhand during an earlier step of the protocol. Some exonuclease activity may result from residual DNA polymerase I used for the cDNA synthesis. Therefore, two, rather than one, phenol extractions are required to purify the cDNA. For micro-SAGE users, the SDS within the protocol may be helpful for inactivation of exonucleases. In any case, if the Nla III site is absent, the library must be remade taking care to make sure all the purifications are precise.
8. The quantity and purity of the ditags can both affect the ligation reaction and ultimate insert size of the pZERO clones. In general, 500 ng to 1 μg of ditags formed from ~300 PCR reactions is needed to get sufficient ditags for good ligation and concatemer formation. Even though the ditag is gel purified, linker material may still contaminate the ditags. Although others have overcome this problem by using an additional purification based on biotinylating primers 1 and 2, we have found that extra purifications create more problems than they solve. Using proper care not to overload the gels or to run the gels too quickly, and cutting out the ditag precisely should result in a nearly pure ditag product. Finally, it is important to make sure that the ends of the pZERO are intact for good cloning efficiency. Freshly cut pZERO should be used if there is a problem with little or no inserts.

References

1. Velculescu, V. E., Zhang, L., Vogelstein, B., and Kinzler, K. W. (1995) Serial analysis of gene expression. *Science* **270**, 484–487.
2. Lal, A., Sui, I.-M., and Riggins, G. (1999) Serial analysis of gene expression: probing transcriptomes for molecular targets. *Curr. Opin. Mol. Ther.* **1**, 720–726.
3. Madden, S. L., Wang, C. J., and Landes, G. (2000) Serial analysis of gene expression: from gene discovery to target identification. *Drug Discov. Today* **5**, 415–425.
4. Velculescu, V. E., Vogelstein, B., and Kinzler, K. W. (2000) Analysing uncharted transcriptomes with SAGE. *Trends Genet.* **16**, 423–425.
5. Powell, J. (2000) SAGE. The serial analysis of gene expression. *Meth. Mol. Biol.* **99**, 297–319.
6. Datson, N. A., van der Perk-de Jong, J., van den Berg, M. P., de Kloet, E. R., and Vreugdenhil, E. (1999) MicroSAGE: a modified procedure for serial analysis of gene expression in limited amounts of tissue. *Nucleic Acids Res.* **27**, 1300–1307.
7. Virlon, B., Cheval, L., Buhler, J. M., Billon, E., Doucet, A., and Elalouf, J. M. (1999) Serial microanalysis of renal transcriptomes. *Proc. Natl. Acad. Sci. USA* **96**, 15286–15291.
8. Ye, S. Q., Zhang, L. Q., Zheng, F., Virgil, D., and Kwiterovich, P. O. (2000) miniSAGE: gene expression profiling using serial analysis of gene expression from 1 microg total RNA. *Anal. Biochem.* **287**, 144–152.
9. St Croix, B., Rago, C., Velculescu, V., et al. (2000) Genes expressed in human tumor endothelium. *Science* **289**, 1197–1202.
10. Lal, A., Lash, A. E., Altschul, S. F., et al. (1999) A public database for gene expression in human cancers. *Cancer Res.* **59**, 5403–5407.
11. Lash, A. E., Tolstoshev, C. M., Wagner, L., et al. (2000) SAGEmap: a public gene expression resource. *Genome Res.* **10**, 1051–1060.
12. Boon, K., Osório, E. C., Greenhut, S. F., et al. (2002) An anatomy of normal and malignant gene expression. *Proc. Natl. Acad. Sci. USA,* **55**, 11, 287–11, 292.
13. Velculescu, V. E., Madden, S. L., Zhang, L., et al. (1999) Analysis of human transcriptomes. *Nat. Genet.* **23**, 387–388.
14. Velculescu, V. E., Zhang, L., Zhou, W., et al. (1997) Characterization of the yeast transcriptome. *Cell* **88**, 243–251.
15. Zhang, L., Zhou, W., Velculescu, V. E., et al. (1997) Gene expression profiles in normal and cancer cells. *Science* **276**, 1268–1272.
16. Hibi, K., Liu, Q., Beaudry, G. A., et al. (1998) Serial analysis of gene expression in non-small cell lung cancer. *Cancer Res.* **58**, 5690–5694.
17. van den Berg, A., Visser, L., and Poppema, S. (1999) High expression of the CC chemokine TARC in Reed-Sternberg cells. A possible explanation for the characteristic T-cell infiltratein Hodgkin's lymphoma. *Am. J. Pathol.* **154**, 1685–1691.
18. Nacht, M., Ferguson, A. T., Zhang, W., et al. (1999) Combining serial analysis of gene expression and array technologies to identify genes differentially expressed in breast cancer. *Cancer Res.* **59**, 5464–5470.
19. Hough, C. D., Sherman-Baust, C. A., Pizer, E. S., et al. (2000) Large-scale serial analysis of gene expression reveals genes differentially expressed in ovarian cancer. *Cancer Res.* **60**, 6281–6287.
20. Xu, L. L., Shanmugam, N., Segawa, T., et al. (2000) A novel androgen-regulated gene, PMEPA1, located on chromosome 20q13 exhibits high level expression in prostate. *Genomics* **66**, 257–263.
21. Loging, W. T., Lal, A., Siu, I. M., et al. (2000) Identifying potential tumor markers and antigens by database mining and rapid expression screening. *Genome Res.* **10**, 1393–1402.
22. Zhou, W., Sokoll, L. J., Bruzek, D. J., et al. (1998) Identifying markers for pancreatic cancer by gene expression analysis. *Cancer Epidemiol. Biomarkers Prev.* **7**, 109–112.

23. Hibi, K., Westra, W. H., Borges, M., Goodman, S., Sidransky, D., and Jen, J. (1999) PGP9.5 as a candidate tumor marker for non-small-cell lung cancer. *Am. J. Pathol.* **155**, 711–715.
24. Madden, S. L., Galella, E. A., Zhu, J., Bertelsen, A. H., and Beaudry, G. A. (1997) SAGE transcript profiles for p53-dependent growth regulation. *Oncogene* **15**, 1079–1085.
25. Polyak, K., Xia, Y., Zweier, J. L., Kinzler, K. W., and Vogelstein, B. (1997) A model for p53-induced apoptosis. *Nature* **389**, 300–305.
26. He, T.C., Sparks, A. B., Rago, C., et al. (1998) Identification of c-MYC as a target of the APC pathway. *Science* **281**, 1509–1512.
27. Hermeking, H., Rago, C., Schuhmacher, M., et al. (2000) Identification of CDK4 as a target of c-MYC. *Proc. Natl. Acad. Sci. USA* **97**, 2229–2234.
28. Boon, K., Caron, H. N., van Asperen, R., et al. (2001) N-myc enhances the expression of a large set of genes functioning in ribosome biogenesis and protein synthesis. *EMBO. J.* **20**, 1383–1393.

30

Differential Display Techniques to Identify Tumor Suppressor Gene Pathways

Siham Biade and Maureen E. Murphy

1. Introduction

Differential display is a reverse-transcription polymerase chain reaction (RT-PCR) technique that was introduced in 1992 by Liang and Pardee *(1)*. In this technique, messenger RNA 3′ termini are amplified using an anchored oligo-dT primer and a series of arbitrary 13-mers; a single round of cDNA synthesis is followed by PCR amplification in the presence of radiolabeled nucleotide, resulting in the amplification of subsets of mRNAs expressed in the cell. This complex PCR product is resolved on a denaturing polyacrylamide gel; following autoradiography, gene expression is detected as a ladder of cDNAs that differ in size by a single nucleotide. Theoretically, the use of different sets of primers from both directions in discrete RT-PCR reactions allows for the examination of the majority of expressed genes within the cell. This chapter will focus first on a brief review of the use of differential display to identify p53 target genes, as well as secondary tumor suppressor genes important for tumor progression. This will be followed by an overview of the methodology of differential display, along with recent advances in this technique; a more comprehensive analysis can be found in *Methods in Molecular Biology, Volume 85: Differential Display Methods and Protocols*, edited by Liang and Pardee *(2)*. An outline of the methodology involved in differential display is provided in **Fig. 1**, and individual steps are described below; typically these steps follow the protocol described by Liang and Pardee *(1,2)*, but in some instances changes have been made and noted.

1.1. p53-Induced Genes: cyclin G, ei24, and Siah

The first p53 response gene identified by differential expression methodology was p21/waf1, which was cloned as a p53-induced gene by subtractive hybridization **(3)**. This gene was later shown to be essential for p53-dependent growth arrest **(4)**. Since that time, several groups have utilized differential display in efforts to identify other targets of p53 influential in p53-dependent growth arrest and apoptosis. For example, Okamoto and Beach **(5)** identified cyclin G as a novel transcriptional target of p53. In

From: *Methods in Molecular Biology, Vol. 222: Tumor Suppressor Genes: Pathways and Isolation Strategies*
Edited by: Wafik S. El-Deiry © Humana Press Inc., Totowa, NJ

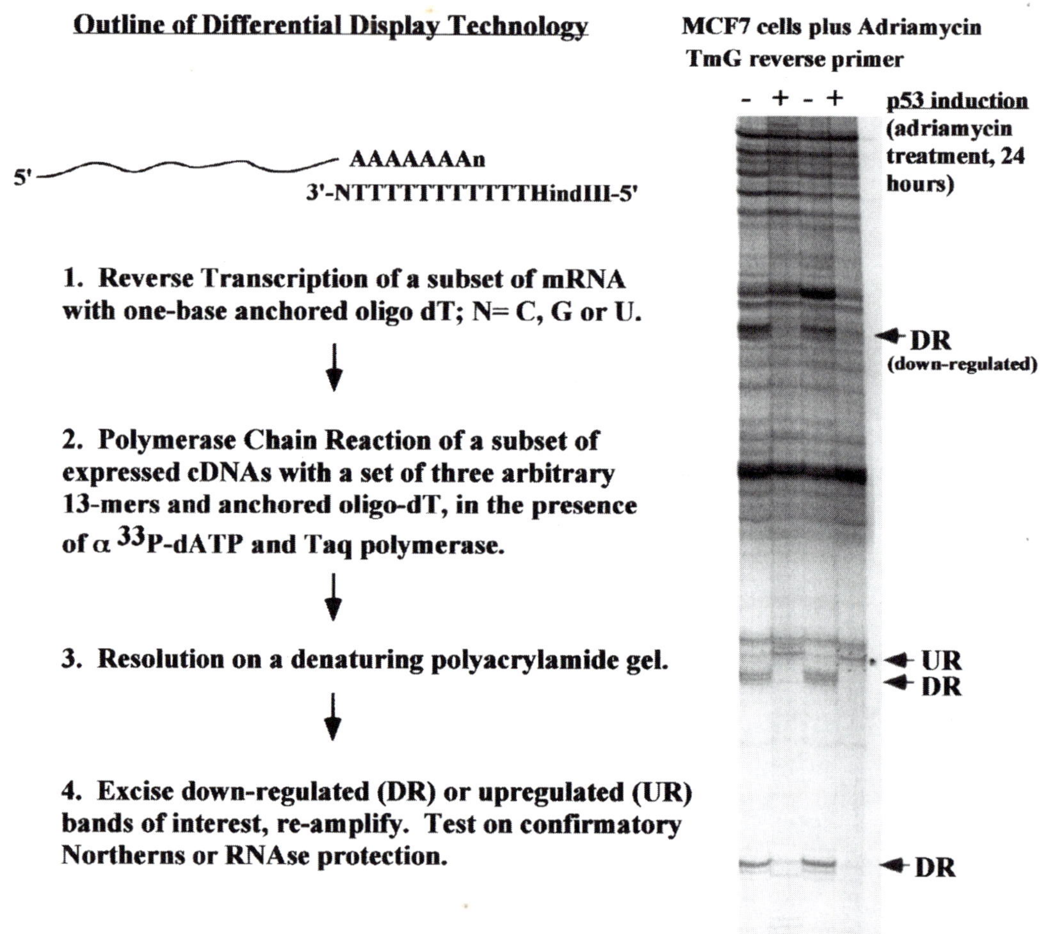

Fig. 1. Outline of the relevant steps in the standard differential display protocol. Shown as a representative figure is a differential display gel from mRNA isolated from the MCF-7 human breast carcinoma cell line, before and after treatment for 24 h with the DNA-damaging agent adriamycin (0.5 μg/mL, final concentration), which induces the activity of the p53 tumor suppressor protein. Samples were performed and run in duplicate; candidate p53 upregulated (UR) and downregulated (DR) cDNAs are denoted.

cells treated with IR, as well as those containing an inducible p53 protein, this gene was found to be upregulated 10-fold in a p53-dependent manner **(5)**. Confirmation that this was a direct target gene of p53 was made following the identification of p53-consensus binding sites in the cyclin G promoter **(5,6)**. While it seemed initially counterintuitive for p53 to be upregulating a positive regulator of the cell cycle, cyclin G overexpression was shown to enhance cell death induced by cisplatin **(7)** and to facilitate apoptosis induced by TNFα, retinoic acid, and serum starvation **(8)**. How cyclin G functions in apoptosis is currently unclear, although it may do so in part by binding and modifying the function of the regulatory subunit of protein phosphatase 2A (PP2A) **(9)**.

ei24 was cloned as a gene induced following etoposide treatment of NIH 3T3 cells ***(10)***. The rationale for elucidating genes induced by etoposide was that the induction of apoptosis by etoposide in this cell line was known to require p53, as well as new RNA synthesis. The *ei24* gene was found to have a p53-consensus binding element in its promoter region, and overexpression of ei24 alone was sufficient to induce apoptosis ***(11)***. ei24 is predicted to encode a novel protein with six membrane-spanning domains; interestingly, ei24-induced apoptosis can be inhibited by overexpression of bcl-x_L, indicating that ei24 may play a role in the death pathway mediated by the mitochondria/ cytochrome c pathway ***(11)***. *Siah* was uncovered in a large-scale differential display screen for p53-regulated genes ***(12)***. In *Drosophila* this gene functions downstream of the sevenless receptor, and is required for specification of the R7 photoreceptor cells in the *Drosophila* eye. Unlike cyclin G and ei24, a p53-response element has not to date been identified in the *siah* gene, indicating that it may be indirectly regulated by p53. However, like cyclin G and ei24, overexpression of SIAH protein directly promotes apoptosis ***(13,14)***. Interestingly, it may do so via its interaction with Pw1/Peg3, which was also identified by differential display as a p53-induced gene ***(15)***. The *siah* gene was shown to encode a RING finger-containing protein that most likely plays a role in protein turnover, stimulating the degradation of proteins such as c-myb and the kinesin Kid ***(16,17)***.

1.2. p53-Repressed Genes: Map4, oncoprotein 18 (stathmin), presenilin-1

That p53 has a role in apoptosis that is separable from its transactivation function is supported by several lines of evidence. For example, mutation of amino acids 22 and 23 in the p53 transactivation domain eliminates the transactivation function of this protein ***(18)***, but not its ability to induce apoptosis ***(19)***. Further, p53-dependent apoptosis has been shown to occur in the presence of inhibitors of new RNA and protein synthesis ***(20,21)***, supporting the existence of a mechanism for p53-dependent apoptosis that is independent of transactivation. Such a mechanism was implicated by the finding that inhibitors of apoptosis, such as BCL-2, E1B-19k, and the Wilms tumor suppressor WT-1, can inhibit p53-dependent apoptosis and transcriptional repression, but do not affect the transactivation function of p53 ***(22–24)***. Tumor-derived mutant forms of p53 were found that were able to induce growth arrest, but not apoptosis; these were demonstrated to retain the ability to transactivate p53-induced genes, but not to repress gene expression ***(25)***. Finally, deletion of the proline-rich domain of p53 was shown to impair the ability of p53 to induce apoptosis, and to repress gene expression ***(26,27)***. The combined data present a compelling case for a role of p53-dependent transcriptional repression of gene expression to the ability of this protein to induce apoptosis.

Several groups capitalized on these data and used differential display to identify genes showing decreased expression following p53 induction. One of the first studies identified *Map4*, which encodes a microtubule-associated protein, as a p53-repressed gene in a cell line containing temperature-sensitive p53 that underwent programmed cell death ***(28)***. Later these analyses were refined to include a comparison between cells with temperature-sensitive p53 that underwent p53-dependent apoptosis (Vm10 cells) with the same cells transfected with the adenovirus E1B-19k gene (Vm10-19k). E1B-19k, an adenoviral homolog of BCL-2, inhibits p53-dependent apoptosis and transcriptional repression, and instead p53-dependent growth arrest occurs. An example of differential display data obtained from this study is depicted in **Fig. 2**; as is evident from the figure, genes upregulated during p53-dependent apoptosis (Vm10, 32 degrees) are identically upregulated during p53-dependent growth arrest (Vm10-19k, 32 degrees). In contrast, genes downregulated during p53-dependent apoptosis are not downregulated when apoptosis is inhib-

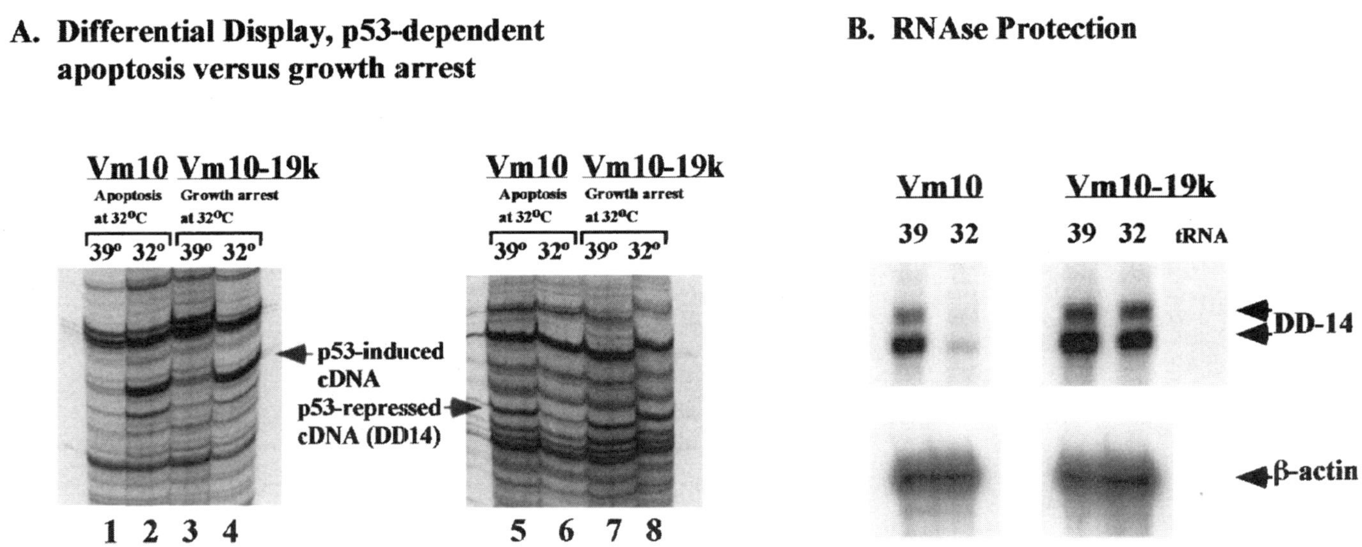

Fig. 2. **(A)** Differential display allows for the identification of gene expression differences between p53-dependent apoptosis (Vm10 cells, at 32°C) and growth arrest (Vm10-19k cells at 32°C). Both cell lines contain temperature-sensitive p53 protein, which exists in a mutant conformation at 39°C, and a wild-type (active) conformation at 32°C. Vm10-19K cells have been stably transfected with the adenovirus E1B-19k gene, which protects from p53-dependent apoptosis, and instead undergo p53-dependent growth arrest (see **ref.** ***24, 30,*** and ***31***). As evident from the figure, genes downregulated following p53 induction in apoptotic cells (*lane 6*, arrow) are not downregulated when cells are protected from apoptosis by E1B-19k *(lane 8)*. **(B)** RNAse protection analysis of the DD14 cDNA isolated in **(A)**. The DD14 antisense cDNA probe detects two RNA species, due to differential poly-adenylation of the mRNA encoded by DD-14, Map4 (*see* **ref.** ***30***). This gene is downregulated by p53 is cells that undergo apoptosis (Vm10), but not in cells protected from apoptosis (Vm10-19k).

ited by E1B-19k, and growth arrest ensues (**Fig. 2**, arrow). These data are consistent with earlier findings that E1B-19k can inhibit p53's transcriptional repression function, and the refinement of this study allowed for the better identification of candidate genes specifically repressed by wild-type p53. In studies such as this one, candidate p53-repressed genes began to be isolated; these included *oncoprotein 18 (stathmin)* ***(29)***, which may play a role in G2/M arrest by p53 ***(30)***, as well as *presenilin-1* ***(12)***. Significantly, repression of either Map4 or presenilin-1 expression was shown to accelerate apoptosis, verifying that the downregulation of these genes plays a role in the induction of apoptosis by p53 ***(28,31)***.

Recently, p53-dependent transcriptional repression of the *Map4* gene was shown to require histone deacetylases (HDACs); further, complexes between p53 and HDAC-1, mediated by a direct interaction between p53 and the corepressor protein Sin3, can be found in the cell ***(32)***. The contribution of this complex, and of p53-dependent transcriptional repression, to apoptosis induction by this protein was verified by findings that the HDAC inhibitor Trichostatin A could inhibit p53-dependent apoptosis ***(32)***. Further, the Sin3-binding domain of p53 maps to the proline-rich region of this protein ***(33)***, which was previously shown to be essential for p53-dependent apoptosis induction, but dispensable for transactivation ***(26)***. In sum, the identification of p53-repressed genes by differential display provided the necessary tools to dissect the mechanism whereby p53 functions as a transcriptional repressor, and should facilitate the elucidation of the contribution of this activity to apoptosis induction.

1.3. Other p53-Induced Genes: Wig-1, rCOP1, p53 RR2

Differential display continues to be useful in the identification of p53-regulated genes. Recently, this technique was used to identify Wig-1, a novel zinc finger-containing protein that is upregulated by p53 ***(34)***, as well as rCop-1, a member of the cysteine-rich growth regulator family that is frequently lost in transformed cells ***(35)***. More recently, Tanaka and colleagues identified p53R2, which encodes a protein with similarity to the ribonucleotide reductase small subunit. Interestingly, this p53-induced gene appears to play a role in DNA repair, such as would occur following p53 induction by DNA-damaging agents ***(36)***. A summary of these p53-regulated genes, and their postulated roles in growth arrest and apoptosis, is presented in **Fig. 3**.

2. Materials

1. CsCl cushion solution: 5.7 *M* CsCl/0.1 *M* EDTA, pH 7.5.
2. Guanidine solution: 6 *M* guanidine HCl/10 m*M* dithiothreital (DTT)/25 m*M* NaOAc, pH 5.0.
3. RNA ethanol wash solution: 20/9/1 100% ethanol/dH_2O/2 *M* NaOAc, pH 5.
4. 6% Acrylamide/7 *M* urea sequencing gel.
5. 1× TBE buffer, pH 8.3–8.9: 0.089 *M* Tris-borate/0.089 *M* boric acid/0.05 *M* EDTA, pH 8.0.
6. All other components can be found in the RNAimage kit (GenHunter Corporation).

3. Methods (*see* Notes 1–4)

The generation of high-quality RNA is essential for differential display, and can be achieved through any of a number of standard protocols; however, the use of cesium chloride step gradients is recommended, and this protocol is listed below ***(37)***. The opti-

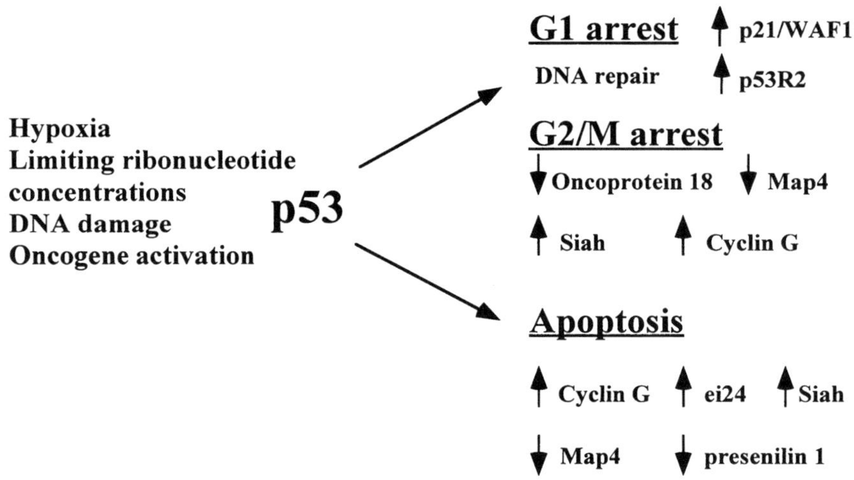

Fig. 3. Summary of the identity and postulated functions of the p53-regulated genes identified by differential display. The exception in this figure is p21/waf1, which plays a paramount role in p53-dependent G1 arrest but was identified by subtractive screening. Upward arrows denote upregulation, downward arrows denote downregulation.

cal density 260/280 ratio of the RNA should range from 1.8 to 2.0, and the RNA should be analyzed on a denaturing formaldehyde gel to assess intactness and purity. Typically the ratio of 28S to 18S RNA should be between 5:1 and 10:1, and there should be no evidence for smearing of low-molecular-weight RNA (less than 500 bp).

3.1. RNA Isolation

1. Place 2 mL of Cesium chloride solution (5.7 *M* CsCl, 0.1 *M* EDTA, pH 7.5, filter-sterilized and autoclaved) into a sterilized polyallomer tube of 9/16 × 3.5 in in diameter. Mark the level of CsCl with a marker, and mark the sample number underneath this line (bottom of tube).
2. Suspend cells in 10 mL of guanidine HCl solution with vigorous pipetting, so that the solution is completely homogeneous and there are no visible clumps. For each milliliter of packed cell volume (cells that are harvested freshly or cells frozen at –80°C), use one polyallomer tube and 10 mL of guanidine HCl. Layer this onto the CsCl cushion by carefully running the solution gently along the side of the polyallomer tube, to avoid mixing with the CsCl.
3. Place tubes in ultracentrifuge buckets, and weigh each, along with the cap; add guanidine HCl such that all tubes weigh the same, to within 0.05 g. Spin in a swinging-bucket ultracentrifuge rotor at 51,000 *g* for 18–24 h at 22°C.
4. The next day, invert each tube to empty its contents and keep the tubes inverted, such that the RNA, which is pelleted on the bottom of the tube, does not come into contact with the guanidine solution along the sides of the tube. Keeping the tubes upside down, cut with scissors at the marked CsCl line and discard the top of the tube.
5. Resuspend the RNA pelleted in the bottom of the tube in 500 µL of sterile dH_2O, pipetting vigorously while scraping the bottom of the tube with the pipet tip. Add this to a sterile microfuge tube, then rinse the bottom of the polyallomer tube with 50 µL of sterile dH_2O and add this to the first microfuge tube. Add 50 µL of 2 *M* NaOAc, pH 5, mix well, and add

1 mL of sterile 100% ethanol. Shake vigorously and place in –80°C freezer for 20 min. Spin in microfuge for 20 min at 4°C, then decant supernatant.

6. Add 1 mL of a 20/9/1 solution of ethanol/dH_2O/2 *M* NaOAc, pH 5. Invert carefully so as to not disturb the pellet, and spin at 4°C for 5 min. Decant. Repeat.
7. Add 1 mL of 100% ethanol, invert gently to wash the pellet, and decant. Spin sample for 20 s and pipet off remaining ethanol. Let pellet air-dry for 5 min. Add 50 µL of dH_2O per milliliter of packed cells used for the isolation, and freeze pellet at –80°C overnight.
8. The next day, thaw tubes on ice for 30 min, and spin in a microfuge for 20 s to aid in resuspending the pellet. Pipet the solution thoroughly to resuspend the RNA and measure the absorbance at 260 and 280 nm.

3.2. cDNA Synthesis, PCR Amplification

1. cDNA synthesis is commonly performed using one of three one-base anchored oligo-dT primers. For each oligo-dT primer, a reaction is carried out with 200–2000 ng of CsCl-purified RNA, to which 1× room-temperature (RT) buffer, 20 µ*M* dNTP mix, and 0.20 µM each oligo-dT primer is added, as per the manufacturer's protocol (RNAimage kit, GenHunter). Samples are preincubated for 5 min at 65°C, followed by 10 min at 37°C to anneal the primers. Then 2–5 U of MMLV reverse transcriptase (SuperScript II, Life Technologies) are added to the reaction, which is then incubated for 50 min at 37–42°C.
2. The reaction is extracted with an equal volume of 1/1 phenol/chloroform, and ethanol-precipitated in the presence of 10 µg yeast tRNA or glycogen as carrier, followed by a wash step in 1 mL of 70% ethanol. The cDNA-RNA hybrid is resuspended in 30 µL dH_2O, and 1–5 µL is added to a 100-µL reaction containing 1× PCR buffer, 2 µ*M* dNTPs, 0.2 µ*M* oligo-dT anchored primer, 1 µL α-^{33}P-dATP, 0.2 µ*M* arbitrary forward primer set, and 1 µL (1–5 U) Taq polymerase (for example, Hot Tub polymerase, Amersham-Pharmacia).
3. The PCR is run for 40 cycles (30 s at 94°C, 2 min at 40°C, 30 s at 72°C), followed by a single extension cycle for 5 min at 72°C and a 4°C soak.

3.3. Resolution and Isolation of Differentially Expressed cDNAs

1. Formamide loading dye is added to an aliquot of each PCR reaction, and samples are boiled and resolved on 6% denaturing polyacrylamide gels (7 *M* urea, 1× TBE). The gel is dried on Whatman paper, wrapped in plastic wrap, and radiolabeled cDNA fragments are revealed by autoradiography. It is imperative to carefully mark the orientation of the film with the gel in as many ways possible, for example, with fluorescent stickers (GloGos, Stratagene).
2. Bands of interest are excised with a clean razor blade, soaked in a microfuge tube in 100 µL of dH_2O for 10 min at room temperature, and boiled for 15 min. The sample is centrifuged to pellet the acrylamide and Whatman paper, and the eluted DNA is ethanol-precipitated in the presence of glycogen or yeast tRNA as carrier.
3. For each fragment isolated, a second round of PCR is performed under the same conditions previously described, except without the inclusion of radiolabeled nucleotide. The cDNA fragment is then isolated from a 1.5% agarose gel; although this fragment can be directly utilized for Northern hybridization from RNA used for the differential display, it is recommended that PCR products are first cloned, as several different cDNA fragments are often coamplified in this procedure.
4. Cloning can be performed using either TOPO TA cloning (Invitrogen) or the PCR-TRAP cloning system (GenHunter). Individually cloned PCR products should be subjected to sequence analysis and to confirmatory Northern or RNAse protection analyses to identify the cDNA that corresponds to the differentially regulated signal.

4. Notes

1. The cloning and confirmation of differentially regulated genes is the most time-consuming aspect of this protocol. Therefore, it is wise to ensure complete commitment to the isolation of each differentially regulated cDNA on the denaturing gel before pursuing this step. This can best be done by analyzing multiple time points of treated cells, at least two independent samples of RNA, and by analyzing multiple different cell lines treated with the agent of choice, relative to identically treated control cell lines. The relative ease and cost-effectiveness of the differential display technique facilitates the use of such multiple controls and time points.
2. Interesting modifications of this technique have emerged that enable one to narrow down target genes of interest; for example, "motif" primers have been used as forward primers, corresponding to conserved sequences present in zinc finger or leucine zipper domains of proteins ***(38)***.
3. Alternative extraction procedures include those utilizing RNAzol (Life Technologies), or RNApure (GenHunter). However, in order to avoid artifacts and false positives, it is essential that the RNA be free from contaminating genomic DNA. 10–100 µg of total RNA is subject to DNAse I (Pharmacia) treatment for 30 minutes at 37°C, followed by phenol/chloroform extraction and ethanol precipitation in the presence of 0.3 *M* NaOAc.
4. Recent improvements in differential display methodology have included improved primer design, the use of modified reverse transcriptases, and the introduction of automated technologies utilizing fluorescently labeled primers; these improvements have increased the sensitivity and reproducibility of the technique, while at the same time reducing the rate of false positive results.

References

1. Liang, P. and Pardee, A. B. (1992) Differential display of eukaryotic messenger RNA by means of the polymerase chain reaction. *Science* **257**, 967–971.
2. Liang, P. and Pardee, A. B. (1997) *Methods in Molecular Biology, Volume 85: Differential Display Methods and Protocols*. (Liang, P. and Pardee, A. B., eds.). Humana Press, Totowa, NJ, pp. 3–11.
3. El-Deiry, W. S., Tokino T., Velculescu V. E., et al. (1993) WAF1, a potential mediator of p53 tumor suppression. *Cell* **75**, 817–825.
4. Deng, C., Zhang, P., Harper, J. W., Elledge, S. J., and Leder, P. (1995) Mice lacking p21/waf1/cip1 undergo normal development, but are defective in G1 checkpoint control. *Cell* **82**, 675–684.
5. Okamoto, K. and Beach, D. (1994) Cyclin G is a transcriptional target of the p53 tumor suppressor protein. *EMBO J.* **13**, 4816–4822.
6. Zauberman, A., Lupo, A., and Oren, M. (1995) Identification of p53 target genes through immune selection of genomic DNA: the cyclin G gene contains two distinct p53 binding sites. *Oncogene* **10**, 2361–2366.
7. Smith, M. L., Kontny, H. U., Bortnick, R., and Fornace, A. J. (1997) The p53-regulated cyclin G gene promotes cell growth: p53 downstream effectors cyclin G and Gadd45 exert different effects on cisplatin toxicity. *Exp. Cell Res.* **230**, 61–68.
8. Okamoto, K. and Prives, C. (1999) A role of cyclin G in the process of apoptosis. *Oncogene* **18**, 4606–4615.
9. Okamoto, K., Kamibayashi, C., Serrano, M., Prives, C., Mumby, M. C., and Beach, D. (1996) p53-dependent association between cyclin G and the B' subunit of protein phosphatase 2A. *Mol. Cell. Biol.* **16**, 6593–6602.
10. Lehar, S. M., Nacht, M., Jacks, T., Vater, C. A., Chittenden, T., and Guild, B. C. (1996) Identification and cloning of ei24, a gene induced by p53 in etoposide-treated cells. *Oncogene* **12**, 1181–1187.

11. Gu, Z., Flemington, C., Chittenden, T., and Zambetti, G. P. (2000) ei24, a p53 response gene involved in growth suppression and apoptosis. *Mol. Cell. Biol.* **20**, 233–241.
12. Amson, R. B., Nemani, M., Roperch, J. P., et al. (1996) Isolation of 10 differentially expressed cDNAs in p53-induced apoptosis: activation of the vertebrate homologue of the drosophila seven-in-absentia gene. *Proc. Natl. Acad. Sci. USA* **93**, 3953–3957.
13. Roperch, J. P., Lethrone, F., Prieur S., et al. (1999) SIAH-1 promotes apoptosis and tumor suppression through a network involving the regulation of protein folding, unfolding, and trafficking: identification of common effectors with p53 and 21Waf1. *Proc. Natl. Acad. Sci. USA* **96**, 8070–8073.
14. Matsuzawa, S., Takayama, S., Froesch, B. A., Zapata, J. M., and Reed, J. C. (1998) p53-inducible human homologue of *Drosophila* seven-in-absentia (Siah) inhibits cell growth: suppression by Bag-1. *EMBO J.* **17**, 2736–2747.
15. Relaix, F., Wei, X., Li, W., et al. (2000) Pw1/Peg3 is a potential cell death mediator and cooperates with Siaha in p53-mediated apoptosis. *Proc. Natl. Acad. Sci. USA* **97**, 2105–2110.
16. Tanikawa, J., Ichikawa-Iwata, E., Kanei-Ishii, C., et al. (2000) p53 suppresses the c-myb-induced activation of heat shock transcription factor 3. *J. Biol. Chem.* **275**, 15578–15585.
17. Germani, A., Bruzzoni-Giovanelli, H., Fellous, A., Gisselbrecht, S., Varin-Blank, N., and Calvo, F. (2000) SIAH-1 interacts with alpha-tubulin and degrades the kinesin Kid by the proteasome pathway during mitosis. *Oncogene* **19**, 5997–6006.
18. Lin, J., Chen, J., Elenbaas, B., and Levine, A. J. (1994) Several hydrophobic amino acids in the p53 amino-terminal domain are required for transcriptional activation, binding to mdm-2 and the adenovirus E1B 55-kD protein. *Genes Dev.* **8**, 1235–1246.
19. Haupt, Y., Rowan, S., Shaulian, E., Vousden, K. H., and Oren, M. (1995) Induction of apoptosis in HeLa cells by trans-activation-deficient p53. *Genes Dev.* **9**, 2170–2183.
20. Caelles, C., Helmberg, A., and Karin, M. (1994) p53-dependent apoptosis in the absence of transcriptional activation of p53-target gene. *Nature* **370**, 220–223.
21. Wagner, A. J., Kokontis, J. M., and Hay, N. (1994) Myc-mediated apoptosis requires wild type p53 in a manner independent of cell cycle arrest and the ability of p53 to induce p21waf1/cip1. *Genes Dev.* **8**, 2817–2830.
22. Sabbatini, P., Chiou, S-K., Rao, L., and White, E. (1995a) Modulation of p53-mediated transcriptional repression and apoptosis by adenovirus E1B 19K protein. *Mol. Cell. Biol.* **15**, 1060–1070.
23. Shen, Y. and Shenk, T. E. (1994) Relief of p53-mediated transcriptional repression by the adenovirus E1B-19K protein or the cellular bcl2 protein. *Proc. Natl. Acad. Sci. USA* **91**, 8940–8944.
24. Maheswaran, S., Englert, C., Bennett, P., Heinrich, G., and Haber, D. A. (1995) The WT1 gene product stabilizes p53 and inhibits p53-mediated apoptosis. *Genes Dev.* **9**, 2143–2156.
25. Ryan, K. M. and Vousden, K. H. (1998) Characterization of structural p53 mutants which show selective defects in apoptosis but not cell cycle arrest. *Mol. Cell. Biol.* **18**, 3692.
26. Walker, K. K. and Levine, A. J. (1996) Identification of a novel p53 functional domain that is necessary for efficient growth suppression. *Proc. Natl. Acad. Sci. USA* **93**, 15335–15340.
27. Sakamuro, D., Sabbatini, P., White, E., and Prendergast, G. C. (1997) The polyproline region of p53 is required to activate apoptosis but not growth arrest. *Oncogene* **15**, 887–898.
28. Murphy, M., Hinman, A., and Levine, A. J. (1996) Wild type p53 negatively regulates the expression of a microtubule-associated protein. *Genes Dev.* **10**, 2971–2980.
29. Ahn, J., Murphy, M., Kratowicz, S., Wang, A., Levine, A. J., and George, D. L. (1999) Down-regulation of the stathmin/Op18 and FKBP25 genes following p53 induction. *Oncogene* **18**, 5954.
30. Johnsen, J. I., Aurelio, O. N., Kwaja, Z., et al. (2000) p53-mediated negative regulation of stathmin/Op18 expression is associated with G2/M cell cycle arrest. *Int. J. Cancer* **88**, 685–691.

31. Roperch, J. P., Alvaro, V., Prieur, S., et al. (1998) Inhibition of presenilin 1 expression is promoted by p53 and p21WAF1 and results in apoptosis and tumor suppression. *Nat. Med.* **4**, 835–838.
32. Murphy, M., Ahn, J., Walker, K. K., et al. (1999) Transcriptional repression by wild-type p53 utilizes histone deacetylases, mediated by interaction with mSin3a. *Genes Dev.* **13**, 2490–2501
33. Zilfou, J., Hoffman, W. H., Sank, M., George, D. L., and Murphy, M. The co-repressor mSin3a interacts with the proline-rich domain of p53 and protects p53 from proteasome-mediated degradation. *Mol. Cell. Biol.* **21**, 3974–3985.
34. Varmeh-Ziaie, S., Okan, I., Wang, Y., et al. (1997). Wig-1, a new p53-induced gene encoding a zinc-finger protein. *Oncogene* **15**, 2699–2704.
35. Zhang, R., Averboukh, L., Zhu, W., et al. (1998) Identification of rCop-1, a new member of the CCN protein family as a negative regulator for cell transformation. *Mol. Cell. Biol.* **18**, 6131–6141.
36. Tanaka, H., Arakawa, H., Yamaguchi, T., et al. (2000) A ribonucleotide reductase gene involved in a p53-dependent cell cycle checkpoint for DNA damage. *Nature* **404**, 42–49.
37. Liu, C. P., Slate, D. L., Gravel, R., and Ruddle, F. H. (1979) Biological detection of specific mRNA molecules by microinjection. *Proc. Natl. Acad. Sci. USA* **76**, 4503–4506.
38. Donahue, P. J., Hsu, D. K. W., and Winkles, J. (1997) *Methods in Molecular Biology, Volume 85: Differential Display Methods and Protocols*. (Liang, P. and Pardee, A. B., eds.) Humana Press, Totowa, NJ, pp. 25–35.

31

Detection of Mismatch Repair Gene Expression in Urologic Malignancies

Fredrick S. Leach

1. Introduction

Genitourinary malignancies were diagnosed in over 250,000 U.S. men and women in the year 2000, and carcinomas of the prostate, kidney and urothelium accounted for over 20% of all adult malignancies ***(1)***. In the twenty-first century, oncologists and molecular biologists will be challenged to expand and advance diagnostic and treatment options for genitourinary malignancies. In order to accomplish these goals, an increasing armamentarium of molecular reagents will be necessary for analysis of molecular pathogenesis. Identification and characterization of genes important for initiation and progression of urologic malignancies should facilitate development of clinically useful reagents as prognostic and metastatic markers.

An essential feature of carcinogenesis is accumulation of genetic alterations ultimately resulting in unregulated cell proliferation, immortalization, invasion, and metastasis ***(2)***. Since genetic alterations precede overt histologic manifestation, detection of genetically altered cells or altered expression of gene products is a rationale approach for assessment of urologic malignancies. Mismatch repair (MMR) is a highly conserved process that suppresses genetic instability in procaryotic and eucaryotic cells ***(3–6)***. Human MMR genes encode redundant and interacting proteins that detect and correct mismatched nucleotides formed primarily during DNA replication. Inherited inactivating mutations in MMR genes are responsible for a cancer predisposition syndrome known as hereditary nonpolyposis colon cancer (HNPCC) ***(7–13)***. HNPCC patients are predisposed to colorectal cancers and a variety of other tumors including genitourinary cancers ***(14–16)***. Tumors lacking MMR gene expression exhibit an enhanced mutation rate that can be detected using microsatellite analysis of matched normal and tumor genomic DNA ***(17–20)***. High frequency of microsatellite instability (MSI) in tumors and cell lines is practically pathognomonic for MMR deficiency. This degree of MSI is almost always due to mutations affecting both copies of a MMR gene ***(21–25)***.

The prototype mismatch repair gene, *hMSH2* (human mut-S homolog 2), has been investigated in urologic malignancies using MSI analysis ***(26–30)*** and standard immunologic techniques ***(31,32)***. In this chapter, an analysis of MSH2 expression in urologic

From: *Methods in Molecular Biology, Vol. 222: Tumor Suppressor Genes: Pathways and Isolation Strategies*
Edited by: Wafik S. El-Deiry © Humana Press Inc., Totowa, NJ

malignancies will be used to illustrate these two methods. Microsatellite analysis of matched normal and tumor genomic DNA and immunohistochemical analysis of MSH2 in prostate tissue will be described in the following sections.

2. Materials

1. Polyacrylamide.
2. α^{32}P-dCTP, 10.0 mCi/mL.
3. Primer Map Pairs (Research Genetics, Huntsville, AL).
4. Ab-2 (FE11) hMSH2 monoclonal antibody (Calbiochem).
5. 10× polymerase chain reaction (PCR) buffer (containing $MgCl_2$), AmpliTaq-1 DNA polymerase (Perkin Elmer).
6. 10× dNTP mix: 2 m*M* dGTP, 2 m*M* dATP, 2 m*M* TTP, 0.025 m*M* dCTP.
7. Paraffin-embedded surgical specimens for analysis.
8. Biotinylated goat anti-mouse IgG (H + L) (Pierce, Rockford, IL).
9. Antibody detection kits (Vector, Burlingame, CA; or DAKO, Carpenteria, CA).
10. Xylenes, ethanol, hematoxylin and eosin (H&E), glass staining trays and racks.
11. Autodewaxer (Research Genetics, Huntsville, AL).
12. 10× Antigen retrieval (Biogenex, San Ramon, CA).
13. PCR machines.
14. Sequencing gel rigs and power supplies.
15. Cryostat (optional).
16. Automated slide stainer for immunohistochemical analysis (optional).

3. Methods

The methods described below outline microsatellite instability (MSI) analysis of genomic DNA previously prepared from fresh or paraffin-embedded matched normal and malignant urologic tissue. Immunohistochemical staining for MSH2 using formalin-fixed paraffin-embedded tissues derived from nonmalignant and malignant urologic tissues will also be described in detail.

3.1. Microsatellite Analysis

Several variations of microsatellite analysis have been described, depending on the type of microsatellite (CA, AT, triplet) or mononucleotide tract being amplified and the investigator's preference for using radioactive- or fluorescent-labeled primers (*see* **Note 1**). The method described here incorporates radioactive dCTP into the PCR product and is faster and easier to perform, since a separate procedure for primer labeling and purification is not required. However, this method requires prior knowledge that the PCR product contains at least one deoxycytidine. In addition, this method could potentially bias the assay for detection of larger repeats, due to higher specific activity in these products. Similarly, the lower concentration of dCTP in the reaction may influence the PCR reaction and the normal PCR conditions may require modification. Finally, free radioactive nucleotides pass into the bottom reservoir of the gel rig and therefore liquid radioactive waste management is desirable. Despite these concerns, this protocol for MSI analysis has worked well for many different tissue types and many different investigators.

3.1.1. Genomic DNA for Analysis

Careful preparation of high-quality DNA may be the most important factor for successful detection of MSI in urologic and nonurologic malignancies. Many reagents are commercially available for genomic DNA preparation, and detailed reagent specific protocols are provided (*see* **Note 2**). In some cases, a cryostat may be necessary to assist in the identification and isolation of tumor cells and minimize contamination with normal cellular components. The frozen tumor block in OCT compound (Tissue Tek) can be "sculpted" to obtain sections that are relatively tumor pure (confirmed by H&E staining of the section) and carefully dissecting away nontumor tissue from the block prior to preparative sectioning. Assessment of the mounted tumor block before, during, and after preparative sectioning for DNA ensures that the tissue is enriched for tumor cells. Since only a few micrograms of DNA are required in PCR-based analyses, a small amount of pure specimen is preferable to a large amount of a mixed specimen. In adenocarcinoma of the prostate, laser capture microdissection may be necessary for precise dissection of normal and malignant nuclei prior to DNA purification ***(33)***.

3.1.2. Microsatellite Analysis

1. Set up a 10 μL PCR reaction using 10× PCR buffer and 10× dNTP mix.
2. Combine in equal volumes the forward and reverse Map Pair primers into a tube labeled "primer mix" and use 1 μL in the PCR reaction (*see* **Note 3**).
3. Aliquot 10–30 ng of normal or tumor DNA into a reaction tube or well.
4. Add 0.5 μCi per reaction of α-^{32}P-dCTP to the PCR reaction mix and 0.5 U per reaction of Taq-1 polymerase.
5. Typical settings for MSI analysis using Research Genetics Map Pairs are:
   a. 95°C for 30 s.
   b. 95°C for 30 s; 55°C for 60 s; 72°C for 60 s.
   c. 72°C for 5 min.
   d. 4°C hold (optional).
   Step 2 is repeated for 27–40 cycles (*see* **Note 4**).
6. Add 3 vol of a denaturing loading buffer (95% deionized formamide, 0.05% bromophenol blue, 0.05% xylene cyanole, and 20 mM NaOH) to the reactions (final volume 40 lL).
7. Denature the PCR products (85–958C for 5–10 min) and load 2.5–10 lL on an 8% polyacrylamide sequencing gel (*see* **Note 5**).

An autoradiogram of a typical MSI analysis in normal and malignant prostate tissue is shown in **Fig. 1**. Preparation of genomic DNA differs depending on the source (fresh vs archival tissue). In addition, many primers that amplify using DNA prepared from fresh tissue may not amplify using DNA prepared from paraffin embedded tissue (*see* **Note 6**).

3.2. MSH2 Immunohistochemistry

Immunohistochemical detection of MMR proteins has been described in different histologic tissues ***(26,27,34–36)***. The monoclonal antibody Ab-2 (FE-11) has been successfully used in paraffin-embedded urologic tumors and is directed to the carboxy terminal portion of the hMSH2 protein product ***(34)***. Although other antibodies have been used in archival materials, the Ab-2 (FE-11) antibody was used exclusively for this protocol.

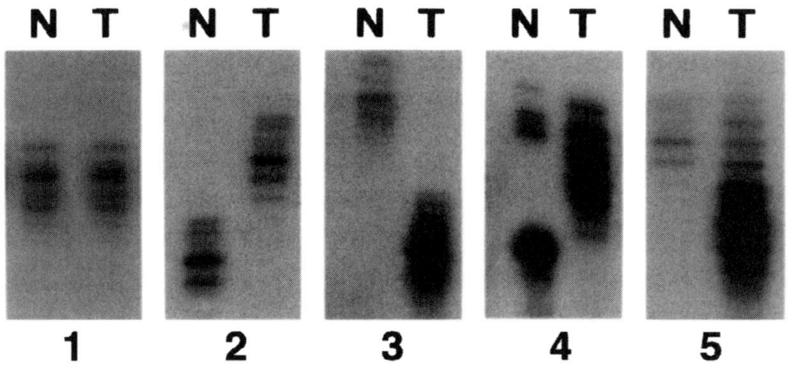

Fig. 1. Microsatellite analysis using normal (N) and tumor (T) genomic DNA from laser-captured microdissected nuclei from paraffin-embedded prostate cancer specimens. An example of no detectable microsatellite instability is shown in panel 1; an increase shift in the homozygous normal alleles is shown in panel 2; a decrease shift seen in the homozygous normal alleles is shown in panel 3; an increase shift in the smaller normal allele is shown in panel 4; a decrease shift in one of the homozygous alleles is shown in panel 5. Examples shown used primer Map Pairs described in text.

3.2.1. Tissues for Analysis

This protocol has been used successfully in paraffin-embedded tissue from bladder ***(26)***, prostate ***(37)***, testis, and kidney (Velasco and Leach, unpublished) carcinomas. However, slight modifications have been used to optimize the staining depending on histology and the detection kit used (*see* **Note 7**). Examples of immunohistochemistry using formalin-fixed paraffin-embedded prostate specimens are shown in **Fig. 2**.

3.2.2. Immunohistochemistry Protocol

1. Obtain pathologic blocks of the urologic tissue of interest. H&E stains should be available and reviewed prior to beginning any immunohistochemical analysis (*see* **Note 8**).
2. Obtain 6-µm-thick sections of interest. Be sure to obtain extra unstained sections for unexpected repeat experiments, positive and negative controls.
3. The sections should be processed on charged slides, heated to 65°C for 30 min and cooled to room temperature prior to immunohistochemical analysis.
4. Deparaffinize the sections by soaking the slides in Autodewaxer solution (Research Genetics, Huntsville, AL) and microwaving for 4 min at maximal power.
5. Cool slides to room temperature, then place in 3 changes of xylenes for a total of 30 min.
6. Rehydrate the sections by soaking in 2 changes of 100% ethanol; 2 changes of 95% ethanol; once in 70% ethanol; once in 50% ethanol; and finally distilled water for 5 min at room temperature each step.
7. Antigen retrieval is carried out by placing the slides in 1× antigen retrieval and microwaving at high power for a total of 5–7 min with 30-s pulses (30 s on and 30 s off). Actual microwaving is only 2.5–3.5 min, and this prevents the solution from boiling over and/or evaporating too quickly.
8. Let the slides cool to room temperature for 15–20 min, then wash for a total of 30 min in phosphate-buffered saline (PBS) without calcium or magnesium. The slides can be stored overnight at 4°C in PBS at this point.
9. Treat slides with proteinase K at 12.5–25-µg/mL in PBS for 5–30 min at room temperature (*see* **Note 9**).

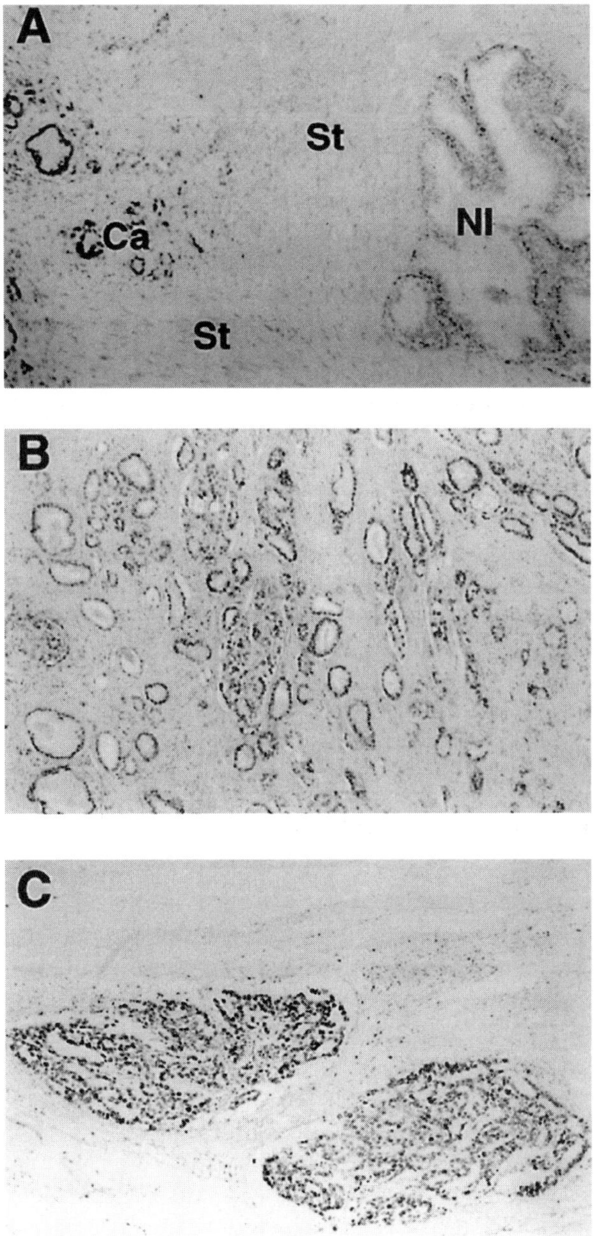

Fig. 2. Immunohistochemical analysis of MSH2 expression in benign and malignant prostate tissue. **(A)** Moderate to high-staining nuclei seen in the cancer cells (Ca) shown to the left of the stroma (St) separating cancer from the benign prostate gland seen to the right (NI). Basal cell staining can be seen in the benign gland. **(B)** Low nuclear staining seen in prostate cancer cells relative to the stromal staining. **(C)** Moderate staining seen in cribriform pattern prostate cancer. All specimens were counterstained with methyl green.

10. Block endogenous peroxidase activity by incubating slides in methanol containing 0.3% hydrogen peroxide. This solution should be made fresh for each experiment.
11. Rinse slides for a total of 30 min in 3 changes of PBS.
12. Block slides with the 50% goat serum in PBS for 1 h at room temperature.

13. Add the MSH2 Ab-2 monoclonal antibody at 2–4 µg/mL for at least 2 h at room temperature or overnight in a humidified chamber at room temperature (*see* **Note 10**).
14. Wash for total of 30 min in PBS for 3 changes at room temperature.
15. Add secondary antibody, biotinylated goat anti-mouse IgG (1-mg/mL) diluted 1/200 in 50% goat serum for 45–60 min at room temperature.
16. Wash for a total of 30 min in 3 changes of PBS.
17. Develop with DAB using any commercially available detection kit that uses conjugated streptavidin-peroxidase substrates (Vector or DAKO). Addition of nickel gives a black signal that is preferable for methyl green (Vector) nuclear counterstaining (optional).
18. Dehydrate the slides in 70% ethanol, 2 changes of 95% ethanol, 2 changes of 100% ethanol for 5 min each, then 2 changes of xylenes for 5 min each.
19. Mount the slides using Permount (Fisher Scientific) and cover slips (Corning).
20. Score the immunohistochemical staining using independent observers (*see* **Note 11**).

4. Notes

1. Microsatellite instability assays can also be performed using primers tagged with ^{32}P or a florescent dye. The radiolabeled primers can be used in standard PCR reactions and autoradiography. Florescent primers eliminate radioactivity but require special equipment, software, and experience for analysis and interpretation.
2. Commercial kits for preparation of high-quality DNA are available. The source of DNA (blood, fresh tissue, or formalin fixed paraffin-embedded) is an important variable as to which reagents to use. A basic laboratory handbook can provide protocols and general principles for DNA extraction from mammalian cells and tissue. Preparation of DNA from blood requires separation of the leukocytes from red blood cells prior to purification (Gentra Systems, Minneapolis, MN). Preparation of DNA from paraffin is somewhat empiric and has been described previously ***(33,37)***.
3. Map Pairs from Research Genetics are supplied at 20 µ*M*. Equal volumes of the forward and reverse primers are combined to form a "primer mix" for the microsatellite locus of interest. Alternatively, 0.5 µL of each primer could be independently added to the PCR reaction.
4. The number of cycles used for PCR amplification depends on the amount and quality of DNA present in the reaction. In general, the fewest number of cycles that give a robust, reliable, and reproducible signal should be used. DNA from fresh tissue may require only 27 cycles, while DNA from formalin-fixed paraffin-embedded laser-captured prostate specimens required up to 40 cycles (**Fig. 1**).
5. The amount of PCR reaction loaded onto the gel is determined empirically. In general, the smaller the amount loaded the better. The remainder of the sample can be stored at –20°C for at least 7 d.
6. Research genetics Map Pairs that have been used successfully in MSI analysis of DNA prepared from archival prostate specimens are D3s1283, D3s1293, D9s66, and D9s113.
7. Conjugated streptavidin–peroxidase complexes have worked the best. Ultrasensitive kits may require lowering the amount of primary antibody or incubation times to decrease background staining. Some specimens, such as from bladder, tend to have high background in the normal tissue. Shorter incubation times for primary and secondary antibodies, decreased amount of primary antibody, meticulous wash cycles, and decreased signal development time will enhance signal to noise.
8. Be sure that the sections to be analyzed contain tumor confirmed by a pathologist if this is the goal of the analysis.

9. The amount and length of time for this step is empirical. For the analysis shown, proteinase K (GibcoBRL) was used at 12.5 μg/mL for 15 min at room temperature.
10. For most procedures, the specimen is incubated overnight with the primary antibody; however, incubations as short as 2 h can be used, especially when using high-sensitivity detection systems and when high background exists.
11. In this protocol, MSH2 staining in urologic malignancies is graded relative to the stromal staining. A scale of 0–4+ is used for staining intensity, with 0 being no nuclear staining seen in tumor cells. 0–1+, 1+, or 1–2+ is low staining, similar to or slightly greater than the stromal cells; 2+ or 2–3+ is moderate staining that is significantly greater than the stromal cell staining; while 3+, 3–4+, or 4+ staining is high staining and is substantially more intense than stromal cell staining. The staining pattern using this antibody is fairly uniform (75% or greater of the tumor cells stain at a given intensity level and less than 10% of the tumor cells stain in absent staining) once conditions are optimized. The intensity scale allows for some subjectivity, although agreement among three observers was high in one study ***(37)***. Absent to low-staining should be grouped together, and moderate and high-staining specimens considered together to compensate for subjectivity and experimental variability. Using this grading system, a clinically useful grouping was identified and absent to low-staining specimens correlated with detection of MSI in prostate cancer ***(37)***.

Acknowledgments

The author thanks Y. Gathwright for helpful suggestions, reagents, and resources in the immunohistochemical analysis.

References

1. Greenlee, R. T., Murray, T., Bolden, S., and Wingo, P. A. (2000) Cancer statistics, 2000. *CA Cancer J. Clin.* **50**, 7–33.
2. Fearon, E. R. and Vogelstein B. (1990) A genetic model for colorectal tumorigenesis. *Cell* **61**, 759–767.
3. Kolodner R. D. and Marsischky, G. T. (1999) Eukaryotic DNA mismatch repair. *Curr. Opin. Genet. Dev.* **9**, 89–96.
4. Buermeyer, A. B., Deschenes, S. M., Baker, S. M., and Liskay, R. M. (1999) Mammalian DNA mismatch repair. *Annu. Rev. Genet.* **33**, 533–564.
5. Modrich, P. and Lahue, R. (1996) Mismatch repair in replication fidelity, genetic recombination, and cancer biology. *Annu. Rev. Biochem.* **65**, 101–133.
6. Leach, F. S. (2001) Molecular genetics of urothelial malignancies: mismatch repair genes as markers for detection and prognosis? *Urol. Oncol.* **6**, 211–219.
7. Nicolaides, N. C., Papadopoulos, N., Liu, B., et al. (1995) Mutations of two PMS homologues in hereditary nonpolyposis colon cancer. *Nature* **371**, 75–80.
8. Liu, B., Parsons, R. E., Hamilton, S. R., et al. (1994) hMSH2 mutations in hereditary non-polyposis colorectal cancer kindreds. *Cancer Res.* **54**, 4590–4594.
9. Papadapoulos, N., Nicolaides, N. C., Wei, Y. F., et al. (1994) Mutations of a mutL homolog in hereditary colon cancer. *Science* **263**,1625–1629.
10. Bronner, C. E., Baker, S. M., Morrison, P. T., et al. (1994) Mutation in the DNA mismatch repair gene homologue hMLH1 is associated with hereditary non-polyposis colon cancer. *Nature* **368,** 258–261.
11. Leach, F. S., Nicolaides, N. C., Papadopoulos, N., et al. (1993) Mutations of a mutS homolog in hereditary nonpolyposis colon cancer. *Cell* **75**, 1215–1225.

12. Fishel, R., Lescoe, M. K., Rao, M. R. S., et al. (1993) The human mutator gene MSH2 and its association with hereditary nonpolyposis colon cancer. *Cell* **75,** 1027–1038.
13. Papadopoulos, N. and Lindblom, A. (1997) Molecular basis of HNPCC: mutations of MMR genes. *Hum. Mutat.* **10,** 89–99.
14. Watson, P. and Lynch, H. T. (1993) Extracolonic cancer in hereditary nonpolyposis colorectal cancer. *Cancer* **71,** 677–685.
15. Lynch, P. M., Smyrk, T. C., Watson, P., et al. (1993) Genetics, natural history, tumor spectrum and pathology of hereditary nonpolyposis colorectal cancer: an updated review. *Gastroentrerology* **104,** 1535–1549.
16. Lynch, H. T., Ens, J. A., and Lynch, J. F. (1990) The Lynch syndrome II and urologic malignancies. *J. Urol.* **143,** 24–28.
17. Strand, M., Prolla, T. A., Liskay, R. M., and Petes, T. D. (1993) Destabilization of tracts of simple repetitive DNA in yeast by mutations affecting DNA mismatch repair. *Nature* **365,** 274–276.
18. Aaltonen, L. A., Peltomaki, P., Leach, F. S., et al. (1993) Clues to the pathogenesis of familial colorectal cancer. *Science* **260,** 812–816.
19. Thibodeau, S. N., Bren, G., and Schaid, D. (1993) Microsatellite instability in cancer of the proximal colon. *Science* **260,** 816–819.
20. Ionov, Y., Peinado, M. A., Malkhosyan, S., Shibata, D., and Perucho, M. (1993) Ubiquitous somatic mutations in simple repeated sequences reveal a new mechanism for colonic carcinogenesis. *Nature* **363**, 558–561.
21. Thibodeau, S. N., French, A. J., Roche, P. C., et al. (1996) Altered expression of hMSH2 and hMLH1 in tumors with microsatellite instability and genetic alterations in mismatch repair genes. *Cancer Res.* **56**, 4836–4840.
22. Liu, B., Nicolaides, N. C., Markowitz, S., et al. (1995) Mismatch repair gene defects in sporadic colorectal cancers with microsatellite instability. *Nat. Genet.* **9,** 48–55.
23. Borresen, A. L., Lothe, R. A., Meling, G. I., et al. (1995) Somatic mutations in the hMSH2 gene in microsatellite unstable colorectal carcinomas. *Hum. Mol. Genet.* **4,** 2065–2072.
24. Boyer, J. C., Umar, A., Risinger, J. I., et al. (1995) Microsatellite instability, mismatch repair deficiency and genetic defects in human cancer cell lines. *Cancer Res.* **55,** 6063–6070.
25. Leach, F. S., Velasco, A., Hsieh, J-T., Sagalowsky, A. I., and McConnell, J. D. (2000) The human mismatch repair gene hMSH2 is mutated in the prostate cancer cell line LNCaP. *J. Urol.* **164,** 1830–1833.
26. Leach, F. S., Hsieh, J-T., Molberg, K., Saboorian, M. H., McConnell, J. D., and Sagalowsky, A. I. (2000) Expression of the human mismatch repair gene hMSH2: a potential marker for urothelial malignancy. *Cancer* **88**, 2333–2341.
27. Jin, T-X., Furihata, M., Yamasaki, I., et al. (1999) Human mismatch repair gene (hMSH2) product expression in relation to recurrence of transitional cell carcinoma of the urinary bladder. *Cancer* **85**, 478–484.
28. Gonzalez-Zulueta, M., Ruppert, J. M., Tokino, K., et al. (1993) Microsatellite instability in bladder cancer. *Cancer Res.* **53**, 5620–5623.
29. Watanabe, M., Shiraishi, T., Muneyuki, T., et al. (1998) Allelic loss and microsatellite instability in prostate cancer in japan. *Oncology* **55**, 569–574.
30. Dahiya, R., Lee, C., McCarville, J., Hu, W., Kaur, G., and Deng, G. (1997) High frequency of genetic instability of microsatellites in human prostatic adenocarcinoma. *Int. J. Cancer* **72,** 762–767.
31. Egawa, S., Uchida, T., Suyama, K., et al. (1995) Genomic instability of microsatellite repeats in prostate cancer: relationship to clinicopathological variables. *Cancer Res.* **55,** 2418–2421.
32. Shin, K-H., Ku, J-L., Kim, W-H., et al. (2000) Establishment and characterization of seven human renal cell carcinoma cell lines. *B.J.U. Int.* **85,** 130–138.

33. Zhuang, Z., Bertheau, P., Emmert-Buck, M. R., et al. (1995) A microdissection technique for archival DNA analysis of specific cell populations in lesions < 1 mm in size. *Am. J. Pathol.* **146**, 620–625.
34. Leach, F. S., Polyak, K., Burrell, M., et al. (1996) Expression of the mismatch repair gene hMSH2 in normal and neoplastic tissues. *Cancer Res.* **56,** 235–240.
35. Chen, Y., Wang, J., Fraig, M. M., et al. (2001) Defects of DNA mismatch repair in human prostate cancer. *Cancer Res.* **61,** 4112–4121.
36. Wilson, T. M., Ewel, A., Duguid, J. R., et al. (1998) Differential cellular expression of the human MSH2 repair enzyme in small and large intestine. *Cancer Res.* **55,** 5146–5150.
37. Velasco, A., Hewitt, S. M., Albert, P. S., et al. (2002) Differential expression of the mismatch repair gene hMSH2 in malignant prostate tissue is associated with cancer recurrence. *Cancer*, in press.

Index